FULTON-MONTGOMERY COMMUNITY COLLEGE   DUPL

3 0388 00117505 1

DATE DUE

D1466005

Avoiding Dangerous Climate Change

Editor in Chief

Hans Joachim Schellnhuber

Co-editors

Wolfgang Cramer, Nebojsa Nakicenovic, Tom Wigley, Gary Yohe

THE EVANS LIBRARY
FULTON-MONTGOMERY COMMUNITY COLLEGE

CAMBRIDGE
UNIVERSITY PRESS

CAMBRIDGE UNIVERSITY PRESS
Cambridge, New York, Melbourne, Madrid, Cape Town, Singapore, São Paulo

CAMBRIDGE UNIVERSITY PRESS
The Edinburgh Building, Cambridge CB2 2RU, UK

Published in the United States of America by Cambridge University Press, New York

www.cambridge.org
Information on this title: www.cambridge.org/9780521864718

© Cambridge University Press, 2006

This publication is in copyright. Subject to statutory exception
and to the provisions of relevant collective licensing agreements,
no reproduction of any part may take place without
the written permission of Cambridge University Press.

First published 2006

Editorial project management by Frameworks
www.frameworks-city.ltd.uk

Typeset by Charon Tec Ltd, Chennai, India
www.charontec.com

Printed in the United Kingdom at the University Press, Cambridge

A catalogue record for this book is available from the British Library

ISBN: 13 978-0-521-86471-8 hardback
ISBN: 10 0-521-86471-2 hardback

Cambridge University Press has no responsibility for the persistence or
accuracy of URLs for external or third-party internet websites referred to in
this publication, and does not guarantee that any content on such websites is,
or will remain, accurate or appropriate.

CONTENTS

FOREWORD

10 DOWNING STREET
LONDON SW1A 2AA

The Rt Hon Tony Blair, MP
UK Prime Minister

Climate change is the world's greatest environmental challenge. It is now plain that the emission of greenhouse gases, associated with industrialisation and economic growth from a world population that has increased six-fold in 200 years, is causing global warming at a rate that is unsustainable.

That is why I set climate change as one of the top priorities for the UK's Presidency of the G8 and the European Union in 2005.

Early in the year, to enhance understanding and appreciation of the science of climate change, we hosted an international meeting at the Hadley Centre in Exeter to address the big questions on which we need to pool the best available answers:

'What level of greenhouse gases in the atmosphere is self-evidently too much?' and 'What options do we have to avoid such levels?'

It is clear from the work presented that the risks of climate change may well be greater than we thought. At the same time it showed there is much that can be done to avoid the worse effects of climate change.

Action now can help avert the worst effects of climate change. With foresight such action can be taken without disturbing our way of life.

The conference provided a scientific backdrop to the G8 summit. At the Gleneagles meeting the leaders of the G8 were able to agree on the importance of climate change, that human activity does contribute to it and that greenhouse gas emissions need to slow, peak and reverse. All G8 countries agreed on the need to make 'substantial cuts' in emissions and to act with resolve and urgency now.

There was agreement to a new Dialogue on Climate Change, Clean Energy and Sustainable Development between G8 and other interested countries with significant energy needs. This process will allow continued discussion of the issues around climate change and measures to tackle it and help create a more constructive atmosphere for international negotiations on future actions to reduce emissions.

This book will serve as more than a record of another conference or event. It will provide an invaluable resource for all people wishing to enhance global understanding of the science of climate change and the need for humanity to act to tackle the problem.

Tony Blair

MINISTERIAL ADDRESS BY Rt Hon MARGARET BECKETT, MP

It is a great pleasure for me to meet so many distinguished climate scientists and in such an impressive new building, which among other things houses the Hadley Centre.

At the time of the Hadley Centre's inception in 1990 the IPCC was in its infancy and the climate change convention had not even been born! Since then it has become one of the world's leading institutes for climate research.

In 1990 carbon dioxide levels were 354 parts per million – now they are at around 377 parts per million and still rising. Since 1990 global temperatures have increased by about 0.2°C and the ten warmest years in the global record have occurred. Absolute temperature records for the UK were broken in 2003 as we passed the 100°F mark.

What the non-specialists have always wanted to know is whether these effects really were connected. In 1990 the first assessment of the IPCC could not unequivocally show that the observed rise in temperatures was linked to increasing greenhouse gases and not just natural variation, even though it was consistent with modelled projections. But by 2001 the IPCC was able to say that 'there is new and stronger evidence that most of the warming observed over the last 50 years is attributable to human activities'.

You are all familiar with the IPCC projections of warming over this century of between about 1.5°C and almost 6°C due to increased greenhouse gases. No doubt they will be refined further but what is clear is that temperatures will go on rising. Indeed, I understand that the warming expected over the next few decades is virtually unavoidable now. Even in this timeframe we may expect significant impacts and so we need to act now to ensure that we limit the scale of warming in the future to avoid the worst effects.

Recent events show that even wealthy modern societies struggle with extreme events, and developing societies are particularly vulnerable to catastrophe. Extreme weather events can be costly, not only in both in human lives and suffering but also in terms of sheer economics. The flooding which swept Europe in 2002 not only caused 37 deaths but cost US$16 billion in direct costs; the European heat-wave in 2003 led to 26,000 premature deaths and US$13.5 billion in direct costs.

Such events can be expected to become more frequent as a result of climate warming. And there are some signs that extremes are increasing in scale and frequency. Recent work published by Hadley Centre has shown that the risk of extreme warmth, such as that of the summer of 2003 over Europe, is now four times greater than 100 years ago and that that increased risk is due to the elevated levels of greenhouse gases in the atmosphere.

The Climate Change Convention's objective, 'to stabilise greenhouse gases in the atmosphere at levels which avoid dangerous anthropogenic climate change', is a protection standard for the global climate, analogous to national and international environmental standards for air quality or critical loads for sulphur or nitrogen.

But for climate operationalising that objective is no mean feat because responsibility is shared across the world. Common, even though differentiated. All countries contribute to the problem to varying degrees but no one country can solve the problem by acting alone. So an international approach is essential. Defining how much climate change is too much is a political, as well as a scientific, question but one which needs to be guided by the best objective information that science can give. That is why we have called this conference. When he announced it in September, the Prime Minister posed these questions, 'What level of greenhouse gases in the atmosphere is self-evidently too much? What options do we have to avoid such levels?' I hope that your discussions here will help society consider these questions.

We need to begin a serious debate to understand how much different levels of climate change will affect the world as a whole, specific regions and particular sectors of society. How fast will change occur and, more significantly, how can we avoid the worst effects? We may not be able to do much to reduce climate change over the next few decades, but what we do now will affect how much and how quickly climate changes. That is why we also need this meeting to look at possible solutions. We in the UK have already committed ourselves to a 60% reduction in carbon dioxide emissions by 2050. We urge others to commit themselves to take comparable steps.

But we should not underestimate the scale of the task. Since 1990, global emissions of CO_2 alone have increased by 20%. By 2010 without the Kyoto Protocol emissions could have risen to 30% above 1990 levels. Nothing less than a radical change in how we generate and use energy will be needed and there will not be one solution but a whole portfolio of measures. Kyoto, which only has targets for developed countries, will shave some 2-3% off the projected emissions. That is very much a first step; but it provides the opportunity to try novel approaches such as giving carbon a value that can be traded to ensure the most economical ways of reducing emissions. The clean development mechanism provides a novel way to slow the growth in developing country emissions whilst at the same time providing resources and new technologies which will aid development.

By comparison to the potential cost of damage due to climate change, the cost of long-term global action to tackle climate change is likely to be short-term and relatively modest. But the level of such costs depends above all on clear long-term signals from government. International action can provide the clarity and confidence that business needs to invest, and to unleash the power of markets to create a low carbon future – both in the developed world and in emerging economies such as China and India where there is such a strong demand for new energy investment.

The UK experience demonstrates that decarbonisation need not be damaging to economic growth. Between 1990 and 2003 our greenhouse gas emissions fell by around 14% while our GDP rose by 36% over the same period.

As the Prime Minister said last week, we need to involve the world's largest current and future emitters in tackling climate change. Also businesses can and must play an absolutely central role in delivering a low carbon economy. To do so industry and investors need the long-term signals to incentives investment in new technology. This is why a clear scientific picture is essential and why your work here is so important.

So what is next? We can all play a part in dealing with the problem but Governments must provide leadership and be prepared to drive change. In Buenos Aires in December, the world took a first small step to looking at what we do beyond 2012, the end of the Kyoto period. This will be a long road but it will help enormously to have at our disposal science which has addressed the questions that this meeting will address, that shows clearly the risks of delay and too little action, and shows us very clearly what the options are to achieve stabilisation. I very much hope that this conference will send a clear message to leaders and decision makers about the scale, the urgency and the necessity of the task before us, that it will encourage more scientists to explore the issues raised and that it will provide through your papers and deliberations helpful guidance to our G8 presidency and important input to the 4th assessment report of the IPCC.

This meeting provides a tremendous opportunity for you as scientists to influence the debate and to help the world to move to a sustainable future and to avoid the worst effects of anthropogenic climate change. I wish you well in your deliberations.

Hadley Centre, Exeter, 1 February 2005

PREFACE

The Meaning and Making of This Book

The International Symposium on Stabilisation of Greenhouse Gas Concentrations, Avoiding Dangerous Climate Change, (ADCC) took place, at the invitation of the British Prime Minister Tony Blair and under the sponsorship of the UK Department for Environment, Food and Rural Affairs (Defra), at the Met Office, Exeter, United Kingdom, on 1–3 February 2005. The conference attracted over 200 participants from some 30 countries. These were mainly scientists, and representatives from international organisations and national governments.

The conference offered a unique opportunity for the scientists to exchange views on the consequences and risks presented to the natural and human systems as a result of changes in the world's climate, and on the pathways and technologies to limit GHG emissions and atmospheric concentrations. The conference took as read the conclusions of the IPCC Third Assessment Report (TAR) that climate change due to human actions is already happening, and that without actions to reduce emissions climate will continue to change, with increasingly adverse effects on the environment and human society.

In particular the scientists were asked to address the following questions:

- What are the key impacts – on regions, sectors, and the world as a whole – of different potential levels of anthropogenic climate change?
- What would such levels of climate change imply in terms of greenhouse gas stabilisation levels and emission pathways required to achieve these levels?
- What technological options are there for implementing these emission pathways, taking into account costs and uncertainties?

By all standards (topicality of contributions, novelty of results, quality of presentations, intensity of discussions) and all accounts (feedback from participants, media coverage, stakeholder reactions and reflections, reverberations in the scientific community), the ADCC Conference was a highly successful event. As a consequence, the conveners were urged by numerous individuals and organisations to summarise the ground covered during the meeting in a self-contained book that makes the pertinent results conveniently accessible to a wider audience. In order to satisfy this demand, Defra established an international Editorial Board (EB) and launched an energetic review and production process.

This book consolidates the scientific findings presented at the Conference and is a resource intended to inform the international debate on what constitutes dangerous climate change. The message coming out of the book is clear – that climate change is happening, that impacts of the change are likely to be more serious than previously thought, and that there are already technological options that can be used to ultimately stabilise the concentration of greenhouse gases in the atmosphere at appropriate levels.

The conference did not attempt to identify a single level of greenhouse gas concentrations to be avoided. The intricacies of climate change prohibit the identification of one single atmospheric concentration that can avoid dangerous levels of climate change on the basis of scientific evidence alone. Indeed consideration of the question requires value judgments by societies and international debate. The conference does however go some way to providing the scientific evidence that could inform such a debate. There is a clear difference between presentation and interpretation of evidence. Scientific evidence is generally restricted to revealing (i) causal aspects of the climate change problem; (ii) the characters, magnitudes and interrelations of the values at stake; and (iii) the potential costs and benefits of the available response strategies. It would be expecting too much of the scientific community to act as the arbiter of society's preferences as reflected in the valuation metrics actually employed and the decision processes actually implemented.

The process of putting together this book has spared no pains in ensuring the scientific quality and credibility of the material presented. All contributions had to survive a four-fold filtering and amendment procedure. Firstly, the submissions to the conference in response to the 2004 open call for papers as well as about ten invited keynotes were scrutinized by the International Scientific Steering Committee on an extended-abstract basis. Secondly, the invited and selected presentations were intensively discussed by the Conference itself and in numerous individual conversations, providing the authors with numerous valuable suggestions and criticisms. Thirdly, all the presenters were invited by the EB in the spring of 2005 to submit an amended version of their Conference contribution that took into account comments from the participants and was restructured for inclusion in this book. Finally all the re-submissions (whether originally invited or selected) were subjected to independent peer review as the basis for a final acceptance or rejection decision by the EB. This process also allowed for some amendment by the authors of their original papers in the light of the reviewers' comments.

We feel that the outcome was well worth the efforts of hundreds of experts, stakeholders and staff involved in this enterprise. We would like to express our deep gratitude to all those involved and in particular to the referees for their invaluable reviews and to the authors of the papers for delivering under brutal time constraints.

The resulting material is organised in seven sections that span all aspects of the problem, starting with climate system analysis and ending with an assessment of the technological portfolio needed for global warming containment. We hope that this book will make a significant contribution to the scientific and policy debates on the ultimate rationale for and level of climate protection, in terms of breadth of coverage, topicality, scientific quality and relevance.

Hans Joachim Schellnhuber (Chair)
Wolfgang Cramer
Nebojsa Nakicenovic
Tom Wigley
Gary Yohe
(Editorial Board)

Dennis Tirpak
(Chair of the International Scientific Steering Committee)

ACKNOWLEDGEMENT

The Editorial Board and Defra would like to express their deep gratitude to all 56 peer-reviewers for their contribution to this work.

SECTION I

Key Vulnerabilities of the Climate System and Critical Thresholds

INTRODUCTION

As a result of anthropogenic greenhouse gas emissions, key components of the climate system are being increasingly stressed. The primary changes in climate and sea level will be relatively slow and steady (albeit much faster than anything previously experienced by mankind). However, superimposed on these trends, there may well be abrupt and possibly irreversible changes that would have far more serious consequences. The main areas of concern here are the large ice sheets in Greenland and Antarctica, and the ocean's thermohaline circulation. The papers in this chapter focus on these areas.

In their introductory paper, Schneider and Lane present a conceptual overview of 'dangerous' climate change issues, noting the difficulty in defining just what 'dangerous' means. They also highlight the different, but complementary, roles that scientists and policymakers play in this complex arena. In particular, they introduce the notion of Type I errors (exaggerated precautionary action based on ultimately unfounded concerns) and Type II errors (insufficient hedging action, delaying measures while waiting for the advent of overwhelming evidence). Schneider and Lane suggest ways out of these dilemmas using recently developed probabilistic methods.

Rapley focuses on the Antarctic ice sheet and its relationship with sea level. He presents new data-based results on the stability of the West Antarctic Ice Sheet and on the overall mass balance of Antarctica. The melting of the ice shelves, such as Larsen B, which has been continuously present since the last glacial period, may be leading to a speed up of some glaciers, by a factor of 2–6, in a 'cork out of bottle' effect. These processes need to be incorporated in advanced ice-sheet models. The extents to which anthropogenic warming or natural variability are contributing to these changes is unknown but many of the changes are consistent with the expected effects of human activities.

The paper by Lowe and co-authors addresses the Greenland ice sheet. If the Greenland ice sheet melted completely, this would raise global average sea level by around 7 metres – so the probability of such melting and the timescale over which it might occur is an important issue. Lowe and co-authors report on a model ensemble experiment based on the finding that local warming of more than 2.7°C would cause the ice sheet to contract. Using a range of models and emissions scenarios leading to CO_2 stabilisation between 450 ppm and 1000 ppm, the study shows that, even with stabilisation at 450 ppm, 5% of the cases lead to a complete and irreversible melting of the ice sheet. Although complete melting would take place over millennia, there would be an accelerated contribution to sea level rise compared with projections given in the IPCC Third Assessment Report.

A package of three papers is dedicated to the stability of the North Atlantic Thermohaline Circulation (THC). Schlesinger and co-authors present a novel assessment based on probability distributions for crucial system parameters and a spectrum of possible policy interventions. Their results quantify both the probability of a THC collapse in the absence of policy, and the effects of different policies on this probability. Challenor and co-authors present similar results for the probability of a THC collapse, based on a large ensemble study using a statistically-based representation of a medium-complexity climate model. Both of these papers suggest that the likelihood of a THC collapse before 2100 could be higher than suggested by previous studies. However, both papers employ simple models so their quantitative results must be treated cautiously – their main contributions are in demonstrating methods for producing probabilistic results. Wood and co-authors show from a model simulation that the cooling effect of a hypothetical rapid THC shutdown in 2049 would more than outweigh global warming in and around the North Atlantic. They demonstrate the feasibility of using ensembles of AOGCMs to quantify the likelihood of THC collapse, noting that no AOGCM in the IPCC TAR or since has shown a shutdown by 2100. They note that further modelling experiments and observational data are essential for more robust answers.

Turley and co-authors review data showing the marked acidification (pH reduction) of the oceans due to the build up of atmospheric carbon dioxide. As atmospheric concentrations continue to increase, so too will acidification, and this in turn may result in drastic changes in marine ecosystems and biogeochemical cycling. Thus, even in the absence of substantial climate change, the oceans may suffer serious damage, providing yet another reason to be concerned about continuing increases in CO_2 emissions.

The papers presented in this section illustrate why the term 'global warming' is inadequate to describe the changes we can expect in the Earth System. We should focus not only on temperature, but also on anticipated shifts (perhaps rapid) in the full range of climate variables, their variability and their extremes; and also on the direct oceanic consequences of atmospheric CO_2 concentration increases. Further, we need to quantify uncertainties arising from uncertainties in future emissions and in climate models, as far as possible, in probabilistic terms. Some of the papers in this section make initial attempts to do this. Addressing climate change will involve balancing uncertainties in both future change and the consequences of policy actions, and understanding the dangers associated with delayed action.

Our understanding of the Earth System is still incomplete and models of the climate system clearly need to be improved. For example, while we have a good sense of how much sea level would rise if the Greenland ice sheet were to disappear, we do not fully understand the thresholds that might lead to such a dramatic effect, nor the time frame over which this might happen. Similarly, while our most physically detailed and realistic models, AOGCMs, indicate that a shutdown of the THC is unlikely, at least by 2100, new analyses presented here using simpler models give somewhat greater cause for concern. A better understanding of the probability of dangerous interference with the climate system requires improved understanding of and quantitative estimates of the thresholds and 'tipping points' explored by the papers in this section.

CHAPTER 1

Avoiding Dangerous Climate Change

Rajendra Pachauri
Presentation given to the Exeter Conference, February 2005

This conference comes at a time when both scientific research in the field of climate change and public policy are waiting for vital inputs. There is a pressing need to provide objective scientific information to assist the process of decision-making in the field.

I am going to talk about the kind of framework within which we need to look at the whole issue of what constitutes dangerous interference with the climate system. This is not a trivial question. The Framework Convention on Climate Change, which was negotiated with a great deal of effort, highlighted the provisions of Article-2 which raises the issue of dangerous levels of anthropogenic emissions and the impacts of human actions on climate change. What I would like to submit is that this is no doubt a question that must be decided on the basis of a value judgment. What is dangerous is essentially a matter of what society decides. It is not something that science alone can decide. But, science certainly can provide the inputs for facilitating that decision. I would like to highlight some cardinal principles which I suggest are important in arriving at a framework and in arriving at what constitutes dangerous. The first, of course, is universal human rights. We need to be concerned with the rights of every society. Every community on this Earth should be able to exist in a manner that they have full rights to decide on. So, therefore, what I would like to highlight is the importance of looking at the impacts of climate change on every corner of the globe and on every community, because we cannot ignore some as being irrelevant to this decision and they certainly have to be part of the larger human rights question that we or most societies today subscribe to.

The next issue that I would like to highlight is the needs of future generations and sustainable development. Climate change is at the heart of sustainable development. If we are going to leave a legacy that essentially creates a negative force for future generations and their ability to be able to meet their own needs then we are certainly not moving on the path of sustainable development. Now, science can provide a basis for this perspective by assessing the impacts and the damage that climate change at different levels can create and, more particularly, the socio-economic dimensions of these impacts. This is an area where I must say that the scientific community has not done enough. And, that is largely because we generally find that social scientists have not really got adequately involved in researching on issues of climate change.

There are several questions which I am sure will come up for discussion in this conference. Setting an explicit threshold for a dangerous level of climate change – how valid is that? You have to start somewhere and I am sure there is no perfect measure, there is no perfect datum on the basis of which you could decide what is dangerous. But this is a question that needs to be answered. Of course, we must also understand that if we fix a certain threshold then reaching that threshold depends to a significant extent on initial conditions. You could have a place that is severely stressed as a result of a variety of factors, where even a slight change in the climate could take you over the threshold. These baseline or initial conditions are extremely important to define and understand. Then we need to look at the marginal impacts and the damage that climate change causes. This requires an assessment of the extent of climate change that is likely to take place and the marginal impacts associated with it. At the same time, we need to determine the costs of the impacts. Of course, when we are dealing with human lives, the classical models of economics will not apply. We need to have some other basis by which we can value the kind of human dimensions that would be involved in assessing impacts. We need to look at irreversibility and the feasibility of appropriate adaptation measures; where is it that you can adapt to a certain level of climate change and thereby tolerate it without really making any stark or major difference to the way we live?

And where is it that we need to seriously consider irreversibility? When we talk about irreversibility, it is not merely issues related to our day-to-day business. It has to do with slow processes that could damage coral reefs; it has to do with various ecosystems across the globe, which may not have an immediate and obvious implication or significance for our day-to-day living but would certainly prove significant over a period of time. And we necessarily have to look at mitigation options; we cannot isolate the impacts question from what is possible from the mitigation point of view. For example, in the UK we have seen a drastic reduction in emissions accompanied by an extremely robust and healthy rate of growth, which gives us an indication of the economic dimensions of mitigation measures. We need to assess these under different conditions and define what the mitigation options would be in the future. Therefore, to sum up what I have said – we need to assess the issue of danger in terms of dangerous for whom (because there is an equity dimension involved), and dangerous by when.

Even if we were to bring about very deep cuts in emissions today, we know that there is an enormous inertia in the system which will result in continuation of climate change for a long time to come. There are intergenerational issues too. We also have to look at plausible adaptation scenarios. Some measures of adaptation can be implemented immediately, others would take a substantial period of time and they would also take a substantial expenditure of effort, finance and other inputs. And, similarly, we need plausible mitigation scenarios. On the basis of these, perhaps we may be able to define in a balanced way actions that would be required.

Now, some practical questions that I am sure will be discussed in the conference. Can a target of increase in temperature capture the limit of what is dangerous? Undoubtedly, that is just one indicator; there are several dimensions to what is dangerous. Of course, we need some measures by which we can decide on a course of action. Is a temperature target the best way to define it? That is the question that I think needs to be answered. Do we have a scientific rationale for setting this target? And, if so, how can we provide its underlying basis? This is where the scientific community really has an enormous responsibility to understand the framework within which this decision would have to be taken and then try to fill in the gaps with adequate and objective scientific knowledge that would assist the politician and the decision maker.

This is where I would like to highlight the character of the IPCC. The IPCC is required to review and assess policy relevant research; i.e. not be policy prescriptive, but policy relevant! And, relevance has to be based on our perception of the decision-making framework and the kinds of issues that become part of policy. Then we can perhaps address in an objective and scientific manner what would assist that system of decision-making. Can a global-mean temperature target, for example, represent danger at the local level? I would mention the importance of looking across the globe and seeing what the impacts would be for different communities and different locations. And, how do we determine a concentration level for GHGs? Where is it that we draw the limit? And what is the trajectory that we require to achieve stabilization because we are not dealing with a static concept, we are not talking about reaching a certain level at a particular point of time. The path by which you reach that particular level is critically important and that necessarily needs to be defined.

Now some issues of initial conditions. Here I will pick out a combination of results from the Third Assessment Report and a few other assessments available in the literature. We know that the global-mean surface temperature has increased by about 0.7°C over the last century. We know that there has been a decrease in Arctic sea ice extent by 10 to 15% and in thickness by 40%; and a decrease in Arctic snow cover area by some 10% since satellite observations started in 1960. We know about the damage to the coral reefs and that the 1990s was likely the warmest decade of the millennium.

In assessing what is dangerous we have to look at every aspect of the impacts on health, agriculture, water resources, coastal areas, species and natural assets. Of course, in coastal areas, natural disasters will take place. We can certainly warn communities against them if we have adequate and effective warning systems. But we must also understand that natural disasters are going to take place no matter what. If climate change is going to exacerbate conditions, which would enhance the severity of the impacts, then that adds another responsibility that the global community has to accept. In Mauritius, a couple of weeks ago, there was the major UN conference involving the small island developing states. In discussions with several people there, I heard an expression of fear based on the question: suppose a tsunami such as that of December 26 were to take place in 2080 and suppose the sea level was a foot higher, can you estimate what the extent of damage would be under those circumstances? Hence, I think when we talk about dangerous it is not merely dangers that are posed by climate change *per se*, but the overlay of climate change impacts on the possibility of natural disasters that could take place in any event.

Another issue that I would like to highlight is the issue of dangerous for whom. There are several studies none of which I am going to endorse, but I just want to put these forward as examples – the work of Norman Myers, for instance. He wrote about the possibility of 150 million environmental refugees by the year 2050. Numbers are not important, but I would like to highlight the issue that we need to look at. What is likely to happen as a result of sea level rise and agricultural changes to human society in different parts of the globe, for instance, in the form of refugees? Bangladesh, which as you know is a low-lying country is particularly vulnerable to sea level rise and the impacts that this would bring. Egypt is another country that would lose 12–15% of its alluvial land, and so on. Consequently, we really need a cataloging of all the impacts that are likely to take place. Science should be able to at least attempt the quantification of what these impacts are likely to be for different levels of climate change. This might help decision makers focus on how to deal with the whole issue.

When we discuss dangerous for whom, then there is also the question of extreme events. The IPCC Third Assessment Report clearly identified that the number of disasters of hydro-meteorological origin have increased significantly, along with an increase in precipitation in the mountains accompanied by melting of glaciers, increased incidence of floods, mud slides, and severe land slides. There is a fair amount of data now available on this, particularly in parts of Asia; large areas with high population densities are susceptible to floods, droughts and cyclones as in Bangladesh and India.

I would now like to highlight some of the social implications of the impacts that are likely to happen related to extreme weather or climatic events. Here I would like to underline the fact that demographic and socio-economic factors can amplify the dangers. There has been an upward trend in weather related losses over the last 50 years linked to socio-economic factors; population growth, increased wealth, urbanization in vulnerable areas, etc. These are trends that are going to continue. If we have to define dangerous then this changing baseline must be considered. Dangerous must be assessed on the basis of scenarios that are consistent with the changes that we already see, for instance, in migration, demographics, and in incomes. All of these in essence define the initial conditions that I mentioned earlier on. We also need to understand the operation of financial services such as insurance in defining the behaviour of societies, in defining where people are likely to settle, because these things are intimately linked with perceptions of the damages – climate-related damages – that might occur over a period of time.

Now the question is, can we adapt to irreversible changes? Can science give us some answers on this? You certainly can adapt to changes like deforestation because we have the means by which we can carry out aforestation, by which we can plant trees in areas wherever deforestation is taking place. But can we bring back the loss of biodiversity which is taking place? Issues of this nature need to be defined because all of this becomes an important part of the package on what is dangerous. In fact, we know that in the 20th century especially during an El Niño event there has been a major impact on coral reef bleaching. Worldwide increase in coral reef bleaching in 1997–98 was coincidental with high water temperatures associated with El Niño. Will future such occurrences be irreversible?

Other examples include the frequency and severity of drought, now fairly well documented in different parts of Africa and Asia. Duration of ice cover of rivers and lakes has decreased by about 2 weeks over the 20th century in mid and high latitudes of the northern hemisphere. Arctic sea ice extent, as I mentioned earlier, decreased by 40% in recent decades in late summer to early autumn and decreased by 10 to 15% since the 1950s in spring and summer. And temperate glaciers are receding rapidly in different parts of the globe.

We also need to look at climate change and its relationship to possible singular events; such as a shutdown of the ocean's thermohaline circulation or rapid ice losses in Greenland or Antarctica. Here, of course, science has a long way to go, but it is a challenge for the scientific community to be able to establish if there is likely to be a relationship between these possible singular events and the process of climate change that we are witnessing. Such events could lead to very high magnitude impacts that could overwhelm our response strategies.

We need to put some of these possible impacts into a framework with an economic perspective where they are translated into the impacts on numbers of people in specific geographical areas. This is a challenge that requires scientists not only to look at the geophysical impacts of climate change, but also start looking at the socio-economic implications. The inertia of the climate system must also be taken into account. Even if we were to stabilize the concentrations of CO_2 and other greenhouse gases today, the inertia in the system can carry the impacts of climate change, particularly sea level rise, through centuries if not a millennium. Indeed, sea level rise could continue for centuries after global-mean temperature was effectively stabilized, complicating the issue of choosing a single metric to defining a dangerous interference threshold.

Even if we are going to think in terms of a temperature target, this necessarily requires that we look at the relationship between emissions, concentrations, and the temperature response. Related to this would be all the other issues that I have put before you in terms of the impacts of climate change as they relate to the global-mean temperature response, particularly adaptation issues. Adaptation strategies can be planned or anticipatory. I highlight the importance of looking at adaptation measures because they need to be considered in defining what is dangerous. If you cannot adapt to a particular change and yet it is likely to have a very harmful impact, then clearly it could be dangerous; but if you can adapt to it without serious consequences then it certainly is not dangerous. We need to define, therefore, adaptation measures within choices including planned and anticipatory as well as autonomous and reactive.

On the mitigation side, we often take a very narrow view of costs and economics of mitigation. We must look at a holistic assessment of mitigation measures and identify measures where there are several co-benefits including those related to goals for sustainable development (in economic, equity, and environmental terms). Then, of course, there is a whole range of so-called no regrets measures that also need to be identified. And the key linkages between mitigation and development are numerous. So, in assessing mitigation costs and options it is absolutely essential that we look at the whole gamut of associated benefits and costs as well.

In addressing the need for assessing the issue of value judgments we must try to see that we create value in terms of scientific information and analysis. But, once again, I would like to emphasize that the decision itself has to be based on a collective assessment by the global community on what they are willing to accept. However, let me repeat that decisions would have to be guided by certain principles, principles that must look at the rights of every community on this globe and at some of the intergenerational implications of climate change (because what may not be dangerous today could very well turn out to be dangerous fifty years from now). It would be totally irresponsible if, as a species, we ignore that reality. So, there is before us a huge agenda for the scientific community. In this context we need to understand the framework within which decisions have to be made. It is my hope that in the Fourth Assessment Report of the IPCC we will be able to provide information through which some of the holes, in the form of uncertainties or unknowns that affect decision-making, can be filled up effectively.

CHAPTER 2

An Overview of 'Dangerous' Climate Change

Stephen H. Schneider and Janica Lane
Stanford University, Stanford, California

ABSTRACT: This paper briefly outlines the basic science of climate change, as well as the IPCC assessments on emissions scenarios and climate impacts, to provide a context for the topic of key vulnerabilities to climate change. A conceptual overview of 'dangerous' climate change issues and the roles of scientists and policy makers in this complex scientific and policy arena is presented, based on literature and recent IPCC work. Literature on assessments of 'dangerous anthropogenic interference' with the climate system is summarized, with emphasis on recent probabilistic analyses. Presenting climate modeling results and arguing for the benefits of climate policy should be framed for decision makers in terms of the potential for climate policy to reduce the likelihood of exceeding 'dangerous' thresholds.

2.1 Introduction

Europe's summers to get hotter... The Arctic's ominous thaw... Study shows warming trend in Alaskan Streams... Lake Tahoe Warming Twice as Fast as Oceans. Global Warming Seen as Security Threat... Global warming a bigger threat to poor... Tibet's glacier's heading for meltdown... Climate change affects deep sea life... UK: Climate change is costing millions. These are just a few of the many headlines related to climate change that crossed the wires in 2004 and they have elicited widespread concern even in the business community. 2004 is thought to have been the fourth warmest year on record and the worst year thus far for weather-related disaster claims – though the devastation in the US Gulf Coast from intense hurricanes in the summer of 2005 could well set a new record for disaster spending. Munich Re, the largest reinsurer in the world, recently stated that it expects natural-disaster-related damages to increase 'exponentially' in the near future and it attributes much of these damages to anthropogenic climate change. Thomas Loster, a climate expert at Munich Re, says: 'We need to stop this dangerous experiment humankind is conducting on the Earth's atmosphere'.

'Dangerous' has become something of a cliché when discussing climate change, but what exactly does it mean in that context? This paper will explore some basic concepts in climate change, how they relate to what might be 'dangerous', and various approaches to characterizing and quantifying 'dangerous anthropogenic interference [DAI] with the climate system' [70]. It will also outline and differentiate the roles of scientists and policymakers in dealing with dangerous climate change by discussing current scientific attempts at assessing elements of dangerous climate change and suggesting ways in which decision makers can translate such science into policy. It will state explicitly that determination of 'acceptable' levels of impacts or what constitutes 'danger' are deeply normative decisions, involving value judgments that must be made by decision makers, though scientists and policy analysts have a major role in providing analysis and context.

2.2 Climate Change: A Brief Primer

We will begin by stressing the well-established principles in the climate debate before turning to the uncertainties and more speculative, cutting-edge scientific debates. First, the greenhouse effect is empirically and theoretically well-established. The gases that make up Earth's atmosphere are semi-transparent to solar energy, allowing about half of the incident sunlight to penetrate the atmosphere and reach Earth's surface. The surface absorbs the heat, heats up and/or evaporates liquid water into water vapor, and also re-emits energy upward as infrared radiation. Certain naturally-occurring gases and particles – particularly clouds – absorb most of the infrared radiation. The infrared energy that is absorbed in the atmosphere is re-emitted, both up to space and back down towards the Earth's surface. The energy channeled towards the Earth causes its surface to warm further and emit infrared radiation at a still greater rate, until the emitted radiation is in balance with the absorbed portion of incident sunlight and the other forms of energy coming and going from the surface. The heat-trapping 'greenhouse effect' is what accounts for the $\sim 33°C$ difference between the Earth's actual surface air temperature and that which is measured in space as the Earth's radiative temperature. Nothing so far is controversial. More controversial is the extent to which non-natural (i.e. human) emissions of greenhouse gases have contributed to climate change, how much we will enhance future disturbance, and what the consequences of such disturbance could be for social, environmental, economic, and other systems – in short, the extent to which human alterations could risk DAI.

It is also well-known that humans have caused an increase in radiative forcing. In the past few centuries, atmospheric carbon dioxide has increased by more than 30%. The reality of this increase is undeniable, and virtually all climatologists agree that the cause is human activity, predominantly the burning of fossil fuels. To a lesser extent, deforestation and other land-use changes and industrial and agricultural activities like cement production and animal husbandry have also contributed to greenhouse gas buildups since 1800. [One controversial hypothesis ([58]) asserts that atmospheric concentrations of carbon dioxide (CO_2) and methane (CH_4) were first altered by humans thousands of years ago, resulting from the discovery of agriculture and subsequent technological innovations in farming. These early anthropogenic CO_2 and CH_4 emissions, it is claimed, offset natural cooling that otherwise would have occurred.]

Most mainstream climate scientists agree that there has been an anomalous rise in global average surface temperatures since the time of the Industrial Revolution. Earth's temperature is highly variable, with year-to-year changes often masking the overall rise of approximately 0.7°C that has occurred since 1860, but the 20th century upward trend is obvious, as shown in Figure 2.1. Especially noticeable is the rapid rise at the end of the 20th century.

For further evidence of this, Mann and Jones, 2003 [33]; Mann, Bradley and Hughes, 1998 [32]; and Mann, Bradley and Hughes, 1999 [31] have attempted to push the Northern Hemisphere temperature record back 1,000 years or more by performing a complex statistical analysis involving some 112 separate indicators related to temperature. Although there is considerable uncertainty in their millennial temperature reconstruction, the overall trend shows a gradual temperature decrease over the first 900 years, followed by a sharp upturn in the 20th century. That upturn is a compressed representation of the 'real' (thermometer-based) surface temperature record of the last 150 years. Though there is some ongoing dispute about temperature details in the medieval period (e.g. [72]), many independent studies confirm the basic picture of unusual warming in the past three decades compared to the past millennium [73].

It is likely that human activities have caused a discernible impact on observed warming trends. There is a high correlation between increases in global temperature and increases in carbon dioxide and other greenhouse gas

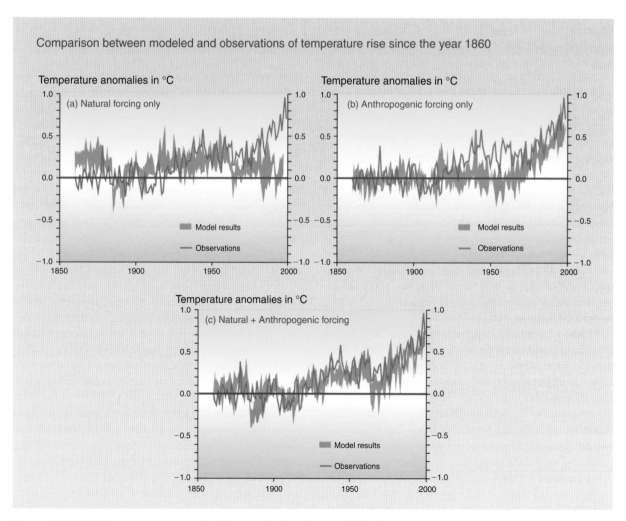

Figure 2.1 Explaining temperature trends using natural and anthropogenic forcing.
Source: IPCC, 2001d.

concentrations during the era, from 1860 to present, of rapid industrialization and population growth. As correlation is not necessarily causation, what other evidence is there about anthropogenic CO_2 emissions as a direct cause of recent warming? Hansen et al. (2005) [18] offer considerable data to suggest that there is currently an imbalance of some $0.85 \pm 0.15\,\text{W/m}^2$ of extra heating in the Earth-atmosphere system owing to the heat-trapping effects of greenhouse gas build-ups over the past century. If accepted, this new finding would imply that not only has an anthropogenic heat-trapping signal been detected in observational records, but that the imbalance in the radiative heating of the Earth-atmosphere system implies that there is still considerable warming "in the bank", and that another 0.6°C or so of warming could be inevitable even in the unlikely event that greenhouse gas concentrations were frozen at today's levels [76].

Other evidence can be brought to bear to show human influences on recent temperatures from a variety of sources, such as the data summarized in Figure 2.1. The Figure suggests that the best explanation for the global rise in temperature seen thus far is obtained from a combination of natural and anthropogenic forcings. Although substantial, this is still circumstantial evidence. However, many recent 'fingerprint analyses' have reinforced these conclusions (i.e. [60], [20], [48], [55], and [59]). Most recently, Root et al. (2005) [54] have shown that the timing of biological events like the flowering of trees or egg-laying of birds in the spring are significantly correlated with anthropogenically-forced climate, but only weakly associated with simulations incorporating only natural forcings. This same causal separation is illustrated in Figure 2.1 comparing observed thermometer data and modeled temperature results for natural, anthropogenic, and combined forcings. (Root et al. came to these results using the HadCM3 model, the same model used to obtain the results depicted in Figure 2.1.) Since plants and animals can serve as independent 'proxy thermometers', these findings put into doubt suggestions that errors in instrumental temperature records due to urban heat island effects as well as claims that satellite-derived temperatures do not support surface warming – the satellite-derived temperature trend dispute apparently has been largely resolved in mid-2005 by a series of reports reconciling lower atmospheric warming in models, balloons and satellite temperature reconstructions. These and other anthropogenic fingerprints in global climate system variables and temperature trends represent an overwhelming preponderance of evidence. In our opinion, results from 30 years of research by the scientific community now convincingly suggest it is fair to call the detection and attribution of human impacts on climate a well-established conclusion.

2.3 Climate Change Scenarios

Since the climate science and historical temperature trends show highly likely direct cause-and-effect relationships, we must now ask how climate may change in the future.

Scientists, technologists, and policy analysts have invested considerable effort in constructing 'storylines' of plausible human demographic, economic, political, and technological futures from which a range of emissions scenarios can be described, the most well-known being the Intergovernmental Panel on Climate Change's (IPCC) Special Report on Emissions Scenarios (SRES), published in 2000 [38]. One grouping is the A1 storyline and scenario family, which describes a future world of very rapid economic growth, global population that peaks in mid-century and declines thereafter and, in several variations of it, the rapid introduction of new and more efficient technologies. Major underlying themes are convergence between regions, capacity-building, and increased cultural and social interactions, with a substantial reduction in regional differences in per capita income. A1 is subdivided into A1FI (fossil-fuel intensive), A1T (high-technology), and A1B (balanced), with A1FI generating the most CO_2 emissions and A1T the least (of the A1 storyline, and the second lowest emissions of all six marker scenarios). But even in the A1T world, atmospheric concentrations of CO_2 still near a doubling of preindustrial levels by 2100.

For a contrasting vision of the world's social and technological future, SRES offers the B1 storyline, which is (marginally) the lowest-emissions case of all the IPCC's scenarios. The storyline and scenario family is one of a converging world with the same global population as A1, peaking in mid-century and declining thereafter, but with more rapid change in economic structures towards service and information economies, which is assumed to cause a significant decrease in energy intensity. The B1 world finds efficient ways of increasing economic output with less material, cleaner resources, and more efficient technologies. Many scientists and policymakers have doubted whether a transition to a B1 world is realistic and whether it can be considered equally likely when compared to the scenarios in the A1 family. The IPCC did not discuss probabilities of each scenario, making a risk-management framework for climate policy problematic since risk is probability times consequences (e.g. see the debate summarized by [14]). Figure 2.2 is illustrative of the SRES scenarios.

2.4 Climate Change Impacts

After producing the SRES scenarios, the IPCC released its Third Assessment Report (TAR) in 2001, in which it estimated that by 2100, global average surface temperatures would rise by 1.4 to 5.8°C relative to the 1990 level. While warming at the low end of this range would likely be relatively less stressful, it would still be significant for some 'unique and valuable systems' [25] – sea level rise of concern to some low-lying coastal and island communities and impacts to Arctic regions, for example. Warming at the high end of the range could have widespread catastrophic consequences, as a temperature change of 5–7°C on a globally-averaged basis is about the difference between an ice age and an interglacial – and over a period

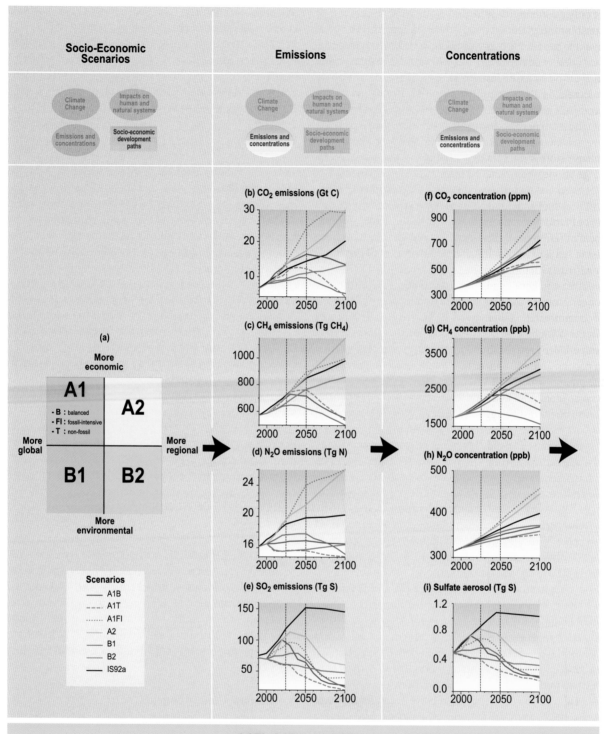

A1FI, A1T, and A1B

The A1 storyline and scenario family describes a future world of very rapid economic growth, global population that peaks in mid-century and declines thereafter, and the rapid introduction of new and more efficient technologies. Major underlying themes are convergence among regions, capacity-building, and increased cultural and social interactions, with a substantial reduction in regional differences in per capita income. The A1 scenario family develops into three groups that describe alternative directions of technological change in the energy system. The three A1 groups are distinguished by their technological emphasis: fossil intensive (A1FI), non-fossil energy sources (A1T), or a balance across all sources (A1B) (where balanced is defined as not relying too heavily on one particular energy source, on the assumption that similar improvment rates apply to all energy supply and end use technologies).

Figure 2.2 SRES emissions scenarios.
Source: IPCC, 2001d.

of only a century [7]. If the IPCC's projections prove reasonable, the global average rate of temperature change over the next century or two will exceed the average rate sustained over the last century, which is already greater than any seen in the last 10,000 years [65].

Based on these temperature forecasts, the IPCC has produced a list of likely effects of climate change, most of which are negative (see [25]). These include: more frequent heat waves (and less frequent cold spells); more intense storms (hurricanes, tropical cyclones, etc.) and a surge in weather-related damage; increased intensity of floods and droughts; warmer surface temperatures, especially at higher latitudes; more rapid spread of disease; loss of farming productivity in many regions and/or movement of farming to other regions, most at higher latitudes; rising sea levels, which could inundate coastal areas and small island nations; and species extinction and loss of biodiversity. On the positive side, the literature suggests longer growing seasons at high latitudes and the opening of commercial shipping in the normally ice-plagued Arctic. Weighing these pros and cons is the normative (value-laden) responsibility of policy-makers, responding in part, of course, to the opinions and value judgments of the public, which will vary from region to region, group to group, and individual to individual.

The IPCC also suggested that, particularly for rapid and substantial temperature increases, climate change could trigger 'surprises': rapid, nonlinear responses of the climate system to anthropogenic forcing, thought to occur when environmental thresholds are crossed and new (and not always beneficial) equilibriums are reached. Schneider et al. (1998) [66] took this a step further, defining 'imaginable surprises'– events that could be extremely damaging but which are not *truly* unanticipated. These could include a large reduction in the strength or possible collapse of the North Atlantic thermohaline circulation (THC) system, which could cause significant cooling in the North Atlantic region, with both warming and cooling regional teleconnections up- and downstream of the North Atlantic; and deglaciation of polar ice sheets like Greenland or the West Antarctic, which would cause (over many centuries) many meters of additional sea level rise on top of that caused by the thermal expansion from the direct warming of the oceans [61].

There is also the possibility of *true* surprises, events not yet currently envisioned [66]. However, in the case of true surprises, it is still possible to formulate 'imaginable conditions for surprise'—like rapidly-forced climate change, since the faster the climate system is forced to change, the higher the likelihood of triggering abrupt nonlinear responses (see page 7 of [27]). Potential climate change and, more broadly, global environmental change, faces both types of surprise because of the enormous complexities of the processes and interrelationships involved (such as coupled ocean, atmosphere, and terrestrial systems) and our insufficient understanding of them individually and collectively (e.g. [21]).

Many systems have been devised for categorizing climate change impacts. IPCC (2001b) [25] has represented impacts as 'reasons for concern', as in Figure 2.3, below.

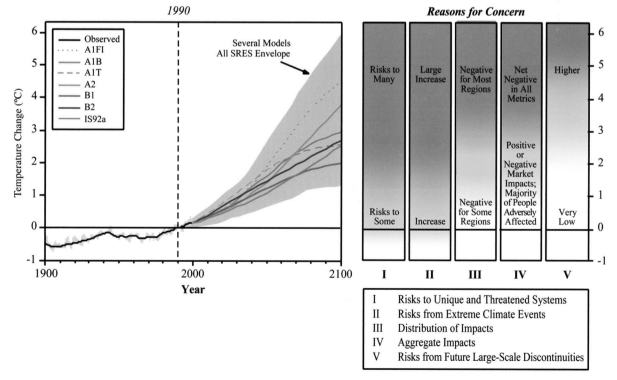

Figure 2.3 IPCC reasons for concern about climate change impacts.
Source: IPCC, 2001b.

These impacts are: risks to unique and threatened systems; risks associated with extreme weather events; the distribution of impacts (i.e. equity implications); aggregate damages (i.e. market economic impacts); and risks of large-scale singular events (e.g. 'surprises'). Leemans and Eickhout (2004) [30] have also suggested including risks to global and local ecosystems as an additional reason for concern, though this could be partially represented under the first reason for concern. The Figure, also known as the 'burning embers diagram', shows that the most potentially serious climate change impacts (the red colors on the Figure) typically occur after only a few degrees Celsius of warming.

Parry et al.'s (2001) [49] 'millions at risk' work suggests another approach. These authors estimate the additional millions of people who could be placed at risk as a result of different amounts of global warming. The risks Parry et al. focus on are hunger, malaria, flooding, and water shortage. Similarly, the 2002 Johannesburg World Summit on Sustainable Development (WSSD) came up with five key areas to target for sustainable development: water, energy, health, agriculture, and biodiversity (WEHAB). These categories, with the addition of coastal regions (as proposed by [49]), are also well-suited to grouping climate change impacts [51].

In looking at climate impacts from a justice perspective, Schneider and Lane (2005) [63] propose three distinct areas in which climate change inequities are likely to be significant: inter-country equity, intergenerational equity, and inter-species equity. (Schneider and Lane and others have also suggested intra-national equity of impacts.) Another justice-oriented impacts classification scheme is Schneider et al.'s (2000) [64] 'five numeraires': market system costs in dollars per ton Carbon (C); human lives lost in persons per ton C; species lost per ton C; distributional effects (such as changes in income differentials between rich and poor) per ton C; and quality of life changes, such as heritage sites lost per ton C or refugees created per ton C. Lane, Sagar, and Schneider (2005) [29] propose examining not just absolute costs in each of the five numeraires, but *relative* costs as well in some of them:

> *…we should consider market-system costs relative to a country's GDP, species lost relative to the total number of species in that family, etc. Expressing impacts through the use of such numeraires will capture a richer accounting of potential damages and could help merge the often-disparate values of different groups in gauging the seriousness of damages. In other cases, such as human lives lost, we believe that the absolute measure remains more appropriate.*

It is our strong belief that such broad-based, multi-metric approaches to impacts categorization and assessment are vastly preferable to focusing solely on market categories of damages, as is often done by traditional cost-benefit analyses. One-metric aggregations probably underestimate the seriousness of climate impacts. Evidence for

this was gathered by Nordhaus (1994a) [41], who surveyed conventional economists, environmental economists, atmospheric scientists, and ecologists about estimated climate damages. His study reveals a striking cultural divide across the natural and social scientists who participated in the study. Conventional economists surveyed suggested that even extreme climate change (i.e. 6°C of warming by 2090) would not likely impose severe economic losses, implying it is likely to be cheaper to emit more in the near term and worry about cutting back later, using additional wealth gained from near-term emitting to fund adaptation later on. Natural scientists estimated the total economic impact of extreme climate change, much of which they assigned to non-market categories, to be 20 to 30 times higher than conventional economists' projections. In essence, the natural scientists tended to respond that they were much less optimistic that humans could invent acceptable substitutes for lost climatic services (see [57]).

Because they typically measure only market impacts, traditional cost-benefit analyses (CBAs) are often considered skewed from a distributional equity perspective. In a traditional CBA, the ethical principle is not even classical Benthamite utilitarianism (greatest good for the greatest number of people), but an aggregated market power form of utilitarianism (greatest good for the greatest number of dollars in benefit/cost ratios). Thus, an industrialized country with a large economy that suffered the same biophysical climate damages as an unindustrialized nation with a smaller economy would be considered to have suffered more by virtue of a larger GDP loss and would, in the aggregate-dollars-lost metric, be more important to 'rescue' and/or rehabilitate, if possible.

Even more problematic, what if an industrial northern country experienced a monetary gain in agriculture and forestry from global warming due to longer growing seasons, while at the same time – as much of the literature suggests – less-developed southern countries suffered from excessive heating that amounted to a monetary loss of the same dollar value as the gain in the north? This could hardly be viewed as a 'neutral' outcome despite a net (global) welfare change of zero (derived from summing the monetary gain in the north and the loss in the south). Very few would view a market-only valuation and global aggregation of impacts in which the rich get richer and the poor get poorer as a result of climate change as an ethically neutral result.

Under the framework of the five numeraires and other systems that rely on multiple metrics, the interests of developing countries and the less privileged within nations would be given a greater weight on the basis of the threats to non-market entities like biodiversity, human life, and cultural heritage sites. Take the example of Bangladesh: Assume that rising sea levels caused by climate change lead to the destruction of lives, property, and ecosystems equivalent to about 80% of the country's GDP. While the losses would be indisputably catastrophic for Bangladesh, they would amount to an inconsequential 0.1% of global

GDP (see Chapter 1 of [25]), causing a market-aggregation-only analysis to classify the damage as relatively insignificant, though a reasonable interpretation of many would be that such a loss clearly qualifies as DAI—what Mastrandrea and Schneider (2005) [35] labeled as "stakeholders metrics". Those considering multiple numeraires would argue that this is clearly unfair, as the loss of life, degraded quality of life, and potential loss of biodiversity in Bangladesh are at least as important as aggregate market impacts.

2.5 Dangerous Climate Change

But what exactly is 'dangerous' climate change? The term was legally introduced in the 1992 United Nations Framework Convention on Climate Change (UNFCCC), which calls for stabilization of greenhouse gases to 'prevent dangerous anthropogenic interference with the climate system' [70]. The Framework Convention further suggests that: 'Such a level should be achieved within a time frame sufficient

- to allow ecosystems to adapt naturally to climate change;
- to ensure that food production is not threatened and;
- to enable economic development to proceed in a sustainable manner'.

While it seems that some of the impacts of climate change discussed thus far suggest that dangerous levels of climate change may occur, the UNFCCC never actually defined what it meant by 'dangerous'.

Many metrics for defining dangerous have been introduced in recent years, and most focus on the consequences (impacts) of climate change outcomes. From an equity perspective, it can be argued that any climate change that has a greater impact on those who contributed the least to the problem is less just and thus arguably more dangerous—and could have repercussions that extend beyond environmental damages (to security, health, and economy, for example). Along similar lines, some scientists defined 'dangerous anthropogenic interference' at the 10th Conference of the Parties (COP10) in Buenos Aires in December 2004 by assessing the key vulnerabilities with regard to climate change. In the IPCC TAR, 'vulnerability' was described as a consequence of exposure, sensitivity, and adaptive capacity (Glossary, [25]). The notion of key vulnerabilities was derived partly from the discussion on 'concepts of danger' that occurred at the European Climate Forum's (ECF) symposium on 'Key vulnerable regions and climate change' in Beijing in October 2004 and was presented at COP 10. The ECF symposium identified three concepts of danger:

- **Determinative dangers** are, on their own, enough to define dangerous levels of climate change. The ECF's list of determinative dangers resulting from climate change include: circumstances that could lead to global and unprecedented consequences, extinction of 'iconic' species or loss of entire ecosystems, loss of human cultures, water resource threats, and substantial increases in mortality levels, among others.
- **Early warning dangers** are dangers already present in certain areas that are likely to spread and worsen over time with increased warming. These dangers could include Arctic Sea ice retreat, boreal forest fires, and increases in frequency of drought, and they could become determinative over time or taken together with other dangers.
- **Regional dangers** are widespread dangers over a large region, most likely related to food security, water resources, infrastructure, or ecosystems. They are not considered determinative, as they are largely confined to a single region [12].

Dessai et al. (2004) [10] also focus on vulnerabilities as an indicator of dangerous climate change. They have separated definitions of danger into two categories: those derived from top-down research processes and those derived from bottom-up methods. The more commonly used top-down approach determines physical vulnerability based on hierarchical models driven by different scenarios of socio-economic change, whereas the bottom-up approach focuses on the vulnerability and adaptive capacity of individuals or groups, which leads to social indications of potential danger like poverty and/or lack of access to healthcare, effective political institutions, etc.

In working drafts of the IPCC Fourth Assessment Report [23], interim definitions and descriptions of 'key vulnerabilities' are framed as follows. Key vulnerabilities are a product of the exposure of systems and populations to climate change, the sensitivity of those systems and populations to such influences, and the capacity of those systems and populations to adapt to them. Changes in these factors can increase or decrease vulnerability. Assessments of key vulnerabilities need to account for the spatial scales and timescales over which impacts occur and the distribution of impacts among groups, as well as the temporal relationship between causes, impacts, and potential responses. No single metric can adequately describe the diversity of key vulnerabilities. Six objective and subjective criteria are suggested for assessing and defining key vulnerabilities:

- Magnitude
- Timing
- Persistence and reversibility
- Likelihood and confidence
- Potential for adaptation
- Importance of the vulnerable system.

Some key vulnerabilities are associated with 'systemic thresholds' in either the climate system, the socio-economic system, or coupled socio-natural systems (e.g. a collapse of the West Antarctic Ice Sheet or the cessation of sea ice touching the shore in the Arctic that eliminates a major prerequisite for the hunting culture of indigenous people in the region). Other key vulnerabilities can be associated with 'normative thresholds', which are defined by groups concerned with a steady increase in adverse impacts caused by an increasing magnitude of climate change (e.g. a magnitude of sea level rise no longer considered acceptable by low-lying coastal dwellers).

While scientists have many ideas about what vulnerabilities may be considered dangerous, it is a common view of most natural and social scientists that it is not the direct role of the scientific community to define what 'dangerous' means. Rather, it is ultimately a political question because it depends on *value judgments* about the relative importance of various impacts and how to face climate change-related risks and form norms for defining what is 'unacceptable' [62, 36]. In fact, the notion of key vulnerabilities itself is also a value judgment, and different decision makers at different locations and levels are likely to perceive vulnerabilities and the concept of 'dangerous' in distinct ways.

Dessai et al. (2004) [10] explain the juxtaposition of science and value judgment by assigning two separate definitions for risk – internal and external. External risks are defined via scientific risk analysis of system characteristics prevalent in the physical or social worlds. Internal risk, on the other hand, defines risk based on the individual or communal perception of insecurity. In the case of internal risk, in order for the risk to be 'real', it must be experienced. Of course, these two definitions are intertwined in complex ways. Decision-makers' perceptions of risk are partly informed by the definitions and guidance provided by scientific experts, and societal perceptions of risk may also play a role in scientific research.

2.6 The Role of Science in Risk Assessment

Ultimately, scientists cannot make expert value judgments about what climate change risks to face and what to avoid, as that is the role of policy makers, but they can help policymakers evaluate what 'dangerous' climate change entails by laying out the elements of *risk*, which is classically defined as ***probability x consequence***. They should also help decision-makers by identifying thresholds and possible surprise events, as well as estimates of how long it might take to resolve many of the remaining uncertainties that plague climate assessments.

There is a host of information available about the possible consequences of climate change, as described in our discussion of the SRES scenarios and of the impacts of climate change, but the SRES scenarios do not have probabilities assigned to them, making risk management difficult. Some would argue that assigning probabilities to scenarios based on social trends and norms should not be done (e.g. [15]), and that the use of scenarios in and of itself derives from the fact that probabilities can't be analytically estimated. In fact, most models do not calculate

objective probabilities for future outcomes, as the future has not yet happened and 'objective statistics' are impossible, in principle, before the fact. However, modelers can assign subjective confidence levels to their results by discussing how well established the underlying processes in a model are, or by comparing their results to observational data for past events or elaborating on other consistency tests of their performance (e.g. [14]). It is our belief that qualified assessment of (clearly admitted) subjective probabilities in every aspect of projections of climatic changes and impacts would improve climate change impact assessments, as it would complete the risk equation, thereby giving policy-makers some idea of the likelihood of threat associated with various scenarios, aiding effective decision-making in the risk-management framework. At the same time, confidence in these difficult probabilistic estimates should also be given, along with a brief explanation of how that confidence was arrived at.

2.7 Uncertainties

A full assessment of the range of climate change consequences *and* probabilities involves a cascade of uncertainties in emissions, carbon cycle response, climate response, and impacts. We must estimate future populations, levels of economic development, and potential technological props spurring that economic development, all of which will influence the radiative forcing of the atmosphere via emissions of greenhouse gases and other radiatively active constituents. At the same time, we must also deal with the uncertainties associated with carbon cycle modeling, and, equally important, confront uncertainties surrounding the climate sensitivity – typically defined as the amount that global average temperature is expected to rise for a doubling of CO_2.

Figure 2.4 shows the 'explosion' that occurs as the different elements of uncertainty are combined. This should not be interpreted as a sign that scientists cannot assign a high degree of confidence to *any* of their projected climate change impacts but, rather, that the scope of possible consequences is quite wide. There are many projected effects, on both global and regional scales, that carry high confidence estimates, but the Figure suggests that there still are many more impacts to which we can only assign low confidence ratings and others that have not yet been postulated – i.e. 'surprises' and irreversible impacts.

One other aspect of Figure 2.4 needs mentioning: Current decision-makers aware of potential future risks might introduce policies to reduce the risks over time – also known as 'reflexive' responses – which would be equivalent to a feedback that affects the size of the bars on Figure 2.4 merely because the prospects for risks created precautionary responses. That possibility is partly responsible for the attitudes of some who are reluctant to assign probabilities – even subjective ones – to the components of Figure 2.4. If no probabilities are associated

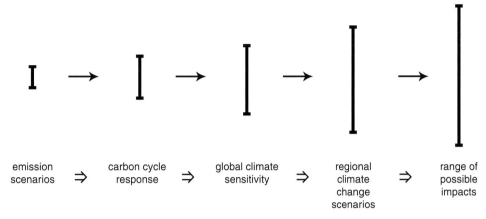

emission scenarios $\Rightarrow$ carbon cycle response $\Rightarrow$ global climate sensitivity $\Rightarrow$ regional climate change scenarios $\Rightarrow$ range of possible impacts

Figure 2.4 Explosion of uncertainty.
Source: Modified after R.N. Jones, Climatic Change 45, 403–419, 2000, and the 'cascading pyramid of uncertainties' in S.H. Schneider, in Social Science Research and Climatic Change: An Interdisciplinary Appraisal, ed. R.S. Chen et al., 9–15, 1983.

with scenarios, however, then the problem still remains of how decision makers should weigh climate risks against other pressing social issues competing for limited resources that could be directed towards a host of social needs.

Various classification schemes have been generated to categorize different types of uncertainties prevalent in scientific assessment (e.g. [79], [20], [66], [39], [56], [11], [34]). In the discussions among authors in the AR4, one classification scheme for uncertainties includes the following categories: lack of scientific knowledge, natural randomness, social choice, and value diversity [23].

The plethora of uncertainties inherent in climate change projections clearly makes risk assessment difficult. In this connection, some fear that actions to control potential risks could produce unnecessary loss of development progress, especially if impacts turned out to be on the benign side of the range. This can be restated in terms of Type I and Type II errors. If governments were to apply the precautionary principle and act now to mitigate risks of climate change, they would be said to be committing a Type I error if their worries about climate change proved unfounded and anthropogenic greenhouse gas emissions did not greatly modify the climate and lead to dangerous change. A Type II error would be committed if serious climate change did occur, yet insufficient hedging actions had been taken as a precaution because uncertainty surrounding the climate change projections was used as a reason to delay policy until the science was 'more certain'.

Researchers, understandably, often are wary of Type I errors, as they are the ones making the projections and do not like to be responsible for actions that turn out to be unnecessary. Decision-makers, and arguably most individuals, on the other hand, might be more worried that dangerous outcomes could be initiated on their watch (Type II error), and thus may prefer some hedging strategies. Most individuals and firms buy insurance, clearly a

Type II error mitigation strategy. Determining levels of climate change that, if reached, would constitute Type II errors can provide decision makers with guidance on setting policy goals and avoiding both Type I and Type II errors. However, as there will almost never (freezing point of water being an obvious exception) be near certainty regarding specific thresholds for specific dangerous climate impacts, such assessment must involve probabilistic analyses of future climate change. With or without information on such thresholds, whether Type I or Type II errors become more likely (i.e., whether we choose to be risk-averse) is necessarily a function of the policymaking process.

2.8 Vulnerability Measurements

The climate science community has been asked to provide decision makers with information that may help them avoid Type II errors (e.g. avoid DAI). In the ongoing AR4 discussions mentioned above, one way to attempt this is through studies providing quantitative measures of key vulnerabilities. In contemplating quantitative values for *human* vulnerabilities, studies have addressed monetary loss [42, 43, 16, 28] and a wide range of population-related metrics, including loss of life [77], risk of hunger as measured by the number of people who earn enough to buy sufficient cereal grains [50], risk of water shortage as measured by annual per capita water availability [3], mean number of people vulnerable to coastal flooding [40], number of people prone to malaria infection or death [69, 71] and number of people forced to migrate as a result of climate change [9].

Non-human quantitative analyses have also been performed. These have calculated potential numbers of species lost [68], numbers of species shifting their ranges [48, 55] and absolute or relative change in range of species or habitat type. Leemans and Eickhout (2004) [30] note that after 1–2°C of warming most species, ecosystems, and landscapes have limited capacity to adapt. Rates of climate change also influence adaptive capacity of social and (especially) natural systems.

Another quantitative measure of vulnerability is the five numeraires, discussed above, as it encompasses both human and non-human metrics of impacts. Each numeraire may be reported separately, or they can be aggregated. Any aggregation should be accompanied by a 'traceable account' of how it was obtained [37].

2.9 Thresholds

Another important step toward achieving the goal of informing decision-makers is identifying climate thresholds or limits. One classification scheme lists three categories of threshold relevant in the context of Article 2 of the UNFCCC: systemic (natural) thresholds, normative (social) impact thresholds, and legal limits. A systemic threshold is a point at which 'the relationship between one or more forcing variables and a valued system property becomes highly negative or nonlinear' [23]. Normative thresholds have been divided into two categories by Patwardhan et al. (2003) [51]. Type I normative thresholds are 'target values of linear or other "smooth" changes that after some point would lead to damages that might be considered "unacceptable" by particular policy-makers' [51]. Type II normative thresholds are 'linked directly to the key intrinsic processes of the climate system itself (often nonlinear) and might be related to maintaining stability of those processes or some of the elements of the climate system' [51]. Examples are presented in Table 2.1 below. Legal limits are policy constraints like environmental

standards placed upon certain factors that are thought to play a part in unfavorable outcomes. They can be influenced by normative thresholds, as well as cost and other factors. [Please note, Types I & II 'thresholds' are not the same as Types I & II 'errors' referred to above.]

Extensive literature relating to Type II thresholds, also referred to as Geophysical and Biological Thresholds, has arisen in recent years. The literature has attempted to incorporate Type II thresholds into integrated assessment and decision-making, both on global scales (e.g. [1], [6], [78], [62], [21], [8], [61]) and on regional scales (e.g. [53]). The next step involves associating specific climate parameters with thresholds. For example, O'Neill and Oppenheimer (2002) [44] have given values of carbon dioxide concentration and global temperature change that they believe may be associated with Type II thresholds corresponding to the disintegration of the West Antarctic Ice Sheet (WAIS), collapse of thermohaline circulation, and widespread decline of coral reefs. Oppenheimer and Alley (2004) [46] also proposed a range of threshold values for disintegration of the WAIS, and Hansen (2004) [17] and Oppenheimer and Alley (2005) [45] discuss quantification of thresholds for loss of WAIS and Greenland ice sheets. Due to large uncertainties in models and in the interpretation of paleoclimatic evidence, a critical issue in all of the above studies is whether the values selected correspond to well-established geophysical or biological thresholds or simply represent best available, subjective judgments about levels or risk.

Type I thresholds, perhaps more accurately called socioeconomic limits, generally do not involve the large-scale discontinuities implied in the word 'threshold', with an exception being the collapse of an atoll society due to climate-change-induced sea level rise [9]. Again, there is

Table 2.1 Proposed numerical values of 'Dangerous Anthropogenic Interference'.

Vulnerability	Global Mean Limit	References
Shutdown of thermohaline circulation	3°C in 100 yr 700 ppm CO_2	O'Neill and Oppenheimer (2002) [44] Keller et al. (2004) [28]
Disintegration of West Antarctic Ice Sheet (WAIS)	2°C, 450 ppm CO_2 2–4°C, <550 ppm CO_2	O'Neill and Oppenheimer (2002) [44] Oppenheimer and Alley (2004, 2005) [45, 46]
Disintegration of Greenland ice sheet	1°C	Hansen (2004) [17]
Widespread bleaching of coral reefs	>1°C	Smith et al. (2001) [67] O'Neill and Oppenheimer (2002) [44]
Broad ecosystem impacts with limited adaptive capacity (many examples)	1–2°C	Leemans and Eickhout (2004) [30], Hare (2003) [19], Smith et al. (2001) [67]
Large increase of persons-at-risk of water shortage in vulnerable regions	450–650 ppm	Parry et al. (2001) [49]
Increasingly adverse impacts, most economic sectors	>3–4°C	Hitz and Smith (2004) [22]

Source: Oppenheimer and Petsonk, 2005 [47].

extensive literature on Type I thresholds. Many studies view climate change impacts in terms of changes in the size of vulnerable populations, typically as a result of climate-change-induced food shortages, water shortages, malaria infection, and coastal flooding (e.g. [4], [5], [49], [50]).

We present a simple example as another approach to the problem of joint probability of temperature rise to 2100 and the possibility of crossing 'dangerous' warming thresholds. Instead of using two probability distributions, an analyst could pick a high, medium, and low range for each factor. For example, a glance at the cumulative probability density function of Andronova and Schlesinger (2001) [2] – included in Figure 2.5, below – shows that the 10th percentile value for climate sensitivity is $1.1°C$ for a doubling of CO_2. $1.1°C$ is, of course, below the $1.5°C$ lower limit of the IPCC's estimate of climate sensitivity and the temperature projection for 2100. But this 10th percentile value merely means that there is a 10% chance that the climate sensitivity will be $1.1°C$ or less, i.e. a 90% chance climate sensitivity will be $1.1°C$ or higher. The 50th percentile result, i.e. the value that climate sensitivity is as likely to be above as below, is $2.0°C$. The 90th percentile value is $6.8°C$, meaning there is a 90% chance climate sensitivity is $6.8°C$ or less, but there

is still a very uncomfortable 10% chance it is even higher than $6.8°C$ – a value well above the 'top' figure in the IPCC range for climate sensitivity ($4.5°C$).

Using these three values ($6.8°C$, $2.0°C$, and $1.1°C$) for high, medium, and low climate sensitivity can produce three alternative projections of temperature over time (using a simple mixed-layer climate model), once an emissions scenario is given. In the example below, these three climate sensitivities are combined with two of the SRES storylines: the fossil-fuel intensive scenario (A1FI) and the high-technology scenario (A1T), where development and deployment of advanced lower carbon-emitting technologies dramatically reduces long-term emissions. These make a good comparison pair since they almost bracket the high and low ends of the six SRES representative scenarios' range of cumulative emissions to 2100. Further, since both are for the 'A1 world', the only major difference between the two is the technology component – an aspect decision-makers have the capacity to influence via policies and other measures. Therefore, asking how different the projected climate change to 2100 is for the two different scenarios is a very instructive exercise in exploring in a partial way the likelihood of crossing 'dangerous' warming thresholds. Of course, as has been emphasized often by us (e.g. see [35] and [36]), the quantitative results of this highly-aggregated, simple model are not intended to be taken literally but, rather, the results can be used to compare the relative temperature projections using different climate sensitivities and thus the framework is intended to be taken seriously.

We will use a conservative (high) estimate of $3.5°C$ above 2000 levels for this 'dangerous' threshold since $3.5°C$ was the highest number projected for the 2100 temperature rise in the IPCC's Second Assessment Report (SAR) and because the IPCC Working Group II TAR suggested that after 'a few degrees', many serious climate change impacts could be anticipated. However, $3.5°C$ is a very conservative number, since the IPCC noted that some 'unique and valuable' systems could be lost at warmings any higher than $1–1.5°C$. In essence, the 'threshold' for what is 'dangerous' depends not only on the probabilities of factors like climate sensitivity and adaptive capacity, but on value judgments as to what is acceptable given any specific level of warming or damage – and who suffers the damage or pays the adaptation costs. Figure 2.6, below, presents the results.

The most striking feature of both Figures 2.6A and 2.6B (A is for the A1FI scenario and B the A1T) is the top 90th percentile line, which rises very steeply above the other two lines below it. This is because of the peculiar shape of the assumed probability density function for climate sensitivity in the cumulative probability density function – it has a long tail to the right due to the possibility that aerosols have been holding back not-yet-realized heating of the climate system.

This simple pair of Figures shows via a small number of curves the amount of temperature change over time for

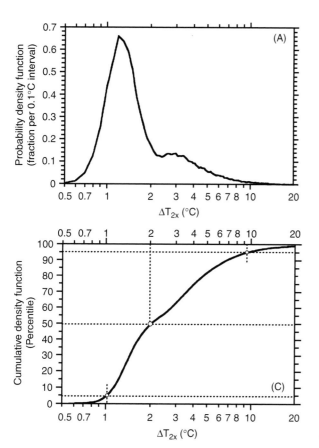

Figure 2.5 Probability density function (A) and cumulative density function (C).
Source: Andronova and Schlesinger, 2001.

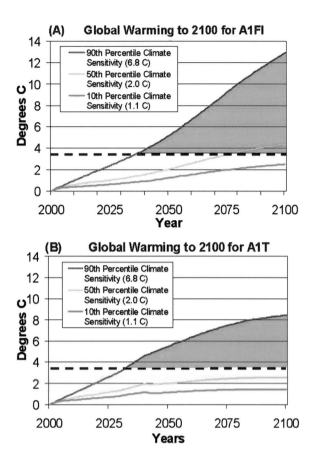

Figure 2.6 Three climate sensitivities and two scenarios.
Source: Unpublished research, posted only on Stephen
Schneider's Web site, http://stephenschneider.stanford.edu

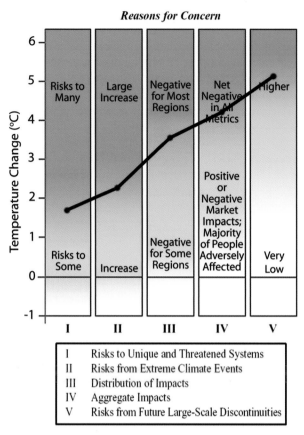

Figure 2.7 An adaptation of the IPCC (2001b) 'Reasons for
Concern' figure from [36], with the thresholds used to
generate their CDF for DAI (black line). The IPCC figure
conceptualizes five reasons for concern, mapped against
global temperature increase. As temperature increases, colors
become redder, indicating increasingly widespread and/or
more severe negative impacts.
Source: Mastrandrea and Schneider, 2004.

three climate sensitivity probabilities (10th, 50th, and
90th percentile). However, it does not give *probabilities*
for the emissions scenarios themselves; only two are
used to 'bracket' uncertainty, and, thus, no joint probability
can be gleaned from this exercise. The problem with
this is that the likelihood of threshold-crossing occurrences
is quite sensitive to the particular selection of scenarios
and climate sensitivities used. This adds urgency
to assessing the relative likelihood of each such entry
(scenario and sensitivity) so that the joint distribution has
a meaning consistent with the underlying probabilistic
assessment of the components. Arbitrary selection of
scenarios or sensitivities will produce conclusions that
could easily be misinterpreted by integrated assessors and
policymakers as containing expert subjective probabilistic
analysis when, in fact, they do not until a judgment is formally
made about the likelihood of each storyline or
sensitivity.

Such joint probability analyses are the next step. A group
at MIT has already made an effort at it (see [74]), as have
Wigley (2004) [75], Rahmstorf and Zickfeld (2005) [52],
and Mastrandrea and Schneider (2004) [36]. We will
summarize here Mastrandrea and Schneider (2004) [36],
which estimates the probability of DAI and the influence
of climate policy in reducing the probability of DAI.

2.10 Climate Science and Policy Crossroads

In defining their metric for DAI, Mastrandrea and
Schneider estimate a cumulative density function (CDF)
based on the IPCC's 'burning embers' diagram by marking
each transition-to-red threshold and assuming that the
probability of 'dangerous' change increases cumulatively
at each threshold temperature by a quintile, as shown by
the thick black line in Figure 2.7. This can be used as a
starting point for analyzing 'dangerous' climate change.

From Figure 2.7, Mastrandrea and Schneider identify
2.85°C as their median threshold for 'dangerous' climate
change, which may still be conservative. Mastrandrea and
Schneider apply this median 2.85°C threshold to three
key parameters – climate sensitivity, climate damages,
and the discount rate – all of which carry high degrees of
uncertainty and are crucial factors in determining the
policy implications of global climate change. To perform
these calculations, they use Nordhaus (1994b) [42] DICE
model because it is well known and is a relatively simple

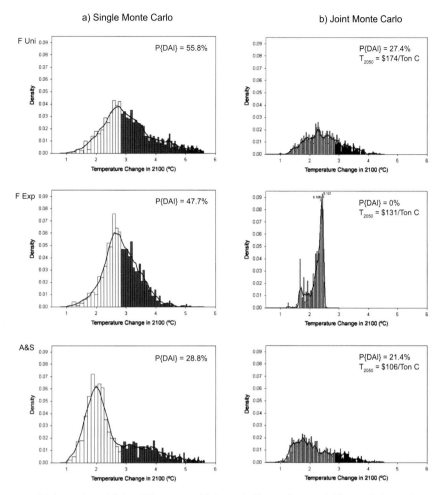

Figure 2.8 Climate sensitivity-only and joint (climate sensitivity and climate damages) Monte Carlo analyses. *Source*: Mastrandrea and Schneider, 2004.

Notes: Panel A displays probability distributions for each climate sensitivity distribution for the climate sensitivity-only Monte Carlo analyses with zero damages. Panel B displays probability distributions for the joint (climate sensitivity and climate damage) Monte Carlo analyses. All distributions indicate a 3-bin running mean and the percentage of outcomes above the median threshold of 2.85°C for 'dangerous' climate change (P{'DAI'}), and the joint distributions display carbon taxes calculated in 2050 (T_{2050}) by the DICE model using the median climate sensitivity from each climate sensitivity distribution and the median climate damage function for the joint Monte Carlo cases. Comparing the joint cases with climate policy controls, b), to the climate sensitivity-only cases with negligible climate policy controls, a), high carbon taxes reduce the potential (significantly in two out of three cases) for DAI. (However, this case uses a PRTP of 0%, implying a discount rate of about 1%. With a 3% PRTP – a discount rate of about 6% – this carbon tax is an order of magnitude less, and the reduction in DAI is on the order of 10%. See the supplementary on-line materials of Mastrandrea and Schneider, 2004 [36] for a full discussion.)

and transparent integrated assessment model (IAM), despite its limitations. Using an IAM allows for exploration of the impacts of a wide range of mitigation levels on the potential for exceeding a policy-relevant threshold such as DAI. Mastrandrea and Schneider focus on two types of model output: (i) global average surface temperature change in 2100, which is used to evaluate the potential for DAI; and (ii) 'optimal' carbon taxes.

They begin with climate sensitivity. The IPCC estimates that climate sensitivity ranges between 1.5°C and 4.5°C but it has not assigned subjective probabilities to the values within or outside of this range, making risk analysis difficult. However, recent studies – many of which have produced climate sensitivity distributions wider than the IPCC's 1.5°C to 4.5°C range, with significant

probability of climate sensitivity above 4.5°C – are now available. Mastrandrea and Schneider use three such probability distributions: the combined distribution from Andronova and Schlesinger (2001) [2], and the expert prior (F Exp) and uniform prior (F Uni) distributions from Forest et al. (2001) [13]. They perform a Monte Carlo analysis sampling from each climate sensitivity probability distribution separately, without applying any mitigation policy, so that all variation in results will be solely from variation in climate sensitivity. The probability distributions they produce show the percentage of outcomes resulting in temperature increases (above current levels) above their 2.85°C 'dangerous' threshold (Figure 2.8A).

Mastrandrea and Schneider's next simulation is a joint Monte Carlo analysis looking at temperature increase in

2100 with climate policy, varying both climate sensitivity *and* the climate damage function, their second parameter (Figure 2.8B). For climate damages, they sample from the distributions of Roughgarden and Schneider (1999) [57], which produce a range of climate damage functions both stronger and weaker than the original DICE function. As shown, aside from the Andronova and Schlesinger climate sensitivity distribution, which gives a lower probability of DAI under the single (climate sensitivity-only) Monte Carlo analysis, the joint runs show lower chances of dangerous climate change as a result of the more stringent climate policy controls generated by the model due to the inclusion of climate damages. Time-varying median carbon taxes are over \$50/Ton C by 2010, and over \$100/Ton C by 2050 in each joint analysis. Low temperature increases and reduced probability of 'DAI' are achieved if carbon taxes are high, but because this analysis only considers one possible threshold for 'DAI' (the median threshold of 2.85°C) and assumes a relatively low discount rate (about 1%), these results cannot fully describe the relationship between climate policy controls and the potential for 'dangerous' climate change. They are given to demonstrate a framework for probabilistic analysis, and, as already emphasized, the highly model-dependent results are not intended to be taken literally.

Because the analysis above only considers Mastrandrea and Schneider's median threshold (DAI[50‰]) of 2.85°C, Mastrandrea and Schneider continue their attempt to characterize the relationship between climate policy controls and the potential for 'dangerous' climate change by carrying out a series of single Monte Carlo analyses varying climate sensitivity and using a *range* of fixed damage functions, rather than just the median case. For each damage function, they perform a Monte Carlo analysis sampling from each of the three climate sensitivity distributions discussed above. They then average the results for each damage function, which gives the probability of DAI at a given 2050 carbon tax under the assumptions described above, as shown in Figure 2.9. Each band in the Figure corresponds to optimization around a different percentile range for the 'dangerous' threshold CDF, with a lower percentile from the CDF representing a lower temperature threshold for DAI. At any DAI threshold, climate policy 'works': higher carbon taxes lower the probability of future temperature increase, and thus reduce the probability of DAI. For example, if climate sensitivity turns out to be on the high end and DAI occurs at a relatively low temperature like 1.476°C (DAI[10‰]), then there is nearly a 100% chance that DAI will occur in the absence of carbon taxes and about an 80% chance it will occur even if carbon taxes were \$400/ton, the top end of Mastrandrea and Schneider's range. If we inspect the median (DAI [50‰]) threshold for DAI (the thicker black line in Figure 2.9), we see that a carbon tax by 2050 of \$150–\$200/Ton C will reduce the probability of 'DAI' to nearly zero, from 45% without climate policy controls (for a 0% pure rate of time preference (PRTP), equivalent to a discount rate of about 1%).

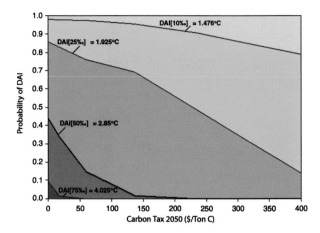

Figure 2.9 Carbon taxes in 2050 and the probability of DAI. *Source*: Mastrandrea and Schneider, 2004.

Notes: Each band represents a different percentile range for the DAI threshold CDF—a lower percentile from the CDF representing a lower temperature threshold for DAI. At any threshold, climate policy controls significantly reduce the probability of DAI. At the median DAI threshold of 2.85°C (the thicker black line above), a 2050 carbon tax of >\$150/Ton C is necessary to virtually eliminate the probability of DAI.

While Mastrandrea and Schneider's results using the DICE model do not provide us with confident quantitative answers, they still demonstrate three very important issues: (1) that DAI can vary significantly, depending on its definition; (2) that parameter uncertainty will be critical for all future climate projections; and (3) most importantly for this volume on the benefits of climate stabilization policies, that climate policy controls (i.e. 'optimal' carbon taxes in this simple framework) can significantly reduce the probability of dangerous anthropogenic interference. This last finding has considerable implications for introducing climate information to policy-makers. We agree with Mastrandrea and Schneider that presenting climate modeling results and arguing for the benefits of climate policy should be framed for decision makers in terms of the potential for climate policy to reduce the likelihood of exceeding a DAI threshold – though we have argued that no such single threshold can be stated independent of the value systems of the stakeholders who name it.

2.11 The Fundamental Value Judgments

Despite the uncertainties surrounding climate change probabilities and consequences, policy-makers must still produce value judgments about what climate change risks to face and what to avoid. They must use all expert information available to decide how to best allocate a pool of limited resources to address avoiding potential DAI versus improving healthcare or reforming education or a host of other worthy causes. It is our personal value judgment that hedging against first-decimal-place odds of DAI is prudent, and we hope that as climate science progresses and more information is available to policy

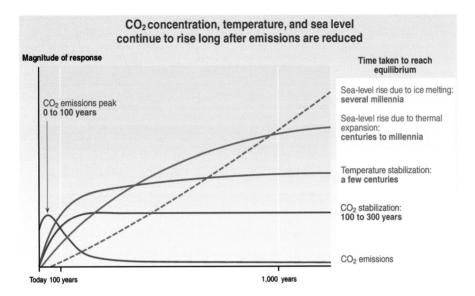

Figure 2.10 Carbon dioxide concentration, temperature, and sea level rise.
Source: IPCC, 2001d.

makers, they will be more willing to risk Type I errors in the climate change arena and will enact effective abatement and adaptation measures. This view is partly supported by Figure 2.10, which suggests that human actions over the next few generations can precondition climatic changes and impacts over the next millennium.

Figure 2.10 shows a 'cartoon' of effects that can play themselves out over a millennium, even for decisions taken within the next century. Such very long-term potential irreversibilities (significant increases in global annual average surface temperature, sea level rise from thermal expansion and melting glaciers, etc.) that the Figure depicts are the kinds of nonlinear events (exceeding Type II thresholds) that would likely qualify as 'dangerous anthropogenic interference with the climate system' [36, 44, 7]. Whether a few dominant countries and/or a few generations of people demanding higher material standards of living and consequently using the atmosphere as an unpriced waste dump to more rapidly achieve such growth-oriented goals is 'ethical' is a value-laden debate that will no doubt heat up as greenhouse gas buildups grow.

REFERENCES

1. Alley, R.B., J. Marotzke, W.D. Nordhaus, J.T. Overpeck, D.M. Peteet, R.A. Pielke Jr., R.T. Pierrehumbert, P.B. Rhines, T.F. Stocker, L.D. Talley, J.M. Wallace, 2003: "Abrupt Climate Change", Science 299(5615): 2005–2010.
2. Andronova, N.G. and M.E. Schlesinger, 2001: "Objective Estimation of the Probability Density Function for Climate Sensitivity", Journal of Geophysical Research 106: 22,605–22,612.
3. Arnell, N.W., 2005: "Climate change impacts on river flows in Britain: the UKCIP02 scenarios", Journal of the Chartered Institution of Water and Environmental Management. In press.
4. Arnell, N.W., 2000: "Thresholds and response to climate change forcing: the water sector", Climatic Change 46: 305–316.

5. Arnell, N.W., R. Nicholls, M.J.L. Livermore, S.R. Kovats, P. Levy, M.L. Parry, and S. Gaffin, 2004: "Climate and socio-economic scenarios for climate change impacts assessments: characterising the SRES storylines", Global Environmental Change 14: 3–20.
6. Azar, C. and K. Lindgren, 2003: "Catastrophic events and stochastic cost-benefit analysis of climate change", Climatic Change 56: 245–255.
7. Azar, C. and H. Rodhe, 1997: "Targets for Stabilization of Atmospheric CO_2", Science 276: 1818–1819.
8. Baranzini, A., M. Chesney, and J. Morisset, 2003: "The impact of possible climate catastrophes on global warming policies", Energy Policy 31(8): 691–701.
9. Barnett, J. and W.M. Adger, 2003: "Climate Dangers and Atoll Countries", Climatic Change 61: 321–337.
10. Dessai, S., W.N. Adger, M. Hulme, J. Köhler, J.P. Turnpenny, and R. Warren, 2004: "Defining and experiencing dangerous climate change", Climatic Change 64(1–2): 11–25.
11. Dessai, S. and M. Hulme, 2003: "Does climate adaptation policy need probabilities?" Climate Policy 4: 107–128.
12. European Climate Forum and Potsdam Institute for Climate Impact Research, 2004: "What is dangerous climate change?" Initial results of a [Oct. 27–30, 2004 Beijing] symposium on key vulnerable regions, climate change, and Article 2 of the UNFCCC. Presented at the 10th Conference of the Parties, Buenos Aires, 14 December. Available online at: http://www.european-climate-forum.net/pdf/ECF_beijing_results.pdf.
13. Forest, C.E., P.H. Stone, A.P. Sokolov, M.R. Allen, and M.D. Webster, 2001: "Quantifying Uncertainties in Climate System Properties with the Use of Recent Climate Observations", Science 295: 113–117.
14. Giles, J., 2002: "When Doubt Is a Sure Thing", Nature 418: 476–478.
15. Grubler, A. and N. Nakicenovic, 2001: "Identifying Dangers in an Uncertain Climate", Nature 412: 15.
16. Hammitt, J.K. and A.I. Shlyakhter, 1999: "The Expected Value of Information and the Probability of Surprise", Risk Analysis 19(1): 135–152.
17. Hansen, J., 2004: "Defusing the global warming time bomb", *Scientific American* 290(3): 68–77.
18. Hansen, J., L. Nazarenko, R. Ruedy, M. Sato, J. Willis, A. Del Genio, D. Koch, A. Lacis, K. Lo, S. Menon, T. Novakov, J. Perlwitz,

G. Russell, G.A. Schmidt, and N. Tausnev, 2005: "Earth's Energy Imbalance: Confirmation and Implications", Science 308: 1431–1435.

19. Hare, W., 2003: "Assessment of knowledge on impacts of climate change-contribution to the specification of Art. 2 of the UNFCCC", background report to the German Advisory Council on Global Change (WBGU) Special Report 94.

20. Helton, J.C. and D.E. Burmaster, 1996: "Treatment of aleatory and epistemic uncertainty in performance assessments for complex systems", Reliability Engineering & System Safety 54: 91–94.

21. Higgins, P.A.T., M.D. Mastrandrea, and S.H. Schneider, 2002: "Dynamics of Climate and Ecosystem Coupling: Abrupt Changes and Multiple Equilibria", Philosophical Transactions of the Royal Society of London (Series B-Biological Sciences) 357: 647–655.

22. Hitz, S. and J. Smith, 2004: "Estimating global impacts from climate change", *Global Environmental Change* 14(3): 201–218.

23. IPCC, 2005: Chapter 19 of Working Group II's contribution to the IPCC Fourth Assessment Report. First Order Draft.

24. IPCC, 2001a: "The Scientific Basis, Summary for Policy Makers – Contribution of Working Group I to the Third Assessment Report of the Intergovermental Panel on Climate Change", ed. J.T. Houghton, Y. Ding, D.J. Griggs, M. Noguer, P.J. van der Linden, X. Dai, K. Maskell, and C.A. Johnson. Cambridge: Cambridge University Press, 881 pp.

25. IPCC, 2001b: Impacts, Adaptation, and Vulnerability – Contribution of Working Group II to the IPCC Third Assessment Report, ed. J.J. McCarthy, O.F. Canziani, N.A. Leary, D.J. Dokken, K.S. White. Cambridge: Cambridge University Press, 1032 pp.

26. IPCC, 2001c: "Synthesis Report 2001- Contribution of Working Groups I, II, and III to the Third Assessment Report of the Intergovernmental Panel on Climate Change", ed. R.T. Watson and the Core Writing Team. Cambridge: Cambridge University Press, 397 pp.

27. IPCC, 1996: "IPCC Second Assessment – Climate Change 1995: Summaries for Policymakers of the three Working Group reports", ed. B. Bolin et al. Cambridge: Cambridge University Press, 73 pp.

28. Keller, K., B.M. Bolker, and D.F. Bradford, 2004: "Uncertain climate thresholds and optimal economic growth", Journal of Environmental Economics and Management 48(1): 723–741.

29. Lane, J., A. Sagar, and S.H. Schneider, 2005: "Equity implications of climate change impacts and policy", Tiempo 55 (April): 9–14.

30. Leemans, R. and B. Eickhout, 2004: "Another reason for concern: regional and global impacts on ecosystems for different levels of climate change", Global Environmental Change (Part A) 14: 219–228.

31. Mann, M.E., R.S. Bradley, and M.K. Hughes, 1999: "Northern Hemisphere Temperatures During the Past Millennium: Inferences, Uncertainties, and Limitations", Geophysical Research Letters 26(6): 759.

32. Mann, M.E., R.S. Bradley, and M.K. Hughes, 1998: "Global-scale temperature patterns and climate forcing over the past six centuries", Nature, 392: 779–787.

33. Mann, M.E. and P.D. Jones, 2003: "Global Surface Temperatures Over the Past Two Millennia", Geophysical Research Letters 30(15): 5-1-5-4.

34. Manning, M. and M. Petit, 2005: "A Concept Paper for the AR4 Cross Cutting Theme: Uncertainties and Risk." Suggestions for IPCC AR4.

35. Mastrandrea, M.D. and S.H. Schneider, 2005: "Probabilistic Integrated Assessment of 'Dangerous' Climate Change and Emissions Scenarios", this volume, Chapter 27, 253–264.

36. Mastrandrea, M.D. and S.H. Schneider, 2004: "Probabilistic Integrated Assessment of 'Dangerous' Climate Change", Science 304: 571–575.

37. Moss, R.H. and S.H. Schneider, 2000: "Uncertainties in the IPCC TAR: Recommendations to Lead Authors for More Consistent Assessment and Reporting", Guidance Papers on the Cross Cutting Issues of the Third Assessment Report of the IPCC, ed. R. Pachauri, T. Taniguchi, and K. Tanaka. Geneva, Switzerland: World Meteorological Organization, 33–51.

38. Nakicenovic, N. et al., 2000: IPCC Special Report on Emissions Scenarios, ed. N. Nakicenovic and R. Swart. Cambridge: Cambridge University Press, pp. 570.

39. New, M. and M. Hulme, 2000: "Representing uncertainty in climate change scenarios: a Monte-Carlo approach", Integrated Assessment 1: 203–213.

40. Nicholls, R.J., 2004: "Coastal flooding and wetland loss in the 21st century: changes under the SRES climate and socio-economic scenarios", Global Environmental Change 14(1): 69–86.

41. Nordhaus, W.D., 1994a: "Expert Opinion on Climatic Change", American Scientist 82: 45–51.

42. Nordhaus, W.D., 1994b: Managing the Global Commons: The Economics of Climate Change. Cambridge, MA: MIT Press, 213 pp.

43. Nordhaus, W.D. and J. Boyer, 2000: Warming the world: Economic models of global warming. Cambridge, MA: MIT Press, 232 pp.

44. O'Neill, B.C. and M. Oppenheimer, 2002 "Climate change – dangerous climate impacts and the Kyoto protocol", Science 296 (5575): 1971–1972.

45. Oppenheimer, M. and R.B. Alley, 2005: "Ice Sheets, Global Warming, and Article 2 of the UNFCCC – An Editorial Essay", Climatic Change 68: 257–267.

46. Oppenheimer, M. and R.B. Alley, 2004: "The West Antarctic Ice Sheet and Long Term Climate Policy", Climatic Change 64: 1–10.

47. Oppenheimer, M. and A. Petsonk, 2005: "Article 2 of the UNFCCC: Historical Origins, Recent Interpretations", Climatic Change, in press.

48. Parmesan, C. and G. Yohe, 2003: "A Globally Coherent Fingerprint of Climate Change Impacts Across Natural Systems", Nature 421: 37–42.

49. Parry, M., N. Arnell, T. McMichael, R. Nicholls, P. Martens, S. Kovats, M. Livermore, C. Rosenzweig, A. Iglesias, and G. Fischer, 2001: "Millions at risk: defining critical climate change threats and targets", Global Environmental Change (Part A) 11(3): 181–183.

50. Parry, M., 2004: "Global impacts of climate change under the SRES scenarios", Global Environmental Change 14(1): 1.

51. Patwardhan, A., S.H. Schneider, and S.M. Semenov, 2003: "Assessing the Science to Address UNFCCC Article 2: A concept paper relating to cross cutting theme number four." Suggestions for IPCC AR4. Available at: http://www.ipcc.ch/activity/cct3.pdf

52. Rahmstorf, S., and K. Zickfeld, 2005: Thermohaline Circulation Changes: A Question of Risk Assessment – An Editorial Review Essay", Climatic Change 68: 241–247.

53. Rial, J.A., R.A. Pielke Sr., M. Beniston, M. Claussen, J. Canadell, P. Cox, H. Held, N. De Noblet-Ducoudré, R. Prinn, J.F. Reynolds, and J.D. Salas, 2004: "Nonlinearities, feedbacks and critical thresholds within the Earth's climate system", Climatic Change 65: 11.

54. Root, T.L., D.P. MacMynowski, M.D. Mastrandrea, and S.H. Schneider, 2005: "Human-Modified Temperatures Induce Species Changes: Combined Attribution", Proceedings of the National Academy of Sciences 102(21): 7465–7469.

55. Root, T.L., J.T. Price, K.R. Hall, S.H. Schneider, C. Rosenzweig, and J.A. Pounds, 2003: "Fingerprints of Global Warming on Animals and Plants", Nature 421: 57–60.

56. Rotmans, J. and M.B.A. van Asselt, 2001: "Uncertainty in integrated assessment modelling: A labyrinthic path", Integrated Assessment 2: 43–55.

57. Roughgarden, T. and S.H. Schneider, 1999: "Climate Change Policy: Quantifying Uncertainties for Damages and Optimal Carbon Taxes", Energy Policy 27: 415–429.

58. Ruddiman, W.F., 2003: "The Anthropogenic Greenhouse Era Began Thousands of Years Ago", Climatic Change 61: 261–293.

59. Santer, B.D., R. Sausen, T.M.L. Wigley, J.S. Boyle, K. AchutaRao, C. Doutriaux, J.E. Hansen, G. A. Meehl, E. Roeckner, R. Ruedy, G. Schmidt, and K.E. Taylor, 2003: "Behavior of Tropopause Height and Atmospheric Temperature in Models, Reanalyses, and Observations: Decadal Changes", Journal of Geophysical Research 108 (D1): 1-1-1-22.

60. Santer, B.D., T.M.L. Wigley, and P.D. Jones, 1993: "Correlation Methods in Fingerprint Detection Studies", Climate Dynamics 8: 265–276.

61. Schneider, S.H., 2004: "Abrupt non-linear climate change, irreversibility and surprise", Global Environmental Change 14: 245–258.

62. Schneider, S.H. and C. Azar, 2001: "Are Uncertainties in Climate and Energy Systems a Justification for Stronger Near-term Mitigation Policies?" Proceedings of the Pew Center Workshop on The Timing of Climate Change Policies, ed. E. Erlich, 85–136. Washington D.C., 11–12 October 2001.

63. Schneider, S.H. and J. Lane, 2005: "Dangers and thresholds in climate change and the implications for justices", in Justice in Adaptation to Climate Change, ed. W.N. Adger. In press.

64. Schneider, S.H., K. Kuntz-Duriseti, and C. Azar, 2000: "Costing Non-linearities, Surprises and Irreversible Events", Pacific and Asian Journal of Energy 10(1): 81–106.

65. Schneider, S.H. and T.L. Root, 2001: "Climate Change: Overview and Implications for Wildlife", in Wildlife Responses to Climate Change: North American Case Studies ed. S.H. Schneider and T.L. Root. Washington D.C.: Island Press, 1–56.

66. Schneider, S.H., B.L. Turner, and H. Morehouse Garriga, 1998: "Imaginable Surprise in Global Change Science", Journal of Risk Research 1(2): 165–185.

67. Smith, J.B. et al., 2001: "Vulnerability to Climate Change and Reasons for Concern: A Synthesis", in Climate Change 2001: Impacts, Adaptation, and Vulnerability, ed. J.J. McCarthy et al. Cambridge: Cambridge University Press, 913–967.

68. Thomas, C.D. et al., 2004: "Extinction risk from climate change", Nature 427: 145–148.

69. Tol, R.S.J. and H. Dowlatabadi, 2002: "Vector-borne diseases, development, and climate change", Integrated Environmental Assessment 2: 173–181.

70. United Nations, 1992: United Nations Framework Convention on Climate Change. Bonn, Germany: United Nations, 33 pp.

71. van Lieshout, M., R.S. Koavates, M.T.J. Livermore, and P. Martens, 2004: "Climate change and malaria: analysis of the SRES climate and socio-economic scenarios", Global Environmental Change 14: 87–99.

72. von Storch, H., E. Zorita, J. Jones, Y. Dimitriev, F. González-Rouco, and S. Tett, 2004: "Reconstructing Past Climate from Noisy Data", Science 306: 679–682.

73. Wahl, E.R. and C. Ammann: "Robustness of the Mann, Bradley, Hughes Reconstruction of Surface Temperatures: Examination of Criticisms Based on the Nature and Processing of Proxy Climate Evidence", Climatic Change, in revision.

74. Webster, M., C. Forest, J. Reilly, M. Babiker, D. Kicklighter, M. Mayer, R. Prinn, M. Sarofim, A. Sokolov, P. Stone, and C. Wang, 2003: "Uncertainty Analysis of Climate Change and Policy Response", Climatic Change 61: 295–320.

75. Wigley, T.M.L., 2004: "Choosing a stabilization target for CO_2", Climatic Change 67: 1–11.

76. Wigley, T.M.L., 2005: "The climate change commitment", Science 307: 1766–1769.

77. World Health Organization, 2003: Climate Change and Human Health – Risks and Responses, ed. A.J. McMichael, D.H. Campbell-Lendrum, C.F. Corvalán, K.L. Ebi, A. Githeko, J.D. Scheraga and A. Woodward. France, 250 pp.

78. Wright, E.L. and J.D. Erickson, 2003: "Incorporating catastrophes into integrated assessment: science, impacts, and adaptation", Climatic Change 57: 265.

79. Wynne, B., 1992: "Uncertainty and environmental learning: Reconceiving science and policy in the preventive paradigm", Global Environmental Change 2: 111–127.

CHAPTER 3

The Antarctic Ice Sheet and Sea Level Rise

Chris Rapley

British Antarctic Survey, High Cross, Cambridge, United Kingdom

ABSTRACT: In its 2001 Third Assessment Report the Intergovernmental Panel on Climate Change (IPCC TAR) concluded that the net contribution of the Antarctic ice sheet to global sea level change would be a modest gain in mass because of greater precipitation. The possibility of a substantial sea level rise due to instability of the West Antarctic Ice Sheet (WAIS) was considered to be very unlikely during the 21st Century. Recent results from satellite altimeters reveal growth of the East Antarctic Ice Sheet north of 81.6 deg S, apparently due to increased precipitation, as predicted. However, a variety of evidence suggests that the issue of the stability of the West Antarctic Ice Sheet should be revisited.

3.1 Antarctica

Antarctica is the fifth largest continent and is the Earth's highest, windiest, coldest, and driest land mass. Its surface is 99.7% covered by a vast ice sheet with an average thickness of $\sim 2\,km$ and a total volume of $\sim 25\,M\,km^3$. The weight of the ice depresses the Earth's crust beneath it by $\sim 0.8\,km$, and, were it to melt, global sea level would rise $\sim 57\,m$.

Two hundred million years ago, Antarctic temperatures were some 20°C warmer than today and the land was vegetated. The Antarctic ice sheet first formed ~ 40 million years ago (Zachos et al., 2001), apparently as a result of a global cooling linked with the shifting arrangement of the continents. The ice sheet became permanent ~ 15 million years ago following the opening of the oceanic gateways that created the circumpolar Southern Ocean. Since that time the Antarctic ice volume has waxed and waned in response to periodic variations in the Earth's orbit. Evidence from marine sediments shows that there have been 46 cycles of growth and decay over the last 2.5 million years. Ice-core data from the last 900,000 years show a periodicity of $\sim 100\,k$ years.

Contemporary snow accumulation over the continent has a (negative) global sea level equivalent (SLE) equal to $\sim 5\,mm/y$. The snowfall is concentrated mainly around the coast, with the Antarctic Peninsula, the region extending northwards towards South America, having the highest accumulation. The ice sheet is dome-shaped, and the central plateau is an extreme desert, with precipitation less than 50 mm/y water-equivalent.

The snow accumulation is offset by ice returned to the ocean. The ice sheet deforms and flows under its own weight, with most of the flow being channeled into ice 'streams', especially at the margin. Thirty-three major basins are drained by ice streams with flow rates that depend on the ice thickness, slope, and the friction at the base. These range from $\sim 10\,m/y$ in the interior to $\sim 1\,km/y$ at the coast. As the ice lifts off the bedrock and begins to float, it displaces a weight of water equal to the part previously above sea level, thereby raising global sea level. The floating ice extends into 'shelves' with thicknesses ranging from hundreds to thousands of metres. The ice shelves fringe approximately 80% of the Antarctic coastline, and the two largest, the Ronne-Filchner and Ross, each exceed the area of France. The ice is ultimately lost through a combination of basal melting and iceberg calving. The former process is highly sensitive to ocean temperature, the latter to air temperature and the occurrence of surface melting, especially if this results in a catastrophic mechanical collapse (as happened to the Larsen B ice shelf in 2002).

Estimates of the mass balance of the ice sheet are derived (i) by aggregating sparse data on input and output and differencing the two, (ii) from measurements of changes in surface topography (and hence ice volume) using data from laser or radar altimeter instruments mounted on aircraft and satellites, or (iii) from estimates of the mass of the ice sheet derived from sensitive spaceborne gravimeters. The mass balance uncertainties are of order $\pm 20\%$, and are complicated by the detailed nature of the observational challenges and differences in behaviour over geographic regions and time.

A particular issue concerns the stability of the West Antarctic Ice Sheet (WAIS). Much of the WAIS rests on bedrock below sea level (as deep as $\sim 2\,km$), with the possibility that a combination of accelerated flow and hydrostatic lift might cause a runaway discharge. Although it contains $\sim 10\%$ of the overall Antarctic ice volume, the WAIS corresponds to only $\sim 7\%$ of the equivalent SLE, or $\sim 5\,m$. This is because much of it is already grounded below sea level. Nevertheless, even a small percentage ice loss would have a significant impact on the millions of people and major infrastructure located on low-lying coastal regions worldwide. Mercer (1978) suggested that the WAIS might collapse as a result of human-induced global warming, a suggestion largely disputed and discounted, based on the results from prevailing glacier models. An issue is whether or not the ice shelves act as buttresses, impeding the flow of the ice streams which feed them. Mercer suggested that a progressive southward wave of ice shelf disintegrations along the coast of the Antarctic Peninsula followed by related glacier accelerations could be a prelude to WAIS collapse.

3.2 The IPCC Third Assessment Report (IPCC TAR)

Based on the evidence available at the time (Church et al., 2001), the IPCC TAR Working Group 1 (WG1) report concluded:

> '... loss of grounded ice (from the WAIS) leading to substantial sea level rise ... is now widely agreed to be very unlikely during the 21st century, although its dynamics are still inadequately understood, especially for projections on longer time-scales.'
>
> (WG1 Technical Summary; p. 74 in Houghton et al., 2001), and

> 'Current ice dynamic models suggest that the West Antarctic ice sheet could contribute up to 3 metres to sea level rise over the next 1,000 years, but such results are strongly dependent on model assumptions regarding climate change scenarios, ice dynamics and other factors.'
>
> (WG1 Summary for Policymakers; p. 17 in Houghton et al., 2001)

More generally, the IPCC TAR considered the Antarctic ice sheet overall to be a net minor player in the contemporary 1.8 mm/y mean sea level rise, and in its projections for accelerated rise over the next century. It stated: *'The Antarctic ice sheet is likely to gain mass because of greater precipitation ...'* (WG1 Technical Summary; p. 74 in Houghton et al., 2001), and it estimated the magnitude of the contribution in the period 1990 to 2100 to be −0.17 m to +0.02 m relative to a total projected rise of 0.11 to 0.77 m. We could characterise the IPCC view of the Antarctic as a 'slumbering giant'.

3.3 Results since the IPCC 2001 Assessment

Since the publication of the IPCC TAR, a number of important new results have been reported:

(i) Bamber et al. (2000) used satellite synthetic aperture radar (SAR) data to reveal that the complex network of ice stream tributaries extends much deeper into the interior of the Antarctic ice sheet, with consequences for the modelled or estimated response time of the ice sheet to climate forcing.

(ii) Shepherd et al. (2001) using satellite altimeter data detected significant thinning of the Pine Island Glacier in the Amundsen Sea Embayment (ASE) of West Antarctica which could only be accounted for by accelerated flow. They pointed out the relevance to the issue of WAIS stability.

(iii) Bamber and Rignot (2002) analysed surface velocities of the Pine Island and Thwaites glaciers derived from satellite-born interferometric synthetic aperture radar data and concluded that the Thwaites glacier had recently undergone a substantial change in its flow regime.

(iv) Joughin and Tulaczyk (2002) used satellite synthetic aperture radar (SAR) data to demonstrate an overall slowing down and thickening of the WAIS ice streams feeding the Ross ice shelf.

(v) Rignot and Thomas (2002) provided a comprehensive review of the mass balance of the Greenland and Antarctic ice sheets and concluded that the WAIS exhibited strong regional differences, but was discharging ice overall. Uncertainties in the data for East Antarctica left them unable to determine the sign of its mass balance. They commented on the rapidity with which substantial changes can occur.

(vi) De Angelis and Skvarca (2003) and Scambos et al. (2004) used satellite imagery to show that the collapse of ice shelves on the eastern Peninsula had resulted in acceleration of the feed glaciers, demonstrating that the ice shelves provided a restraining force as Mercer had speculated.

(vii) Thomas et al. (2004) used aircraft and satellite laser altimeter data to provide a comprehensive summary of the state of discharge from the Pine Island, Thwaites and Smith glaciers of the ASE. They showed that glacier thinning rates near the coast of the ASE in 2002–2003 were much larger than observed during the 1990s, revealing a substantial imbalance and an estimated 0.24 mm/y contribution to sea level rise.

(viii) Cook et al. (2005) used over 200 historical aerial photographs dating from 1940 to 2001 and more than 100 satellite images from the 1960s onwards to show that, of 244 glaciers on the Antarctic Peninsula, 87% have retreated over the past 61 years, and that the pattern of retreat has moved steadily southward over that period. They noted the likely connection between this behaviour and the strong warming trend seen in the Peninsula surface air temperature data.

(ix) Davis et al. (2005) show that radar altimetry measurements indicate that the East Antarctic Ice Sheet interior north of 81.6 deg S increased in mass by 45 ± 7 billion tons per year between 1992 and 2003. Comparisons with contemporaneous meteorological model snowfall estimates suggest that the gain in mass was associated with increased precipitation. A gain of this magnitude is enough to slow sea level rise by 0.12 ± 0.02 mm/y. They note that: '*Although both observations are consistent with the IPCC prediction for Antarctica's likely response to a warming climate ... the results have only sparse coverage of the coastal areas where recent dynamic changes may be occurring. Thus the overall contribution of the Antarctic Ice Sheet to global sea level change will depend on the balance between mass changes on the interior and those in coastal areas.*'

3.4 Summary

(a) The East Antarctic Ice Sheet is growing, apparently as a result of increased precipitation, as predicted by the IPCC TAR.
(b) The Antarctic ice in the Peninsula is responding strongly to the regional climatic warming.
(c) The extension of ice stream tributaries deep into the ice sheet interior might allow for more rapid drainage than had previously been appreciated.
(d) The disintegration of ice shelves can result in a significant acceleration of the feed glaciers, although it is not known yet whether this can be sustained.
(e) The Amundsen Sea Embayment region of the WAIS is exhibiting strong discharge, which, if sustained over the long-term, could result in a greater contribution to sea level rise than accounted for in the IPCC projections.

These new insights suggest that the issue of the contribution of Antarctica to global sea level rise needs to be reassessed. We could characterise the situation as 'giant awakened?'

Since relevant observational data remain sparse and since even the best numerical models of the ice sheet are unable simultaneously to represent the known retreat since the end of the last ice age and its current behaviour, it is recommended that an intensive programme of internationally coordinated research focussed on the issue should be carried out. This should exploit the opportunities provided by existing space initiatives such as NASA's ICESat and the European Space Agency's CryoSat satellite (due for launch in October 2005), the ongoing relevant national and international research programmes, and especially research activities being planned under the auspices of the International Polar Year 2007–2008 (Rapley et al., 2004). A good start has been made by joint NASA/Chilean flights out of Punta Arenas in 2002 and 2004, which showed that many of the Amundsen Sea glaciers flow over deeper bedrock than earlier thought, and that recent thinning rates are larger than those based on earlier measurements. Also relevant is joint fieldwork carried out in the 1995 field season by the British Antarctic Survey and University of Texas. This work, made possible by major US logistics, acquired 100,000 km of flight lines of radio echo sounding data

covering approximately 30% of the WAIS centred over the area that is currently active. Once analysed, the data will provide valuable new knowledge about the internal and basal state and basal topography of the WAIS, which should allow important progress on the issue of its stability.

In the meantime, the question of what would constitute a dangerous level of climatic change as regards the contribution of Antarctica to global mean sea level remains unknown.

REFERENCES

Bamber, J.L., Vaughan, D.G. and Joughin, I., 2000: Widespread complex flow in the interior of the Antarctic Ice Sheet. *Science*, **287**, 1248–1250.

Bamber, J. and Rignot, E., 2002: Unsteady flow inferred for Thwaites Glacier and comparison with Pine Island Glacier, West Antarctica. *Journal of Glaciology*, **48**, 161, 237–246.

Church, J.A., Gregory, J. M., Huybrechts, P., Kuhn, M., Lambeck, K., Nhuan, M.T., Qin, D. and Woodworth, P.L., 2001: Changes in Sea Level. In, Houghton, J.T., Ding, Y., Griggs, D.J., Noguer, M., van den Linden, P.J., Dai, X., Maskell, K. and Johnson, C.A. (eds.), *Climate Change 2001: The Scientific Basis*. Cambridge University Press, Cambridge, UK, 583–638.

Cook, A.J., Fox, A.J., Vaughan, D.G. and Ferrigno, J.G., 2005: Retreating glacier fronts on the Antarctic Peninsula over the past half-century. *Science*, **308**, 541–544.

Davis, C.H.Y., Li, Y., McConnell, J.R., Frey, M.M. and Hanna, E., 2005: Snowfall-driven growth in East Antarctic Ice Sheet mitigates recent sea-level rise. *Science*, **308**, 1898–1901.

De Angelis, H. and Skvarca, P., 2003: Glacier surge after ice shelf collapse. *Science*, **299**, 1560–1562.

Houghton, J.T., Ding, Y., Griggs, D. J., Noguer, M., van den Linden, P. J., Dai, X., Maskell, K. and Johnson, C. A. (eds.), *Climate Change 2001: The Scientific Basis*. Cambridge University Press, Cambridge, UK, pp. 881.

Joughin, I. and Tulaczyk, S., 2002: Positive mass balance of the Ross Ice streams, West Antarctica. *Science*, **295**, 476–480.

Mercer, J.H., 1978: West Antarctic ice sheet and CO_2 greenhouse effect: a threat of disaster. *Nature*, **271**, 321–325.

Rapley, C., Bell, R., Allison, I., Bindschadler, R., Casassa, G., Chown, S., Duhaime, G., Kotlyakov, V., Kuhn, M., Orheim, O., Pandey, P., Petersen, H., Schalke, H., Janoscheck, W., Sarukhanian, E. and Zhang, Z. 2004: A Framework for the International Polar Year 2007–2008. International Council of Science.

Rignot, E. and Thomas, R.H., 2002: Mass balance of polar ice sheets. *Science*, **297**, 1502–1506.

Scambos, T.A., Bohlander, J.A., Shuman, C.A. and Skvarca, P., 2004: Glacier acceleration and thinning after ice shelf collapse in the Larsen B embayment, Antarctica. *Geophysical Research Letters*, **31**, L18402, doi: 10.1029/2004GL0260670.

Shepherd, A., Wingham, D.J., Mansley, J.A.D., and Corr, H.F.J., 2001: Inland thinning of Pine Island Glacier, West Antarctica. *Science*, **291**, 862–864.

Thomas, R., Rignot, E., Casassa, G., Kanagaratnam, P., Acuna, C., Akins, T., Brecher, H., Frederick, E., Gogineni, P., Krabill, W., Manizade, S., Ramamoorthy, H., Rivera, A., Russell, R., Sonntag, J., Swift, R., Yungel, J. and Zwally, J., 2004: Accelerated sea-level Rise from West Antarctica. *Science*, **306**, 255–258.

Zachos, J., Pagani, M., Sloan, L., Thomas, E. and Billups, K., 2001: Trends, rhythms, and aberrations in global climate 65 Ma to present. *Science*, **292**, 686–693.

CHAPTER 4

The Role of Sea-Level Rise and the Greenland Ice Sheet in Dangerous Climate Change: Implications for the Stabilisation of Climate

Jason A. Lowe[1], Jonathan M. Gregory[1,2], Jeff Ridley[1], Philippe Huybrechts[3,4], Robert J. Nicholls[5] and Matthew Collins[1]

[1] *Hadley Centre for Climate Prediction and Research, Met Office, Exeter, UK*
[2] *Centre for Global atmospheric Modelling, Department of Meteorology, University of Reading, UK*
[3] *Alfred-Wegener-Institut für Polar- und Meeresforschung, Bremerhaven, Germany*
[4] *Department Geografie, Vrije Unversiteit Brussel, Belgium*
[5] *School of Civil Engineering and the Environment, University of Southampton, UK*

ABSTRACT: Sea level rise is an important aspect of future climate change because, without upgraded coastal defences, it is likely to lead to significant impacts. Here we report on two aspects of sea-level rise that have implications for the avoidance of dangerous climate change and stabilisation of climate.

If the Greenland ice sheet were to melt it would raise global sea levels by around 7 m. We discuss the likelihood of such an event occurring in the coming centuries. The results suggest that complete or partial deglaciation of Greenland may be triggered for even quite modest stabilisation targets. We also examine the time scales associated with sea-level rise and demonstrate that long after atmospheric greenhouse gas concentrations or global temperature have been stabilised coastal impacts may still be increasing.

4.1 Introduction

Sea level is reported to have risen during the 20th century by between 1 and 2 mm per year and model predictions suggest the rise in global-mean sea level during the 21st century is likely to be in the range of 9–88 cm (Church et al., 2001). It is also well known that there has been considerable growth in coastal populations and the value of assets within the coastal zone during the 20th century, and this may continue in the future. Consequently, there is a concern that future increases in sea level will lead to sizeable coastal impacts (Watson et al., 2001). The issue of sea-level rise in dangerous climate change has also recently been discussed by Oppenheimer and Alley (2004) and Hansen (2005).

The main causes of increased global average sea level during the 21st century are likely to be thermal expansion of the ocean, melting of small glaciers, and the melting of the Greenland and Antarctic ice sheets (Church et al., 2001). Thermal expansion and the melting of small glaciers are expected to dominate, with Greenland contributing a small but positive sea-level rise, which may be partly offset by a small and negative contribution from Antarctica. This negative contribution results from an increase in precipitation over Antarctica, which is assumed to more than offset small increases in melting during the 21st century. With further warming the Antarctic ice sheet is likely to provide a positive sea-level rise contribution, especially if the West Antarctic Ice Sheet (WAIS) becomes unstable. Beyond the 21st century the changes in the ice sheets and thermal expansion are expected to be make the largest contributions to increased sea level.

In this work we concentrate on two issues associated with sea-level rise. First, how likely is it that the Greenland ice sheet will undergo complete or significant partial deglaciation during the coming centuries, thus providing a large additional sea-level rise? Second, what are the time scales of sea-level rise, especially those associated with thermal expansion and Greenland deglaciation, and what are the consequences of the time scales for mankind?

4.2 Models and Climate Change Scenarios

Results are presented from a range of physical models, including: simple climate models; complex climate models with detailed representation of the atmosphere, ocean and land surface; and a high-resolution model of the Greenland ice sheet.

A small number of long simulations have been performed with the coupled ocean-atmosphere general circulation climate model, HadCM3. This is a non flux-adjusted coupled model with an atmospheric resolution of $2.5° \times 3.75°$ and 19 levels in the atmosphere. The ocean is a 20 level rigid-lid model with a horizontal resolution of $1.25° \times 1.25°$ and 20 levels. More details of the model and its parameterisations are given by Pope et al. (2000) and Gordon et al. (2000).

Recently, we used this model to simulate around 1000 years for an experiment in which atmospheric carbon dioxide concentration was increased from a pre-industrial level of approximately 285 ppm at 2% compound per annum, then stabilised after 70 years at four times the pre-industrial value for the remainder of the simulation. An increase in atmospheric carbon dioxide to four times pre-industrial atmospheric carbon dioxide corresponds to a radiative forcing of around $7.5\,\mathrm{Wm^{-2}}$, which is comparable to the $6.7\,\mathrm{Wm^{-2}}$ increase in forcing between years 2000 and 2100 for the SRES A2 scenario and $7.8\,\mathrm{Wm^{-2}}$ for SRES A1FI (IPCC 2001, Appendix 2). In a second simulation, HadCM3 was coupled to a 20 km resolution dynamic ice sheet model (Ridley et al., 2005; Huybrechts et al., 1991) and used to simulate more than 3000 years of ice sheet evolution. Importantly, the coupling method allowed changes in climate to influence the evolution of the ice sheet and changes in the ice sheet to feedback on the climate, affecting its subsequent evolution.

We have also made a number of additional simulations using a large number of slightly different but plausible versions of HadCM3. These models used a simplified slab ocean, which responds to radiative forcing changes much faster than the ocean in the fully coupled model, allowing estimates of equilibrium response to be made relatively quickly. For this work we used an ensemble of 129 simulations in which atmospheric carbon dioxide levels were first prescribed at pre-industrial levels ($1\times\mathrm{CO_2}$) and then doubled ($2\times\mathrm{CO_2}$). In both the $1\times\mathrm{CO_2}$ and $2\times\mathrm{CO_2}$ phases the simulations were run until they first reached an equilibrium and then for a further 20 years.

Like other models, the Hadley Centre climate model contains a number of parameters that may be modified within a sensible range. In this work, there is one ensemble member in which model parameters and parameterisation schemes take their standard values (Pope et al., 2000), with the exception of the use of a prognostic sulphur cycle model component. In the remaining 128 ensemble members, perturbations were made simultaneously to these standard values for a range of important model parameters. The choice of parameters perturbed and the effects of perturbations on global mean equilibrium climate sensitivity are described in Murphy et al. (2004) and Stainforth et al. (2005).

The precise algorithm for generating the perturbations is complex but, briefly, the ensemble was designed on the basis of linear statistical modelling to produce a range of different magnitude climate sensitivities while maximising the chance of high-fidelity model base climates and exploring as much of the model parameter space as possible. More details are given in Webb et al. (2005), together with an assessment of cloud feedbacks in the ensemble. A method for producing probability density functions of future climate change predictions is to first run the ensemble of simulations to generate a frequency distribution and second to give a relative weight to each ensemble member based on some assessment of its 'skill' in simulating the forecast variable of interest. The details of the correct way of doing this are still subject to considerable debate and

require much further work, particularly when addressing the question of regional climate change as we do here. We therefore limit ourselves to the production of frequency distributions. The consequences of this for the use of these results in a formal risk assessment are discussed in Section 3. A further limitation is that our model ensemble is based on a single climate model and we have not attempted to account for results from other climate models. However, we do note that the range of climate sensitivities produced by the 129 member ensemble are not inconsistent with those published in other studies (e.g. Frame et al., 2005) which tend to use simple models and a range of different observational constraints.

Finally, we have used simple model formulations in which both temperature change and sea-level rise are represented using Green's functions. The Green's functions are taken as the sum of two exponential modes derived from the 1000 year HadCM3 stabilisation experiment without an ice sheet. Predictions were made with the simple model by convolving either the temperature Green's function or sea-level rise Green's function with an estimate of the radiative forcing. These simple models have only been used here to extend more complex Hadley Centre model results further into the future or to scale to alternative emissions scenarios.

4.3 Likelihood of a Deglaciation of Greenland

If the Greenland ice sheet were to melt completely, it would raise global average sea level by around 7 m (Church et al., 2001). Without upgraded sea defences this would inundate many cities around the world. There are also concerns that the fresh water from Greenland could help trigger a slow-down or collapse of the ocean thermohaline circulation[1] (Fichefet et al., 2003). This could lead to a significant cooling over much of the northern hemisphere (Vellinga and Wood, 2002).

The Greenland ice sheet can only persist if the loss of ice by ablation and iceberg discharge is balanced by accumulation. Under present day conditions the two loss terms are each roughly half the accumulation. If the accumulation were greater than the sum of the loss terms then the ice sheet would grow. However, in a warmer climate it is expected that the increase in ablation will outweigh the increase of accumulation. Under these circumstances, the ice sheet will shrink. For a small warming, the ice sheet could still evolve towards a new equilibrium by reducing its rate of iceberg calving and/or obtaining a different geometry that reduces ablation sufficiently to counterbalance the initial increase of the surface melting. However, as reported in the IPCC's third assessment report (Church et al., 2001), based on Huybrechts et al. (1991; see also Oerlemans, 1991; Van de

---

[1] The ocean thermohaline circulation plays a role in the transport of large amounts of heat from the tropics to high latitudes.

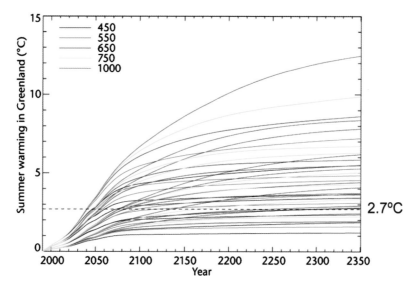

Figure 4.1 Predicted warming for various CO_2 stabilisation levels (purple, 450 ppm; light blue, 550 ppm; green, 650 ppm; yellow, 750 ppm; red, 1,000 ppm). Scenarios involving higher carbon dioxide concentrations stabilize later. The threshold for deglaciation is shown as a dotted line.

Wal and Oerlemans, 1994), for a mean temperature rise of 2.7°C the ablation is predicted to increase beyond the accumulation. Since the ice sheet can not have a negative discharge, this represents the temperature above which the ice sheet can no longer be sustained and will retreat in-land, even if the calving rate were to be reduced to zero.

Alternative thresholds could have been defined, such as the temperature rise leading to a particular loss of Greenland ice by a particular time. Huybrechts and De Wolde (1999) showed that for a local Greenland temperature rise of 3°C the ice sheet would lose mass equivalent to around 1 m of global mean sea-level rise over 1000 years and that the rate of sea-level rise at the end of the 1000-year simulation remained sizeable. In their 5.5°C warming scenario the sea-level rise contribution from Greenland over 1000 years was around 3 m. Thus, we believe that above the chosen temperature threshold a significant Greenland ice loss will occur, although we acknowledge that for warming that is close to the threshold the warming may either not lead to complete deglaciation or that a complete deglaciation may take much longer than a millennium. In Ridley et al., (2005) and Section 4 of this article the ice loss for a high forcing scenario is reported.

Gregory et al. (2004) used the simple MAGICC climate model (Wigley and Raper, 2001), with a range of climate sensitivity and heat uptake parameters to look at the warming over Greenland for a range of greenhouse gas emission scenarios that lead to stabilisation of atmospheric carbon dioxide at levels between 450 ppm and 1000 ppm. The emissions of other greenhouse gas species followed the SRES A1B scenario up to 2100 and were then stabilised. The climate model parameters and the relationship between global mean warming and local

warming over Greenland were estimated from the more complex models used in the IPCC third assessment (Church et al., 2001). When annual mean warming was considered, all but one of the model simulations led to a warming above the 2.7°C threshold by approximately 2200. When uncertainty in the threshold and only summer seasonal warming were considered, 69% of the model versions led to the threshold being exceeded before 2350 (Figure 4.1). This use of summer only warming is more appropriate because little melting occurs during the cold winter months.

We have recently attempted to re-examine this issue using the 'perturbed parameter ensemble' of Hadley Centre complex climate models (described in Section 2). For each ensemble member the carbon dioxide stabilisation level that would lead to a Greenland temperature rise equal to the threshold for deglaciation is estimated, assuming a logarithmic relationship between stabilisation carbon dioxide concentration and equilibrium temperature increases. We also make the assumption that the ratio of the summer warming over Greenland to global mean warming and the climate sensitivity will remain constant for a given model over a range of climate forcing and temperature rise.

The orange curve in Figure 4.2 shows a smoothed frequency distribution of the stabilisation carbon dioxide levels that lead to a local Greenland warming of 2.7°C and, thus, a complete or partial Greenland deglaciation being triggered. The red and green curves are the carbon dioxide stabilisation levels that would lead to warmings of 2.2°C and 3.2°C respectively, which represents uncertainty in the value of the deglaciation threshold. The vertical bars show the raw data to which the orange curve was fitted. The results suggest that even if carbon dioxide levels are stabilised below 442 ppm to 465 ppm then 5%

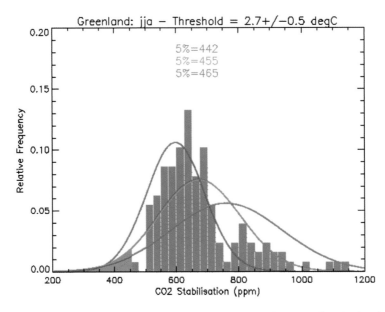

Figure 4.2 Predicted CO_2 stabilisation levels that lead to the local Greenland warming exceeding the threshold for deglaciation, and ±0.5°C of this amount. The raw results are shown as bars for the central threshold case and the curves are a fit to the raw results.

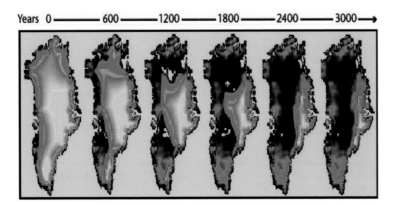

Figure 4.3 Predicted change in the ice sheet volume following a quadrupling of atmospheric CO_2. Red and yellow indicate thick ice while green and blue indicate thin (or no) ice.

of our plausible model simulations will still lead to a complete or partial deglaciation. A stabilisation level of 675 ppm would lead to 50% of our model versions exceeding 2.7°C. At this level, however, the uncertainty in the value of the threshold becomes more important and, when this is taken into account, the carbon dioxide concentration level that leads to 50% of the model version reaching the deglaciation threshold varies between 600 ppm to 750 ppm.

It is important to emphasize that because the 'perturbed parameter ensemble' technique is still in its infancy and we have not attempted to apply a weighting to the frequency distribution of carbon dioxide stabilisation levels, so this result can not be taken as a formal probability density function or definitive estimate of the risk of collapse. Rather, we have used the ensemble to illustrate the method whereby such a risk may be estimated. To that end, our results are likely to be a credible first attempt at linking the collapse of the Greenland ice

sheet to a particular stabilisation level using a perturbed parameter approach with complex climate models.

4.4 Timescales of sea level response

Having established that even for quite modest carbon dioxide stabilisation levels the Greenland ice sheet might become deglaciated, we now discuss the time scales over which this might occur. For a pessimistic, but plausible, scenario in which atmospheric carbon dioxide concentrations were stabilised at four times pre-industrial levels (Section 2) a coupled climate model and ice sheet model simulation predicts that the ice sheet would almost totally disappear over a period of 3,000 years, with more than half of the ice volume being lost during the first millennium (Figure 4.3). The peak rate of simulated sea-level rise was around 5 mm/year and occurred early in the simulation. These results are discussed more fully by Ridley

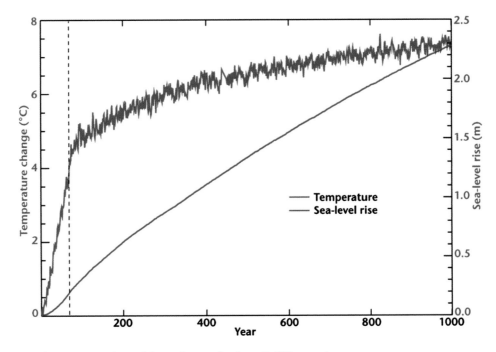

Figure 4.4 Simulated temperature rise and thermal expansion for a 4×CO$_2$ experiment.

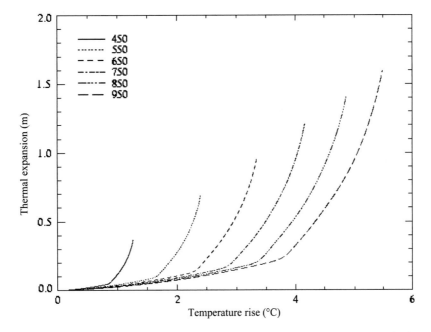

Figure 4.5 Simulated temperature rise and thermal expansion for a range of stabilisation levels. The stabilisation of atmospheric carbon dioxide takes place 70 years into the experiment following a linear increase.

et al. (2005) who also note that in the Hadley Centre climate model, the freshwater provided by the melting of Greenland ice had a small but noticeable effect on the model's ocean circulation, temporarily reducing the thermohaline circulation by a few per cent. However, this was not enough to lead to widespread northern hemisphere cooling.

A further issue associated with the loss of ice from Greenland is that of reversibility. If the climate forcing were returned to pre industrial levels once the ice sheet

had become totally or partially ablated could the ice sheet eventually reform? If not, when would the point of no return be reached? The studies of Lunt et al. (2004) and Toniazzo et al. (2004) offer conflicting evidence on whether a fully-ablated ice sheet could reform, and this is an active area of current research.

In the parallel HadCM3 experiment without an ice sheet the thermal expansion was estimated and also found to make a considerable sea-level rise contribution over millennial time scales (Figure 4.4). The timescale associated

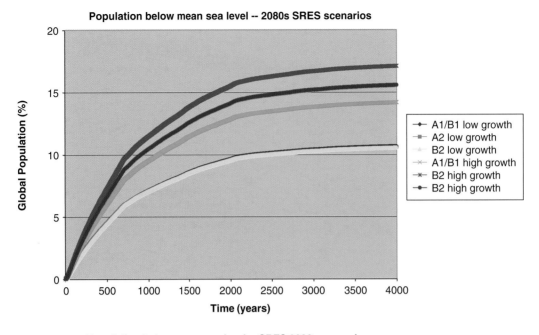

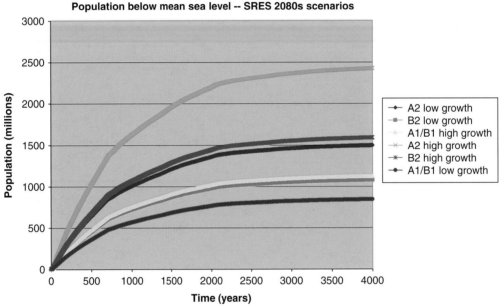

Figure 4.6 (a) Exposed population and (b) percentage of world population exposed to Greenland deglaciation and the thermal expansion from a stabilisation level of four time pre-industrial values.

with the thermal expansion component of sea-level rise depends strongly on the rate at which heat can be transported from near the surface into the deep ocean. The thermal expansion response time in the Hadley Centre coupled climate model was found to be greater than 1000 years, which is much longer than the time needed to stabilise temperature (the global average surface temperature rise for the same experiment is also shown in Figure 4.4). Using the simple Green's function model formulations for thermal expansion and temperature rise, tuned to the HadCM3 results, we have constructed a set of curves showing the time dependent relationship between the two quantities for a range of different carbon dioxide stabilisation

levels. These curves were generated for scenarios in which the carbon dioxide was increased linearly over 100 years then fixed at the stabilisation levels.

Figure 4.5 shows that during the period of rapidly-increasing carbon dioxide concentration, the sea-level rise and temperature both increase and there is an approximately linear relationship between them. However, once the carbon dioxide concentration has stabilised, the differing time scales affecting surface temperature and sea-level rise become important and the gradient of the curves increases significantly.

Taken together, the Greenland deglaciation and the thermal expansion results show that sea level is likely to continue rising long after stabilisation of atmospheric carbon dioxide, agreeing with earlier studies, such as Wigley (1995). Changes in the WAIS are also likely to provide an important contribution to future multi-century increases in sea level. However, we can not yet comment with any degree of confidence on the time scales of Antarctic ice sheet collapse. A review of expert opinions (Vaughan and Spouge, 2002) suggested this is not thought likely to occur in the next 100 years, although recent work (Rapley, this volume) suggests the Antarctic ice sheet may make a sizeable contribution to sea-level rise earlier than previously thought.

4.5 Consequences of these Results for Mankind

A detailed assessment of impacts is beyond the scope of this paper. However, it is instructive to add together the Greenland and thermal expansion sea-level rise estimates and consider the potential exposure of people to this rise. The thermal expansion estimate for the first 1000 years is from HadCM3 but this is extended using the simple Green's function climate model formulation. The exposure is based on population estimates for the 2080s, when they are expected to have increased significantly compared to the present situation. The base data comes from the CIESIN PLACE database (http://:www.ceisin.org), and this is transformed using the SRES scenarios, including different growth rates for coastal areas (see Nicholls, 2004).

Figure 4.6 shows the population that is exposed based on absolute numbers and as a proportion of the global population estimates in the 2080s. While this is translating changes over 4000 years, the potential scale of impacts is evident. Within 500 years, the exposed population could be in the range of 300–1000 million people, rising to 800 to 2400 million people at the end of the simulation. This is 10–17% of the world's population, and represents the number of people who would need to be protected or relocated. Nicholls and Lowe (this volume) have extended the calculation to include a contribution from the WAIS but acknowledge that this term is likely to be even more uncertain than the contribution from Greenland.

4.6 Conclusions

Simulations of the Greenland ice sheet and ocean thermal expansion have highlighted several issues that are relevant to the stabilisation of climate at a level that would avoid dangerous changes. In particular:

● Complete or partial deglaciation of Greenland may be triggered for even quite modest stabilisation targets.
● Sea level is likely to continue rising for more than 1000 years after greenhouse gas concentrations have been stabilised, so that with even a sizeable mitigation effort adaptation is also likely to be needed.

We are currently addressing the question of whether the Greenland deglaciation is irreversible or whether, if greenhouse gas concentrations were reduced, the ice sheet could be regrown. If it can recover, we also need to establish the greenhouse gas levels that would permit this to occur. Finally, we note that there is a large uncertainty on sea-level rise predictions, especially those made for times beyond the 21st century.

REFERENCES

Church, J.A., Gregory, J.M., Huybrechts, P., Kuhn, M., Lambeck, K., Nhuan, M.T., Qin, D. and Woodworth, P.L. Changes in Sea Level. In: Houghton, J.T., Ding, Y., Griggs, D.J., Noguer, M., van der Linden, P.J. and Xiaosu, D. (eds.) Climate Change 2001. The Scientific Basis. Cambridge University Press, Cambridge, 639–693, 2001.

Fichefet, T., Poncin, C., Goosse, H., Huybrechts, P., Janssens, I., Le Treut, H. Implications of changes in freshwater flux from the Greenland ice sheet for the climate of the 21st century. Geophysical Research Letters, 30(17), 1911, doi: 10.1029/2003GL017826, 2003.

Frame, D.J., Booth, B.B.B., Kettleborough, J.A., Stainforth, D.A., Gregory, J.M., Collins, M. and Allen, M.R. Constraining climate forecasts. Geophysical Research Letter, 32(9), L09702, doi: 10.1029/2004GL022241, 2005.

Gordon, C., Cooper, C., Senior, C.A., Banks, H., Gregory, J.M., Johns, T.C., Mitchell, J.F.B. and Wood, R.A. The simulation of SST, sea ice extents and ocean heat transports in a version of the Hadley Centre coupled model without flux adjustments. Climate Dynamics, 16, 147–168, 2000.

Gregory, J.M., Huybrechts, P. and Raper, S.C.B. Threatened loss of the Greenland ice-sheet. Nature 428, 616, 2004.

Hansen, J.E.A. slippery slope: How much global warming constitutes "dangerous anthropogenic interference"? Climatic Change, 68, 269–279, 2005.

Huybrechts, P., Letreguilly, A. and Reeh, N. The Greenland Ice Sheet and greenhouse warming. Paleogeography, Paleoclimatology, Paleoecology (Global and Planetary Change Section), 89, 399–412, 1991.

Huybrechts, P. and de Wolde, J. The dynamic response of the Greenland and Antarctic ice sheets to multiple-century climatic warming. Journal of Climate, 12, 2169–2188, 1999.

IPCC. 'Climate Change 2001: The Scientific Basis. Contribution of Working Group I to the Third Assessment Report of the Intergovernmental Panel on Climate Change', Houghton, J.T., Ding, Y., Griggs, D.J., van der Linden, P.J., Dai, X., Maskell, K. and Johnston, C.A. (eds.), Cambridge University Press, Cambridge, UK and New York, NY, USA, pp. 881, 2001.

Murphy, J.M., Sexton, D.M.H., Barnett, D.N., Jones, G.S., Webb, M.J., Collins, M. and Stainforth, D.A. Quantification of modelling uncertainties in a large ensemble of climate change simulations. Nature, 430, 768–772, 2004.

Nicholls, R.J. Coastal flooding and wetland loss in the 21st Century: Changes under the SRES climate and socio-economic scenarios. Global Environmental Change, 14, 69–86, 2004.

Nicholls, R.J. and Lowe, J.A. Climate Stabilisation and Impacts of Sea-Level Rise. This volume.

Oerlemans, J. The mass balance of the Greenland ice sheet: sensitivity to climate change as revealed by energy balance modelling. Holocene, 1, 40–49, 1991.

Oppenheimer, M. and Alley, R.B. Ice sheets, global warming, and Article 2 of the UNFCCC. Climatic Change, 68, 257–267, 2005.

Pope, V.D., Gallani, M.L., Rowntree, P.R. and Stratton, R.A. The impact of new physical parameterisations in the Hadley Centre climate model – HadAM3, Climate Dynamics, 16, 123–146, 2000.

Rapley, C. The Antarctic ice sheet and sea level rise. This volume.

Ridley, J.K., Huybrechts, P., Gregory, J.M. and Lowe, J.A. Elimination of the Greenland ice sheet in a high CO_2 climate. Journal of Climate, 18, 3409–3427, 2005.

Stainforth, D.A., Aina, T., Christensen, C., et al. Uncertainty in predictions of the climate response to rising levels of greenhouse gases. Nature, 433, 403–406, 2005.

Van de Wal, R.S.W. and Oerlemans, J. An energy balance model for the Greenland ice sheet. Glob. Planetary Change, 9, 115–131, 1994.

Vaughan D.G. and Spouge, J.R. Risk estimation of collapse of the West Antarctic Ice Sheet. Climatic Change 52, 65–91, 2002.

Vellinga M. and Wood R.A. Global climatic impacts of a collapse of the Atlantic thermohaline circulation. Climatic Change 54, 251–267, 2002.

Watson, R.T. and the Core Writing Team (eds.). Climate Change 2001: Synthesis Report, Cambridge University Press, Cambridge, pp. 396, 2001.

Webb M.J., Senior, C.A., Williams, K.D., et al. On the contribution of local feedback mechanisms to the range of climate sensitivity in two GCM ensembles. Submitted to Climate Dynamics.

Wigley, T.M.L. and Raper, S.C.B. Interpretation of high projections for global-mean warming. Science 293, 451–454, 2001.

Wigley, T.M.L. Global-mean temperature and sea level consequences of greenhouse gas concentration stabilization. Geophysical Research Letters, 22, 45–48, 1995.

CHAPTER 5

Assessing the Risk of a Collapse of the Atlantic Thermohaline Circulation

Michael E. Schlesinger[1], Jianjun Yin[2], Gary Yohe[3], Natalia G. Andronova[4], Sergey Malyshev[5] and Bin Li [1]

[1]Climate Research Group, Department of Atmospheric Sciences, University of Illinois at Urbana-Champaign
[2]Program in Atmospheric & Oceanic Sciences, Princeton University
[3]Department of Economics, Wesleyan University
[4]Department of Atmospheric, Oceanic and Space Sciences, University of Michigan
[5]Department of Ecology and Evolutionary Biology, Princeton University

ABSTRACT: In this paper we summarize work performed by the Climate Research Group within the Department of Atmospheric Sciences at the University of Illinois at Urbana-Champaign (UIUC) and colleagues on simulating and understanding the Atlantic thermohaline circulation (ATHC). We have used our uncoupled ocean general circulation model (OGCM) and our coupled atmosphere-ocean general circulation model (AOGCM) to simulate the present-day ATHC and how it would behave in response to the addition of freshwater to the North Atlantic Ocean. We have found that the ATHC shuts down 'irreversibly' in the uncoupled OGCM but 'reversibly' in the coupled AOGCM. This different behavior of the ATHC results from different feedback processes operating in the uncoupled OGCM and AOGCM. We have represented this wide range of behaviour of the ATHC with an extended, but somewhat simplified, version of the original model that gave rise to the concern about the ATHC shutdown. We have used this simple model of the ATHC together with the DICE-99 integrated assessment model to estimate the likelihood of an ATHC shutdown between now and 2205, both without and with the policy intervention of a carbon tax on fossil fuels. For specific subjective distributions of three critical variables in the simple model, we find that there is a greater than 50% likelihood of an ATHC collapse, absent any climate policy. This likelihood can be reduced by the policy intervention, but it still exceeds 25% even with maximal policy intervention. It would therefore seem that the risk of an ATHC collapse is unacceptably large and that measures over and above the policy intervention of a carbon tax should be given serious consideration.

5.1 Introduction

The Atlantic thermohaline circulation (ATHC) is driven by temperature (thermo) and salt (haline) forcing over the ocean surface (Stommel, 1961). The ATHC currently transports poleward about 1 petawatt (10^{15} W) of heat, that is, a million billion Watts. Since human civilization currently uses 10 terawatts of energy (10^{13} W), the heat transported by the ATHC could run 100 Earth civilizations. Conversely, 1% of the heat transported by the ATHC could supply all of humanity's current energy use. As a result of this enormous northward heat transport, Europe is up to 8°C warmer than other longitudes at its latitude, with the largest effect in winter. It is this comparatively mild European climate, as well as the inter-related climates elsewhere, that has given concern about the possible effect of a collapse of the ATHC, in terms of political and economic instability (Gagosian, 2003, Schwartz and Randall, 2003) and the onset of an ice age (Emmerich, 2004). Public concern has also been expressed in the novel 'Forty Signs of Rain' (Robinson, 2004) – the first book in a trilogy about a human-induced 'stall' of the ATHC – with an opposing view expressed in the novel 'State of Fear' (Crichton, 2004).

Why would the ATHC collapse? There are two threads of evidence that suggest this possibility. One is based on modeling and the other is drawn from paleoclimate evidence. The first model of the ATHC was developed by Henry Stommel (1961), which is the simplest possible model to study the dynamical behavior of the ATHC. In this very simple model, heat and salt are transported from an equatorial box to a polar box, with each box taken to have its own temperature and salinity. The direction of the net transport is the same regardless of whether the circulation is clockwise (viewed from Europe toward North America) as for the present-day ATHC configuration or counterclockwise – a reversed ATHC. Many years later Barry Saltzman (2002) simplified the model to consider only salt transport. He took the temperature difference between the boxes as being constant and extended the model to include salt transport by the non-THC motions in the ocean – the wind-driven gyre circulation and eddies akin to weather disturbances in the atmosphere.

As freshwater is added to the polar box in the Stommel-Saltzman (S-S) model the ATHC intensity weakens because the density of the polar box decreases, leading to a reduction in the density differential between the equatorial box and the polar box. As increasing amounts of freshwater are added, the intensity continues to decrease, but only to a point. At this threshold or bifurcation point, this continuous behavior ceases and is replaced by a non-linear abrupt change to a counterclockwise reversed ATHC (RTHC). Further addition of freshwater enhances the intensity of this RTHC. More importantly, a reduction of the freshwater addition does not cause the circulation to return to the bifurcation point from which it came. Rather, it weakens the RTHC. Eventually, if the freshwater addition is reduced sufficiently, another bifurcation point is reached such that the ATHC abruptly restarts. This irreversible behavior of the ATHC in the S-S model results in hysteresis – a change in the system from one stable equilibrium to another and then back along a different path.

Why should there be an additional freshwater addition to the North Atlantic Ocean? The surface air temperature of central Greenland has been reconstructed as a function of time from about 15,000 years ago to the present based on the isotopic composition of an ice core that was drilled in the Greenland ice sheet (Alley et al., 1993, Taylor et al., 1997, Alley, 2000). The reconstruction shows a rise in surface air temperature at the end of the last Ice Age nearly 15,000 years ago followed by a return to Ice Age conditions thereafter for about 2000 years. During this episode, an Arctic plant called Dryas *Octopetala* arrived in Europe, hence the appellation Younger Dryas. Additional evidence that the Younger Dryas was global in extent has been provided by terrestrial pollen records, glacial-geological data, marine sediments, and corals (e.g. Chinzei et al., 1987, Atkinson et al., 1987, Alley, 2000, McManus, 2004). This evidence of abrupt cooling in the North Atlantic and Europe has been taken as being due to a slowdown or collapse of the ATHC. This ATHC slowdown/shutdown appears to have occurred as the meltwater stored in Lake Agassiz from the retreating Laurentide ice sheet on North America, which had previously flowed to the Gulf of Mexico via the Mississippi River, instead flowed out either the St. Lawrence waterway to the North Atlantic Ocean (Johnson and McClue, 1976, Rooth, 1982, Broecker, 1985, Broecker et al., 1988, Broecker et al., 1989, Broecker, 1997, Alley, 1998, Teller et al., 2002, Broecker, 2003, Nesje et al., 2004, McManus et al., 2004) or to the Arctic Ocean via the Mackenzie River and then to the North Atlantic Ocean (Tarasov and Peltier, 2005), thereby freshening it sufficiently to slow down or halt the ATHC.

So the ATHC has apparently slowed or shut down in the past. Might it do so in the future as a result of global warming? The ATHC intensity simulated by 9 AOGCMs for a scenario of future IS92a greenhouse gas emissions (IS92a, Leggett et al., 1992) slows down for all models but one (Cubasch et al., 2001, Figure 9.21). As the world warms, both precipitation (P) and evaporation (E) increase over the North Atlantic, but the difference (P − E) also increases there. Freshwater is thereby added to the ocean. Both the surface ocean freshening and warming reduces the density of the surface water and thus its ability to sink (Manabe and Stouffer, 1994).

In the AOGCM simulations of a greenhouse-gas (GHG)-induced slowdown or shutdown of the ATHC, the resulting climate change is due to both the increased concentrations of GHGs and to the ATHC change. However, the magnitude of GHG-induced climate change required to slowdown

or shutdown the ATHC is highly uncertain. Thus, it is desirable to separate the ATHC-induced climate change from the GHG-induced climate change so that they can subsequently be combined to address a series of critical questions. Suppose the ATHC begins to slowdown for a change of global-mean surface air temperature of x°C due to increased concentrations of GHGs: (1) What would the resulting climate changes look like? (2) What would the impacts of those changes look like? and (3) What near-term policies are robust against the uncertainty of an ATHC slowdown/shutdown (Lempert and Schlesinger, 2000)?

We began a program of research in 1999 that would allow us to answer the first of these questions by simulating the slowdown and shutdown of the ATHC using our AOGCM. We performed our ATHC-shutdown simulations first with our uncoupled ocean GCM (OGCM) and then with it coupled to our atmospheric GCM. Like all other simple models (Rahmstorf, 1995, Ganopolski and Rahmstorf, 2001, Schmittner and Weaver, 2001, Titz et al., 2002, Prange et al., 2002, Schmittner et al., 2002, Rahmstorf, 1995) beginning with that of Stommel (1961), the OGCM simulated an irreversible ATHC shutdown. By way of contrast, though, the AOGCM simulated a reversible ATHC shutdown, as found by all AOGCMs (Schiller et al., 1997, Manabe and Stouffer, 1999, Rind et al., 2001, Vellinga et al., 2002) other than by Manabe and Stouffer (1988). Below we describe this finding, comparing for the first time a single uncoupled and coupled OGCM, and note that the S-S model can reproduce not only the irreversible ATHC shutdown, but also the reversible ATHC shutdown. We shall also discuss some of the climate changes induced by the ATHC collapse simulated by our AOGCM. Subsequently, we will use the S-S model with wide-ranging behavior to examine how to reduce the risk of an ATHC collapse.

5.2 Simulations of the ATHC Shutdown with the UIUC OGCM and AOGCM

The zonally integrated meridional circulation in the Atlantic Ocean simulated by the UIUC coupled atmosphere/ocean general circulation model (AOGCM) in its control simulation for present-day conditions is shown in Figure 5.1. The ocean currents simulated by the AOGCM in the upper (0–1000 m) and deep (1000–3000 m) Atlantic Ocean are shown in Figure 5.2. A longitude-depth cross-section of currents at 30°N and 50°N is shown in Figure 5.3.

Below we describe the freshwater perturbation experiments that we have performed with our OGCM and AOGCM, discuss the climate changes induced by a collapse of the ATHC, and describe how the S-S model is capable of simulating a range of ATHC shutdown behavior, from an irreversible collapse to a reversible one.

5.2.1 *Freshwater Perturbation Experiments*

The freshwater perturbation experiments with the uncoupled OGCM were performed by very slowly increasing and then decreasing the external freshwater addition to the North Atlantic between 50°~70°N latitudes (Rahmstorf, 1995). The freshwater perturbation changes at a rate of 0.2 Sv (Sv = $10^6 \, m^3/sec$) per 1000 years. Although the setup of the experiment is a transient run, the ATHC is always in quasi-equilibrium with the external freshwater forcing due to the extremely slow change of the freshwater perturbation flux. To facilitate comparison with the AOGCM simulations, several steady-state runs with fixed freshwater perturbations were also carried out using the uncoupled OGCM.

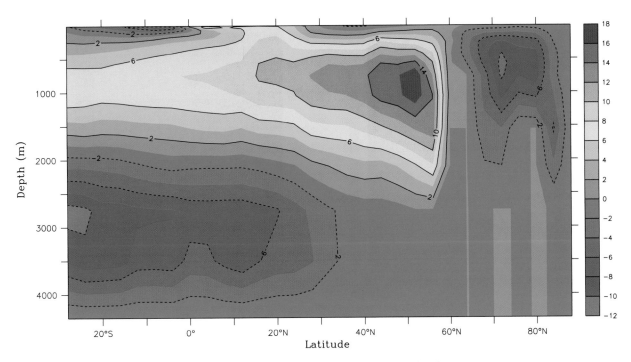

Figure 5.1 Zonally integrated meridional streamfunction simulated by the UIUC AOGCM.

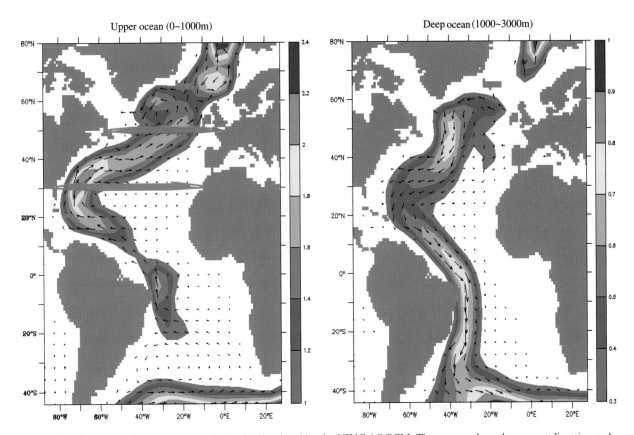

Figure 5.2 Plan view of the ocean currents (cm/s) simulated by the UIUC AOGCM. The vectors show the current direction and the contours indicate the velocity. The arrows in the left panel show the locations of the longitude-depth cross-sections in Fig. 5.3.

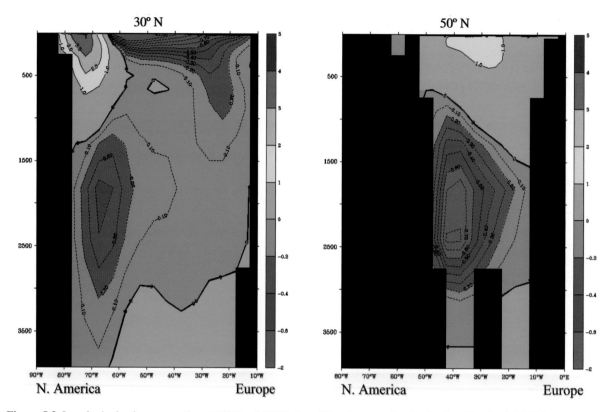

Figure 5.3 Longitude-depth cross-section at 30°N and 50°N of meridional current (cm/s) simulated by the AOGCM.

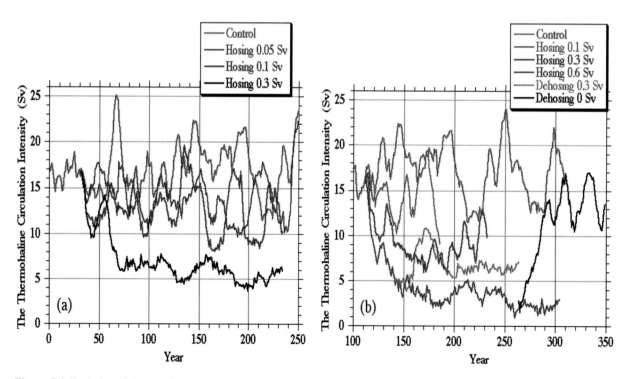

Figure 5.4 Evolution of the meridional mass streamfunction in the AOGCM hosing and dehosing simulations. (a) The experiments starting from the 30th year of the control; (b) the experiments starting from the 110th year of the control.

The set of AOGCM simulations was performed for fixed freshwater addition ('hosing') and removal ('dehosing') rates over the same latitude band in the North Atlantic as for the OGCM-only simulations (Figure 5.4). Two groups of freshwater perturbation experiments were carried out to test the response of the ATHC. The first group included three 'hosing' experiments starting from the 30th year of the control run. Perturbation freshwater fluxes of 0.05, 0.1 and 0.3 Sv were uniformly input into the perturbation region in separate experiments. The 110th year of the control run was chosen as the initial condition for the second group. This group consisted of three 'hosing' experiments (0.1, 0.3 and 0.6 Sv) and two 'dehosing' experiments. The two 'dehosing' experiments started from the shutdown state of the ATHC induced by the 0.6 Sv freshwater addition, and included a moderate reduction of the perturbation flux from 0.6 to 0.3 Sv and the total elimination of the 0.6 Sv freshwater addition.

The strength of the ATHC simulated by the uncoupled OGCM with boundary conditions of prescribed heat and freshwater fluxes from the atmosphere has a pronounced hysteresis loop in which the ATHC, after shutdown, can be restarted only after the freshwater addition is eliminated and changed into a freshwater extraction (Figure 5.5a). Three equilibria of the ATHC coexist under the present-day freshwater forcing. Points a and e correspond to two active ATHC modes, while point c is an inactive ATHC mode. The different intensity between points a and e is caused by the switch-on (point e) and switch-off (point a) of deep convection in the Labrador Sea. Points b and d are thresholds along the hysteresis curves. Beyond these critical points, the ATHC undergoes a rapid transition between the active and inactive modes. All of these features indicate a remarkable nonlinearity of the ATHC in the ocean-only model, which results from the domination by the positive feedbacks in the ATHC system. This irreversibility of the ATHC shutdown, if true, would warrant the use of precaution in formulating climate policy.

In contrast, the strength of the ATHC simulated by the AOGCM does not have a hysteresis loop when the freshwater added to the North Atlantic is increased until shutdown occurs and is then reduced (Figure 5.5b). Instead, once the freshwater addition is reduced from its shutdown value, the ATHC restarts. Furthermore, the relation between the ATHC intensity and the change in freshwater addition is roughly linear throughout the entire range of freshwater addition. Moreover, the freshwater addition required to shut down the ATHC is much larger for the AOGCM than for the uncoupled OGCM.

Why does the ATHC behave differently in the uncoupled OGCM and the AOGCM? Yin (2004) and Yin et al. (2005) investigated this question and found different feedback processes operating in the uncoupled OGCM and AOGCM. After the shutdown of the ATHC, a reversed cell develops in the upper South Atlantic in the uncoupled OGCM. This ATHC reversal cannot occur in the AOGCM simulation. The reversed cell transports a large amount of salt out of the Atlantic basin and facilitates the decrease of the basin-averaged salinity in the Atlantic, thereby stabilizing the 'off' mode of the ATHC in the uncoupled OGCM. In contrast, the salinity increases in the Caribbean in the AOGCM simulation of the ATHC shutdown because the intertropical convergence zone shifts from the Northern Hemisphere into the Southern Hemisphere, thereby decreasing the precipitation over the

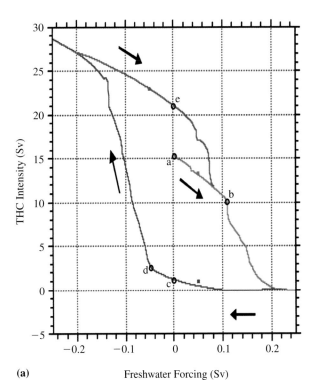

(a)   Freshwater Forcing (Sv)

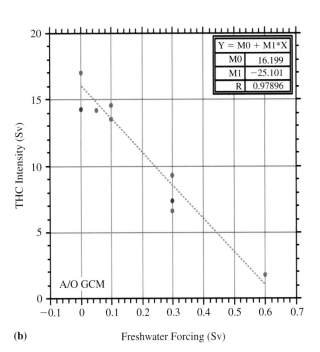

(b)   Freshwater Forcing (Sv)

Figure 5.5 The stability diagrams of the ATHC established by the uncoupled OGCM and the coupled AOGCM. (a) The OGCM with prescribed surface heat and salinity fluxes; (b) The AOGCM (50-year mean). Red, blue and green colors represent the increase in freshwater addition, the subsequent decrease in freshwater addition after the ATHC is shut down, and the following increase in freshwater addition. The origin of the x axis represents the 'present-day' freshwater flux. The rectangles indicate the equilibrium runs with the uncoupled OGCM. The red points in (b) with the same freshwater forcing come from the two simulation groups. The red dashed line is the linear fit based on the red points.

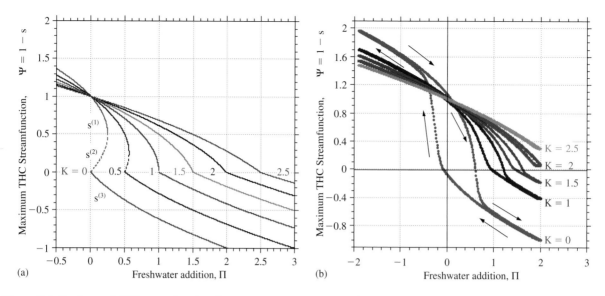

Figure 5.6 Maximum ATHC streamfunction Ψ versus freshwater addition Π in the S-S model for K from 0 to 2.5: (a) equilibrium and (b) hosing-dehosing simulation.

Caribbean. The resulting more-dense salty water is then transported poleward by the gyre circulation in the North Atlantic. This acts as a negative feedback on the ATHC shutdown which works both to make it more difficult to shut down the ATHC – a larger freshwater addition is required than in the uncoupled OGCM – and to help restart the ATHC when the freshwater, which has been added to shut down the ATHC, is reduced. This negative feedback cannot exist in the uncoupled OGCM simulations because of the need therein to prescribe boundary conditions in the atmosphere.

5.2.2 *Climate Changes Induced by an ATHC Shutdown*

In the 0.6 Sv hosing experiment simulated by the AOGCM, the clockwise meridional circulation of the control run is eliminated. A clockwise circulation near 15°N latitude at the surface remains due to the wind-driven upwelling and downwelling. The ocean currents in the upper (0–1000 m) and deep (1000–3000 m) Atlantic Ocean simulated by the AOGCM of the control run both collapse in the 0.6 Sv hosing simulation. The counter-clockwise Antarctic Bottom Water (AABW) circulation centered near 3000 m that is caused by water sinking off the West Antarctic coast is barely influenced by the shutdown of the ATHC in the North Atlantic.

The January and July surface air temperatures resulting from the ATHC shutdown in the 0.6 Sv simulation are lower over the U.S. midwest, Greenland, the North Atlantic Ocean and Europe, with larger cooling in winter than in summer. Interestingly, strong warming occurs over Alaska and the Palmer Peninsula in Northern Hemisphere and Southern Hemisphere winter, respectively. If such a simulated warming were to occur, it would likely harm the Alaskan permafrost and the West Antarctic ice sheet that is grounded on the ocean floor.

5.2.3 *Simulation of the ATHC Shutdown by a Simple Model*

As noted in the Introduction, it was the simple two-box model proposed by Stommel (1961) that raised the first alert that the ATHC could collapse irreversibly if sufficient freshwater were added there to reach its threshold bifurcation point. Here we describe how this model, as generalized by Saltzman (2002), can simulate not only an irreversible ATHC collapse, as obtained by all simple models, but also the reversible ATHC shutdown described above which is obtained by most AOGCMs (Yin, 2004). The calibration of the S-S model is described in the Appendix.

The ATHC simulated by the S-S model exhibits sharply-different behavior for different values of the ratio of the transport coefficient K for the gyre circulation and eddies to that for the ATHC. For K = 0 (the case examined by Stommel (1961)) there is an unstable equilibrium circulation connecting two stable equilibrium circulations; one displays sinking in high latitudes and upwelling in low latitudes while the other moves in the opposite direction (Figure 5.6a). As K increases from zero to unity, the range examined by Saltzman (2002), the region of the unstable equilibrium shrinks. Larger values of freshwater addition are required to weaken the ATHC intensity to any particular value. When K takes the value of unity, the unstable equilibrium circulation disappears, and the two stable equilibrium circulations merge. In this case the flow between the two boxes is the combination of wind-driven flow and ATHC flow. The contribution of the wind-driven flow to the poleward salinity transport is significant. As K is increased above unity – a case examined by Yin (2004) and Yin et al. (2005) – still larger values of freshwater addition are required to weaken the ATHC to any particular intensity, and the discontinuity in slope between the two stable circulations decreases. The curve gradually approaches a straight line with increasing K. In this case, the contribution of the non-THC flow to the mass exchange dominates that of the thermohaline flow.

When the S-S model is run in a hosing–dehosing simulation like that of the OGCM and AOGCM, the result for K = 0 shows the classical hysteresis loop of the Stommel model (Figure 5.6b). Much weaker hysteresis is obtained for K = 1, and it is shifted toward larger values of freshwater addition. As K increases upward from unity the slopes of the two stable modes approach each other and the hysteresis disappears at about K = 2.5. This behavior is quite similar to the transition from the hysteresis loop simulated by the uncoupled OGCM to the single curve simulated by the coupled AOGCM.

5.3 Assessing the Likelihood of a Human-Induced ATHC Collapse

We are now in a position to ask, 'How likely is a collapse of the Atlantic thermohaline circulation?', and if not highly unlikely, 'How can we reduce the risk of an ATHC shutdown?' To show how the significance of these questions might be investigated, and to offer some answers expressed in terms of the relative likelihood of ATHC collapse, we use the S-S model together with a simple Integrated Assessment Model, the Dynamic Integrated Climate Economy (DICE) model. DICE was developed by Bill Nordhaus (1991) to simulate a wide range of possibilities that an assessment of the more complicated process-based models cannot now exclude from the realm of possibility. More specifically, we use DICE-99 (Nordhaus and Boyer, 2001) to drive an ensemble of S-S model simulations across a range of future temperature trajectories that are themselves uncertain, given our current estimates of the range of climate sensitivity.

DICE-99 uses a reduced-form submodel (called by some the IPCC-Bern model) to calculate time-dependent GHG concentrations, radiative forcings, and change in global-mean surface air temperature from a base-case of greenhouse-gas emissions. For the latter, the climate sensitivity – the change in the equilibrium global-mean surface air temperature due to a doubling of the pre-industrial CO_2 concentration, ΔT_{2x} – must be prescribed. For this we use the probability density function (pdf) calculated by Andronova and Schlesinger (2001) from the observed record of surface air temperature from 1856 to 1997, as discretized by Yohe et al. (2004). Because simple climate models have simulated an irreversible ATHC shutdown, akin to K = 0 in the S-S model, while our and other AOGCMs simulate a reversible ATHC shutdown akin to K = 2.5 in the S-S model, we take K in the S-S model to be uncertain with a uniform pdf between these values. To close the problem, we specify the (non-dimensional) amount of freshwater added to the North Atlantic, $\Pi(t)$, as a function of the change in global-mean surface air temperature simulated by DICE-99, $\Delta \overline{T}(t)$.

Results from simulations by our atmospheric GCM coupled to a 60 m deep mixed-layer ocean model for several different radiative forcings (Schlesinger et al., 2000) suggest the linear relationship,

$$\Pi(t) = \alpha[\Delta\overline{T}(t) - \Delta\overline{T}_c(t)]H[\Delta\overline{T}(t) - \Delta\overline{T}_c],$$

where

$$H(x) = \begin{cases} 0 \text{ if } x < 0 \\ 1 \text{ if } x \geq 0 \end{cases}$$

is the Heavyside step function and α is the 'hydraulic sensitivity'. The Heavyside step function is introduced to prevent any freshwater addition until a critical temperature change is reached, $\Delta\overline{T}_c$. As noted in the Appendix, we treat both α and $\Delta\overline{T}_c$ as uncertain independent quantities with uniform pdfs between 0.2 and 1.0 (1/°C) and between 0 and 0.6°C, respectively (Yohe et al., 2005).

The policy instrument within DICE is a tax on the carbon content of fossil fuels, from an initial tax of $10 a ton of carbon (tC) – about 5 cents a gallon of gasoline – to $100 per tC – about 6 pence per liter of petrol. This carbon tax rises through time at the then prevailing interest rate that is determined by the model. The tax can be considered as economic 'shorthand' for a wide range of possible policy interventions such as the Clean Development Mechanism and Joint Implementation.

We now address the question, 'How likely is a collapse of the Atlantic thermohaline circulation?' For the base-case CO_2 emission from 2005 to 2205 and $\Delta T_{2x} = 3°C$, the likelihood of an ATHC shutdown obtained over the uniform probability distributions for K, α and $\Delta\overline{T}_c$ rises monotonically to 4 in 10 in 2100 and 65 in 100 in 2200 (Figure 5.7(d)).

Having found that the collapse of the Atlantic thermohaline circulation is not highly unlikely, we now address the question, 'How can we reduce the risk of an ATHC shutdown?' Policy intervention in the form of a carbon tax (Figure 5.7): (1) reduces CO_2 emissions to zero, earlier the larger the initial tax; (2) causes the CO_2 concentration to peak and then decrease as the carbon sinks begin to dominate the declining CO_2 emissions, earlier the larger the initial tax; and (3) causes the global-mean surface temperature change to peak and then decrease in response to the declining CO_2 concentration, to lower values the larger the initial tax. As a result, mitigation can cause the likelihood of an ATHC shutdown to peak, with lower maximum probabilities (MP) associated with larger initial taxes.

We now consider MP as a function of the initial tax in 2005 (IT) contingent on (Figure 5.8): (a) climate sensitivity, ΔT_{2x}; (b) the critical temperature threshold for the input of freshwater into the North Atlantic, $\Delta\overline{T}_c$; (c) the hydraulic sensitivity, α; and (d) the ratio of the salt transport by the non-THC oceanic motions to that by the ATHC, K. Each of these likelihoods is obtained over the probability distributions of the three non-contingent quantities. For example, for the contingency on ΔT_{2x}, the likelihood is calculated

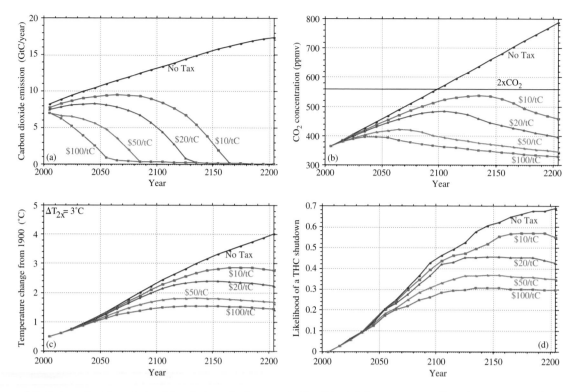

Figure 5.7 Carbon dioxide emission (a) and atmospheric concentration (b), global-mean near-surface air temperature change (c), and the likelihood of an ATHC shutdown (d) versus time for different initial taxes.

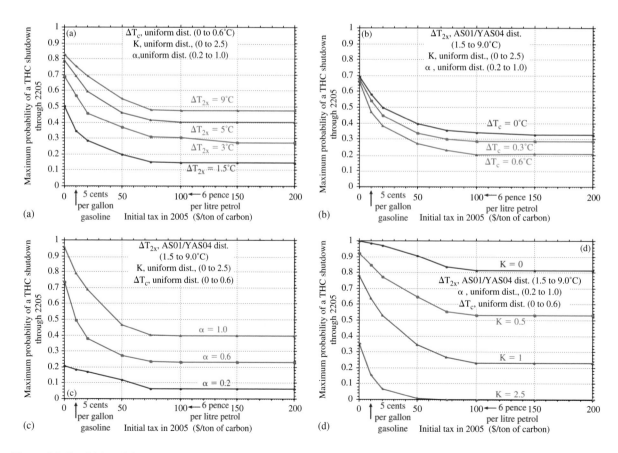

Figure 5.8 Sensitivity of the maximum probability of an ATHC shutdown versus carbon tax to climate sensitivity, ΔT_{2x} (a); threshold temperature, $\Delta \bar{T}_c$ (b); hydraulic sensitivity α (c); and ratio of the non-THC transport of salinity to the ATHC transport, K (d).

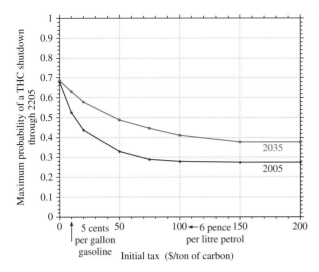

Figure 5.9 Maximum probabilities of a collapse of the ATHC between 2005 and 2205 are plotted against various carbon taxes initiated in either 2005 or 2035. Once they are imposed, the taxes increase over time at the endogenously determined rate of interest derived by DICE-99. The probabilities were computed across a complete sample of scenarios defined by spanning all sources of uncertainty.

over the probability distributions for $\Delta\bar{T}_c$, K and α. It is found that MP decreases with increasing IT, but the rate of decrease slows to zero when IT reaches \$100/tC. Also, the MP for any IT is most sensitive to K; that is, whether the shutdown of the ATHC is irreversible (small K) or reversible (large K). The MP–IT relationship is also sensitive to the uncertainty in hydraulic sensitivity, α, and climate sensitivity, ΔT_{2x}, but less so than to the uncertainty in K. Lastly, the MP–IT relationship is relatively insensitive to the uncertainty in the threshold, $\Delta\bar{T}_c$.

MP as a function of IT beginning in 2005 (Figure 5.9), obtained over the probability distributions of all four uncertain quantities K, α, ΔT_{2x} and $\Delta\bar{T}_c$, is reduced from a 65-in-100 occurrence for no initial tax to a 28-in-100 occurrence for an initial tax of \$100/tC. If the tax were initiated 30 years later in 2035, then the \$100/tC tax would reduce the 65-in-100 likelihood to a 42-in-100 likelihood, and a \$200/tC tax somewhat further to a 38-in-100 occurrence. We also found the expected value of global warming required to shutdown the ATHC is 2.3°C (Figure 5.10).

5.4 Conclusion

We have used, of necessity, very simple models of the Earth's climate system, within DICE-99, and of the Atlantic thermohaline circulation, the S-S model. Note, though, that the latter contains the original Stommel model (for K = 0) that gave rise to the concern about the possible collapse of the ATHC. Accordingly, one should take the quantitative results with caution.

This caution notwithstanding, one cannot but be taken by the finding that in the absence of any policy intervention to

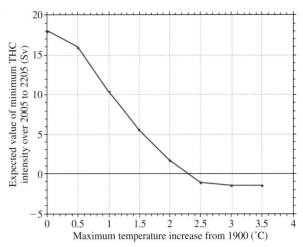

Figure 5.10 Expected value of the minimum ATHC intensity over 2005–2205 versus global-mean temperature increase from 1990.

slow the emission of greenhouse gases, uncertainty in our understanding of ATHC processes supports a greater than 50% likelihood of an Atlantic THC collapse. Furthermore, even with a carbon tax, this uncertainty supports a likelihood of an ATHC collapse in excess of 25%. Such high probabilities are worrisome. Of course, they should be checked by additional modelling studies. Nonetheless, simulations based on simple models do identify major sensitivities and thus provide guidance for these future studies. If further work produces similar results, it would indicate that the risk of an ATHC collapse is unacceptably large. In this case, measures over and above the policy intervention of a carbon tax should be given serious consideration.

Acknowledgements

This material is based upon work supported by the National Science Foundation under Award No. ATM-0084270. Any opinions, findings, and conclusions or recommendations expressed in this publication are those of the authors and do not necessarily reflect the views of the National Science Foundation. The authors express their gratitude to Tom Wigley and two anonymous referees for constructive comments on the earlier draft of this paper. GY also acknowledges the support of B. Belle. Remaining errors, of course, reside with the authors.

APPENDIX

Calibration of the Stommel-Saltzman Model

The governing equation of the Stommel-Saltzman (S-S) 2-box ocean model for nondimensional variables is

$$\frac{ds}{dt^*} = \Pi - |1 - s|s - Ks, \tag{5.1}$$

where s is the difference in salinity between the equatorial and polar boxes, t* is time, Π is the freshwater addition, and K is the ratio of the transport coefficient for the gyre circulation and eddies (denoted k_ϕ) to that for the ATHC (denoted k_ψ). The K term was absent from the original Stommel model and was taken to be as large as unity by Saltzman. The maximum streamfunction of the ATHC is

$$\Psi = k_\psi \mu_T \delta T^* (1 - s), \tag{5.2}$$

where μ_T is the thermal volume expansion coefficient, and δT^* is the temperature difference between the equatorial and polar boxes, taken to be constant.

We calibrated the S-S model so that it is about as sensitive to a freshwater addition as the University of Illinois at Urbana-Champaign (UIUC) coupled atmosphere-ocean general circulation model (AOGCM), which requires a freshwater addition of 0.6 Sv (10^6 m^3/sec) between 50°N to 70°N in the Atlantic to shut down the ATHC [Yin (2004); Yin et al. (2005)]. From Equation (5.2), an ATHC shutdown ($\Psi = 0$) requires s = 1. From the steady-state version of Equation (5.1), the latter condition requires a dimensionless freshwater addition of $\Pi = K$. The corresponding dimensional freshwater addition is F = $\beta\Pi$ = βK, where β is a conversion coefficient. The largest value of K we consider is K = 2.5, which is the value required by the S-S model to reproduce the reversible ATHC shutdown simulated by the UIUC AOGCM [Yin (2004); Yin et al. (2005)]. Taking F = 0.6 Sv for K = 2.5 yields β = 0.24 Sv.

Schlesinger et al. (2000) report results from simulations by the UIUC atmospheric GCM coupled to a 60 m deep mixed-layer ocean model for several different radiative forcings that suggest a linear relationship between freshwater addition, Π, and global-mean temperature change, $\Delta\bar{T}$,

$$\Pi(t) = \alpha[\Delta\bar{T}(t) - \Delta\bar{T}_c]H[\Delta\bar{T}(t) - \Delta\bar{T}_c], \tag{5.3}$$

where

$$H(x) = \begin{cases} 0 \text{ if } x < 0 \\ 1 \text{ if } x \geq 0 \end{cases} \tag{5.4}$$

is the Heavyside step function and α is the 'hydraulic sensitivity'. The Heavyside step function is introduced to prevent any freshwater addition until a critical temperature change, $\Delta\bar{T}_c$, is reached.

Substituting Equation. (5.3) into F = $\beta\Pi$ and solving for α yields

$$\alpha = \frac{F}{\beta[\Delta\bar{T} - \Delta\bar{T}_c]H[\Delta\bar{T} - \Delta\bar{T}_c]} \tag{5.5}$$

If we assume that $\Delta\bar{T} - \Delta\bar{T}_c = 2.5°C$ for F = 0.6 Sv, then α = 1.0 (°C)$^{-1}$ for β = 0.24 Sv. The values of α and $\Delta\bar{T}_c$ are highly uncertain, though. Accordingly, we took these quantities to have uniform probability distributions between 0.2 and 1.0 (°C)$^{-1}$ (in increments of 0.2) for α

and between 0.0°C and 0.6°C (in 0.1 degree increments) for $\Delta\bar{T}_c$.

Finally, the S-S model translates freshwater addition to flow in the ATHC. Yin (2004) and Yin et al. (2005) show that this depends critically on the ratio of salinity transports by the gyre/eddies and the ATHC, represented by K. A uniform prior ranging from 0.0 through 2.5 (in six increments of 0.5) was chosen based on the study by Yin (2004) and Yin et al. (2005) which showed that the S-S model with K = 0 (the original Stommel model) reproduced the irreversible ATHC shutdown simulated by the uncoupled UIUC ocean general circulation model, while the S-S model with K = 2.5 reproduced the reversible ATHC shutdown simulated by the coupled UIUC atmosphere-ocean general circulation model.

The likelihood of any specific combination of climate sensitivity, $\Delta\bar{T}_c$, α, and K thus equaled ($\pi_i/210$), where π_i represents the likelihood of the various climate sensitivities.

REFERENCES

Alley, R.B. (1998) Palaeoclimatology: Icing the North Atlantic. *Nature*, **392**, 335–337.

Alley, R.B. (2000) The Younger Dryas cold interval as viewed from central Greenland. *Quaternary Sci. Rev.*, **19**, 213–226.

Alley, R.B., Meese, D.A., Shuman, C.A., Gow, A.J., Taylor, K.C., Grootes, P.M., White, J.W.C., Ram, M., Waddington, E.D., Mayewski, P.A. and Zielinski, G.A. (1993) Abrupt increase in Greenland snow accumulation at the end of the Younger Dryas event. *Nature*, **362**, 527–529.

Andronova, N.G. and Schlesinger, M.E. (2001) Objective estimation of the probability density function for climate sensitivity. *J. Geophys. Res.*, **106**, 22, 605–22, 612.

Atkinson, T.C., Briffa, K.R. and Coope, G.R. (1987) Seasonal temperatures in Britain during the past 22,000 years, reconstructed using beetle remains. *Nature*, **325**, 587–592.

Broecker, W.S. (1985) Does the ocean-atmosphere system have more than one stable mode of operation? *Nature*, **315**, 21–26.

Broecker, W.S. (1997) Thermohaline Circulation, the Achilles Heel of Our Climate System: Will Man-Made CO$_2$ Upset the Current Balance? *Science*, **278**, 1582–1588.

Broecker, W.S. (2003) Does the Trigger for Abrupt Climate Change Reside in the Ocean or in the Atmosphere? *Science*, **300**, 1519–1522.

Broecker, W.S., Andree, M., Wolfi, W., Oeschger, H., Bonani, G., Kennett, J. and Peteet, D. (1988) The chronology of the last deglaciation: Implications to the cause of the Younger Dryas event. *Paleoceanography*, **3**, 1–19.

Broecker, W.S., Kennett, J.P., Flower, B.P., Teller, J., Trumbore, S., Bonani, G. and Wolfi, W. (1989) The routing of Laurentide ice-sheet meltwater during the Younger Dryas cold event. *Nature*, **341**, 318–321.

Chinzei, K., Fujioka, K., Kitazato, H., Koizumi, I., Oba, T., Oda, M., Okada, H., Sakai, T. and Tanimura, Y. (1987) Postglacial environmental change of the Pacific Ocean off the coasts of central Japan. *Marine Micropaleontology*, **11**, 273–291.

Crichton, M. (2004) *State of Fear*, New York, HarperCollins, pp. 603.

Cubasch, U., *et al.* 2001: Projections of future climate change. In: *Climate Change 2001: The Scientific Basis. Contribution of Working Group I to the Third Assessment Report of the Intergovernmental Panel on Climate Change* [Houghton, J.T., et al., Eds.]. Cambridge University Press, Cambridge, UK, pp. 525–582.

Emmerich, R. (2004) The Day After Tomorrow. 20th Century Fox, http://www.foxhome.com/dayaftertomorrow/

Gagosian, R.B. (2003) Abrupt Climate Change: Should We Be Worried? Woods Hole, MA, Woods Hole Oceanographic Institution, http://www.whoi.edu/institutes/occi/currenttopics/climatechange_wef.html

Ganopolski, A. and Rahmstorf, S. (2001) Rapid changes of glacial climate simulated in a coupled climate model. *Nature, 409*, 153–158.

Johnson, R.G. and McClue, B.T. (1976) A model for northern hemisphere continental ice sheet variation. *Quaternary Research, 6*, 325–353.

Leggett, J., Pepper, W.J. and Swart, R.J., 1992: Emissions scenarios for the IPCC: An update. (In) *Climate Change 1992: The Supplementary Report to the IPCC Scientific Assessment* [Houghton, J.T., et al., Eds.]. Cambridge University Press, Cambridge, UK, pp. 71–95.

Lempert, R.J. and Schlesinger, M.E. (2000) Robust Strategies for Abating Climate Change. *Climatic Change, 45*, 387–401.

Manabe, S. and Stouffer, R.J. (1994) Multiple-century response of a coupled ocean-atmosphere model to an increase of atmospheric carbon dioxide. *J. Climate, 7*, 5–23.

Manabe, S. and Stouffer, R.J. (1988) Two Stable Equilibria of a Coupled Ocean-Atmosphere Model. *J. Climate, 1*, 841–866.

Manabe, S. and Stouffer, R.J. (1999) Are two modes of thermohaline circulation stable? *Tellus, 51A*, 400–411.

McManus, J.F. (2004) Palaeoclimate: A great grand-daddy of ice cores. *Nature, 429*, 611–612.

McManus, J.F., Francois, R., Gherardi, J.-M., Keigwin, L.D. and Brown-Leger, S. (2004) Collapse and rapid resumption of Atlantic meridional circulation linked to deglacial climate change. *Nature, 428*, 834–837.

Nesje, A., Dahl, S.O. and Bakke, J. (2004) Were abrupt late glacial and early-Holocene climatic changes in northwest Europe linked to freshwater outbursts to the North Atlantic and Arctic Oceans? *The Holocene, 14*, 299–310.

Nordhaus, W.D. (1991) To slow or not to slow? *Economic Journal, 5*, 920–937.

Nordhaus, W.D. and Boyer, J. (2001) *Warming the World – Economic Models of Global Warming,* MIT Press, Cambridge, MA.

Prange, M., Romanova, V. and Lohmann, G. (2002) Influence of vertical mixing on the thermohaline hysteresis: analyses of an OGCM. *J. Phys. Oceanogr., 33*, 1707–1721.

Rahmstorf, S. (1995) Bifurcations of the Atlantic thermohaline circulation in response to changes in the hydrological cycle. *Nature, 378*, 145–149.

Rind, D., Demenocal, P., Russell, G., Sheth, S., Collins, D., Schmidt, G. and Teller, J. (2001) Effects of glacial meltwater in the GISS coupled atmosphere-ocean model. *J. Geophys. Res., 106*, 27, 335–27, 353.

Robinson, K.S. (2004) *Forty Signs of Rain,* New York, Bantam Dell, pp. 358.

Rooth, C. (1982) Hydrology and ocean circulation. *Prog. Oceanogr., 11*, 131–139.

Saltzman, B. (2002) *Dynamical Paleoclimatology: Generalized Theory of Global Climate Change,* San Diego, Academic Press.

Schiller, A., Mikolajewicz, U. and Voss, R. (1997) The stability of the North Atlantic thermohaline circulation in a coupled ocean-atmosphere general circulation model. *Climate Dynamics, 13*, 325–347.

Schlesinger, M.E., Malyshev, S., Rozanov, E.V., Yang, F., Andronova, N.G., Vries, B.D., Grübler, A., Jiang, K., Masui, T., Morita, T., Penner, J., Pepper, W., Sankovski, A. and Zhang, Y. (2000) Geographical Distributions of Temperature Change for Scenarios of Greenhouse Gas and Sulfur Dioxide Emissions. *Technological Forecasting and Social Change, 65*, 167–193.

Schmittner, A. and Weaver, A.J. (2001) Dependence of multiple climate states on ocean mixing parameters. *Geophys. Res. Lett., 28*, 1027–1030.

Schmittner, A., Yoshimori, M. and Weaver, A.J. (2002) Instability of glacial climate in a model of ocean-atmosphere-cryosphere system. *Science, 295*, 1489–1493.

Schwartz, P. and Randall, D. (2003) An abrupt climate change scenario and its implications for United States national security, http://pubs.acs.org/cen/topstory/8209/pdf/climatechange.pdf

Stommel, H.M. (1961) Thermohaline convection with two stable regimes of flow. *Tellus, 13*, 224–230.

Tarasov, L. and Peltier, W.R. (2005) Arctic freshwater forcing of the Younger Dryas cold reversal. *Nature, 435,* 662–665.

Taylor, K.C., Mayewski, P.A., Alley, R.B., Brook, E.J., Gow, A.J., Grootes, P.M., Meese, D.A., Saltzman, E.S., Severinghaus, J.P., Twickler, M.S., White, J.W.C., Whitlow, S. and Zielinski, G. (1997) The Holocene-Younger Dryas Transition Recorded at Summit, Greenland. *Science, 278*, 825–827.

Teller, J.T., Leverington, D.W. and Mann, J.D. (2002) Freshwater outbursts to the oceans from glacial Lake Agassiz and their role in climate change during the last deglaciation. *Quaternary Science Reviews, 21*, 879–887.

Titz, S., Kuhlbrodt, T., Rahmstorf, S. and Feudel, U. (2002) On freshwater-dependent bifurcation in box model of the interhemispheric thermohaline circulation. *Tellus, 54A*, 89–98.

Vellinga, M., Wood, R.A. and Gregory, J.M. (2002) Processes governing the recovery of a perturbed thermohaline circulation in HadCM3. *J. Climate, 15*, 764–780.

Yin, J. (2004) The Reversibility/Irreversibility of the Thermohaline Circulation after its Shutdown: Simulations from a Hierarchy of Climate Models. *Atmospheric Sciences.* Urbana, University of Illinois at Urbana-Champaign.

Yin, J., Schlesinger, M.E., Andronova, N.G., Malyshev, S. and Li, B. (2005) Is a Shutdown of the Thermohaline Circulation Irreversible? *J. Geophys. Res.,* (submitted).

Yohe, G., Andronova, N. and Schlesinger, M. (2004) To Hedge or Not Against an Uncertain Climate Future. *Science, 306*, 416–417.

Yohe, G., Schlesinger, M.E. and Andronova, N.G. (2005) Reducing the Risk of a Collapse of the Atlantic Thermohaline Circulation. *Integrated Assessment,* (submitted).

CHAPTER 6

Towards a Risk Assessment for Shutdown of the Atlantic Thermohaline Circulation

Richard Wood[1], Matthew Collins[1], Jonathan Gregory[1,2], Glen Harris[1] and Michael Vellinga[1]
[1]*Hadley Centre for Climate Prediction and Research, Met Office, Exeter, UK*
[2]*NERC Centre for Global Atmospheric Modelling, Reading, UK*

ABSTRACT: The possible shutdown of the Atlantic Ocean Thermohaline Circulation (THC) has attracted considerable attention as a possible form of dangerous climate change. We review evidence for and against three common assertions, which imply that THC shutdown could pose particular problems for adaptation: first, associated climate changes would be in the opposite direction to those expected from global warming; secondly, such changes could be rapid (timescale one or two decades); and thirdly the change could be irreversible. THC shutdown is generally considered a high impact, low probability event. Assessing the likelihood of such an event is hampered by a high level of modelling uncertainty. One way to tackle this is to develop an ensemble of model projections which cover the range of possible outcomes. Early results from a coupled GCM ensemble suggest that this approach is feasible.

Many scientific challenges remain before we can provide robust estimates of the likelihood of THC shutdown, or of 'THC-safe' stabilisation pathways. However, recent developments in ensemble climate projection and in observations provide the prospect of real progress on this problem over the next 5–10 years.

6.1 Review of Current Knowledge

Here we provide a brief, non-comprehensive review of current thinking on some of the key scientific questions concerning the future of the Atlantic THC.

6.1.1 *Impact of the THC on Climate*

The THC, or more precisely the meridional overturning circulation (MOC), transports around 10^{15} W of heat northwards in the North Atlantic [1]. This heat is lost to the atmosphere northwards of about 24°N, and represents a substantial heat source for the extratropical northern hemisphere climate. The impact of this heat transport on the atmosphere has been estimated using coupled climate models. The THC can be artificially suppressed in such models by adding large amounts of fresh water to the North Atlantic to stop deep water formation there [e.g. 2,3,4]. The resulting climate response varies in detail between models, but robust features include substantial cooling of the northern hemisphere (strongest in regions close to the North Atlantic) and major changes in precipitation, particularly in regions bordering the tropical Atlantic. Modelled impacts of THC shutdown on net primary production of carbon by terrestrial vegetation are shown in Figure 6.1. General cooling and drying of the Northern Hemisphere results in a reduction of 11% in hemispheric primary production. Regionally, changes are larger and in some regions current vegetation types become unsustainable, leading to large scale ecosystem change [5]. A shutdown of the THC may be expected to have substantial impacts on sea level. In a recent study using an intermediate-complexity climate model [6], an artificially-induced THC shutdown resulted in global sea level rise of order 10 cm per century due to buildup of heat in the deep ocean. Furthermore, there was a more rapid dynamical response resulting in a sea level rise of up to 50 cm around the North Atlantic margins, with a compensating fall distributed over the rest of the ocean. Similar magnitudes of signal are seen in the HadCM3 study shown here [7].

While downscaling of the impacts of rapid THC shutdown from global models to local scale has not been widely performed as yet, and model estimates vary in detail, there is sufficient evidence that the impacts of such a rapid shutdown would be substantial. Figure 6.2 shows the modelled effect on surface temperature of a hypothetical (and here artificially-induced) rapid THC shutdown in 2049, after following the IS92a scenario of global warming up to that point [7, 8]. We see that around the North Atlantic, the cooling effect of the THC change more than outweighs the effects of global warming, leading to a net cooling relative to the pre-industrial climate in those regions. In the UK, for example, winter temperatures are comparable to those typical of the 'Little Ice Age' of the 17th and 18th Centuries. It should be stressed that this is a 'what if?' scenario, and the model does not predict that this would actually occur.

6.1.2 *Rapid Climate Changes*

A number of palaeoclimatic records point to the occurrence of rapid changes in the past. Particular events, which have been argued to show spatial coherence over a

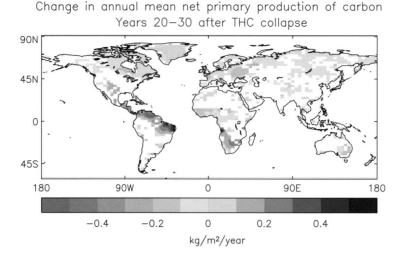

Figure 6.1 Change in net primary productivity (kg carbon per m² per year) when the THC is artificially turned off in the HadCM3 climate model, from [4]. Reductions are seen over Europe (−16%), Asia (−10%), the Indian subcontinent (−36%) and Central America (−106%). The latter figure implies that present vegetation types would become unsustainable and large- scale ecosystem adjustment could be expected [5]. At the point in the model run shown (the third decade after the artificial fresh water was introduced) the meridional overturning circulation has recovered to about 30% of its strength in the control run.

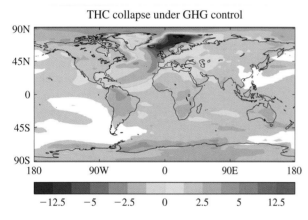

Figure 6.2 Change in surface air temperature (°C) relative to pre-industrial (1860s) values, in a HadCM3 experiment in which the THC is artificially turned off in 2049, after following the IS92a greenhouse gas emission scenario up to that point, from [8]. Note that this is a 'what if?' scenario; the model does not actually predict a THC shutdown at that time. Values shown are for the first decade after the artificial fresh water perturbation. The meridional overturning has about 18% of its strength in the pre-industrial control run and about 25% of its strength in the unperturbed IS92a run (see [7] for more details).

wide region, include the Dansgaard-Oeschger events during glacial periods, and, more recently, the so-called '8.2 kbp cold event', seen in Greenland ice cores and other proxies. These events appear to have timescales of decades, and their amplitudes are well in excess of variability seen in the later Holocene (last 8000 years). A *prima facie* case has been made for a link between these events and major reorganisations of the THC. See [9] for a review of the palaeoclimatic evidence of such events.

Modelling evidence also shows that the internal dynamics of the atmosphere-ocean-sea ice system may include the possibility of large changes occuring on a decadal timescale, not directly related to any climatic forcing. This has been seen both in rapid fluctuations during the recovery of the THC after a fresh water pulse [10] and in a more localised rapid cooling event arising spontaneously in a model control run with fixed forcing [11].

6.1.3 *Can the Present THC Exhibit Multiple Equilibria and Rapid Change?*

The climatic state of the late Holocene (last few thousand years) is substantially different from the state during glacial or early post-glacial periods, when ice sheets and sea ice covered much of the northern high latitudes, resulting in a geographically different ice-albedo feedback and the potential for substantial fresh water input to the North Atlantic through ice melt. Since there is no evidence of any order (1) changes in the THC over the past 8000 years at least (i.e. changes of magnitude similar to the current magnitude of the THC), it needs to be asked whether the present (and likely future) climate states do in fact have the potential for THC shutdown.

Many simpler climate models, ranging from the box model of [12] to climate models of intermediate complexity [13, 14], suggest that the present climate state may possess an alternative mode of operation with the THC weaker or absent. In many such studies increased greenhouse gas forcing can take the system beyond some threshold, after which only the 'THC off' state is stable. In that case, even if greenhouse gas forcing is returned to present day values, the THC remains off. Once the threshold is passed, the THC shutdown is effectively irreversible. Since the evidence for such hysteresis behaviour is largely based on simpler models, it is important to ask whether such bistable behaviour exists in the most comprehensive climate models used to make climate projections (GCMs).

The computational cost of coupled GCMs prohibits a complete exploration of the hysteresis curve. Experimentation has therefore concentrated on applying a temporary perturbation (usually a fresh water flux) to the models, in order to turn off the THC. In most cases when the perturbation is removed, the THC recovers, implying that a stable 'THC off' state has not been found in that model (though it may nevertheless exist) [15–17]. However, a stable 'THC off' state has been demonstrated in two GCMs [16, 18]. A number of factors have been proposed as influencing the stability of the 'off' state, including ocean mixing [16], atmospheric feedbacks through wind stress [15] and the hydrological cycle [15, 17, 19]. At present, it is not possible to say definitively from these model studies whether the present day THC is bistable, or whether there is a threshold beyond which irreversible shutdown would occur. It is also worth noting that in many of the model experiments used to show bistable THC states, the transition between states occurs on a slow advective timescale (centuries) rather than on a rapid (decadal) timescale. Thus, the issues of rapid and irreversible change, though related, are distinct.

6.1.4 *Model Projections of the Future THC*

The current state of uncertainty in modelling the future behaviour of the THC can be illustrated by comparing the THC response of a number of different climate general circulation models (GCMs) used in the IPCC 3rd Assessment Report, under a common greenhouse gas forcing scenario ([20], see Figure 9.21). Under this scenario, the models suggest changes in the maximum strength of the overturning circulation, ranging from a slight strengthening to a weakening of around 50%. It is notable that none of the GCMs suggests a complete THC shutdown in the 21st century. It should be noted that none of the GCM results used in [20] fully include the effects of melting of the Greenland ice sheet, which may be expected to add extra fresh water to the North Atlantic, and so further weaken the THC. Two recent studies have explored the impact of Greenland melt on the THC [21, 22]; in [22] the impact is weak, in [21] it is somewhat larger, but in neither case is a complete shutdown seen.

Why is there so much uncertainty in modelling the response of the THC to increasing greenhouse gases? In the above study, even two models that showed a similar THC change could be shown to obtain that response for different reasons, dominated in one case by thermal forcing and in the other by fresh water forcing ([20], Figure 9.22). The difficulty arises because the THC response is likely to be the net result of a number of positive and negative feedbacks. Different feedbacks dominate in different models, and to obtain the correct net

outcome it may be necessary to model each of the key feedbacks quite accurately. A further difficulty is likely to arise because simplified models that do show the possibility of the THC crossing a threshold suggest that, near the threshold, predictability becomes very poor, i.e. even if we could accurately determine that the THC was near a threshold, it could be difficult to predict the timing of a shutdown (e.g. [23], [24]).

In the present state of scientific knowledge it is not possible to identify a 'safe' CO_2 stabilisation level that would prevent THC shutdown. While the history of the past 8000 years suggests that the late Holocene THC is rather stable, there is no clear consensus from modelling work as to whether there is currently an alternative 'THC off' state, and hence a (remote) possibility of the THC switching to that state as a result of some random climate fluctuation. A variety of simpler models suggests that the THC has a bistable structure with some threshold beyond which only a weak THC state is stable, but there is disagreement among the models about the location of the current climate relative to the threshold [25]. Further, there is currently no clear understanding about whether and how fast the THC approaches the threshold as greenhouse gas forcing increases. Progress is being made towards answering these questions (e.g. see Section 2), but this can only be achieved through a programme of painstaking analysis of model processes, linked with use of appropriate observations to constrain possible responses.

As we work towards defining 'THC-safe' CO_2 stabilisation levels in future it will be important to consider stabilisation pathways as well as just the final stabilised concentrations. In particular the *rate* of CO_2 increase, as well as the final concentration, may determine the outcome. For example, in an intermediate-complexity climate model it was shown that for a given stabilisation level, a faster approach to that level was more likely to result in irreversible THC shutdown [14] and a GCM study found that a faster approach to the stabilisation level resulted in a weaker minimum overturning rate [26]. In the latter study, however, the overturning recovered slowly once CO_2 was stabilised.

6.1.5 *Summary: Where Are We Now?*

Comprehensive GCM climate projections suggest a slowdown of the THC in response to global warming over the next century, in the range 0–50%. The amount of THC change is likely to be an important factor in determining the magnitude of warming throughout the Northern hemisphere. No GCMs have shown a complete shutdown, or a net cooling over land areas. Hence a shutdown during the 21st century must be regarded as unlikely. Nonetheless, a range of theoretical, modelling and palaeoclimate studies shows that large, rapid changes are a possibility that needs to be taken seriously.

To produce a risk assessment for THC shutdown requires an understanding of both the impacts of a shutdown and the probability of occurrence. The evidence of 1.1 above points to substantial impacts (although these have not been assessed in detail). However, little can currently be said about the probability, except that it is subjectively considered low during the 21st century, based on the results of Section 1.4. To work towards a more quantitative probabilistic assessment, including information about 'safe' stabilisation levels, requires further development of models and methods. Some promising progress has recently been made towards this goal, and this is described in Section 2 below.

6.2 Towards Quantifying and Reducing Uncertainty in THC Projections

6.2.1 *Understanding What Drives THC Changes*

The first step to reducing uncertainty is to understand the processes that contribute to the wide range of THC responses currently seen in models. A recent international initiative under the auspices of the Coupled Model Intercomparison Project (CMIP) addresses this goal by analysing a number of climate models, all subject to a number of standardised forcing experiments. Figure 6.3 shows the roles of heat and water forcing in the response of the THC to a compound 1% p.a. CO_2 increase, across this range of models, based on [27]. The large variation in the forcing processes is apparent, although it can be seen that in all models except one the heat forcing dominates the fresh water forcing over the timescale of this experiment (the caveat, discussed above, that Greenland meltwater is not fully taken into account in the

models, also applies here). More detailed analysis is required to obtain a full picture of the processes determining the THC response in each model (e.g. [28]), but we can expect this research eventually to allow a good understanding to be developed of why the model responses are so different. This in turn will suggest targeted observational constraints than can be used to determine how much weight to give to particular model's' THC projections, and suggest specific priorities for model development.

6.2.2 *Probabilistic Estimation of the Future THC*

Some uncertainty will inevitably remain and in order to obtain some form of objective assessment of the likelihood of major THC changes, it will be necessary to sample the range of possible model outcomes more systematically than is possible using the few model runs shown in [20] or in Figure 6.3. Recent progress has been made in this area by generating 'perturbed physics' model ensembles (e.g. [29, 30, 31]). An ensemble of models is generated by varying a set of model parameters within a defined range. The parameter settings are chosen from a prior distribution based on expert judgement about reasonable allowable ranges. Climate projections made using each ensemble member may then be weighted according to some chosen set of observational constraints [30], or the ensemble may be allowed to evolve in such a way as to improve the goodness of fit to the observations [29, 31].

Studies to date have used either highly simplified models [29, 31] or atmosphere-only GCMs coupled to 'thermal slab' oceans [30]. Here we demonstrate the feasibility of generating a coupled GCM ensemble that can exhibit a range of THC responses to a given forcing. We use an

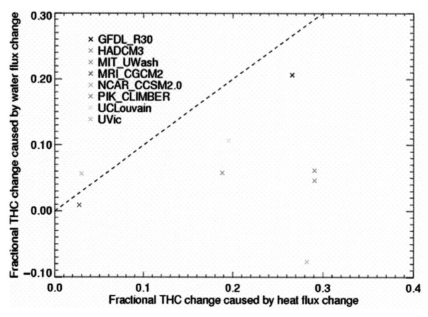

Figure 6.3 Contributions of changes in thermal and fresh water forcing to the total THC change, following a 1% per annum CO_2 increase up to four times the initial concentration, in a range of climate models. Changes are expressed as a fraction of the THC strength in the control run. The dashed line divides the regions where thermal and fresh water forcing dominate. Data derived from [27], courtesy of the CMIP co-ordinated experiment on THC stability.

existing ensemble of atmosphere-slab ocean model runs using the HadAM3 atmospheric model [30] to generate a set of atmospheric model parameters that are likely to result in a range of different THC responses, based on detailed analysis of the coupled model HadCM3 (with standard parameter settings) [28]. An ensemble of coupled models is thus produced, and a range of THC responses can be seen. The problem of climate drift in the coupled models is overcome by one of two methods: either flux adjustment, or pre-selection of parameter settings to minimise climate drift without using flux adjustment. The latter pre-selection is made by only allowing parameter settings that give an accurate global heat budget in the atmosphere-slab ocean ensemble.

In the standard HadCM3 model, the THC weakening in response to CO_2 increase is limited by a tropical fresh water feedback [28]. Warming of the tropical oceans results in an intensification of the hydrological cycle, including an increase of evaporation from the tropical Atlantic. Much of this water is transported away from the Atlantic by the trade wind circulation and falls into the Pacific catchment. Thus the tropical Atlantic becomes saltier, and this salty anomaly is transported by the ocean circulation to the subpolar North Atlantic, where it helps to maintain deep water formation. The intensity of this evaporative feedback varies quite widely in the ensemble of atmosphere-slab ocean integrations with doubled CO_2, leading us to hypothesise that by selecting parameter settings on the basis of the atmosphere-slab integrations we can generate an ensemble of coupled integrations that have stable control (constant CO_2) climates, yet which show a range of THC responses.

Early results show that a range of THC responses can be produced, in models whose control runs have minimal climate drift. For example an ensemble member has been produced whose climate drifts are similar to those in the standard HadCM3 model, but which has a significantly greater THC weakening in response to 1% p.a. CO_2 increase at the time of CO_2 doubling. The greater THC response is consistent with a weaker evaporative feedback (as described above) in the corresponding atmosphere-slab ocean run. The ensemble is now being expanded to cover as wide a region of parameter space as possible, thus allowing a plausible range of THC behaviour to be quantified. Both flux adjusted and non-flux adjusted ensembles will be explored, since it could be argued that climate drift may be a result of small model errors and imbalances that do not impact on the THC response. Hence one might argue that by insisting on non-drifting models one may not sample the full range of possible responses. On the other hand, it has been suggested that use of flux adjustments may distort the stability properties of the THC [32, 33].

The longer-term goal is to incorporate a range of models into such studies (in order to explore and transcend any constraints due to the structural features of different models). This should include a spectrum of models, including appropriately formulated but computationally cheaper models to allow thorough exploration of a wide parameter space (including a plausible range of stabilisation scenarios). This will allow for the first time an objective estimate of the likelihood of major THC change and identification of 'safe' stabilisation pathways. However, the difficulties of reaching such a goal should not be underestimated. Two specific issues will need to be addressed:

i. The choice of observational constraints used to weight the ensemble members may be critical in determining the shape of the resulting probability distributions. This has been demonstrated in [31], where different choices of observational constraints resulted in either a significant or a near-zero probability of THC shutdown. To address this issue we will need to develop a process-based understanding of the role of specific observables in THC stability.

ii. While simplified models will be valuable in exploring parameter space and developing methods, they inevitably involve a choice to omit certain processes that may be crucial to THC stability. The results must therefore be used with caution. It will be important to develop the idea of a 'traceable' spectrum of models, in which the simpler models include (albeit in highly parameterised form) all processes that have been shown to be important for the THC response in the more comprehensive models. The processes in the comprehensive models must in turn be evaluated against observations, as discussed in (i) above. If such traceability cannot be established then there is no demonstrable link between the simpler model and the real (observed) world.

6.3 Summary and Prospects

The currently very high level of modelling uncertainty makes accurate projection of the future of the THC difficult, beyond the rather vague statement that complete shutdown is 'unlikely' over the next century. Methods of probabilistic climate projection are in their infancy and quantifying the relatively low probability of THC shutdown will be particularly challenging. But recent progress in ensemble methods, along with some exciting new observational developments (e.g. continuous monitoring of the MOC at 26°N [34, 35]) suggests that real progress can be made towards providing broad limits on 'THC-safe' stabilisation pathways. If we can make and sustain the 'right' observations (and we need to determine what these are: see e.g. [36]), and focus model developments on those processes that currently contribute to the large differences among models, we can expect uncertainty to reduce substantially over the next decade.

Acknowledgements

This work was supported under Defra contract number PECD 7/12/37. We thank James Murphy for valuable discussions.

REFERENCES

1. Bryden, H.L. and Imawaki, S., 2001: Ocean heat transport. In *Ocean Circulation and Climate*, ed. Siedler G., Church J. and Gould J., Academic Press, pp. 715.

2. Schiller, A., Mikolajewicz, U. and Voss, R., 1997: The stability of the thermohaline circulation in a coupled ocean-atmosphere general circulation model. *Clim. Dyn.* **13**, 325–347.

3. Manabe, S. and Stouffer, R.J., 1999a: The role of the thermohaline circulation in climate *Tellus* **51A**, 91–109.

4. Vellinga, M. and Wood, R.A., 2002: Global climatic impacts of a collapse of the Atlantic thermohaline circulation. *Climatic Change* **54**, 251–267.

5. Higgins, P. and Vellinga, M., 2004: Ecosystem responses to abrupt climate change: teleconnections, scale and hydrological cycle. *Climatic Change* **64**, 127–142.

6. Levermann, A., Griesel, A., Hofmann, M., Montoya, M. and Rahmstorf, S., 2005: Dynamic sea level changes following changes in the thermohaline circulation. *Clim. Dyn,* **24**, 347–354.

7. Vellinga, M. and Wood, R.A., 2005: Impacts of thermohaline circulation shutdown in the twenty-first century. *Climatic Change* (submitted).

8. Wood, R.A, Vellinga, M. and Thorpe, R., 2003: Global Warming and THC stability. *Phil. Trans Roy. Soc. A* **361**, 1961–1976.

9. Alley, R.B., 2003: Palaeoclimatic insights into future climate challenges. *Phil. Trans. Roy. Soc. A* **361**, 1831–1850.

10. Manabe, S. and Stouffer, R.J., 1995: Simulation of abrupt climate change induced by freshwater input to the North Atlantic Ocean. *Nature* **378**, 165–167.

11. Hall, A. and Stouffer, R.J., 2001: An abrupt climate event in a coupled ocean-atmosphere simulation without external forcing. *Nature* **409**, 171–174.

12. Stommel, H., 1961: Thermohaline convection with two stable regimes of flow. *Tellus* **13**, 224–230

13. Rahmstorf, S. and Ganopolski, A., 1999: Long term global warming scenarios computed with an efficient coupled climate model. *Climatic Change* **43**, 353–367.

14. Stocker, T.F. and Schmittner, A., 1997: Influence of CO_2 emission rates on the stability of the thermohaline circulation. *Nature* **388**, 862–865.

15. Schiller, A., Mikolajewicz, U. and Voss, R., 1997: The stability of the thermohaline circulation in a coupled ocean-atmosphere general circulation model. *Clim. Dyn.* **13**, 325–347.

16. Manabe, S. and Stouffer, R.J., 1999b: Are two modes of thermohaline circulation stable? *Tellus* **51A**, 400–411.

17. Vellinga, M., Wood, R.A. and Gregory, J.M., 2002: Processes governing the recovery of a perturbed thermohaline circulation in HadCM3. *J. Climate* **15**, 764–780.

18. Rind, D., deMenocal, P., Russell, G., Sheth, S. D., Schmidt, G., and Teller, J., 2001: Effects of glacial meltwater in the GISS coupled atmosphere-ocean model. Part 1. North Atlantic Deep Water response. *J. Geophys. Res.* **106**, 27335–27353.

19. Rahmstorf, S., 1996: On the freshwater forcing and transport of the Atlantic thermohaline circulation. *Clim. Dyn.* **12**, 799–811.

20. Cubasch, U., Meehl, G.A., Boer, G.J., Stouffer, R.J., Dix, M., Noda, A., Senior, C.A., Raper, S., and Yap, K.S., 2001: Projections of future climate change. *In Climate Change 2001: The Scientific Basis. Contribution of Working Group 1 to the Third Assessment Report of the Intergovernmental Panel on Climate* (J.T. Houghton et al., editors), Cambridge University Press, 525–582.

21. Fichefet, T., Poncin, C., Goose, H., Huybrechts, P., Janssens, I. and Le Treut, H., 2003: Implications of changes in freshwater flux from the Greenland ice sheet for the climate of the 21st century, *Geophys. Res. Lett,* **30**(17), 1911, doi:10.1029/2003GL017826.

22. Ridley, J., Huybrechts, P., Gregory, J.M., Lowe, J., 2005: Future changes in the Greenland ice sheet: a 3000 year simulation with a high resolution ice sheet model interactively coupled to an AOGCM, *Journal of Climate* (submitted).

23. Wang, X., Stone, P.H. and Marotzke, J., 1999: Global thermohaline circulation, Part I: Sensitivity to atmospheric moisture transport. *J. Climate* **12**, 71–82.

24. Knutti, R. and Stocker, T.F., 2002: Limited predictability of the future thermohaline circulation close to an instability threshold. *J. Climate* **15**, 179–186.

25. Rahmstorf, S., Crucifix, M., Ganopolski, A., Goosse, H., Kamenkovich, I., Knutti, R., Lohmann, G., Marsh, R., Mysak, L.A., Wang, Z. and. Weaver, A.J., 2005: Thermohaline circulation hysteresis: A model intercomparison. *Geophys. Res. Lett.* (submitted).

26. Stouffer, R.J. and Manabe, S., 1999: Response of a coupled ocean-atmosphere model to increasing atmospheric carbon dioxide: Sensitivity to the rate of increase. *J. Climate* **12**, 2224–2237.

27. Gregory, J.M., Dixon, K.W., Stouffer, R.J., Weaver, A.J., Driesschaert, E., Eby, M., Fichefet, T., Hasumi, H., Hu, A., Jungclaus, J.H., Kamenkovich, I.V., Levermann, A., Montoya, M., Murakami, S. Nawrath, S., Oka, A., Sokolov, A.P. and Thorpe, R.B., 2005: A model intercomparison of changes in the Atlantic thermohaline circulation in response to increasing atmospheric CO_2 concentration. *Geophys. Res. Lett.* **32**, L12703, doi: 10.1029/2005GL023209.

28. Thorpe, R.B., Gregory, J.M., Johns, T.C., Wood, R.A. and Mitchell, J.F.B., 2001: Mechanisms determining the Atlantic thermohaline circulation response to greenhouse gas forcing in a nonflux-adjusted coupled climate model. *J. Climate* **14**, 3102–3116.

29. Annan, J.D., Hargreaves, J.C., Edwards, N.R. and Marsh, R., 2005: Parameter estimation in an intermediate complexity earth system model using an ensemble Kalman filter. *Ocean Modelling* **8**, 135–154.

30. Murphy, J.M., Sexton, D.M.H., Barnett D.N., et al., 2004: Quantification of modelling uncertainties in a large ensemble of climate change simulations. *Nature* **430**, 768–772.

31. Hargreaves, J.C. and Annan, J.D., 2005: Using ensemble prediction methods to examine regional climate variation under global warming scenarios. *Ocean Modelling* (in press).

32. Marotzke, J. and Stone, P.H., 1995: Atmospheric transports, the thermohaline circulation, and flux adjustments in a simple coupled model. *J. Phys. Oceanogr.* **25**, 1350–1364.

33. Dijkstra, H.A. and Neelin, J.D., 1999: Imperfections of the thermohaline circulation: multiple equilibria and flux correction. *J. Climate* **12**, 1382–1392.

34. Hirschi, J., Baehr, J., Marotzke, J., Stark, J., Cunningham, S., and Beismann, J.-O., 2003: A monitoring design for the Atlantic meridional overturning circulation. *Geophys. Res. Lett,* **30**, 66.1–66.4, doi:10.1029/2002GL016776.

35. Baehr, J., Hirschi, J., Beismann, J-O. and Marotzke, J., 2004: Monitoring the meridional overturning circulation in the North Atlantic: a model-based array design study. *J. Mar. Res.* **62**, 283–312.

36. Vellinga, M. and Wood, R.A., 2004: Timely detection of unusual change in the Atlantic meridional overturning circulation. *Geophys Res. Lett,* **31**, L14203, doi:10.1029/2004GL020306.

CHAPTER 7

Towards the Probability of Rapid Climate Change

Peter G. Challenor, Robin K.S. Hankin and Robert Marsh
National Oceanography Centre, Southampton, University of Southampton, Southampton, Hampshire, UK

ABSTRACT: The climate of North West Europe is mild compared to Alaska because the overturning circulation in the Atlantic carries heat northwards. If this circulation were to collapse, as it appears to have done in the past, the climate of Europe, and the whole Northern Hemisphere, could change rapidly. This event is normally classified as a 'low probability/high impact' event, but there have been few attempts to quantify the probability. We present a statistical method that can be used, with a climate model, to estimate the probability of such a rapid climate change. To illustrate the method we use an intermediate complexity climate model, C-GOLDSTEIN combined with the SRES illustrative emission scenarios. The resulting probabilities are much higher than would be expected for a low probability event, around 30–40% depending upon the scenario. The most probable reason for this is the simplicity of the climate model, but the possibility exists that we may be at greater risk than we believed.

7.1 Introduction

Northwest Europe is up to 10°C warmer than equivalent latitudes in North America because a vigorous thermohaline circulation transports warm water northwards in the Atlantic basin (Rind et al., 1986). However, due to increasing concentrations of CO_2 in the atmosphere, this circulation could slow markedly (Cubasch et al., 2001) or even collapse (Rahmstorf and Ganopolski, 1999). The climatic impact of such a change in the ocean circulation would be severe, especially in Europe (Vellinga and Wood, 2002), but with worldwide consequences, and could happen on a rapid time scale. It is important therefore that we assess the risk of such a collapse in the thermohaline circulation (Marotzke, 2000). Recent studies have developed and adopted a probabilistic approach to address the climate response to rising levels of greenhouse gases (Wigley and Raper, 2001; Allen and Stainforth, 2002; Stainforth et al., 2005). However, to our knowledge, no study has yet addressed the probability of substantial weakening of the overturning circulation and the implied rapid climate change. In this paper we present a statistical technique that can be used to estimate the probability of such a rapid climate change using a model of the climate and illustrate it with a model of intermediate complexity.

7.2 A Method for Calculating Probabilities of Climate Events

Most modern climate models are deterministic: given a set of inputs they always give the same results on a given hardware platform. There are two standard ways to introduce an element of randomness and hence to make probabilistic predictions. The first is to use the internal, chaotic variability of the model. The initial conditions are varied by a small amount and an ensemble of model runs is performed. This method is widely used in weather forecasting. This is suitable for problems where the initial conditions are the important factor for predictability, predictability of the first kind. However, for long-range climate forecasting we believe we have predictability of the second kind where it is the boundary conditions that matter. In this case the perturbations need to be made on the boundary conditions. In our case these are the model parameters. A numerical model of the climate system contains a number of parameters, the 'true' value of which is unknown. If we represent our ignorance of these parameters in probabilistic terms we can propagate this uncertainty through the numerical model and hence produce a probability density function of the model outputs. This is the method we will use in this paper.

In essence, our method is to sample from a specified uncertainty distribution for the model input parameters, run the model for this combination of inputs and compute the output. This process is repeated many thousands of times to build up a Monte Carlo estimate of the probability density of the output. This type of Monte Carlo method is too computationally expensive for practical use; even intermediate complexity climate models such as C-GOLDSTEIN (Edwards and Marsh, 2005) are not fast enough to allow us to carry out such calculations with the required degree of accuracy. To overcome this problem we introduce the concept of an *emulator*. An emulator is a technique in which Bayesian statistical analysis is used to furnish a statistical approximation to the full dynamical model. In preference to a neural network (Knutti et al., 2003), we follow Oakley and O'Hagan (2002) and use a Gaussian process to build our emulator.

This has the advantage that is easier to understand and interpret, and every prediction comes with an associated uncertainty estimate. This means that the technique can reveal where the underlying assumptions are good and where they are not. Our emulators run about five orders of magnitude faster than a model such as C-GOLDSTEIN.

Full mathematical details of Gaussian processes and the Bayesian methods we use to fit them to the data are given in Oakley and O'Hagan (2002). The basic process of constructing and using an emulator is as follows:

1. For each of the parameters of the model, specify an uncertainty distribution (a 'prior') by expert elicitation and thereby define a prior pdf for the parameter space of the model.
2. We generate a set of parameter values that allow us to span the parameter space of these prior pdfs and run the climate model at each of these points to provide a calibration dataset of predicted MOC strength.
3. Estimate the parameters of the emulator using the calibration dataset using the methods given in Oakley and O'Hagan (2002).
4. Sample a large number (thousands) of points from the prior pdf.
5. Evaluate the emulator at each of these points. The output from the emulator then gives us an estimate of pdf of the variable being emulated from which we can calculate statistics such as the probability of being less than a specified value.

Ideally, in step 2 we would use an ensemble of model runs that spanned the complete parameter space of the model. However, as dimensionality increases this becomes difficult, and a factorial design soon requires an impractically large number of model runs. We therefore use the latin hypercube design (McKay et al., 1979), which requires us to specify in advance the number of model runs we can afford, in our example below this is 100. The range of each parameter is split up into this number of intervals of equal probability according to the uncertainty distribution of the input parameters. Our experience is that this distribution should be longer tailed than the input distribution used for the Monte Carlo calculations: the emulator is, along with all such estimation techniques, poor at extrapolation but good at interpolation so we want model runs out in the tails of the distribution to minimise the amount of extrapolation the emulator is called upon to do. For step 4, the order of the values of each parameter is now shuffled so that there is one and only one value in each of the equiprobable interval of each parameter (that is, the marginal distribution is unchanged), but the points are randomly scattered across multi-dimensional parameter space.

A Gaussian process is the extension of a multivariate Gaussian distribution to infinite dimension. For full mathematical details of Gaussian processes and the Bayesian methods we use to fit them to the data see Oakley and O'Hagan (2002). A Gaussian process is given by the sum of two terms: a deterministic, or mean, part and a stochastic part. The mean part can be considered as a general trend while the stochastic part is a local adjustment to the data. There is a trade-off between the variation explained by the mean function and the stochastic part. Following Oakley and O'Hagan (op. cit) we specify *a priori* that the mean function has a simple form (linear, in our case) with unknown parameters. The stochastic term in the Gaussian process is specified in terms of a correlation function. We use a Gaussian shape for the correlation function. This is parameterised by a correlation matrix. The elements of this matrix give the smoothness of the resulting Gaussian process. For simplicity we use a diagonal matrix, setting the off-diagonal terms to zero. These correlation scales cannot be estimated in a fully Bayesian way so are estimated using cross-validation. An alternative approach is to use regression techniques to model the mean function in a complex way. This means that the stochastic term is much less important and may make problems such as non-stationarity less important; for a non-climate example where this is done see Craig et al. (2001). Gaussian process emulators specified in this way are perfect interpolators of the data and it can be shown that any smooth function can be expressed as a Gaussian process.

It is important to specify the uncertainty distributions of the model inputs/parameters in step 2 carefully. In our case we elicit the information from experts, in this case the model builders and tuners. Our method was to request reasonable lower and upper limits for each parameter and interpret these as fifth and ninety-fifth percentiles of a log normal distribution. Because of the importance of the input distributions a sensitivity analysis was carried out to identify important input parameters; step 4 was repeated with doubled standard deviation for those parameters (see below for details). It is difficult to elicit the full joint input distribution so we have elicited the marginals and assumed that the inputs are independent. This assumption is almost certainly wrong and needs to be tested in further work. More complex elicitation methods (see the review by Garthwaite et al., 2005) need to be considered.

7.3 An Illustration: Emulating the MOC Response to Future CO$_2$ Forcing in C-GOLDSTEIN

To illustrate the methods described above we estimate the probability of the collapse of the thermohaline circulation under various emission scenarios using an intermediate complexity climate model. The climate model we use is C-GOLDSTEIN (Edwards and Marsh, 2005). This is a global model comprising a 3-D frictional-geostrophic ocean component configured in realistic geometry, including bathymetry, coupled to an energy-moisture balance model of the atmosphere and a thermodynamic model of sea ice. We use *a priori* independent log-normal distributions for 17 model parameters (Table 7.1). For 12 of the parameters, we use the distributions derived in an objective

Table 7.1 Mean value and standard deviation for each model parameter.

Parameter*	Mean	St. Dev.
Windstress scaling factor	1.734	0.1080
Ocean horizontal diffusivity (m^2s^{-1})	4342	437.9
Ocean vertical diffusivity (m^2s^{-1})	5.811e−05	1.428e−06
Ocean drag coefficient ($10^{-5}s^{-1}$)	3.625	0.3841
Atmospheric heat diffusivity (m^2s^{-1})	3.898e+06	2.705e+05
Atmospheric moisture diffusivity (m^2s^{-1})	1.631e+06	7.904e+04
'Width' of atmospheric heat diffusivity profile (radians)	1.347	0.1086
Slope (south-to-north) of atmospheric heat diffusivity profile	0.2178	0.04215
Zonal heat advection factor	0.1594	0.02254
Zonal moisture advection factor	0.1594	0.02254
Sea ice diffusivity (m^2s^{-1})	6786.0	831.6
Scaling factor for Atlantic-Pacific moisture flux (x 0.32 Sv)	0.9208	0.05056
Threshold humidity, for precipitation (%)	0.8511	0.01342
'Climate sensitivity'[†] (CO_2 radiative forcing, Wm^{-2})	6.000	5.000
Solar constant (Wm^{-2})	1368	3.755
Carbon removal e-folding time (years)	111.4	15.10
Greenland melt rate due to global warming[‡]	0.01(Low)	0.005793
(Sv/°C)	0.03617 (High)	

* The first 15 parameters control the background model state. The first 12 of these have been objectively tuned in a previous study, while the last three (threshold humidity, climate sensitivity and solar output) are specified according to expert elicitation. The last two parameters control transient forcing (CO_2 concentration and ice sheet melting). Italics show the parameters that exert particular control on the strength of the overturning and which we varied in our experiment. For these parameters, the standard deviation was doubled in the cases with high uncertainty.

[†] The climate sensitivity parameter, ΔF_{2x}, determines an additional component in the outgoing planetary long-wave radiation according to $\Delta F_{2x}\ln(C/350)$, where C is the atmospheric concentration of CO_2 (units ppm). Values for ΔF_{2x} of 1, 6 and 11 Wm^{-2} yield 'orthodox' climate sensitivities of global-mean temperature rise under doubled CO_2 of around 0.5, 3.0 and 5.5 K, respectively.

[‡] We used two mean values of the Greenland melt rate parameter (see main text).

tuning exercise (Hargreaves et al., 2004). For the others we elicited values from one of the model authors (Marsh) using the method described above. We specify particularly high variance for climate sensitivity, in line with recent results (Stainforth et al., 2005). We thus account for uncertainty in the model parameters, but not in the model physics (so called 'structural' uncertainty).

To generate our emulator as described above we need an ensemble of model runs to act as our 'training set'. We use an ensemble of 100 members in a latin hypercube design. We first 'spin up' the climate model for 4000 years to the present day (the year 2000, henceforth 'present day') in an ensemble of 100 members that coarsely samples from a range of values for fifteen key model parameters (see Table 7.1); the remaining two parameters are only used for simulations beyond the present day. Following 3800 years of spin-up under pre-industrial CO_2 concentration, the overturning reaches a near-equilibrium state in all ensemble members (see Figure 7.1). For the last 200 years of the spin-up, we specify historical CO_2 concentrations (Johnston, 2004), leading to slight (up to 5%) weakening in the overturning circulation. After the complete 4000-year spin-up we have 100 simulations of the current climate and the thermohaline circulation. Figure 7.2 shows fields of mean and standard deviation in surface temperature. The mean temperature field is

similar to the ensemble–mean obtained by Hargreaves et al. (2005). The standard deviations reveal highest sensitivity to model parameters at high latitudes, especially in the northern hemisphere, principally due to differences (between ensemble members) in Arctic sea ice extent. We obtain an ensemble of present day overturning states, with ψ_{max} in the plausible range 12–23 Sv for 91 of the ensemble members (see Figure 7.1). The overturning circulation collapsed in the remaining nine members after the first 1000 years. Since we know that the overturning is not currently collapsed, we remove these from further analysis. This is a controversial point that we will return to in the discussion. We then specify future anthropogenic CO_2 emissions according to each of the six illustrative SRES scenarios (Nakicenovic and Swart, 2000) (A1B, A2, B1, B2, A1FI, A1T), to extend those simulations with a plausible overturning to the year 2100.

In extending the simulations over 2000–2100, we specify the SRES CO_2 emissions scenarios and introduce two further parameters (the last two parameters in Table 7.1) that relate to future melting of the Greenland ice sheet and the rate at which natural processes remove anthropogenic CO_2 from the atmosphere. The rate of CO_2 uptake is parameterised according to an e-folding timescale that represents the background absorption of excess CO_2 into marine and terrestrial reservoirs. This timescale can be

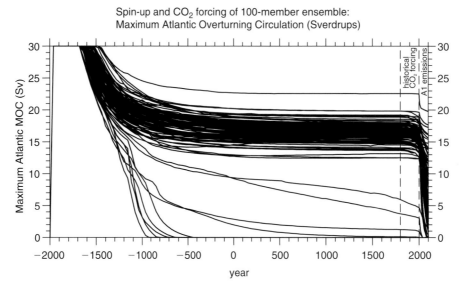

Figure 7.1 Spin-up of the Atlantic MOC, including CO_2 forcing from 1800.

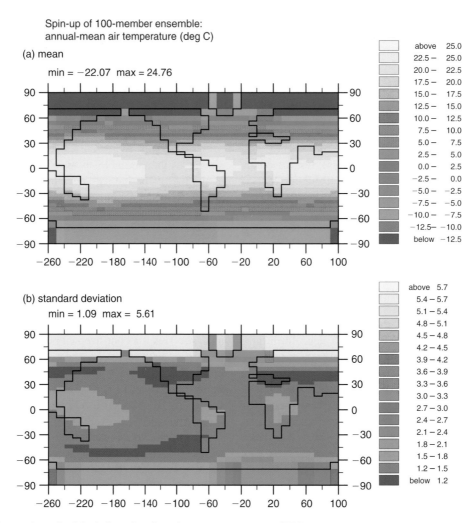

Figure 7.2 Mean and standard deviation of surface air temperature at year 2000.

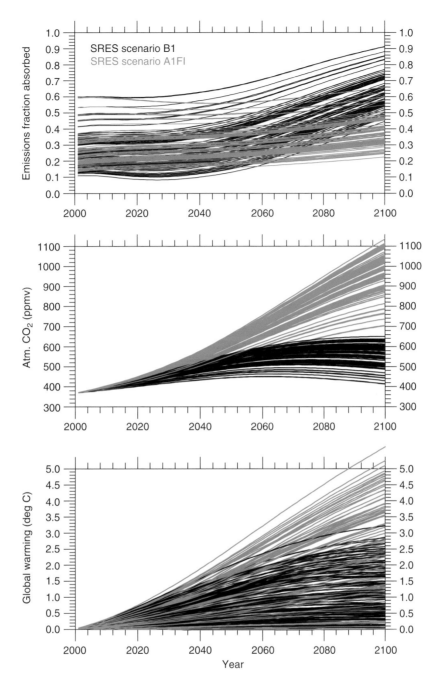

Figure 7.3 Time series of emitted CO_2 uptake, atmospheric CO_2 concentration and temperature rise over 2000–2100, under scenarios B1 and A1FI.

roughly equated with a fractional annual uptake of emissions. Timescales of 50, 100 and 300 years equate to fractional uptakes of around 50%, 30% and 10% respectively (see Figure 7.3, top panel), spanning the range of uncertainty in present and future uptake (Prentice et al., 2001). For each emissions scenario, a wide range of CO_2 rise is obtained, according to the uptake timescale (see Figure 7.3, middle panel). This in turn leads to a wide range of global-mean temperature rise, which is further broadened by the uncertainty in climate sensitivity (see Figure 7.3, bottom panel). The freshwater flux due to melting of the Greenland ice sheet is linearly proportional to

the air temperature anomaly relative to 2000 (Rahmstorf and Ganopolski, 1999). This is consistent with evidence that the Greenland mass balance has only recently started changing (Bøggild et al. 2004). Over the range chosen for this parameter (combined with the uncertainty in emissions and climate sensitivity), the resultant melting equates to sea level rise by 2100 mostly in the range 0–30 cm (see Figure 7.4), consistent with predictions obtained with a complex ice sheet model (Huybrechts and de Wolde, 1999).

As a consequence of the applied forcing, ψ_{max} declines to varying degrees, in the range 10–90% in the case of the

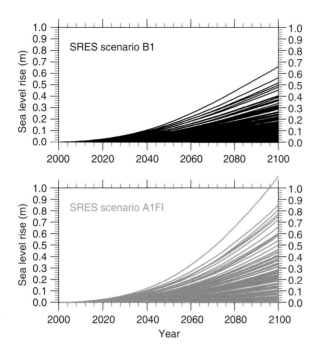

Figure 7.4 Sea level rise due to Greenland melting over 2000–2100, under scenarios B1 and A1FI.

A1FI scenario (see Figure 7.1). The range of MOC weakening is compatible with that suggested by IPCC (2001) AOGCM results. At 2100, the IPCC AOGCMs cover a range of $+2$ to $-14\,\mathrm{Sv}$ with 9 model runs. The range for our 91 run ensemble is -1 to -17 with 90% between -2 and -15. Under the B1 scenario, the regional impact of this MOC slow-down is a local cooling in the Atlantic (see Figure 7.5, upper panel), also the location of highest standard deviation (Figure 7.5, lower panel), due to wide variation in the extent of slow-down. In several extreme cases (not clear from the ensemble-mean temperature change) of substantial slow-down, North Atlantic cooling under B1 exceeds 5°C. Under the A1FI scenario, global warming is amplified and the effect of MOC slow-down is to locally cancel warming (Figure 7.6, upper panel), and highest standard deviations are found in the Arctic (Figure 7.6, lower panel) due to disappearance of Atlantic sector Arctic sea ice cover in some ensemble members.

Using the model results for each SRES scenario at 2100, we build a statistical model (emulator) of ψ_{max} as a function of the model parameters. A separate emulator is built for each emissions scenario. We then use these six emulators, coupled with probability densities of parameter uncertainty, to calculate the probability that ψ_{max} falls below $5\,\mathrm{Sv}$ by 2100 using Monte Carlo methods. We use a sample size of 20,000 for all our Monte Carlo calculations. An initial, one-at-a-time, sensitivity analysis shows that the four most important parameters are: (1) sensitivity to global warming of the Greenland Ice Sheet melt rate, providing a fresh water influx to the mid-latitude North Atlantic that tends to suppress the overturning; (2) the rate at which anthropogenic CO_2 is removed from the atmosphere; (3) climate sensitivity (i.e., the global

warming per CO_2 forcing); (4) a specified Atlantic-to-Pacific net moisture flux which increases Atlantic surface salinity and helps to support strong overturning. We perform a number of experiments calculating the probability of substantial slow-down of the overturning under variations in the values of these parameters and their uncertainties.

For each SRES scenario, we show in Table 7.2 the probability of substantial reduction in Atlantic overturning for five uncertainty cases. Each case is split into low and high *mean* Greenland melt rate, as this has been previously identified as a particularly crucial factor in the thermohaline circulation response to CO_2 forcing (Rahmstorf and Ganopolski, 1999). The probabilities in Table 7.2 are much higher than expected: substantial weakening of the overturning circulation is generally assumed to be a 'low probability, high impact' event, although 'low probability' tends not to be defined in numerical terms. Our results show that the probability is in the range 0.30–0.46 (depending on the SRES scenario adopted and the uncertainty case): this could not reasonably be described as 'low'. Even with the relatively benign B2 scenario we obtain probabilities of order 0.30, while with the fossil fuel intensive A1FI we obtain even higher probabilities, up to a maximum of 0.46.

Our probabilities are clearly less sensitive to the uncertainty case than to the SRES scenario. Increasing the mean Greenland melt rate from 'low' to 'high' increases only slightly the chance of shutdown in the circulation, probably because even the low melt rate already exceeds a threshold value (for substantial weakening of the overturning rate). The dependence of probability on parameter uncertainty is unclear, but any increase in uncertainty will broaden the distribution of the overturning strength and should theoretically lead to a higher proportion less than $5\,\mathrm{Sv}$. While in some cases this is reflected in a slightly higher probability under higher parameter uncertainty (as expected), in other cases the probabilities are slightly lower. By comparing estimates from our sample of 20,000 between sub-samples of size 1,000 we estimate the standard error of our probability estimates to be about 0.01. If we had simple binomial sampling we would expect a standard error of about 0.05. We believe this difference in error comes from the correlation between estimates of the output. How much of this correlation comes from C-GOLDSTEIN and how much from the emulation process needs to be investigated. These error estimates imply that most of the random variation in our estimates is due to uncertainty coming from the fact that our emulation is not perfect, although some may also be caused by complex positive and negative feedbacks in the climate model.

7.4 Conclusions and Discussion

We have described a method that can be used to estimate the probability of a substantial slow-down in the Atlantic thermohaline circulation and a consequent rapid climate

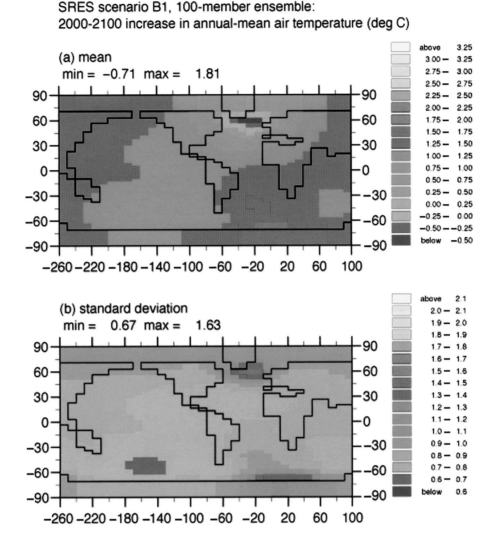

Figure 7.5 Mean and standard deviation of air temperature change in 2100 (relative to 2000) under scenario B1.

change. To illustrate the method we have applied it to an intermediate complexity climate model, C-GOLDSTEIN. The results we obtained were surprising. The probabilities we estimate are much higher than our expectations. *A priori* we expected to obtain probabilities of the order of a few percent or less. The probabilities in Table 7.2 are order 30–40%. There are a number of possible explanations for these differences. Our statistical methodology may be somewhat flawed, the model we have used could be showing unusual behaviour or our *a priori* ideas (and the current consensus) could be wrong. Let us consider each in turn.

The first possibility is that there is a problem with our statistical methodology. The basic method is sound but in our implementation we have made some assumptions and compromises that may influence our results. For example, we have assumed that the input distributions for our parameters are independent of each other and we have discarded the nine runs where the circulation collapsed during spin up. Both of these decisions could have altered our estimated probabilities of collapse. A more

thorough elicitation of the input distributions and better sensitivity analysis will enable us to address the problems of specifying input distributions in future work. Moving on to the nine runs that collapsed during the spin up: from measurements we know that the current strength of the Atlantic overturning circulation is in the range 15–20 Sv. When we performed the spin-up, nine of our runs produced current day climates with the overturning circulation approximately zero. We therefore infer that the parameter values used in these runs are not possible. We simply ignored these runs when we built the emulator. This is not correct. When we perform our Monte Carlo simulation we will still be sampling from these regions with parameter sets that we know do not generate the present day climate. Because we discarded those runs, the emulator will interpolate across this region from adjacent parts of parameter space. It is likely that these will themselves have collapsed in 2100 so we may well be overestimating the probability of collapse by including this region. A better procedure would be to build an emulator for the present day and to map out those parts of

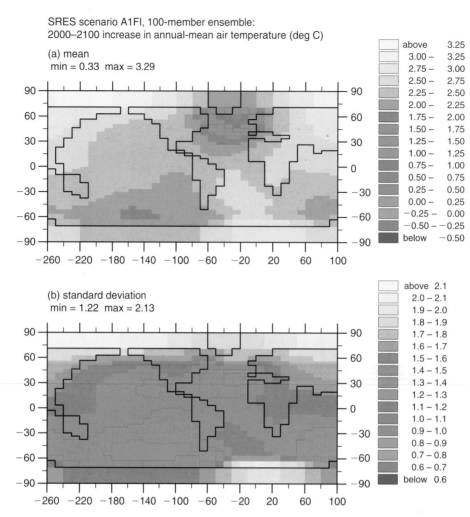

Figure 7.6 As Figure 7.5, under scenario A1FI.

parameter space that result in a collapsed present day circulation. This region could then be set to have zero probability in the input distribution before carrying out the Monte Carlo simulations. This discussion leads us to consider more widely how we might include data in our procedure. The methodology for doing this is explained in Kennedy and O'Hagan (2001).

The second possibility is that the circulation in C-GOLDSTEIN is much more prone to collapse than reality. An intermediate complexity model must by necessity include many assumptions and compromises. A consensus view is that, compared to AOGCMs, the overturning circulation in such models is generally considered more prone to the collapse. However, no one has yet managed to fully explore the behaviour of the overturning circulation across the parameter space of an AOGCM. As discussed above, the spread of our ensemble is not dissimilar to the variation across the set of AOGCMs used by the IPCC. This gives us some confidence that the response of C-GOLDSTEIN's overturning is not very different from the AOGCMs.

The final possibility is that the current consensus is wrong and that the probability of a collapse in the overturning circulation is much higher than believed. There has

been little previous work attempting to quantify the probability. Schaeffer et al. (2002) using ECBilt-CLIO, a different intermediate complexity model, state that 'for a high IPCC non-mitigation emission scenario the transition has a high probability', but they do not quantify what they mean by 'high'. Most model runs investigating the collapse of the overturning circulation, such as CMIP, are run at the most likely value for the parameters and therefore approximately at the 50% probability level so would not detect probabilities of collapse of less than 50%. We should, therefore, at least consider the possibility that the current consensus is wrong and that the probability of a shutdown in the overturning circulation is higher than presently believed. However, the most likely reason for our high probabilities is the model we have used is too simple and has omitted important aspects of the climate system. We caution against giving our results too much credence at this stage. However, we believe that our results do show that it is important that quantitative estimates of dangerous, even if unlikely, climate changes can be made. Our calculations need to be repeated with other models and in particular our statistical methodology needs to be extended to make it viable for use with AOGCMs.

Table 7.2 Probability of Atlantic overturning falling below 5 Sv by 2100.

Uncertainty Case	SRES scenario					
	A1B	A2	B1	B2	A1FI	A1T
default uncertainty						
Case 1a	0.37	0.38	0.31	0.32	0.43	0.32
Case 1b	0.38	0.40	0.30	0.31	0.46	0.31
doubled uncertainty in climate sensitivity						
Case 2a	0.37	0.38	0.33	0.33	0.43	0.33
Case 2b	0.39	0.40	0.31	0.32	0.46	0.32
doubled uncertainty in Atlantic-Pacific moisture flux						
Case 3a	0.37	0.38	0.32	0.33	0.43	0.33
Case 3b	0.40	0.40	0.30	0.30	0.46	0.32
doubled uncertainty in CO_2 uptake						
Case 4a	0.38	0.38	0.31	0.32	0.44	0.33
Case 4b	0.38	0.39	0.31	0.31	0.44	0.32
doubled uncertainty in Greenland melt rate						
Case 5a	0.37	0.38	0.31	0.32	0.43	0.32
Case 5b	0.38	0.39	0.30	0.32	0.45	0.32

In Case 1, 'default uncertainty' refers to the standard deviations for all 17 parameters in Table 7.1. In Cases 2–5, 'doubled uncertainty' refers to twice the standard deviation on an individual parameter (italics in Table 7.1). In each case, 'a' ('b') indicates low (high) mean Greenland melt rate.

Acknowledgements

We thank Jonathan Rougier and Tony O'Hagan for discussions and two anonymous referees for their helpful comments. This work was supported by the 'RAPID' directed research programme of the UK Natural Environment Research Council, and by the Tyndall Centre for Climate Change Research.

REFERENCES

Allen, M.R. and D.A. Stainforth. (2002) Towards objective probabilistic climate forecasting. *Nature,* **419**, 228.

Bøggild, C.E., Mayer, C., Podlech, S., Taurisano, A., and Nielsen, S., 2004. Towards an assessment of the balance state of the Greenland Ice Sheet. *Geological Survey of Denmark and Greenland Bulletin* **4**, 81–84.

P.S. Craig, Goldstein, M., Rougier, J.C. and Seheult, A.H., 2001. Bayesian forecasting for complex systems using computer simulators, *Journal of the American Statistical Association,* **96**, 717–729.

Cubasch, U., Meehl, G.A. and 39 others, 2001. Projections of future climate change. *Climate Change 2001: The Scientific Basis. Contribution of Working Group I to the Third Assessment Report of the Intergovernmental Panel on Climate Change,* J. T. Houghton, et al. Eds., Cambridge University Press, Cambridge, UK, 525–582.

Edwards, N.R. and Marsh, R., 2005. Uncertainties due to transport-parameter sensitivity in an efficient 3-D ocean-climate model. *Clim. Dyn.,* **24**, 415–433.

Garthwaite, P.H., Kadane, J.B. and O'Hagan, A., 2005. Statistical methods for eliciting probability distributions. *Journal of the American Statistical Association,* **100**, 680–701.

Hargreaves, J.C., Annan, J.D., Edwards, N.R. and Marsh, R., 2004. An efficient climate forecasting method using an intermediate complexity Earth system model and the ensemble Kalman Filter. *Clim. Dyn.,* **23**, 745–760.

Huybrechts, P. and de Wolde, J., 1999. The dynamic response of the Greenland and Antarctic ice sheets to multiple-century climatic warming. *J. Climate,* **12**, 2169–2188.

Johnston, W.R., 2004. Historical data relating to global climate change. Available from http://www.johnstonsarchive.net/environment/co2table.html

Kennedy, M.C. and O'Hagan, A., 2001. Bayesian calibration of computer models *Journal of the Royal Statistical Society Series B – Statistical Methodology,* **63**, 425–450.

Knutti, R., Stocker, T.F., Joos, F. and Plattner, G.-K., 2003. Probabilistic climate change projections using neural networks. *Clim. Dyn.,* **21**, 257–272.

McKay, M.D., Beckman, R.J., and Conover, W.J., 1979. A comparison of three methods for selecting values of input variables in the analysis of output from a computer code. *Technometrics,* **21**, 239–245.

Marsh, R., Yool, A., Lenton, T.M., Gulamali, M.Y., Edwards, N.R., Shepherd, J.G., Krznaric, M., Newhouse, S. and Cox, S.J., 2004. Bistability of the thermohaline circulation identified through comprehensive 2-parameter sweeps of an efficient climate model. *Clim. Dyn.,* **23**, 761–777.

Marsh, R., De Cuevas, B.A., Coward, A.C., Bryden, H.L and Alvarez, M., 2005. Thermohaline circulation at three key sections in the North Atlantic over 1985–2002. *Geophys. Res. Lett.,* **32**, doi:10.1029/2004GL022281.

Marotzke, J., 2000. Abrupt climate change and thermohaline circulation: Mechanisms and predictability. *Proceedings of the National Academy of Sciences of the United States of America,* **97**, 1347–1350.

Nakicenovic, N. and Swart, R. (eds.), 2000. *Special Report on Emissions Scenarios.* Cambridge University Press, Cambridge, UK, 612 pp.

Oakley, J. and O'Hagan, A., 2002. A Bayesian inference for the uncertainty distribution of computer model outputs. *Biometrika,* **89**, 769–784.

Prentice, I.C. and 60 others, 2001. The Carbon Cycle and Atmospheric Carbon Dioxide. *Climate Change 2001: The Scientific Basis. Contribution of Working Group I to the Third Assessment Report of the Intergovernmental Panel on Climate Change,* Houghton, J.T. et al. Eds., Cambridge University Press, Cambridge, UK, 183–237.

Rahmstorf, S. and Ganopolski, A., 1999. Long-term global warming scenarios computed with an efficient coupled climate model. *Clim. Change,* **43**, 353–367.

Rind, D., Peteet, D., Broecker, W., McIntyre, A. and Ruddiman, W., 1986. The impact of cold North Atlantic sea surface temperatures on climate: Implications for the Younger Dryas cooling (11–10 k). *Clim. Dyn.,* **1**, 3–33.

Stainforth, D.A., Aina, T., Christensen, C. and 13 others, 2005. Uncertainty in predictions of the climate response to rising levels of greenhouse gases. *Nature,* **433**, 403–406.

Schaeffer, M., Selten, F.M., Opsteegh, J.D. and Goosse, H., 2002. Intrinsic limits to predictability of abrupt regional climate change in IPCC SRES scenarios. *Geophys. Res. Lett.,* **29**, doi:10.1029/2002GL015254.

Vellinga, M. and Wood, R.A., 2002. Global climate impacts of a collapse of the Atlantic thermohaline circulation. *Climatic Change,* **54**, 251–267.

Wigley, T.M.L. and Raper, S.C.B., 2001. Interpretation of high projections for global-mean warming, *Science,* **293**, 451–454.

CHAPTER 8

Reviewing the Impact of Increased Atmospheric CO_2 on Oceanic pH and the Marine Ecosystem

C. Turley, J.C. Blackford, S. Widdicombe, D. Lowe, P.D. Nightingale and A.P. Rees

Plymouth Marine Laboratory, Prospect Place, Plymouth

ABSTRACT: The world's oceans contain an enormous reservoir of carbon, greater than either the terrestrial or atmospheric systems. The fluxes between these reservoirs are relatively rapid such that the oceans have taken up around 50% of the total carbon dioxide (CO_2) released to the atmosphere via fossil fuel emissions and other human activities in the last 200 years. Whilst this has slowed the progress of climate change, CO_2 ultimately results in acidification of the marine environment. Ocean pH has already fallen and will continue to do so with certainty as the oceans take up more anthropogenic CO_2. Acidification has only recently emerged as a serious issue and it has the potential to affect a wide range of marine biogeochemical and ecological processes. Based on theory and an emerging body of research, many of these effects may be non-linear and some potentially complex. Both positive and negative feedback mechanisms exist, making prediction of the consequences of changing CO_2 levels difficult. Integrating the net effect of acidification on marine processes at regional and basin scales is an outstanding challenge that must be addressed via integrated programs of experimentation and modelling. Ocean acidification is another argument, alongside that of climate change, for the mitigation of anthropogenic CO_2 emissions.

8.1 Introduction

The 1999 EU Energy Outlook to 2020 suggests that, despite anticipated increases in energy generation from renewable sources, up to 80% will still be accounted for by fossil fuels. On current trends, CO_2 emissions could easily be 50% higher by 2030. Already about 50% of anthropogenic CO_2 has been taken up by the oceans [1] and thus the oceans have been acting as a buffer, limiting atmospheric CO_2 concentrations. CO_2 in the atmosphere is relatively inert but when dissolved in seawater it becomes highly reactive and takes part in a range of chemical, physical, biological and geological reactions, some of which are predictable while some are more complex. Warming of the oceans will only have a small direct impact on the rate of oceanic uptake via changes in the solubility of CO_2. However, the oceans' capacity to absorb more CO_2 decreases as they take up CO_2.

Of all the predicted impacts attributed to this inevitable rise in atmospheric CO_2 and the associated rise in temperature (e.g. large-scale melting of ice sheets, destabilisation of methane hydrates, sea level rise, slowdown in the North Atlantic thermohaline circulation) one of the most pressing is the acidification of surface waters through the absorption of atmospheric CO_2 and its reaction with seawater to form carbonic acid [2, 3].

Predictions of atmospheric CO_2 concentrations, due to the unrestricted release of fossil fuel CO_2, by 2100 are 700 ppm [4] and by 2300 are 1900 ppm [3, 5] (based on median scenarios). This would equate to a decrease in surface ocean pH of 0.3 and 0.8 pH units from pre-industrial

levels respectively [2, 3]. The top-end prediction of 1000 ppm CO_2 by 2100 would equate to a pH decrease of 0.5 units which is equivalent to a threefold increase in the concentration of hydrogen ions [5]. While climate change has uncertainty, these geochemical changes are highly predictable. Only the timescale and thus mixing scale length are really under debate. Such dramatic changes in ocean pH have probably not been seen for millions of years of the Earth's history [6, Figure 8.1].

8.2 Global Air–Sea Fluxes of Carbon Dioxide

There has been an increase in atmospheric carbon dioxide from 280 ppm in AD1800 to 380 ppm at the present day. This increase is due to a supply of anthropogenic CO_2 to the atmosphere which is currently estimated at 7 GtC yr^{-1} [4]. The observed annual increase in atmospheric CO_2 represents 3.2 GtC yr^{-1}, the balance being removed from the atmosphere and taken up by the oceans and land. There is now generally good agreement that the ocean absorbs 1.7 ± 0.5 GtC yr^{-1} [4]. Note that the rate-limiting step in the long-term oceanic uptake of anthropogenic CO_2 is not air-sea gas exchange, but the mixing of the surface waters with the deep ocean [7]. Whilst the ocean can theoretically absorb 70–80% of the projected production of anthropogenic CO_2, it would take many centuries to do so [8].

There is also a large natural annual flux of CO_2 between the ocean and the atmosphere of almost 90 GtC yr^{-1} that, pre-1800, was believed to be almost in balance. This

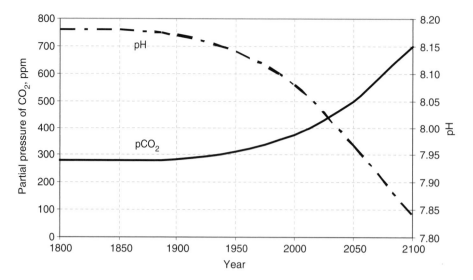

Figure 8.1 The past and projected change in atmospheric CO$_2$ and seawater pH assuming anthropogenic emissions are maintained at current predictions (redrawn from Zeebe and Wolf-Gladrow 2001).

huge influx and efflux is due to a combination of marine productivity and particle sinking (the biological pump) and ocean circulation and mixing (the solubility pump). Phytoplankton growth consumes dissolved inorganic carbon (DIC) in the surface seawater causing an undersaturation of dissolved CO$_2$ and uptake from the atmosphere. The re-equilibration time for CO$_2$ is slow (typically several months) due to the dissociation of CO$_2$ in seawater (see below). Ocean circulation also results in air-sea exchange of CO$_2$ as the solubility of CO$_2$ is temperature dependent. Warming decreases the solubility of CO$_2$ and promotes a net transfer of CO$_2$ to the atmosphere, whereas cooling results in a flux from the atmosphere to the ocean. Anthropogenic CO$_2$ modifies the flux from the solubility pump as CO$_2$ availability does not normally limit biological productivity in the world's oceans.

However, the observation that the net oceanic uptake of anthropogenic CO$_2$ is only about 2% of the total CO$_2$ cycled annually across the air-sea interface ought to be of major concern. The significant perturbations arising from this small change in flux imply that the system is extremely sensitive. Any resulting changes in the biogeochemistry of the mixed layer could have a major impact on the magnitude (or even sign) of the total CO$_2$ flux and hence on the Earth's climate [9].

8.3 The Carbonate System

The chemistry of carbon dioxide in seawater has been the subject of considerable research and has been summarized by Zeebe and Wolf-Gladrow [2]. Dissolved inorganic carbon can be present in any of 4 forms, dissolved carbon dioxide (CO$_2$), carbonic acid (H$_2$CO$_3$), bicarbonate ions (HCO$_3^-$) and carbonate ions (CO$_3^{2-}$). Addition of CO$_2$ to seawater, by air–sea gas exchange due to increasing CO$_2$

in the atmosphere, leads initially to an increase in dissolved CO$_2$ (equation 8.1). This dissolved carbon dioxide reacts with seawater to form carbonic acid (equation 8.2). Carbonic acid is not particularly stable in seawater and rapidly dissociates to form bicarbonate ions (equation 8.3), which can themselves further dissociate to form carbonate ions (equation 8.4). At a typical seawater pH of 8.1 and salinity of 35, the dominant DIC species is HCO$_3^-$ with only 1% in the form of dissolved CO$_2$. It is the relative proportions of the DIC species that control the pH of seawater on short to medium timescales.

$$CO_{2(atmos)} \Leftrightarrow CO_{2(aq)} \tag{8.1}$$

$$CO_2 + H_2O \Leftrightarrow H_2CO_3 \tag{8.2}$$

$$H_2CO_3 \Leftrightarrow H^+ + HCO_3^- \tag{8.3}$$

$$HCO_3^- \Leftrightarrow H^+ + CO_3^{2-} \tag{8.4}$$

It is also important to consider the interaction of calcium carbonate with the inorganic carbon system. Calcium carbonate (CaCO$_3$) is usually found in the environment either as calcite or less commonly aragonite. Calcium carbonate dissolves in seawater forming carbonate ions (CO$_3^{2-}$) which react with carbon dioxide as follows:

$$CaCO_3 + CO_2 + H_2O \Leftrightarrow Ca^{2+} + CO_3^{2-} + CO_2 + H_2O$$
$$\Leftrightarrow Ca^{2+} + 2HCO_3^- \tag{8.5}$$

This reaction represents a useful summary of what happens when anthropogenic carbon dioxide dissolves in seawater. The net effect is removal of carbonate ions and production of bicarbonate ions and a lowering in pH. This in turn will encourage the dissolution of more calcium carbonate. Indeed, the long-term sink for anthropogenic CO$_2$ is dilution in the oceans and reaction with carbonate sediments.

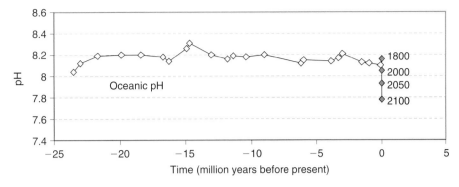

Figure 8.2 Past (white diamonds, data from Pearson and Palmer, 2000) and contemporary variability of marine pH (grey diamonds with dates). Future predictions are model derived values based on IPCC mean scenarios.

As can clearly be seen above, formation of calcite (the reverse of equation 8.5) actually produces CO_2.

Seawater at current pH levels is highly buffered with respect to carbon dioxide and has a great capacity to absorb carbon dioxide, as most of the CO_2 added will rapidly be converted to bicarbonate ions. It can be shown that if the atmospheric CO_2 levels doubled, dissolved CO_2 would only rise by 10%, with most of the remaining 90% being converted to bicarbonate ions. However, if bicarbonate ions increase, then the equilibrium of reaction 3 will be forced forwards and hence the pH of the seawater will be reduced. This is of great importance both for seawater chemistry and for the buffering capacity of seawater as it reduces the ability of seawater to buffer further CO_2 increases [2]: i.e. as the partial pressure of carbon dioxide increases the buffering capacity of seawater decreases.

The mean pH of seawater has probably changed by less than 0.1 units over the last several million years [6, Figure 8.2]. Since the start of the Industrial Revolution (circa 1800), the release of anthropogenic CO_2 to the atmosphere and subsequent flux into the surface oceans has already led to a decrease in the pH of oceanic surface waters of 0.1 unit [10, 5]. The same calculations show that the current rate of increase in atmospheric CO_2 concentration (15 ppm/decade) will cause a decrease in pH of 0.015 units/decade [11]. Globally, oceanic surface water pH varies over a range of 0.3 pH units, due to changes in temperature and seasonal CO_2 uptake and release by biota. However, the current surface ocean pH range is nearly distinct from that assumed for the inter-glacial period and the predicted pH for 2100 is clearly distinct from that of the pre-industrial period (Figure 8. 2). In some sense therefore the marine system is accelerating its entry into uncharted territory. Whilst species shifts and adaptation of physiology and community structure might maintain the system's gross functionality over longer timescales, the current rates of environmental change are far more rapid than previously experienced. We do not know if marine organisms and ecosystems will be able to adapt at these timescales.

8.4 Ecosystem Impacts

Although studies looking at ecosystem response are in their infancy, reduced pH is a potent mechanism by which high CO_2 could affect marine biogeochemistry [5, 12, 13]. The changes to the carbonate chemistry of the system [14, 15] may affect plankton species composition and their spatial or geographical distribution [16], principally by inhibiting calcifying organisms such as coccolithophores, pteropods, gastropods, foraminifera and corals in waters with high CO_2 [5]. Reduced calcification in cultures of two species of coccolithophores has been observed when grown at 750 ppm CO_2 [17]. Other non-calcifying organisms may grow in their place and impact the structure and processes occurring in the whole ecosystem. The main calcifiers in the ocean are the planktonic microalgae, coccolithophores [18], which secrete calcite platelets called liths. These organisms can form massive blooms, often of 100,000s km². They play an important role in the global carbon cycle through the transport of calcium carbonate to the marine sediments. Coccolithophores are also a major producer of dimethyl sulphide (DMS) which may have a role in climate regulation via the production of cloud condensation nuclei [19]. A reduction in the occurrence of the coccolithophore blooms that occur in large areas of the global oceans could lead to a reduced flux of DMS from the oceans to the atmosphere and hence further increases in global temperatures via cloud changes. International efforts to examine the impacts of high CO_2 in more natural enclosed seawater systems (mesocosms) with blooms of coccolithophores shows that calcification, growth rates and exudation can be affected by high CO_2 and this has implications on biogeochemical cycling, carbon export and food web dynamics [20, 21]. Over long timescales calcium carbonate is the major form in which carbon is buried in marine sediments, hence species composition is intimately linked to the strength of the biological pump and carbon burial in sediments [22, 23].

The effect of high CO_2 on tropical coral reefs has received particular attention [24, 25, 26] because calcification rates in corals (which secrete a more thermodynamically stable form of $CaCO_3$, aragonite) decline under elevated CO_2 conditions. Predictions are that coral calcification rates may decrease by 21–40% over the period 1880–2065 in response to changes in atmospheric CO_2 concentrations [27, 28, 29]. Reduction in coral calcification can result in declining coral cover and loss of the reef environments [25]. Coral reefs are essentially oases of high productivity such that they produce 10–12% of the fish caught in the tropics and 20–25% of the fish caught by developing nations [30]. The sea contributes about 90% of the animal protein consumed by many Pacific Island countries.

Calcification rates respond not only to carbonate saturation state, but also to temperature, nutrients, and light. It has been argued that increasing temperature, at least in corals, may invoke a biological response that leads to higher calcification rates in the short term. This might offset the impact of declining carbonate ion concentrations [31]. Although there is concern over these studies [5, 25] they do show the importance of looking at the impacts synergistically.

Extensive cold water corals have been discovered in the last decade in many of the world's oceans that may equal or even exceed the coverage of the tropical coral reefs [32]. A decrease in the depth below which aragonite dissolves, due to reduced carbonate ion concentrations, may make these ecosystems particularly vulnerable [33]. This effect will be greatest in the higher latitudes and impact calcifying organisms that live there [5]. For instance, pteropods are the dominant calcifiers in the Southern Ocean and are an important part of the Antarctic food web and ecosystem [33].

The availability of marine nutrients, necessary for primary production, is affected by pH. The form of both phosphorus and nitrogen, the key macro nutrients, are pH sensitive; acidification provoking a reduction in the available form of phosphate (PO_4^{3-}) and a decrease in ammonia (NH_3) with respect to ammonium (NH_4^+), changing the energetics of cellular acquisition. A second consequence of low pH may be the inhibition of microbial nitrification [34] with a resulting decrease in the oxidised forms of nitrogen (e.g. NO_3^-). As a result we may see a decrease in the NO_3^- dependant denitrification process which removes nitrogen from the marine system in the form of nitrogen gas. The resulting build-up of marine nitrogen (mainly as NH_4^+) may trigger eutrophication effects.

The solubility (and availability) of iron, an important micro-nutrient, is likely to increase with acidification, perhaps increasing productivity in some remote ocean basins that are currently iron limited. The net effect of these processes is likely to change the nutrient availability to phytoplankton, impacting species composition and distribution and consequently the rate of carbon cycling in the marine system. Changes to the phytoplankton community structure are likely to affect the organisms that prey on phytoplankton, including economically important species [35, 36, 37].

If the environmental CO_2 concentration is high (equivalent to three-fold increases in atmospheric CO_2 relative to pre-industrial), fish and other complex animals are likely to have difficulty reducing internal CO_2 concentrations, resulting in accumulation of CO_2 and acidification of body tissues and fluids (hypercapnia) [38]. The effects of lower level, long term increases in CO_2 on reproduction and development of marine animals is unknown and of concern. High sensitivity to CO_2 is shown by squid (Cephalopods), because of their high energy and oxygen demand for jet propulsion, with a relatively small decrease in pH of 0.25 having drastic effects (reduction of c. 50%) on their oxygen carrying capacity [39].

Experiments, using CO_2 concentration beyond that expected to be seen in the next few hundred years, have shown that decreased motility, inhibition of feeding, reduced growth, reduced recruitment, respiratory distress, decrease in population size, increased susceptibility to infection, shell dissolution, destruction of chemosensory systems and mortality can occur in high CO_2/low pH waters in the small range of higher organisms tested to date, many of which are shellfish [5]. However, further experiments are required to investigate the impacts of the CO_2 and pH levels relevant to ocean uptake of anthropogenic CO_2.

Juvenile forms of shellfish may be less tolerant to changes in pH than adults. Indeed, greater than 98% of the mortality of settling marine bivalves occurs within the first few days or weeks after settling. This is thought to be in part due to their sensitivity to the carbonate saturation state at the sediment-water interface [40]. The higher seawater CO_2 concentrations that will occur in the future may therefore enhance shell dissolution and impact recruitment success and juvenile survival.

The average carbonate saturation state of benthic sediment pore waters could decline significantly, inducing dissolution of carbonate phases within the pore-water-sediment system [14]. Further, the benthic sediment chemistry of shallow coastal seas exhibits a delicate balance between aerobic and anaerobic activity which may be sensitive to varying pelagic CO_2 loads. In short, marine productivity, biodiversity and biogeochemistry may change considerably as oceanic pH is reduced through oceanic uptake of anthropogenic CO_2.

Changes that may occur in the same time frame as increased seawater CO_2 and reduced pH, include increased seawater temperature, changes in the supply of nutrients to the euphotic zone through stronger water column stratification, changes in salinity, and sea-level rise. There are likely to be synergistic impacts on marine organisms and ecosystems. There is surprisingly little research on the potential impact of a high CO_2 ocean on marine organisms and ecosystems let alone the impact this might have when combined with other climate-induced changes. This needs to be redressed. Whilst about 28 million people are employed in fishing and aquaculture with a global fish trade of US\$53,000 million [30], the marine environment provides other valuable services [41] and its existence and diversity is treasured. As the oceans play a key role in the Earth's life support system, it would seem that a better understanding of the impacts of high CO_2 on the marine environment and consideration of mitigation and stabilization choices is worthy of substantial investment.

8.5 International Recognition

The global scientific community is increasingly concerned about the impacts of a high CO_2 ocean. This community includes the International Global Biosphere Programme (*IGBP*), the Scientific Committee on Oceanic Research (*SCOR*), the Commission on Atmospheric Chemistry and Global Pollution (*CACGP*) and the International Council for Science (*ICSU*). A SCOR and IOC-funded International Science Symposium held at UNESCO, Paris on 10–12 May 2004, *Symposium on the Ocean in a High-CO₂ World*, brought together scientists working in this area for the first time. The scientific consensus has been summarised in the report *Priorities for Research on the Ocean in a High-CO₂ World* [42] and the overwhelming conclusion was that there is an urgent need for more research in this area. The Royal Society formed an international working group to report on ocean acidification and published on 30 June 2005 [5]. Commissions and conventions that are policy instruments for the protection of our seas (such as the OSPAR (Oslo–Paris) Commission and the London Convention) have held workshops on the environmental impact of placement of CO_2 in geological structures in the maritime area and recognise the significance of ocean acidification caused by uptake of anthropogenic CO_2 as a strong argument, along with climate change, for global mitigation of CO_2 emission. A report to Defra, summarising the current knowledge of the potential impact of ocean acidification (by direct uptake or by release from sub-seabed geological sequestration) concluded that there was a need for urgent research to help inform government of the potential impact of both ocean uptake of anthropogenic CO_2 and its release from maritime sea bed geological structures [43].

8.6 Conclusions

This paper outlines only a few of the potential effects that higher CO_2 may have on the marine system. Many other processes are pH sensitive: for example, changes in pH also have the potential to disrupt metal ion uptake causing symptoms of toxicity, and intra-cellular enzymatic reactions are also pH sensitive [5]. Given continued CO_2 emissions, further marine acidification is inevitable and effects on the marine ecosystem are likely to be measurable. Whilst many of the effects are nominally negative, some could be considered positive. How these may balance out is unknown. The scientific community is far from being able to predict accurately the impact of acidification on the oceans and whether an appreciable decline in resource base may occur. We also need to address the key question of whether marine organisms and ecosystems have the ability to adapt to the predicted changes in CO_2 and pH. Ocean acidification will occur within the same time scales as other global changes associated with climate impacts. These also have much potential to alter marine biogeochemical cycling.

Modelling techniques provide an important mechanism for resolving whole system impact. Indeed, several researchers cite the need for integrated modelling studies [e.g. 35]. The problem is multi-disciplinary. We need to integrate atmosphere, hydrodynamic and ecosystem modellers, to build on experimental knowledge, and require significantly more system measurements in order to validate models. UK and international momentum is building towards this challenge and many of the required collaborations are being forged. However, the provision of manpower, computer, experimental and observational resources still needs to be addressed. Mitigation of CO_2 emissions will decrease the rate and extent of ocean acidification [5]. This is another powerful argument to add to that of climate change for reduction of global anthropogenic CO_2 emissions.

REFERENCES

1. Sabine, C.L., Feely, R.A., Gruber, N., Key, R.M., Lee, K., Bullister, J.L., Wanninkhof, R., Wong, C.S., Wallace, D.W.R., Tilbrook, B., Millero, F.J., Peng, T.H., Kozyr, A., Ono, T. and Rios, A.F., 2004. The oceanic sink for anthropogenic CO_2. *Science*, **305**: 367–371.
2. Zeebe, R.E. and Wolf-Gladrow, D.A., 2001. CO_2 in seawater: equilibrium, kinetics and isotopes. Elsevier Oceanography Series. **65**: pp. 346.
3. Caldeira, K. and Wickett, M.E., 2003. Anthropogenic carbon and ocean pH. *Nature*, **425**, p. 365.
4. IPCC, 2001. *Climate Change 2001: The Scientific Basis. Contribution of Working Group I to the Third Assessment Report of the Intergovernmental Panel on Climate Change* [Houghton, J.T., Ding, Y., Griggs, D.J., Noguer, M., van der Linden, P.J., Dai, X., Maskell, K., and Johnson C.A. (Eds.)]. Cambridge University Press, Cambridge, United Kingdom and New York, NY, USA, pp. 881.
5. The Royal Society, 2005. Ocean acidification due to increasing atmospheric carbon dioxide. Document 12/05 Royal Society: London.
6. Pearson, P.N. and Palmer M.R., 2000. Atmospheric carbon dioxide concentrations over the past 60 million years. *Nature*, **406**, 695–699.
7. Sarmiento, J.L. and Sundquist, E.T., 1992. Revised budget for the oceanic uptake of anthropogenic carbon-dioxide. *Nature*, **356**, 589–593.
8. Maier-Reimer, E. and Hasselmann, K., 1987. Transport and storage of CO_2 in the ocean – An inorganic ocean-circulation carbon cycle model. *Climate Dynamics*, **2**, 63–90.
9. Nightingale, P.D. and Liss P.S., 2003. Gases in seawater. In, *The oceans and marine geochemistry. A Treatise on Geochemistry, Vol. 6* [H. Elderfield, Ed.]. Elsevier.
10. Haughan, P.M. and Drange, H., 1996. Effects of CO_2 on the ocean environment. *Energy Conservation and Management*, **6–8**, 1019–1022.

11. Ormerod, W.G., Freund, and Smith, A., 2002. Ocean storage of carbon dioxide, 2nd edn. *IEA Greenhouse Gas R&D Programme*, pp. 26.

12. Wolf-Gladrow, D.A., Riebesell, U., Burkhardt, S. and Bijma, J., 1999. Direct effects of CO$_2$ concentration on growth and isotopic composition of marine plankton. *Tellus*, **B51**, 461–476.

13. Knutzen, J., 1981. Effects of decreased pH on marine organisms. *Marine Pollution Bulletin*, **12**, 25–29.

14. Andersson, A. J., Mackenzie, F. T, and Ver, L.M., 2003. Solution of shallow-water carbonates: An insignificant buffer against rising atmospheric CO$_2$. *Geology*, **31**, 513–516.

15. Feely, R.A., Sabine, C.L., Lee, K., Berelson, W., Kleypas, J., Fabry, V.J. and Millero, F.J., 2004. Impact of anthropogenic CO$_2$ on the CaCO$_2$ system in the ocean. *Science*, **305**, 362–366.

16. Boyd, P.W. and Doney, S.C., 2002. Modelling regional responses by marine pelagic ecosystems to global climate change. *Geophysical Research Letters*, **29**, 1806, doi: 10.1029/2001GL014130.

17. Riebesell, U., Zondervan, I., Rost, B., Tortell, P.D., Zeebe, R.E. and Morel, F.M.M., 2000. Reduced calcification of marine plankton in response to increased atmospheric CO$_2$. *Nature*, **407**, 364–367.

18. Holligan, P. M., Fernandez, E., Aiken, J., Balch, W.M., Boyd, P., Burkill, P.H., Finch, M., Groom, S.B., Malin, G., Muller, K., Purdie, D.A., Robinson, C., Trees, C.C., Turner, S.M. and Vanerwal, P., 1993. A Biogeochemical Study of the Coccolithophore, *Emiliania huxleyi*, in the North-Atlantic. *Global Biogeochemical Cycles*, **7**, 879–900.

19. Charlson, R.J., Lovelock, J.E., Andreae, M.O. and Warren, S.G., 1987. Oceanic phytoplankton, atmospheric sulphur, cloud albedo and climate. *Nature*, **326**, 655–661.

20. Engel, A., Zondervan, I., Aerts, K., Beaufort, L., Benthien, A., Chou, L., Delille, B., Gattuso, J.-P., Harly, J., Heemaan, C., Hoffmann, L., Jacquet, S., Nejstgaard, J., Pizay, M.-D., Rochelle-Newall, E.S., Schneider, U., Terbrueggen, A. and Riebesell, U., 2005. Testing the direct effect of CO$_2$ concentration on marine phytoplankton: A mesocosm experiment with the coccolithophorid *Emiliania huxleyi*. *Limnology & Oceanography*, **50**, 493–507.

21. Delille, B., Harlay, J., Zondervan, I., Jacquet, S., Chou, L., Wollast, R., Bellerby, R.G.J., Frankignoulle, M., Borges, A.V., Riebesell, U. and Gattuso, J-P., 2005. Response of primary production and calcification to changes of pCO$_2$ during experimental blooms of the coccolithophorid Emiliania huxleyi. *Global Biogeochemical Cycles*, **19**, GB2023 10.1029/2004GB002318.

22. Archer, D., 1995. Upper Ocean Physics as Relevant to Ecosystem Dynamics – a Tutorial. *Ecological Applications*, **5**, 724–739.

23. Tian, R.C., Vezina, A., Legendre, L., Ingram, R.G., Klein, B., Packard, T., Roy, S., Saenkoff, C., Silverberg, N., Therriault, J.C. and Tremblay, J.E., 2000. Effects of pelagic food-web interactions and nutrient remineralization on the biogeochemical cycling of carbon: a modeling approach. *Deep Sea Research, Part II*, **47**, 637–662.

24. Kleypas, J.A., Buddemeier, R.W., Archer, D., Gattuso, J.-P., Langdon, C. and Opdyke, B.N., 1999. Geochemical Consequences of Increased Atmospheric Carbon Dioxide on Coral Reefs. *Science*, **284**, 118–120.

25. Kleypas, J.A., Buddemeier, R.W. and Gattuso, J.-P., 2001. The future of coral reefs in an age of global change. *International Journal Earth Sciences*, **90**, 426–437.

26. Marubini, F. and Atkinson M. J., 1999. Effects of lowered pH and elevated nitrate on coral calcification. *Marine Ecology Progress*, **188**, 117–121.

27. Langdon, C., Takahashi, T., Marubini, F., Atkinson, M., Sweeney, C., Aceves, H., Barnett, H., Chipman, D. and Goddard, J., 2000. Effect of calcium carbonate saturation state on the calcification rate of an experimental coral reef. *Global Biogeochemical Cycles*, **14**, 639–654.

28. Langdon, C., Broecker, W.S., Hammond, D.E., Glenn, E., Fitzsimmons, K., Nelson, S.G., Peng, T.-H., Hajdas, I. and Bonani, G., 2003. Effect of elevated CO$_2$ on the community metabolism of an experimental coral reef. *Global Biogeochemical Cycles*, **17**, 1011, doi: 10.1029/2002GB001941.

29. Leclercq, N., Gattuso, J.-P. and Jaubert, J., 2000. CO$_2$ partial pressure controls the calcification rate of a coral community. *Global Change Biology*, **6**, 329–334.

30. Garcia, S.M. and De Leiva Moreno, I., 2001. Global overview of marine fisheries. Report of the Conference on Responsible Fisheries in the Marine Ecosystem (ftp://ftp.fao.org/fi/DOCUMENT/reykjavik/Default.htm), Reykjavik, Iceland, 1– 4 October 2001.

31. Lough, J.M. and Barnes, D.J., 2000. Environmental controls on growth of the massive coral porites. *Journal of Experimental Marine Biology and Ecology*, **245**, 225–243.

32. Freiwald, A., Fossa, J.H., Grehan, A., Koslow, T. and Roberts, J. M., 2004. Cold-water coral reefs: Out of sight no longer out of mind. No. 22 in Biodiversity Series, UNEP-WCMC, Cambridge, UK.

33. Orr, J.C., Fabry, V.J., Aumont, O., Bopp, L., Doney, S.C., Feely, R.A., Gnanadesikan, A., Gruber, N., Ishida, A., Joos, F., Key, R.M., Lindsay, K., Maier-Reimer, E., Matear, R., Monfray, P., Mouchet, A., Najjar, R.G., Plattner, G-K., Rodgers, K.B., Sabine, C.L., Sarmiento, J.L., Schlitzer, R., Slater, R.D., Totterdell, I.J., Weirig, M-F., Yamanaka, Y. and Yool, A., 2005. Anthropogenic ocean acidification over the twenty-first century and its impact on calcifying organisms. *Nature*, **437**, 681–686.

34. Huesemann, M.H., Skillman, A.D. and Creclius, E.A., 2002. The inhibition of marine nitrification by ocean disposal of carbon dioxide. *Marine Pollution Bulletin*, **44**, 142–148.

35. DeAngelis, D. L. and Cushman R. M., 1990. Potential application of models in forecasting the effects of climate changes on fisheries. *Transactions of the American Fisheries Society*, **119**, 224–239.

36. Frank, K.T., Perry, R.I. and Drinkwater, K.F., 1990. Predicted response of Northwest Atlantic invertebrate and fish stocks to CO$_2$-induced climate change. *Transactions of the American Fisheries Society*, **119**, 353–365.

37. Nejstgaard, J.C., Gismervik, I. and Solberg, P.T., 1997. Feeding and reproduction by Calanus finmarchicus and microzooplankton grazing during mesocosm blooms of diatoms and the coccolithophore *Emiliania huxleyi*. *Marine Ecology Progress*, **147**, 197–217.

38. Portner, H.O., Langenbuch, M. and Reipschlager, A., 2004. Biological impact of elevated ocean CO$_2$ concentrations: lessons from animal physiology and Earth history. *Journal of Oceanography*, **60**, 705–718.

39. Portner, H.O. and Reipschlager, A., 1996. Ocean dispersal of anthropogenic CO$_2$: physiological effects on tolerant and intolerant animals. In, *Ocean storage of CO$_2$. Environmental Impact.* [Omerod, B. and Angel, M., Eds.] MIT and IEA Greenhouse Gas ROD Programme, Cheltenham, Boston, pp. 57–81.

40. Green, M.A., Jones, M.E., Boudreau, C.L., Moore, R.L. and Westman, B.A., 2004. Dissolution mortality of juvenile bivalves in coastal marine deposits. *Limnology & Oceanography*, **49**, 727–734.

41. Costanza, R.R., d'Arge, R., de Groot, R., Farber, S., Grasso, M., Hannon, B., Limburg, K., Naeem, S., O'Neill, R.V., Paruleo, J., Raskin, R.G., Sutton, P. and van den Belt, M.,1997. The value of the world's ecosystem services and natural capital. *Nature*, **387**, 253–260.

42. Priorities for Research on the Ocean in a High-CO$_2$ World. http://ioc.unesco.org/iocweb/co2panel/Docs/Research%20Priorities%20Report-Final.pdf

43. Turley, C., Nightingale, P., Riley, N., Widdicombe, S., Joint, I., Gallienne, C., Lowe, D., Goldson, L., Beaumont, N., Mariotte, P., Groom, S., Smerdon, G., Rees, A., Blackford, J., Owens, N., West, J., Land, P. and Woodason, E., 2004. Literature Review: Environmental impacts of a gradual or catastrophic release of CO$_2$ into the marine environment following carbon dioxide capture. DEFRA: MARP 30 (ME2104) 31 March 2004.

SECTION II

General Perspectives on Dangerous Impacts

INTRODUCTION

There are evidently different approaches towards a common scientific understanding of the notion 'dangerous climate change'. The approach highlighted in Section I tries to identify key elements of the Earth System that might be altered ('activated', 'switched', 'tipped') – possibly abruptly and irreversibly – by anthropogenic global warming. In a sense, this is the search for potential 'knock-out criteria' inspiring the public debate on climate protection. The approach introduced in this section is less elegant but certainly not less relevant: instead of focussing on one or two geophysical watershed events (like the collapse of the West Antarctic Ice Sheet), the entire range and diversity of potential climate change impacts on natural and human systems is considered. This exercise is driven by the hope that, for all the complexities involved, certain structures might emerge in impact space that allow telling 'dangerous' from 'innocuous'. For instance, going along the global mean temperatures axis, there may be sections where individual negative impacts tend to cluster or change character collectively.

The section presents several general perspectives on this very approach that will be underpinned by a wealth of concrete and detailed studies as presented in Chapters 9–11.

Izrael and Semenov develop some fundamental thoughts on the various quantitative components that should be taken into account when addressing the 'D Question'. They refer, on the one hand, to critical thresholds and vulnerabilities of the planetary system as discussed in Section I, yet underline, on the other hand, the importance of calculating the residual damages associated with any given stabilization level. The paper argues that humankind's burning of the entire fossil fuel pool would not cause dramatic atmospheric changes in the very long run (ten thousand of years), yet would bring about pernicious interference with civilization at the secular/millennium scale. The authors propose tentative limits of temperature rise, namely 2.5°C above pre-industrial level for the globe and 4°C for the Arctic. Sea-level rise should be limited to 1m overall.

By way of contrast, Yamin et al. suggest that there are many levels of potentially dangerous anthropogenic interference, given the complexity of climate change impacts and the multiple scales at which they are felt. It may be

desirable to establish a goal which stabilises concentrations at as low a level as feasible and which can be revisited in the light of improvements in scientific understanding, the capacity to reduce emissions or as values change. This should recognise that impacts below the goal may still be dangerous and will need to be the focus of adaptation. To be broadly accepted and meaningful, any process to determine a target should be as transparent as possible and incorporate public values and perceptions. The authors conclude their discourse by musing on an alternative approach to UNFCCC-Article 2, which would re-direct the debate away from 'dangerous' climate change in favour of identifying 'tolerable' levels.

The Warren contribution, finally, is an heroic effort of aggregating all possible impacts information from the pertinent literature. This paper may be seen as the bottom-up counterpart to the top-down approach adopted by Izrael and Semenov. Through several tables and appendices, a general (but, of course, preliminary) picture is constructed that sketches the distribution of impacts in response to increasing levels of global warming. The emerging pattern is still far too weak to be conclusive, yet confirms the IPCC TAR assessment that a multitude of damaging effects will be triggered by a 2–3°C temperature increase.

Several additional points worth mentioning in this introduction are either made in the section papers or were raised in the pertinent plenary discussions at the Exeter conference. First, the scientific assessment of climate change risks needs to take into account both gradual and discontinuous processes, the interactions between them, and the synergistic effects of climate change and other human-induced stresses. Second, as the planet warms, societies will also be changing. New technologies will emerge, ground-breaking discoveries will be made and population structures and distributions will alter. These dynamics will, in turn, transform the adaptive capacities of communities at all scales and, thereby, the character of dangers faced. Third, the notion of resilience is a key element of the analysis. For instance, climate change will expose more people to infection by malaria, but the increment is probably small in relation to the total number at risk. A resilient society, with excellent public health measures containing malaria, will be able to cope.

CHAPTER 9

Critical Levels of Greenhouse Gases, Stabilization Scenarios, and Implications for the Global Decisions

Yu. A. Izrael and S. M. Semenov
Institute of Global Climate and Ecology

ABSTRACT: Critical values for greenhouse gas concentrations and global surface temperature can be obtained through either cost-benefit analysis of mitigation cost and residual damage to climate and socio-economic systems, or investigations of critical thresholds for climate change for key vulnerable elements of the systems. The scientific basis for the estimation of such critical values has not yet been completely developed, although intensive studies in this field are being carried out worldwide. The Earth's climate system has natural variations observed on millennium and century scales. They are driven, in particular, by solar and orbital factors interacting with the climate system of the Earth. Anthropogenic perturbations of the climate system are to be assessed against this baseline. The ability of humans to influence the CO_2 amount in the atmosphere in the long-term perspective is very limited, because the world ocean has a huge capacity to accumulate carbon, As follows from calculations with a simple linear model, even if all the known commercially-efficient resources of fossil fuels are used, the associated asymptotic CO_2 level will be substantially lower than at present. However, transition values may be much higher and cause serious damage to vulnerable earth systems and socio-economic systems. A set of concentration trajectories to be assessed in the analysis of 'safe' global stabilization scenarios for emissions should not only include monotonic ones, but also so-called 'overshoot' trajectories allowing concentrations to exceed the target value for a while. Analysis of uncertainties is absolutely crucial for correct establishment of critical values for greenhouse gas concentrations and global surface temperature.

The global CO_2 concentration ranged from 180 to 300 ppmv over the past 400,000 years (Barnola et al., 2003). It varied roughly within a 270–290 ppmv interval over the last 1000 years in the pre-industrial era to 1860 and thus was practically stable (Climate Change 2001, 2001a, p. 185). Since the middle of the 19th century, CO_2 concentration has been increasing rapidly (Climate Change 2001, 2001a, p. 201) and exceeds 370 ppmv at present.

Regional natural variations of surface temperature are large on a century scale. For example, as paleodata from Vostok station (Antarctica) show, in the last millennium 200 and 400 years ago, a temperature rise of roughly 0.5–1.5°C emerged, developed, and ended within approximately 100 years (Petit et al., 2000; Semenov, 2004b). These events were caused by natural factors, most probably by solar and orbital factors interacting with the non-linear climate system of the Earth. Anthropogenic emissions of greenhouse gases raising their concentrations in the atmosphere undoubtedly lead to the enhancement of the greenhouse effect and a respective increase in global mean surface temperature. However, this increase will be against the baseline determined by natural variations of global climate, which is not completely understood yet.

The unprecedented (for the last 400,000 years) rise in atmospheric CO_2 since the 1850s and a discernible increase in global surface temperature (0.6 ± 0.2°C) in the 20th century, usually associated with the anthropogenic enhancement of the greenhouse effect, were the major reasons for the development and adoption of the United Nations Framework Convention on Climate Change (UNFCCC) in 1992 aiming at **stabilization**, i.e. keeping greenhouse gas concentrations below a certain constant 'not dangerous' level. However, until now no inter-governmental decision on a **particular level** has been taken, and its nature still remains unclear. Working group II of the IPCC has included the investigation of such potential levels in its outline for the Fourth Assessment Report (to be issued in 2007).

The economic analysis of stabilization scenarios for 1000, 750, 650, 550 and 450 ppmv of CO_2 as stabilization targets showed that stabilization is not free of charge for the world community. In particular, for 450 ppmv, this may cost as much as \$3.5–17.5 trillion in 1990 prices over 100 years (Climate Change 2001, 2001c, p. 119). Although some publications have shown that this level of spending will have little effect on worldwide GDP growth over a 100-year timescale (Azar and Schneider, 2002), the potential efficiency of such non-negligible 'investments' should be properly analysed using the cost-benefit approach. The framework could be outlined as follows.

It is usually assumed that with no emission control, certain climate-change damage to the Earth's systems and socio-economic systems will occur. The likely extent of the damage appears to be substantial, at least comparable to the mitigation costs. Otherwise, there would be no reason for any control measures. In this connection, one can consider emission reduction scenarios, the implementation of which prevents a certain part of the damage.

However, some residual part remains. If a special set of emission control scenarios is considered, namely stabilization scenarios (where CO_2 concentration approaches a certain target level), this residual part is probably monotonically increasing with the stabilization level.

A reasonable stabilization target value could be found by ensuring equilibrium between the marginal STABILIZATION COST and the climate-change caused RESIDUAL DAMAGE associated with a given stabilization level (adaptations are taken into account). In other words, the following criterion can be employed:

{STABILIZATION COST + RESIDUAL DAMAGE}

should be minimal. Of course, discounting coefficients are to be applied as needed in calculating both components of the criterion. This approach is illustrated in Figure 9.1. A value c_0 is the lowest stabilization level under consideration. A function characterizing RESIDUAL DAMAGE is the sum of partial damage functions characterizing climate-change caused damage for different recipients. A partial damage function is just a respective response function if the response is expressed in monetary equivalent.

While costing methodologies for emission control programs are available (although some refinements are evidently needed), less attention has been paid to the assessment of residual damage. The IPCC TAR (Climate Change 2001, 2001b) characterized major actual and potential effects of climate change. This was made for certain sectors and regions. Unfortunately, the global estimate has not been obtained even for the globally-aggregated metrics/numeraires proposed in (Schneider et al., 2000), namely, for market impacts, human lives lost, biodiversity loss, distributional impacts, and quality of life. Thus, at present the information on actual stabilization costs is much more certain than on residual damage, and

assessing the latter in aggregated terms and finally in monetary equivalent is still a priority research task.

The stabilization cost and residual damage can be assumed to be concave functions of the stabilization level, monotonically decreasing and monotonically increasing with the level, respectively (Semenov, 2004b, pp. 122–124). This ensures, in particular, that their sum reaches a unique minimum. In our illustrative example this point is c_{opt}. If a component of residual damage was missing in the analysis (e.g. the component associated with some element of the socio-economic system) and it can also be described by a monotonically increasing partial damage function, the actual point of minimum will shift to the left of that found using incomplete information on the components of the total damage (to the left of c_{opt} ppmv in our example). Thus, 'optimal' values for the stabilization level produced by the proposed procedure are to be considered as majorizing (upper) estimates of actual optimal values. This estimate will decrease as the new components of the damage are involved in the analysis.

While assessing different damage functions, it is expedient to investigate carefully those associated with large-scale key vulnerabilities (Patwardhan et al., 2003), i.e. the large-scale key elements of the Earth's system or socio-economic systems that are both highly sensitive to climate change and have a limited adaptation capacity (like some physical elements of the climate system, for example, West Antarctic and Greenland Ice Sheets, Thermohaline Circulation (O'Neill and Oppenheimer, 2002, etc.)). Their damage functions have the potential for a strong non-linear behaviour, namely, the abrupt rise near a certain threshold c_{thr} (like line 4 in Figure 9.1). In this case, the optimal stabilization level should not exceed the threshold, otherwise such an interference with the climate system may result in a nearly-infinite magnitude of the damage. Thus, thresholds of this kind could also serve as the majorizing estimates of and temporary upper limits for the optimal stabilization level. Recently, a set of such thresholds for global surface temperature has been presented in (Corfee-Morlot and Höhne, 2003). The concept of critical thresholds for the anthropogenic impact on the climate system and biosphere was initially proposed in (Izrael, 1983) and recently developed in (Izrael, 2004).

Since the IPCC began, (IPCC XVIII Session, Wembley, UK, 24–29 September, 2001) its deliberations of key vulnerabilities in connection with the scientific basis of UNFCCC Article 2, many potential stabilization levels for atmospheric CO_2 concentration associated with different critical thresholds for climatic parameters have been investigated in the scientific literature. They vary widely, mostly from 450 to 700 ppmv of CO_2 (see e.g. (Swart et al., 2002; Izrael and Semenov, 2003; O'Neill and Oppenheimer, 2002, 2004)). However, it should be emphasized that such levels are to be considered as medium-term target values for CO_2 concentration (over centuries) rather than actual asymptotic levels (over millennia). Indeed, the current amount of carbon available

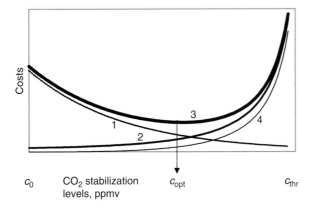

Figure 9.1 Stabilization target value for CO_2: (1) stabilization cost as a function of stabilization level; (2) residual climate-change caused damage increasing with the level; (3) – their sum {STABILIZATION COST + RESIDUAL DAMAGE} as a function of the level; (4) – residual damage associated with a key vulnerable element of the Earth's system or the socio-economic system.

for fossil fuel combustion is estimated at 1643 Gt(C) (Putilov, pp. 61–65; Semenov, 2004b, p. 113). This includes oil, gas, and coal (commercially efficient coal fields only). According to (Brovkin et al., 2002, pp. 86–9), in the pre-industrial time when a distribution of carbon among the atmosphere, terrestrial reservoirs, and the ocean was near equilibrium, the total amount of exchangeable C was 40,851 Gt(C), while the atmosphere contained 600 Gt(C). If for a rough estimate the non-linearity of the global carbon cycle is ignored, the immediate burning of all current resources of fossil fuels (1643 Gt(C)) will lead asymptotically to the enrichment of the atmosphere with $1643 \cdot (600/40,851) \approx 24$ Gt(C). This corresponds to about 11 ppmv of carbon dioxide.

CO_2 concentration has been varying within a 270–290 ppmv interval over the past 1000 years (Climate Change 2001, 2001a, p. 185), which gives a range for the 'pre-industrial equilibrium value'. The additional 11 ppmv of CO_2 may shift the equilibrium concentration to 281–301 ppmv. Such values were typical of the first decade of the 20th century, and from the authors' point of view they cannot be qualified as 'dangerous'.

However, transition values may have such a potential. To illustrate this, the transition curve and respective perturbation of global surface temperature are plotted in Figure 9.2 (all resources of fossil fuels are used at the beginning of 2000, and then anthropogenic emissions of all types are stopped). This figure and the next one are drawn using results of calculations made with a model of minimal complexity. The model allows the computation of anthropogenic perturbations of the global CO_2 cycle and respective perturbations of global surface temperature (Izrael and Semenov, 2003; Semenov, 2004a, 2004b; Izrael and Semenov, 2005). As can be seen from Figure 9.2, the global mean surface temperature will in this case

exceed the pre-industrial value by 3°C over 2050–2200 and 1°C over 2050–3000. Many recent studies have qualified such exceedances as at least 'suspicious' with respect to risks of large-scale singularities, the increasing frequency of extreme weather events, monetary or economic welfare losses for some regions, and so forth (see, for example, (Corfee-Morlot and Höhne, 2003)). This also appears to be true for the rates of temperature increase by 2100.

Once a stabilization level for greenhouse gas concentration in the atmosphere (i.e., the target value for the next few centuries) is adopted, one should investigate opportunities to reach it. A first attempt to develop pathways from the present CO_2 concentration to different constant future levels was undertaken in (Enting et al., 1994, pp. 75–76). Polynomial approximation was used to construct so-called S350 and S750 profiles. Later on, this approach was developed in (Wigley et al., 1996) where the well-known WRE-profiles were proposed. These concentration profiles were then transformed into respective stabilization scenarios through inverse modelling using the Bern-CC (Joos et al., 1996, 2001) and ISAM (Kheshgi, 2004) models. The major limitation of these profiles is their monotonic behaviour, i.e. stabilization level is reached through a monotonic increase in CO_2 concentration starting from the present one.

Actually, monotonic behaviour is not a necessary assumption, and the concentration may exceed the target value for a while. Such 'overshoot' concentration trajectories have been recently investigated in a series of publications (Kheshgi, 2004; O'Neill and Oppenheimer, 2004; Wigley, 2004; Semenov, 2004b; Izrael and Semenov, 2005; Kheshgi et al., 2005). They may give additional, somewhat more realistic, stabilization scenarios to be considered in the development of climate policy.

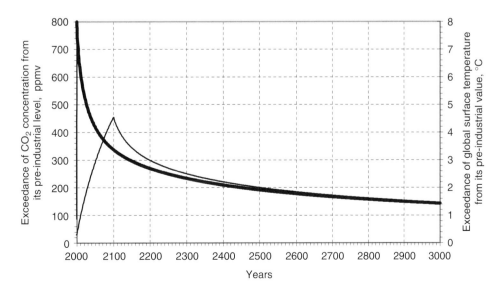

Figure 9.2 Changes in CO_2 concentration (thick line) and global mean surface temperature (thin line) under a hypothetical scenario: all known resources of gas, oil, and coal (commercially-efficient coal fields only) are used at once at the beginning of 2000, and then anthropogenic emissions of all types are stopped (Izrael and Semenov, 2005, p.10).

Perhaps the simplest type of stabilization scenarios could be associated with the implementation of two programs for the reduction in global CO_2 emissions. They are labelled as $BC\_T_{st}\_T_{imp}$ and $LU\_T_{st}\_T_{imp}$. Letters BC and LU indicate which type of CO_2 emissions is reduced, namely, emissions associated with fuel burning and cement production or with changes in land use and land management, respectively. Each of them is characterized by a certain year T_{st} at which a stabilization program begins and by a certain characteristic time T_{imp} (years) for the implementation of the program; 'st' and 'imp' are the abbreviations for 'start' and 'implementation', respectively. No emission control measures are taken before T_{st}. In each year beyond T_{st}, the total amount of emissions is reduced by a certain factor, namely, by a factor of $\exp(1/T_{imp})$. Thus, the initial emission rate (i.e., in year T_{st}) will decrease by factor of $e \approx 2.71$ over T_{imp} years.

In applications, the efficiency of a stabilization program with respect to its effect on the climate system is to be evaluated quantitatively. The means for such an analysis have not yet been completely developed, although many in-depth studies in this field have already been carried out, e.g. (Toth et al., 2002). In this paper, for a preliminary analysis, we will use the following criterion: a given exceedance of atmospheric CO_2 concentration from its pre-industrial level is considered undesirable ('dangerous') if it is greater than 300 ppm in 2000–3000 on average. The rationales for such a criterion are as follows. A long-term increase by 300 ppm in CO_2 concentration above the pre-industrial level leads to a long-term increase in mean surface temperature of about 3.0°C above the pre-industrial value (Izrael and Semenov 2003, p. 613). Such an increase, if it takes place during a period longer than that over which the Earth's climate system can reach the equilibrium (1000 years or more), leads to undoubtedly negative outcomes, in particular, to the complete melting of the Greenland ice sheet (Climate Change 2001, 2001a, p.17) with multiple regional climatic and ecological consequences.

The numbers characterising the temperature change in response to CO_2 increase given above require a short explanation. A long-term increase in surface temperature T caused by a given increase in the long-term CO_2 concentration is most commonly described through the so-called 'equilibrium climate sensitivity'. This parameter is defined as a change ΔT from the pre-industrial value associated with a doubling of the pre-industrial CO_2 level in equilibrium (Climate Change 2001, 2001a, p. 789). This parameter is produced by mathematical models of the climate system. Since the model constants are not known precisely and the climate system itself has a stochastic component in its evolution in time, the model estimates of climate sensitivity have uncertainties. A range from 1.5 to 4.5°C is commonly used for quantifying the climate sensitivity: see, for example, (Kheshgi et al., 2005, p. 219). The value '3°C/300 ppmv (CO_2)' mentioned in the previous paragraph corresponds to about 2.8°C for a

doubling of the pre-industrial CO_2 level and thus is practically at the center of the range. The latter estimate was produced by a highly aggregated model of the greenhouse effect (Izrael and Semenov, 2003) based upon the IPCC data on the Earth's energy budget and radiative forcing.

Using the minimal complexity model described in (Izrael and Semenov, 2003; Semenov, 2004a; 2004b; Izrael and Semenov, 2005), we have calculated atmospheric CO_2 concentrations in 2000–3000 corresponding to the simultaneous implementation of stabilization programs $BC\_T_{st}\_T_{imp}$ and $LU\_T_{st}\_T_{imp}$. The year T_{st} was chosen identical for both programs, while T_{imp} might be different. A series of values were considered for T_{st}, namely, from 2012 to 2112 with a 10 year time step; T_{imp} varied from 100 to 1000, and a 20 year time step was applied, which corresponds to an annual reduction in CO_2 emissions from 0.1 to 1%. For each T_{st}, Figure 9.3 shows maximum permissible values for the implementation time T_{imp} for programs of reduction in industrial emissions (i.e., BC-emissions) and respective rates of its annual reduction (%). In the calculations, the land-atmosphere net flux of CO_2 associated with changes in land use and land management was assumed to be annually reduced by 0.1%.

Results of computations of atmospheric CO_2 concentration and global surface temperature (exceedance from the pre-industrial values) in 2000–3000 under two 'opposite' scenarios of those described above are presented in Figure 9.4:

1. the simultaneous implementation of programs $BC\_$ 2012_340 and LU_2012_1000, which implies the annual reduction of 0.29% and 0.1% in both types of emissions, respectively, starting from 2012;
2. the simultaneous implementation of programs $BC\_$ 2112_120 and LU_2112_1000, which implies that both types of emissions are reduced annually by 0.83% and 0.1%, respectively, starting from 2112.

What is actually more expedient, namely, to postpone the reduction in emissions for 100 years (and then reduce them more rapidly as compared with lower rates of emission reduction required if the reduction programs were started immediately) or to start reductions in 2012, should be properly investigated using, in particular, the temperature magnitudes and rates of its change shown in Figure 9.4. Key vulnerabilities of a geophysical, ecological, social, and economic nature should be widely involved in such an analysis. The analysis has also to include the estimation of uncertainties.

It should be emphasized that 'knowing' the uncertainty is absolutely crucial for the establishment of critical limits for climate change and for long-term greenhouse gas concentration levels (Patwardhan et al., 2003). Assume that the upper limit for an increase above the pre-industrial value in long-term mean surface temperature for a region is estimated at $\Delta_0 T$ (see illustrative

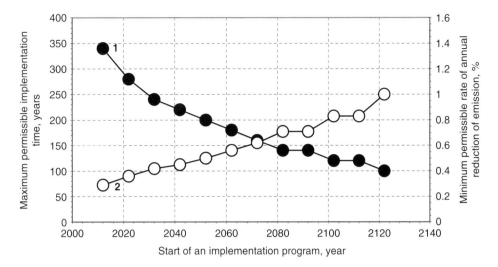

Figure 9.3 Maximum permissible values for the implementation time T_{imp} (curve 1) and corresponding minimum permissible values for annual reduction in global industrial emission (curve 2) for different initial years of the implementation of stabilization programs; T_{imp} for the global land-atmosphere net flux associated with changes in land use and land management is 1000 years (corresponds to the 0.1% annual reduction in emissions) (Izrael and Semenov, 2005, p. 11).

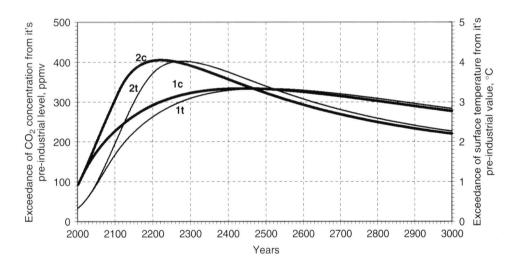

Figure 9.4 Changes in CO_2 concentration (thick lines 1c and 2c) and global mean surface temperature (thin lines 1t and 2t) under two scenarios: (1) annual, starting from 2012, reduction of 0.29% in global emission associated with fuel burning and cement production, while the land-atmosphere net-flux associated with changes in land use and land management is annually reduced by 0.1%; (2) annual, starting from 2112, reduction by 0.83% in global emission associated with fuel burning and cement production, while the land-atmosphere net-flux associated with changes in land use and land management is annually reduced by 0.1% (Izrael and Semenov, 2005, p. 12).

Figure 9.5). Keeping the actual rise in long-term mean temperature below this limit implies a high confidence in the stability of some key element of the climate system, for example, the Greenland ice sheet. In this case, $\Delta_0 T$ is approximately 3°C according to the IPCC TAR (Climate Change 2001, 2001a, p. 17). This deterministic case is shown by the 'step-like' curve 1 in Figure 9.5. However, any models used in such assessments cannot be absolutely precise. This inevitably results in the uncertainty of $\Delta_0 T$ quantified by probability P of the event: if the long-term increase in mean temperature exceeds ΔT,

the chosen key element is assumed losing stability with a probability greater than $P(\Delta T)$. This stochastic case is shown by the 'smooth' curve 2 in Figure 9.5. Assume that $\Delta_1 T \approx 1°C$ and $\Delta_2 T \approx 5°C$ are the lower and upper 90% confidence limits for $\Delta_0 T$. If $\Delta T < \Delta_1 T$, the critical threshold will not be exceeded with probability 0.9. If $\Delta T > \Delta_2 T$, the critical threshold will be exceeded with probability 0.9. In this example, the range from about 1 to 5°C is a zone of uncertainty (see Figure 9.5).

The size of such a zone of uncertainty can be reduced through obtaining new knowledge and data only. This

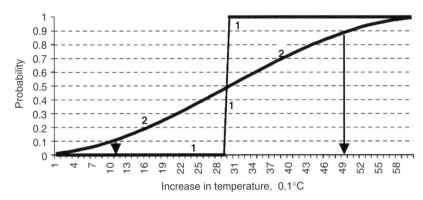

Figure 9.5 The long-term increase in surface temperature and probability of an 'undesirable' event: (1) deterministic case; (2) stochastic case (Semenov, 2004b, p. 139).

requires more assessments, research, monitoring and modelling activity. However, which value is to be chosen in this example – the lower or the upper one? Those who prefer a precautionary approach will choose the lower one, while the upper value is to be chosen by sceptics. Actually the whole probabilistic distribution should be investigated and ultimately taken into account in the establishment of the critical limit.

Concluding remarks

For the achievement of the main goal of the UN FCCC, a wide range of stabilization emission scenarios should be explored, and those ensuring that the concentration trajectory is kept within the safe corridor should be selected for ultimate adoption by policy-makers. This scientific work is in progress. The authors hope that some thoughts presented above will be useful in this connection. The work may be finished by 2007 when the Fourth Assessment Report (AR4) of the IPCC is issued. The AR4 has the analysis of key vulnerabilities and risks to the climate system among its major topics.

However, along with this scientific process, the stabilization target values for greenhouse gas concentrations and scenarios for achieving them have already become a subject of vigorous discussions by policy-makers and the public. In this connection it would be expedient to have some tentative limits for climate change for the 21st century proposed by scientists and based upon scientific expertise even if the studies are still in progress and numbers are very preliminary. This may help to discuss the issue in a more balanced manner.

The numbers below present the authors' expert opinion on the upper limits (majorizing estimates) for permissible values of some climate parameters for the 21st century:

- CO_2 concentration: 550 ppmv;
- A rise in surface temperature above the pre-industrial level: 2.5°C for the globe and 4°C for the Arctic;
- Sea level rise (above the pre-industrial level): 1 m.

We hope that in the very near future the world research community will produce scientifically-based levels for greenhouse gas concentrations in the atmosphere which could be presented to policy-makers for further deliberations. This will allow different countries to advance their national expertise for climate policy and to develop reasonable actions in the implementation of the UN FCCC and protocols to it.

REFERENCES

1. Azar C., Schneider S.H. 2002. Are the Economic Costs of Stabilizing the Atmosphere Prohibitive? Ecological Economics 42: 73–80.
2. Barnola J.M., Raynaud D., Lorius C., Barkov N. I. 2003. Historical CO_2 record from the Vostok ice core. In Trends: A Compendium of Data on Global Change. Carbon Dioxide Information Analysis Center, Oak Ridge National Laboratory, U.S. Department of Energy, Oak Ridge, Tenn., U.S.A., <http://cdiac.esd.ornl.gov/>
3. Brovkin V., Bendtsen J., Claussen M., Ganopolski A., Kubatzki C., Petoukhov V., Andreev A. 2002. Carbon cycle, vegetation, and climate dynamics in the Holocene: Experiments with the CLIMBER-2 model. *Global Biogeochemical Cycles*, Vol. 16, No. 4, 1139, doi: 10.1029/2001GB001662, 86–1 – 86–20.
4. Climate Change 2001. 2001a. The scientific basis. Contribution of Working Group I to the Third Assessment Report of the Intergovernmental Panel of Climate Change. (Houghton J. T., Ding Y., Griggs D. J., et al., editors). Cambridge University Press, 881 pp.
5. Climate Change 2001. 2001b. Impacts, Adaptation and Vulnerability. Contribution of Working Group II to the Third Assessment Report of the Intergovernmental Panel of Climate Change. (McCarthy J.J., Canziani O.F., Leary N.A., et al., editors). Cambridge University Press, 1032 pp.
6. Climate Change 2001, 2001c. Synthesis Report. Contribution of Working Group I, II, and III to the Third Assessment Report of the Intergovernmental Panel of Climate Change. (Watson R.T. and the Core Writing Team, editors). Cambridge University Press, 2001, 397 pp.
7. Corfee-Morlot J., Höhne N. 2003. Climate change: long-term targets and short-term commitments. *Global Environmental Change* 13, pp. 277–293.
8. Enting I.G., Wigley T.M.L., Heimann M. 1994. Future emissions and concentrations of carbon dioxide: key ocean/atmosphere/land analyses. CSIRO Tech. Pap. No 31.
9. Izrael Yu. A. 1983. Ecology and control of the natural environment. Kluwer, 400 pp.
10. Izrael Yu. A., Semenov S. M. 2003. Example calculation of critical limits for greenhouse gas content in the atmosphere using a

minimal simulation model of the greenhouse effect. *Doklady Earth Sciences* (English Translation of Doklady Akademii nauk, Vol. 390, No 4, May-June 2003, pp. 533–536), Volume 390, Number 4, pp. 611–614.

11. Izrael Yu.A. 2004. On the concept of dangerous anthropogenic interference with the climate system, and abilities of the biosphere. *Meteorology and Hydrology*, N 4, pp. 30–37. (in Russian)

12. Izrael Yu.A., Semenov S.M. 2005. Calculations of a change in CO_2 concentration in the atmosphere for some stabilization scenarios of global emission with use of minimal complexity model. *Meteorology and Hydrology*, N 1, pp. 5–13. (in Russian)

13. Joos F., Bruno M., Fink R., Stocker T.F., Siegenthaler U., Le Quéré C., Sarmiento J.L. 1996. An efficient and accurate representation of complex oceanic and biospheric models of anthropogenic carbon uptake. *Tellus*, 48B: 397–417.

14. Joos F., Prentice I.C., Sitch S., Meyer R., Hooss G., Plattner G.K., Gerber S., Hasselmann K. 2001.Global warming feedbacks on terrestrial carbon uptake under the Intergovernmental Panel on Climate Change (IPCC) emission scenarios. *Global Biogeochemical Cycles*, 15, 891–908.

15. Kheshgi H.S. 2004. Evasion of CO_2 injected into the ocean in the context of CO_2 stabilization. Energy 29, pp. 1479–1486.

16. Kheshgi H.S., Smith S.J., Edmonds J.A. 2005. Emissions and atmospheric CO_2 stabilization: long-term limits and paths. Mitigation and Adaptation Strategies for Global Change, 10: 213–220.

17. O'Neill, B.C., Oppenheimer M. 2002. Climate change – dangerous climate impacts and the Kyoto protocol. *Science* 296 5575: 1971–1972.

18. O'Neill B.C., Oppenheimer M. 2004. Climate change impacts are sensitive to the concentration stabilization path, *Proc. Nat. Acad. Sci.*, 101, 16411–16416.

19. Patwardhan A., Schneider S.H., Semenov S.M. 2003. Assessing the science to address UNFCCC Article 2. <http://www.ipcc.ch/>, 13 p.

20. Petit J.R., Raynaud D., Lorius C., Jouzel J., Delaygue G., Barkov N.I., Kotlyakov V.M. 2000. Historical isotopic temperature record from the Vostok ice core. In Trends: A Compendium of Data on Global Change. Carbon Dioxide Information Analysis Center, Oak Ridge National Laboratory, U.S. Department of Energy, Oak Ridge, Tenn., U.S.A. <http://cdiac.esd.ornl.gov/>

21. Putilov V.Ya. (ed.: 2003. Ecology of Power Engineering. Published by Moscow Energy Engineering Institute, Moscow. (in Russian)

22. Schneider, S., Kuntz-Duriseti, K., Azar, C. 2000. Costing nonlinearities, surprises and irreversible events. *Pacific and Asian Journal of Energy*, 10, pp. 81–106.

23. Semenov S.M. 2004a. Modeling of anthropogenic perturbation of the global CO2 cycle. *Doklady Earth Sciences* (English Translation of Doklady Akademii nauk, Vol. 398, No 6, pp. 810–814), Vol. 399, Number 8, pp. 1134–1138.

24. Semenov S.M. 2004b. Greenhouse Gases and Present Climate of the Earth. Moscow, Publishing Center 'Meteorology and Hydrology', 175 pp. (in Russian)

25. Swart R., Mitchell J., Morita T., Raper S. 2002. Stabilization scenarios for climate impact assessment. *Global Environmental Change*, 12, pp. 155–165.

26. Toth, F.L., T. Bruckner, H.-M. Füssel, M. Leimbach, G. Petschel-Held, H.J. Schellnhuber. 2002. Exploring Options for Global Climate Policy: A New Analytical Framework. Environment 44(5): 22–34.

27. Wigley T.M.L, Richels R., Edmonds J.A. 1996. Economic and environmental choices in the stabilization of atmospheric CO_2 concentrations. *Nature*, 379, 242–245.

28. Wigley, T.M.L. 2004. Choosing a stabilization target for CO_2. *Climatic Change* 67: 1–11.

CHAPTER 10

Perspectives on 'Dangerous Anthropogenic Interference'; or How to Operationalize Article 2 of the UN Framework Convention on Climate Change

Farhana Yamin, Joel B. Smith and Ian Burton[1]

10.1 Introduction

Science forms the backbone of the international climate change regime. The negotiation and entry into force of the 1992 UN Framework Convention on Climate Change (UNFCCC) in only four years, was due, in large part, to the strong international scientific consensus on the need for a convention – the draft elements of which were appended to the first scientific assessment report produced by the Intergovernmental Panel on Climate Change (IPCC) in 1990 (Bodansky, 1993). Although more circumspect in terms of policy recommendations, the IPCC's Second and Third Assessment Reports generated significant momentum for the negotiations leading to the 1997 Kyoto Protocol and decisions subsequently adopted by the UNFCCC Conference of the Parties (COP) in 2001, the Marrakesh Accords, that enabled the Protocol's entry into force in February 2005.

What contribution will the Exeter conference and the Fourth Assessment Report (FAR), scheduled for completion in 2007, make to future climate policy? An important focus of attention for scientists and policy makers in the coming decade will most likely be on making operational sense of Article 2 of the Convention: avoidance of dangerous anthropogenic interference with the climate system. The crucial science/policy issues would thus be how Article 2 relates to future efforts (under the UNFCCC, Kyoto and/or a new legal instrument) to prevent climate change (mitigation) as well as how it provides policy guidance for dealing with adverse impacts and potential beneficial opportunities (adaptation).

This paper does not attempt to provide an answer to what constitutes dangerous anthropogenic interference with the climate system. Instead, it reviews some of the various perspectives on Article 2 that have emerged over the last 15 years of negotiations, as ways have been sought to arrive at a common understanding. It then offers an assessment of the current situation of a dangerous change in climate. The paper aims to catalyze future science/policy discussions by providing an overview of the main approaches to Article 2 and some sense of history about how changing science/policy considerations have created

challenges for climate science and policy. It then focuses on three issues germane to the evolution and operation of Article 2 that have been, in our view, relatively neglected in climate literature related to Article 2, namely: the categorization of climate change in terms of timing, scale and types of impacts; the role of adaptation; and the development of a new process of global decision-making or negotiations that can accommodate divergent human values.

We conclude by suggesting that the categorization of climate impacts (geophysical, biophysical, human health and wellbeing) and the scale at which impacts are assessed are critical for determining what may be a 'dangerous' level of climate change. To date, the scientific community has been given insufficient guidance about scale and categorization issues in policy processes. Unless remedied, the resulting lacunae will, by default, be filled by scientists resorting to familiar mental frameworks and unexplained values and preferences which may or may not accord with the perception, values and framework of policy-makers or broader publics. This would not likely lead to effective implementation of Article 2 given that a number of levels of dangerous anthropogenic interference (DAI) can legitimately be chosen for the purposes of climate policy.

Our conclusions about the way in which values are interlaced with 'technical' issues in a unique way in climate change suggest that process issues are of critical concern, particularly in terms of who makes decisions and the values embedded in those decisions. The setting of any climate goal or target, long or short term, should be the result of informed dialogue between researchers, negotiators, and the public. Thus, a crucial part of the next phase of the climate science/policy nexus is development of a process which can enable a full and open discussion on Article 2 and lead to a consensus and resolution on shorter term aspects of climate policy such as targets and timetables.

10.2 Perspectives on Article 2

The definition or framing of a problem plays an important part in shaping subsequent institutional and political responses, including which kind of knowledge will be considered relevant for devising solutions. Climate change was identified as a problem by scientists and came to be framed as an international environmental problem. Even though climate change profoundly implicates economic, social and political developments which are the responsibility

[1] Farhana Yamin, Fellow at the Institute of Development Studies, University of Sussex, UK. Joel Smith, Vice President, Stratus Consulting, USA. Ian Burton, Scientist Emeritus, Meteorological Service of Canada & Emeritus Professor, University of Toronto, Canada.

of treasuries and economic and planning ministries, the initial framing meant ministries of the environment were typically given lead responsibilities over climate change.

10.2.1 *Environmental Standard Approaches*

Although core economic and development actors are now beginning to take a more active interest in climate change, the basic architecture of the climate regime reflects and shapes institutional and policy responses most familiar to those engaged in environmental science and policy. The underlying framework of Article 2, for example, draws on an environmental standards-based approach to setting a long-term goal for stabilization of atmospheric greenhouse gas concentrations. It also draws on approaches to the setting of environmental and health safety standards, as the three criteria mentioned in Article 2 – food security, sustainable development and ecosystem adaptation – aim to protect and promote human wellbeing.

The environmental standards approach typically involves the specification of a standard based on certain policy goals and criteria that are accepted as worthy of protection. In the case of contaminants such as toxic substances, carcinogens or bacteria the maximum level or amounts of the contaminants in air, water, or food is specified. Emissions that result in exposures at or above these maximum levels are typically prohibited. The standards are based on evidence based on scientific studies and often interpreted by expert advisory bodies, who often relate the amount of the contaminant to the impact or the response as in dose-response curves for example. The criteria enable decisions to be made about acceptable levels of risk. The actual choice of acceptable levels can involve comparative risk information (how high is this risk compared with other socially accepted risks?), and risk-benefit information (how much benefit is being gained and would be foregone if regulations were to be imposed that limited use or access?). The standards are periodically reviewed and revised in the light of new scientific evidence.

10.2.2 *Acid Rain and Ozone: Precedents or Problems?*

The environmental standards approach features in many domestic and international environmental regimes. Importantly for climate change, this approach had been successfully deployed in two international environmental regimes dealing with the atmosphere that were influential precedents or models for those negotiating the UNFCCC and later on, Kyoto itself. The two regimes were the acid rain regime, comprising the 1972 Long Range Transboundary Air Pollution Convention and now eight protocols dealing with specific air pollutants, and the ozone regime, comprising the 1985 Vienna Convention for the Protection of the Ozone Layer and the 1987 Montreal Protocol on Substances that Deplete the Ozone Layer (Andersen and Madhava Sarma, 2002; Benedict, 1991; Sands, 1991). An important feature of both regimes is that the long-term goal of protecting the atmosphere is to be reached through the adoption of short-term, legally binding quantitative targets that are reviewed and revised in response to changing scientific, technical and other relevant information.

The acid rain, ozone and climate change regime also share the 'framework convention/protocol' approach to standard setting. The basic feature of this is to institutionalize an iterative policy cycle presided over by a conference or meeting of the parties, which is able to promulgate more detailed rules through the adoption of decisions or other legal instruments negotiated periodically in the light of evolving scientific information provided by independent scientists. This means decisions do not have to be made in an all-or-nothing fashion that might bog down things for decades or else result in bad decisions being made that cannot easily be reversed. In combination, these factors help explain why Article 2 of the UNFCCC was drafted in a way which provides less guidance than both policy makers and scientists now want. They also explain why it has not been further elaborated as originally intended by climate negotiators, with the focus shifting instead to the more manageable task of agreeing short-term emission reductions targets of the kind set out in Kyoto.

The success of the environmental standards approach rests, however, on a number of characteristics of the issues in question that are arguably not applicable to climate change. First, an environmental standards approach is typically based on the determination of a level of exposure to a pollutant above which would cause injury or mortality. The level of danger can be based on testing how animals or humans react when exposed to different levels of pollutants. Once such a relationship is established, it can be applied anywhere geographically. Climate change is more complex, partly because the exposure is not to a pollutant but to a characteristic of the environment, namely the climate itself. Different people, societies, and ecosystems will not be affected in the same way by the same change in climate. A 1°C increase in temperature could be harmful to some species and societies and could benefit others. Furthermore, it is difficult to prescribe diverse climate impacts with the same degree of detail and confidence. By setting up a framework that requires a large amount of impacts knowledge to be well circumscribed and certain before preventative or even precautionary action can be justified, the environmental standards approach may have set up more unhelpful obstacles for climate policy than might otherwise have existed.

Second, environmental standards approaches tend not to have to grapple with the issue of adaptation. For example, the adaptation options in the face of the impacts of acid rain are very limited. Some liming of lakes to reduce their acidity has been tried but this addresses only a small part of the impacts of acid rain so was not seen as a large part of the solution. Some suggested that staying out of the sun, wearing a hat and sunscreen lotion would be sufficient to offset much of the risk related to ozone depletion before acceptance that such adaptations would not solve the problem. Although complex issues of equity and of capacity may arise, in the case of climate change the opportunities for adaptation measures to reduce impacts are potentially much larger, and in many cases could prove effective in shifting the level that might be considered dangerous. For other cases, such as impacts on ecosystems and on the poor, the limited capacity for adaptation would result in little or no change to the level of climate change that might be considered dangerous.

Third, the environmental standards approach works well when the scope and size of the decision-making process is limited and well circumscribed. The major source of acid rain is the power generation sector which is clearly under national jurisdiction. Likewise, the production of ozone-depleting substances which in any case is confined to a handful of countries. By contrast, the sources of climate 'pollutants' are virtually all of humanity. These sources are spread across virtually all sectors of economic activity and are widely distributed across the planet. The challenge of then deciding which sectors and sources to regulate, how to do so, and then enforcing regulations over millions of sources for the next 50–100 years is frankly unprecedented – not just in the arena of environmental but of international affairs. The hugely complex, multi-level decision-making processes emerging under climate change are justifiably regarded as groundbreaking in international affairs.

10.2.3 *Values, Science and Politics*

Fourth, whilst value judgments are ultimately always implicated in the setting of standards, in most domestic and international environmental policy-making to date, controversy over values has tended to be relatively limited in scope and/or has been settled fairly early on as part and parcel of the conditions of regulatory action being undertaken. In the international arena, defining values is complex given the sovereign equality of states and the lack of a central authority to force closure over value-related disputes and to compel enforcement over agreed ones. Because value disputes could lead to negotiating impasse, negotiators often take great care to frame disagreements in technical, more issue-specific neutral terms. More often than not in the international environmental context, closure over diverging values is reached by a powerful nation or group of nations showing moral leadership through implementation of significant domestic action – as happened in the case of acid rain by the '30 per cent club' countries and in the case of the ozone regime by the USA. Other countries are then compelled to follow suit for a mix of reasons: they want to do the right thing (or at least not lose face), switch out of obsolete technologies and/or fear incurring the wrath of the more powerful states.

In spite of the intractable nature of diverging values, priorities and perspectives, the ratification of the Convention by 189 countries and of the Kyoto Protocol by 155 countries highlights the fact that a measure of consensus exists among states as regards basic values, the nature of the problem structure and practical responses to climate change. This favors an iterative cycle of policy focused on emission reductions targets, led in the early stages by developed countries, with technological and financial assistance for needy developing countries to decarbonize development and also to cope with climate impacts. Nevertheless, it remains the case that values and approaches differ markedly on many issues germane to future climate policy, including what future action (if any) should be taken, by whom and how short term efforts to mitigate and adapt to climate change relate to the ultimate objective set out in Article 2. On these issues, the guidance that is provided by Article 2 and other existing principles and rules is indeterminate.

Thus, an environmental standards approach, as embodied in Article 2, poses challenges for implementation. The remainder of the paper addresses how these challenges can be overcome.

10.3 Categorization of Climate Change: Impacts, Scale and Timing

For future discussions of Article 2, we believe it would be useful to separate out the three fundamental elements to determining a dangerous level of climate change: what is dangerous, to whom is it dangerous and how much is dangerous? These elements raise questions about values that determine the types of impacts selected as relevant for policy. They also raise questions about the extent to which it is possible or desirable to aggregate different kinds of impacts under a common metric in terms of deciding what is deemed to be significant in policy terms.

10.3.1 *What is Dangerous?*

Three broad types of adverse impacts can be identified that can be used to define what is a 'dangerous' level of climate change: geophysical impacts, biophysical impacts, and impacts on human health and wellbeing.

Geophysical impacts could be large-scale change in the Earth's physical processes such as breakdown of the Thermohaline Circulation or disintegration of the Greenland or West Antarctic Ice Sheets. Essentially these are impacts that either have widespread implications for society or nature, or are so valued that their occurrence is deemed unacceptable.

Biophysical impacts could be loss of valuable ecosystems such as coral reefs or arctic ecosystems, or loss of valuable species. The loss of ecosystems or species falls into the latter category (noting that they can have socio-economic impacts as well). This category can be linked to the ecosystems clause in Article 2, but also to sustainable development.

Human health and wellbeing addresses direct impacts to humanity. It includes impacts on individual and public health (e.g. heat waves, floods, infectious diseases), impacts on key sectors of the economy such as agriculture (which is the only specific societal impact mentioned in Article 2) as well as on the economy as a whole. Net economic impacts, e.g. retarding development, would also be within the economics category, as would inundation of low-lying coastal communities or small island states by sea-level rise, flooding, drought, loss of cultures, loss of sovereignty, or increased displacement leading to internal and external refugees.

10.3.2 *To Whom is it Dangerous?*

One aspect of Article 2 that is not clear from the literature and has received less attention is at what scale Article 2 is to be interpreted. Is it to be interpreted to apply only to impacts that are global in nature, such as disintegration of the WAIS? Does it apply to impacts that while more limited in immediate effect, might have global importance, such as destruction of a valuable ecosystem? Alternatively, could Article 2 be applied at a finer scale, perhaps to limited geographic impacts which may only be of high importance to a region, country, province, or even a village? (see e.g. Dessai et al., 2004).

Defining dangerous at the *global scale* implies that there is a process for achieving a global consensus on what is dangerous. This process could be based on avoiding impacts that are widespread, such as a collapse of the WAIS or a runaway greenhouse effect. Alternatively, it could be the development of a consensus on avoiding impacts that while perhaps not directly affecting all or even many, are deemed unacceptable. Severe harm to coral reefs, loss of arctic ecosystems, and loss of some small island states may be examples.

The global scale implies that we collectively reach an agreement on defining such a level of danger. The 'burning embers' diagram from the IPCC TAR (Smith et al., 2001) was an approach to define options for identifying globally unacceptable outcomes.

Use of a global scale approach might imply, however, that adverse impacts at less than a global scale may not be deemed to be dangerous. For example, loss of species or reduction in agricultural production in some regions may be not found to be dangerous for the planet as a whole. Or it might be considered not dangerous if the losses to one region are offset by the gains to another.

The second option or scale is at the *regional level*. The concept is generally meant to apply to nations with common vulnerabilities, such as small island nation states or sub-Saharan African states. They are likely to face common adverse impacts of climate change such as inundation of low lying areas and possible loss of existence in the case of small island states, or increased drought, famine, or spread of infectious disease in the case of sub-Saharan African states. Their specific vulnerabilities may be masked in an assessment of impacts carried out at the global scale.

Likewise, for the third option of scale which concerns the level of governance: *the nation-state*. Each country would determine what level of adverse impacts would be considered dangerous. This could be based on impacts within its territory or other impacts outside its territory that it considers particularly important or valuable. So, a reduction in a nation's agricultural production or ability to be self-sufficient in food production could be deemed dangerous – even if net global agricultural production is rising. It also could mean that a change that is judged globally to be dangerous at a particular level of climate change, such as rapid sea level rise or loss of ecosystems, might be judged to be dangerous within the state at a lower level of climate change.

The fourth scale option for the determination of what is dangerous is at a *local* (e.g. village) or even individual level. A shift in agricultural competitiveness could undermine a village's livelihood. So too, a small rise in sea level might threaten existence of the village.

It is most likely that the use of a framework that has a finer scale of decision making, e.g. national rather global or local/individual rather than national, would imply the definition of dangerous at a lower level of climate change. Indeed it may be that at finer scales, almost any change in climate would be determined to be dangerous. That is because even small changes in climate can be or already may be dangerous at the village or individual level. For example, warming of the Arctic has already adversely affected some indigenous communities (ACIA, 2004). Application of a governance scale approach to determining what is dangerous may well result in selection of different levels of climate change being deemed dangerous. Some countries may find that a very small change in climate results in adverse outcomes that are determined to be dangerous. Others may find it takes a higher level of change in climate to result in what is deemed to be a dangerous outcome. The existence of other stressors may mean that some villages or communities would probably find an even smaller level of climate change to be dangerous. We expect such differences to arise not just from differences in impacts within different countries but also differences in how impacts are perceived or valued.

The issue of scale highlights that the process of agreeing on what is dangerous is not clear from Article 2. It is very likely that individual countries or communities will determine what they regard as 'dangerous' before the UNFCCC COP does. Because the application of scale and process could result in very different outcomes for different countries and communities, the salient issue will be deciding how to deal with this diversity in policy terms which we discuss below under the section on decision-making.

10.3.3 *How Much is Dangerous?*

Article 2 implies that a dangerous level of climate change will be determined based on definition of an unacceptable outcome, i.e., a 'dangerous' outcome. This may be best met if climate change results in the crossing of a threshold which is widely perceived as unacceptable. Thresholds may be associated with discrete events such as destruction of an ecosystem, extinction of a species, decline in economic production, or state change in the climate system. Some of the impact categories are quite consistent with thresholds, particularly the geophysical and biophysical categories, whereas human health and wellbeing may face a steady increase in many adverse impacts with higher concentrations of greenhouse gases. Some impacts, such as global agricultural production, may be marginal or positive at a relatively small level of climate change and negative at larger levels (Gitay et al., 2001; ECF/Potsdam Report, 2004). For continuous impacts such as sea level rise, where adverse impacts increase monotonically with the level of climate change, it may be difficult to discern an unambiguous threshold.

It should also be noted that, as applied in numerous cases of controlled substances such as DDT and CFCs, the impact on ecosystems, and indirectly on human health, has been considered sufficiently risky that a standard of zero tolerance has been adopted. National governments and in some cases the international community have decided that no amount of these substance can be tolerated, and total bans have been enacted and enforced. If greenhouse gases are viewed in this light then the zero tolerance level would be that at which no new adverse effects occur above the baseline level. This might be the natural background level of GHGs or the pre-industrial level.

An important issue in climate science/policy is how scientists can provide objective, value-free information to policy makers about the myriad impacts climate change may bring. A crucial issue here is whether relevant information about different kinds of impacts can be usefully aggregated for the purposes of policy. This means deciding on how impacts will be categorized and how they can be counted. Conventional cost-benefit analysis (CBA) approaches aim to provide information by quantifying different impacts in monetary terms and comparing these with the costs of taking action to prevent climate change impacts (e.g. Nordhaus and Boyer, 2001; Tol, 2003). Alternative approaches, termed 'sustainability' or 'tolerable windows' approaches, highlight the incommensurate nature of climate change impacts (e.g. numbers of people at risk, ecosystems put under stress and welfare losses) and insist that attempts by scientists to aggregate impacts under a common metric can be difficult for policy purposes (Parry, 1996; Bruckner et al., 1999; Azar and Schneider, 2001; Patwardhan, Schneider and Semenov, 2003; Grassl et al., 2003; Leemans and Eickout, 2004; Jacoby, 2004). Both types of approaches have in common an attempt to separate issues of value from the purposes of scientific assessment: in the tolerable windows approach it is up to policy-makers to assign value to different types of impacts to be avoided whilst in CBA the values are embedded in a host of assumptions made about how and which things are counted, compared and discounted.

Although the application of approaches such as CBA or tolerable windows do not to us appear to have generated a consensus on what is a dangerous level of climate change, it is also becoming clear there is a desire to move beyond the 'impacts are incommensurable and it is up to policy-makers, not scientists, to decide how which are important' type approach. Prime Minister Blair's exhortation to the Exeter Conference that scientists should identify a level that is 'self-evidently too much' is a clear challenge to scientists to say something more than climate impacts are like apples and oranges and it is up to policy makers to choose between them. The attention recently accorded to the adaptation needs of least developed countries (LDCs) within the UNFCCC also signals growing sophistication of the international process in being able to move beyond a 'one size fits all' approach to climate policy. Moving beyond this point will require policy-makers to consider what weight they may wish to give to particular types of impacts and to select scales that merit particular attention in terms of scientific assessments. In turn, this will require scientists to better explain the relationships between different kinds of impacts, scale and their timing.

10.4 Adaptation

Article 2 focuses on preventing dangerous interference but within a timeframe that allows ecosystems to adapt naturally, and food production and economic development not to be threatened or disrupted. The Convention also contains extensive provisions on adaptation. These provisions reflect the fact that climate change differs from other environmental problems in that there may be much room for adaptation. This means the calculus of 'dangerous' cannot be made simply on the grounds of impacts and their consequences. There are the 'gross impacts' that have been the subject of much research and comment as reported in successive IPCC reports. Then there are the 'net impacts' which are the impacts that will remain after adaptation. The term 'vulnerability' encompasses consideration of the capacity of a system to adapt to climate change (Smit et al., 2001). But there are few studies that examine what might be achieved by adaptation or that estimate the limits to or costs of adaptation. It has been widely assumed that the impacts of climate change in the absence of mitigation will be so great that adaptation will be of no avail. While this is probably true at the more extreme levels of climate change, it is also clear that for at least some sectors and countries a moderate degree of warming, adaptation, if effectively deployed, can substantially reduce impacts. In other cases, by increasing resilience, adaptation can 'buy time' so that there will be a delay in reaching any level that might be considered to be dangerous.

As touched upon earlier, adaptation has not received as much attention in the UNFCCC process as mitigation in the early years of the climate regime (Yamin and Depledge, 2004). During the Kyoto negotiations, many viewed focusing on adaptation as a response strategy as simply a way out for rich developed countries to avoid making politically difficult decisions about mitigation. While in economic theory it is possible to construct graphs which suggest that adaptation and mitigation are alternatives and that a balance of the two would form an optimum strategy (e.g. Fankhauser, 1996; Wilbanks et al., 2003), the practicalities are different. The decisions about adaptation and mitigation are made by different players in different jurisdictions, and there is no authority that can choose how much of each is to be deployed. The likelihood is that for the foreseeable future, not enough will be done on the mitigation side nor on the adaptation side.

Perspectives on adaptation are now shifting. With the Kyoto architecture firmly, if not universally, in place until 2012, the consensus is shifting towards giving greater attention to the role of adaptation (Yamin, 2005). But it should be noted that many developed and developing countries remain somewhat wary of adaptation being a focal point of the climate policy, although for quite different reasons than in the past: the costs and effectiveness of adaptation have not been established (Hitz and Smith, 2004; Corfee-Morlot and Agrawala, 2004). Moreover, there are bigger knowledge gaps about future climate impacts than about patterns of current emissions, and adaptation options are highly localized and solutions more deeply context-specific. Additionally, unlike mitigation efforts that can be focused on large emitter or upstream activities that can be easily regulated, to be effective, an adaptation instrument would have to engage the agency of billions of individuals and thousands of communities – something which international processes are not good at doing. All these considerations make forging an international agreement on adaptation as difficult, if not more so, than negotiating mitigation commitments. Because adaptation will play a more central role in increasing resilience to climate variability, climate change science will have to do far more work on defining what impacts can be avoided or reduced through adaptation and which cannot.

10.5 The Process of Decision-Making

In this section we focus on the elaboration of Article 2, the framing of climate change as an environmental problem and climate related decision-making processes.

10.5.1 *The Role of UNFCCC Institutions and the IPCC*

In traditional standard setting, the experts or scientists have often played a major role, and while the final choice of standard has been made at the political level, the authority of science has been such that it has substantially contributed to the development of standards. This pattern, followed to a large extent in the acid rain and ozone regimes, has only been partly possible in the climate regime

for two reasons; one of relevance to science-based policy processes generally, and the second related to considerations specific to the climate regime. The generic factors are demands by the lay public and stakeholder organizations for increased accountability and transparency in scientific research and related processes of standard setting, particularly those dealing with issues of public health and safety. We address these concerns and possible ways to meet them in more detail below.

The more specific factors concern the political sensitivity of UNFCCC Parties: governments do not want international scientists prescribing policy in areas vital to national security and development such as energy, food and transport, all of which are implicated in climate change. These sensitivities explain why international decision-making processes related to the scientific and technical aspects of the climate regime are uniquely structured and function very differently to those found in the acid rain and ozone regimes – even if at first sight they appear in name to be quite similar.

The political and legal authority to interpret and further elaborate the provisions of the Convention rests with its supreme decision-making body, the Conference of the Parties – and no one else. The COP cannot, of course, stop individual or groups of countries or others from coming up with their own interpretations of what counts as dangerous. If they have scientific credibility and engage political imaginations, these views may, over time, influence and guide COP thinking, but they will not be legally or politically authoritative until the COP makes its determination.

If the COP decides to elaborate Article 2 – a big 'if' – the COP has distinct legal, institutional and scientific advantages but also some drawbacks. With 190 governments now Parties, the COP has near-universal representation and with that comes the legitimacy to make decisions. Compared with other international regimes, the COP also has well-funded scientific institutions to furnish it with scientific advice. The main organ for such advice is the Convention's Subsidiary Body for Scientific and Technological Advice (SBSTA), established pursuant to Article 9 of the Convention. The semi-annual meetings of SBSTA draw up draft recommendations for consideration by the COP largely on the basis of scientific input provided by the IPCC.

In terms of drawbacks, for a number of political and institutional reasons, the COP and SBSTA get bogged down in scientific issues confronting the regime. It has taken four years, for example, to agree that SBSTA's agenda should include consideration of the policy implications of the scientific work published in the IPCC 2001 Third Assessment Report. Unlike the acid rain and ozone regime whose scientific bodies are limited in number and comprise independent scientists selected principally for their expertise, SBSTA is an open-ended body, comprising representatives from all Parties. Whilst many delegates are selected for their scientific backgrounds, others are known primarily for their diplomatic skills and political acumen. This means SBSTA is too large and sometimes too politicized a body to deliberate in much detail on complex scientific findings put forward by the IPCC (Yamin and Depledge, 2004). An additional problem is that because the political stakes for countries are so high, and many fear future scientific findings might catalyze momentum for new commitments, the scientific rigor and independence of the IPCC itself is coming under strain.

These difficulties may explain why suggestions have been made that a 'top-down' approach to determining Article 2, focused on the formal UNFCCC negotiations process, is unlikely to make much progress (Pershing and Tudela, 2003). Others have suggested that perhaps parallel private efforts, informed by the IPCC, with some government participation may be able to make more headway in generating the basis of consensus (Oppenheimer and Petsonk, 2004).

Given the dynamic and complex nature of climate change and the changing state of scientific knowledge, such alternative processes, however, may have their own problems which have not been given adequate attention in the literature. Top-down global policy-making processes, for example, whether focused on the COP or undertaken by a panel of distinguished international experts (as has usually been the case for many environmental and health and safety standards), might ignore or underplay the significance of impacts that occur if assessment is undertaken at finer scales such as at the household or community level. As discussed earlier, scale has a crucial bearing on the determination of dangerous. Given the wider demands for accountability and transparency, top-down, expert-led processes might also be critiqued in terms of legitimacy and long-term effectiveness.

10.5.2 *The Role of Stakeholders*

In the climate change context, the failure of international processes to take into account the social or individual perceptions of danger has, for example, led to calls for a more 'bottom-up' approach to give more weight to stakeholder perceptions of dangers (Desai et al., 2004). A top-down decision on a dangerous level of greenhouse gas concentrations that is inconsistent with bottom-up views of danger is also less likely to be successfully carried through to implementation.

Procedural issues of how the COP can consider Article 2, and how COP processes relate to determinations of 'dangerous' agreed at the national level by individual Parties or by sub-national entities or informal processes, thus need to be thought through in greater detail. This begs broader questions about the involvement of stakeholders and civil society in processes that weigh up public risks and benefits.

The importance of stakeholder and civil society involvement in decision-making processes relating to environmental and health-related risks is becoming increasingly recognized for legitimacy and long-term effectiveness (Yamin, 2001; Millstone, 2004). In many parts of the world, the role of science and scientific judgment in policy processes compared with that of stakeholders or civil society has been criticized (Wildavsky, 1979; Jasanoff and Wynne, 1997; Stirling, 2001, 2003). Recent international controversies with the scientific assessment of the risks and benefits of genetically modified crops, BSE and hormone-treated beef have demonstrated that risk assessment and risk management is not a matter that can be left to scientists and policy-makers alone when fundamental values and choices are at stake (Millstone, 2004).

To ensure effective, more legitimate policy-making, the involvement of stakeholders in decision processes is becoming standard practice in many countries. Although terminologies vary between jurisdictions, the concept of 'risk assessment policy' has emerged to cover the process prior to assessment during which issues that carry fundamental normative implications are mutually agreed between policy-makers and stakeholders.

Risk assessment policy focuses on the purpose of risk assessment and the context in which that assessment is to be carried out by technical experts (May, 2000). Agreed guidance is provided to the experts who are to undertake the assessment on matters relating to the scope, scale and distribution of risks or potential impacts to be assessed. What weights should be given to different risks and benefits, what kind of evidence is counted and discounted, what level of proof is required and whether trade-offs between impacts and benefits of different kinds is deemed appropriate, and if so, how it is to be made explicit, are considerations for policy-makers. These framing issues have a large impact on outcomes but in the past have not been openly addressed in policy processes. The trend now is for such matters to be determined in advance by risk policy-makers with input and agreement of those with an interest (May, 2000).

But how do we decide who has an interest for the purposes of Article 2? The myriad impacts of climate change appear at first sight to make the problem of stakeholder involvement intractable. As the above discussion on categorization of impacts, scale and timing shows, there can be few people if any that are not in some way at risk from climate change. And there are future generations to consider. How can sufficient stakeholder involvement be developed at the international and global level for citizens of the world to feel that their concerns have been taken into account? In the next section we present some suggestions for how the process of elaborating Article 2 might be advanced by the scientific and policy community with greater transparency and legitimacy.

10.6 New Directions for Defining Long Term Goals for the Convention

10.6.1 *Is Universal Agreement on 'Dangerous' Possible?*

Adverse impacts, particularly at a fine scale, happen at virtually all levels of climate change. Indeed, some may argue that we have already exceeded a dangerous level of climate change by adversely affecting some species (see e.g. some species of penguins in Antarctica (Kaiser, 1997), toads in Central America (Pounds, 2001), or extreme climate events (Stott et al., 2004)). The problem is that there is no intuitively obvious level of climate change to accept what is happening as 'dangerous.'

Reaching such agreement internationally may appear today to be an insurmountable task. But policy-makers exercise collective judgment on a daily basis – which means nothing more complicated than that they adopt a normative course of action on the basis of the facts presented to them, however incomplete and imperfect. The universal acceptance of norms of a fundamentally normative character is not uncommon in international affairs. It is certainly within the realm of possibility then that a consensus can be achieved on a dangerous level of climate change, particularly if that level is defined at a relatively low level to ensure that all possible dangers have been taken into account.

Normally in international negotiations, the task of agreement is simplified if dangers that affect everyone are addressed. But perversely in climate change, the setting of a dangerous target may be complicated by the difficulty of achieving it, particularly if it is set at a relatively low degree of change in climate. This is because costs rise with lower stabilization targets; indeed they can rise substantially with relatively low targets (Metz et al., 2001). To be sure, Article 2 does not consider the costs or feasibility of holding greenhouse gas concentrations at the level that would avoid dangerous impacts. But the Convention and Protocol are based on the principle of common but differentiated responsibilities which puts a greater share of abatement burden on developed countries, as well as mandating that they provide developing countries with financial resources for adaptation. In these circumstances, limiting climate change to very low levels that would avoid all impacts that could be considered dangerous may be practically infeasible or unacceptably costly to developed countries.

Given these complexities and difficulties with trying to apply Article 2 in a manner that would cover all aspects of what may constitute dangerous and to incorporate different scales, we suggest an alternative framing of Article 2 that may be more likely to result in practical guidance for the climate regime.

10.6.2 *An Alternative Approach to Article 2*

Rather than trying to find what level of climate change is dangerous and implicitly what level below the dangerous level is 'safe', perhaps we should consider whether asking ourselves to define a single level of climate change which can be termed 'dangerous' is indeed the right question. As we have argued in this paper, many different levels of climate change can, with legitimacy, be conceived as dangerous. Yet most of the literature on Article 2 presumes

that ultimately a single level would need to be selected as the basis for guiding global climate policy and that the main purpose of doing so would be to set mitigation goals (O'Neill and Oppenheimer, 2002, 2004).

It may be more practical, in fact, to identify levels of climate change that might be deemed "tolerable" by the full range of stakeholders and interests affected by climate change. Focusing on 'tolerable' levels of climate change as a way of defining the long-term goal for the UNFCCC shifts the focus away from scientists making expert judgments about 'dangerous' on the basis of crucial, but generally unexplained assumptions about the choice of scale to be applied. The tolerable approach implies that there may be no one 'safe' level of climate change: whatever level of climate change is selected as being tolerable would likely have adverse impacts at lower levels. Those adverse impacts would either be absorbed or dealt with in some other way, such as adaptation, but in each case the focus of attention is on whether the stakeholders and interests affected by climate impacts find them to be (in) tolerable.

Use of an approach focusing on a tolerable level of climate change might allow for the consequences of limiting change to such a level to be factored in. Costs and feasibility of mitigation might be factors in defining a tolerable level of climate change. It also recognizes that risks exist even below the level considered tolerable and continued efforts should be made to further reduce GHG concentrations.

By advocating the use of a 'tolerable' approach to Article 2, we are not suggesting that the Convention needs to be amended in any formal sense. Our aim is to challenge the current framing of Article 2 so that we engage more explicitly with issues dealing with the choice of scale, the full range of response strategies including adaptation and crucially of focusing on the process of reaching agreement. These issues tend to be hidden or side stepped when Article 2 is framed in terms of defining what *impacts* are dangerous. We note that the Tolerable Windows Approach (TWA; Bruckner et al., 1999) addressed developing acceptable emissions control pathways to avoid a dangerous level of climate change as defined by global policy-makers. Our approach instead focuses on the process for selecting the dangerous level. Thus TWA could be complementary or even a part of the discussion on what impacts are tolerable.

10.6.3 *Assessing Politically Defined Long Term Goals*

An additional suggestion for approaching Article 2 concerns how information is organized to help with the assessment of different kinds of impacts. Section 2 identified three categories which can be considered to be dangerous based on Article 2 (geophysical impacts, biophysical impacts, and impacts on human health and wellbeing). Effectively, there is a fourth category, which we label as 'political'. This policy-based category involves making a judgment that impacts that occur above the achievement of a long-term stabilization target are 'dangerous.' Such a target could be expressed, for example, in terms of GHG concentrations, changes to mean global temperature or mean sea level rise. These proposals imply selection of a target defining 'dangerous' based on political judgment. Such targets are informed by studies on impacts of climate change, but some of them may or may not take feasibility considerations into account. Such an approach is typified by the European Union which has proposed limiting the increase in mean global climate to 2°C above pre-industrial (EU Council, 1996; see also the in-depth discussion in Grassl et al., 2003).

It seems to us that this 'political' category may be a more likely one to be eventually applied. For all the effort to define a long-term goal of the Convention based on avoiding particular outcomes, what has been emerging from the political arena is the use of relatively more arbitrary goals. This is due, in part, to the fact that the purpose of Article 2 is to give broad guidance to the climate change regime and this 'agenda-setting' function must necessarily involve a political choice.

10.6.4 *Strengthening the Process to Elaborate Article 2*

We have argued that the process used to come to agreement on the long-term goal for the UNFCCC may be as important as the goal itself. We have emphasized the need to find explicit ways of incorporating stakeholders and civil society at a global level in the policy process on climate change.

On a national scale, it is likely that dedicated stakeholder meetings may be utilized as they are for other types of consultations. Internationally, meetings may be impractical or may not suffice. For climate change, international practices to engage stakeholders need further elaboration. Two possibilities suggest themselves: direct and indirect. The most direct approach is to involve organized civil society in the shape of non-governmental groups (NGOs) in a pro-active fashion. This happens to some extent in that such groups as environmental NGOs and industry associations, are already present as observers at COP negotiations. They are often permitted to go beyond 'observing' by making interventions and by lobbying national delegations. The evolution of COP negotiation to provide greater involvement opportunities is a step in the right direction and evidence of the recognition that stakeholders are important and should be heard (Gupta, 2003; Ott, 2004). One obvious defect of the present system is that the civil society voices are not evenly represented from across the globe with fewer Southern NGOs being present due to funding constraints (Yamin, 2001). We suggest this aspect could be remedied by greater attention to funding and representation.

The indirect approach could work by trying to survey the public's perceptions of dangerous climate. This could be done through polls, random sample interviews with representatives of individuals or groups on issues defined by the COP. Such polls could be organized by the Secretariat itself or provided by NGOs. Such solicitation has not been tried before. But systematic information of this kind might be given a more official backing (and a more legitimate role) within the UNFCCC process. Use of a transparent and open process is quite consistent with application of the 'political' category of setting a long-term goal for the UNFCCC. Indeed, it has the virtue of recognizing the importance of process in developing a consensus. It should be informed by scientific and other analyses, but may not necessarily be a mechanical application of the outcome of those analyses.

A process to elaborate Article 2 will take years. But the resulting dialogue may produce a more legitimate consensus on what the long-term goals for the UNFCCC should be or at least better define differences in perceptions around the world in ways that do not damage the credibility of the international process. It should be noted that apart from a very brief set of discussions that took place in the run up to Kyoto, issues about the long-term objective of the Convention have not been given discussion time in the COP process. The time is surely ripe for Parties and stakeholders to submit their views on the merits of elaborating Article 2 and their substantive views on what constitutes a determination of dangerous anthropogenic interference with the climate system.

10.7 Conclusions

We agree that a new modality for global decision-making is struggling to emerge, and that climate change is the guinea pig or leading experiment through which it is being developed (Kjellen, 2004). There is no scientific basis for determining a single level at which danger can be said to begin, and if anything like a global consensus can eventually emerge it seems likely that it will do so from a process that takes a more nuanced view of the distribution of impacts, the scale and timing of the consequences, and gives full recognition to the role of adaptation. The new modality also involves innovations in the pattern and practice of global governance such that stakeholders and scientists take their place in the negotiations, and in which, above all, credence and weight is given to the diversity of values which impinge onto the climate debate.

It is difficult to see now how such an evolving regime could ever arrive at a single value of dangerous. But there are many precedents in international affairs when humanity has agreed on fundamental normative principles and rules. The climate change regime has defied critics in terms of reaching the measure of universality it currently has, and those involved in negotiations show remarkable capacity to re-invent the international process so it can better overcome the challenges it faces (Grubb and Yamin, 2001). As with other international issues, tentative first steps may lie more in the direction of each country and each stakeholder or interest group taking their own action to define dangerous. The process of sharing such understandings through an enhanced international process would then emerge. Drawn out and uncertain as it is, perhaps such a process offers a more transparent, legitimate and ultimately more effective path forward to operationalizing Article 2.

REFERENCES

ACIA. (2004). *Impacts of a Warming Arctic, Arctic Climate Impacts Assessment*. New York: Cambridge University Press.

Andersen, S. and Madhava Sarma, K. (2002). Protecting the Ozone Layer. In *The United Nations History*. UNEP. 513.

Azar, C. and Schneider, S. (2001). Are Uncertainties in Climate and Energy Systems a Justification for Stronger Near-term Mitigation Policies? Paper, Prepared for the Pew Center on Global Climate Change, Washington: USA.

Bodansky, D. (1993). The United Nations Framework Convention on Climate Change: *A Commentary, Yale Journal of International Law*. **18** (2), 451–58.

Benedict, R. (1998) *Ozone Diplomacy, New Directions in Safeguarding the Planet,* Harvard University Press.

Bruckner, T., Petschel-Held, G., Toth, F.L., Fussel, H-M., Helm, C., Leimbach, M. and Schellnhuber, H.-J. (1999) Climate change decision-support and the tolerable windows: Approach, Environmental Modeling and Assessment **4**, 217–234.

Corfee-Morlot, J. and Agrawala, S. (ed.) (October, 2004). The Benefits of Climate Policy, *Global Environmental Change,* **14** (3).

Dessai, S., Adger, W. N., Hulme, M., Turnpenny, J. Köhler, J. and Warren, R. (2004). Defining and Experiencing Dangerous Climate Change, *Climatic Change,* **64**, 11–25.

European Climate Forum (ECF) and Potsdam Institute, Report on the Beijing Symposium on Article 2, September, 2004, http://www.european-climate-forum.net/pdf/ECF_beijing_results.pdf

EU Council. (1996), 1939th Council, Environment, Luxembourg, 25th June 1996.

Fankhauser, S. (1996). The Potential Costs of Climate Change Adaptation, In Smith, J.B., Bhatti, N., Menzhulin, G., Benioff, R., Budyko, M., Campos, M., Jallow, B. and Rijsberman, F. (eds.). *Adapting to Climate Change*. New York: Springer.

Gitay, H., Brown, S., Easterling, W. and Jallow, B. (2001). Ecosystems and Their Goods and Services, in McCarthy, J. et al., eds *Climate Change 2001, Impacts, Adaptation and Vulnerability, Contribution of Working Group II to the Third Assessment Report of the Intergovernmental Panel on Climate Change*, Chapter 5, Cambridge: University Press.

Grassl, H. et al. (2003). Climate Protection Strategies for the 21st Century. Kyoto and Beyond. WBGU Special Report, Berlin: WBGU.

Grubb, M. and Yamin, F. (2001). Climate collapse at The Hague, what happened, why and where do we go from here? *International Affairs,* **2**, 261–277.

Gupta, J., Berk, M. and Van Asselt, H. (2003), Defining dangerous: report of the annex I workshop on Article 2 of the Climate Convention (HOT WDI), Report number W-03/35, and Netherlands, Institute for Environmental Studies.

Hitz, S. and Smith, J.B. (2004). Estimating Global Impacts from Climate Change, *Global Environmental Change,* **14** (3): 201–218.

Jacoby, H. (2004). Informing Climate Policy Given Incommensurable Benefits Estimates. *Global Environmental Change,* **14** (3).

Jasanoff, S. and Wynne, B. (1997). Science and Decision-making, in Malone, E. and Rayner, S. (eds) *Human Choice and Climate Change,* **1**, The Societal Framework, Battelle Press.

Kaiser, J. (1997). Is Warming Trend Harming Penguins? *Science* **276**: 1790.

Kjellen, B. (2004). Pathways to the Future, The New Diplomacy for Sustainable Development, in Yamin, F. (ed.), *Climate Change and Development*. Brighton: Institute of Development Studies, IDS Bulletin, **35** (3), 1–10.

Leemans, R. and Eickout, B. (2004). Analysing Changes in Ecosystems for Different Levels of Climate Change. *Global Environmental Change* **14** (3), 219–28.

May, R. (2000). The use of scientific advice in policy making, OST, UK

NAS (National Academy of Sciences), 1979. *Carbon Dioxide and Climate: A Scientific Assessment*. Washington, D.C: National Academy Press.

Metz, B., Davidson, O., Swart, R. and Pan, J. (2001). Climate Change 2001, *Mitigation*. New York: Cambridge University Press. 752.

Millstone, et al. (2004). Science in Trade Disputes related to potential risks, comparative case studies, JRC, ESTO and IPTS.

Nordhaus, W. and Boyer, J. (2000). *Roll the DICE Again, Economic Modeling of Climate Change*. Cambridge, MA, MIT Press.

O'Neill, B.C. and Oppenheimer, M. (2002). Dangerous Climate Impacts and the Kyoto Protocol. *Science* **296**, 1971–72.

O'Neill, B.C. and. Oppenheimer M. (2004). *Climate Change Impacts Sensitive to Path to Stabilization. Proc. Nat. Acad. Sci.,* **101**, 411–416.

Oppenheimer, M. and Petsonk, A. (2004), UNFCCC, Historical Origins, *Recent Interpretations,* Climatic Change, **2**.

Ott, K. et al. (2004). Reasoning Specifications of Climate Protection, Europaische Akademie GmbH, prepared for the Federal Environmental Agency of Germany, Report No. UBA-FB202 41 252.

Parry, M., Carter, T. and Hulme, M. (1996). What is dangerous climate change*?, Global Environmental Change,* **6** (1), 1–6.

Patwardhan, A., Schneider, S. and Semenov, S (September 2003). Key vulnerabilities including issues relating to Article 2, a concept paper prepared for the first scoping meeting of the IPCC Workshop on Fourth Assessment Report, Potsdam. 1–4, 2003.

Patwardhan, A. et al. Assessing the Science to address Article 2 UNFCCC: A concept paper relating to cross cutting theme four, http://ipcc.ch/activity/cct.pdf

Pounds J.A. (2001). Climate and Amphibian Declines. *Nature* **410**, 639–40.

Pershing, J. and Tudela, F. (2003). Developing Long Term Targets, Aldy, et al (eds), Beyond Kyoto, *Advancing the International Effort Against Climate Change,* USA: Pew Centre.

Sand, P. (1990). *Lessons Learned in Global Environmental Governance*, Washington DC: World Resources Institute.

Schneider, S. (2001). What is dangerous climate change? *Nature,* **411**, 17–19.

Smith, J.B., Schellnhuber, H-J., Mirza, M.Q., Fankhauser, S., Leemans, R., Lin, E., Ogallo, L., Pittock, B., Richels, R., Rosenzweig, C., Safriel, U., Tol, R.S.J., Weyant, J., and Yohe, G. (2001). Vulnerability to Climate Change and Reasons for Concern: A Synthesis. In: McCarthy, J., Canziana, O., Leary, N., Dokken, D. and White, K. (eds) *Climate Change (2001). Impacts, Adaptation, and Vulnerability*. New York: Cambridge University Press, 913–67.

Smit, B., Pilifosova, O., Burton, I., Challenger, B., Huq, S., Klein, R. and Yohe, G. (2001). Adaptation to Climate Change in the Context of Sustainable Development and Equity. In McCarthy, J., Canziana, O., Leary, N., Dokken, D. and White, K. (eds) *Climate Change (2001) Impacts, Adaptation, and Vulnerability*. New York: Cambridge University Press, 877–912.

Stirling, A. (2001). Rethinking risk, application of a novel technique to GM crops', *Technology, Innovation & Society* **18** (1), 1–23.

Stirling, A. (2003). Risk, uncertainty and precaution, some instrumental implications from the social sciences, in Berkhout, F., Leach, M. and Scoones, I (ed.), *Negotiating Environmental Change, Some Perspectives from Social Sciences,* Edward Elgar, Massachusetts.

Stott, P.A., Stone, D.A. and Allen, M.R. (2004). Human contribution to the European heatwave of 2003. *Nature* **432**, 610–14.

Tol, R.S.J. (2003). Is the Uncertainty about Climate Change too Large for Expected Cost-Benefit Analysis. *Climatic Change* **56**: 265–89.

Wilbanks, T.J., Kane, S.M., Leiby, P.N., Perlack, R.D., Settle, C., Shogren, J.F. and Smith, J.B. (2003). Possible Responses to Global Climate Change, Integrating Mitigation and Adaptation. *Environment,* **45** (5): 28–38.

Wildavsky, A. (1979). *Speaking Truth To Power.* Boston: Little, Brown.

Yamin, F. and Depledge, J. (2004). *The International Climate Change Regime, A Guide to Rules, Institutions and Procedures,* Cambridge: Cambridge University Press.

Yamin, F. (2001). NGOs and International Environmental Law, A critical evaluation of their roles and responsibilities, *Review of European Community and International Environmental Law,* **10** (2) 149–162.

Yamin, F. (2005). The European Union and future climate policy: Is mainstreaming adaptation a distraction or part of the solution? *Climate Policy* **5**.

CHAPTER 11

Impacts of Global Climate Change at Different Annual Mean Global Temperature Increases

Rachel Warren

Tyndall Centre for Climate Change Research, School of Environmental Sciences, University of East Anglia, Norwich

ABSTRACT: Based on peer-reviewed literature, climate change impacts on the earth system, human systems and ecosystems are summarised for different amounts of annual global mean temperature change (ΔT) relative to pre-industrial times. Temperature has already risen by $\Delta T = 0.6°C$ and effects of climate change are already being observed globally. At $\Delta T = 1°C$ world oceans and Arctic ecosystems are damaged. At $\Delta T = 1.5°C$ Greenland Ice Sheet melting begins. At $\Delta T = 2°C$ agricultural yields fall, billions experience increased water stress, additional hundreds of millions may go hungry, sea level rise displaces millions from coasts, malaria risks spread, Arctic ecosystems collapse and extinctions take off as regional ecosystems disappear. Serious human implications exist in Peru and Mahgreb. At $\Delta T = 2–3°C$ the Amazon and other forests and grasslands collapse. At $\Delta T = 3°C$ millions at risk to water stress, flood, hunger and dengue and malaria increase and few ecosystems can adapt. The thermohaline circulation could collapse in the range $\Delta T = 1–5°C$, whilst the West Antarctic Ice Sheet may commence melting and Antarctic ecosystems may collapse. Increases in extreme weather are expected.

11.1 Introduction

This paper reports the results of literature-review based assessment of the impacts of climate change on the earth system, on human systems and on ecosystems for different changes in annual global mean temperature change with respect to pre-industrial times (ΔT). It summarises observed changes which have either been directly attributed to, or are at least consistent with the expected effects of, climate change at $\Delta T = 0.6°C$. It continues with predictions of the impacts of potential further temperature change of $\Delta T = 1, 2$ and $3°C$ or larger increases in annual mean global temperature. A summary table reports the main findings. Detailed information and an extensive reference list are provided in the tables A to H given in the Appendix. The policy context is to allow assessment of the benefits of stabilisation of greenhouse gases at different levels in the atmosphere, since this will alter the probabilities of reaching the different levels of temperature change. The summary table and tables A to G in the Appendix allow different potential temperature changes to be associated with their respective likely impacts.

11.2 Methodology

A literature search was made to assess pertinent impacts of climate change on all sensitive systems. These references were scanned for specific information about thresholds in temperature change/sea level rise or rates of temperature change/sea level rise above which adverse consequences could be expected, taking note of the climate scenario and GCM used in any quantitative study of impacts, together with any assumptions about adaptation. In quantitative analysis the following methods were used to tabulate impact thresholds for the tables.

11.2.1 *Harmonisation of Reference Point for Temperature*

Information from studies was converted to the same pre-industrial reference point for temperature, noting that pre-industrial temperature is approximately 0.6°C below present day temperatures (IPCC 2001); that the mean 1961–1990 temperature is approximately 0.3°C below present day (*pers. comm.*); and using Table 11.1 below, taken from Parry *et al.* (2004) showing the HadCM3

Table 11.1 Projected changes in global mean temperature relative to the 1961–1990 mean (0.3°C is then added to the figures to convert to the pre-industrial reference point required in the tables).

Year	IS92a	A1F1	A2a	A2B	A2C	B1	B2a	B2b
2020s	1.1	0.99	0.86	0.93	0.88	0.84	0.91	0.91
2050s	2.06	2.26	1.92	1.89	1.85	1.4	1.56	1.66
2080s	3.00	3.97	3.21	3.28	3.32	2.06	2.35	2.40

Table 11.2 Had CM2 ensemble and HadCM3 values of ΔT (Global temperature change relative to pre-industrial temperature) (taken from Hulme _et al._ 1999).

MODEL	1961–1990	2020s	2050s	2080s
IS92A	0.3	1.3	2.0	2.7
HADCM2	0.3	1.5	2.4	3.4
HADCM3	0.3	1.4	2.4	3.4

Table 11.3 Population scenarios in SRES (taken from Parry _et al._ 2004).

	IS92a	A1	B1	A2	B2
2025	~8200	7926	7926	8714	8036
2050	~9800	8709	8709	11778	9541
2075	~15200	7914	7914	14220	10235

simulations of global mean temperature changes for different SRES scenarios.

11.2.2 _Upscaling_

Whilst some of the literature relates impacts directly to global mean temperature rises, many studies give only local temperature rises, and hence a global temperature rise had to be inferred. Owing to the limited resources of this study, where upscaling information was not provided in source literature, HadCM2/3 only was used to upscale from local to global temperature changes, using temperature trajectories from HadCM2/3 outputs (Table 11.2) taken from (Hulme _et al._ 1999). However the tables report, where possible, the GCM used in the source literature (frequently HadCM2 or HadCM3), since this affects the relationship between local and global temperature change assumed in the study, as well as the associated precipitation changes.

11.2.3 _Population/Socio-Economic Scenarios_

Climate change impacts on the human system are, not surprisingly, strongly affected by the future development pathway of the human system, which affects the stock at risk and its vulnerability. Since impacts depend in a highly non-linear manner on population and population is not the only driver for climate impacts, no scaling was attempted between different socio-economic scenarios. However it is important that the reader should take into account, on perusing the impacts tables, in particular Tables B and C detailing human system impacts (in the Appendix), and the summary table, the very different population projections used in the various scenarios (Table 11.3).

11.2.4 _Adaptation_

In general adaptation is treated superficially and inconsistently in the literature, and assumptions made are often poorly documented. Hence reference is made verbally in the tables to indicate if any adaptation is taken into account in the studies.

11.2.5 _Linkage Between Temperature and Sea Level Rise_

Information related to the impacts of sea level rise on human systems and ecosystems were presented separately to temperature impacts, although there is intentionally some overlap in the information presented in sea level

Table 11.4 Global annual mean temperature rise (ΔT) since pre-industrial times and sea level rise in HadCM3 (taken from Parry _et al._ 1999).

ΔT (HadCM3)	Year	Sea Level rise (total) relative to 1961–1990
0.6	1990	2.7 cm
1.5	2020	12.1
2.4	2050	24.1
3.4	2080	39.8

rise and temperature Tables (B and C of the Appendix for human systems and D and F of the Appendix for ecosystems). Some estimates of millions at risk (Nicholls _et al._ 1999, Parry _et al._ 1999) due to sea level rise were related to temperature rise using Table 11.4, taken from Parry _et al._ 1999, which is based on a simulation from HadCM3. This ignores the fact that sea level rise will continue increasing even if temperature ceases to rise. Thus in the full tables in the Appendix, sea level rise and temperature effects are quoted separately.

11.3 Results

Table 11.5 summarises the observed changes consistent with or attributed to the effects of climate change, and continues with a summary of the impacts of climate change which have been predicted for different levels of global annual mean temperature rise (ΔT). These impacts are further detailed in Tables A to H of the Appendix. Table A summarises impacts upon the earth system, Tables B and C summarise impacts of temperature and sea level rise on human systems, whilst Tables D and F summarise impacts of temperature and sea level rise on ecosystems. Tables E and G report impacts of different rates of temperature change and sea level rise upon ecosystems, whilst Table H indicates the impacts of ocean acidification.

11.3.1 _Observed Changes Attributed to, or Consistent with, Climate Change_

To date, annual mean global temperature has increased by $\Delta T = 0.6°C$ relative to pre-industrial times and is increasing at a rate of 0.17°C/decade. The summary Table 11.5, and Tables A to G of the Appendix all show that the effects of this fairly small climate change are already

Table 11.5 Summary of Climate Change Impacts on the Earth System, Human Systems and Ecosystems.

Global Average Surface Temperature rise above pre-industrial (ΔT, °C)	Region	IMPACTS: Note that impacts are cumulative (that is those accruing at $\Delta T = 2$°C are additional to those accruing at $\Delta T = 1$°C) except for the agricultural sector
		OBSERVED CHANGE
0.6	GLOBE	Sea level increasing at 1.8 mm/yr; glaciers retreating worldwide; changes in rate and seasonality of streamflows; 80% of 143 studies of phenological, morphological and distributional changes in species show changes in direction consistent with expected response to climate change e.g. spring advanced 5 days, losses in alpine flora; increase in extreme rainfall patterns causing drought and flood; substantial and increasing damage due to extreme weather events partly due to climatic factors, particularly in small islands.
	Arctic	Local temperature rise of 1.8°C; damage to built infrastructures due to melting permafrost; accelerating sea ice loss now at $0.36 \pm 0.05 \times 10^6$ km^2/10 yr.
	Antarctic	Collapse of ice shelves; changes in penguin populations.
	Africa	Abrupt change in regional rainfall caused drought & water stress, food insecurity and loss of grassland in the Sahel.
	Americas	Extinction of Golden Toad in Central America.
	Europe	N shifts in plankton distribution in N Sea, likely to have caused observed decline in sand eels and hence breeding failure of seabirds; changes in fish distributions; extreme heat & drought in 2003 which caused 25,000 deaths has been attributed to anthropogenic climate change.
1C	Globe	Oceans continue to acidify, with unknown consequences for entire marine ecosystem; 80% loss of coral reefs due to climate-change induced changes in water chemistry and bleaching; potential disruption of ecosystems as predators, prey and pollinators respond at different rates to climatic changes and damage due to pests and fire increases; 10% ecosystems transformed, variously losing between 2 to 47% of their extent, loss cool conifer forest; further extinctions in cloud forests; increase in heatwaves and associated mortality, decrease in cold spells and associated mortality, further increase in extreme precipitation causing drought, flood, landslide, likely to be exacerbated by more intense El Niño; increased risk malaria & dengue; rise in insurance prices and decreased availability of insurance; 18–60 million additional millions at risk to hunger and 20 to 35 million ton loss in cereal production depending on socioeconomic scenario, GCM and realisation of CO_2 fertilisation effect; 300–1600 additional millions suffer increase in water stress depending on socioeconomic scenario and GCM.
	Arctic	only 53% wooded tundra remains stable.
	Africa	Decreases in crop yields e.g. barley, rice estimated ~10%; significant loss of Karoo the richest floral area in world; increased risk of death due to flooding; southern Kalahari dunefield begins to activate.
	Americas	Serious drinking water, energy and agricultural problems in Peru following glacier melt; increased risk death due to flooding; increased crop yields in N America in areas not affected by drought if C fertilisation occurs.
	Europe, Russia	Increased crop yields if C fertilisation occurs in areas not affected by drought; increased drought in steppes, Mediterranean causing water stress and crop failure.
	Australia	Extinctions in Dryandra forest; Queensland rainforest 50% loss endangering endemic frogs & reptiles.
1.5°C	GLOBE	Onset of melt of Greenland ice sheet causing eventual additional sea level rise of 7m over several centuries.
2°C	GLOBE	Threshold above which agricultural yields fall in developed world; 1.0 to 2.8 billion people experience increase in water stress depending on socioeconomic scenario and GCM model used; 97% loss of coral reefs; sea level rise and cyclones displace increasing numbers (12–26 million, less those protected by adaptation schemes) of people from coasts;

(continued)

Table 11.5 (*contd*)

Global Average Surface Temperature rise above pre-industrial (ΔT, °C)	Region	IMPACTS: Note that impacts are cumulative (that is those accruing at $\Delta T = 2$°C are additional to those accruing at $\Delta T = 1$°C) except for the agricultural sector
		additional millions at risk to malaria particularly in Africa and Asia, depending on socioeconomic scenario; 16% global ecosystems transformed: ecosystems variously lose between 5 and 66% of their extent; -12 to $+220$ additional millions at risk to hunger and 30–180 million ton loss global cereal production depending on socioeconomic scenario, GCM and realisation of CO_2 fertilisation effect.
	Arctic	Destruction of Inuit hunting culture; total loss of summer Arctic sea ice; likely extinction polar bear, walrus; disruption of ecosystem due to 60% lemming decline; only 42% existing Arctic tundra remains stable, high arctic breeding shorebirds & geese in danger, common mid-arctic species also impacted.
	Antarctic	Potential ecosystem disruption due to extinction of key molluscs.
	Africa	Large scale displacement of people (climate refugees from low food security, poverty and water stress) in Mahgreb as rainfall declines by at least 40%; all Kalahari dunefields begin to activate;
	Americas	Vector borne disease expands poleward e.g. 50% increase in malarial risk in N America; extinction of many Hawaiian endemic birds and impacts on salmonid fish;
	Europe, Russia	Tripling of bad harvests increasing Russian inter-regional political tensions;
	Asia	1.8 to 4.2 billion experience decrease in water stress (again depending on socioeconomic scenario and GCM model used) but largely in wet season and not in arid areas; vector borne disease increases poleward; 50% loss of Chinese boreal forest; 50% loss of Sundarbans wetlands in Bangladesh;
	Australia	Risk of extinctions accelerates in N Australia, e.g. Golden Bowerbird; 50% loss Kakadu wetland;
1–5°C	GLOBE	Expert judgements and models predict increasing probability of complete THC collapse in this range; predictions of 50% collapse probability range from 2 to 5°C.
2–3°C	GLOBE	Conversion of vegetation carbon sink to source; collapse of Amazon rainforest; 0.9–3.5 billion additional persons suffer increased water stress.
	Africa	80% Karoo lost endangering 2800 plants with extinctions
		Loss Fynbos causing extinction of endemics
		5 S African parks lose >40% animals
		Great Lakes wetland ecosystems collapse
		Fisheries lost in Malawi
		Crop failures of 75% in S Africa
		All Kalahari dunefields may be mobile threatening sub-Saraharan ecosystems and agriculture
	Americas	Maples threatened in N American temperate forest
	Australia	Total loss Kakadu wetlands and Alpine zone
	Asia	Large impacts (desertification, permafrost shift) on Tibetan plateau; complete loss Chinese boreal forest, food production threatened in S
3°C	GLOBE	Few ecosystems can adapt; 50% nature reserves cannot fulfil conservation objectives; 22% ecosystems transformed; 22% loss coastal wetlands; ecosystems variously lose between 7 and 74% of their extent; 65 countries lose 16% agricultural GDP even if CO_2 fertilisation assumed to occur; irrigation requirements increase in 12 of 17 world regions; 17–18% increase in seasonal *and* perennial potential malarial transmission zones exposing 200 to 300 additional people; overall increase for all zones 10%; 50–60% world population exposed to dengue compared to 30% in 1990; 25 to 40 additional millions displaced from coasts due to sea level rise, less those protected by adaptation schemes; -20 to $+400$ additional millions at risk to hunger and 20–400 million tonne loss global cereal

(*continued*)

Table 11.5 (*contd*)

Global Average Surface Temperature rise above pre-industrial (ΔT, °C)	Region	IMPACTS: Note that impacts are cumulative (that is those accruing at $\Delta T = 2$°C are additional to those accruing at $\Delta T = 1$°C) except for the agricultural sector
		production depending on socioeconomic scenario and realisation of CO_2 fertilisation effect; 1200 to 3000 additional millions suffer increase in water stress depending on socioeconomic scenario and GCM.
	Africa	70–80% of those additional millions at risk from hunger are located in Africa
	Americas	50% loss world's most productive duck habitat; large loss migratory bird habitat
	Europe, Russia	Alpine species near extinction; 60% species lost from Mediterranean region; high fire risk in Mediterranean region; large loss migratory bird habitat
	Asia	Chinese rice yields fall by 10–20% or increase by 10–20% if CO_2 fertilisation is realised
	Australia	50% loss eucalypts; 24% loss suitable (80% loss original) range endemic butterflies.
2–4.5°C	Antarctic	Potential to trigger melting of the West Antarctic Ice Sheet raising sea levels by a further 5 to 6 m i.e. 60 to 120 cm/century
	Africa	Crop failure rises by 50–75% in S Africa
4°C	GLOBE	Entire regions out of agricultural production; -30 to $+600$ additional millions at risk to hunger; 25% increase in potential malarious zones: 40% increase in seasonal zones and 20% decrease in perennial zones; timber production increases by 17%; probability of thermohaline shutdown at or above 50% according to many experts; 44% loss taiga, 60% loss tundra.
	Australia	Out of agricultural production; total loss alpine zone.
	Africa	70 to 80% of those additional millions at risk from hunger are in Africa.
	Europe	38% European alpine species lose 90% range
	Russia	5–12% drop in production including 14–41% in agricultural regions.

being observed across the world, from the Arctic to the Tropics, from the oceans to the mountains, in the earth system, in ecosystems and in human systems. Across the globe species are changing their phenology and geographical distribution in a direction consistent with their expected response to climate change. Glaciers are melting throughout most of the world, the ocean has already acidified by 0.1 pH units, unprecedented heat waves are causing episodes of mortality in large cities, and drought is intensifying in many regions. The first extinction which is likely to be attributable to climate change has already occurred, that of the Golden Toad in the cloud forest covering mountaintops in Costa Rica. All the tables show observed changes in response to existing climate change. Some of these have been directly attributed to anthropogenic climate change through rigorous calculations, such as the unusually warm European summer temperatures of 2005 (Stott *et al.* 2004), and sets of observed phenological changes (Root *et al.* 2005) whilst all are consistent in direction with its expected effects (for example, that warmer sea surface temperatures would lead to increasing destructiveness of tropical cyclones (Emanuel 2005)).

11.3.2 *Impacts at DT = 1°C of Global Annual Mean Temperature Rise Since Pre-industrial Times*

A temperature rise of only $\Delta T = 1$°C since pre-industrial – that is only a further 0.4°C above today's – would cause additional climate impacts. Of great concern at this temperature rise is that oceans would continue to acidify, with completely unknown consequences for the entire marine ecosystem of our planet, through damage to marine calcifying organisms such as corals and calcareous plankton. Secondly, the planet's coral reefs would be subject to damage due to bleaching and changed ocean chemistry (also resulting from climate change). At $\Delta T = 1$°C, it is predicted that 10% of the global ecosystems would be transformed losing between 2% and 47% of their extent, whilst in the Arctic, where temperatures are currently rising at 0.46°C/decade, much greater local temperature rises are predicted, which would lead to losses of tundra, sea ice, and associated impacts on fauna such as high arctic breeding birds and polar bears.

In Peru, at $\Delta T = 1$°C, the continued melting of glaciers are expected to cause serious drinking water, energy and agricultural problems. In Africa crop yields are predicted to begin to decline, whilst in Europe and North America, CO_2 fertilisation could increase crop yields and high latitudes would become more suitable for cultivation. However, recent evidence (Royal Society, 2005) shows that CO_2 fertilisation is lower in the field than in the laboratory, and is significantly offset by yield losses due to the predicted increasing frequency in extreme weather (e.g. a day or an hour of extreme heat) and exposure to rising levels of tropospheric ozone, also a greenhouse gas, even if soil-nutrient and water availability remain constant under climate change. Once changes in precipitation are taken into account yields may fall further. Species extinctions are predicted in Australian Dryandra forest, and the Queensland rainforest may shrink by 50%.

11.3.3 *Impacts at ΔT = 1.5°C of Global Annual Mean Temperature Rise Since Pre-industrial Times*

Coral reefs in the Indian Ocean are not expected to survive above a temperature rise of ΔT = 1.4°C. Of perhaps even greater concern is the potential to trigger irreversible melting of the Greenland ice sheet at a local temperature rise of 2.7°C, matching a global ΔT of 1.5°C (only 0.9°C above today's temperatures), a process that results in an eventual 7 m sea level rise over and above that caused by thermal expansion of the oceans, and potentially, causing an additional sea level rise of 0.75 m as soon as 2100 (Hansen 2005).

11.3.4 *Impacts at ΔT = 2°C of Global Annual Mean Temperature Rise Since Pre-industrial Times*

At ΔT = 2°C, all of the impacts seen at 1°C global temperature rise would already have occurred. As temperature continues to increase a wide range of further impacts would occur in both ecosystems and human systems. The impacts are thus cumulative, since damages generally increase with temperature, except in the case of agriculture where there are initially benefits in some developed regions for small temperature rises. However at ΔT = 2°C agricultural yields would begin to fall in the developed world. Thus with the exception of the agricultural sector, all of the impacts which are listed for ΔT = 2°C would be additional to those already experienced at ΔT = 1°C. At ΔT = 2°C it is predicted that 97% of coral reefs would be gone and 16% of the global ecosystems would be transformed, losing between 5% and 66% of their extent. Approximately one to three billion people would experience an increase in water stress, the range reflecting the consequences of different socio-economic futures, as well as the use of different GCM models to predict regional climate changes. In Asia millions would theoretically experience a decrease in water stress, but this decrease would occur in the wet season when the additional water would need to be stored for use in the dry season and might cause floods. Sea level rise and cyclones would displace millions from the world's coastlines, and malaria risks would increase northwards and up mountainsides particularly in Africa, Asia, and Latin America, the extent of which depends on the socioeconomic future of these regions. Global cereal production would fall leading to a rise in food prices, exposing from between 12 million less to 220 million more people to the risk of hunger – the range reflecting the aforementioned uncertainty in the realisation of the theoretical benefits of CO_2 fertilisation, as well as differing potential socioeconomic futures. In the Arctic, ecosystem disruption is predicted owing to complete loss of summer sea ice whilst only 42% of the tundra would remain stable. This would destroy the unique Inuit hunting culture, cause the extinction of the polar bear and large losses in global populations of birds and local populations of lemmings. Meanwhile in the Antarctic, key molluscs are predicted to become extinct with damaging ramifications for the rest of the Antarctic ecosystem. In Africa, severe problems would occur in the Mahgreb where increased drought, hence poverty and hunger, are expected to create the world's first climate refugees. The expected mobilisation of dunes in the Kalahari Desert would also displace human populations (Thomas *et al.* 2005). Meanwhile in Russia, inter-regional political tension would be aggravated by an expected tripling of bad harvests due to drought. Peru and the Mahgreb emerge as the two regions where it is known that the effects of climate change are expected to be very serious for human society at only relatively small global temperature rises of up to ΔT = 2°C. At ΔT = 2°C there would be high risks of extinctions of frogs, reptiles and the Golden Bowerbird in a shrinking Australian Queensland rainforest, and of endemic Hawaiian birds. The famous Kakadu wetland in Australia and the Chinese boreal forest would lose 50% of their extent.

At ΔT = 2.5°C both Kakadu and the Chinese boreal forest would be completely lost. Eighty per cent of the South African Karoo would also be lost threatening 2800 endemic plants with extinction, and the South African Fynbos would also be lost and along with it its endemic species. Five famous South African safari parks would lose over 40% of their animals; the Great Lakes wetland ecosystems in Africa would collapse, and along with it the fisheries on which local people (for example in Malawi) depend. At this temperature all Kalahari dunefields may be mobile threatening sub-Saharan ecosystems and agriculture (Thomas *et al.* 2005). The Tibetan Plateau would experience large-scale melting of permafrost and desertification. Mangroves are known not to be able to withstand more than a 45 cm sea level rise in Asia, which is in the middle of the IPCC (2001) range for 2100.

It is known that if the world continues to warm, feedbacks in the climate system would cause a shift in the terrestrial carbon cycle. Currently, carbon on land is acting as a sink for CO_2, helping to buffer some of the effects of anthropogenic climate change. If CO_2 concentrations soar this sink would become a source, owing to increased soil respiration, further exacerbating climate change.

This is predicted to occur between $\Delta T = 2$ to 3°C of global mean temperature rise, and will cause widespread loss of forests and grasslands including the Amazon rainforest, which would undergo a transition to savannah, with massive implications for local populations and for global biodiversity, as well as the global carbon cycle.

11.3.5 *Impacts at 3°C of Global Annual Mean Temperature Rise Since Pre-industrial Times*

At a global temperature rise of $\Delta T = 3$°C, many additional impacts in human and natural systems would occur over and above those predicted for $\Delta T = 2$°C. Few ecosystems can adapt to such a large temperature rise: 22% of them would be transformed losing 7% to 74% of their extent whilst 50% of nature reserves could not fulfil their conservation objectives. Much larger losses in global cereal production than are predicted at $\Delta T = 2$°C would cause further food price rises and expose potentially 400 additional million people, largely in Africa, to hunger (or potentially 20 million less, the range reflecting the aforementioned uncertainty in the realisation of the theoretical benefits of CO_2 fertilisation, as well as differing potential socio-economic futures). Even with full CO_2 fertilisation, 65 countries would lose 16% of their agricultural GDP. Globally irrigation requirements would increase in 12 of 17 world regions whilst one to three billion people would experience an increase in water stress. As at $\Delta T = 2$°C, in Asia millions would experience a theoretical decrease in water stress, but the same caveats as above apply.

At 3°C, 50–60% of the world's population would be exposed to dengue fever (compared with 30% currently).

At 3°C, 50% of the world's most productive duck habitat would be lost, 50% Australian eucalypts would vanish, and very substantial range losses would occur for many species, for example Australia's endemic butterflies. Most alpine species in Europe would be near extinction. High fire risks would occur in the Mediterranean and 60% of its species would be lost.

Between $\Delta T = 2$ to 4.5°C there would be the potential (according to expert judgement) to trigger the melting of the West Antarctic Ice Sheet, which has recently proved to be less stable than was previously thought. This would induce further sea level rise of 5 to 6 m, implying potentially up to 75 cm or more by 2100 (Hansen 2005).

If global temperature rise reached $\Delta T = 4$°C, whole regions, including the entirety of Australia, would be forced out of agricultural production. Many experts judge that there would be a greater than 50% chance of a breakdown of the thermohaline circulation at this temperature, although a range of $\Delta T = 1$ to 5°C is given by various researchers. Up to approximately 600 million people could be at risk of hunger, and losses of tundra would reach 60% and taiga 44%.

The reader may find plots of 'impact guardrails' used in an integrated assessment study known as the 'tolerable windows approach' (Toth *et al.* 2003a, 2003b) useful to complement some of the information presented in the tables. These plots show impact guardrails that indicate (i) the percentages of ecosystems worldwide (agricultural areas excluded) that would undergo a change in biome and (ii) the changes in crop performance, for various increases in annual global mean temperature and CO_2 concentrations. The impact guardrails for biome shifts are based on Leemans and Eickhout, 2003, quoted in the accompanying tables, and the plots may be found in Fussel *et al.* (2003).

11.3.6 *Important Trends Not Associated with a Particular Temperature Increase*

In addition some general statements may be made about gradual changes which accrue as global mean temperature increases. Increases in the magnitude/frequency of extreme weather, wildfires, and outbreaks of pests and diseases are expected with climate change. Oceans are predicted to continue to acidify as temperature rises. The precise relationship between temperature rise and ocean acidification depends on the climate sensitivity, because acidification is related to CO_2 enrichment of the oceans rather than directly to temperature. Climate sensitivity is thought to lie with in the range of 1.6°C to 11.5°C but the precise value is not yet known (Stainforth *et al.* 2005). Coral reefs and calcifying plankton would be at risk from ocean acidification potentially altering the marine food chain and the ecosystem service that the ocean provides (Appendix Table H). Unpredictable ecological changes would also occur on sea and land as climate changes if predators and prey become decoupled, which could occur if they have differing phenological/geographical/physiological responses to climate change (Burkett *et al.* 2005, Price 2002). Reductions in sea ice in the Antarctic are likely to have contributed to the dramatic 80% declines in krill observed since 1970 (Gross 2005) with penguin populations already affected, and particularly if climate change shifts the Antarctic Circumpolar Current, krill could suffer further and the ecosystem could collapse. Climate change is expected to cause the deglaciation of the Himalayan region, which would adversely affect the hydrology of Indian region and disrupting agriculture, in an analogous situation to that of Peru at smaller temperature rises. There is also an expectation of monsoon disruption (Zickfeld *et al.,* 2005). Whilst the effect of climate change on El Niño remains unclear, at high CO_2 concentrations the globe would feature permanent El Niño.

As sea level rises and storm surges become more frequent, the risk of inundation of small island states would increase. Sea level rises of 1 m (at the highest end of the range predicted by IPCC for the year 2100) would expose millions of people to flood, inundation and storm, displace coastal and small island residents from their homes, and require the construction of large protective barriers for some cities (e.g. the Thames Barrier would need to be upgraded). Higher sea level rises of 2 m and above would inundate many of the world's large cities and obliterate major deltaic areas such as Bangladesh, the Nile, the Yangtze and the Mekong. Large sea level rises would occur by 2300 particularly if Greenland and West Antarctic Ice Sheets melt, whilst Hansen (2005) believes that a 2 m sea level rise is possible by 2100 in the absence of greenhouse gas mitigation due to instability of these ice sheets. Tables C, E and G of the Appendix detail the predicted impacts of various sea level rises.

The rate of climate change is also important, with ecosystems now predicted to be able to withstand a temperature increase of only 0.05°C/decade, much slower than the current rate and hugely slower than the current rate near the poles. Other estimates suggest a limit of 0.1°C/decade, such that the current rate of over 0.4°C/decade in the Arctic is considered sufficient to cause serious ecosystem disruption. The faster the rate of change, the greater the damage to an ecosystem since this reduces the time it has to adapt to the higher temperatures. Similarly human systems would be damaged to a greater extent for faster temperature rises, since there would be less time to adapt.

Faster rates of change also make adaptation more difficult for human societies, owing to the reduced adaptation times required. Rapid adaptation would be most difficult where adaptive capacities are low, for example in many developing countries. As Tol & Downing *et al.* (2004) show, the distribution of impacts across human systems is expected to be strongly skewed, with the worst impacts being experienced in the developing world and by poor sectors of society. Overall, the faster the rate of change, the less damage in human systems can be avoided through adaptation.

At the earth system scale, as temperature continues to rise, the risk for the potential release of methane from melting tundra and clathrates from shallow seas would increase. Such a release of methane would trigger a strong amplification of the greenhouse effect, greatly exacerbating the existing climate change.

Table 11.5 summarises the impacts of climate change predicted for different levels of temperature rise, which are further detailed in Tables A to G of the Appendix. The study could not encompass a complete survey of the literature, or a rigorous treatment of how adaptation is included, whilst some regions/human systems/ecosystems may not feature in the literature. Hence the tables provide a guide to the major known impacts. They show that many of the impact levels are affected by socio-economics.

For example, stocks and risk, populations and adaptive capacities of human society determine the magnitude of impacts on human systems. Impacts on ecosystems would act in conjunction with other human stresses

such as development of wilderness areas and water use, for example Nicholls *et al.* 1999 suggests that if sea levels rise by ~40 cm in the 2080s, 22% of coastal wetlands would be lost, but 70% would be lost when expected human destruction is also considered. Similarly, direct climate impacts on freshwater ecosystems are expected to be dwarfed by indirect impacts as climate change enormously increases water stress in large areas of the world.

11.4 Conclusion

The literature reveals that through phenological and distributional change of species, glacier melt, and unprecedented heat waves, the effects of climate change are already being felt throughout the world although annual global mean temperatures have thus far risen by just 0.6°C relative to pre-industrial times. At $\Delta T = 1$°C temperature rise oceans acidify, coral reefs and Arctic ecosystems would be damaged, whilst at $\Delta T = 1.5$°C the Greenland Ice Sheet is predicted irreversibly to melt. At $\Delta T = 1$°C agricultural yields would begin to fall and additional billions of people would experience an increase in water stress, hundreds of millions may go hungry, whilst sea level rise would displace millions from the world's coastlines, malaria risks would spread, Arctic ecosystems would collapse as summer sea ice vanishes, and species extinctions would begin to take off as other regional ecosystems are lost. Serious implications for humans would exist in Peru and the Mahgreb where climate refugees would be expected. Between $\Delta T = 2$ and 3°C the Amazon is predicted to collapse along with other forests and grasslands. At $\Delta T = 3$°C additional millions at risk to water stress, flood, hunger and malaria and dengue would increase further whilst few ecosystems would be able to adapt causing many extinctions. Many experts believe the thermohaline circulation could collapse for global annual mean temperature changes of between $\Delta T = 1$ to 5°C, the temperature threshold being influenced by the rate as well as the absolute level of temperature change. The West Antarctic Ice Sheet may also begin to lose mass between $\Delta T = 2$ to 4.5°C, and the Antarctic ecosystem may collapse through krill declines. Increases in the frequency/intensity of extreme weather and possibly El Niño are expected as climate changes.

Acknowledgements

The author is grateful to Defra for funding this study, to Prof. Terry Root for permission to use her database of observed ecosystem changes, and to Dr. Jeff Price for proof-reading the manuscript. This paper builds on the existing Hare 2003 review and the author wishes to acknowledge the particular value of this work.

APPENDIX

Table A Observed Changes and Impacts of Climate Change on the Earth System at different levels of global mean annual temperature rise, ΔT, relative to pre-industrial times.

ΔT	[CO_2] ppm	Year in which impact occurs	Impacts to the earth system	Region affected	Source
			OBSERVED CHANGE		
0.6	378	2004	Annual average temperature has risen by 0.6°C	Globe	IPCC 2001
0.6		2004	Temperature has risen by 1.8°C; could rise by 10°C by 2100	Arctic	ACIA 2004
0.6		2004	Sea surface temperature increased by 0.6°C ± 0.1°C	Globe e.g. N Sea where 0.5°C rise in 15 years	IPCC 2001, EEA 2004
0.6		2005	Index of potential destructiveness of hurricanes has increased since 1970s (closely correlated with sea surface temperature rise)	Globe	Emanuel 2005
0.6		2004	90% globe's glaciers retreating since 1850 (not attributed)	Globe e.g. Alps where 70–90% mass loss (30–40% since 1980), Peru	EEA 2004, Street & Melnikov 1990
0.6		2004	Increased freshwater flux from Arctic rivers appears to be already 20% of what would cause shutdown of THC	Northern and Western Europe (to Arctic Ocean)	ECF 2004
0.6		2004	Arctic sea ice reduced by 15–20%	Arctic	ACIA 2004
0.6		2004	Arctic sea ice extent decreased by $0.30 \pm 0.03 \times 10^6 km^2/10$ yr from 1972 through 2002, but by $0.36 +/- 0.05 \times 10^6 km^2/10$ yr from 1979 through 2002, indicating an acceleration of 20% in the rate of decrease.	Arctic	Cavalieri *et al.* 2003
0.6		2004	3.7 ± 1.6°C warming/century observed	Antarctic Peninsula	Vaughan *et al.* 2003
0.6		2004	N hemisphere snow cover decreased by 10% since 1966	N hemisphere	EEA 2004
0.6		1846–1995	Lake & river ice: Average freeze dates 5.8 days/century later, and breakup dates 6.5 days/century earlier	N hemisphere	Magnuson *et al.* 2000
0.6		2004	Measured spring snowpack decreased in Alps and Pyrenees	Switzerland, Spain	Lopez-Moreno 2005
0.6		2004	Measured spring snowpack declined, (not attributed) correlated with rising temperature/declined precipitation	Cascades & N California, USA	Mote 2005
0.6		2004	Bottom melt rates of Antarctic glaciers increase by 1 m/year for each 0.1°C rise in ocean temperature	Antarctic	Rignot & Jacobs 2002
0.6		2004	Some evidence that savannaisation of parts of Amazon triggered by land use change interacting with warming	Amazon	ECF 2004

(continued)

Table A (*contd*)

ΔT	[CO$_2$] ppm	Year in which impact occurs	Impacts to the earth system	Region affected	Source
0.6		2004	Greenland ice sheet losing mass (not attributed)	Greenland	Rignot & Jacobs 2002
0.6		2004	West Antarctic Ice Sheet losing mass overall	Antarctic	Rignot & Jacobs 2002
0.6		2004	Larsen B ice shelf collapse; subsequent ice discharge from land (not attributed)	Antarctic	Rignot *et al.* 2004
0.6		2004	Increase in global sea level of 1.8 mm/year: about 50% of this caused by melting of terrestrial ice (remainder from thermal expansion of water), of which 0.4 mm/yr from non-polar glaciers, 0.4 mm/yr from Greenland, estimated 0.2 mm/yr from West Antarctic Ice Sheet	Globe	Thomas *et al.* 2004a
0.6		2004	Green biomass increased by 12% (not attributed)	Europe	EEA 2004
			PREDICTED CHANGE		
0.7		2015	Africa's last tropical glacier on Kilimanjaro lost (not attributed)		Thompson *et al.* 2002
1.5–1.6		over a few centuries	Onset of complete melting of Greenland ice: when complete 7 m of additional sea level rise or additional 75 cm by 2100	All coastal regions; many world cities inundated	Gregory *et al.* 2004, Hansen 2005
2–3	At approx CO$_2$ dbling		Collapse of Amazon rainforest, forest replaced by savannah: enormous consequences for biodiversity and human livelihoods	S America, also globe	Cox *et al.* 2004, Betts 2005
2 to 3	~550 ppm inevitable at some point		Conversion of terrestrial carbon sink to carbon source, due to temperature-enhanced soil and plant respiration overcoming CO$_2$-enhanced photosynthesis. Resulting in desertification of many world regions as there is widespread loss of forests and grasslands, and accelerating warming through a feedback effect	Global	Cox *et al.* 2000, Cox 2005, ECF 2004
Any			Release of C to atmosphere due to deterioration of ecosystems at rapid rates of temperature change	Global	Neilson 1993
	Double		Net primary production increases by 10%	Globe	Betts 2005
	Double		Runoff increases by 12%	Globe	Betts 2005
2.3		2100	Collapse of thermohaline circulation: a maximum likelihood analysis gives a shutdown probability of 4 in 10 for climate sensitivity of 3°C (and climate sensitivity could lie between 1.5 and 11°C)	Globe; cooling NW Europe, warming Alaska and Antarctic, decreasing rainfall in S America	Schlesinger 2005
1			Kalhari dune activation commences	Africa	Thomas *et al.* 2005

(*continued*)

Table A (*contd*)

ΔT	[CO_2] ppm	Year in which impact occurs	Impacts to the earth system	Region affected	Source
1–3		2100	Collapse of thermohaline circulation cooling N hemisphere and altering precipitation patterns affecting fisheries, ecosystems, agriculture: expert opinion probability "a few percent"	Northern and Western Europe	Rahmstorf in ECF 2004
2		2100	Probability of collapse exceeds 50% (taking into account range for climate sensitivity of 1.5 to 11°C)		Schlesinger 2005
2			All Kalahari dunes active	Africa	Thomas *et al.* 2005
2–4	700	2100	THC collapse		O'Neill & Oppenheimer 2002
4	750	2200	THC collapses permanently for CO_2 concentration increases of 1%/year (current value) or slows recovering to a 15% weakened state		Stocker & Schmittner 1997
2–4.5			Potential to trigger melting of the West Antarctic Ice Sheet raising sea levels by a further 5 to 6 m or up to 75 cm by 2100	Globe	ECF 2004, Hansen 2005
4–5			Expert opinion: probability of thermohaline shutdown up to or above 50%	Northern and Western Europe	Rahmsdorf in ECF 2004
			THC collapse, Greenland Ice Sheet melt and West Antarctic Ice Sheets may interact in ways that we have not begun to understand		Discussed at conference
			Potential release of methane from melting tundra and clathrates from shallow seas	Globe, especially Arctic: feedback accelerates warming	IPCC 2001
		2100	Acidification of the oceans, pH falls by up to 0.4: may disrupt marine ecosystem functioning, in turn reducing buffering capacity of oceans (positive feedback)	World oceans	IPCC 2001, Blackford 2005, Archer 1995
		2250	Acidification, pH falls by 0.77	World oceans	IPCC 2001, Blackford 2005
			Increased variability in summer monsoons exacerbating flood/drought damage	Asia, Australia	IPCC 2001, Gordon *et al.* 2005, Lal *et al.* 2002, Zickfeld *et al.* 2005
	16×CO_2		Permanent El Niño	Globe	Navarra 2005

Table B Observed and predicted impacts of climate change on human systems at different levels of global mean annual temperature rise, ΔT, relative to pre-industrial times.

ΔT	Year	Population scenario	Impact to human systems m.a.r. = additional millions of people at risk than would be the case in absence of climate change	GCM used	Region affected	Source
			OBSERVED IMPACTS			
≤0.6	1967 onward		Abrupt change in regional rainfall pattern causing food insecurity, water stress (not attributed)		Sahel	Dore 2005
0.6	2004		Extreme weather is causing substantial and increasing damage partly due to climatic factors (not attributed)		Globe	IPCC 2001
0.6	2004		Increase in severity and frequency of extreme events in tropical small island states (not attributed)		Small islands	Krishna *et al.* 2000, Trotz 2002, Hay *et al.* 2003
0.6	2004		Changes in stream flows, flood and drought observed (e.g. earlier peak runoff) (not attributed)		Europe, Russia, N America, Sahel, Peru, Brazil, Colombia	ECF 2004
0.6	2004		High temperatures of 2003 summer in Europe attributed to anthropogenic cause with confidence of 1 in 8; likelihood of such events doubled by human influence.		Europe	Stott 2004
0.6	2004		Heat wave associated with unusual 2003 summer caused 14802 deaths in France, and approximately 25000 in Europe			WHO 2004
0.6	2000		Since 1970, number people affected by drought increased from 0 to 35 million (not attributed)		Southern Africa	ECF 2004
0.6			Increased frequency and intensity of drought (not attributed)		S. Africa, Sahel, Asia, SW Australia	IPCC 2001, ECF 2004
0.6			Decline in growing season rainfall		Ethiopia	Royal Society 2005
0.6			Damage to infrastructure/buildings as permafrost melts		Alaska	ACIA 2004
0.6			Inuit affected by changing animal distributions/abundance		Alaska	ACIA 2004
0.6			Increased cloud amount, annual precipitation, and heavy precipitation events (not attributed)		Mid- and high-latitudes N hemisphere	IPCC 2001, Dore 2005
0.6			Lake and river ice duration reduced by 2 weeks; onset averaged 5.8 days per 100 years earlier and breakup averaged 6.5 days per 100 years earlier (not attributed)		Mid- and high-latitudes N hemisphere	Dore 2005, Magnuson *et al.* 2000
0.6	2004		Water stress increase associated with drying & warming (not attributed)		Australia	ECF 2004

Temp	Year	Predicted/observed change	Region	Reference
0.6		Rainfall decline in W hemisphere, subtropics, E equatorial region observed, consistent with more frequent El Niño-like conditions	S hemisphere especially 5 Andean countries	ECF 2004
	—	**PREDICTED CHANGES**		
0.6	2000	Climate change has been *modelled* (not observed) to have caused the loss of 150,000 lives and 5.5 million days-of-life-lost/yr since 1970	Globe	McMichael *et al.* 2004
≥0.6		Increase in frequency and length heatwaves typified by those occurring in Paris, 2003, and Chicago, 1995, causing elevated mortality rates in elderly/urban poor, risk crop damage, stress to livestock, increased cooling demand	All land areas	IPCC 2001, Meehl and Tebaldi 2004
≥0.6		Decreased cold days in twentieth century. Higher minimum temperatures, reducing cold-relatedmortality. Increased risk to some crops, decreased to others, reduced heating demand. Extended range of some pests and disease vectors	Almost all land areas	IPCC 2001, Tol 2002
≥0.6		Increased summer drying over continents likely, decreasing crop yields, damaging buildings, decreasing water resources and increasing forest fire	Continental interiors	IPCC 2001
≥0.6		Increase in magnitude/frequency of precipitation events, very likely: causing floods, landslides, avalanche, increased soil erosion (not attributed)		IPCC 2001
≥0.6		More intense El Niño, increasing strength of associated droughts/floods likely, decreasing agricultural productivity and hydro-power potential, causing water stress	S America, Australia	IPCC 2001
≥0.6	2025	Water quality degraded	Some regions	IPCC 2001
≥0.6		Melting permafrost disrupts built infrastructure and destabilises slopes causing landslides	Arctic	IPCC 2001
≥0.6	2025	Increased energy demand for summer cooling demand and decreased winter heating demand very likely	Europe, N America	IPCC 2001
≥0.6	2025	Market sector losses likely in many developing countries, mixture of gains and losses in developed countries	Globe	IPCC 2001

(*continued*)

Table B (*contd*)

ΔT	Year	Population scenario	Impact to human systems m.a.r. = additional millions of people at risk than would be the case in absence of climate change	GCM used	Region affected	Source
≥0.6			Large scale damage to infrastructure and threat to human lives		Caribbean & tropical small island states: increased magnitude & frequency of extreme weather events	IPCC 2001
≥0.6			As above		Himalayas: glacier lake outbursts	IPCC 2001
≥0.6			As predator-prey and plant-pollinator relationships disconnect in shifting ecosystems, leading to extinctions of pollinators and pest-predators, agricultural crops lose key pollinators and pests increase in many areas, reducing yields		Globe	Burkett *et al.* 2005, Price 2002
≥0.6			Rainfall decline, loss of glaciers predicted; *serious drinking water; energy generation and agriculture problems, adaptation may not be economically feasible.* In 20 years glaciers below 5500 m will have disappeared causing hydropower problems		Peru	ECF 2004
0.8	2030	S550	Malarial risk increased by factor 1.27, dengue by 1.3		N America	McMichael *et al.* 2004
0.8	2030	S550	Risk of death due to flooding increased by 1.44		W Africa	McMichael *et al.* 2004
0.8	2030	S550	Risk of death due to flooding increased by 3.58		C/S America	McMichael *et al.* 2004
0.8–2.6	2050		Higher market impact likely in developing countries, fewer losses and more gains in developed countries		Globe	IPCC 2001
0.8–2.6	2050		Increased insurance prices and reduced availability of insurance very likely		Globe	IPCC 2001
1	2020	IS92a + S750	240 mar from water stress	HadCM2	Globe	Arnell *et al.* 2002
1			10% decrease barley yield		Uruguay	Gitay *et al.* 2001
1			6–10% decrease rice yield		S Asia	ECF 2004
1	2020	–	Disbenefit to agriculture		Less developed world	Hare 2003
1	2020	–	Benefit to agriculture		Developed world	Hare 2003
1.1	2025	B1:2882 or 37% population under water stress if no cc	400 additional mar from water stress under climate change;1819 m with decrease in water stress[1]	HadCM3	Globe	Arnell 2004

Temp	Year	Scenario	Impact	Model	Region	Reference
1.2	2025	A2: 3320 or 39% population under water stress if no cc	615–1660 or 500–915 (5 GCMs) mar from water stress 1385–1893 or 1140–2423 (5 GCMs) m with decrease in water stress[1]	5 GCMs	Globe	Arnell 2004
1.2	2025	B2: 2883 (36%) population under water stress if no cc	508–592 (HadCM3) or 374–1183 (5GCMs) mar from water stress 1651–1937 or 1261–2202 (5 GCMs) m with decrease in water stress[1]	5 GCMs	Globe	Arnell 2004
1.3	2025	A1FI: 2882 (37%) population under water stress if no cc	829 mar from water stress 649 m with decrease in water stress[1]	HadCM3	Globe	Arnell 2004
1.0			300–1600 additional millions suffer increase in water stress depending on socioeconomic scenario and GCM[1]		Globe	Arnell 2004
1.0			18–60 additional millions at risk of hunger depending on socioeconomic scenario and GCM		Globe	Parry et al. 2004
1.3	–	–	Food price rise begins		Globe	Hare 2003
1.3	2060	–	21% rise in timber production for 2045–2095; 30% rise by 2095–2145 (Temp assumed to be stable in 2060)	Hamburg GCM	Globe	Sohngen et al. 2001
1.3	2050	S550	Risk of death due to flooding increased by 1.48		W Africa	McMichael et al. 2004
1.3	2050	S550	Risk of death due to flooding increased by 3.76		C/S America	McMichael et al. 2004
1.3	2030	S750	Malarial risk increased by factor 1.33, dengue by 1.33		N America	McMichael et al. 2004
1.3	2050	IS92a + S550	160–220 mar from malaria	HadCM2	Globe	Parry et al. 2001
1.3	2050	IS92a + S550	5 mar from hunger	HadCM2	Less developed	Parry et al. 2001, Hare 03
1.3	2080	IS92a + S450	400 mar from water stress	HadCM2	Globe	Parry et al. 2001
1.3	2080	IS92a + S450	150 mar from malaria	HadCM2	Globe	Parry et al. 2001
1.4	2050		Shorelines behind bleached coral reefs now vulnerable to storm damage; damage and tourism loss could lead to $140–420 million loss in Caribbean alone.	HadCM2	Caribbean, Indian Ocean, small island states	ECF 2004
1.4	2020		Irrigation requirements increase in 11 out of 17 world regions as result of climate change	HadCM2	Globe	Döll 2002
1.4–5.8	2100		High market impacts likely in developing countries, net losses in developed countries			IPCC 2001
1.5	2080	IS92a + S450	165 mar from malaria	HadCM2	Globe	Parry et al. 2001, Hare 2003
1.5 with 8% increase in precipitation			Farm values increase by between $188–311 billion		USA	Mendelsohn et al. 1996[2]

(continued)

Table B (contd)

ΔT	Year	Population scenario	Impact to human systems m.a.r. = additional millions of people at risk than would be the case in absence of climate change	GCM used	Region affected	Source
1.5	2025		Increase in water stress in Africa & S America; decrease in Europe and N America		Africa, S America; Europe, N America	Vorosmarty et al. 2000
2–4	2055–2085		Increase in water stress in Mediterranean, C & S Africa, Europe, C & S America. Decreases in SE Asia[1]			Arnell 2004
1.5	Any		$5.3–5.4 billion losses in dryland agriculture	8% increase in precipitation assumed	USA	Schlenker et al. 2004
1.5–2°C	–	–	Poor farmers' income declines in this range		Less developed	Hare 2003
1.6	2030		Malarial risk increased by factor 1.51		N America	McMichael et al. 2004
1.6	2030	S550	Risk of death due to flooding increased by 1.64		W Africa	McMichael et al. 2004
1.6	2030	S550	Risk of death due to flooding increased by 4.64		C/S America	McMichael et al. 2004
1.7	2030		Winter yield increases or decreases by 30–40% depending on GCM used to model precipitation changes		USA	Tubiello et al. 2002
1.7	2030		Maize yield changes by −30% to +20% depending on degree to which CO_2 fertilisation is realised[3]		USA Great Plains	Tubiello et al. 2002
1.7	2055	B2: 3988 (42%) population under water stress if no cc	1020–1057 (HadCM3) or 670–1538 (5 GCMs) mar from water stress 2407–2623 or 1788–3138 (5 GCMs) m with decrease in water stress[1]	5 GCMs	Globe	Arnell 2004
1.75	2055	B1: 3400 (39%) population under water stress if no cc	988 mar (HadCM3) from water stress 2359 m with decrease in water stress[1]	HadCM3	Globe	Arnell 2004
1.8	2025	A2	0.05 diarrhoeal incidence per capita per year		Globe	Hijioka et al. 2002
1.8	1200	S550	International tourism flows negatively impacted		S Europe, Caribbean, SE Asia	Viner 2005, IPCC 2001
1.8–2.6	2050		Large scale displacement of people (climate refugees from low food security, poverty and water stress)	Rainfall decrease of 40% simulated by most GCMs included	Mahgreb (N Africa) and Sahel	ECF 2004

Temperature increase (°C)	Year	Scenario / Model	Impact	Region	Reference
1.8–2.6	2050		40% rainfall decline from 1961–1990 average (in all GCMs)	Africa Mahgreb	ECF 2004
1.9	2050	5 GCMs; A2: 3320 (39%) population under water stress if no cc	1620–1973 (HadCM3) or 1092–2761 (5 GCMs) mar from water stress 2804–3813 or 1805–4286 (5 GCMs) m with decrease in water stress[1]	Globe	Arnell 2004
Any			Increase in magnitude of cyclones likely, increasing risks to human life, infectious disease epidemics, coastal erosion and damaging coastal infrastructure, coral reefs and mangroves	Tropical & sub-tropical regions	IPCC 2001
Any	Any		River flood hazard increase	Europe	IPCC 2001
Any	Any		Drought, reduced water supplies for irritation, and increases in crop pests/diseases	All regions	IPCC 2001, Rosegrant & Cline 2003
Any	Any		Sea level rise and cyclones displace several million people from coasts	Tropical Asia	IPCC 2001
Any	Any	5 GCMs	Runoff increase in N but decrease in arid areas; however in N may not be in useful season	Asia	IPCC 2001
Any	Any		Vector borne disease expands poleward	Latin America and Asia	IPCC 2001
Not known			Loss of sovereignty of small island states and countries with large low lying deltaic regions	Small low-lying islands	ECF 2004
Not known			Regional conflict over water supplies or food supplies	Nile, parts of Russia	ECF 2004
Not specified			Deglaciation of Himalayan region affects hydrology of Indian region, disrupting agriculture	Nepal, India	ECF 2004
2	—		Threshold above which agricultural yields fall	EU, Canada, USA, Australia Russia	Hare 2003
2			Double/triple frequency of bad harvests leading to inter-regional political tension		ECF 2004
2			Destruction of Inuit hunting culture	Arctic	ECF 2004
2			1.0 to 2.8 billion people experience increase in water stress depending on socioeconomic scenario and GCM model used	Globe	Arnell 2004
2			Wheat yield decrease	S Asia	ECF 2004
2			Maize yield 15% decrease	Uruguay	Gitay et al. 2001
2–2.5			Food production threatened	Southern Africa, S Asia, parts of Russia	ECF 2004

(continued)

Table B (*contd*)

ΔT	Year	Population scenario	Impact to human systems m.a.r. = additional millions of people at risk than would be the case in absence of climate change	GCM used	Region affected	Source
2–2.5			Fisheries impacted		NW Africa, E African lakes	ECF 2004
2–2.5			Fishery damage removes primary protein source for 50% of population		Malawi	ECF 2004
2–2.5			Combined effects of precipitation changes, floods, droughts, reducing crop yields leading to significant risk commercial & subsistence of up to 80% crop failure		Southern Africa	ECF 2004
2–3			Kalahari dune activation threatens Sub-Saharan agriculture and ecosystems		Africa	Thomas *et al.* 2005
1–3 (not known)	2050–2100		Dry season water security loss & complete loss glaciers		W China	ECF 2004
2–3	2050–2100	A1B:	Increase in magnitude/frequency of precipitation: causing high flood damage		Japan	Emori 2005
~2	Range over SRES scenarios	Any	220 to (−12) additional mar from hunger depending on whether CO_2 fertilisation is included	HadCM3	Globe	Parry *et al.* 2004
2.1	2080	IS92a + S750	2.3–3.0 bar from water stress	HadCM2	Globe	Parry *et al.* 2001
2.3	2050	IS92a	26 mar from coastal flood (i.e. a doubling of the 26 mar in absence of climate change)	HadCM2	Globe especially S & SE Asia	Parry *et al.* 2001, IPCC 2001, Nicholls 2004
2.3	2050	IS92a	180–230 mar from malaria	HadCM2	Globe	Parry *et al.* 2001
2.3	2050	IS92a	23–25 mar from malaria	HadCM2	Globe	Rogers & Randolph, 2000
2.3	2050	IS92a	10% loss in maize production equivalent to losses of $2bn/yr		Africa & Latin America	Jones & Thornton 2003
2.3	2100		30–70% loss snow pack losing 13–30% water supply		California	Hayhoe 2005
2.3	2080	IS92a > S1000	230–270 mar from malaria	HadCM2	Globe	Parry *et al.* 2001
2.3	2080	IS92a > S1000	33 mar from hunger	HadCM2	Less developed	Parry *et al.* 2001, Hare 2003
2.36	2080	B1	4–8% increase in mar of hunger (4–8 million)	HadCM3 CSIRO NCAR CGCM2	Globe	Fischer *et al.* 2001
2.3–2.7	2080	B1/B2	5% fall in cereal production yield	HadCM3	Globe	Parry *et al.* 2004
2.36	2080	B1	2–3 mar from coastal flood	HadCM3	Globe	Nicholls 2004

Temp	Year	Scenario	Impact	GCM	Region	Reference
2.36	2080	B1	10–40 mar from hunger	HadCM3		Parry et al. 2004
2.36	2050	IS92a, unmitigated	7 mar of hunger	HadCM2	Less developed	Parry et al. 2001, Hare 2003
2.36	2080	B1	250 more mar 1 month exposure to malaria 153 less mar 3 month exposure to malaria	HadCM3	Globe	Van Lieshout et al. 2004
2.36	2080	B1	1 less mar 1 month exposure to malaria 3 month exposure to malaria		W Africa	Van Lieshout et al. 2004
2.36	2080	B1	38 more mar 1 month exposure to malaria 21 more mar 3 month exposure to malaria		SubSaharan Africa	Van Lieshout et al. 2004
2.36	2080	B1	18 more mar 1 month exposure to malaria 41 less mar 3 month exposure to malaria[4]		Latin & South America	Van Lieshout et al. 2004
2.36	2080	B1	134 more mar 1 month exposure to malaria 2 less mar 3 month exposure to malaria		West Asia	Van Lieshout et al. 2004
2.36	2080	Constant population	15.5 million additional person months exposure	HadCM3	Africa	Tanser et al. 2003
2.36	2085	B1:2860 (37%) population under water stress if no cc	1135 mar water stress increase 1732 m with decrease in water stress[1]	HadCM3	Globe	Arnell 2004
2.36	2085	B2:4530 (45%) population under water stress if no cc	1196–1535 (HadCM3) or 867–2015 (5 GCMs) mar water stress[1] 2791–3099 or 2317–3460 (5 GCMs) m with decrease in water stress	5 GCMs	Globe	Arnell 2004
2.5–3			Rice yields reduced 10–20% (no CO$_2$ fertilisation) (or change by −10% to 20% assuming total CO$_2$ fertilisation)		China	ECF 2004
2.5 to 4	–		Crop failure rise from 50 to 75%		S Africa	ECF 2004
2.56	2055	A1FI:3400 (39%) population under water stress if no cc	1136 mar (HadCM3) from water stress 2364 m with decrease in water stress[1]	HadCM3	Globe	Arnell 2004
2.6	–		Rapid increase in flooding damaging agriculture and endangering life		Bangladesh	ECF 2004
2.6 and −20% precipitation			5 to 30% loss rice/wheat yields putting food security at risk		Indian subcontinent	ECF 2004
2.7	2060		Increase of 265 million or decrease of 84 million from reference level of 641 million in 1960, at risk of hunger in developing countries as cereal production falls	GISS	Globe	Rosenzweig et al. 1995

(continued)

Table B (*contd*)

ΔT	Year	Population scenario	Impact to human systems m.a.r. = additional millions of people at risk than would be the case in absence of climate change	GCM used	Region affected	Source
2.7	2080	B2	by 4 to 9%, whilst production increases by 2 to 11% in developed countries	HadCM3	Globe	Fischer et al. 2001
2.7	2080	B2	15% increase in millions at risk of hunger, includes CO_2 fertilisation (40 mar)	HadCM3	Globe	Nicholls 2004
2.7	2080	B2	16–27 mar from coastal flood	HadCM3	Globe	Van Lieshout et al. 2004
2.7	2080	B2	307 mar 1 month exposure to malaria 31 mar 3 month exposure to malaria	HadCM3	W Africa	Van Lieshout et al. 2004
2.7	2080	B2	2 less mar 1 month exposure to malaria 8 less mar 3 month exposure to malaria	HadCM3	SubSaharan Africa	Van Lieshout et al. 2004
2.7	2080	B2	67 more mar 1 month exposure to malaria 51 more mar 3 month exposure to malaria	HadCM3	Latin & S America	Van Lieshout et al. 2004
2.7	2080	B2	66 more mar 1 month exposure to malaria 66 less mar 3 month exposure to malaria[4]	HadCM3	West Asia	Van Lieshout et al. 2004
2.7	2080	B2	159 more mar 1 month exposure to malaria 62 more mar 3 month exposure to malaria	HadCM3		Van Lieshout et al. 2004
2.7	2080	B2	−15 to 200 mar from hunger (range due to CO_2 fertilisation inclusion or not)	HadCM2/3		Parry et al. 2004
3	–	–	65 countries lose 16% agricultural GDP, includes CO_2 fertilisation	HadCM3 CSIRO CGCM2 NCAR	Less developed	Fischer et al. 2001
3	–	2070	Irrigation requirements increase in 12 of world's 17 regions	HadCM3 (also ECHAM 4)	Globe	Doll 2002
3	IS92a	2090	Massive reduction in extreme rainfall return periods for the UK	HadCM2/HadRCM	UK	Huntingford et al. 2003
3	IS92a With own population	2085	Proportion of world population exposed to dengue fever increases from 30% in 1990 to 50–50% in 2085	HadCM3/2 ECHAM4 CCSR/NIES, CGCMA1/2	Globe	Hales et al. 2002
~3	Range over SRES scenarios	Any	400 to (−20) additional mar from hunger depending on whether CO_2 fertilisation is included	HadCM3	Globe	Parry et al. 2004

Temp	Year	Scenario	Impact	GCM	Region	Reference
3			17–18% increase in seasonal AND perennial potential malarial transmission zones; overall increase for all zones 10%	HadCM2/3	Globe	Martin & Lefevre 1995
3–4			Loss in farm income between 9 and 25%		Indian subcontinent	ECF 2004
3–4			Wheat yield decline of up to 34%		Indian subcontinent	ECF 2004
3.1	2090	No evolving baseline: fixed at 1990 world	19% fall in cereal supply without farm level adaptation, 4% with; falls to zero allowing for trade, changes in demand and land use changes to provide new cropland	OSU	Globe	Darwin et al. 1995[4]
3 with 25% less rain			Maize and potato yields increase		Chile	Fischer et al. 2001
3 with −25% less rain			Wheat and grape yields fall		Norte Chico,	Fischer et al. 2001
3 with 8% higher precipitation			Farm values increase by between $227–403 billion		USA	Mendelsohn et al. 1996[2]
3.3	2070–2100	IS92a 710 ppm	Increase in cropland suitability of estimated 16% Average 4 GCMs if three agree		N Hemisphere	Ramankutty et al. 2002
3.3	2070–2100	IS92a 710 ppm	Small decrease in cropland suitability Average 4 GCMs if 3 agree		Tropics	Ramankutty et al. 2002
3.3	2080	IS92a	75–100 mar from hunger		Globe	Parry et al. 2001
3.3	2080	IS92a	80 mar from coastal flooding (only 14 million at risk in absence of climate change)	HadCM2	Globe	Parry et al. 2001, Nicholls 2004
3.3	2080	IS92a unmitigated	280–330 mar from malaria	HadCM2	Globe	Parry et al. 2001
3.3	2080	—	560–1350 thousand at risk from coastal flooding	HadCM2	Caribbean	Parry et al. 1999
3.3	2080	IS92a, unmitigated	3.1–3.5 bar from water stress	HadCM2	Globe	Parry et al. 2001
3.3	2080	IS92a unmitigated	Coastal flooding several times worse than in 1990		Globe	Arnell et al. 2002
3.3–6.3			5–12% drop in country's production; 14–41% in agricultural regions		Russia	ECF 2004
3.55	2085	A2:8065 (57%) population under water stress if no cc	2583–3210 (HadCM3) or 1560–4518 (5 GCMs) water stress 4688–5375 or 3372–5375 (5 GCMs) m with decrease in water stress[1]	5 GCMs	Globe	Arnell 2004

(continued)

Table B (contd)

ΔT	Year	Population scenario	Impact to human systems m.a.r. = additional millions of people at risk than would be the case in absence of climate change	GCM used	Region affected	Source
3.55	2080	A2	29–50 mar from coastal flood	HadCM3	Globe, especially S/SE Asia, Africa, Mediterranean, and small islands of Indian & Pacific Oceans	Nicholls 2004
3.55	2080	A2	416 mar 1 month exposure to malaria 141 less mar 3 month exposure to malaria	HadCM3	W Africa	Van Lieshout et al. 2004
3.55	2080	A2	1 less mar 1 month exposure to malaria 25 less mar 3 month exposure to malaria	HadCM3	SubSaharan Africa	Van Lieshout et al. 2004
3.55	2080	A2	38 more mar 1 month exposure to malaria 21 more mar 3 month exposure to malaria		Latin & S America	Van Lieshout et al. 2004
3.55	2080	A2	47 less mar 1 month exposure to malaria 211 less mar 3 month exposure to malaria[4]	HadCM3	West Asia	Van Lieshout et al. 2004
3.55	2080	A2	299 more mar 1 month exposure to malaria 16 more mar 3 month exposure to malaria			
3.55	2080	Constant population	23.2 million additional person months exposure	HadCM3	Africa	Tanser et al. 2003
3.55	2080	A2	600 mar from hunger (-30 CO_2 ff)			Parry et al. 2004
3.55	2080	A2	15% increase in number at risk from hunger (120 million), includes CO_2 fertilisation	HadCM3 CSIRO	Globe	Fischer et al. 2001
3.55	2055	A2	0.1 diarrhoeal incidence per capita per year		Globe	Hijioka et al. 2002
3.55	2060	–	Global timber production increases by 17% (2045–2095) and 28% (2095–2145). Temperature is at equilibrium in 2060.	UIUC	Globe	Sohngen et al. 2001
4.3	2060	10.2 billion people (UN medium population estimates, similar to IS92a)	$+11$ to -33% change in wheat yields (depending on CO_2 fertilisation included/not); $+16$ to -57% change in soy -15 to -31% change in maize -2 to -12% change in rice cereal price rise of -17 to 145% -13 to 58% increase in numbers at risk of hunger	GISS GFDL HadCM2	Globe, if no adaptation	Rosenzweig et al. 1995
4.3	2090	No evolving baseline:	23% fall in cereal supply without farm level adaptation, 4.4% with;	GFDL	Globe	Darwin et al. 1995[5]

Temp	Year	Scenario	Impact	Model	Region	Reference
4.3		fixed at 1990 world	falls to zero allowing for trade, changes in demand and land use changes to provide new cropland	GISS GFDL HadCM2	Globe, with farm-level adaptation	Rosenzweig et al.1995
4.3	2060	10.2 billion people	−2 to 19% increase in numbers at risk of hunger			
4.3	2080	A1FI	Increase of 26% in mar of hunger (28 million), includes CO_2 fertilisation	HadCM3 NCAR CSIRO CCCma	Globe	Fischer 2001[5]
4.3	2080	A1FI	227 mar 1 month exposure to malaria 100 mar 3 month exposure to malaria	HadCM3	Globe	Van Lieshout et al. 2004
4.3	2080	A1FI	13 less mar 1 month exposure to malaria 46 less mar 3 month exposure to malaria	HadCM3	W Africa	Van Lieshout et al. 2004
4.3	2080	A1FI	44 more mar 1 month exposure to malaria 49 more mar 3 month exposure to malaria	HadCM3	SubSaharan Africa	Van Lieshout et al. 2004
4.3	2080	A1FI	179 more mar 1 month exposure to malaria 23 more mar 3 month exposure to malaria	HadCM3	West Asia	Van Lieshout et al. 2004
4.3	2080	A2FI	26 less mar 1 month exposure to malaria 111 less mar 3 month exposure to malaria[4]		Latin & S America	Van Lieshout et al. 2004
4.3	2080	Constant population	28.2 million additional person months exposure	HadCM3	Africa	Tanser et al. 2003
4.3	2085	A1:2080 (37%) population under water stress if no cc	1256 mar water stress 1818 m with decrease in water stress[1]	HadCM3	Globe	Arnell 2004
4.3	2080	A1FI	7–10 mar from coastal flood	HadCM3	Globe	Nicholls 2004
4.3	2080	A1FI	300 mar from hunger (30 CO_2 ff)	HadCM3	Globe	Parry et al. 2004
4.3	–	–	Entire regions out of production		Australia, S Africa, parts of S Asia	Hare 2003
4.3/3.6	2080	A1/A2	10% fall in cereal production	HadCM3	Globe	Parry et al. 2004
4.5	2090	No evolving baseline: fixed at 1990 world	30% fall in cereal supply without farm level adaptation, 6% with; falls to zero allowing for trade, changes in demand and land use changes to provide new cropland	GISS	Globe	Darwin et al. 1995[6]
4.5			25% increase in potential malarious zones; 40% increase in seasonal zones and 20% decrease in perennial	HadCM2/3	Globe	Martin & Lefevre 1995

(continued)

Table B (*contd*)

ΔT	Year	Population scenario	Impact to human systems m.a.r. = additional millions of people at risk than would be the case in absence of climate change	GCM used	Region affected	Source
5.5			30% increase in potential malarial transmission zones; 55% increase in seasonally affected zones and 40% reduction in perennially affected zones	HadCM2/3	Globe	Martin & Lefevre 1995
5.5	2090	No evolving baseline: fixed at 1990 world	23% fall in cereal supply without farm level adaptation, 2.4% with; falls to zero allowing for trade, changes in demand and land use changes to provide new cropland	UKMO	Globe	Darwin 1995[4]

[1] Arnell (2004) shows that although under climate change more watersheds move out of the water stressed category than into it, *the increases in runoff generally occur in high flow seasons, and thus will not alleviate water stress unless this water is stored*, and indeed, increased flooding in the wet season, rather than reduced water stress, may result. Secondly *the watersheds where rainfall increases are in limited areas of the world only, but these happen to be populous, that is mainly SE Asia*.

[2] This result is based on the hedonic method, which uses the spatial difference in bio-economics of agriculture between warm and cold regions to predict the consequences of increasing temperatures in present-day cold regions to those of present-day warm regions, thus assuming that changes in time and space are equivalent, that systems immediately just to a new stable state so that there is no consideration of time-dependence, and only annual average regional temperatures are considered, so changes and seasonal variability in temperature or rainfall are not considered (Schneider 1997). The author does not think that these assumptions are credible. It also assumes that precipitation measures the water supply for crops and that future changes in production costs will be capitalised in land values in the same way that past production costs were capitalised in past land values, both of which are problematic assumptions for the area of study, the USA, where large areas of cropland are irrigated, and construction of new water systems would be very much more costly than continued operations of existing ones. Using a hedonic model tied to a national data set of farmland values that combines both dryland and irrigated farming counties is likely to be questionable both on econometric grounds, because it combines what we expect to be two heterogeneous equations with different variables and different coefficients into a single regression, and also on economic grounds, since we expect it to understate future capital costs, especially those borne by farmers, in the areas that will need additional surface water irrigation due to the effects of climate change. (Schlenker 2004).

[3] Full CO$_2$ fertilisation effects assume no yield reductions due to potential changes in soil nutrients, pollinator scarcity, pest outbreaks and food quality that are associated with climate change.

[4] The decreased risk of malaria is Latin & S America is due to reductions in precipitation predicted by HadCM3 for this region. For further regional detail see Van Lieshout *et al.* 2004. C. Thomas *et al.* (2004) suggest that the increases in exposure to malaria in Africa are largely in regions where existing risks occur before 2050, whilst after 2050 new exposure in highland areas of Africa occurs.

[5] Fischer highlight the fundamental role of SRES scenario choice in influencing additional millions at risk. Under the A2 scenario, the increase in millions at risk due to climate change is very significant, whilst the increase risk is smaller under the other three scenarios. However, note that full benefits of CO$_2$ fertilisation are assumed in this study. Without this assumed benefit, more significant risks would be found for the other scenarios, as found by Parry 2001. None of the agricultural studies consider the impacts of extreme weather events on crop production, and only the Parry study provides any insight on the effects of rates of change of climate. All studies consider farm level adaptation.

[6] The Darwin study predicts large impacts of climate change, but puts forward the view that adaptations and economic processes, together with land use change can largely offset these impacts. It also does not consider impacts of extreme events or rates of change of climate. To offset the impacts in the UKMO model in 2090, a 15% increase in world cropland is considered necessary, including a doubling of the area farmed in Canada. Such large scale conversion of previously uncultivated land would increase the stresses on ecosystems.

Table C Observed and predicted impacts of sea level rise on human systems.

Sea-level rise above 1961–1990 average (m)	Year in which this occurs	Population scenario	Impacts to human systems	Region affected	Source
0.0	Present day	Present day	46 million people are exposed to storm surge flooding at present		Hoozemans *et al.* 1993, Baarse 1995
0.3	2050	IS92a	26 mar from coastal flood (i.e. a doubling of the 26 million in absence of climate change)	HadCM2	Parry *et al.* 2001, Nicholls 2004
0.4	2140	S550 (stabilisation) in IS92a	45 mar coastal flooding (compared to 3 million in absence of climate change)	HadCM2	Nicholls 2004
0.46	2140	S750 (stabilisation) in IS92a	60 mar coastal flooding (compared to 3 million in absence of climate change)	HadCM2	Nicholls 2004
0.5	If occurred present day	Present day	Sea level rise causes number of people exposed to storm surge flooding to 92 million per year		Hoozemans *et al.* 1993, Baarse 1995
0.5	2080	IS92a	80 mar from coastal flooding (only 14 million at risk in absence of climate change)	HadCM2	Parry *et al.* 2001, Nicholls 2004
0.58	2110	IS92a	Additional 140 mar coastal flooding (only 3 million at risk in absence of climate change)	HadCM2	Nicholls 2004
0.75	2140	IS92a	Additional 160 mar coastal flooding (only 1 million at risk in absence of climate change)	HadCM2	Nicholls 2004
1.0	If occurred present day	Present day	Sea level rise causes number of people exposed to storm surge flooding to almost triple to 118 million per year		Hoozemans *et al.* 1993, Baarse 1995
1.0			$1000 billion damage due to sea level rise	Global	Fankhauser 1995
1		–	Additional 2 m people and additional 55 trillion yen of assets exposed to tides, requiring protection barriers of between 2.8 and 3.5 m high	Japan	Harasawa 2005
1.0	2100		Damages due to the 1:1000 year flood increase from zero to £25 billion (we are currently protected by the Thames barrier against the 1:1000 year flood) for constant population	London if Thames Barrier not upgraded	Hall 2005
Any	Any		Population displaced	Nile delta	IPCC 2001
Any	Any		Population displacement & livelihood impacts due to inundation and coastal erosion	Banjul, Gambia Lagos, Nigeria, Gulf of Guinea, Senegal	IPCC 2001
2.0			$2000 billion damage due to sea level rise	Globe	Fankhauser 1995
Above 2 m	2300		Widespread loss of many of the world's largest cities, widespread loss coastal and deltaic areas including Bangladesh, Nile, Yangtze, Mekong	Globe	ECF 2004; Oppenheimer & Alley 2004; Hansen 2005

Table D Observed and Predicted Impacts of Climate Change upon Ecosystems at different levels of global mean annual temperature rise, ΔT, relative to pre-industrial times.

ΔT	Year in which this occurs	Impacts to unique and threatened ecosystems	Region affected	GCM used where known	Source
		OBSERVED CHANGE			
0.6	2004	Analysis of 143 studies of species which showed changes in phenology, morphology, range of abundance shows that 80% of the changes are in the direction consistent with the expected physiological response to climate change	All regions	N/A	Root *et al.* 2003, Root *et al.* 2005, Parmesan & Yohe 2003.
0.6	2004	50 species of frogs & toads locally extinct in area, including global extinction of Golden Toad	Monteverde, Costa Rica	N/A	Pounds *et al.* 1999
0.6	2005	Oceans have acidified by 0.1 pH units since preindustrial times	All oceans		Caldeira & Wickett 2003
0.6	2004	Changes in tree growth rates, increase in fire/pest outbreaks, permafrost melting causing collapse of trees and creation of new wetlands	Arctic boreal forest	N/A	ACIA 2004
0.6	2004	Decline in growth of white spruce as summers warm	Alaska	N/A	ACIA 2004
0.6	2004	Northward spread of spruce budworm	Alaska	N/A	ACIA 2004
0.6	2004	Spruce bark beetle infestations spread	Alaska, Canada	N/A	ACIA 2004
0.6	2004	Area of forest burnt by fires in Russia has doubled in 1990s	Russia	N/A	ACIA 2004
0.6	2004	Condition of polar bears declines; polar bear cub births decline	Hudson Bay	N/A	ACIA 2004
0.6	2004	90% decline in Ivory Gull	Canada	N/A	ACIA 2004
0.6	1989–2001	Declines in caribou of approx. 3.5% /year	Canada, Alaska, Greenland	N/A	ACIA 2004
0.6	2004	Algae at base of marine food chain underwent shifts in community composition	Beaufort Sea	N/A	ACIA 2004
0.6	1965–2004	Loss of grassland & acacia, loss of flora/fauna, shifting sands (not attributed)	Sahel	N/A	ECF 2004
0.6	1979–2004	Chinstrap penguins (ice-phobic) increased 400% whilst ice-dependent Adelie decreased 25%	West Antarctic (where T rise 4 to 5°C since 1954)	N/A	Fraser and Patterson 1997, Smith 1999
0.6	2004	Vascular plant range increases	Antarctica	N/A	Smith 1994
0.6	2004	Decline of Rockhopper Penguins correlated to sea surface temperature	S Ocean	N/A	Cunningham & Moors 1994
0.6	2004	Birds nesting earlier	Finland	N/A	Jarvinen 1989
0.6	2004	Birds nesting earlier	Germany	N/A	Ludwichowski 1997
0.6	2004	Earlier migrant arrival	Slovak Republic	N/A	Sparks & Bravslavska 2001
0.6	2004	Earlier egg-laying	N America, Europe, Australia	N/A	Winkel & Hudde 1997, Schiegg *et al.* 2002, Crick & Sparks 1999,

(continued)

Table D (*contd*)

ΔT	Year in which this occurs	Impacts to unique and threatened ecosystems	Region affected	GCM used where known	Source
					Oglesby & Smith 1995, Mickelson *et al.* 1992
0.6	2004	Earlier emergence of butterflies 1883–1993	UK	N/A	Sparks & Yates 1997
0.6	2004	Poleward migration of plants; disappearance of species from S Europe	Europe/	N/A	EEA 2004
0.6	2004	Advanced spring phenology	Asia	N/A	Yoshino & Ono 1996, Kai *et al.* 1996
0.6	2004	Spring phenology advanced by 5 days e.g. tree flowering, leaf unfolding, egg-laying date of birds, emergence date of insects, hatching date of birds, spring arrival of birds.	All regions. Specifically Europe, Asia, North America	N/A	Root *et al.* 2003
0.6		Advanced bird migration	Germany	N/A	Huppop & Huppop 2003
0.6	2004	Advanced arrival of birds, leaf unfolding and flowering	Spain	N/A	Penuelas *et al.* 2002
0.6	2004	Growing season lengthened 11 days	Europe	N/A	Gitay *et al.* 2001
0.6	2004	N movement of warm water plankton of 1000 km in only 40 years	E Atlantic	N/A	Richardson & Schoeman 2004
0.6	2004	Major reorganisation of plankton ecosystems: Change in plankton distribution; increasing phytoplankton biomass; extension of the seasonal growth period; N shift of zooplankton	North Sea, Pacific Ocean	N/A	EEA 2004, Richardson & Schoeman 2004, Mackas *et al.* 1998
0.6	2004	Severe decrease in sandeel abundance likely due to reorganisation of plankton above	North Sea	N/A	Arnott & Ruxton 2002
0.6	2004	Large scale breeding failure of seabirds likely due to decline of sandeels above	UK	N/A	Lanchbery 2005
0.6	2004	Dramatic change in community composition of UK marine fish	English & Bristol Channels	N/A	Hawkins 2005
0.6	2004	Decreased alpine flora, migration to higher altitudes	Japan, Europe	N/A	Harasawa 2005, EEA 2004
0.6	2004	Altered distribution of trees, butterflies, birds, insects	Japan	N/A	Harasawa 2005
0.6	2004	Northward movement of cold-water fish	Bering Sea	N/A	ACIA 2004
0.6		50% of Southern Ocean krill stocks are found in SW Atlantic sector, where their density has declined by 80% since the 1970s, probably as a result of decreasing sea-ice extent; a huge drop was observed in 2004	Antarctic	N/A	Gross 2005
0.6		Range change in native trees	New Zealand	N/A	Wardle & Coleman 1991
0.6		Range shift in birds	Central America	N/A	Pounds *et al.* 1999
0.6		Density change in reptiles	Central America	N/A	Pounds *et al.* 1999

(*continued*)

Table D (*contd*)

ΔT	Year in which this occurs	Impacts to unique and threatened ecosystems	Region affected	GCM used where known	Source
0.6		Advance in spring phenology of birds and trees	N America	N/A	Bradley *et al.* 1999
0.6		Advance in flowering of plants	N America	N/A	Abu-Asab *et al.* 2001
0.6		Advance in spring phenology of grasses	N America	N/A	Chuine *et al.* 2000
0.6		Range shift and density change in intertidal invertebrates, zooplankton and fish	English Channel	N/A	Southward *et al.* 1995
0.6		Mammal range shifts	North America	N/A	Frey 1992
0.6		Bird density changes	California	N/A	Sydeman *et al.* 2001
0.6		Fish, bird and flowering plant phenology advances	Estonia	N/A	Ahas 1999
0.6		Bird phenological advances	Russia	N/A	Minin 1992
0.6		Salmon return rate changes	Japan	N/A	Ishida *et al.* 1996
0.6		Amphibian arrival and spawning advances	UK	N/A	Beebee 1995
0.6	2004	Mammal spring phenology advances	USA	N/A	Inouye *et al.* 2000
≥0.6	2004	Climate change impacts such as rising sea levels, sea-surface temperatures, droughts and storms are adding to threats to 18 endangered/vulnerable/ threatened birds	Globe, particularly coastal areas/low-lying islands	N/A	BTO (unpublished)
		PREDICTED CHANGE			
≥0.6		Since ecosystem species do not shift in concert as climate changes, predator-prey and pollinator-plant relationships are disrupted, leading to many extinctions and pest outbreaks	Globe		Burkett *et al.* 2005, Price 2002
≥0.6		Cloud forest ecosystems continue to shift to higher elevations, causing further extinctions of endemic species over and above the frogs mentioned previously	Tropical mountainous areas e.g. Central & S America, Borneo, Africa		Still *et al.* 1999
≥0.6		More pronounced ecosystem disturbance by fire/pests	Globe		Gitay *et al.* 2001
≥0.6		Cod populations may increase off Greenland, whilst N shrimp will decrease	Greenland		ACIA 2004
≥0.6		Increased overwinter survival of resident and wintering birds	Europe		EEA 2004
≥0.6		Northward extensions in ranges of European butterflies	Europe		EEA 2004
≥0.6		Increased drought in the Sahel would cause many local fauna and flora to disappear	Sahel		ECF 2004
≥0.6		Decreased survival of long distance migrants crossing Sahel as climate change is predicted to increase drought; global effects if long-distance migrants suffer phenological miscuing	Eurasia Globe		Berthold 1990
≥0.6		Increased ecosystem disturbance by pest/disease,	Globe, especially in		Gitay *et al.* 2001, Hare 2003, ECF

(*continued*)

Table D (*contd*)

ΔT	Year in which this occurs	Impacts to unique and threatened ecosystems	Region affected	GCM used where known	Source
			Boreal forest, Australia, California		2004
<1		Coral reefs at high risk	Caribbean, Indian Ocean, Great Barrier Reef		Hoegh-Guldberg 1999
<1		Loss in extent of Australia's most biodiverse region, the Queensland World Heritage Rainforest	N Australia		Hilbert *et al.* 2001
<1		Loss in extent of Karoo, the richest floral area in world	S Africa	HadCM2 HADGGAX50 (CO_2 doubling)	Rutherford *et al.* 1999
<1		Risk extinction of vulnerable species in Dryandra forest	SW Australia		Pouliquen-Young & Newman 1999
<1		Range losses begin for animal species in S Africa, and Golden Bowerbird in Australia	S Africa, Australia	HadCM2 HadCM3 **	Rutherford *et al.* 1999, Hilbert *et al.* 2004
Not known		Snow leopards at risk	Russia		ECF 2004
1		Coral reefs 82% bleach including Great Barrier Reef	Globe, i.e. Australia, Caribbean, Indian Ocean		Hoegh-Guldberg 1999
1		10% Global Ecosystems transformed; only 53% wooded tundra remains stable, loss cool conifer forest. Ecosystems variously lose between 2 to 47% of their extent.	Globe	5 GCMS: HadCM2GFDL ECHAM4 CSIROMK2 CGCM1	Leemans & Eickhout 2003
1	2050	50% loss highland rainforest, range losses of endemics and 1 of these extinct	Queensland Australia	Sensitivity study covered range of precipitation outcomes	Hilbert *et al.* 2001, Williams *et al.* 2003
1.3	2020 IS92a	Risk extinction of Golden Bower bird: at 1°C local temperature rise habitat reduced by 50%	Australia	Not specified[7]	Hilbert *et al.* 2004
1.4		Extinction of coral reefs	Indian Ocean		Sheppard 2003
1.4		>50% loss Kakadu	Australia	HadCM2/3	Hare 2005
1–2		Risks for many ecosystems	Globe		Leemans & Eickhout 2003
1–2		Many eucalypts out of range	Australia		Hughes *et al.* 1996
1–2		Large impacts to salmonid fish	N America	Range of GCMs	Hare 2005 based on Keleher & Rahel 1996
1–2		Significant loss Alpine zone	Australia		Busby 1988
1–2	2050	Severe loss of extent of Karoo	S Africa	HadCM2 HADGGAX50 (CO_2 doubling)	Rutherford *et al.* 1999
1–2		Risk extinction frogs/mammals (40% loss World Heritage Rainforest area)	Australia's most		Williams *et al.* 2003

(*continued*)

Table D (*contd*)

ΔT	Year in which this occurs	Impacts to unique and threatened ecosystems	Region affected	GCM used where known	Source
			biodiverse region (Queensland wet tropics)		
1–2		Loss of aerobic capacity, potential for local extinction of key mollusc species from the Southern Ocean at local T rise of 2°C.	Antarctic		Peck *et al.* 2004
1–2		Moderate stress Alpine zone	Europe		Hare 2005
1–2		Severe damage to Arctic ecosystem	Arctic		ACIA 2004
1–2		60% loss lemming (for local T rise 4°C) affecting whole ecosystem, including snowy owl	Arctic	GISS GCM;	Kerr & Packer 1998
1.5	2050 (SRES B1)	18% all species extinct	Globe		Thomas *et al.* 2004b[8]
2		Coral reefs 97% bleached	Globe		Hoegh-Guldberg 1999
2	2100	Total loss Arctic summer ice, high risk of extinction of polar bears, walrus, seals, whole ecosystem stressed	Arctic		ACIA 2004
2		16% global ecosystems transformed: ecosystems variously lose between 5 and 66% of their extent		5 GCMs: HadCM2GFDL ECHAM4 CSIROMK2 CGCM1	Leemans & Eickhout 2003
2		Further ecosystem disturbance by fire & pests	Globe		IPCC 2001
2		50% loss of Sundarbans wetlands	Bangladesh	HadCM2/3 to convert local T to global	Hare 2005, Qureshi & Hobbie, 1994, Smith *et al.* 1998
2		Only 42% existing Arctic tundra remains stable	Arctic		Folkestad 2005
2		Millions of the world's shorebirds nest in Arctic, from the endangered Spoon-billed Sandpiper to the and very common Dunlin and would lose between 10% and 45% of breeding area; high arctic species most at risk	Globe		Folkestad 2005
2		Millions of Geese e.g. White-fronted and endangered Red-breasted Goose lose up to 50% breeding area	N hemisphere		Folkestad 2005
2		60% N American wood warblers ranges contract, whilst only 8% expand, such that between 4 and 13 (34%) (range allows for uncertainty in precipitation change) reach "vulnerable" conservation status		Sensitivity analysis	J.T. Price (unpublished)
2		Severe damage (590% loss) to boreal forest	China		Ni 2001, Hare 2003
2		>50% salmonid fish habitat loss	N America	Range of GCMs	Hare 2005 based on Keleher & Rahel 1996

(*continued*)

Table D (*contd*)

ΔT	Year in which this occurs	Impacts to unique and threatened ecosystems	Region affected	GCM used where known	Source
2	2050 IS92a	Transformation of ecosystems e.g. 32% of plants move from 44% European area with potential extinction of endemics/ specialists	N Europe		ECF 2004, Bakkenes *et al.* 2002
2		High risk extinctions of forest mammals; inflexion point at which extinction rates take off	Australia (Queensland)		Williams *et al.* 2003
~2		Cloud forest regions lose hundreds of metres of elevational extent	Central America, tropical Africa & Indonesia	GENESIS GCM $2\times CO_2$	Still *et al.* 1999
2		Extinctions of endemics such as Hawaiian honeycreeper birds	Hawaii		Benning *et al.* 2002
2		Loss of 9%–62% mammal species from mountainous areas	USA Great Basin	Not specified[1]	Hannah *et al.* 2002
~2		Loss of forest wintering habitat of Monarch butterfly	Mexico	CCC GFDL	Villers-Ruiz & Trejo-Vasquez 1998
2.2	A1F1	15–37% species extinct	Globe		Thomas *et al.* 2004b[2]
2.3	2050 IS92a	High risk extinction of Golden Bowerbird: at 2C local temperature rise habitat reduced by 90% and at 3C by 96% to $37\,km^2$	Australia	Not specified[1]	Hilbert *et al.* 2004
2.4	2055 IS92a	Large range loss animals & risk extinctions of 11% species	Mexico	HadCM2 HADGGA×50 (CO_2 doubling)	Peterson *et al.* 2002
2.4	2050 IS92a	Succulent Karoo fragmented and reduced to 20% of area, threatening 2800 plants with extinction; 5 S African parks lose >40% animals	S Africa	HadCM2 HADGGA×50 (CO2 doubling)	Rutherford *et al.* 1999, Hannah *et al.* 2002
2.4		66% animals lost from Kruger; 29 endangered species lose >50% range; 4 species becomes locally extinct	S Africa	HadCM2	Erasmus *et al.* 2002, Hare 2005
2–2.5		Fish populations decline strongly with drought, wetland ecosystems dry and disappear	Malawi, African Great Lakes		ECF 2004
2–3		Amazon collapse	S America, Globe		Cox *et al.* 2004
2–3		Total loss Kakadu	Australia	HadCM2/3	Hare 2005
2–3		Extinctions of alpine flora	New Zealand		Halloy & Mark 2003
2–3		Large impacts eg permafrost shifts N by 1 to 2 degrees latitude, acceleration of desertification	Tibetan plateau	HadCM2 500 ppm CO_2	Ni 2000
2.5	2050	Extinctions 10% endemics in Fynbos hotspot for plant biodiversity; 51–65% loss of Fynbos area.	S Africa	HadCM2 CSM	Midgley *et al.* 2002
2.5		Complete loss alpine zone	Australia		Hare 2005 based on Pouliquen-Young & Newman 1999

(*continued*)

Table D *(contd)*

ΔT	Year in which this occurs	Impacts to unique and threatened ecosystems	Region affected	GCM used where known	Source
2–2.5		Cold temperate forest e.g. maple (responsible for New England fall colours) at risk	USA		ECF 2004
2.6	2100	20–70% loss (average 44%) migratory & wintering shorebird habitat at 4 major sites	USA coasts		Galbraith *et al.* 2002, Hare 2005
3		Few ecosystems can adapt to temperature increases of 3°C and above	Globe		Lemans & Eickhout 2003
3	2080 IS92a	Increase of fire frequency converting forest and macquis to scrubland, increased vulnerability to pests	Mediterranean	HadCM3 for T; reduced low and increased high intensity rainfall events	Mouillot *et al.* 2002
3		50% all nature reserves cannot fulfil their conservation objectives	Globe	5 GCMs: Had-CM2GFDLLR ECHAM4 CSIROMK2 CGCM1	Leemans & Eickhout 2003
3		Risk extinction of 90% Hawaiian honeycreeper birds	Hawaii		Benning *et al.* 2002
3	2100	Risk of loss of up to 60% species	Europe especially Southern		ECF 2004
3		Complete loss of Chinese boreal forest ecosystem	China		Ni 2001
3		Large loss migratory bird habitat	Baltic, USA, Mediterranean	HadCM3IPCC 2001 IS92a sea level scenario	Nicholls *et al.* 1999, Najjar *et al.* 2000
3 (2.8–3.6)	2050	50% loss world's most productive duck habitat in prairie pothole region 38% HadCM3; 54% GFDL; others 0–100% but 11 of 12 simulations show losses, even if precipitation increases	USA	GFDL Had-CM2 Other GCM ranges covered via sensitivity analysis	Sorenson *et al.* 1998
3		22% global ecosystems transformed: ecosystems variously lose between 7 and 74% of their extent	Globe	Range of GCMs (via IMAGE)	Leemans & Eickhout 2003
3		Alpine species near extinction	Europe	Explored range of regional climate outcomes	Bugmann 1997
3		50% loss eucalypts	Australia		Hughes *et al.* 1996
3.3	2050	>50% range loss (and 80% current range loss) of 24 latitudinally restricted endemic butterflies	Australia	Median of 10 GCMs	Beaumont & Hughes 2002
3.3		77% loss low tundra	Canada		Neilson *et al.* 1997
3.4		22% loss coastal wetlands	Globe	HadCM2 HadCM3	Nicholls *et al.* 1999
3.8		60% loss tundra ecosystem	Globe		Neilson *et al.* 1997
3.8		44% loss taiga ecosystem	Globe		Neilson *et al.* 1997
4		38% European alpine species lose 90% range	Europe		Hare 2005

(continued)

Table D (*contd*)

ΔT	Year in which this occurs	Impacts to unique and threatened ecosystems	Region affected	GCM used where known	Source
5.3	2100	Average 79% loss at 4 key sites for migratory & wintering shorebird habitat (2C SF Bay)	USA coasts		Galbraith *et al.* 2002, Hare 2005

References in bold appear in this volume.

[7]The literature gives only the effects of local temperature rises, hence the author has used Hulme *et al.* 1999's presentation of HadCM2 and HadCM3 scenarios to convert from local to global temperature rise, in which the IS92a scenario is simulated (see temperature table in accompanying "methodology" section).

[8]Thomas *et al.*(2004) has been subject to debate (Thuiller *et al.* 2004; Harte *et al.* 2004; Buckley & Roughgarden 2004; Thomas *et al.* reply 2004c). Potential biases include (i) overestimation due to questions related to the validity of the particular application of the species-area relationship used, though Thomas *et al.* contest this in their reply (ii) over or under estimation due to the use of a common formula for all species, since sparsely distributed species will be more vulnerable (iii) the potential effects of methodological uncertainty concerning niche models (iv) the validity of the relation between range reduction and extinction likelihood (v) underestimation due to ignoring genetic adaptation to climate at the population level. It has been suggested that endemics-area relationships might better be used. What is clear is that climate change and land use change together place enormous threats to biodiversity in the twenty-first century.

Table E Predicted Impacts of Rate of Temperature Change upon Ecosystems.

Rate of Temperature rise above pre-industrial	Population scenario	Impacts to unique and threatened ecosystems	Region affected	Source
0.6°C over 20th century; now 0.17 +/− 0.05°C/decade		Fastest rise of millennium	Globe	IPCC 2001
0.05°C/decade		Proposed threshold to protect ecosystems		Leemans & van Vliet 2005
0.1°C/decade		Threshold above which ecosystems are damaged	Globe	Vellinga & Swart 1991
0.1°C/decade		50% of ecosystems can adapt; forest ecosystems impacted first	Globe	Leemans & Eickhout 2003
General remark		Warming may require migration rates much faster than those in post-glacial times & therefore has potential to reduce biodiversity through selection for mobile/opportunistic species		Malcolm *et al.* 2002; using 7 climate scenarios from GFDL and HadCM2
General remark		Ecosystem response lags behind equilibrium, hence vulnerability to pests, diseases, fire is high, this is worse for higher rates of change	Globe	IPCC 2001, Leemans & Eickhout 2003
0.3°C/decade		30% ecosystems can adapt; ecosystem response lags behind equilibrium, vulnerability to pests, diseases, fire is high	Globe	Leemans & Eickhout 2004
0.4°C/decade		All ecosystems rapidly deteriorate, disturbance regimes, low biodiversity, aggressive opportunistic species dominate globe: resulting in release of carbon to the atmosphere	Globe	Leemans & Eickhout 2003, Neilson 1993
0.46°C/decade		Current rate in Arctic (1977–2003)		Folkestad 2005

Table F Predicted Impacts of Different Levels of Sea Level Rise upon on Ecosystems.

Sea-level rise above 1961–1990 average (cm)	Year	Matching Temperature increase range (TAR) for this time period	Impacts to unique and threatened ecosystems	Region affected	Source
2.7	2004	0.6		Globe	Parry *et al.* 1999
3–14	2025	0.4–1.1	Loss of some coastal wetlands likely, increased shoreline erosion, saltwater intrusion into coastal aquifers	Globe	IPCC 2001
30	Any		57% sandy beaches eroded	Asia	Harasawa 2005
5–32	2050	0.8–2.6	More extensive loss coastal wetlands, further shore erosion		IPCC 2001
34			20–70% loss of key bird habitat at 4 major sites	USA	Galbraith *et al.* 2002
34			Large loss migratory bird habitat	Baltic, Mediterranean	Nicholls *et al.* 1999, Najjar *et al.* 2000
45	Any	Any	Mangroves cannot survive 45 cm sea level rise	Asia	Harasawa 2005
9–88	2100	1.4–5.8	More extensive wetland loss, further erosion of shorelines		IPCC 2001
100	Any	Any	90% sandy beaches eroded	Asia	Harasawa 2005
40	2080	3.4 (particular GCM scenario used)	5–22% world's coastal wetlands lost	Globe	Nicholls *et al.* 1999
100	2100	5.8[9]	25–55% world's coastal wetlands lost	Globe	Hoozemans *et al.* 1993
300–500	2300	3	With 3C temperature rise this will occur by 2300 even if Greenland and WA ice sheets do not melt	Globe	ECF 2004
300–500	2300	3	Widespread loss coastal and deltaic areas including Bangladesh, Nile, Yangtze, Mekong	Globe	ECF 2004

[9] Volume assuming upper range of IPCC temperature matches upper range of IPCC sea level rise.

Table G Predicted Impacts of Different Rates of Sea Level Rise upon on Ecosystems.

Rate of sea-level rise	Status	Impacts to unique and threatened ecosystems	Region affected	Source
1 to 2 mm/yr Between 0.8 and 3 mm/year	Observed in twentieth century		Globe	IPCC 2001
Between 0.8 and 3 mm/yr	Observed in twentieth century		Europe	EEA 2004
5 mm/yr		Coastal erosion, loss of coastal ecosystem such as mangroves and coral reefs thus destroying natural coastal defences; saltwater intrusion, dislocation of people, increased risk to storm surge, this being especially problematic in small island states	Globe, particularly Asia, N America, Latin America, and small island states.	IPCC 2001

(*continued*)

Table G (*contd*)

Rate of sea-level rise	Status	Impacts to unique and threatened ecosystems	Region affected	Source
6 mm/yr	Prediction	Wetlands lost	New England	Hare 2005 based on Donnelly & Bertness 2001

Note to observed changes reported in Table A to G: Not all of the observed changes are directly attributed to anthropogenic climate change. They are listed because they are changes which are consistent with the patterns of change predicted to result from anthropogenic climate change.

Table H Predicted Effects of Climate-change-induced Acidification on the Oceans.

[CO_2]	Ocean pH	Impacts to marine ecosystems	Source
265	8.2	Marine biogeochemistry altered, disrupting carbonate chemistry and altering plankton composition	Riebesell *et al.* 2000
750	7.82	Calcifying organisms at risk: Replacement of coccilithiphores, gastropods & formanifera by non-calcifying organisms	Turley *et al.* 2005
Not known		Calcifying organisms at risk: Corals growth rates reduced by up to 40% by 2065	Langdon *et al.* 2000, Leclercq *et al.* 2000
Not known		Impacts on plankton grazers including economically important species such as shellfish and fish.	Turley *et al.* 2005

REFERENCES

Abu-Asab, M.S., Peterson, P.M., Shetler, S.G., and Orli, S.S.: 2001, 'Earlier plant flowering in spring as a response to global warming in the Washington DC area', *Biodiversity and Conservation* 10, 597–612.

Ahas, R.: 1999, 'Long-term phyto-, ornitho- and ichthyophenological time-series analyses in Estonia', *International Journal of Biometeorology* 42, 119–123.

Archer, D.: 1995, 'Upper ocean physics as relevant to ecosystem dynamics – a tutorial'. *Ecological Applications* 5, 724–739.

Arctic Climate Impact Assessment (ACIA): 2004, 'Impacts of a warming Arctic'. Cambridge University Press, Cambridge, U.K.

Arnell, N.W.: 2004, 'Climate change and global water resources: SRES emissions and socio-economic scenarios', *Global Environmental Change* 14, 31–52.

Arnell, N.W., Cannell, M.G.R., Hulme, M., Kovats, R.S., Mitchell, J.F.B., Nicholls, R.J., Parry, M.L., Livermore, M.T.J. and White, A.: 2002, 'The consequences of CO_2 stabilisation for the impacts of climate change', *Climatic Change* 53, 413–446.

Arnott, S.A., and Ruxton, G.D.: 2002, 'Sandeel recruitment in the North Sea: demographic, climatic and trophic effects', *Mar. Ecol. Prog. Ser.* 238, 199–210.

Bakkenes, M., Alkemade, J.R.M., Ihle, F., Leemans, R., and Latour, J.B.: 2002, 'Assessing effects of forecasted climate change on the diversity and distribution of European higher plants for 2050', *Global Change Biology* 8, 390–407.

Baarse, G.: 1995, 'Development of an Operational Tool for Global Vulnerability Assessment (GVA)-Update of the Number of People at Risk Due to Sea Level Rise and Increased Flooding' Institute for Environmental Studies of the Vrije Universiteit, Amsterdam (The Netherlands), http://www.dinas-coast.net/.

Beaumont, L.J., and Hughes, L.: 2002, 'Potential changes in the distributions of latitudinally restricted Australian butterfly species in response to climate change', *Global Change Biology* 8, 954–971.

Beebee, T.J.C.: 1995, 'Amphibian breeding and climate', *Nature* 374, 219–220.

Benning, T.L., LaPointe, D., Atkinson, C.T., and Vitousek, P.M.: 2002, 'Interactions of climate change with biological invasions and land use in the Hawaiian Islands: Modeling the fate of endemic birds using a geographic information system', *PNAS* 99, 14246–14249.

Berthold, P.: 1990, 'Patterns of avian migration in light of current global greenhouse effects: a central European perspective', *Acta Congr Int Ornithol* 20, 780–786.

Betts, R., 'Impacts of Doubled-CO_2 Climate Change', *Avoiding Dangerous Climate Change Conference*, Exeter, UK, www.stabilisation2005.com

Bradley, N.L., Leopold, A.C., Ross, J., and Huffaker, W.: 1999, 'Phenological changes reflect climate change in Wisconsin', *PNAS* 96, 9701–9704.

Buckley, L.B., and Roughgarden, J.: 2004, 'Biodiversity conservation: effects of climate change and land use', *Nature* 430, 35.

Bugmann, H.: 1997, 'Sensitivity of forests in the European Alps to future climatic change', *Climate Research* 8, 35–44.

Burkett, V., Wilcox, D.A., Stottlemeyer, R., Barrow, W., Fagre, D., Baron, J., Price, J., Nielsen, L., Allen, C.D., Peterson, D.L., Ruggerone, G., and Doyle, T.: 2005, 'Nonlinear dynamics in ecosystem responses to climatic change: case studies and policy implications', *Ecological Complexity* (in press).

Busby, J.R.: 1988, 'Potential implications of climate change on Australia's flora and fauna', in Pearman, G.I. (ed.), *Greenhouse: Planning for Climate Change*, CSIRO Division of Atmospheric Research, Melbourne, pp. 387–388.

Caldeira, K., and Wickett, M.E.: 2003, 'Anthropogenic carbon and ocean pH', *Nature* 425, 365.

Cavalieri, D.J., Parkinson, C.L., and Vinnikov, K.Y.: 2003, '30-Year satellite record reveals contrasting Arctic and Antarctic decadal sea ice variability', *Geophysical Research Letters* 30, no. 18, 1970. DOI: 10.1029/2003GL018031.

Chuine, I., Cambon, G., and Comtois, P.: 2000, 'Scaling phenology from the local to the regional level: advances from species-specific phenological models', *Global Change Biology* 6, 943–952.

Cox, P.: 2005, 'Conditions for Sink-to-Source Transitions and Runaway Feedbacks from the Land Carbon Cycle', this volume.

Cox, P.M., Betts, R.A., Collins, M., Harris, P.P., Huntingford, C., and Jones, C.D.: 2004, 'Amazonian forest dieback under climate-carbon cycle projections for the 21st century', *Theoretical and Applied Climatology* 78, 137–156.

Cox, P.M., Betts, R.A., Jones, C.D., Spall, S.A., and Totterdell, I.J.: 2000, 'Acceleration of global warming due to carbon-cycle feedbacks in a coupled climate model', *Nature* 408, 184–187.

Crick, H.Q.P., and Sparks, T.H.: 1999, 'Climate change related to egg-laying trends', *Nature* 399, 423–424.

Cunningham, D.M., and Moors, P.J.: 1994, 'The decline of the Rockhopper Penguin *Eudyptes chrysocome* at Campbell Is, S Ocean', Emu 94, 27–36.

Darwin, R., Tsigas, M., Leqandrowski, J., and Raneses, A.: 1995, *World Agriculture and Climate Change: Economic Adaptations*. Agricultural economic report number 703, US Dept of Agriculture.

Döll, P.: 2002, 'Impact of climate change and variability on irrigation requirements: A global perspective', *Climatic Change* 54, 269–293.

Donnelly, J.P., and Bertness, M.D.: 2001, 'Rapid shoreward encroachment of salt marsh cordgrass in response to accelerated sea-level rise', *PNAS* 98, 14218–14223.

Dore, M.: 'Dangerous Climate Change and Kyoto Mechanisms on Adaptations for Least Developed Countries'. *Avoiding Dangerous Climate Change Conferenc*e, Exeter, UK, www.stabilisation2005.com

Emanuel, K.: 2005, 'Increasing destructiveness of tropical cyclones over the past 30 years', *Nature* 436, 686–688.

Emori, S.: 'Japan as a Possible Hot Spot of Flood Damage in Future Climate Illustrated by High-Resolution Climate Modeling Using the Earth Simulator', *Avoiding Dangerous Climate Change Conferenc*e, Exeter, UK, www.stabilisation2005.com

European Environment Agency (EEA): 2004, 'Indicators of Europe's Changing Climate', (Prepared by Thomas Voigt, Jelle van Minnen, Markus Erhard, David Viner, Robert Koelemeijer, Marc Zebisch) EEA Copenhagen.

Erasmus, B.F.N., Van Jaarsfeld, A.S., Chown, S.L., Kshatriya, M., and Wessels, K.J.: 2002, 'Vulnerability of S African animal taxa to climate change', *Global Change Biology* 8, 679–693.

European Climate Forum (ECF): 2004, 'What is dangerous climate change? Initial results of a symposium on Key Vulnerable Regions, Climate Change and Article 2 of the UNFCCC', held at Beijing, 27–30 October 2004, and presented at Buenos Aires, 14 Dec 2004.

Fankhauser, S.: 1995, 'Valuing Climate Change: The Economics of the Greenhouse', London, Earthscan.

Fisher, G., Shah, Mahendra, and van Velthuizen, H.: 2002, 'Climate Change and Agricultural Vulnerability', special report prepared by the International Institute for Applied Systems Analysis under UN contract agreement no. 1113 as contribution to the World Summit on Sustainable Development, Johannesburg 2002. IIASA, Laxenburg, Austria.

Folkestad, T.: 2005, 'Evidence and Implications of Dangerous Climate Change in the Arctic', this volume.

Frey, J.K.: 1992, 'Response of a mammalian faunal element to climatic changes', *Journal of Mammalogy* 73, 43–50.

Fraser, W.R., and Patterson, D.L.: 1997, 'Human disturbance and long-term changes in Adelie penguin populations: a natural experiment at Palmer Station, Antarctica', in Battaglia, B., Valencia, J., and Walton, D. (eds), *Antarctic Communities: Species, Structure, Survival*. Cambridge University Press, Cambridge, pp. 445–452.

Fussel, H.-M., Toth, F.L., van Minnen, J.G., and Kaspar, F.: 2003, 'Climate Impact Response Functions as Impact Tools in the Tolerable Windows Approach', *Climatic Change* 56, 91–117.

Galbraith, H., Jones, R., Park, R., Clough, J., Herrod-Julius, S., Harrington, B., and Page, G.: 2002, 'Global climate change and sea level rise: Potential losses of intertidal habitat for shorebirds', *Waterbirds* 25, 173–183.

Gitay, H., Brown, S., Easterling, W., and Jallow, B.: 2001, 'Chapter 5: Ecosystems and Their Goods and Services', *Climate Change 2001: Impacts, Adaptation and Vulnerability*, Cambridge University Press, Cambridge, UK, pp. 237–342.

Gordon, L.J., Steffen, W., Jonsson, B.F., Folke, C., Falkenmark, M., and Johannessen.missing initial: 2005: 'Human modification of global water vapour flow from the land surface', PNAS 102, pp. 7612–7617.

Gregory, J.M., Huybrechts, P., and Raper, S.C.: 2004, 'Climatology threatened loss of the Greenland ice sheet', *Nature* 428, 616.

Gross, L.: 2005: 'As the Antarctic pack ice recedes, a fragile ecosystem hangs in the balance', *Plos Biology* 3, 557–561.

Hales, S., deWet, N., Maindonald, J., and Woodward, A.: 2002, 'Potential effect of population and climate changes on global distribution of dengue fever: an empirical model', *Lancet* 360, 830–834.

Hall, J., 'Tidal Flood Risk in London Under Stabilisation Scenarios', *Avoiding Dangerous Climate Change Conferenc*e, Exeter, UK, www.stabilisation2005.com

Halloy, S.R.P., and Mark, A.F.: 2003, 'Climate change effects on alpine plant biodiversity: A New Zealand perspective on quantifying the threat', *Arctic Antarctic and Alpine Research* 35, 248–254.

Hannah, L., Midgley, G.F., Lovejoy, T., Bond, W.J., Bush, M., Lovett, J.C., Scott, D., and Woodward, F.I.: 2002, 'Conservation of biodiversity in a changing climate', *Conservation Biology* 16, 264–268.

Hansen, J.E.: 2005, 'A slippery slope: How much global warming constitutes "dangerous anthropogenic interference?"' *Climatic Change* 68, 269–279.

Harasawa, H.: 2005, 'Key vulnerabilities and Critical Levels of Impacts on East and Southeast Asia', this volume.

Hare, W. 2003, '*Assessment of Knowledge on Impacts of Climate Change:– Contribution to the specification of Art. 2 on the UNFCCC*', WGBU, Berlin 2003.

Hare, B.: 2005, 'Relationship between Increases in Global Mean Temperature and Impacts on Ecosystems, Food Production, Water and Socio-Economic Systems', this volume.

Harte, J., Ostling, A., Green, J.L., and Kinzig, A.: 2004, 'Biodiversity conservation: climate change and extinction risk', *Nature* 430, 37.

Hawkins, S., 'Regional climatic warming drives long-term changes in British marine fish', *Avoiding Dangerous Climate Change Conferenc*e, Exeter, UK, www.stabilisation2005.com

Hay, J., Mimura, N., Campbell, J., Fifita, S., Koshy, K., McLean, R., Nakalevu, T., Nunn, P., and de Wet, N.: 2003, '*Climate Variability and Change and Sea-level Rise in the Pacific Islands Region: A Resource book for policy and decision makers, educators and other stakeholders*', Apia, Samoa, South Pacific Region.

Hayhoe, K.: 'Regional Assessment of Climate Impacts on California under Alternative Emissions Pathways – Key Findings and Implications for Stabilisation, this volume.

Hijioka, Y., Takahashi, K., Matsuoka, Y., and Harasawa, H.: 2002, 'Impact of global warming on waterborne diseases', *Journal of Japan Society on Water Environment* 25, 647–652.

Hilbert, D.W., Bradford, M., Parker, T., and Westcott, D.A.: 2004, 'Golden bowerbird (*Prionodura newtonia*) habitat in past, present and future climates: predicted extinction of a vertebrate in tropical highlands due to global warming', *Biological Conservation* 116, 367–377.

Hilbert, D.W., Ostendorf, B., and Hopkins, M.S.: 2001, 'Sensitivity of tropical forests to climate change in the humid tropics of north Queensland', *Austral Ecology* 26, 590–603.

Hoegh-Guldberg, O.: 1999, 'Climate change, coral bleaching and the future of the world's coral reefs', *Marine and Freshwater Research* 50, 839–866.

Hoozemans, F.M.J., Marchand, M., and Pennekamp, H.A.: 1993, '*Sea Level Rise: A Global Vulnerability Assessment-Vulnerability Assessments for Population, Coastal Wetlands and Rice Production on a Global Scale*', 2nd revised edition, Delft Hydraulics and Rijkswaterstaat, Delft and The Hague, The Netherlands, xxxii + 184 pp.

Hughes, L., Cawsey, E.M., and Westoby, M.: 1996, 'Climatic range sizes of Eucalyptus species in relation to future climate change', *Global Ecology and Biogeography Letters* 5, 23–29.

Hulme, M., Mitchell, J., Ingram, W., Lowe, J., Johns, T., New, M., and Viner, D.: 1999, 'Climate change scenarios for global impacts studies', *Global Environmental Change* 9, S3–S19.

Huntingford, C., Jones, R.G., Prudhomme, C., Lamb, R., Gash, J.H.C. and Jones, D.A.: 2003, 'Regional climate-model predictions of extreme rainfall for a changing climate', *Quarterly Journal of the Royal Meteorological Society* 129, 1607–1621 Part A.

Huppop, O., and Huppop, K.: 2003, 'North Atlantic Oscillation and timing of spring migration in birds', *Proc Royal Soc London B* 270, 233–240.

Inouye, D.W., Barr, B., Armitage, K.B., and Inouye, B.D.: 2000, 'Climate change is affecting altitudinal migrants and hibernating species', *PNAS* 97, 1630–1633.

Intergovernmental Panel on Climate Change (IPCC): 2001, '*Climate Change 2001: A synthesis report*', Cambridge University Press, 397 pp.

Ishida, Y., Welch, D.W., and Ogura, M.: 1996, 'Potential influence of North Pacific sea-surface temperature on increased production of chum salmon (*Oncorhynchus keta*) from Japan', in Omasa, K., Kai, K., Taoda, H., Uchijima, Z., and Yoshino, M. (eds.) Climate change and Plants in East Asia. Springer.

Jarvinen, A.: 1989, 'Patterns and causes of long-term variation in reproductive traits of the Pied Flycatcher *Ficedula hypoleuca* in Finnish Lapland', *Ornis Fennica* 66, 24–31.

Jones, P.G., and Thornton, P.K.: 2003, 'The potential impacts of climate change on maize production in Africa and Latin America in 2055', *Global Environmental Change* 13, 51–59.

Kai, K., Kainuma, M., and Murakoshi, N.: 1996, 'Effects of global warming on the phenological observation in Japan', in Omasa, K., Kai, K., Taoda, H., and Uchijima, Z. (eds.) Climate Change and Plants in East Asia. Springer.

Keleher, C.J., and Rahel, F.J.: 1996, 'Thermal limits to salmonid distributions in the Rocky Mountain region and potential habitat loss due to globalwarming: A geographic information system (GIS) approach', *Transactions of the American Fisheries Society* 125, 1–13.

Kerr, J., and Packer, L.: 1998, 'The impact of climate change on mammal diversity in Canada', *Environmental Monitoring and Assessment* 49, 263–270.

Krishna, R.P.F., Lefale, M., Sulliva, E., Young, J., Pilon, C., Shulz, G., and Clarke, G.: 2000, 'Pacific Meteorological Services: meeting the challenges', *South Pacific Regional Environment Programme SPRP*. Apia, Samoa.

Lal, M., Nozawa, T., Emori, S.: 2002, 'Future Climate Change: Implications for India summer monsoon and its variability', *Current Science* 81, 1196–1207.

Lanchbery, J.: 2005, 'Climate change-induced Ecosystem Loss and its Implications for Greenhouse Gas Concentration Stabilisation', this volume.

Langdon, C., Takahisi, T., Sweeney, C., Chipman, D., Goddard, J., Marubini, F., Aceves, H., Barnett, H., and Atkinson, M.J.: 2000, 'Effect of calcium carbonate saturation state on the calcification rate of an experimental coral reef', *Global Biogeochemical Cycles* 14, 639–654.

Leclercq, N., Gattuso, J-P., and Jaubert, J.: 2000, 'CO_2 partial pressure controls the calcification rate of a coral community', *Global Change Biology* 6, 329–334.

Leemans, R., and Eickhout, B.: 2003, '*Analysing ecosystems for different levels of climate change*', Report to OECD Working Party on Global and Structural Policies ENV/EPOC/GSP 5/FINAL.OECD.

Lopez-Moreno, J.I.: 2005, 'Recent Variations of Snowpack Depth in the Central Spanish Pyrenees', *Arctic Antarctic and Alpine Research* 37, 253–260.

Leemans, R., and Eickhout, B.: 2004, 'Another reason for concern: regional and local impacts on ecosystems for different levels of climate change', *Global Environmental Change* 14, 219–228.

Ludwichowski, I.: 1997, 'Long-term changes of wing-length, body mass and breeding parameters in first-time breeding females of goldeneyes (*Bucephala clangula clangula*) in northern Germany', Vogelwarte 39, 103–116.

Mackas, D.L., Goldblatt, R., and Lewis, A.G.: 1998, 'Interdecadal variation in developmental timing of *Neocalanus plumchrus* populations at Ocean Station P in the subarctic North Pacific', *Canadian Journal of Fisheries and Aquatic Science* 55, 1878–1893.

Magnuson, J.J., Robertson, D.M., Benson, B.J., Wynne, R.H., Livingstone, D.M., Arai, T., Assel, R.A., Barry, R.G., Card, V., Kuusisto, E., Granin, N.G., Prowse, T.D., Stewart, K.M., and Vuglinski, V.S.: 2000, 'Historical trends in lake and river ice cover in the Northern Hemisphere', *Science* 289, 1743–1746.

Malcolm, J.R., Markham, A., Neilson, R.P., and Garaci, M.: 2002, 'Estimated migration rates under scenarios of global climate change', *Journal of Biogeography* 29, 835–849.

Martin, P.H., and Lefevre, M.G.: 1995, 'Malaria and Climate: Sensitivity of malaria potential transmission to climate', *Ambio* 24, 200–207.

McMichael, A.J., Campbell-Lendrum, D., Kovats, R.S., Edwards, S., Wilkinson, P., Edmonds, N., Nicholls, N., Hales, S., Tanser, F.C., Le Sueur, D., Schlesinger, M., and Andronova, N.: 2004, '*Climate Change*', In *Disease due to Selected Major Risk Factors. Vol 2; Comparative Quantification of Health Risks: Global and Regional Burden* Ezzaoti, M., Lopez, A.D., Rogers, A., Murray, C.J. (eds) WHO, Geneva; Ch 20.

Meehl, G.A., and Tebaldi, C.: 2004, 'More intense, more frequent and longer lasting heat waves in the 21st century', *Science*, 305, 994–7.

Mendelsohn, R., Nordhaus, W., and Shaw, D.: 1996: 'Climate Impacts on Aggregate Farm Value: Accounting for Adaptation', *Agricultural and Forest Meteorology*, 80, 55–66.

Mickelson, M.J., Dann, P., and Cullen, J.M.: 1992, 'Sea temperature in Bass Straits and breeding success of the Little Penguin *Eudyptula minor* at Philip Island, SE Australia', *Emu* 91, 355–368.

Midgley, G.F., Hannah, L., Millar, D., Rutherford, M.C., and Powrie, L.W.: 2002, 'Assessing the vulnerability of species richness to anthropogenic climate change in a biodiversity hotspot', *Global Ecology and Biogeography* 11, 445–452.

Minin, A.A.: 1992, 'Spatio temporal variability of starting dates of some phenological phenomena in birds on the Russian plain', *Byulleten' Moskovskogo Obshchestva Ispytatelei Prirody Otdel Biologicheskii* 97, 28–34.

Mote, P., 'Observed Hydrologic Consequences of Climate Change in Western North America', *Avoiding Dangerous Climate Change Conference*, Exeter, UK, www.stabilisation2005.com

Mouillot, F., Rambal, S., and Joffre, R.: 2002, 'Simulating climate change impacts on fire frequency and vegetation dynamics in a Mediterranean-type ecosystem', *Global Change Biology* 8, 423–437.

Najjar, R.G., Walker, H.A., Anderson, P.J., Barron, E.J., Bord, R.J., Gibson, J.R., Kennedy, V.S., Knight, C.G., Megonigal, J.P., O'Connor, R.E., Polsky, C.D., Psuty, N.P., Richards, B.A., Sorenson, L.G., Steele, E.M., and Swanson, R.S.: 2000, 'The potential impacts of climate change on the mid-Atlantic coastal region', *Climate Research* 14, 219–233.

Navarra, A., 'Impact of Increased CO_2 Levels on Interannual Tropical Variability', *Avoiding Dangerous Climate Change Conference*, Exeter, UK, www.stabilisation2005.com

Neilson, R.P.: 1993, 'Vegetation redistribution: a possible biosphere source of CO_2 during climatic change', *Water, Air and Soil Pollution*, 70, 659–673.

Neilson, R.P., Prentice, I.C., Smith, B., Kittel, T., and Viner, D.: 1997, 'Simulated changes in vegetation distribution under global warming', in Dokken, D.J. (eds.), *The Regional Impactions of Climate Change. An Assessment of Vulnerability*, Cambridge University Press, New York, p. 439.

Ni, J.: 2000, 'A simulation of biomes on the Tibetan Plateau and their responses to global climate change', *Mountain Research and Development* 20, 80–89.

Ni, J.: 2001, 'Carbon Storage in Terrestrial Ecosystems of China: Estimates at Different Spatial Resolutions and Their Responses to Climate Change', *Climatic Change* 49, 339–358.

Nicholls, R.J., Hoozemans, F.M.J., and Marchand, M.: 1999, 'Increasing flood risk and wetland losses due to global sea-level rise: regional and global analyses', *Global Environmental Change-Human and Policy Dimensions* 9, S69–S87.

Nicholls, R.J.: 2004, 'Coastal flooding and wetland loss in the 21st century: Changes under the SRES climate and socio-economic scenarios', *Global Environmental Change* 14, 69–86.

Oglesby, T.T., and Smith, C.R.: 1995, '*Climate change in the Northeast*', in LaRoe, E.T., Farris, G.S., Puckett, C.E., Doran, P.D., and Mac, M.J. (eds) *Living Resources: A Report to the Nation on the Distribution, Abundance and Health of U.S. Plant, Animals and Ecosystems*. DOI, National Biological Service, Washington D.C.

O'Neill, B.C., and Oppenheimer, M.: 2002, 'Dangerous climate impacts and the Kyoto Protocol', *Science* 296, 1971–1972.

Oppenheimer, M., and Alley, R.B.: 2004, 'The West Antarctic Ice Sheet and long term climate policy', *Climatic Change* 64, 1–10.

Parmesan, C., and Yohe, G.: 2003, 'A globally coherent fingerprint of climate change impacts across natural systems', *Nature* 421, 37–42.

Parry, M., Arnell, N., McMichael, T., Nicholls, R., Martens, P., Kovats, S., Livermore, M., Rosenzweig, C., Iglesias, A., and Fischer, G.: 2001, 'Millions at risk: defining critical climate change threats and targets', *Global Environmental Change* 11, 181–183.

Parry, M., Rosenzweig, C., Iglesias, A., Fischer, G., and Livermore, M.: 1999, 'Climate change and world food security: a new assessment', *Global Environmental Change* 9, S51–S68.

Parry, M.L., Rosenzweig, C., Iglesias, A., Livermore, M., and Fischer, G.: 2004, 'Effects of climate change on global food production under SRES emissions and socio-economic scenarios', *Global Environmental Change* 14, 53–67.

Peck, L.S., Webb, K.E., and Bailey, D.M.: 2004, 'Extreme sensitivity of biological function to temperature in Antarctic marine species', *Functional Ecology* 18, 625–630.

Penuelas, J., Filella, I., and Comas, P.: 2002, 'Changed plant and animal life cycles from 1952 to 2000 in the Mediterranean region', *Global Change Biology* 8, 531–544.

Peterson, A.T., Ortega-Huerta, M.A., Bartley, J., Sanchez-Cordero, V., Soberon, J., Buddemeier, R.H., and Stockwell, D.R.B.: 2002, 'Future projections for Mexican faunas under climate change scenarios', *Nature* 416, 626–629.

Pouliquen-Young, O. and Newman, P.: 1999, '*The Implications of Climate Change for Land-Based Nature Conservation Strategies.*' Perth, Australia, Australian Greenhouse Office, Environment Australia, Canberra, and Institute for Sustainability and Technology Policy, Murdoch University: 91.Final Report 96/1306.

Pounds, J.A., Fogden, M.P.L., and Campbell, J.H.: 1999 'Biological response to climate change on a tropical mountain', *Nature* 398, 611–615.

Price, J.: 2002, 'Climate change, birds and ecosystems – why should we care?' in Rapport, D.J., W.L. Lasley, D.E. Rolston, N.O. Nielsen, C.O. Qualset and A.B. Damania (eds). *Managing for Healthy Ecosystems*. Lewis Publishers, Boca Raton, Florida, pp. 465–469.

Qureshi, A., and Hobbie, D.: 1994, 'Climate change in Asia'. Manila, Asian Development Bank. Cited by World Bank 2000. Chapter 2 *Potential Impacts of climate change in Bangladesh.*

Ramankutty, N., Foley, J.A., Norman, J., and McSweeney, K.: 2002, 'The global distribution of cultivable lands: current patterns and sensitivity to possible climate change', *Global Ecology and Biogeography* 11, 377.

Richardson, A.J., and Schoeman, D.S.: 2004, 'Climate impact of plankton ecosystems in the Northeast Atlantic', *Science* 305, 1609–1612.

Riebesell, U., Zonderban, I., Rost, B., Tortell, P.D., Zeebe, R.E., and Morel, F.M.M.: 2000, 'Reduced calcification of marine plankton in response to increased atmospheric CO_2', *Nature* 407, 364–7.

Rignot, E., and Jacobs, S.S.: 2002, 'Rapid Bottom-melting widespread near Antarctic ice-sheet grounding lines', *Science* 2020–2023.

Rignot, E., Casassa, G., Gogineni, P., Krabill, W., Rivera, A., and Thomas, R.: 2004, 'Accelerated ice discharge from the Antarctic Peninsula following collapse of the Larsen B ice shelf', *Geophysical research letters* 31, L18401.

Rogers, D.J., and Randolph, S.E.: 'The global spread of malaria in a future, warmer world', *Science* 289, 1763–1766.

Root, T.L., Price, J.T., Schneider, S.H., Rosenzweig, C., and Pounds, J.A.: 2003, 'Fingerprints of global warming on wild animals and plants', *Nature* 421, 57–60.

Root, T.L., MacMynowski, D.P., Mastrandrea, M.D., and Schneider, S.H.: 2005, 'Human modified temperatures induce species changes: Joint attribution', *PNAS* 102, 7465–7469.

Rosegrant, M.W., and Cline, S.A.: 2003, 'Global Food Security: Challenges & Policies', *Science* 302, 1917–1922.

Rosenzweig, C., Parry, M., and Fischer, G.: 1995, 'World food supply', in *As climate changes: International Impacts and Implications*, Strzepek, K.M., and Smith, J.B. (eds), Cambridge University Press, Cambridge, UK, pp. 27–56.

Royal Society: 2005, 'Food Crops in a Changing Climate: Report of a Royal Society Discussion held in April 2005', *Policy Document* 10/05, June 2005. ISBN 0 85403 615 6 (available at www.royal-soc.ac.uk).

Rutherford, M.C., Midgley, G.F., Bond, W.J., Powrie, L.W., Musil, C.F., Roberts, R., and Allsopp, J.: 1999, '*South African Country Study on Climate Change*'. Pretoria, South Africa, Terrestrial Plant Diversity Section, Vulnerability and Adaptation, Department of Environmental Affairs and Tourism.

Schiegg, K., Pasinelli, G., Walters, J.R., and Daniels, S.J.: 2002, 'Inbreeding and experience affect response to climate change by endangered woodpeckers', *Proc. Royal Soc. London B* 269, 1153–1159.

Schlesinger, M.E., Yin, E.J., Yohe, G., Andronova, N.G., Malyshev, S. and Li, B.: 2005, 'Assessing the Risk of a Collapse of the Atlantic Thermohaline Circulation', this volume.

Schneider, S.: 1997, 'Integrated assessment modelling of global climate change: transparent rational tool for policy making or opaque screen for hiding value-laden assumptions?' *Environmental Modelling & Assessment* 2, 229–249.

Schlenker, W., Hanemann, M., and Fisher, A.: 2004, '*Will US Agriculture Really Benefit from Global Warming? Accounting for Irrigation in the Hedonic Approach*', CUDARE working paper, University of California, Berkeley. Paper 941 http://repositories.cdlib.org/are_ucb/941.

Sheppard, C.R.C.: 2003, 'Predicted recurrences of mass coral mortality in the Indian Ocean', *Nature* 425, 294–297.

Smith, J.B., Rahman, A., Haq, S., and Mirza, M.Q.: 1998, '*Considering Adaptation to Climate change in the sustainable development of Bangladesh. World Bank Report*'. Washington DC, World Bank.

Smith, R.C.: 1999, 'Marine ecosystem sensitivity to historical climate change in the Antarctic Peninsula', *Bioscience* 49, 393–404.

Smith, R.I.L.: 1994, 'Vascular plants as bioindicators of regional warming in Antarctica', *Oecologica* 99, 322–328.

Sohngen, B., Mendelsohn, R., and Sedjo, R.: 2001, 'A global model of climate change impacts on timber markets', *Journal of Agricultural and Resource Economics* 26, 326–343.

Sorenson, L.G., Goldberg, R., Root, T.L. and Anderson, M.G.: 1998, 'Potential effects of global warming on waterfowl populations breeding in the Northern Great Plains', *Climatic Change* 40, 343–369.

Southward, A.J., Hawkins, S.J., and Burrows, M.T.: 1995, 'Seventy years' observations of changes in distribution and abundance of zooplankton and intertidal organisms in the western English Channel in relation to rising sea temperature', *Journal of Thermal Biology* 20, 127–155.

Sparks, T.H., and Braslavska, O.:2001, 'The effects of temperature, altitude and latitude on the arrival and departure dates of the swallow *Hirundo rustica* in the Slovak Republic', *International Journal of Biometeorology* 45, 212–216.

Sparks, T.H., and Yates, T.J.: 1997, 'The effect of spring temperature on the appearance dates of British butterflies 1883–1993', *Ecogeography* 20, 368–374.

Stainforth, D., Aina, T., Christensen, C., Collins, M., Faull, N., Frame, D.J., Kettleborough, J.A., Knight, S., Martin, A., Murphy, J.M., Piani, C., Sexton, D., Smith, L.A., Spicer, R.A., Thorpe, A.J., and Allen, M.R.: 2005, 'Uncertainty in predictions of the climate response to rising levels of greenhouse gases', *Nature*, 433, 403–406.

Still, C.J., Foster, P.N., and Schneider, S.H.: 1999 'Simulating the effects of climate change on tropical montane cloud forests', *Nature* 398, 608–10.

Stocker, T.F., and Schmittner, A.: 1997 'Influence of CO_2 emission rates on the stability of the thermohaline circulation', *Nature* 388, pp. 862–865.

Street, R.B., and Melnikov, P.I.: 1990, 'Seasonal snow cover, ice and permafrost', in *Climate Change: The IPCC Impacts Assessment*. (eds. Izrael, Y.A., Hashimoto, M. and McTegart, W.J.), Australian Government Publishing Service, Canberra. pp. 1–46.

Stott, P.A., Stone, D.A., and Allen, M.R.: 2004 'Human contribution to the European heatwave of 2003', *Nature* 432, 610–614.

Sydeman, W.J, Hester, M.M., Thayer, J.A., Gress, F., Martin, P., and Buffa, J.: 2001, 'Climate change, reproductive performance and diet composition of marine birds in the southern California Current system, 1969–1997', *Progress in Oceanography* 49, 309–329.

Tanser, F.C., Sharp, B., and Le Sueur, D.: 2003, 'Potential effect of climate change on malaria transmission in Africa', *Lancet* 362, 1792–1798.

Thomas, C.J., Davies, G., and Dunn, C.E.: 2004, 'Mixed picture for changes in stable malaria distribution with future climate in Africa', *Trends in Parasitology* 20, 216–220.

Thomas, D.S.G., Knight, M., and Wiggs, G.S.F.: 2005, 'Remobilization of southern African desert dune systems by twenty-first century global warming', *Nature* 435, 1218–1222.

Thomas, R., Rignot, E., Cassase, G., Kanagaratnam, P., Acuna, C., Akins, T., Brecher, H., Frederick, E., Gogineni, P., Krabill, W., Manizade, S., Ramamoorthy, H., Rivera, A., Russell, R., Sonntag, J., Swift, R., Yungel, J., and Zwally, J.: 2004a, 'Accelerated Sealevel Rise from West Antarctica', *Science* 306, 255–258.

Thomas, C.D., Cameron, A., Green, R.E., Bakkenes, M., Beaumont, L.J., Collingham, Y.C., Erasmus, B.F.N., de Siqueira, M.F., Grainger, A., Hannah, L., Hughes, L., Huntley, B., van Jaarsveld, A.S., Midgley, G.F., Miles, L., Ortega-Huerta, M.A., Peterson, A.T., Phillips, O.L., and Williams, S.E.: 2004b, 'Extinction risk from climate change', *Nature* 427, 145–148.

Thomas, C.D., Williams, S.E., Cameron, A., Green, R.E., Bakkenes, M., Beaumont, L.J., Collingham, Y.C., Erasmus, B.F.N., de Siqueira, M.F., Grainger, A., Hannah, L., Hughes, L., Huntley, B., van Jaarsveld, A.S., Midgley, G.F., Miles, L., Ortega-Huerta, M.A., Peterson, A.T., and Phillips, O.L.: 2004c, 'Biodiversity conservation: Uncertainty in predictions of extinction risk/Effects of changes in climate and land use/Climate change and extinction risk (reply)', *Nature* 430, 37.

Thompson, L.G., Mosley-Thompson, E., Davis, M.E., Henderson, K.A., Brecher, H.H., Zagorodnov, V.S., Mashiotta, T.A., Lin, P.-N., Mikhalenko, V.N., Hardy, D.R., and Beer, J.: 2002, 'Kilimanjaro ice core records: evidence of Holocene climate change in tropical Africa', *Science* 298, 589–593.

Thuiller, W., Araujo, M.B., Pearson, R.G., Whittaker, R.J., Brotons, L., and Lavorel, S.: 2004, 'Biodiversity conservation: uncertainty in predictions of extinction risk', *Nature* 430, 34.

Tol, R., 2002: 'Estimates of the damage costs of climate change: part I: Benchmark estimates', *Environmental and Resource Economics* 21, 47–73 and part II: Dynamic estimates 21, 135–160.

Tol, R.S.J., Downing, T.E., Kuik, O.J., and Smith, J.B.: 2004, 'Distributional aspects of climate change impacts', *Global Environmental Change* 3, 259–272.

Toth, F.L., Bruckner, T., Fussel, H.-M., Leimbach, M., and Petschel-Held, G.: 2003a, 'Integrated assessment of long-term climate policies: part I – Model Presentation', *Climatic Change* 56, 37–56.

Toth, F.L., Bruckner, T., Fussel, H.-M., Leimbach, M., and Petschel-Held, G.: 2003a, Integrated assessment of long-term climate policies: part II – Model Results and Uncertainty Analysis', *Climatic Change* 56, 57–72.

Trotz, U.O.: 2002 *'Disaster Reduction and Adaptation to Climate Change – A CARICOM Experience'*, Havana, Cuba.

Tubiello, F.N., Rosenzweig, C., Goldberg, R.A., Jagtap, S., and Jones, J.W.: 2002, 'Effects of climate change on U.S. crop production: simulation results using two different GCM scenarios. Part I: Wheat, potato, maize and citrus', *Climate Research* 20, 259–270.

Turley, C. and Blackford, J.: 2005, 'Reviewing the impact of Increased Atmospheric CO_2 on Oceanic pH and the Marine Ecosystem', this volume.

Van Lieshout, M., Kovats, R.S., Livermore, M.T., Martens, P.: 2004, 'Climate change and malaria: Analysis of the SRES climate and socio-economic scenarios', *Global Environmental Change* 14, 87–99.

van Vliet, A. and Leemans, R.: 2005, 'Rapid species' responses to changes in climate require stringent climate protection targets', this volume.

Vaughan, D.G., Marshall, G.J., Connolley, W.M., Parkinson, C., Mulvaney, R., Hodgson, D.A., King, J.C., Pudsey, C.J., and Turner, J.: 2003, 'Recent rapid regional climate warming on the Antarctic Peninsula', *Climatic Change* 60, pp. 243–274.

Vellinga, P. and Swart, R.J.: 1991, 'The greenhouse marathon: A proposal for a global strategy', *Climatic Change*, 18, 7–12.

Villers-Ruiz, L., and Trejo-Vasquez, I.: 1998, 'Impacts of climate-change on Mexican forests and natural protected areas', *Global Environmental Change* 8, 141–157.

Viner, D., 'The Implications of Greenhouse Gas Stabilisation for International Tourism Flows', *Avoiding Dangerous Climate Change Conferenc*e, Exeter, UK, www.stabilisation2005.com

Vorosmarty, C.J., Green, P., Salisbury, J., and Lammers, R.B.: 2000, 'Global water resources: vulnerability from climate change and population growth', *Science* 289, 284–288.

Wardle, P., and Coleman, M.C.: 1991, 'Evidence for rising upper limits of four native New Zealand forest trees', *New Zealand Journal of Ecology* 15, 13–152.

World Health Organisation (WHO), 2004, *'Heat Waves: Risks and Responses'*, Health and Global Environmental Change Series, No. 2, WHO Regional Office for Europe, Denmark.

Williams, S.E., Bolitho, E.E., and Fox, S.: 2003, 'Climate change in Australian tropical rainforests: an impending environmental catastrophe', *Proc. Roy. Soc. London B*, 270, 1887–1892.

Winkel, W. and Hudde, H.:1997, 'Long-term trends in reproductive traits of tits (*Parus major, P. caeruleus*) and pied flycatchers (*Ficedula hypoleuca)'*, *Journal of Avian Biology* 28, 187–190.

Yoshino, M., and Ono, H.P.: 1996, 'Variations in plant phenology affected by global warming', in Omasa, K., Kai, K., Toado, H., Uchijima, Z., and Yoshino, M. (eds) *Climate Change and Plants in East Asia*. Springer-Verlag.

Zickfeld, K., Knopf, B., Petoukhov, V., and Schellnhuber, H.J.: 2005, 'Is the Indian summer monsoon stable against global change?', *Geophysical Research Letters* 32, L15707, 5 pp.

SECTION III

Key Vulnerabilities for Ecosystems and Biodiversity

INTRODUCTION

This section considers impacts of recent climate change on the carbon cycle and ecosystems. The literature on numerous observed changes in ecosystems contains overwhelming evidence for their attribution to recent climate change – although rates and processes differ, depending on the nature of the organisms involved. Feedbacks from changes in vegetation and soils to the carbon cycle and climate change are now increasingly better understood, and the papers demonstrate both the importance of tropical forests in this context and recent advances in the assessment of the possible saturation of the land biosphere carbon sink.

Van Vliet and Leemans note first that the number of studies published in the literature now provides substantial evidence of ecosystems changes caused by recent climate change; while only 21 papers were available to the IPCC Third Assessment Report (TAR) there now are over 1000. They emphasise that studies focusing on species-specific responses provide higher sensitivity in depicting impacts than earlier impact assessments focusing on shifts of entire biomes. The paper includes a summary of widespread and immediate phenological changes, species-range shifts and food-web responses. This literature covers insects, birds, pathogens, lichens and trees, all affected by climate change. They also note that many ecosystems respond more strongly to changes in extreme weather events than to average climate. Their concluding recommendation is that, in order to avoid significant ecosystem damage, climate change should be limited to 1.5°C above pre-industrial levels with a rate of less than 0.5°C per century.

Lanchbery argues that, on the basis of ecological effects and the observed inability of some natural ecosystems to adapt, atmospheric concentrations of greenhouse gases can be considered to be already too high. He points out alterations to species ranges, ecosystem loss and the unpredictability of subsequent impacts arising from changes in

one key species. He then introduces recent work in the North Sea on seabird populations, and notes that climate impacts on plankton abundance may have resulted in a substantial reduction in sandeel numbers – a key feed species for many seabirds. This shortage has been independently indicated by Danish sandeel fisheries where 2003/4 catches were half the typical catch. In his conclusion, Lanchbery shows that achievement of a stabilisation target of 2°C above pre-industrial levels clearly implies heavy damage for many species and ecosystems, but that higher levels of warming would lead to much greater damage.

Lewis et al. discuss the role of tropical forests in the global carbon cycle. They show on the basis of observations (particularly permanent plot studies), how the remaining forests currently act as an important sink of about 1.2 Pg C a^{-1}, while ongoing deforestation is a very important source or more than 2 Pg C a^{-1}. They then demonstrate that the remaining forests are unlikely to retain their sink strength. They cite a number of processes that could turn these forests into a source, mainly due to changing physiological or other functional conditions under high CO_2, but also due to increasing drought or fire. These changes could rapidly amplify current CO_2 concentrations and hence climate change.

Cox et al. present an analysis of the possible transition from carbon sink to carbon source in the terrestrial biosphere. They note that carbon cycle feedbacks have been an important consideration in developing the newest generations of GCMs which now include the key processes of photosynthesis, respiration and vegetation dynamics, as well as their responses to changes in CO_2 and climate. There is still uncertainty in the relevant parameters, but there is a significant probability of shift from carbon sink to source in the terrestrial environment before the year 2100 under business as usual emissions scenarios. Beyond this, they consider the question of whether the critical positive feedbacks might reach a level where 'runaway conditions' would appear. This instability is found to be unlikely to occur within a foreseeable future.

CHAPTER 12

Rapid Species' Responses to Changes in Climate Require Stringent Climate Protection Targets

Arnold van Vliet & Rik Leemans
Environmental Systems Analysis Group, Wageningen UR, The Netherlands

ABSTRACT: Widespread ecological impacts of climate change are visible in most ecosystems. Plants and animals respond immediately to the ongoing changes. Responses significantly differ from species to species and from year to year. Traditional impact studies that focus on average climate change at the end of this century and long-term range shifts of biomes, correctly estimate the direction of these ongoing changes but not the magnitude. More recent studies using species and population specific models show more widespread impacts but also do not reproduce the full extent of observed changes. Impacts and vulnerability assessment therefore likely underestimate responses, especially at the lower levels of climate change. Over the last decades extreme weather has changed more markedly than average weather and ecosystems have responded more rapidly to this more complex set of changes than the average climate change in most climate scenarios. This can explain the unexpected rapid appearance of ecological responses throughout the world.

Tighter political climate protection targets are therefore needed to cope with the greater vulnerability of species and ecosystems. Based on current understanding of the response of species and ecosystems, and extreme weather events, we propose that efforts be made to limit climate change to maximally 1.5°C above pre-industrial levels and limit the rate of change to less than 0.5°C per century.

12.1 Introduction

The history of the Earth's climate has been characterized by many changes. But the extent and the rate of current climate change now exceeds most natural variation. Most of this climate change is attributable to human activities, in particular to the increase in the atmospheric concentrations of greenhouse gases. IPCC [1] concluded that 'an increasing body of observations gives a collective picture of a warming world and other changes in the climate system.' Climate change already has resulted in considerable impacts on species and ecosystems, human health and society [2–6].

As a response to the threats posed by these climate change impacts, the United Nations Framework Convention on Climate Change (UN-FCCC) was established. Its objective is to realize stabilization of greenhouse gas concentrations at a level that would prevent dangerous anthropogenic interference with the climate system. Such a level should be achieved, among others, within a time frame sufficient to allow ecosystems to adapt naturally to climate change (i.e. Article 2, the objective of UN-FCCC). Although some UN-FCCC members proposed clear climate protection targets, these were never seriously discussed within the UN-FCCC. Europe, for example, aims to limit climate change to 2°C, while the Alliance of Island States insisted on a maximum sea-level rise target of 30 cm. IPCC clearly demonstrated that a global mean increase in average surface temperature of more than 1 to 2°C leads to rapidly increasing risks for adverse impacts on ecosystems [the 'Reasons for Concern' or 'Burning Embers' diagram in 7]. In its own assessment, the UN Convention on Biological Diversity (UN-CBD) reviewed IPCC's evidence [8, 9]. They concluded that a climate change beyond 2°C was unacceptable for ecosystems and biodiversity. This was recently reaffirmed by the Millennium Ecosystem Assessment [10].

Responses of ecosystem represent complex phenomena that generally have multiple causal agents. While many trends in impacts are consistent with climate change trends, a statistically rigorous attribution of impacts to climate change is often impossible because long-term observations on weather and climate and impacts are rarely collected simultaneously. Observation of a specific response seems anecdotal but all responses put together start to corroborate clearer proof. The analysis and mapping of the few studies available to IPCC [i.e. IPCC's global map of observed responses by 7] led to the conclusion that 'recent regional climate changes, particularly temperature increases, have already affected many physical and biological systems'. Over the last few years, reports on observed impacts on climate change have increased enormously. Recently, Lovejoy and Hannah [6] evaluated the observed responses of many species and stated that 80% of these changes could be explained by climatic change.

In this paper we present additional examples of observed ecological responses to climate change. We focus on the Netherlands because long-term trends in many ecological monitoring networks for plants, amphibians and reptiles, birds, lichens, insects, spiders, etc. are available. These trends were recently analyzed [e.g. 11, 12–18] and compiled in a popular publication by Roos et al. [19]. Additional examples are added to illustrate that these Dutch responses are not exceptional. For example, Parmesan and Galbraith [20], Root et al. [21] and Lovejoy and Hannah [6] provide similar compilations for North America. These examples are not, however, intended to be exhaustive. Then we compare these responses with expected changes derived from traditional impacts assessments based on models and scenarios. One of the problems with such a comparison is that these impact assessments apply large climate changes (more than 2°C warming), while the observed responses result from a less than a 1°C warming. Another problem is that impact assessments aggregate ecosystems into coarse units, while the observed responses show that each species display unique responses locally. Despite these limitations, we comment on the disagreements and discuss the consequences for defining climate protection targets by policy makers.

12.2 Observed Changes in Climate

Reconstructed temperatures over the last 1000 years indicate that the 20th century climate change is the largest and exceeds by far all natural climate variations during this period [22, 23]. In addition, direct measurements show that the 1990s are the warmest decade of the century. This rapid warming has continued during the first years of the 21st century. The increase in global temperatures has resulted mainly from an accompanying smaller increase in the frequency of much above normal temperatures. Klein Tank [24] recently analyzed European patterns of climate change and concluded: 'Although there have been obvious changes in the mean climate, most of the observed ongoing climate change can be attributed to changes in the extremes'. His analysis showed statistically significant and non-trivial changes in extremes: fewer cold extremes, more heat waves, smaller diurnal and seasonal ranges, more precipitation that come mostly in intense showers. He further concluded that larger extremes should be expected in the future, often aggravated by systematic interactions. Such an effect is illustrated by the exceptionally hot summer in Europe in 2003. These high temperatures were caused by a lack of soil moisture and evaporation, which amplified the warming [24].

12.3 Impacts of the Observed Climate Change

The first signs that such climate change caused obvious changes in ecosystems comes from high latitudes and alpine systems. Anisimov [25] was among the first to analyze long-term data for Russia and Siberia and concluded that permafrost was thawing. Such melting actually began in the middle of the 19th century and climate change has accelerated more over the last decades in all Polar Regions than in any other region of the world. The Arctic Climate Impact Assessment [26] provided well-documented evidence of all these changes in permafrost, ice thickness and ice cover and the subsequent negative impacts on polar ecosystems. Similar trends are reported from Antarctica [e.g. 27].

Glaciers are also retreating almost everywhere in the world. The last ice of the glacier on Mount Kilimanjaro, for example, will likely melt before 2020 [28]. This threatens unique alpine ecosystems, local biodiversity and runoff volumes. Similar trends are observed for most other glaciers [29]. The accelerated melting of glaciers, permafrost, ice and snow cover will alter the hydrology of many rivers. Water availability downstream could be threatened and adversely impact the livelihoods of many people [10].

Climatic change has also increased the length and intensity of summer drought in many regions. This has increased the susceptibility of ecosystems to fires. Over the last decade fire frequencies increased in many regions. For example, fires burned up to 810,000 hectares of rainforest land in Indonesia [30], including almost 100,000 hectares of primary forest and parts of the already severely reduced habitat of the Kalimantan Orang Utan.

Since the seventies, satellites have been used to monitor changes in the environment. Myneni et al. [31] analyzed such data to detect a climate change over land in the Northern hemisphere. From their data for 1981 to 1991 they found surprisingly large changes over many regions. They detected an earlier greening of vegetation in spring of up to ten days and a later decline of a few days in autumn. These changes indicate a longer growing season to which vegetation growth and phenology immediately responds [32]. Such phenomena have also been observed elsewhere [e.g. 33, 34].

One of the most obvious early indicators of ecological impacts is therefore phenological change. Phenology deals with the times of annual recurring natural events like flowering, leaf unfolding, fruit ripening, leaf coloring and fall, migration, and spawning, and can be observed by easy means everywhere. Many phenological networks that monitor the timing of life cycle events have been established [35]. The records go back hundreds of years and most are still expanding. These networks now help us to assess long-term changes. In the Netherlands, for example, systematic phenological observations were made from 1869 till 1968. In 2001 this Dutch network was successfully revived under the name 'Nature's Calendar' (http://www.natuurkalender.nl). Since then, thousands of volunteer observers have submitted their own phenological observations on plants, birds and insects. Many species groups have showed significant changes in the timing of their own life cycle events [c.f. Figure 12.1 and 36, 37–39].

Other studies highlight the intricate linkages between species. The long-term observations made on the Pied flycatcher [13, 40], for example, revealed that although

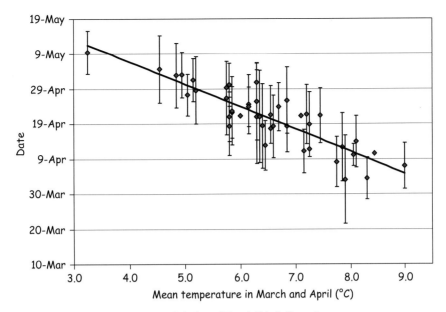

Figure 12.1 Relation between spring temperature and timing of Dutch Birch flowering.

the Pied flycatcher advanced its egg laying date by seven days, the main food source for their young, caterpillars of the Winter moth, appear 14 days earlier than they did in the past. Timing mismatches develop, which rapidly reduces the breeding success of the Pied flycatcher. With the complexity of food webs in natural systems, it is highly likely that many more problems will emerge.

The global distributions of plants and animals are primarily limited by climate and locally, mainly by soil properties, topography and land use. The climate change indicator report of the European Environmental Agency (EEA) [41] concludes that over the past decades a northward extension of many plant species has been observed in Europe. In Western Europe, warmth-demanding plant species have become more abundant compared with 30 years ago [e.g. 15]. Despite the increase in abundance of warmth-demanding plants, a remarkably small decline in the presence of traditionally cold-tolerant species is observed. The location of tree lines and growth has also recently changed [e.g. 42, 43, 44].

Endemic species have been replaced by more general species in many mountain regions due to a number of factors, including climate change [34]. Higher temperatures and longer growing seasons appear to have created suitable conditions for plant species that have migrated upward and which now compete with endemic species. It is expected that species with a high migration capacity have the ability to quickly change their geographic distribution. Recent changes in the Dutch lichen flora provide such an example [14]. Since the end of the 1980s Mediterranean and tropical species have been increasing. Lichen species with a boreo-montane distribution are decreasing.

Increasing evidence also indicates whole food webs in marine systems are undergoing major changes [45–47]. Some zooplankton species have shown a northward shift

of up to 1000 km. These shifts have taken place southwest of the British Isles since the early 1980s and, from the mid 1980s, in the North Sea. The diversity of colder temperate, sub-Arctic and Arctic species has decreased. Furthermore, a northward extension of the ranges of many warm-water fish species in the same region has occurred. Most of the warm-temperate and temperate species have migrated northward by about 250 km per decade, which is much faster than the migration rates expected in terrestrial ecosystems.

Coral reefs are the most diverse marine ecosystem. Mass coral bleaching and mortality has affected the world's coral reefs with increasing frequency and intensity since the late 1970s. Mass bleaching events are triggered by small increases (+1 to 3°C above mean maximum) in water temperature [e.g. 48, 49]. The loss of living coral cover (e.g. 16% globally in 1998, an exceptionally warm year) is resulting in an as yet unspecified reduction in the abundance of a myriad other species.

Insects also have the ability to quickly respond to climate change. This is illustrated by the rapid recent northward expansion of the mountain pine beetle in Canada. Data from the Canadian Forestry Center shows a large increase in the number of infestations occurring in areas that were historically climatically unsuitable [50]. The mountain pine beetle population has doubled annually in the last several years, causing mortality of pine trees across two million hectares of forest in British Colombia in 2002 alone. These large-scale pest infestations have large economic impacts. Another range change that is becoming a societal problem is the northward expansion of the oak processionary caterpillar in the Netherlands [12]. After the first observations in 1991 in the southern part of the Netherlands, it advanced its distribution range to the mid-Netherlands. This southern European species requires warm conditions. The caterpillars are a concern to human health because of the many stinging hairs that can cause rashes in skin and bronchial tubes.

All the above examples show that recent changes in climate have caused significant ecological impacts everywhere in the world. The changes observed should be seen in the context of a global climate change, expressed as a mean average temperature increase in temperature of approximately 0.5°C [1].

12.4 Are these Responses Consistent with Expected Changes?

The ecological impacts of climate change are now observed in many places and many of those changes were not anticipated. The question that immediately arises is 'Were these changes expected to happen so fast and with such a magnitude?' To address this question we evaluate how future impacts of climate change have been determined.

Most of the traditional impacts assessments have used two components. First, scenarios for a gradually changing average climate were produced [based on the outcome of climate models 51]. Second, these scenarios were applied to drive models that simulate possible responses. Applying this approach is straightforward and potential impacts of different systems are established [see for example: 52]. Most of these impact assessments are done for doubled CO_2 conditions or even larger levels of climate change (i.e. more than 2°C in global mean warming). Most studies ignored transient responses (and thus the rate of change) and only indicated potential final responses. However, despite these obvious limitations, the majority of impacts assessments during the last two decades used this static approach. Emanuel et al. [53], for example, were among the first to use this approach. They showed that climate change would have large impacts on the distribution of ecosystems and concluded that 35% of all the world's ecosystems would change under a doubled-CO_2 climate. Their pioneering result can still be compared favourably with recent studies based on more advanced models [e.g. 54]. Of course, the more recent studies have added more spatial detail, used dynamic models, more realistic species and ecosystem responses but the magnitude of impacts has not changed much.

Nowadays transient climate-change scenarios are more commonly used. These studies generally show little response during the first few decades, then an accelerated response, followed by a levelling off after a century. Still, the simulated impacts replicate those of the equilibrium approaches. Leemans and Eickhout [55] used a simple transient scenario approach to calculate whether vegetation can adapt to the simulated changes over a century. For a 0.5°C warming 5% of the terrestrial vegetation changed. This increased to 10% and 22% with a warming of respectively 2 and 3°C. At a warming of 1°C in 2100, only 50% of the affected ecosystems were able to adapt.

With increasing rates of climate change, the adaptation capacity rapidly declines. Their study indicated that with a warming over 0.1°C per decade, most ecosystems would definitely not adapt naturally, as required by the objective of the UN Framework Convention on Climate Change.

One of the problems with all these approaches, however, is the unrefined aggregation of the unit of analysis. Generally, only between 10 and 30 biomes are distinguished. Changes start at biome margins and rarely affect whole biomes. Using such highly aggregated models conceals many relevant impacts at the local scale. Several studies have used species models instead of biome models [56–58]. All these studies showed many more subtle impacts in many more regions than just along margins of biomes. In fact, they all indicated much larger adverse impacts (i.e. 30–50%) using species than Leemans and Eickhout [55] did using biomes. This means that earlier impact studies, as assessed by IPCC [52], underestimate projected future impact levels. However, the species-based studies also show relative smaller impacts at lower levels of climate change [as depicted in the maps provided by 59]. Over time impacts seem to accelerate or increase exponentially. These impact levels closely follow the exponential increase of global mean temperature in the used IPCC scenarios.

Most of the changes that we observed over the last decade are consistent with the directions of the projected impacts. However, many of the changes that we are experiencing seem to occur faster than indicated by all impact studies. The observed changes indicate that almost all species [e.g. 80% in 6] and not just a fraction of all species, respond immediately and extensively. Our overall impression from the above review of observed responses is that they are more widespread and appear more rapidly than impact studies suggest. Note that the observed mean climate change still closely follows the simulated trends in the IPCC scenarios.

When we link the observed responses to observed changes in weather patterns, most seem to be directly caused by extreme events, such as high temperatures early in the season, warmer and wetter winters and dry summers. Generally responses to these extreme changes are pronounced. For example, the early budding and leafing in the Netherlands in 2004 and 2005 were clearly caused by unexpectedly high temperatures in early February. Also the emergence of subtropical lichen species is clearly encouraged by more frequent hot and dry summers and mild winters. Klein Tank's conclusion [24] that extreme weather events contribute most to recently observed climate change, explains why ecological impacts are becoming so abundant over the last decade. Ecosystems respond most rapidly and vigorously especially to these events, which lead to higher impact levels in the earliest phases of climate change. Other authors have indicated similar presumptions [3, 60]. Unfortunately, extreme events are rarely considered in the most model based impact studies. This is an obvious reason to underestimate expected ecological

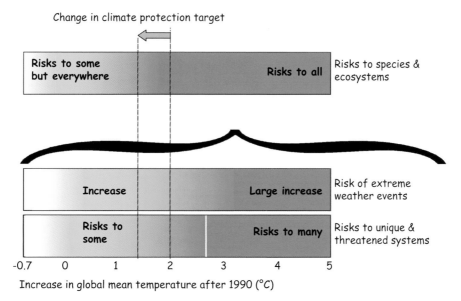

Figure 12.2 A new 'Burning Ember' that combines the reasons for concern 'extreme weather events' and 'unique & threatened systems'.

impacts. Scientifically, there is thus an urgent need to test the impacts models against the observed changes to quantify the actual underestimation. Another obvious improvement of impacts studies is to include changes in extremes in the scenarios. Both will make model based impact studies more realistic.

Species, communities, landscapes, ecosystems and biomes are probably much more vulnerable than is commonly appreciated. With continued climate change over the coming decades, natural responses of species and ecosystems (c.f. Article 2) will not be adequate for survival, and many ecosystems will rapidly become depauperated [6].

Most of the observed responses that we list stem from European studies. Some argue that the milder and wetter winters over the last decade were due to an anomalous North-Atlantic Oscillation (NAO). The responses therefore are not attributable to anthropogenic climatic change. Recent research, however, concludes that the NAO is one of the surface components of the Northern Hemisphere Oscillation (NHO). The NHO is expected to change, especially in the winter, due to anthropogenic changes [61]. This alters the NAO in the way it is observed. It is therefore likely that the anomalous NAO will continue for several decades [62], thus contributing to a more rapid climate change in the European winters. This consequently also leads to more impacts.

Many have argued that the observed changes show that species and ecosystems are resilient and can cope with climate change. Unfortunately, it is not as simple as it seems. The continued climate change trend pushes many species into conditions that they have never experienced. This increases stress. Such stressed and degraded systems can be rapidly replaced by better-adapted ones. That may be true, but degradation generally happens fast (days to decades), while recovery is slow (decades to millennia)

and often constrained by habitat fragmentation, pollution and other stressors [c.f. 5, 10]. This will lead to local diebacks and increased local extinction rates, and opportunistic species with wide ranges and a rapid dispersal will become more abundant, while specialist species with narrow habitat requirements and long lifetimes will decline [8, 9].

12.5 Conclusion: Many More Reasons for Concern

The EU has accepted a climate protection target of maximally 2°C global mean temperature increase since preindustrial times. IPCC [1] indicated that above 2°C warming the risks for adverse impacts rapidly increase. Although IPCC explicitly mentioned that below that level, risks already exist, they judged (at that time) that these risks would be acceptable. By linking observed changes in species and ecosystems with the changes in extreme weather events (two of IPCC's independent 'reasons for concern' in the Burning-Ember diagram), we provide a more consistent correlation of forcing and response (Figure 12.2). Most impact approaches do not precisely estimate the extent of responses [63] and thus provide poor indicators to select climate protection targets. Additionally, the studies using transient scenarios show that not only is the magnitude of climate change important for identifying climate protection targets, but also the rate of change.

We conclude that a target of 2°C warming is too high. Even with small changes, there will be large changes in the frequency and magnitude of extreme events and consequently, unpredictable but devastating impacts to species and ecosystems, even with a moderate climate change (an increase of 1 to 2°C). Defining tight climate protection targets and subsequent emission reduction targets is becoming, more than ever, a must. **Based on our current understanding of responses of species and ecosystems, we propose that efforts be made to limit the increase in global mean surface temperature to maximally 1.5°C above pre-industrial levels and limit the rate of change to less than 0.05°C per decade or 0.5°C per century.**

The maximum of 1.5°C tightens the existing climate protection targets of 2°C. This is necessary because impacts are more widespread, threaten delicate species interactions, and are triggered by the more rapidly occurring changes in extreme events. Together, this creates a strong argument for simultaneously limiting the rate of change to maximally 0.5°C per century.

REFERENCES

1. IPCC, 2001. *Climate change 2001: synthesis report*. Cambridge: Cambridge University Press. pp. 155.

2. Walther, G.-R., E. Post, P. Convey, A. Menzel, C. Parmesan, T.J.C. Beebee, J.-M. Fromentin, O. Hoegh-Guldberg, and F. Bairlein, 2002. *Ecological responses to recent climate change*. Nature 416: 389–395.

3. Parmesan, C. and G. Yohe, 2003. *A globally coherent fingerprint of climate change impacts across natural systems*. Nature 421: 37–42.

4. Root, T.L., J.T. Price, K.R. Hall, S.H. Schneider, C. Rosenzweig, and J.A. Pounds, 2003. *Fingerprints of global warming on wild animals and plants*. Nature 421: 57–60.

5. Hannah, L. and T.E. Lovejoy, 2003. *Climate Change and Biodiversity: Synergistic Impacts*. Advances in Applied Biodiversity Science 4: 1–123.

6. Lovejoy, T.E. and L. Hannah, eds., 2005. *Climate Change and Biodiversity*. Yale University Press: London. pp. 440.

7. Smith, J.B., H.-J. Schellnhuber, M. Qader Mirza, S. Fankhauser, R. Leemans, L. Erda, L.A. Ogallo, B.A. Pittock, R. Richels, C. Rosenzweig, U. Safriel, R.S.J. Tol, J. Weyant, and G. Yohe, 2001. *Vulnerability to Climate Change and Reasons for Concern: A Synthesis*, in *Climate Change 2001. Impacts, Adaptation, and Vulnerability*, J.J. McCarthy, O.F. Canziani, N.A. Leary, D.J. Dokken, and K.S. White, Editors, Cambridge University Press: Cambridge. pp. 913–967.

8. Ad-Hoc Technical Expert Group on Biological Diversity and Climate Change, 2003 *Interlinkages between biological diversity and climate change. Advice on the integration of biodiversity considerations into the Implementation of the United Nations Framework Convention on Climate Change and its Kyoto Protocol*, Secretariat of the Convention on Biological Diversity: Montreal. pp. 157.

9. Gitay, H., A. Suárez, R.T. Watson, O. Anisimov, F.S. Chapin, R.V. Cruz, M. Finlayson, W.G. Hohenstein, G. Insarov, Z. Kundzewicz, R. Leemans, C. Magadza, L. Nurse, I. Noble, J. Price, N.H. Ravindranath, T.L. Root, B. Scholes, A. Villamizar, and X. Rumei, 2002 *Climate change and biodiversity*, Intergovernmental Panel on Climate Change: Geneva. pp. 77.

10. Reid, W.V., H.A. Mooney, A. Cropper, D. Capistrano, S.R. Carpenter, K. Chopra, P. Dasgupta, T. Dietz, A.K. Duraiappah, R. Hassan, R. Kasperson, R. Leemans, R.M. May, A.J. McMichael, P. Pingali, C. Samper, R. Scholes, R.T. Watson, A.H. Zakri, Z. Shidong, N.J. Ash, E. Bennett, P. Kumar, M.J. Lee, C. Raudsepp-Hearne, H. Simons, J. Thonell, and M.B. Zurek, 2005. *Millennium Ecosystem Assessment Synthesis report*. Washington DC: Island Press. pp. 160.

11. van Oene, H., F. Berendse, W.J. Arp, and J.R.M. Alkemade, 2000. *Veranderingen op de Veluwe. Simulatie van veranderingen in ecosysteemprocessen en botanische diversiteit op regionale schaal*. Landschap 65–80: 65–80.

12. Moraal, L.G., G.A.J.M. Jagers op Akkerhuis, and D.C. van der Werf, 2003. *Veranderingen in insectenplagen op bomen: monitoring sinds 1946 maakt trends zichtbaar*. Nederlands Bosbouwkundig Tijdsschrift 74: 29–32.

13. Both, C., A.V. Artemyev, B. Blaauw, R.J. Cowie, A.J. Dekhuijzen, T. Eeva, A. Enemar, L. Gustafsson, E.V. Ivankina, A. Järvinen, N.B. Metcalfe, N.E.I. Nyholm, J. Potti, P.-A. Ravussin, J.J. Sanz, B. Silverin, F.M. Slater, L.V. Sokolov, J.N. Török, W. Winkel, J. Wright, H. Zang, and M.E. Visser, 2004. *Large-scale geographical variation confirms that climate change causes birds to lay earlier*. Proceedings of the Royal Society Biological Sciences Series B 271: 1657–1662.

14. van Herk, C.M., A. Aptroot, and H.F.S.O. van Dobben, 2002. *Long-term monitoring in the Netherlands suggests that lichens respond to global warming*. Lichenologist 34: 141–154.

15. Tamis, W.L.M., M. van't Zelfde, R. van der Meijden, C.G.L. Groen, and H.A.U. De Haes, 2005. *Ecological interpretation of changes in the Dutch flora in the 20th century*. Biological Conservation 125: 211–224.

16. Visser, M.J. and L.J.M. Holleman, 2001. *Warmer springs disrupt the synchrony of Oak and Winter Moth*. Proceeding of the Royal Society, London 268: 289–294.

17. Roos, R., M. Visser, B. van Tooren, and I. Schimmel, 2003. *Themanummer: Klimaatsverandering*. De Levende Natuur 104: 69–133.

18. RIVM, CBS, and Stichting DLO 2003. *Natuurcompendium 2003: Natuur in cijfers*. Utrecht: KNNV Uitgeverij. pp. 494.

19. Roos, R., S. Woudenberg, G. Dorren, and E. Brunner, eds., 2004. *Opgewarmd Nederland*. Stichting Natuurmedia & Uitgeverij Jan van Arkel: Amsterdam. pp. 224.

20. Parmesan, C. and H. Galbraith, 2004 *Observed Impacts of global climate change in the U.S.*, Pew Center on Global Climate Change: Washington DC. pp. 56.

21. Root, T.L., D.P. MacMynowski, M.D. Mastrandrea, and S.H. Schneider, 2005. *Human-modified temperatures induce species changes: Joint attribution*. Proceedings of the National Academy of Science, USA 102: 7465–7469.

22. Mann, M.E., 2002. *Climate reconstruction: the value of multiple proxies*. Science 297: 1481–1482.

23. Moberg, A., D.M. Sonechkin, K. Holmgren, N.M. Datsenko, and W. Karlén, 2005. *Highly variable Northern Hemisphere temperatures reconstructed from low- and high-resolution proxy data*. Nature 433: 613–617.

24. Klein Tank, A., 2004 *Changing temperatures and precipitation extremes in Europe's climate of the 20th century*, Utrecht University: Utrecht. pp. 124.

25. Anisimov, O.A., 1989. *Changing climate and permafrost distribution in the Soviet arctic*. Physical Geography 10: 285–293.

26. Arctic Climate Impact Assessment, 2004. *ACIA Overview report*. Cambridge: Cambridge University Press. pp. 149.

27. Vaughan, D.G., G.J. Marshall, W.M. Connolly, C. Parkinson, R. Mulvaney, D.A. Hodgson, J.C. King, C.J. Pudsey, and J. Turner, 2003. *Recent rapid regional climate warming on the Antarctic Peninsula*. Climatic Change 60: 243–274.

28. Thompson, L.G., E. Mosley Thompson, M.E. Davis, K.A. Henderson, H.H. Brecher, V.S. Zagorodnov, T.A. Mashiotta, P.N. Lin, V.N. Mikhalenko, D.R. Hardy, and J. Beer, 2002. *Kilimanjaro ice core records: Evidence of Holocene climate change in tropical Africa*. Science 298: 589–593.

29. Meier, M.F., M.B. Dyurgerov, and G.J. McCabe, 2003. *The health of glaciers: Recent changes in glacier regime*. Climatic Change 59: 123–135.

30. Page, S.E., F. Siegert, J.O. Rieley, H.-D.V. Boehm, A. Jayak, and S. Limink, 2002. *The amount of carbon released from peat and forest fires in Indonesia during 1997*. Nature 420: 61–65.

31. Myneni, R.B., C.D. Keeling, C.J. Tucker, G. Asrar, and R.R. Nemani, 1997. *Increased plant growth in the northern high latitudes from 1981 to 1991*. Nature 386: 698–702.

32. Lucht, W., I.C. Prentice, R.B. Myneni, S. Sitch, P. Friedlingstein, W. Cramer, P. Bousquet, W. Buermann, and B. Smith, 2002.

Climatic control of the high-latitude vegetation greening trend and Pinatubo effect. Science 296: 1687–1689.

33. White, M.A., R.R. Nemani, P.E. Thornton, and S.W. Running, 2002. *Satellite evidence of phenological differences between urbanized and rural areas of the eastern United States deciduous broadleaf forest.* Ecosystems 5: 260–273.

34. Fagre, D.B., D.L. Peterson, and A.E. Hessl, 2003. *Taking the pulse of mountains: Ecosystem responses to climatic variability.* Climatic Change 59: 263–282.

35. Schwartz, M.D., ed. 2003. *Phenology: An integrative environmental science.* Kluwer Academic Publishers: Dordrecht, Boston, London. pp. 564.

36. Marra, P.P., C.M. Francis, R.S. Mulvihill, and F.R. Moore, 2005. *The influence of climate on the timing and rate of spring bird migration.* Oecologia 142: 307–315.

37. Primack, D., C. Imbres, R.B. Primack, A.J. Miller-Rushing, and P. Del Tredici, 2004. *Herbarium specimens demonstrate earlier flowering times in response to warming in Boston.* American Journal of Botany 91: 1260–1264.

38. Visser, M.E., F. Adriaensen, J.H. van Balen, J. Blondel, A.A. Dhondt, S. van Dongen, C. du Feu, E.V. Ivankina, A.B. Kerimov, J. de Laet, E. Matthysen, R. McCleery, M. Orell, and D.L. Thomson, 2003. *Variable responses to large-scale climate change in European Parus populations.* Proceedings Of The Royal Society Of London Series B-Biological Sciences 270: 367–372.

39. Stefanescu, C., J. Penuelas, and I. Filella, 2003. *Effects of climatic change on the phenology of butterflies in the northwest Mediterranean Basin.* Global Change Biology 9: 1494–1506.

40. Both, C. and M.E. Visser, 2001. *Adjustment to climate change is constrained by arrival date in a long-distance migrant bird.* Nature 411.

41. EEA ETC/ACC, UBA, and RIVM, 2004 *Impacts of Europe's changing climate. An indicator-based assessment*, European Environment Agency.

42. Serreze, M.C., J.E. Walsh, F.S. Chapin, T. Osterkamp, M. Dyurgerov, V. Romanovsky, W.C. Oechel, J. Morison, T. Zhang, and R.G. Barry, 2000. *Observational evidence of recent change in the northern high-latitude environment.* Climatic Change 46: 159–207.

43. Villalba, R., A. Lara, J.A. Boninsegna, M. Masiokas, S. Delgado, J.C. Aravena, F.A. Roig, A. Schmelter, A. Wolodarsky, and A. Ripalta, 2003. *Large-scale temperature changes across the southern Andes: 20th-century variations in the context of the past 400 years.* Climatic Change 59: 177–232.

44. Körner, C. and J. Paulsen, 2004. *A worldwide study of high altitude treeline temperatures.* Journal of Biogeography 31: 713–732.

45. Edwards, M. and A.J. Richardson, 2004. *Impact of climate change on marine pelagic phenology and trophic mismatch.* Nature 430: 881–884.

46. Beaugrand, G., K.M. Brander, J.A. Lindley, S. Souissi, and P.C. Reid, 2003. *Plankton effect on cod recruitment in the North Sea.* Nature 426: 661–664.

47. Frederiksen, M., M.P. Harris, F. Daunt, P. Rothery, and W.S., 2004. *Scale-dependent climate signals drive breeding phenology of three seabird species.* Global Change Biology 10: 1214–1221.

48. Goreau, T.J., 1999. *Weighing up the threat to the world's corals.* Nature 402: 457.

49. Knowlton, N., 2001. *The future of coral reefs.* Proceedings National Academy of Sciences 98: 5419–5425.

50. Anonymous, 2003. *Climate change and Mountain pine beetle range expansion in BC.* The Forestry Chronicle 79: 1025.

51. Carter, T.R., E.L. La Rovere, R.N. Jones, R. Leemans, L.O. Mearns, N. Nakícenovíc, B. Pittock, S.M. Semenov, and J. Skea, 2001. *Developing and applying scenarios,* in *Climate Change 2001. Impacts, adaptation and vulnerability,* J. McCarthy, O.F. Canziani, N. Leary, D.J. Dokken, and K.S. White, Editors, Cambridge University Press: Cambridge. pp. 145–190.

52. McCarthy, J.J., O.F. Canziani, N. Leary, D.J. Dokken, and K.S. White, eds., 2001. *Climate Change 2001. Impacts, adaptation and vulnerability.* Cambridge University Press: Cambridge. pp. 1032.

53. Emanuel, W.R., H.H. Shugart, and M.P. Stevenson, 1985. *Climatic change and the broad-scale distribution of terrestrial ecosystems complexes.* Climatic Change 7: 29–43.

54. Chapin, F.S., III, O.E. Sala, and E. Huber-Sannwald, 2001. *Global biodiversity in a changing environment: Scenario for the 21st century.* New York: Springer Verlag. pp. 372.

55. Leemans, R. and B. Eickhout, 2004. *Another reason for concern: regional and global impacts on ecosystems for different levels of climate change.* Global Environmental Change 14: 219–228.

56. Bakkenes, M., J.R.M. Alkemade, F. Ihle, R. Leemans, and J.B. Latour, 2002. *Assessing effects of forecasted climate change on the diversity and distribution of European higher plants for 2050.* Global Change Biology 8: 390–407.

57. Thomas, C.D., A. Cameron, R.E. Green, M. Bakkenes, L.J. Beaumont, Y.C. Collingham, B.F.N. Erasmus, M.F. de Siqueira, A. Grainger, L. Hannah, L. Hughes, B. Huntley, A.S. van Jaarsveld, G.F. Midgley, L. Miles, M.A. Ortega Huerta, A.T. Peterson, O.L. Phillips, and S.E. Williams, 2004. *Extinction risk from climate change.* Nature 427: 145–148.

58. Thuiller, W., S. Lavorel, M.B. Araujo, M.T. Sykes, and I.C. Prentice, 2005. *Climate change threats to plant diversity in Europe.* Proceedings National Academy of Sciences 102: 8245–8250.

59. Metzger, M.J., R. Leemans, D. Schröter, W. Cramer, and ATEAM Consortium, 2005 *The ATEAM vulnerability mapping tool,* C.T. de Wit Graduate School for Production Ecology and Resource Conservation (PE&RC): Wageningen.

60. Easterling, D.R., G.A. Meehl, C. Parmesan, S.A. Channon, T.R. Karl, and L.O. Mearns, 2000. *Climatic extremes: observations, modelling, and impacts.* Science 289: 2068–2074

61. Hartmann, D.L., J.M. Wallace, V. Limpasuvan, D.W.J. Thompson, and J.R. Holton, 2000. *Can ozone depletion and global warming interact to produce rapid climate change?* Proceedings National Academy of Sciences 97: 1412–1417.

62. KNMI, 2003 *De toestand van het klimaat in Nederland 2003,* Koninklijk Nederlands Meteorologisch Instituut: De Bilt. pp. 32.

63. Hampe, A., 2004. *Bioclimate envelope models: what they detect and what they hide.* Global Ecology & Biogeography 13: 469–476.

CHAPTER 13

Climate Change-induced Ecosystem Loss and its Implications for Greenhouse Gas Concentration Stabilisation

John Lanchbery
Royal Society for the Protection of Birds, The Lodge, Sandy, Bedfordshire, UK

ABSTRACT: The objective of the Climate Change Convention requires that atmospheric concentrations of greenhouse gases should be stabilised at a level which allows ecosystems to adapt naturally to climate change. Yet there is substantial and compelling evidence that the degree of climate change which has already occurred is affecting both species and ecosystems, in many cases adversely. It appears very likely that species will increasingly become extinct and ecosystems will be lost as a result of little further change in the climate. In the context of the objective of the Convention, it can thus be argued that at least some ecosystems are not 'adapting naturally' to climate change and that atmospheric concentrations of greenhouse gases are already too high.

13.1 Introduction

The ultimate objective of the UN Framework Convention on Climate Change requires greenhouse gas concentrations in the atmosphere to be stabilised at a level that would 'prevent dangerous anthropogenic interference with the climate system'. However, policy-makers have consistently failed to decide what 'dangerous' means, in spite of increasing evidence of the likelihood of large and widespread impacts upon both people and wildlife as a result of quite small changes in mean global surface temperature. [1] Consequently, they have also failed to agree on a level at which atmospheric concentrations of greenhouse gases should be stabilised in order to avoid dangerous interference with the climate system.

Yet the second part of the Convention's objective provides guidance as to what 'dangerous' means. It says that 'such a [concentration stabilisation] level should be achieved within a time-frame sufficient to allow ecosystems to adapt naturally to climate change …'. So, if there is evidence that ecosystems will not be able to adapt to a particular mean surface temperature rise, then that increase in temperature should constitute 'dangerous' and atmospheric concentrations of greenhouse gases should be stabilised at a level which avoids the temperature being attained. Whilst all ecosystems will not respond equally to the changing climate, evidence that at least some are already being affected adversely would indicate that dangerous levels are being approached.

There is abundant and increasing evidence that individual species are already being affected by climate change. Indeed, at least one species would appear to have become extinct due to recent, human-induced climate change: the golden toad of Costa Rica. [2] Also, models of species' responses to climate change, particularly in terms of changes in their natural ranges, indicate that species extinctions are likely with quite small further changes in the climate. In addition, model-based studies indicate that unique ecosystems will be lost under medium or even low range warming scenarios, for example, the Succulent Karoo in South Africa. [3] (Many of these studies are summarised by Hare in his paper in this book.)

However, although there is considerable evidence of species having already changed their behaviour as a result of climate change, and a large number of modelled studies which indicate that both species and ecosystem loss is likely in future, there is comparatively little evidence that indicates that ecosystems have failed, or are beginning to fail, to adapt to the degree of climate change that has already occurred. There is some strong evidence of impending ecosystem loss, for example, of coral reefs worldwide and the Succulent Karoo, but actual loss or severe damage is usually forecast. [4] Therefore, whilst most modelled studies are compelling and well reasoned, there is still at least some scope for those that are sceptical about the severity of the impacts of climate change to question either the models or their underlying assumptions.

In this paper, evidence is presented of large-scale ecosystem change in the year 2004 which apparently occurred mainly as a result of climate change. This was an observed, well-recorded event indicating that an ecosystem is failing to adapt to climate change. First, short summaries of evidence for species responses to climate change are given, as background, together with some of the modelled studies referred to above.

13.2 Species Responses to Climate Change

Over the last decade, a host of evidence has been gathered that shows a very strong correlation between changes in the climate and changes in species behaviour. Two recent so-called meta-analyses by Parmesan and Yohe [5] and by Root *et al.*, [6] are instructive because they combine a broad range of results to test whether or not a coherent pattern of correlations between climate change and species behaviour exists across different geographical regions and a wide range of different species.

Parmesan and Yohe's analysis examined the results of 143 studies of 1,473 species from all regions of the world. Of the 587 species showing significant changes in distribution, abundance, phenology, morphology or genetic frequencies, 82% had shifted in the direction expected if they were climate change-induced, i.e. towards higher latitudes or altitudes, or earlier spring events. The timing of spring events, such as egg-laying by birds or flowering by plants, was shown by 61 studies to have shifted earlier by an average 5.1 days per decade over the last half-century, with changes being most pronounced at higher latitudes. The analysis of Root *et al.*, reviewed studies of more than 1,700 species, overlapping with Parmesan and Yohe's, and found similar results: 87% of shifts in phenology and 81% of range shifts were in the direction expected from climate change. These studies give a very high confidence that climate change is already impacting biodiversity.

However, simply because species are affected by climate change does not necessarily mean that the effects will be adverse; some may be beneficial. Neither does it necessarily follow that ecosystems will be threatened or lost. Some changes, however, would be expected to have potentially adverse impacts and one of these is climate change-induced alteration of species ranges. The concept of 'climate space' is often employed to describe where a species range, or potential range, would be if it were determined solely by climate. Whilst ranges are determined by many factors, of which climate is just one important factor, the climate space of a species is helpful in trying to forecast whether a species may be affected by climate change.

If the preferred climate space of a species moves as the climate changes there can be many reasons why the species may be unable move with it; for example, because the underlying geology and flora of the intervening area is different or because it is intensively utilised by human beings. Land-based species which are likely to be unable to move include those that currently inhabit islands or mountain ranges and whose preferred climate space moves to other islands or mountain ranges, or to an ocean. This would not necessarily spell extinction, which would depend on a number of factors including the extent and rate of climate change, but it would make it more likely, especially for endemic species.

Many workers have modelled species' responses to future climate change. Such models typically work on the basis of establishing the preferred climate space for a particular species and then employing models to forecast where that space will be as the climate changes. Whilst species will not necessarily move to fill their future climate

space, the models give a good picture of possible future movement and hence of where movement might be difficult. [7, 8] Recently, a number of workers have focused upon species that are endemic to limited areas that have few, if any, options for movement. [9] For example, Williams *et al.* [10] conducted a study of the Australian Wet Tropics World Heritage Area which is the most biologically rich area in Australia. They assessed the effects of increases in temperature of between 1°C and 7°C on species distribution using bioclimatic modelling based on over 220,000 records. Estimates were made of the change in the core range of each species under different climate scenarios, assuming that species continued to occupy the climate space they currently use. Models for 62 endemic montane (greater than 600 m altitude) species indicated that 1°C warming will result in an average of 40% loss of potential core range, 3.5°C warming a 90% loss and 5°C warming a 97% loss. Warming of 7°C resulted in the loss of all potential core range for all species.

Early in 2004, a number of those who had conducted studies that modelled species' responses to climate change produced a joint paper that assessed the extinction risks for sample regions covering about 20% of the Earth's terrestrial surface, including parts of Australia, Brazil, Europe, Mexico and South Africa. [11] They concluded that '15% to 37% of species in our sample of regions and taxa will be "committed to extinction" as a result of mid-range climate warming scenarios for 2050. Taking the average of the three methods and two dispersal scenarios, minimal climate warming scenarios produce lower projections of species committed to extinction (~18%) than mid-range (~24%) and maximum change (~35%) scenarios.'

13.3 Some Reasons for Concern about Ecosystem Loss

In the context of range changes, ecosystem loss is possible because species will not all move to the same extent or at the same rate as their climate space changes. Any particular ecosystem consists of an assemblage of species, some of which are near the edges of their ranges and others that are not. Those at their range edges will tend to move as their climate space changes whereas those nearer their range centres need not. This differential movement will be exaggerated by opportunistic, robust species tending to move more rapidly and faring better when they do. The composition of ecosystems, and hence the ecosystems themselves, will thus change.

A further concern is that, because species do not act in isolation, changes in one particular species or group of species can affect many others, often in unpredictable ways. For example, a species which is otherwise unaffected by a particular degree of climate change will be radically affected if its source of food changes its range and moves somewhere else. In the next section of this paper, a recent example of this type of occurrence is examined.

13.4 Ecosystem Change in the Northeast Atlantic

Seabirds on the North Sea coast of Britain suffered a large-scale breeding failure in 2004. [12] In Shetland, Orkney and Fair Isle, tens of thousands of seabirds failed to raise any young. The total Shetland population of nearly 7000 pairs of great skuas (*stercorarius skua*) produced only a handful of chicks, and the 1000 or more pairs of arctic skuas (*stercorarius parasiticus*) none at all. Shetland's 24,000 pairs of arctic terns (*sterna paradisaea*) and more than 16,000 pairs of kittiwakes (*larus tridactyla*) have also probably suffered near total breeding failure. This continues a trend (especially in south Shetland) of several years, so much so that some kittiwake colonies are beginning to disappear, despite the fact that the birds are long-lived and can thus survive short-term breeding failures. In Orkney, all of the large arctic tern breeding colonies in the north isles failed. Arctic and great skuas also had a very poor breeding season and numbers of guillemots (*uria aalge*) and kittiwakes were very low.

Whilst the exact cause and extent of the breeding failures is still being investigated, the phenomenon very strongly indicates a widespread food shortage, especially of sandeels, a small fish that forms the staple diet of many UK seabirds. (Five species of sandeels inhabit the North Sea, of which the lesser sandeel, *ammodytes marinus,* is the most abundant and comprises over 90% of sandeel fishery catches). Whilst surface feeders such as terns and kittiwakes might be expected to be disadvantaged by a shortage of sandeels, it is indicative of the probable scale of shortage that deep-diving birds like guillemots (which can dive down to 100 m) also failed to breed in 2004.

A shortage of sandeels is independently indicated by the Danish sandeel fishery which accounts for about 90% of the North Sea catch. In recent years, this fishery has been allocated quotas of around 800,000 to 900,000 tonnes, of which 600,000 to 700,000 tonnes was usually taken. In 2003, however, Denmark undershot its quota significantly, catching only 300,000 tonnes and the 2004 catch is apparently similar. [13] However, whilst the sandeel population has apparently fallen significantly, this does not seem to result solely, or even mainly, from overfishing, in at least some of areas where sea birds' breeding failures have occurred. Shetland has, for example, operated a seabird-friendly sandeel fishing regime for several years. In 2004, the waters around the south of Shetland were closed to sandeel fishing altogether, and a reduced 'Total Allowable Catch' was introduced around the north of Shetland.

It appears likely that climate change has played a significant part in sandeel declines. The temperature of the North Sea is controlled by local solar heating and heat exchange with the atmosphere. [14] The temperature of the North Sea rose by an average of 1.05°C between 1977 and 2001, and in 2001 a very long run of positive temperature anomalies began. In August 2004, the sea surface temperatures peaked at about 2.5°C above the 1971 to 1993 average.

A study of sandeels in the North Sea indicates that their numbers are inversely proportional to sea temperature during the egg and larval stages, and there is further evidence that this is, in turn, linked to plankton abundance around the time of sandeel egg hatching. [15] The same study also indicates that the adverse effect of rising sea temperatures is most marked in the southern North Sea where the lesser sandeel is near the southern limit of its range, leading to the conclusion that the southern limit of sandeel distribution is likely to shift northwards as the sea warms.

Plankton populations in the North Sea have certainly changed. Work by the Sir Alister Hardy Foundation, based on continuous plankton recording over more than four decades, has identified a 'regime shift' in the plankton composition of the North Sea since about 1986. [16] Indeed, the Foundation has recently shown that across the entire Northeast Atlantic sea surface temperature change is accompanied by increased phytoplankton abundance in cooler regions and decreased phytoplankton abundance in warmer regions. [17] They conclude that 'Future warming is therefore likely to alter the spatial distribution of primary and secondary pelagic production, affecting ecosystem services and placing additional stress on already-depleted fish and mammal populations'.

In summary, it would appear that a large-scale change in marine ecosystems is occurring in the North Sea, caused in large part by climate change. The plankton regime has certainly changed and it is hard to find an explanation other than sea temperature rise that adequately accounts for it. Sandeel numbers have declined and a change in sea temperature coupled with a change in the plankton population (also induced by temperature change) seems a likely explanation. Sea bird breeding success was certainly low in 2004, most probably due to the fall in sandeel numbers.

13.5 Implications for Concentration Stabilisation

There is substantial and compelling evidence that the degree of climate change which has already occurred has affected both species and ecosystems, in some cases adversely. It appears very likely that species will increasingly become extinct and that ecosystems will be lost with little further change in the climate. Recent evidence of ecosystem change in the North Sea indicates that at least one major ecosystem is not adapting at all well to the degree of climate change that has already occurred.

In terms of the ultimate objective of the Climate Change Convention, it can thus be argued that atmospheric concentrations of greenhouse gases are already too high. However, atmospheric concentrations will certainly rise from where they are now and so, in the context of this book, the question is at what concentration would it be practical to stabilise greenhouse gas concentrations so as to avoid the worst damage to species and ecosystems. But this is hard to estimate for individual cases or types of cases, because different species and ecosystems will respond at different rates and to different extents to any particular temperature rise and, anyway, temperature, precipitation and other key factors affecting wild species will vary considerably from globally averaged values. Also, species are often critically affected by short spells of high or low temperatures, soil moisture content and a host of other parameters not captured by average values of temperature.

A pragmatic decision based on evidence of avoiding harm to as many species and ecosystems as possible is thus called for. As long ago as 2001, the IPCC gave clear guidance on this matter, as summarised in the so-called 'burning ember' figure in the Third Assessment Report which indicates that risks to unique and threatened systems moves from 'risks to some' to 'risks to many' for a mean global temperature rise of about 2°C. [1] There is now more modelled evidence that supports this finding. For example, the paper by Thomas *et al.*, mentioned earlier, indicates that at least a sixth of species studied in an area covering 20% of the terrestrial surface of the Earth could be 'committed to extinction' as a result of mid-range warming scenarios by 2050. [10] This figure could be as much as one third, according to the authors, and the overall study included many unique ecosystems that could be lost. This level of loss would seem to the author to be unacceptable and would be a clear breach of the Climate Convention's objective to allow systems to adapt naturally.

Whilst such studies are forecasts, not observed evidence, there is an increasing body of evidence that shows that ecosystems are already changing significantly, especially marine ecosystems such as that described here. It appears that although individual forecasts are subject to uncertainty, overall they may well prove reasonably accurate. On balance, therefore, stabilisation of atmospheric concentrations of greenhouse gases at a level that keeps mean global surface temperatures at below 2°C appears necessary if the worst damage to species and ecosystems is to be avoided.

REFERENCES

1. See for example, Figure SPM-2, sometimes known as the 'Burning Ember' diagram: in Smith J. B., Schellnhuber H-J, Qader Mirza M., Fankhauser S., Leemans R., Erda L., Ogallo L.A., Pittock B.A., Richels R., Rosenzweig C., Safriel U., Tol R.S.J., Weyant J., and Yohe G. (2001), Vulnerability to climate change and reasons for concern: a synthesis. Pages 913–967 in McCarthy J.J., Canziani O.F., Leary N.A., Dokken D.J., and White K.S., editors, Climate Change 2001, Impacts, Adaptation, and Vulnerability, Cambridge University Press, UK, 2001.

2. Pounds J.A., Fogden M.L.P. and Campbell J.H. (1999), Biological response to climate change on a tropical mountain, *Nature,* **398,** (6728), 611–615, 15 April 1999.

3. See, Rutherford M.C., Midgley G.F., Bond W.J., Powrie L.W., Musil C.F., Roberts R. and Allsopp J. (1999) South African Country Study on Climate Change. Pretoria, South Africa, Terrestrial Plant Diversity Section, Vulnerability and Adaptation, Department of Environmental Affairs and Tourism, 1999.
   Hannah L., Midgley G.F., Lovejoy T., Bond W.J., Bush M., Lovett J.C., Scott D. and Woodward F.I. (2002) Conservation of Biodiversity in a Changing Climate, *Conservation Biology* **16,** (1), 264–268, 2002.
   Midgley G.F., Hannah L., Millar D., Rutherford M.C. and Powrie L.W. (2002) Assessing the vulnerability of species richness to anthropogenic climate change in a biodiversity hotspot, *Global Ecology and Biogeography,* **11,** (6), 445–452, 2002.

4. See above reference for the Karoo and for coral reefs: Spalding M., Teleki K., and Spencer T. (2001) Climate change and coral bleaching, in Impacts of Climate Change and Wildlife, Eds: Green R.E., Harley M., Spalding M., and Zöckler C., RSPB publications 2001.

5. Parmesan C. and Yohe G. (2003) A globally coherent fingerprint of climate change impacts across natural systems, *Nature,* **421,** (6918), 37–42, 2 January 2003.

6. Root T.L., Price J.T., Hall K.R., Schneider S.H., Rosenzweig C. and Pounds J.A. (2003), Fingerprints of global warming on wild animals and plants, *Nature,* **421,** (6918), 57–60, 2 January 2003.

7. Peterson A.T. *et al.* (2002), Future projections for Mexican faunas under global climate change scenarios, *Nature* **416,** (6881), 626–629, 11 April 2002.

8. For example, Erasmus B.F.N., van Jaarsveld A.S., Chown S.L., Kshatriya M. and Wessels K. (2002), Vulnerability of South African animal taxa to climate change, *Global Change Biology.* **8,** (7), 679–693, July 2002.

9. For example, Midgley G.F., Hannah L., Rutherford M.C. and Powrie L.W. (2002), Assessing the vulnerability of species richness to anthropogenic climate change in a biodiversity hotspot, *Global Ecology and Biogeography,* **11,** (6), 445–451, November 2002.

10. Williams S.E., Bolitho E. E. and Fox, S. (2003), Climate change in Australian tropical rainforests: an impending environmental catastrophe, *Proceedings of the Royal Society of London B.,* **270,** (1527), 1887–1892, 22 September 2003.

11. Thomas C.D. *et al.* (2004), Extinction risk from climate change, *Nature,* **427,** (6970), 145–148, 8 January 2004.

12. The numerical estimates included in this paragraph are provisional figures provided by Euan Dunn of the Royal Society for the Protection of Birds.

13. Proffitt F. (2004), Reproductive failure threatens bird colonies on North Sea coast, *Science,* **305,** (5687), 1090, 20 August 2004.

14. This and all other temperature data included in the paper is from the International Council for the Exploration of the Sea (ICES), specifically their research reports on the North Sea, see The Annual ICES Ocean Climate Status Summary, http://www.ices.dk/marineworld/climatestatus/CRR275.pdf

15. Arnott S.A. and Ruxton G.D. (2002), Sandeel recruitment in the North Sea: demographic, climatic and trophic effects, *Marine Ecology Progress Series,* **238,** 199–210, 8 August 2002.

16. Beaugrand G. (2004), The North Sea regime shift: evidence, causes, mechanisms and consequences, *Progress in Oceanography,* **60,** issues (2/4), 245–262, February/March 2004.

17. Richardson A.J. and Schoeman D.S. (2004), Climate impact of plankton ecosystems in the Northeast Atlantic, *Science,* **305,** (5690), 1609–1612, 10 September 2004.

CHAPTER 14

Tropical Forests and Atmospheric Carbon Dioxide: Current Conditions and Future Scenarios

Simon L. Lewis[1], Oliver L. Phillips[1], Timothy R. Baker[1], Yadvinder Malhi[2] and Jon Lloyd[1]

[1] *Earth & Biosphere Institute, School of Geography, University of Leeds, Leeds*
[2] *School of Geography & the Environment, University of Oxford, Oxford*

ABSTRACT: Tropical forests affect atmospheric carbon dioxide concentrations, and hence modulate the rate of climate change – by being a source of carbon, from land-use change (deforestation), and as a sink or source of carbon in remaining intact forest. These fluxes are among the least understood and most uncertain major fluxes within the global carbon cycle. We synthesise recent research on the tropical forest biome carbon balance, suggesting that intact forests presently function as a carbon sink of approx. $1.2 \, \text{Pg C a}^{-1}$, and that deforestation emissions at the higher end of the reported 1–$3 \, \text{Pg C a}^{-1}$ spectrum are likely. Scenarios suggest that the source from deforestation will remain high, whereas the sink in intact forest is unlikely to continue, and remaining tropical forests may become a major carbon source via one or more of (i) changing photosynthesis/respiration rates, (ii) functional/ biodiversity changes within intact forest, or widespread forest collapse via (iii) drought, or (iv) fire. Each scenario risks possible positive feedbacks with the climate system suggesting that current estimates of the possible rate, magnitude and effects of global climate change over the coming decades may be conservative.

14.1 Introduction

Tropical forests are an important component of the global carbon cycle, as they are relatively extensive, carbon-dense and highly productive. From 1750–2000 global land-use change is estimated to have released approx. 180 Pg C (Pg C = billion tons of carbon) to the atmosphere, 60% from the tropics [1,2], alongside 283 Pg C released from fossil fuel use [3]. Thus tropical forest conversion has released approx. 108 Pg C. Further major carbon additions may be expected, with 553 Pg C residing within remaining tropical forests and soils [4, 5], the equivalent of over 80 years of fossil fuel use at current rates.

The total carbon release from land-use change and fossil fuel use from 1750–2000 has been estimated at 463 Pg C, but the increase in atmospheric CO_2 concentrations has been only 174 Pg C [5]. The remainder has been absorbed into the oceans (approx. 129 Pg C; [6]) and terrestrial ecosystems (approx. 160 Pg C). This 160 Pg C is a potentially transient sink: what if this sink becomes a source? Such a change would radically increase atmospheric CO_2 concentrations, both accelerating the rate, and increasing the magnitude, of climate change.

Understanding the role of the terrestrial tropics as an accelerator or buffer of the rate of climate change via additions and subtractions to the atmospheric CO_2 pool is essential. However, tropical forests are among the least-understood and -quantified major sources of C from deforestation, *and* sources or sinks from intact forest and soil. Below, we assess the state of knowledge regarding these sinks and sources, and sketch a range of future possible scenarios for this important and threatened biome.

14.2 Tropical Forests and the Global Carbon Cycle over the 1990s

14.2.1 *Estimating and Partitioning the Terrestrial Carbon Sink*

Accounting for known annual global carbon fluxes from fossil fuel use and known land-use change, the known additions of carbon to the atmosphere and the known oceanic uptake of carbon show that there must be a residual carbon sink in terrestrial ecosystems. The change in atmospheric CO_2 and emissions from fossil fuel use are known with reasonable precision ($3.2 \pm 0.1 \, \text{Pg C a}^{-1}$ & $6.3 \pm 0.4 \, \text{Pg C a}^{-1}$ respectively [5]). Partitioning of the terrestrial and oceanic fluxes using simultaneous atmospheric measurements of CO_2 and O_2 give the net terrestrial flux as a sink of approx. $1.0 \pm 0.8 \, \text{Pg C a}^{-1}$, (and an oceanic sink of approx. $2.1 \pm 0.7 \, \text{Pg C a}^{-1}$, [6]). Using CO_2 and $^{13}\delta C$ (inverse models) the net terrestrial flux estimates ranges from a sink of 0.8 to $1.4 \, \text{Pg C a}^{-1}$ [5]. Thus terrestrial ecosystems are estimated to be a net sink for carbon, using two independent methods. Assuming land-use change contributes $1.7 \pm 0.8 \, \text{Pg C a}^{-1}$ [5], the residual term, the sink in terrestrial ecosystems is therefore $2.7 \pm 1.1 \, \text{Pg C a}^{-1}$.

Partitioning this global terrestrial sink between northern extratropical and tropical lands, using atmospheric transport models, show that while the terrestrial land-mass as a whole is a sink, tropical regions may be neutral, or a source of C (1.5 ± 1.2 Pg C a^{-1} [7, 8, 9]). This in turn is composed of (1) tropical land-use change (defor-estation), which studies show to be a source of anything between 0.9 Pg C a^{-1} and 3.0 Pg C a^{-1} [2, 8, 10–13], (2) intact forests being, on average, neutral [14, 15], or a modest approx. 1 Pg C a^{-1} [16–18] or major 3 Pg C a^{-1} sink [19, 20], and (3) rivers and wetlands being a source of approx. 0.9 Pg C a^{-1} [21].

14.2.2 *Large or Small Changes Across the Tropics?*

The fluxes of carbon from the tropics are very poorly constrained due to a lack of data and methodological limi-tations. Current evidence, summarised above, suggests two possibilities for the tropics: (1) a large release of car-bon from deforestation, partially offset by a large sink in intact forest, and (2) a smaller release of C from defor-estation with little, if any, sink in intact forest [8, 12, 22].

Differences in carbon flux estimates from deforesta-tion are largely due to contrasting estimates of the rate of deforestation, decisions regarding the average carbon content of a tropical forest [23], and inclusion of all relevant emissions [11]. All aspects are controversial. Two recent studies reporting 'low' deforestation rates and emissions [10, 20, 24] need careful interpretation. The Defries et al. study is based on coarse-resolution (8 km^2) satellite data, calibrated with high-resolution satellite data to identify the smaller clearings not detectable at the coarse scale. Thus, this is likely to be the less reliable than the Achard et al. study. However, the Achard et al. deforestation figures run from 1990–1997, and do not include one of the most important tropical car-bon emission events of the 1990s – the fires associated with the 1997–1998 El Niño Southern Oscillation (ENSO) event. Given 20 million ha that may have burnt [25], releasing possibly 3 Pg C to the atmosphere [25, 26], extrapolating the 1990–1997 results to the 1990s as a whole would underestimate deforestation C emissions rates. Furthermore, depending on the average carbon content estimate selected, fluxes can differ by 50% [11, 23], with the Achard et al. studies utilising lower average carbon content figures than other authors [11, 23]. Lastly, there may be major omissions from the carbon budget, which some authors suggest may double emissions to 2 Pg C a^{-1}, compared to those obtained by Achard et al., using the identical deforestation figures (see [24], and response and counter-responses [10, 11, 27, 28] and see [29] for a detailed discussion).

Two methods have been used to detect whether intact tropical forests are a major sink: forest inventories and micrometeorological techniques (eddy-covariance). Both show sinks [16, 17], but are controversial [15, 30, 31]. Although inventories of *single* well-studied sites have reported no significant carbon sink [32–34], large compil-ations of inventory data from *multiple* sites show that

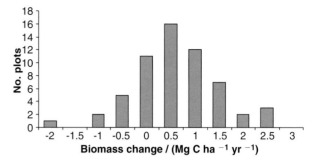

Figure 14.1 Frequency distribution of above-ground biomass change, from 59 × 1 ha long-term monitoring plots from across Amazonia over the 1980's and 1990's (from [18]). Includes corrections for wood density, lianas and small trees. The distribution is normal and shifted to the right of zero. The average increase is significantly greater than zero (0.61 ± 0.22 Mg C ha^{-1} a^{-1}).

most forest plots are increasing in biomass [16, 18], including recent results taking explicit account of high-lighted methodological concerns (Figure 14.1, [18]). If the South American results (0.6 Mg C ha^{-1} a^{-1}) are scaled to the biome (FAO figures [35]), this indicates a total sink within intact tropical forests of approx. 1.2 Pg C a^{-1}. By contrast, on average, the eddy-covariance stud-ies show much larger sinks (1 to 5.9 Mg C ha^{-1} a^{-1} [17, 20, 31]). The differences may be caused by methodolog-ical problems which underestimate night-time fluxes [31], or because inventories include only the fraction of the annual photosynthesis flux into wood production (10–25%), and there may be other sinks, or that carbon may be being transported to rivers, which release the equivalent of 1.2 Mg C ha^{-1} a^{-1} [21, 36].

Two interpretations of the new inventory data (and eddy flux data, see [14]) have been suggested: (1) that the sink is an artefact of the sampling, as most forests increase in biomass, and carbon, most of the time, as forests are naturally affected by rare disturbance events in which they rapidly lose carbon: they then accrue bio-mass and carbon slowly over long periods of time, or (2) that the sink is caused by an increase in net primary pro-ductivity (see [37]). However, if the sink is an artefact of disturbance, then growth fluxes must exceed mortality fluxes within intact forest plots but, on average, there should be no large change in these fluxes over time. By contrast, if the sink is caused by an increase in net pri-mary productivity then the growth flux should increase markedly through time [38].

Inventory data from across South America show that the growth flux is rapidly increasing (Figure 14.2; [37]). Furthermore, the mortality flux is increasing at a similar rate, but lagging the growth (which suggests an increase in inputs of coarse woody debris, which may offset some of the carbon sink, if they are far from equilibrium with the inputs from mortality, however this is unlikely given the long-term changes documented). These results are also replicated on a per stem basis which excludes most

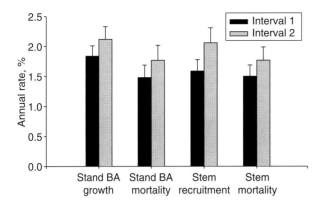

Figure 14.2 Annual rates of stand-level basal area growth, stand-level basal area mortality (correlated with biomass and carbon), Is there an A and B to precede the C??NO! stem recruitment and stem mortality from two consecutive census intervals, each giving the mean from 50 plots from across South America, with 95% CIs (from [37]). The average mid-year of the first and second censuses was 1989 and 1996 respectively. All four parameters show significant increases ($P < 0.05$).

potential measurement errors (Figure 14.2). The large increases ($\sim 2\%\ a^{-1}$) suggest a continent-wide increase in resource availability, increasing net primary productivity, and altering forest dynamics. Time-lag analyses suggest losses from a forest are $\sim 10–15$ years behind the gains, implicating long-term changes in available plant resources [39]. The most obvious candidate increasing resource availability is rising atmospheric CO_2 concentrations, consistent with theoretical, model and experimental results [38, 40–42], possibly coupled with increasing solar radiation [29, 38, 43].

The evidence from multi-site long-term forest monitoring plots, alongside other techniques, suggests that intact tropical forests are a carbon sink. Recent evidence also suggests tropical rivers are significant sources of carbon [21, 36]. This suggests that the 'large source, large sink' option above seems more plausible than the 'low source, no sink' option for tropical forests over the 1990s, and hence that the recent lower estimates of C release from deforestation, of $\sim 1\ Pg\ C\ a^{-1}$ are unlikely, as others have contested on a variety grounds [11, 12, 29]. Overall, this suggests that tropical forests were a highly dynamic component of the global carbon cycle over the 1990s, in terms of being a major source from deforestation, rivers and wetlands and a major sink in intact forest.

Major programs of on-the-ground monitoring of tropical forests, satellite campaigns and ongoing monitoring of the physical, chemical and biological environment across the tropics alongside targeted experimental work (e.g. exposing an entire tropical forest stand to elevated CO_2) will be necessary both to narrow the considerable uncertainty in the two major C fluxes from tropical forests, and also to elucidate their spatial location and causes.

14.3 Future Scenarios

To make predictions about the future, we must understand the drivers of change and how these then percolate through and alter the Earth System. There is great uncertainty at all stages of this predictive process. For example, the drivers of land-use change, in particular deforestation, are a complex mix of political, economic and climatic factors [44]. However, in short we can say with reasonable confidence that the demand for land that is currently tropical forest to be converted to other uses is expected to remain high, keeping carbon emissions high (notably as integration into market economies is the single most important pan-tropical underlying cause of deforestation, [44]). Here we focus solely on interactions and feedbacks between the tropics and changes expected from climate change.

14.3.1 *Photosynthesis/Respiration Changes*

Intact forests will remain a sink while carbon uptake associated with photosynthesis exceeds the carbon efflux from respiration. Under the simplest scenario of a steady rise in forest productivity over time, it is predicted that forests would remain a carbon sink for decades [45, 46]. However, the current increases in productivity, apparently caused by continuously improving conditions for tree growth, cannot continue indefinitely: if CO_2 is the cause, trees are likely to become CO_2 saturated (limited by another resource) at some point in the future. More generally, whatever these 'better conditions for growth' are, forest productivity will not increase indefinitely, as other factors, e.g. soil nutrients, will limit productivity.

Rising temperatures may also cause a reduction in the intact tropical forest sink, or cause forests to become a source in the future. Warmer temperatures increase the rates of virtually all chemical and biological processes in plants and soils (including the enhancement of any CO_2 fertilisation effect), until temperatures reach inflection-points where enzymes and membranes lose functionality. There is some evidence that the temperatures of leaves at the top of the canopy, on warm days, may be reaching such inflection-points around midday at some locations [38]. Canopy-to-air vapour deficits and stomatal feedback effects may also be paramount in any response of tropical forest photosynthesis to future climate change [47, 48].

The relationship between temperature changes and respiration is critical [49]. The first global circulation model (GCM) to include dynamic vegetation and a carbon cycle that is responsive to these dynamic changes, shows that under the 'business as usual' scenario of emissions, IS92a, atmospheric CO_2 concentrations are 900–980 ppmv (parts per million by volume) in 2100, compared to ~ 700 ppmv from previous GCMs [50, 51, 52]. These concentrations depend critically on (1) the alarming dieback of the Eastern Amazon rainforests, caused by climate change-induced drought, and (2) the subsequent release of C from soils. The release of C from soils is critically dependent on the assumed response of respiration to temperature and the modelling of soil carbon [52].

Carbon losses from respiration will almost certainly increase as air temperatures continue to increase. The key question is what form this relationship takes. Carbon gains from photosynthesis cannot rise indefinitely, and will almost certainly asymptote. Thus, the sink in intact tropical forests will diminish and eventually reverse. The major uncertainly is *when* this will occur.

14.3.2 *Functional or Biodiversity Changes*

Subtle functional composition, or biodiversity, changes could plausibly reduce or even reverse the current intact tropical forest C sink. A shift in species composition may be occurring as tree mortality rates have increased by $\sim$3% a^{-1} in recent decades [39, 53], causing an increase in the frequency of tree-fall gaps. This suggests a shift towards light-demanding species with high growth rates at the expense of more shade-tolerant species [38, 54]. Such fast-growing species are associated with lower wood specific gravity, and hence lower volumetric carbon content [55]. A decrease in mean wood specific gravity across Amazonia of just 0.4% a^{-1} would be enough remove the carbon sink effect of 0.6 Mg C ha^{-1} a^{-1}. As mean stand-level wood specific gravity values differ by >20% among Amazonian forests and species values vary 5-fold [56] it is possible that changes in species composition alone could remove or reverse the current sink contribution of tropical forests [54]. We know that forest stands with many fast-growing species that are highly dynamic have lower mean wood specific gravity and hold less above-ground C (Figure 14.3). Whether this plausible scenario will occur within forest stands, and over what timescales, is unknown at present.

In addition, lianas are structural parasites that decrease tree growth and increase mortality, and are disturbance adapted [57]. Thus, the rapid rise in large lianas across

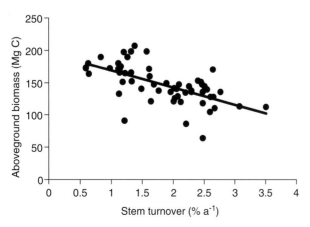

Figure 14.3 Relationship between forest dynamism (stem turnover), and carbon storage (above-ground biomass), from 59 plots from across Amazonia (biomass data from [18], corresponding turnover data from [39]).

Western Amazonia could also turn some surviving forests into a C source over time [58]. None of these functional shifts are present in current GCM models.

14.3.3 *Tropical Forest Collapse: Drought*

Climate change will alter precipitation patterns [4]. There are critical thresholds of water availability below which tropical forests cannot persist and are replaced by savanna systems, often around 1,200–1,500 mm rainfall per annum [59]. Thus, changing precipitation patterns may cause shifts in vegetation from carbon-dense tropical forests, to lower carbon savanna systems, if thresholds are crossed. Such a shift was seen in the first GCM model that included dynamic vegetation and a carbon cycle that is responsive to these dynamic changes, with the Eastern Amazon moving from a tropical forest system, eventually to a desert system [50–52]. However, such a transition was not seen in another 'fully-coupled' GCM model [60]. This is because of the poor agreement between the fully-coupled models on changing precipitation patterns, in terms of locations, durations and magnitudes, and on how soil carbon is modelled [52, 60].

Rainfall has reduced dramatically over the Northern Congo basin over the past two decades [61]. This current drying trend is of unknown cause. These forests are already relatively dry for tropical forests (*ca.* 1,500 mm a^{-1}), and may become savanna if current trends continue, leading to large carbon fluxes to the atmosphere. If the current drying trend is caused by climate change, this could lead to a positive feedback with the climate system exacerbating forest losses and carbon fluxes to the atmosphere.

14.3.4 *Tropical Forest Collapse: Fire*

In terms of climatic interactions, the flammability of a given forest is a key attribute. The hot and dry conditions of El Niño years compared to non-El Niño years partially explains the high incidence of forest burning, and hence partially explains the higher than average atmospheric CO_2 concentrations in these years [26, 62]. The 1997/8 ENSO event coincided with the burning of up to 20 million hectares of tropical forest [25] *and* showed the highest annual increase of atmospheric CO_2 concentrations since direct measurements began [63].

Approximately one-third of Amazonia was susceptible to fire during the much less severe 2001 ENSO period [64]. If droughts, temperatures and ENSO events increase in frequency and severity then the carbon flux from the tropics could rise rapidly in the future, potentially creating a dangerous positive feedback with the climate system.

14.4 Conclusions

While there is considerable uncertainty concerning the future trajectory of the tropical forest biome, (1) continued deforestation will undoubtedly lead to major C additions to the atmosphere, (2) the C sink contribution of remaining intact tropical forests, which we currently think to be ~15% of global fossil fuel emissions, appears unlikely to continue for the rest of this century, and (3) plausible mechanisms have been identified which may turn this biome to a modest or even mega-source of C. A carbon sink of $\sim 1.2\,\text{Pg C a}^{-1}$ occurs with an increase of above-ground C stocks of $\sim 0.5\%\,\text{a}^{-1}$, thus a modest switch to a C source would have serious global implications.

Mechanisms not currently incorporated into GCM models indicate that projected atmospheric CO_2 concentrations of 700–980 ppmv by 2100 under 'business as usual' scenarios (IS92a) may be conservative. This is especially true given the possibility of synergistic interactions, where ongoing forest loss, subsequent habitat fragmentation, more frequent fires, warmer temperatures, rising respiration costs, and periodically severe droughts may combine to dramatically increase C fluxes to the atmosphere, reinforcing the drivers of the increased carbon fluxes. Efforts to limit atmospheric concentrations of CO_2 cannot ignore the danger of substantial net carbon emissions from tropical forests in a globally changing world.

Acknowledgements

Simon L. Lewis and Yadvinder Malhi are supported by Royal Society University Research Fellowships.

REFERENCES

1. DeFries, R.S., Field C.B., Fung. I., Collatz G.J. and Bounoua L. *Combining satellite data and biogeochemical models to estimate global effects of human-induced land cover change on carbon emissions and primary productivity.* Global Biogeochemical Cycles, 1999. **13**(3): p. 803–815.

2. Houghton, R.A. *Revised estimates of the annual net flux of carbon to the atmosphere from changes in land use and land management 1850–2000.* Tellus Series B-Chemical and Physical Meteorology, 2003. **55**(2): p. 378–390.

3. Marland, G., T.A. Boden, and R. J. Andres., *Global, Regional, and National CO_2 Emissions. In Trends: A Compendium of Data on Global Change. Carbon Dioxide Information Analysis Center.* 2003, Oak Ridge National Laboratory: U.S. Department of Energy, Oak Ridge, Tenn., U.S.A.

4. IPCC, *Climate change 2001: the scientific basis.*, in *J. T. Houghton, Y. Ding, D. J. Griggs, M. Noguer, P.J. van der Linden, X. Dai, K. Maskell, C. A. Johnson.* 2001, Cambridge University Press.

5. Prentice, I.C., *The Carbon Cycle and Atmospheric Carbon Dioxide,* in *Climate change 2001: the scientific basis,* IPCC, Editor. 2001, Cambridge University Press.

6. Feely R.A., Sabine C.L., Lee K., Berelson W., Kleypas J., Fabry V.J. and Millero F.J. *Impact of anthropogenic CO_2 on the $CaCO_3$ system in the oceans.* Science, 2004. **305**(5682): p. 362–366.

7. Gurney K.R., Law R.M., Denning A.S., Rayner P.J., Baker D., Bousquet P., Bruhwiler L., Chen Y.H., Ciais P., Fan S., Fung I.Y., Gloor M., Heimann M., Higuchi K., John J., Maki T., Maksutov S., Masarie K., Peylin P., Prather M., Pak B.C., Randerson J., Sarmiento J., Taguchi S., Takahashi T. and Yuen C.W., *Towards robust regional estimates of CO_2 sources and sinks using atmospheric transport models.* Nature, 2002. **415**(6872): p. 626–630.

8. Houghton, R.A., *Why are estimates of the terrestrial carbon balance so different?* Global Change Biology, 2003. **9**(4): p. 500–509.

9. Rodenbeck C., Houweling S., Gloor M. and Heimann M., *CO_2 flux history 1982–2001 inferred from atmospheric data using a global inversion of atmospheric transport.* Atmospheric Chemistry and Physics, 2003. **3**: p. 1919–1964.

10. Achard F., Eva H.D., Mayaux P., Stibig H.J. and Belward A., *Improved estimates of net carbon emissions from land cover change in the tropics for the 1990s.* Global Biogeochemical Cycles, 2004. **18**(2): p. art. no.-GB2008.

11. Fearnside, P.M. and W.F. Laurance, *Tropical deforestation and greenhouse-gas emissions.* Ecological Applications, 2004. **14**(4): p. 982–986.

12. House J.I., Prentice I.C., Ramankutty N., Houghton R.A. and Heimann M., *Reconciling apparent inconsistencies in estimates of terrestrial CO_2 sources and sinks.* Tellus Series B-Chemical and Physical Meteorology, 2003. **55**(2): p. 345–363.

13. DeFries R.S., Houghton R.A., Hansen M.C., Field C.B., Skole D. and Townshend J. *Carbon emissions from tropical deforestation and regrowth based on satellite observations for the 1980s and 1990s.* Proceedings of the National Academy of Sciences of the United States of America, 2002. **99**(22): p. 14256–14261.

14. Saleska, S.R.M., S.D.; Matross, D. M.; Goulden, M.L.;Wofsy,S.C.; da Rocha, H.R.; de Camargo, P.B.; Crill, P.; Daube, B.C.; de Freitas, H.C.; Hutyra,L.; Keller, M.; Kirchhoff, V.; Menton, M.; Munger, J.W.; Pyle, E.H.; Rice, A.H.; Silva, H., *Carbon in Amazon Forests: Unexpected Seasonal Fluxes and Disturbance-Induced Losses.* Science, 2003. **403**: p. 1554–1557.

15. Clark, D.A., *Are tropical forests an important global carbon sink? revisiting the evidence from long-term inventory plots.* Ecological Applications, 2002. **12**: p. 3–7.

16. Phillips O.L., Malhi Y., Higuchi N., Laurance W.F., Nunez P.V., Vasquez R.M., Laurance S.G., Ferreira L.V., Stern M., Brown S., Grace J., *Changes in the carbon balance of tropical forests: evidence from long-term plots.* Science, 1998. **282**: p. 439–442.

17. Grace J., Lloyd J., McIntyre J., Miranda A.C., Meir P., Miranda H.S., Nobre C., Moncrieff J., Massheder J., Malhi Y., Wright I. and Gash J., *Carbon dioxide uptake by an undisturbed tropical rain forest in Southwest Amazonia, 1992 to 1993.* Science, 1995. **270**: p. 778–780.

18. Baker T.R., Phillips O.L., Malhi Y., Almeida S., Arroyo L., Di Fiore A., Erwin T., Higuchi N., Killeen T.J., Laurance S.G., Laurance W.F., Lewis S.L., Monteagudo A., Neill D.A., Vargas P.N., Pitman N.C.A., Silva J.N.M. and Martinez R.V., *Increasing biomass in Amazonian forest plots.* Philosophical Transactions of the Royal Society of London Series B-Biological Sciences, 2004. **359**(1443): p. 353–365.

19. Malhi, Y. and J. Grace, *Tropical forests and atmospheric carbon dioxide.* Trends in Ecology and Evolution, 2000. **15**: p. 332–337.

20. Malhi Y., Nobre A.D., Grace J., Kruijt B., Pereira M.G.P., Culf A. and Scott S., *Carbon dioxide transfer over a Central Amazonian rain forest.* Journal of Geophysical Research, 1998. **103**: p. 31513–31612.

21. Richey, J.E.M., J.M.; Aufdenkampe, A.K.; Ballester, V.M.; Hess, L.H., *Outgassing from Amazonian rivers and wetlands as a large tropical source of CO_2.* Nature, 2002. **416**: p. 617–620.

22. Cramer, W.B., Schaphoff, S. Lucht, W., Smith, B. and Sitch, S. *Tropical forests and the global carbon cycle: impacts of atmospheric carbon dioxide, climate change and rate of deforestation.* Philosophical Transactions of the Royal Society Series B: Biological Sciences, 2004. **359**: p. 331–343.

23. Houghton, R.A., *Aboveground Forest Biomass and the Global Carbon Balance.* Global Change Biology, 2005. **11**: p. 945–958.

24. Achard F., Eva H.D., Stibig H.J., Mayaux P., Gallego J., Richards T. and Malingreau J.P., *Determination of deforestation rates of the world's humid tropical forests.* Science, 2002. **297**(5583): p. 999–1002.

25. Cochrane, M.A., *Fire science for rainforests.* Science, 2003. **421**: p. 913–919.

26. Page S.E., Siegert F., Rieley J.O., Boehm H.D.V., Jaya A. and Limin S., *The amount of carbon released from peat and forest fires in Indonesia during 1997.* Nature, 2002. **420**(6911): p. 61–65.

27. Fearnside, P.M. and W.F. Laurance, *Comment on "Determination of deforestation rates of the world's humid tropical forests".* Science, 2003. **299**(5609): p. U3–U4.

28. Eva H.D., Achard F., Stibig H.J. and Mayaux P., *Response to comment on "Determination of deforestation rates of the world's humid tropical forests".* Science, 2003. **299**(5609): p. U5–U6.

29. Lewis, S.L., *Tropical forests and the Changing Earth System.* Philosophical Transactions of the Royal Society of London Series b-Biological Sciences, 2005. in press.

30. Phillips O.L., Malhi Y., Vinceti B., Baker T., Lewis S.L., Higuchi N., Laurance W.F., Vargas P.N., Martinez R.V., Laurance S., Ferreira L.V., Stern M., Brown S., Grace J., *Changes in growth of tropical forests: evaluating potential biases.* Ecological Applications, 2001.

31. Kruijt B., Elbers J.A., von Randow C., Araujo A.C., Oliveira P.J., Culf A., Manzi A.O., Nobre A.D., Kabat P. and Moors E.J., *The robustness of eddy correlation fluxes for Amazon rain forest conditions.* Ecological Applications, 2004. **14**(4): p. S101–S113.

32. Miller, S.D.G., M. L.; Menton, M.C.; Da Rocha H.R.; De Freitas, H.C.; Silva Figueira, A.M.E.; Dias de Sousa, C.A., *Biometric and micrometerological measurements of tropical forest carbon balance.* Ecological Applications, 2004. **14**: p. S114–S126.

33. Chave J., Condit R., Lao S., Caspersen J.P., Foster R.B., Hubbell S.P., *Spatial and temporal variation of biomass in a tropical forest: results from a large census plot in Panama.* Journal of Ecology, 2003. **91**: p. 240–252.

34. Rice, A.H.P., E.H.; Saleska, S.A.; Hutyra, L.; Palace, M.; Keller, M.; de Camargo, P.B.; Portilho, K.; Marques, D.F.; Wofsy, S.C., *Carbon balance and vegetation dynamics in an old-growth Amazonian forest.* Ecological Applications, 2004. **14**: p. S55–S71.

35. FAO, *State of the World's Forests.* 1999, Rome: FAO.

36. Mayorga, E.A., A.K.; Masiello, C.A.; Krusche, A.M.; Hedges, J.I.; Quay, P.D.; Richey, J.E.; Brown, T.A., *Young organic matter as a source of carbon dioxide outgassing from Amazonian rivers.* Nature, 2005. **436**: p. 538–541.

37. Lewis S.L., Phillips O.L., Baker T.R., Lloyd J., Malhi Y., Almeida S., Higuchi N., Laurance W.F., Neill D.A., Silva J.N.M., Terborgh J., Lezama A.T., Martinez R.V., Brown S., Chave J., Kuebler C., Vargas P.N. and Vinceti B., *Concerted changes in tropical forest structure and dynamics: evidence from 50 South American long-term plots.* Philosophical Transactions of the Royal Society of London Series B-Biological Sciences, 2004. **359**(1443): p. 421–436.

38. Lewis, S.L., Y. Malhi, and O.L. Phillips, *Fingerprinting the impacts of global change on tropical forests.* Philosophical Transactions of the Royal Society of London Series B-Biological Sciences, 2004. **359**(1443): p. 437–462.

39. Phillips O.L., Baker T.R., Arroyo L., Higuchi N., Killeen T.J., Laurance W.F., Lewis S.L., Lloyd J., Malhi Y., Monteagudo A., Neill D.A., Vargas P.N., Silva J.N.M., Terborgh J., Martinez R.V., Alexiades M., Almeida S., Brown S., Chave J., Comiskey J.A., Czimczik C.I., Di Fiore A., Erwin T., Kuebler C., Laurance S.G., Nascimento H.E.M., Olivier J., Palacios W., Patino S., Pitman N.C.A., Quesada C.A., Salidas M., Lezama A.T. and Vinceti B., *Pattern and process in Amazon tree turnover, 1976–2001.* Philosophical Transactions of the Royal Society of London Series B-Biological Sciences, 2004. **359**(1443): p. 381–407.

40. Lloyd, J. and G.D. Farquhar, *The CO_2 dependence of photosynthesis, plant growth responses to elevated atmospheric CO_2 concentrations and their interaction with soil nutrient status.1. General principles and forest ecosystems.* Functional Ecology, 1996. **10**(1): p. 4–32.

41. Long S.P., Ainsworth E.A., Rogers A., Ort D.R., *Rising atmospheric carbon dioxide: Plants face the future.* Annual Review of Plant Biology, 2004. **55**: p. 591–628.

42. Ainsworth, E.A. and S.P. Long, *What have we learned from 15 years of free-air CO_2 enrichment (FACE)? A meta-analytic review of the responses of photosynthesis, canopy.* New Phytologist, 2005. **165**(2): p. 351–371.

43. Wild M., Gilgen H., Roesch A., Ohmura A., Long C.N., Dutton E.G., Forgan B., Kallis A., Russak V. and Tsvetkov A., *From Dimming to Brightening: Decadal Changes in Solar Radiation at Earth's Surface.* Science, 2005. **308**(5723): p. 847–850.

44. Geist, H.J. and E.F. Lambin, *Proximate causes and underlying driving forces of tropical deforestation.* Bioscience, 2002. **52**(2): p. 143–150.

45. Chambers J.Q., Higuchi N., Tribuzy E.S., Trumbore S.E., *Carbon sink for a century.* Nature, 2001. **410**(6827): p. 429–429.

46. Cramer W., Bondeau A., Woodward F.I., Prentice I.C., Betts R.A., Brovkin V., Cox P.M., Fisher V., Foley J.A., Friend A.D., Kucharik C., Lomas M.R., Ramankutty N., Sitch S., Smith B., White A., Young-Molling C., *Global response of terrestrial ecosystem structure and function to CO_2 and climate change: results from six dynamic global vegetation models.* Global Change Biology, 2001. **7**(4): p. 357–373.

47. Sellers P.J., Bounoua L., Collatz G.J., Randall D.A., Dazlich D.A., Los S.O., Berry J.A., Fung I., Tucker C.J., Field C.B. and Jensen T.G., *Comparison of radiative and physiological effects of doubled atmospheric CO_2 on climate.* Science, 1996. **271**(5254): p. 1402–1406.

48. Lloyd J., Grace J., Miranda A.C., Meir P., Wong S.C., Miranda B.S., Wright I.R., Gash J.H.C., and McIntyre J., *A Simple Calibrated Model of Amazon Rain-Forest Productivity Based on Leaf Biochemical-Properties.* Plant Cell and Environment, 1995. **18**(10): p. 1129–1145.

49. Amthor, J.S., *The McCree-de Wit-Penning de Vries-Thornley respiration paradigms: 30 years later.* Annals of Botany, 2000. **86**(1): p. 1–20.

50. Cox P.M., Betts R.A., Collins M., Harris P.P., Huntingford C., Jones C.D., *Amazonian forest dieback under climate-carbon cycle projections for the 21st century.* Theoretical and Applied Climatology, 2004. **78**(1–3): p. 137–156.

51. Cox P.M., Betts R.A., Jones C.D., Spall S.A., Totterdell I.J., *Acceleration of global warming due to carbon-cycle feedbacks in a coupled climate model.* Nature, 2000. **408**: p. 184–187.

52. Jones, C.M., Coleman, C., Cox, P., Falloon, P., Jenkinson, D., Powelson, D., *Global climate change and soil carbon stocks; predictions from two contrasting models for the turnover of organic carbon in soil.* Global Change Biology, 2005. **11**: p. 154–166.

53. Phillips, O.L. and A.H. Gentry, *Increasing turnover through time in tropical forests.* Science, 1994. **263**: p. 954–7.

54. Korner, C., *CO_2 enrichment may cause tropical forest to become carbon sources.* Philosophical Transactions of the Royal Society, in press.

55. West, G.B., J.H. Brown, and B.J. Enquist, *A general model for the structure and allometry of plant vascular systems.* Nature, 1999. **400**: p. 664–667.

56. Baker T.R., Phillips O.L., Malhi Y., Almeida S., Arroyo L., Di Fiore A., Erwin T., Killeen T.J., Laurance S.G., Laurance W.F., Lewis S.L., Lloyd J., Monteagudo A., Neill D.A., Patino S., Pitman N.C.A., Silva J.N.M. and Martinez R.V., *Variation in wood density determines spatial patterns in Amazonian forest biomass.* Global Change Biology, 2004. **10**: p. 545–562.

57. Schnitzer, S.A. and F. Bongers, *The ecology of lianas and their role in forests.* Trends in Ecology & Evolution, 2002. **17**(5): p. 223–230.

58. Phillips O.L., Martinez R.V., Arroyo L., Baker T.R., Killeen T., Lewis S.L., Malhi Y., Mendoza A.M., Neill D., Vargas P.N., Alexiades M., Ceron C., Di Fiore A., Erwin T., Jardim A., Palacios W., Saldias M. and Vinceti B., *Increasing dominance of large lianas in Amazonian forests.* 2002.

59. Salzmann, U. and P. Hoelzmann, *The Dahomey Gap: an abrupt climatically induced rain forest fragmentation in West Africa during the late Holocene.* Holocene, 2005. **15**(2): p. 190–199.

60. Zeng N., Qian H.F., Munoz E., Iacono R., *How strong is carbon cycle-climate feedback under global warming?* Geophysical Research Letters, 2004. **31**(20): p. art. no.-L20203.

61. Malhi, Y. and J. Wright, *Spatial patterns and recent trends in the climate of tropical rainforest regions.* Philosophical Transactions of the Royal Society of London Series B-Biological Sciences, 2004. **359**(1443): p. 311–329.

62. Langenfelds R.L., Francey R.J., Pak B.C., Steele L.P., Lloyd J., Trudinger C.M., Allison C.E., *Interannual growth rate variations of atmospheric CO$_2$ and its delta C-13, H-2, CH$_4$, and CO between 1992 and 1999 linked to biomass burning.* Global Biogeochemical Cycles, 2002. **16**(3): p. art. no.-1048.

63. Cochrane, M.A., *Fire science for rainforests.* Nature, 2003. **421**(6926): p. 913–919.

64. Nepstad D., Lefebvre P., Da Silva U.L., Tomasella J., Schlesinger P., Solorzano L., Moutinho P., Ray D., Benito J.G., *Amazon drought and its implications for forest flammability and tree growth: a basin-wide analysis.* Global Change Biology, 2004. **10**(5): p. 704–717.

CHAPTER 15

Conditions for Sink-to-Source Transitions and Runaway Feedbacks from the Land Carbon Cycle

Peter M. Cox[1], Chris Huntingford[2] and Chris D. Jones[3]

[1]*Centre for Ecology and Hydrology, Winfrith, Dorset, UK*
[2]*Centre for Ecology and Hydrology, Wallingford, Oxon, UK*
[3]*Hadley Centre, Met Office, Fitzroy Road, Exeter, UK*

ABSTRACT: The first GCM climate-carbon cycle simulation indicated that the land biosphere could provide a significant acceleration of 21st century climate change (Cox *et al.* 2000). In this numerical experiment the carbon storage was projected to decrease from about 2050 onwards as temperature-enhanced respiration overwhelmed CO_2-enhanced photosynthesis. Subsequent climate-carbon cycle simulations also suggest that climate change will suppress land-carbon uptake, but typically do not predict that the land will become an overall source during the next 100 years (Friedlingstein *et al.*, accepted). Here we use a simple land carbon balance model to analyse the conditions required for a land sink-to-source transition, and address the question; could the land carbon cycle lead to a runaway climate feedback?

The simple land carbon balance model has effective parameters representing the sensitivities of climate and photosynthesis to CO_2, and the sensitivities of soil respiration and photosynthesis to temperature. This model is used to show that (a) a carbon sink-to-source transition is inevitable beyond some finite critical CO_2 concentration provided a few simple conditions are satisfied, (b) the value of the critical CO_2 concentration is poorly known due to uncertainties in land carbon cycle parameters and especially in the climate sensitivity to CO_2, and (c) that a true runaway land carbon-climate feedback (or linear instability) in the future is unlikely given that the land masses are currently acting as a carbon sink.

15.1 Introduction

Vegetation and soil contain about three times as much carbon as the atmosphere, and they exchange very large opposing fluxes of carbon dioxide with it. Currently the land is absorbing about a quarter of anthropogenic CO_2 emissions, because uptake by plant photosynthesis is outstripping respiration from soils (Houghton *et al.* 1996). However, these opposing fluxes are known to be sensitive to climate, so the fraction of emissions taken up by the land is likely to change in the future. A number of authors have discussed the possibility of the land carbon sink either saturating or reversing (see for example Woodwell and Mackenzie (1995), Lenton and Huntingford (2003)), primarily because of the potential for accelerated decomposition of soil organic matter under global warming (Jenkinson *et al.* 1991). Simple box models of the climate-carbon system have also demonstrated sink-to-source transitions in the land carbon cycle (e.g. Lenton 2000).

The General Circulation Models (GCMs) used to make climate projections have typically neglected such climate-carbon cycle feedbacks, but recently a number of GCM modelling groups have begun to include representations of vegetation and the carbon cycle within their models. The first GCM simulation of this type suggested that feedbacks between the climate and the land biosphere could significantly accelerate atmospheric CO_2 rise and climate change over the 21st century (Cox *et al.* 2000). Subsequent GCM climate-carbon cycle projections also suggest that climate change will suppress land carbon uptake, but typically do not predict that the land will become a carbon source within the simulated period to 2100 (Friedlingstein, accepted).

The terrestrial components used in these first generation coupled climate-carbon cycle GCMs reproduce the land carbon sink as a competition between the direct effects of CO_2 on plant growth, and the effects of climate change on plant and soil respiration. Whilst increases in atmospheric CO_2 are expected to enhance photosynthesis (and reduce transpiration), the associated climate warming is likely to increase plant and soil respiration. Thus there is a battle between the direct effect of CO_2, which tends to increase terrestrial carbon storage, and the indirect effect through climate warming, which may reduce carbon storage.

The outcome of this competition has been seen in a range of dynamic global vegetation models or 'DGVMs' (Cramer *et al.* 2001), each of which simulate reduced land carbon under climate change alone and increased carbon storage with CO_2 increases only. In most DGVMs, the combined effect of the CO_2 and associated climate change results in a reducing sink towards the end of the 21st century, as CO_2-induced fertilisation begins to saturate but soil respiration continues to increase with temperature. This is in itself an important result as it suggests that climate change will suppress the land carbon sink, and therefore lead to greater rates of CO_2 increase and global warming than previously assumed. However, in most models the land carbon cycle remains an overall sink for CO_2, and thus continues to provide a brake on increasing atmospheric CO_2.

The impact of climate change on the land carbon cycle is especially strong in the coupled model projections of Cox *et al.* (2000), leading to the land carbon cycle becoming an overall source of CO_2 from about 2050 onwards (under a 'business as usual' emissions scenario). In this case the land carbon cycle stops slowing climate change, and instead starts to accelerate it by releasing additional CO_2 to the atmosphere. This 'sink-to-source' transition point may be seen as one possible definition of 'dangerous climate change'. In the next section of this chapter we use a transparently simple land carbon cycle model to derive a condition for the critical CO_2 concentration at which the sink-to-source transition will occur. The resulting analytical expression is used to highlight the key uncertainties that contribute to divergences amongst existing DGVM and GCM model projections (section 15.2.1).

Section 15.3 examines the conditions for an even stronger 'runaway' land carbon cycle feedback. In this case the carbon cycle-climate system becomes linearly unstable to an arbitrary perturbation, leading to a release of land carbon to the atmosphere even in the absence of anthropogenic emissions. This state therefore represents not just 'dangerous climate change' but 'rapid climate change' in which the CO_2 increase and climate change are potentially much faster than the rate of anthropogenic forcing of the system. We use the simple model to show that such a runaway feedback is possible in principle (e.g. if the climate sensitivity to CO_2 is very high), but is unlikely given the existence of a land carbon sink in the present day.

15.2 Conditions for Sink-to-Source Transitions in the Land Carbon Cycle

In this section we introduce a very simple terrestrial carbon balance model to demonstrate how the conversion of a land CO_2 sink to a source is dependent on the responses of photosynthesis and respiration to CO_2 increases and climate warming. We consider the total carbon stored in vegetation and soil, C_T, which is increased by photosynthesis, Π, and reduced by the total ecosystem respiration, R:

$$\frac{dC_T}{dt} = \Pi - R \qquad (15.1)$$

where Π is sometimes called Gross Primary Productivity (GPP), and R represents the sum of the respiration fluxes from the vegetation and the soil. In common with many others (McGuire *et al.* 1992, Collatz *et al.* 1991, Collatz *et al.* 1992, Sellers *et al.* 1998), we assume that GPP depends directly on the atmospheric CO_2 concentration, C_a, and the surface temperature, T (in °C):

$$\Pi = \Pi_{max} \left\{ \frac{C_a}{C_a + C_{0.5}} \right\} f(T) \qquad (15.2)$$

where Π_{max} is the value which GPP asymptotes towards as $C_a \to \infty$, $C_{0.5}$ is the 'half-saturation' constant (i.e. the value of C_a for which Π is half this maximum value), and $f(T)$ is an arbitrary function of temperature. We also assume that the total ecosystem respiration, R, is proportional to the total terrestrial carbon, C_T. The specific respiration rate (i.e. the respiration per unit carbon) follows a 'Q10' dependence, which means that it increases by a factor of q_{10} for a warming of T by 10°C. Thus the ecosystem respiration rate is given by:

$$R = r \, C_T q_{10}^{(T-10)/10} \qquad (15.3)$$

where r is the specific respiration rate at T = 10°C. It is more usual to assume separate values of r and q_{10} for different carbon pools (e.g. soil/vegetation, leaf/root/wood), but our simpler assumption will still offer good guidance as long as the relative sizes of these pools do not alter significantly under climate change. Near surface temperatures are expected to increase approximately logarithmically with the atmospheric CO_2 concentration, C_a (Houghton *et al.* 1996):

$$\Delta T = \frac{\Delta T_{2 \times CO_2}}{\log 2} \log \left\{ \frac{C_a}{C_a(0)} \right\} \qquad (15.4)$$

where ΔT is the surface warming, $\Delta T_{2 \times CO_2}$ is the climate sensitivity to doubling atmospheric CO_2, and $C_a(0)$ is the initial CO_2 concentration. We can use this to eliminate CO_2 induced temperature changes from Equation 15.3:

$$R = r_0 \, C_T \left\{ \frac{C_a}{C_a(0)} \right\}^{\mu} \qquad (15.5)$$

where $r_0 \, C_T$ is the initial ecosystem respiration (i.e. at $C_a = C_a(0)$) and the exponent μ is given by:

$$\mu = \frac{\Delta T_{2 \times CO_2}}{10} \frac{\log q_{10}}{\log 2} \qquad (15.6)$$

We can now use Equations 15.1, 15.2 and 15.5 to solve for the equilibrium value of terrestrial carbon, C_T^{eq}:

$$C_T^{eq} = \Pi \left\{ \frac{C_a}{C_a + C_{0.5}} \right\} \left\{ \frac{C_a(0)}{C_a} \right\}^{\mu} \frac{f(T)}{r_0} \qquad (15.7)$$

The land will tend to amplify CO_2-induced climate change if C_T^{eq} decreases with increasing atmospheric CO_2 (i.e. $dC_T^{eq}/dC_a < 0$). Differentiating Equation 15.7 with respect to C_a yields:

$$\frac{dC_T^{eq}}{dC_a} = C_T^{eq} \left[\frac{(1 - \mu_*)}{C_a} - \frac{1}{C_a + C_{0.5}} \right] \qquad (15.8)$$

where

$$\mu_* = \frac{\Delta T_{2 \times CO_2}}{\log 2} \left\{ \frac{\log q_{10}}{10} - \frac{1}{f} \frac{df}{dT} \right\} \qquad (15.9)$$

The equilibrium land carbon storage, C_T^{eq} (Equation 15.7), and the rate of change of equilibrium land carbon with respect to atmospheric carbon dC_T^{eq}/dC_A (Equation 15.8), are plotted in Figures 15.1 and 15.2 for three values of μ_*. For small values of μ_* the equilibrium land carbon increases monotonically over the range of CO_2 concentrations of interest (180–1000 ppmv), implying that the land would act as a carbon sink throughout the 21st century. By contrast, large values of μ_* show a monotonically decreasing land carbon storage with CO_2 concentration, implying a continuous land carbon source, which is at odds with the existence of a current-day land carbon sink. Only for intermediate values of μ_* do we see a turning point in the land carbon storage as a function of CO_2, with a current-day land carbon sink becoming a source before the end of the century (Figure 15.2).

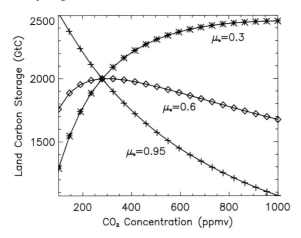

Figure 15.1 Equilibrium land carbon storage, C_T^{eq}, versus atmospheric CO_2 concentration for three values of μ_*. These curves are calculated from Equation 15.7 assuming $C_a(0) = 280$ ppmv, $C_T(0) = 2000$ GtC, $\Pi(0) = 120$ GtC yr^{-1}, $C_{0.5} = 500$ ppmv, and $f(T) = 1$.

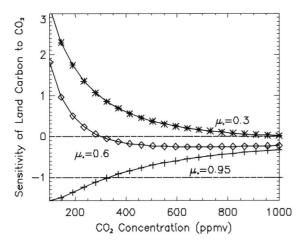

Figure 15.2 Rate of change of equilibrium land carbon with respect to atmospheric carbon, dC_T^{eq}/dC_A, versus atmospheric CO_2 concentration for three values of μ_*. These curves are calculated from Equation 15.8 assuming $C_a(0) = 280$ ppmv, $C_T(0) = 2000$ GtC, $\Pi(0) = 120$ GtC yr^{-1}, $C_{0.5} = 500$ ppmv, and $f(T) = 1$.

The sink-to-source turning point occurs where the rate of change of land carbon storage with CO_2 passes through zero, from positive (carbon sink), to negative (carbon source). From Equation 15.8, the condition for the land to become a source of carbon under increasing CO_2 is therefore:

$$C_a > \frac{1 - \mu_*}{\mu_*} C_{0.5} \qquad (15.10)$$

This means that there will always be a critical CO_2 concentration beyond which the land becomes a source, as long as:

(i) CO_2 fertilisation of photosynthesis saturates at high CO_2, i.e. $C_{0.5}$ is finite.
(ii) $\mu_* > 0$, which requires:
    (a) climate warms with increasing CO_2, i.e. $\Delta T_{2 \times CO_2} > 0$
    (b) respiration increases more rapidly with temperature than GPP, i.e.

$$\frac{\log q_{10}}{10} > \frac{1}{f} \frac{df}{dT} \qquad (15.11)$$

Conditions (i) and (ii)(a) are satisfied in the vast majority of terrestrial ecosystem and climate models. Detailed models of leaf photosynthesis indicate that $C_{0.5}$ will vary with temperature from about 300 ppmv at low temperatures, up to about 700 ppmv at high temperatures (Collatz *et al.* 1991). Although there are differences in the magnitude and patterns of predicted climate change, all GCMs produce a warming when CO_2 concentration is doubled.

There is considerable disagreement over the likely long-term sensitivity of respiration fluxes to temperature, with some suggesting that temperature-sensitive 'labile' carbon pools will soon become exhausted once the ecosystem enters a negative carbon balance (Giardina and Ryan 2000). However, condition (ii)(b) is satisfied by the vast majority of existing land carbon cycle models, and seems to be implied (at least on the 1–5 year timescale) by climate-driven inter-annual variability in the measured atmospheric CO_2 concentration (Jones and Cox (2001), Jones *et al.* 2001).

15.2.1 *Application to the Contemporary Climate*

We therefore conclude that the terrestrial carbon sink has a finite lifetime, but the length of this lifetime is highly uncertain. We can see why this is from our simple model (Equation 15.10). The critical CO_2 concentration is very sensitive to μ_* which is itself dependent on the climate sensitivity, and the difference between the temperature dependences of respiration and GPP (Equation 15.9).

We expect the temperature sensitivity of GPP to vary regionally, since generally a warming is beneficial for photosynthesis in mid and high latitudes (i.e. $df/dT < 0$), but not in the tropics where the existing temperatures are near optimal for vegetation (i.e. $df/dT \leqslant 0$). As a result, we might expect global mean GPP to be only weakly dependent on temperature (df/dT $\approx$ 0), even though there may be significant regional climate effects on GPP through changes in water availability.

Most climate models produce estimates of climate sensitivity to doubling CO_2 in the often-quoted range of 1.5 K to 4.5 K (Houghton *et al.* 1996), but there is now a growing realisation that the upper bound on climate sensitivity is much higher. A recent 'parameter ensemble' of GCM experiments (in which each ensemble member has a different set of feasible internal model parameters) produced model variants with climate sensitivities as high as 11 K (Stainforth *et al.* 2005). In principle it ought to be possible to estimate climate sensitivity by using the observed warming over the 20th century as a constraint. Unfortunately, in practice high climate sensitivities cannot be ruled out owing to uncertainties in the extent to which anthropogenic aerosols have offset greenhouse warming (Andreae *et al.* 2005).

In order to demonstrate the uncertainties in the critical CO_2 concentration we take the conservative 1.5 to 4.5 K range for the global climate sensitivity. Mean warming over land is likely to be a more appropriate measure of the climate change experienced by the land biosphere. We estimate a larger range of $2\,K < \Delta T_{2 \times CO_2} < 7\,K$ because the land tends to warm more rapidly than the ocean (Huntingford and Cox 2000). The sensitivity of ecosystem respiration to temperature, as summarised by the q_{10} parameter, is known to vary markedly amongst ecosystems, but here we require an effective value to represent the climate sensitivity of global ecosystem respiration. Fortunately, anomalies in the growth-rate of atmospheric CO_2, associated with El Niño events (Jones *et al.* 2001), and the Pinatubo volcanic eruption (Jones and Cox 2001), give a reasonably tight constraint on this parameter of $1.5 < q_{10} < 2.5$.

We can therefore derive a range for μ_*, based on plausible values of climate sensitivity over land ($2\,K < \Delta T_{2 \times CO_2} < 7\,K$) and respiration sensitivity ($1.5 < q_{10} < 2.5$). This range of $0.1 < \mu_* < 0.9$, translates into a critical CO_2 concentration which is somewhere between 0.1 and 9 times the half-saturation constant (Equation 15.10). Therefore on the basis of this simple analysis the range of possible critical CO_2 values spans almost two orders of magnitude. Evidently, the time at which the sink-to-source transition will occur is extremely sensitive to these uncertain parameters. This may explain why many of the existing terrestrial models do not reach this critical point before 2100 (Cramer *et al.* (2001), Friedlingstein *et al.* 2005). It is also interesting to note that the 'central estimate' of $q_{10} = 2$, $C_{0.5} = 500$ ppmv, and $\Delta T_{2 \times CO_2} = 4.8$ K (which is consistent with the warming over land in the Hadley Centre coupled model) yields a critical CO_2 value of about 550 ppmv, which is remarkably close to the sink-to-source transition seen in the Hadley Centre experiment.

In the absence of significant non-CO_2 effects on climate change (i.e. assuming that anthropogenic aerosols have approximately offset the warming due to the minor greenhouse gases), we can reduce the uncertainty range further. Under this assumption, critical CO_2 values which are lower than the current atmospheric concentration are not consistent with the observations, since the 'natural' land ecosystems appear to be a net carbon sink rather than a source at this time (Schimel *et al.* 1996). For a typical half-saturation constant of $C_{0.5} = 500$ ppmv this implies that combinations of q_{10} and $\Delta T_{2 \times CO_2}$ which yield values of $\mu_* > 0.6$ are unrealistic. We will return to this point in section 15.3.

We draw two main conclusions from this section. The recognised uncertainties in climate and respiration sensitivity imply a very large range in the critical CO_2 concentration beyond which the land will act as a net carbon source. However, the central estimates for these parameters suggest a real possibility of this critical point being passed by 2100 in the real Earth system, under a 'business as usual' emissions scenario, in qualitative agreement with the results from the Hadley Centre coupled climate-carbon cycle model.

15.3 Conditions for Runaway Feedback from the Land Carbon Cycle

The sensitivity of a system can be defined in terms of the relationship between the forcing of the system (e.g. anthropogenic CO_2 emissions) and its response (e.g. global warming). Rapid or abrupt change is normally associated with responses that are much larger than the forcing, or even independent of it. The latter are typically described as 'instabilities'.

Although a sink-to-source transition in the land carbon cycle would imply an acceleration of climate change, it would not necessarily lead to a sudden change in the Earth System. In this section we examine the necessary conditions for the land carbon-climate system to be linearly unstable at some finite CO_2 concentration. If such a threshold existed, and was crossed, the land would spontaneously lose carbon to the atmosphere, leading to sufficient greenhouse warming to sustain the release even in the absence of anthropogenic emissions. Such instabilities are often termed 'runaway feedbacks' because of their self-sustaining nature.

Even such strong positive feedbacks are ultimately limited by the depletion of reservoirs (e.g. soil carbon), and longer-term negative feedbacks (e.g. uptake of CO_2 by the oceans). In the context of land carbon-climate feedbacks on the century timescale, fast carbon loss from the tropics may completely overwhelm slow carbon uptake in high latitudes, even though in the longer term the biosphere may contain more carbon under high CO_2 conditions. These very different timescales for carbon loss and accumulation mean that the existence of high-carbon storage on the land during hot climates of the past (e.g. the mid-Cretaceous 100 million years ago) does not rule out the possibility of transient runaway instabilities under anthropogenic climate change in the future.

A runaway condition is defined by an instability such that a small perturbation grows exponentially, i.e. a runaway positive feedback requires linear instability (i.e. a feedback gain factor greater than 1). Although Equation 15.10 defines the critical CO_2 concentration for the land carbon cycle to provide a positive feedback, it does not ensure that this feedback is strong enough for a runaway. In order to define the condition for linear instability we rewrite Equation 15.1 in the form:

$$\frac{dC_T}{dt} = \frac{\{C_T^{eq} - C_T\}}{\tau} \qquad (15.12)$$

Here we have used Equation 15.5 to define the timescale, τ, which characterizes the rate at which the terrestrial carbon storage, C_T, approaches its equilibrium value, C_T^{eq},

$$\tau = \frac{1}{r_0} \left\{ \frac{C_a(0)}{C_a} \right\}^{\mu} \qquad (15.13)$$

We consider a perturbation to an initial equilibrium state defined by $C_T = C_T^{eq}(0)$ and $C_A = C_A(0)$, where C_A is the atmospheric carbon content, in GtC, associated with the CO_2 concentration C_a, in ppmv ($C_A = 2.123\, C_a$). A runaway occurs when C_A increases even in the absence of any CO_2 emissions, such that the total carbon in the atmosphere-land-ocean system is conserved:

$$\Delta C_A + \Delta C_T + \Delta C_O = 0 \qquad (15.14)$$

where ΔC_A, ΔC_T and ΔC_O represent perturbations to the carbon in the atmosphere, land and ocean respectively. For simplicity we assume that the ocean takes-up a fraction χ_o of any increase in atmospheric carbon, i.e. $\Delta C_o = \chi_o\, \Delta C_A$, so the carbon conservation Equation becomes:

$$\Delta C_A = -\frac{1}{1 + \chi_o} \Delta C_T \qquad (15.15)$$

Now C_T^{eq} is a function of C_a as described by Equation 15.7, such that:

$$C_T^{eq} \approx C_T^{eq}(0) + \frac{dC_T^{eq}}{dC_A} \Delta C_A \qquad (15.16)$$

Substituting Equations 15.15 and 15.16 into 15.12 yields an Equation for the perturbation to the land carbon:

$$\frac{d\Delta C_T}{dt} = -\frac{\Delta C_T}{\tau} \left\{ 1 + \frac{1}{(1 + \chi_o)} \frac{dC_T^{eq}}{dC_A} \right\} \qquad (15.17)$$

This is a linear Equation with a solution of the form $\Delta C_T = Ke^{\lambda t}$ where λ is the growth-rate of the linear instability,

$$\lambda = -\frac{1}{\tau} \left\{ 1 + \frac{1}{(1 + \chi_o)} \frac{dC_T^{eq}}{dC_A} \right\} \qquad (15.18)$$

The condition for linear instability or 'runaway' is $\lambda > 0$, i.e.:

$$\frac{dC_T^{eq}}{dC_A} < -(1 + \chi_o) \qquad (15.19)$$

This is much more stringent than the condition for positive feedback ($dC_T^{eq}/dC_A < 0$).

Equations 15.7, 15.8 and 15.19 together provide a condition for runaway in terms of the CO_2 concentration (C_a) and parameters associated with the climate change and the carbon cycle response (μ_*, $C_{0.5}$, Π_{max}, r_0, df/dT).

Now we search for the conditions necessary for runaway to occur at any CO_2 concentration, by determining whether the minimum value of dC_T^{eq}/dC_A satisfies Equation 15.19. The minimum value occurs where $d^2C_T^{eq}/dC_a^2 = 0$, so we first differentiate Equation 15.8 with respect to C_a:

$$\frac{d^2C_T^{eq}}{dC_a^2} = \frac{C_T^{eq}}{C_a^2(C_a + C_{0.5})^2} [\mu_*(\mu_* + 1)C_a^2 + 2(\mu_*^2 - 1)C_{0.5}C_a + \mu_*(\mu_* - 1)C_{0.5}^2] \qquad (15.20)$$

The turning points of dC_T^{eq}/dC_a occur where the quadratic equation within the square brackets is zero. The root corresponding to the minimum value (i.e. maximum positive feedback) is given by:

$$\frac{C_a}{C_{0.5}} = \frac{(1 - \mu_*)}{\mu_*} \left\{ 1 + \frac{1}{\sqrt{1 - \mu_*^2}} \right\} \qquad (15.21)$$

Equation 15.21 gives the CO_2 concentration at which the positive feedback from the carbon cycle is strongest. Note that this critical CO_2 concentration is always larger than the critical CO_2 concentration for sink-to-source transition (see Figure 15.3).

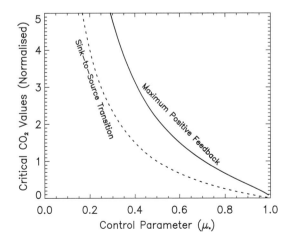

Figure 15.3 The critical CO_2 concentrations beyond which the land becomes an overall source of CO_2 (dashed line), and at which the positive feedback is maximised (continuous line), as a function of the control parameter, μ_*. These curves are calculated from Equations 15.10 and 15.21 respectively.

By substituting Equation 15.21 into Equation 15.8, we can determine the most negative value of dC_T^{eq}/dC_a which represents the strongest positive feedback from the land carbon cycle (Figure 15.4). Only values which satisfy Equation 15.19 are capable of producing a runaway feedback/linear instability, which requires $dC_T^{eq}/dC_a < -1$ even in the absence of ocean carbon uptake (i.e. $\chi_o = 0$). This necessary condition for a runaway land carbon cycle feedback is represented by the horizontal dashed line in Figure 15.4. Note that $\mu_* > 0.9$ is required for runaway.

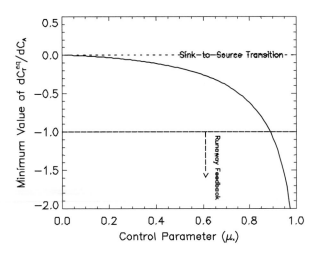

Figure 15.4 The minimum value of dC_T^{eq}/dC_A (i.e. the maximum positive feedback) versus the control parameter, μ_*. Values below the dashed line have the potential to produce a runaway feedback. Other parameters are as listed in the caption to Figure 15.1.

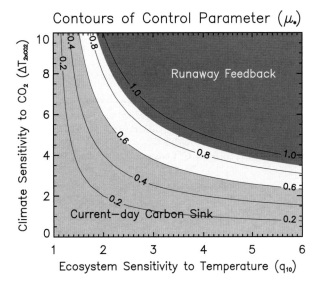

Figure 15.5 Contours of the control parameter, μ_*, versus ecosystem sensitivity to temperature (as summarised by the effective q_{10} parameter), and climate sensitivity to CO_2 ($\Delta T_{2 \times CO_2}$). The region of parameter space where a linear instability or "runaway feedback" is possible is shaded dark grey. The light grey region defines the region of parameter space consistent with a land carbon sink now (in the absence of significant net non-CO_2 effects on climate).

The fact that such a large value of μ_* implies a present-day land carbon source (Figure 15.1), indicates that a land carbon cycle runaway in the future is unlikely given the existence of a current-day land carbon sink. Figure 15.5 shows the separation of the 'Current-day Carbon Sink' and 'Runaway Feedback' regions in the $\{\Delta T_{2 \times CO_2} - q_{10}\}$ parameter space.

15.4 Conclusions

The results from offline dynamic global vegetation models (Cramer *et al.* 2001) and from the first generation coupled climate-carbon cycle GCMs (Friedlingstein *et al.* 2003), suggest that climate change will adversely affect land carbon uptake. In some models this effect is strong enough to convert the current land carbon sink to a source under 21st century climate change (Cox *et al.* 2000). In this paper we have applied a very simple land carbon balance model to produce an analytical expression for the critical CO_2 concentration at which the source-to-sink transition will occur. Beyond this critical point the land carbon cycle accelerates anthropogenic climate change, so this also represents one possible definition of 'dangerous climate change' in the context of the United Nations Framework Convention on Climate Change.

We have shown that the critical CO_2 concentration for such a sink-to-source transition in the land carbon cycle is dependent on a single control parameter (μ_*), which is itself dependent on the climate sensitivity to CO_2 and the sensitivities of photosynthesis and ecosystem respiration to climate. Relatively small changes in these parameters can change the critical CO_2 concentration significantly, helping to explain why most existing terrestrial carbon cycle models do not produce a sink-to-source transition in the 21st century.

We have also used the simple carbon balance model to examine the necessary conditions for a runaway land carbon cycle feedback. A runaway occurs when the gain factor of the (climate-land carbon storage) feedback loop exceeds one, which is equivalent to the condition for the system to be linearly unstable to an arbitrary perturbation. In this case a change in atmospheric CO_2 concentration could occur in the absence of significant anthropogenic emissions, leading to a rapid climate change (i.e. one that is potentially much faster than the anthropogenic forcing that prompted it). We have shown that the condition for such a runaway feedback is much more stringent than the condition for a positive feedback. Furthermore, although a runaway is theoretically possible (e.g. if the climate sensitivity to CO_2 is very high), the simple model indicates that such a strong land carbon source in the future is unlikely given the existence of a land carbon sink now.

Our analysis confirms the importance of reducing the uncertainties in eco-physiological responses to climate change and CO_2 if we are to be forewarned of a possible source-to-sink transition in the land carbon cycle. However, it also highlights the critical nature of uncertainties in the climate sensitivity, which not only determines the magnitude of climate change for a given CO_2, but also influences the strength of the land carbon cycle feedback, and therefore the anthropogenic emissions consistent with stabilisation at the given CO_2 concentration (Jones *et al.*, this volume).

Acknowledgements

This work PMC was supported by the European Commission under the 'CAMELS' project (PMC and CH); the UK Department of the Environment, Food and Regional Affairs, under contract PECD 7/12/37 (CDJ and PMC); and Science Budget funding from the Centre for Ecology and Hydrology (CH).

REFERENCES

Andreae M.O., C.D. Jones and P.M. Cox, 2005: Strong present-day aerosol cooling implies a hot future. *Nature*, **435**, 1187–1190.

Collatz, G.J., J.T. Ball, C. Grivet, and J.A. Berry, 1991: Physiological and environmental regulation of stomatal conductance, photosynthesis and transpiration: A model that includes a laminar boundary layer. *Agric. and Forest Meteorol.*, **54**, 107–136.

Collatz, G.J., M. Ribas-Carbo, and J.A. Berry, 1992: A coupled photosynthesis-stomatal conductance model for leaves of C_4 plants. *Aus. J. Plant Physiol*, **19**, 519–538.

Cox, P.M., C. Huntingford, and R.J. Harding, 1998: A canopy conductance and photosynthesis model for use in a GCM land surface scheme. *J. Hydrology*, **212–213**, 79–94.

Cox, P.M., R.A. Betts, C.D. Jones, S.A. Spall, and I.J. Totterdell, 2000: Acceleration of global warming due to carbon-cycle feedbacks in a coupled climate model. *Nature*, **408**, 184–187.

Cox, P.M., R.A. Betts, M. Collins, C. Harris, C. Huntingford, and C.D. Jones, 2004: Amazon dieback under climate-carbon cycle projections for the 21st century. *Theoretical and Applied Climatology*, **78**, 137–156.

Cramer, W., A. Bondeau, F. Woodward, I. Prentice, R. Betts, V. Brovkin, P. Cox, V. Fischer, J. Foley, A. Friend, C. Kucharik, M. Lomas, N. Ramankutty, S. Sitch, B. Smith, A. White, and C. Young-Molling, 2001: Global response of terrestrial ecosystem structure and function to CO_2 and climate change: results from six dynamic global vegetation models. *Global Change Biol.*, **7**, 357–374.

Friedlingstein, P., J.-L. Defresne, P.M. Cox, and P. Rayner, 2003: How positive is the feedback between climate change and the carbon cycle? *Tellus*, **55B**, 692–700.

Friedlingstein, P., P. Cox, R. Betts, L. Bopp, W. von Bloh, V. Brovkin, S. Doney, M. Eby, I. Fung, B. Govindasamy, J. John, C. Jones, F. Joos, T. Kato, M. Kawamiya, W. Knorr, K. Lindsay, H.D. Matthews, T. Raddatz, P. Rayner, C. Reick, E. Roeckner, K.-G. Schnitzler, R. Schnur, K. Strassmann, A. J. Weaver, C. Yoshikawa, and N. Zeng,

2005: Climate-carbon cycle feedback analysis: Results from the C^4MIP model intercomparison, *J. Climate*, accepted.

Giardina, C., and M. Ryan, 2000: Evidence that decomposition rates of organic carbon in mineral soil do not vary with temperature. *Nature*, **404**, 858–861.

Houghton, J.T., L.G. Meira Filho, B.A. Callander, N. Harris, A. Kattenberg, and K. Maskell, 1996: *Climate Change 1995 – The Science of Climate Change*. Cambridge University Press. 572 pp.

Huntingford, C., and P.M. Cox, 2000: An analogue model to derive additional climate change scenarios from existing GCM simulations. *Clim. Dyn.*, **16**, 575–586.

Jenkinson, D.S., D.E. Adams, and A. Wild, 1991: Model estimates of CO_2 emissions from soil in response to global warming. *Nature*, **351**, 304–306.

Jones, C.D., and P.M. Cox, 2001: Modelling the volcanic signal in the atmospheric CO_2 record. *Global Biogeochem. Cycles*, **15**, 453–466.

Jones, C.D. and P.M. Cox, 2001a: Constraints on the temperature sensitivity of global soil respiration from the observed interannual variability in atmospheric CO_2. *Atmospheric Science Letters*, doi:10.1006/asle.2001.0041.

Jones, C.D., M. Collins, P.M. Cox, and S.A. Spall, 2001: The carbon cycle response to ENSO: A coupled climate-carbon cycle study. *J. Climate*, **14**, 4113–4129.

Jones, C.D., P.M. Cox and C. Huntingford, 2005: Impact of climate-carbon cycle feedbacks on emissions scenarios to achieve stabilisation. *This volume.*

Lenton, T.M., 2000: Land and ocean carbon cycle feedback effects on global warming in a simple Earth system model. *Tellus* **52B**, 1159–1188.

Lenton, T.M. and C. Huntingford, 2003: Global terrestrial carbon storage and uncertainties in its temperature sensitivity examined with a simple model *Global Change Biology* **9**, 1333–1352.

McGuire, A., J. Melillo, L. Joyce, D. Kicklighter, A. Grace, B.M. III, and C. Vorosmarty, 1992: Interactions between carbon and nitrogen dynamics in estimating net primary productivity for potential vegetation in North America. *Global Biogeochem. Cycles*, **6**, 101–124.

Schimel, D., D. Alves, I. Enting, M. Heimann, F. Joos, D. Raynaud, T. Wigley, M. Prather, R. Derwent, D. Enhalt, P. Fraser, E. Sanhueza, X. Zhou, P. Jonas, R. Charlson, H. Rodhe, S. Sadasivan, K. P. Shine, Y. Fouquart, V. Ramaswamy, S. Solomon, J. Srinivasan, D. Albritton, R. Derwent, I. Isaksen, M. Lal, and D. Wuebbles, 1996: Radiative forcing of climate change. In: Houghton, J.T., L.G.M. Filho, B.A. Callander, N. Harris, A. Kattenberg, and K. Maskell (Eds) *Climate Change 1995. The Science of Climate Change*, Cambridge University Press, 65–131.

Sellers, P., D. Randall, C. Collatz, J. Berry, C. Field, D. Dazlich, C. Zhang, and G. Collelo, 1996: A revised land surface parameterisation (SiB2) for atmospheric GCMs. Part I: Model formulation. *Journal of Climate*, **9**, 676–705.

Stainforth, D.A., T. Aina, C. Christensen, M. Collins, D.J. Frame, J.A. Kettleborough, S. Knight, A. Martin, J. Murphy, C. Piani, D. Sexton, L.A. Smith, R.A. Spicer, A.J. Thorpe, and M.R. Allen, 2005: Uncertainty in predictions of the climate response to rising levels of greenhouse gases. *Nature* **433**, 403–406.

Woodwell, G.M., and F.T. Mackenzie, 1995: Biotic Feedbacks in the Global Climatic System – Will the Warming Feed the Warming? *Oxford University Press.*

SECTION IV

Socio-Economic Effects: Key Vulnerabilities for Water Resources, Agriculture, Food and Settlements

INTRODUCTION

In this section the papers focus on the science behind the determination of key magnitudes, rates and aspects of timing related to the estimated effects of climate change.

Patwardhan suggests that key vulnerabilities, as measured in terms of socio-economic outcomes, could provide useful information for countries to arrive at a well-informed judgement about what might be considered as dangerous levels or rates of climate change. He notes that climate change may be either a triggering effect on events which may have been pre-conditioned by other forces, or may be an underlying cause in itself. Even in its causative role, however, climate change most frequently occurs as one part of a long list of stressors. It is, therefore, necessary to consider a quite complex set of interactions between climate and non-climate factors affecting future human and biophysical systems.

An illustration of this point is offered by Arnell. Looking at the effects of climate change on water systems, he identifies three key variables which define socio-economic context: demand (dependant on population and its income level); vulnerability (dependant on income level and governance); and resource supply (in part dependant on climate change). Even without climate change, water stress is expected to increase, especially in Central Asia, North Africa and the drier parts of China. Projected changes in climate are likely to alter the magnitude and timing of this stress, but the manifestation of change will be influenced by socio-economic variables. Arnell explores the effect of different development pathways (as reflected in IPCC SRES projections of population and GDP) on possible future impacts of climate change. Increases in water stress are likely to be higher under an IPCC A2 scenario in comparison with a IPCC B2 scenario, for example, primarily because of higher vulnerability under IPCC A2 and not necessarily because of greater climate forcing.

Hare illustrates results from an expert review of extensive literature across several systems and sectors. He used a four-fold scale of risk (from 'not significant' to 'severe') drawn from estimates of the proportion of damage expected. He concluded that the trigger of large damage (20–50%) varies considerably from one exposure unit to another. In most ecosystems, it appears to be below a warming threshold of 2°C above pre-industrial levels. In other systems, though, large damages may not appear even above 3°C. In general he reiterates a conclusion that can be found in many other studies: up to 1°C warming (measured in terms of an increase in mean global temperature) is likely to be associated with damages in developing countries and with some benefits in developed countries. Beyond 1°C, though, net damages would likely appear and grow in all areas.

Much of the analysis of potential impacts has been derived from modelling studies using input data from GCMs and statistical downscaling. New work in this area has allowed process-based crop models specifically designed to be coupled to GCMs to explore the effects of changes in CO_2, climate and the frequency and/or intensity of extremes (for example, high temperature events that can reduce yields). Challinor, Wheeler, Osborn and Slingo offer the coupling of a processed-based crop model to the Hadley climate model as an illustration of this approach.

The importance of adaptive responses in affecting key vulnerabilities is stressed by Nicholls in his examination of coastal flooding driven by sea-level rise. Considerable differences across estimates of populations at risk were observed across a range of possible futures that assume either constant protection of coasts in opposition to evolving or enhanced protection. Specifically, Nicholls shows how additional risk levels due to climate change might be avoided almost entirely with enhanced protection in a B2 world. This result is consistent with the conclusions reported by Parry and Arnell (see above) that different levels of vulnerability and wealth in various development pathways greatly affect the ability to delay or avoid 'dangerous' effects; and that choice of development pathway can be an effective response to climate change. This possibility is especially relevant because stabilisation cannot avoid all of the additional risk from future flooding due to the 'commitment' to sea-level rise in the ocean system.

CHAPTER 16

Human Dimensions Implications of Using Key Vulnerabilities for Characterizing 'Dangerous Anthropogenic Interference'

Anand Patwardhan and Upasna Sharma
S. J. Mehta School of Management, Powai, Mumbai, India

16.1 Introduction

The ultimate objective of the UN Framework Convention on Climate Change (UNFCCC) is 'the stabilization of greenhouse gas concentrations at levels that would prevent dangerous anthropogenic interference (DAI) with the climate system'. The notion of what may be considered as 'dangerous' is one of the central, and unresolved and contentious, questions in the climate change debate. Article 2 of the UNFCCC provides a set of criteria that help in addressing this question, but practice requires that these criteria be given operational definition.

Some of the criteria in the Convention are, at least to some extent, measurable in terms of biophysical end-points. For example, ecosystem response to climate change may be characterized in terms of variables such as species distribution and abundance or ecosystem structure and function. Even in this natural sphere, however, separating the human dimensions of climate-related issues (such as perception, values and preferences) from their physical effects is a difficult task because ecosystem functions are closely integrated with human activities for both managed and unmanaged ecosystems. It is, though, possible to cast the definition of 'dangerous' in other dimensions that may not be immediately obvious or widely applauded. If one were to consider food production, for example, is vulnerability to be characterized in terms of aggregate output or in terms of food security? The former is easier to measure, but the latter might be more policy-relevant (and far more complex because it includes questions of availability, price and distribution). In any case, the two metrics may not be strongly related.

It is useful to distinguish between biophysical and socio-economic outcomes of climate change end-points when inferences about DAI are being drawn. This is because the role of climate change, and the extent to which an undesirable outcome may be attributed to anthropogenic climate change, may differ considerably based on the outcome being considered. In some cases (such as the coastal impacts of sea level rise), climate change is directly responsible for the eventual outcome. In many other situations, climate change may only play a triggering or precipitating role because the primary causative factors may be socio-economic in nature. When end-points are defined in biophysical terms, a reasonably direct causative link may be drawn between the biophysical stressor (climate change) and the bio-physical end-point. When the end-points being considered are socio-economic, however, climate change (or more generally, biophysical stress) may not be the primary causative factor (or even a causative factor at all).

16.2 Key Vulnerabilities as an Approach for Characterizing DAI

The notion of 'vulnerability' or, more specifically, 'key vulnerability', can be a useful means for accommodating different types of measures and end-points. The term 'vulnerability' has been conceptualized in many different ways by the various research communities addressing the climate change problem. The Third Assessment Report of the IPCC (2001) characterized vulnerability as the consequence of three factors: exposure, sensitivity and adaptive capacity. In broader terms, 'key vulnerability' may be used to describe those interactions between elements of the climate system, climate-sensitive resources and the services where significant adverse outcomes are possible when expressed in terms of ecological, social and/or economic implications.

There may be good reasons why adopting a vulnerability framework might be advantageous. If DAI is to be defined by a socio-political negotiation process, for example, then inputs to this process need to reflect outcomes that can serve as adequate reasons of concern for parties to engage in dialogue and negotiation. Vulnerability of socio-economic systems to climate change can therefore provide useful information for countries as they try to formulate well-informed judgments about what might be considered dangerous.

In using key vulnerabilities for characterizing DAI, the main issue is the link between climate change (biophysical stress) and significant adverse outcomes. As mentioned earlier, it may often be difficult to assume a direct causative link between socio-economic end-points and bio-physical stressors. Human and socio-economic outcomes often manifest themselves as different forms of social disorder such as displacement or migration of people or extreme actions by individuals in response to livelihood insecurity, such as suicides committed by the farmers in India (Reddy, et al., 1998 or Kumar, 2003). These extreme forms of social disorder arise mainly from the disruption or the loss of livelihoods of people, and they are caused by any one or more of a multitude of social-economic and/or political factors. Bio-physical stressors (e.g. climate change) may, in these cases, be simply a triggering to the observed response – the last straw, as it were. It is therefore important to understand the processes that lead to a particular socio-economic endpoint before attributing it to climate change *per se*. In other cases, of course, climate change may be a direct cause of persistent and/or chronic hazard and exposure. My only point is that it is important to make this distinction.

The complex interplay of hazard, exposure and adaptive capacity that underlies vulnerability makes it difficult to draw direct correspondences between outcomes that matter, and levels or rates of climate change; and even more so, levels or rates of change of GHG concentrations or emissions. For some regions and sectors, even a one-degree temperature change may be unacceptable; for others, even a much larger change may be acceptable. This is true not only across individual countries, but also within countries.

16.3 Implications for the Policy Debate

These brief observations lead quite directly to a few obvious, but nonetheless important conclusions that need to be emphasized in any discussion of what is 'dangerous' climate change. First of all, focusing policy attention exclusively (or largely) on the question of setting a stabilization target may actually miss the significance of the link between targets and impacts. In many cases, outcomes that matter may be weakly or indirectly related to concentration (or temperature targets) because of the complexity of the interactions of other sources of stress. It follows that policy responses need to consider adaptation as an integral and distinct part of the portfolio of responses to climate change; complementary, and additional to mitigation. The UNFCCC calls for this, and it makes sense scientifically.

When all is said and done, though, recognizing the link between key vulnerabilities and DAI leads to a dilemma. On the one hand, the practicalities of the policy and negotiation process suggest policy-makers will be engaged only if the research community focuses attention on issues that have high salience. That is, researchers must focus negotiators' attentions on the key vulnerabilities to climate change so that they can reach a shared consensus around what constitutes 'dangerous anthropogenic interference with the climate system'. On the other, focusing on key vulnerabilities can make it extremely difficult to draw direct inferences backwards from undesirable outcomes that need to be avoided. Perhaps, instead of trying to identify a particular target (whether it be global mean temperature change, or CO_2 concentration or whatever), it may be helpful to recognize that preventing dangerous anthropogenic interference is a *process* that needs to be informed by a growing understanding of the consequences of climate change in all of its richness.

A key issue that needs to be addressed for further progress in the area of promoting complementarity across a portfolio of policy responses to 'dangerous' climate change is that of adaptation and adaptive capacity. In the absence of understanding the adaptation baseline, or the extent to which planned and autonomous adaptation would lead to adjustments and coping with regard to climate change, setting of very specific targets becomes problematic.

REFERENCES

Intergovernmental Panel on Climate Change (IPCC), 2001. *Climate Change 2001: Impacts, Adaptation and Vulnerability.* Cambridge University Press. Cambridge (Chapter 18).
Reddy, A.S., Vendantha, S., Rao, B.V., Redd, S.R. and Reddy, Y.V., 1998. 'Gathering agrarian crisis: farmers' suicides in Warangal district (A.P.) India. Citizens Report prepared by Centre for Environmental Studies, Warangal, AP.
Kumar, N.S., 2003. 'Done in by cash crops.' *Frontline*, Vol. 19 (26).

CHAPTER 17

Climate Change and Water Resources: A Global Perspective

Nigel W. Arnell

Tyndall Centre for Climate Change Research, School of Geography, University of Southampton, UK

ABSTRACT: This paper summarises the demographic, economic, social and physical drivers leading to change in water resources pressures at the global scale: climate change is superimposed onto these other drivers. In some parts of the world climate change will lead to reduced runoff, whilst in others it will result in higher streamflows, but this extra water may not be available for use if little storage is available and may appear during larger and more frequent floods.

The actual impacts of climate change on water resource availability (expressed in terms of runoff per capita per watershed) depend not only on the assumed spatial pattern of climate change and, from the 2050s, the assumed rate of climate change, but also on the economic and demographic state of the world. By the 2050s, between 1.1 and 2.8 billion water-stressed people could see a reduction in water availability due to climate change under the most populous future world, but under less populated worlds the numbers impacted could be between 0.7 and 1.2 billion. These impacted populations are largely in the Middle East and central Asia, Europe, southern Africa and parts of central, north and south America.

Climate policies which reduce greenhouse emissions reduce, but do not eliminate, the impacts of climate change. Stabilisation at 550 ppmv (resulting in an increase in temperature since pre-industrial times below the EU's 2°C target), for example, reduces the numbers of people adversely affected by climate change by between 30% and 50%, depending on the unmitigated rate of change and future state of the world. The thresholds of temperature increase, beyond which the impacts of climate change increase markedly, vary between regions.

17.1 Introduction

At present, approximately a third of the world's population lives in countries deemed to be 'water-stressed' (WMO, 1997), where withdrawals for domestic, industrial and agricultural purposes exceed 20% of the available average annual runoff. Around 1 billion people currently lack access to safe drinking water, approximately 250 million people suffer health problems associated with poor quality water, and each year river floods claim thousands of lives. During the course of the 21st century increasing population totals, changing patterns of water use and an increasing concentration of population and economic activities in urban areas are likely to increase further pressures on water resources. Changes in catchment land cover, the construction of upstream reservoirs and pollution from domestic, industrial and agricultural sources have the potential to alter the reliability and quality of supplies. Superimposed onto all these pressures is the threat of climate change.

At the global scale, an increasing concentration of greenhouse gases would lead to an increase in rainfall, largely due to increased evaporation from the oceans. However, due to the workings of the climate system, climate change would mean that whilst some parts of the world – predominantly in high latitudes and some tropical regions – would receive additional rainfall, rainfall in large parts of the world would decrease (see IPCC (2001)).

Climate change therefore has the potential to increase water resource stresses through increasing flood risk in some areas and increasing the risk of shortage in others: some parts of the world may see increased flood risk in one season and increased risk of shortage during another.

"Water resource stress" is difficult to define in practice, and manifests itself in three main, but linked, ways. First, it reflects *exposure* to water-related hazard, such as flood, drought or ill-health. Indicators include the numbers of people flooded or suffering drought each year. These indicators are difficult to model at anything other than the catchment scale, and it is therefore difficult to project global or regional future exposure to water-related hazard, even in the absence of climate change. Secondly, stress can be manifest in terms of *access* to water, as characterised by the widely used measures of access to safe drinking water and access to sanitation. These too are difficult to model, because they depend not only on resource availability but also on local-scale economic, social and political factors limiting access to water supply and sanitation: in most cases, these are arguably much more important in affecting access than the volume of water potentially available. Third, water resources stress can be represented in terms of the *availability* of water, as characterised for example by the amount of water available per person or withdrawals as a percentage of available water. These measures are much easier to model, and therefore project into the future, than the other two groups of measures, although the relationship between "stress" and simple measures of availability is not simple as it is influenced by, for example, water management infrastructure and institutions, governance and aspects of access outlined above.

Over the last decade there have been many catchment-scale studies into the effects of climate change on water resources and flood risk, showing for example how even relatively small changes in average conditions can lead to major changes in the risk of occurrence of extremes. There have, however, been few assessments of the implications of climate change for water resources stresses over a large region or the globe as a whole (see Alcamo *et al.*, 2000; Vorosmarty *et al.*, 2000 and Arnell, 1999; 2004). This paper presents an assessment of the global-scale implications of both climate change and population growth for water resources through the 21st century, using resources per capita as an indicator of water resource availability. It considers the effects of unmitigated emissions (Arnell, 2004) and the implications of policies to limit the rate of increase in global average temperature.

17.2 Data, Models and Projections

The approach followed here basically involves the simulation of current and future river flows by major watershed, the estimation of future watershed populations, and the simple calculation of the amount of water available per person for each watershed (expressed in m³/capita/year).

River flows are simulated at a spatial resolution of 0.5 × 0.5° using a macro-scale hydrological model

driven by gridded climate data (Arnell, 2003). Model parameters are estimated from spatial soil and vegetation data-bases, and whilst the model has not been calibrated against observed river flow data, a validation exercise showed that river flows were simulated "reasonably" well

Table 17.1 Summary of the SRES storylines (IPCC, 2000).

Market-oriented	
A1	**A2**
Very rapid economic growth, global population peaks in mid 21st century, rapid introduction of new technologies. Increased economic and cultural convergence	Very heterogeneous world, with self-reliance and preservation of local identities. Fertility patterns converge slowly, so population growth high. Per capita economic growth and technological change slow and fragmented
Globalised	**Localised**
B1	**B2**
Convergent world with same population as A1. Rapid changes in economic structures towards service economy, with reductions in material intensity and introduction of resource-efficient technologies. Emphasis on global solutions to sustainability issues	Emphasis on local solutions to sustainability issues. Continuously increasing population, with intermediate levels of economic development. Less rapid and more diverse technological change than B1 and A1 storylines.
Community oriented	

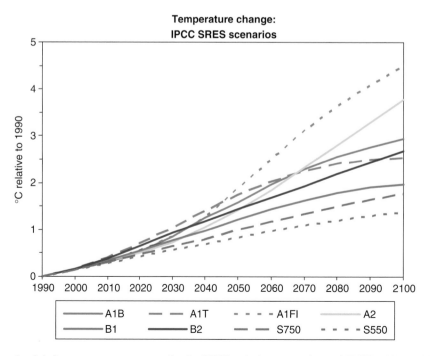

Figure 17.1 Change in global average temperature under the SRES emissions scenarios and IPCC (1997) stabilisation scenarios, assuming average climate sensitivities.

(Arnell, 2003). The model tends to overestimate river flows in dry regions, primarily because it does not account explicitly for evaporation of runoff generated from the surface of the catchment before it reaches a river, or for transmission loss along the river-bed. However, this water may in practice be available for use within the catchment, and gauged river flows may actually underestimate resources available in dry areas. River flows simulated at the 0.5 × 0.5° resolution are summed to estimate total runoff in around 1200 major watersheds, covering the entire ice-free land surface of the world.

The climate scenarios were constructed from simulations with six climate models and four SRES emissions scenarios, held on the IPCC's Data Distribution Centre (www.ipcc-ddc.cru.uea.ac.uk), and as described in IPCC (2001). The four emissions scenarios represent four different possible "storylines" describing the way population, economies, political structures and lifestyles may change during the 21st century (Table 17.1: IPCC, 2000). Figure 17.1 summarises the change in global average temperature under the SRES emissions scenarios.

Figure 17.2 shows the global population totals under the SRES storylines (the A1 and B1 storylines assume the same population changes). Watershed populations were estimated by applying projections at the national level to the 2.5 × 2.5′ resolution 1995 Gridded Population of the World data (CIESIN, 2000), and summing across the watershed (Arnell, 2004).

17.3 Water Resources Stresses in the Absence of Climate Change

Approximately 1.4 billion people currently live in watersheds with less than 1000 m³ of water per person per year (Arnell, 2004[1]), mostly in south-west Asia, the Middle East and around the Mediterranean. Table 17.2 summarises the numbers of people living in such watersheds by 2025, 2055 and 2085, under the three population projections. The increase is greatest under the most populous A2 scenario, which also shows the largest increase in the percentage of total global population living in water-stressed watersheds. Figure 17.3 shows the geographical distribution of water-stressed watersheds in 1995 and 2055.

17.4 The Effect of Climate Change: Unmitigated Emissions

Figure 17.4 shows the percentage change in river runoff by 2055 under the A2 emissions scenario and the six climate models (changes under B2 emissions have similar patterns, but smaller magnitudes). Changes in average annual runoff less than the standard deviation of 30-year

---

[1]A rather arbitrary index of "stress".

mean runoff (calculated from a 240-year long simulation assuming no climate change) are shown in grey in Figure 17.4, and are assumed to be insignificantly different from the effects of natural multi-decadal variability. There is a

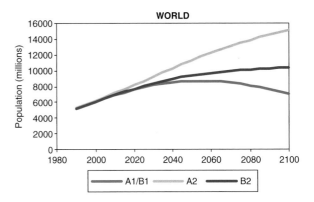

Figure 17.2 Global population totals under the SRES storylines.

Table 17.2 Numbers of people (millions) living in water-stressed watersheds in the absence of climate change (Arnell, 2004).

	A1/B1	A2	B2
1995	–	1368 (24)	–
2025	2882 (37)	3320 (39)	2883 (36)
2055	3400 (39)	5596 (48)	3988 (42)
2085	2860 (37)	8065 (57)	4530 (45)

Percentage of global population in parentheses.

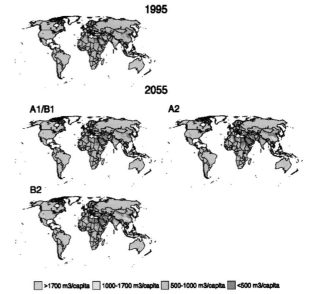

Figure 17.3 Geographical distribution of water-stressed watersheds in 1995 and 2055. Water-stressed watersheds have runoff less than 1000 m³/capita/year (after Arnell, 2004).

Change in average annual runoff: 2050s
A2

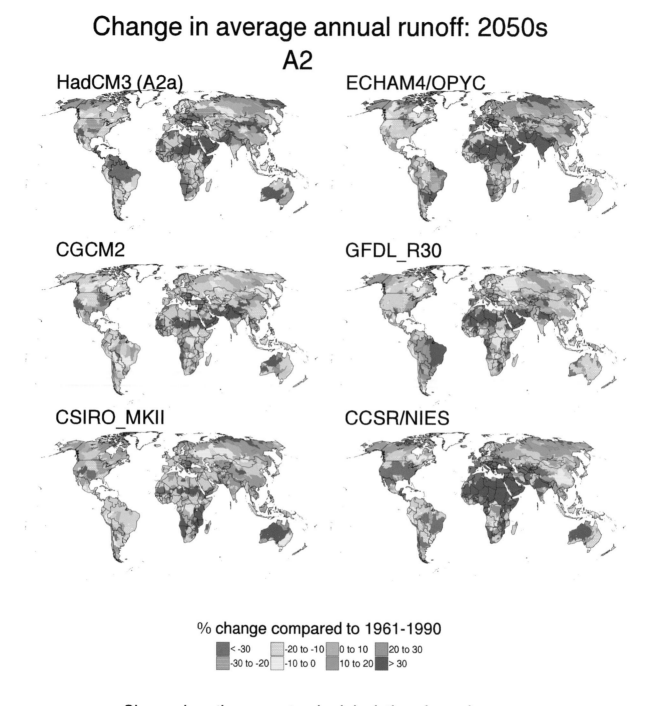

% change compared to 1961-1990

■ < -30	▨ -20 to -10	▨ 0 to 10	▨ 20 to 30
▨ -30 to -20	□ -10 to 0	▨ 10 to 20	■ > 30

Change less than one standard deviation shown in grey

Figure 17.4 Change in watershed runoff by 2055 under the A2 emissions scenario and six climate models (after Arnell, 2003).

broad degree of consistency between the six climate models, with consistent increases in runoff in high latitude North America and Siberia, east Africa and east Asia, and consistent decreases across much of Europe, the Middle East, southern Africa, and parts of both North and Latin America. However, the magnitudes of change vary between climate models, and there are some differences in simulated direction of change in parts of south Asia in particular. Changes in average runoff also only

reveal part of the impact of climate change on water availability: in many parts of the world where precipitation in winter falls as snow, higher temperatures would mean that this precipitation would fall as rain and hence run off rapidly into rivers rather than be stored on the surface of the catchment.

Table 17.3 shows the numbers of people with a simulated increase in water stress under each emissions scenario and climate model. Populations with an "increase

Table 17.3 Numbers of people (millions) with an increase in water stress due to climate change (Arnell, 2004).

	HadCM3	ECHAM4	CGCM2	CSIRO	GFDL	CCSR
2025						
A1	829					
A2	615–1661	679	915	500	891	736
B1	395					
B2	508–592	557	1183	594	374	601
2055						
A1	1136					
A2	1620–1973	1092	2761	2165	1978	1805
B1	988					
B2	1020–1157	885	1030	1142	670	1538
2085						
A1	1256					
A2	2583–3210	2429	4518	2845	1560	3416
B1	1135					
B2	1196–1535	909	1817	1533	867	2015

The range for the HadCM3 model represents the range between ensemble members.

in water stress" are those living in a watershed that becomes water-stressed due to climate change (resources fall below 1000 m³/capita/year) plus those living in already-stressed watersheds that suffer a significant decrease in runoff due to climate change (see Arnell (2004) for a discussion of this index). By the 2020s there is little clear difference between the different population or emissions scenarios, but a large difference in apparent impact of climate change between different climate models: the numbers of people with an increase in water stress vary between 374 and 1661 million. By the 2050s, the effect of the different population totals becomes very clear, with substantially more people adversely affected by climate change under the most populous A2 scenario. Significantly, this is not the emissions scenario with the largest climate change.

Figure 17.5 shows the geographical distribution of the impacts of climate change on water stress by 2055, under the A2 emissions scenario and population distribution. Areas with an increase in stress occur in Europe, around the Mediterranean, parts of the Middle East, central and southern Africa, the Caribbean, and in parts of Latin and North America. It also appears from Figure 17.5 that some watersheds would see a *decrease* in water stress due to climate change, because river flows increase with climate change. However, increasing river flows does not necessarily mean that water-related problems would reduce, because in most cases these higher flows occur during the high flow season. The risk of flooding would therefore increase, and without extra reservoir storage or changes to operating rules water would not be available during the dry season. It is therefore not appropriate to calculate the net effect of apparent decreases and increases in water stress.

17.5 The Effects of Mitigation

The climate simulations described in the previous section assume no policy interventions to reduce the future rate of climate change. Arnell *et al.* (2002) compared the effects of two stabilisation scenarios with unmitigated emissions, as simulated by HadCM2, and concluded that whilst stabilisation at 750 ppmv would not significantly reduce the impacts of climate change, stabilisation at 550 ppmv would have a clearer effect. However, this study used only one set of population projections and just one climate model: it proved difficult to separate out the effects of the different emissions profiles from decade-to-decade variability.

An alternative approach, which eliminates the effect of decade-to-decade variability, is to rescale the pattern of climate change produced by one climate model to different rates of temperature increase, and use this rescaled pattern in the impacts model. This makes the crucial assumption that the pattern of change can be scaled simply, which is reasonable within a relatively small range of temperature changes. This approach was applied in the current study, scaling the patterns of change produced by each of the six climate models by the end of the 21st century to different increases in global average temperature.

Figure 17.6 shows the effect of increasing global temperature on the global total number of people with an increase in water stress, at different time horizons (2025, 2055 and 2085, shown in the top, middle and bottom) and under different population growth scenarios (A1/B1, A2 and B2, shown in the left, middle and right respectively). The curves define six different *climate impact response functions*, constructed from six different climate models. They differ largely because of differences in the spatial pattern of change in precipitation, and hence runoff. An

Change in water resources stress: 2055 A2 world

Watersheds with <1000 m3/capita/year

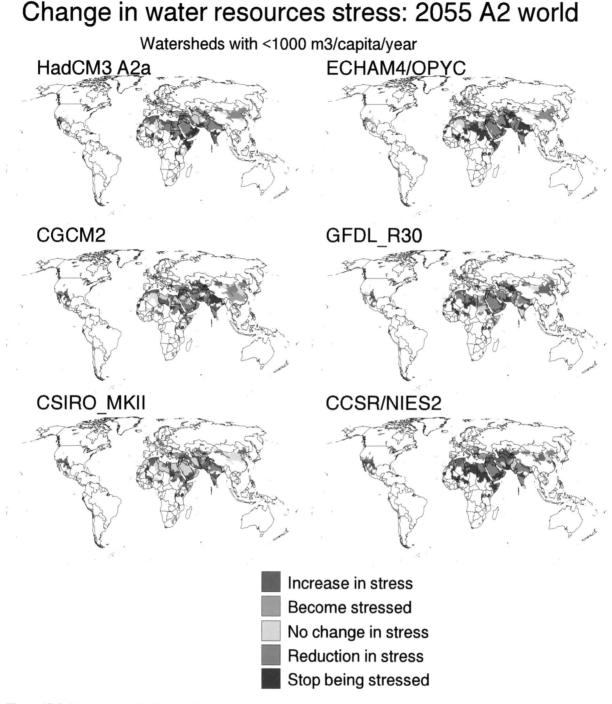

Figure 17.5 Geographical distribution of the impacts of climate change on water stress by 2055, under the A2 emissions and population scenario (after Arnell, 2004).

increase in temperature of 2°C above the 1961–1990 mean by 2055 would lead to increased water stress for between 500 and 1000 million people under the A1/B1 population projection, between 800 and 2200 million under A2, and between 700 and 1100 million under B2, depending on climate model.

The shaded grey area on each panel represents the range in change in global temperature with unmitigated emissions. The dashed vertical line shows the temperature increase with eventual stabilisation at 750 ppmv (S750: IPCC, 1997), and the dotted vertical line shows the temperature increase with stabilisation at 550 ppmv (S550). Stabilisation at 550 ppmv meets the EU's target of restricting the increase in temperature to 2°C above pre-industrial levels (approximately 1.5°C above the 1961–1990 mean).

As a broad approximation, aiming for stabilisation at 550 ppmv appears to reduce the numbers of people with an increase in water stress by between 15 and 25% by

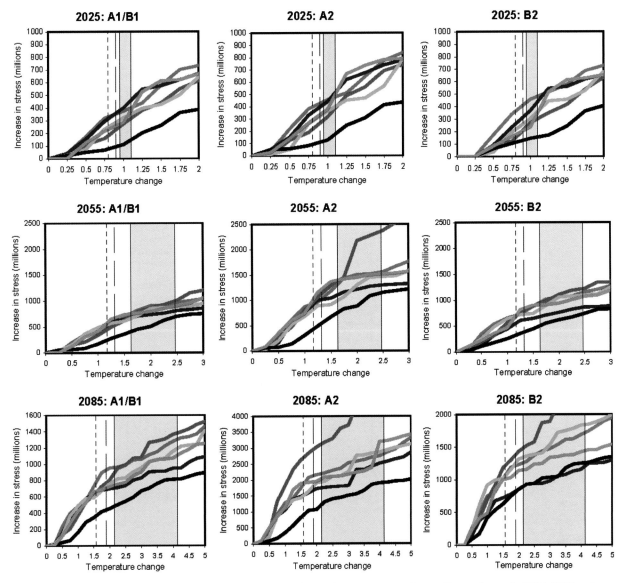

Figure 17.6 Numbers of people living in watersheds with an increasing water stress due to climate change in 2025, 2055 and 2085, with different amounts of global temperature change relative to 1961–1990.

2025, and between 25 and 40% thereafter, but the effect varies with climate model and, to a lesser extent, with assumed population totals: stabilisation appears to have the least effect with the most populous A2 world.

Figure 17.6, however, hides substantial geographic variation, and in many regions the corresponding graphs show more obvious thresholds as increasing climate change pushes additional watersheds into water-stressed conditions and reduces runoff further in stressed watersheds. Figure 17.7 shows the effect of different temperature increases by the 2050s for six key regions, using just the HadCM3 climate model to characterise the spatial pattern of changes in rainfall and temperature for a given global temperature change. In this case, the different lines represent the three different population projections. The effects of stabilisation are, under this climate model, most beneficial in Africa, particularly in central Africa where climate change appears to have little effect until temperatures rise

by 1°C above the 1961–1990 average, and where an increase of 1.5°C results in a step change in impact. Figure 17.7 also demonstrates that the thresholds of increase beyond which climate change has a substantial impact vary between regions.

17.6 Conclusions

A number of drivers – demographic, economic and physical – are stimulating changes in exposure to water-related hazards, access to water and the future availability of water supplies. Climate change is the most important and geographically extensive physical driver, and is likely to lead to changes in both the volume and, importantly, the timing of river flows and groundwater recharge. In some parts of the world runoff is likely to be reduced, and in others increased – but an increase in river flows is not

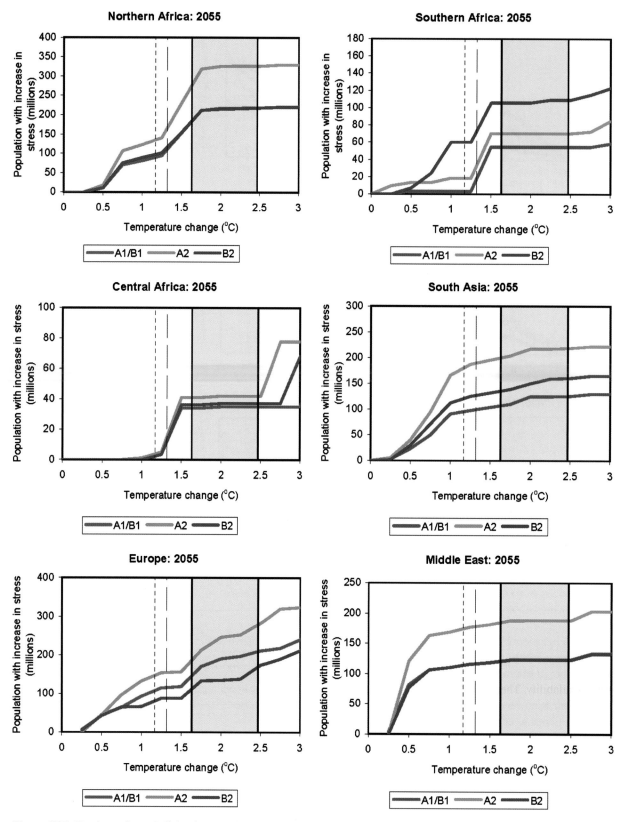

Figure 17.7 Numbers of people living in watersheds with an increasing water stress, by region, in 2055, with different amounts of global temperature change relative to 1961–1990. Changes in temperature and rainfall derived from HadCM3.

necessarily beneficial as there may not be sufficient storage to hold this extra water and it may come in the form of larger and more frequent floods.

The impact of climate change on future water resource availability, however, depends to a very large extent on the future state of the world, and particularly on the numbers of people potentially exposed to water shortage. By the 2050s, nearly twice as many people would be adversely affected by climate change under the most populous of the projections considered in this paper, compared to the projections with the lowest population increase. For a given assumed state of the world, estimates of the magnitude of the impact of climate change also vary with the climate models used to create climate scenarios.

Climate policies which reduce greenhouse emissions reduce the impacts of climate change on water resources stresses, but do not eliminate impacts. Stabilisation at 550 ppmv (resulting in an increase in temperature since pre-industrial times below the EU's 2°C target), for example, reduces the numbers of people adversely affected by climate change by between 30 and 50%, depending on the unmitigated rate of change and future state of the world. The thresholds of temperature increase, beyond which the impacts of climate change increase markedly, vary between regions.

The assessment described in this paper, however, is rather simplistic. It uses very simple indicators of water resource stress – river runoff per capita per watershed – which can be modelled and estimated, at least for defined scenarios. It does not, for example, take into account either geographical variations in the amount of water actually abstracted per capita, or in how abstractions may change over time. It would be expected that areas with large abstractions for irrigation would be more vulnerable to climate change than implied in this study, but unfortunately estimates of future irrigation abstractions are very dependent on assumed future irrigation efficiencies. The study also does not consider impacts on exposure to hazards associated with either too little or too much water, and reveals nothing about how access to safe water may change with climate change: this is as much a function of governance and resources as of changes in water availability. The indicators used also do not help in an assessment of the extent to which adaptation is either feasible or able to offset the effects of climate change. These limitations all point towards the areas requiring further study. It is already technically feasible to estimate the hydrological consequences of a given climate change scenario over a large geographic area, with a reasonable degree of precision. It is, however, necessary to build on these hydrological foundations to understand how the other drivers of water resources stress – demographic, economic, cultural and political – will determine the impacts of climate change for all dimensions of water resources stress, and either constrain or provide opportunities for adaptation.

Acknowledgements

The research described in this paper has been supported by Defra and the Tyndall Centre for Climate Change Research. The author thanks the reviewers for their helpful comments.

REFERENCES

Alcamo, J., Heinrichs, T. & Rösch, T. (2000) *World Water in 2025 – GlobalModeling and Scenario Analysis for the 21st Century.* Report A0002, Center for Environmental Systems Research, University of Kassel, Kurt Wolters Strasse 3, 34109 Kassel, Germany.

Arnell, N.W. (1999) Climate change and global water resources. *Global Environmental Change* 9, S31–S49.

Arnell, N.W. (2003) Effects of IPCC SRES emissions scenarios on river runoff: a global perspective. *Hydrology and Earth System Sciences* 7, 619–641.

Arnell, N.W. (2004) Climate change and global water resources: SRES emissions and socio-economic scenarios. *Global Environmental Change* 14, 31–52.

Arnell, N.W. Cannell, M.G.R., Hulme, M., Kovats, R.S., Mitchell, J.F.B., Nicholls, R.J., Parry, M.L., Livermore, M.T.J. & White, A. (2002) The consequences of CO_2 stabilisation for the impacts of climate change. *Climatic Change* 53: 413–446.

Center for International Earth Science Information Network (CIESIN), Columbia University; International Food Policy Research Institute (IFPRI); and World Resources Institute (WRI). (2000) *Gridded Population of the World (GPW), Version2.* Palisades, NY: CIESIN, Columbia University. Available at *http://sedac.ciesin.columbia.edu/plue/gpw.*

IPCC (1997) *Stabilisation of Atmospheric Greenhouse Gases: Physical, Biological and Socio-economic Implications.* Intergovernmental Panel on Climate Change Technical Paper III. Intergovernmental Panel on Climate Change.

IPCC (2000) *Emissions Scenarios. A Special Report of Working Group II of the Intergovernmental Panel on Climate Change.* Cambridge University Press: Cambridge.

IPCC (2001) *Climate Change 2001: The Scientific Basis.* Contribution of Working Group 1 to the Third Assessment Report of the Intergovernmental Panel on Climate Change. Cambridge University Press: Cambridge.

Vörösmarty, C.J., Green, P.J., Salisbury, J. & Lammers, R.B. (2000) Global water resources: vulnerability from climate change and population growth. *Science* 289, 284–288.

World Meteorological Organisation (1997) *Comprehensive Assessment of the Freshwater Resources of the World.* WMO: Geneva.

CHAPTER 18

Relationship Between Increases in Global Mean Temperature and Impacts on Ecosystems, Food Production, Water and Socio-Economic Systems

Bill Hare
Visiting Scientist, Potsdam Institute for Climate Impact Research

ABSTRACT: This paper attempts to associate different levels of global mean surface temperature increase since pre-industrial and/or sea level rise with specific impacts and risks for species, ecosystems, agriculture, water and socio-economic damages. It is found that that the risks arising from projected human-induced climate change increase significantly and systematically with increasing temperature. Below a 1°C increase the risks are generally low but in some cases not insignificant, particularly for highly vulnerable ecosystems and/or species. Above a 1°C increase risks increase significantly, often rapidly for vulnerable ecosystems and species. In the 1–2°C increase range risks across the board increase significantly, and at a regional level are often substantial. Above 2°C the risks increase very substantially, involving potentially large numbers of extinctions or even ecosystem collapses, major increases in hunger and water shortage risks as well as socio-economic damages, particularly in developing countries.

18.1 Introduction

The ultimate objective of the United Nations Framework Convention on Climate Change specifies in its Article 2 that the stabilization of greenhouse gas concentrations at levels that 'would prevent dangerous anthropogenic interference with the climate system' be achieved 'within a time frame sufficient to allow ecosystems to adapt naturally to climate change, to ensure that food production is not threatened and to enable economic development to proceed in a sustainable manner' [1]. In this paper the relationship between increases in global mean temperature and the latter elements mentioned in Article 2 are explored in order to cast light on the risks posed by climate change.

18.2 Method

An extensive review of the literature on the impacts of climate change on ecosystems and species, food production, water and damages to economic activity has been undertaken and studies analysed to determine relationships between global mean temperature and the risks identified in each work [2][1]. Studies were drawn predominantly from the peer reviewed literature or work that was reviewed for the IPCC Third Assessment Report. Many different baseline climatologies and other climatic assumptions have been used in the literature. For those studies analysed, all climate scenarios used were reduced to a common global mean temperature scale with respect

to the pre-industrial period defined as the 1861–1890 climate using a standard methodology [2]. Regional temperature increases used in scenarios were converted to a range of global mean temperature increases using the MAGICC 4.1/SCENGEN climate model tool [3]. All temperatures referred to in this paper are global mean and with respect to the pre-industrial global mean defined as the 1861–1890 average.

In making the estimates of changes in the metrics used in each study (loss of ecosystem area, changes in food production etc.) for different temperatures, simplifying assumptions needed to be made. Where studies report on a range of scenarios, full use of the diversity of these scenarios has been made to interpolate to different temperature increases than were evaluated in the studies. Often studies report on the effects of only one or a few scenarios, and in these cases extrapolations have been made in a conservative manner. In general most studies do not examine the effects of temperature increases below 2°C above pre-industrial levels, and hence in most cases extrapolations have had to be made to a zero point of damage assumed in reference to the 1961–1990 period. This assumes in effect a slower rate of increase of damage with temperature than if a zero point later in time had been chosen, and vice verse for an earlier zero point for the damage or risk. Particularly for the ecosystem and species analysis, the risks of loss of area or species may be underestimated for temperature increases below 2°C. Modelling work by Leemans and Eickhout [4], as well as the observations of the effects of climate change to date on species and ecosystem [5–8] tends to indicate that present assessments may underestimate the effects of climate change on many species and ecosystems. The annotations to the figures below provide more details on the

---

[1] The work described in this paper is described in substantial detail in this larger report.

assumptions used in each case study and the larger report on which this work is based contains a full description [2].

The results for the ecosystems and species have been mapped onto a risk scale involving five categories of risk: Less than a 5% reduction in area (or other appropriate indicator) is regarded as not significant, a 5–10% reduction is defined as small risk, a 10–20% loss as moderate, a 20–50% loss is defined as large and a severe loss is defined as more than a 50% loss of area or population.

18.3 Results

18.3.1 *Species and Ecosystems*

Discernible effects of climate change that can be attributed to warming over the past century experience have been observed on many species and ecosystems in Europe, North America and Asia [5–8]. One case, the extinction of the Golden Toad, has been attributed to climate change [9, 10]. Unprecedented, widespread coral bleaching has also occurred in the last 10–15 years [11].

Between present temperatures and a 1°C increase, at least three ecosystems appear to be moving into a high risk zone – coral reefs [12, 13], the highland tropical forests in Queensland, Australia [14–16] and the Succulent Karoo in South Africa [17–19]. Increased fire frequency and pest outbreaks may cause disturbance in boreal forests and other ecosystems. There appears to be a risk of extinction for some highly vulnerable species in south-western Australia [20] and to a lesser extent in South Africa. Range losses for species such as the Golden Bower bird in the highland tropical forests of North Queensland, Australia, and for many animal species in South Africa are likely to become small to moderate.

Between 1 and 2°C warming the Australian highland tropical forest, the Succulent Karoo biodiversity hot spot, coral reef ecosystems and some Arctic and alpine ecosystems are likely to suffer large or even severe damage. The Fynbos of South Africa is very likely to experience increased losses. Coral reef bleaching will likely become much more frequent, with slow or no recovery, particularly in the Indian Ocean south of the equator. Australian highland tropical forest types, which are home to many endemic vertebrates, are projected to halve in area in this range. The Australian alpine zone is likely to suffer moderate to large losses [21, 22] and the European Alpine may be experiencing increasing stress [22, 23]. A large to severe loss of Arctic sea ice likely to occur [24] will harm ice-dependent species such as the polar bears and walrus [25]. Increased frequency of fire and insect pest disturbance is likely to cause increasing problems for ecosystems and species in the Mediterranean region [26–29]. Moderate to large losses of boreal forest in China can be expected [30]. Moderate shifts in the range of European plants can be expected and in Australia moderate to large number of Eucalypts may be outside out of their climatic

range [31]. Large and sometimes severe impacts appear possible for some Salmonid fish habitats in the USA [32], the collared lemming in Canada [33], many South African animals and for Mexico's fauna. There is an increasing risk of extinctions in South Africa [34], Mexico [35] for the most vulnerable species and for especially vulnerable highland rainforest vertebrates in North Queensland, Australia. Extinctions in the Dryandra forest of south-western Australia seem very likely [20]. Mid summer ice reduction in the Arctic ocean seems likely to be at a level that would cause major problems for polar bears at least at a regional level.

Between 2 and 3°C warming coral reefs are projected to bleach annually in many regions. At the upper end of this temperature band, the risk of eliminating the Succulent Karoo and its 2800 endemic plants is very high. Moderate to large reductions in the Fynbos can be expected, with the risk of a large number of endemic species extinctions. Australian mainland alpine ecosystems are likely to be on the edge of disappearance, a large number of extinctions of endemic Alpine flora in New Zealand are projected [36] and European alpine systems are likely to be at or above their anticipated tolerable limits of warming with some vulnerable species close to extinction. Severe loss of boreal forest in China is projected and large and adverse changes are also projected for many systems on the Tibetan plateau [37]. Large shifts in the range of European plants seem likely and a large number of Eucalypt species may expect to lie outside of their present climatic range [38]. Moderate to large effects are projected for Arctic ecosystems and boreal forests. Within this temperature range there is a likelihood of the Amazon forest suffering potentially irreversible damage leading to its collapse [39, 40].

Above 3°C, large impacts begin to emerge for waterfowl populations in the Prairie Pothole region of the USA [41]. In the Arctic, the collared lemming range is reduced by 80%, and very large reductions are projected for Arctic sea ice cover particularly in summer that is likely to further endanger polar bears. There seems to be a very high likelihood that large numbers of extinctions would occur among the 65 endemic vertebrates of the highland rainforests of North Queensland, Australia. In Mexico very severe range losses for many animals are projected, as is the case also in South Africa, with Kruger national park projected to lose two thirds of the animals studied.

Results of the analysis of the risks for species and ecosystems are presented graphically in the Appendix at Figures 18.2, 18.3, 18.4 and 18.5 where detailed notes are also given on the data sources and assumptions made.

18.3.2 *Coastal Wetlands*

A key issue is the inertia of sea level rise, which makes the assignment of risk to different temperature levels misleading. Should, for example, sea level rise by 30 cm in the coming decades to a century (threatening Kakadu for example), the thermal inertia of the ocean is such that an ultimate sea level rise of 2–4 times this amount may be inevitable even if temperature stops rising. The prognoses for wetlands in this context is not clear, as many damages are linked to the rate of sea level rise compared to the accretion and/or migratory capacity of the system. A major determinant of the latter will be human activity adjacent to, or in the inland catchments of the wetland system.

Below a 1°C increase[2] the risk of damage is low for most systems. Between 1 and 2°C warming moderate to large losses appear likely for a few vulnerable systems. Of most concern are threats to the Kakadu wetlands of northern Australia [42] and the Sundarbans of Bangladesh [43, 44], both of which may suffer 50% losses at less than 2°C and are both on the UNESCO World Heritage List. Between 2 and 3°C warming, it is likely that the Mediterranean, Baltic and several migratory bird habitats in the US would experience a 50% or more loss [45–47]. It also seems likely that there could be the complete loss of Kakadu and the Sundarbans.

Results of the analysis of the risks for coastal wetlands are presented graphically in the Appendix at Figure 18.1 where detailed notes are also given on the data sources and assumptions made.

18.3.3 *Agriculture and Food Security*

Warming of around 1°C produces relatively small damages when measured from the point of increased risk of hunger and/or under-nourishment over the next century. In this temperature range nearly all developed countries are projected to benefit, whilst many developing countries in the tropics are estimated to experience small but significant crop yield growth declines relative to an unchanged climate [48]. Above this level of change the number of people at risk of hunger increases significantly. Between 2 and 3°C warming the risk of damage begins to increase significantly [49–52]. Whilst developing countries may still gain in this temperature range, the literature indicates that production is finely balanced between the effects of increased temperature and changes in precipitation [53]. 'Drier' climate models show losses in North America, Russia and Eastern Europe whereas 'wetter' models show increases. One study shows rapidly rising hunger risk in this temperature range, with 45–55 million extra people at risk of hunger by the 2080s for 2.5°C warming, which rises to 65–75 million for a 3°C warming [49, 50]. Another study shows that a very large number of people, 3.3–5.5 billion, may be living in countries or regions expected to experience large losses in crop production potential at 3°C warming [54].

For a 3–4°C warming, in one study the additional number at risk of hunger is estimated to be in the range 80–125 million depending on the climate model [50]. In Australia a warming of the order of 4°C is likely to put entire regions out of production, with lesser levels of warming causing moderate to severe declines in the west and the south [55].

At all levels of warming, a large group of the poor, highly vulnerable developing countries is expected to suffer increasing food deficits. It is anticipated that this will lead to higher levels of food insecurity and hunger in these countries. Developed countries will not be immune to large effects of climate change on their agricultural sectors.

18.3.4 *Water Resources*

The number of people living in water stressed countries, defined as those using more than 20% of their available resources, is expected to increase substantially over the next decades irrespective of climate change. Particularly in the next few decades population and other pressures are likely to outweigh the effects of climate change, although some regions may be badly affected during this period. In the longer term, however, climate change becomes much more important. Exacerbating factors such as the link between land degradation, climate change and water availability are in general not yet accounted for in the global assessments.

Around 1°C of warming may entail high levels of additional risk in some regions, particularly in the period to the 2020s and 2050s, with this risk then decreasing due to the increased economic wealth and higher adaptive capacity projected for the coming century. For the 2020s the additional number of people in water shortage regions is estimated to be in the range 400–800 million [50] [56].

Between 1–2°C warming the level of risk appears to depend on the time frame and assumed levels of economic development in the future. One study for the middle of this temperature range has a peak risk in the 2050s at over 1,500 million, which declines to around 500 million in the 2080s [50] [56].

Over 2°C warming appears to involve a major threshold increase in risk. One study shows risk increasing for close to 600 million people at 1.5°C to 2.4–3.1 billion people at around 2.5°C. This is driven by the water demand of mega-cities in India and China in their model. In this study the level of risk begins to saturate in the range of 3.1–3.5 billion additional persons at risk at 2.5–3°C warming [50, 56].

One of the major future risk factors identified is that of increased water demand from mega-cities and large population centres in India and China [50]. It is not clear whether or to what extent additional water resource options would be available for these cities and hence, to what extent this finding is robust. This may have broad implications for environmental flows of water in major rivers of China, India and Tibet should the mega-cities of

---

[2]Impacts at different levels of global mean temperature increase above the 1861–1890 climate state, which is here used as the proxy for the pre-industrial climate.

India and China seek large-scale diversion and impound-ments of flows in the region.

18.3.5 *Socio-Economic Impacts*[3]

For a 1°C warming a significant number of developing countries appear likely to experience net losses, which can range as high as 2% or so of GDP. Most developed countries are likely to experience a mix of damages and benefits, with net benefits predicted by a number of models. For a 2°C warming the net adverse effects projected for developing countries appear to be more consistent and of the order of a few to several percentage points of GDP, depending upon the model. Regional damages for some developing countries and regions, particularly in Africa, may exceed several percentage points of GDP. Above 2°C the likelihood of global net damages increases but at a rate that is quite uncertain. The effects on several developing regions appear to be in the range of 3–5% of GDP for a 2.5–3°C warming, if there are no adverse climate sur-prises. Global damage estimates are in the range of 1–2% of GDP for 2.5–3°C warming, with some estimates increasing substantially with increasing temperature.

18.4 Conclusions

The risks arising from projected human-induced climate change increase significantly with increasing temperature. Below a 1°C global mean increase above pre-industrial levels the risks are low but in some case not insignificant, particularly for highly vulnerable ecosystems. In 2004 global mean temperatures were about 0.7°C above the pre-industrial climate, with this temperature increase already being associated with significant effects on ecosystems and species, as well as substantial damage from the European heat wave of 2003 (Stott, 2004 #12795), (WHO, 2004 #13148). In the 1–2°C increase range risks across the board increase significantly, and at a regional level are often substantial. The risk of large ecosystem damages and losses particularly to coral reefs, as well as large losses of species in some vulnerable regions, seems large in this temperature range. In some regions, particularly in Africa, there is a risk of substan-tial agricultural and water supply damages in this tem-perature range. Above 2°C the risks increase very substantially involving potentially large extinctions or even ecosystem collapses, major increases in hunger and water shortage risks as well as socio-economic damages, particularly in developing countries. Africa seems to be consistently amongst the regions with high to very high projected damages.

The results of this work provide some support for the position adopted by the European Union in 1996 aiming to limit global warming to a global mean increase of 2°C

above pre-industrial levels [59]. It seems clear however that there are substantial risks even below this level of warming, particularly for vulnerable ecosystems, species and regions, which tends to confirm assessments made in the late 1980s [60]. The implications are that for a num-ber of important ecosystems and for some regions there is a large risk that 2°C of global warming will lead to large or severe damages or losses.

18.5 Appendix Figures

18.5.1 *Impacts on Coastal Wetlands*

Notes on Figure 18.1:

1a. Global assessment: high – progressive coastal wet-land loss with increasing warming (22.2% for ca. 3.4°C warming). Based on the Nicholls *et al.* [45] assessment using the high estimate of wetland loss (22.2% in 2100 for around a 3.4°C warming). A lin-ear extrapolation used to calculate 50% loss, which is likely to very much overestimate the temperature at which this would occur.

1b. Global assessment: low – progressive coastal wetland loss with increasing warming (5.7% for ca. 3.4°C warming). As above but for low estimates (5.7% loss by 2100) with linear extrapolation to 50%, which is likely to underestimate the rate at which this would occur.

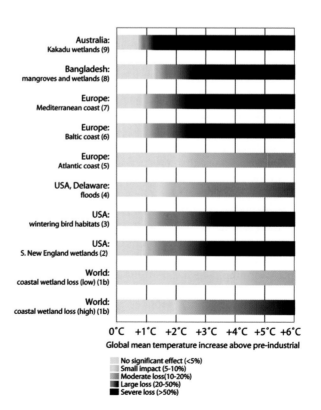

Figure 18.1 Impacts on Coastal Wetlands.

---

[3] See [48] and relevant papers [57, 58].

2. USA: Southern New England – extensive loss of wetlands if sea level rise greater than 6 mm/yr: based on Donnelly and Bertness [61] with assumption that a 5°C increase (3–5°C range) by 2100 is associated with a 6 mm/yr increase in sea level rise and an 80% (extensive) loss of wetlands.

3. USA: loss of important foraging, migratory and wintering bird habitat at four sites (20–70% loss for ca. 2.6°C warming). Based on Galbraith *et al.* [47]. The graph shown is for the average range of losses at the four sites that lose inter-tidal habitat for all warming and sea level rise scenarios – Willapa Bay, Humboldt Bay and northern and southern San Francisco Bay. The average loss at these sites in 2100 for the 2.6°C scenario is 44% (range 26% to 70%) and for 5.3°C is 79% (range 61% to 91%). The latter point is used to scale the average losses with temperature, which increases the temperature slightly for a given loss compared to the 2.6°C scenario. The Delaware Bay site loses 57% of inter-tidal habitat for the 2.6°C (34 cm sea level rise) but gains 20% in the 5.3°C (77 cm sea level rise scenario). Whilst the Bolivar flats site loses significantly by the 2050s for both scenarios (38–81%) it gains by the 2100s for both scenarios.

4. USA: Delaware – loss of 21% for ca. 2.5–3.5°C warming – 100 year floods occurring 3–4 times more frequently. Based on Najjar *et al.* [46] assuming 21% loss at 3.5°C warming with linear extrapolation to 50%. A linear extrapolation used to calculate 50% loss, which is likely to very much overestimate the temperature at which this would occur.

5. European wetlands – Atlantic coast: based on IPCC WGII TAR Table 13–4 which is based on new runs using the models described by Nicholls *et al.* [45], with a linear extrapolation of the high range 17% loss with 4.4°C warming to higher loss rates. This is likely to very much overestimate the temperature at which this would occur.

6. European wetlands – Baltic coast: as above with linear extrapolation of high range 98% loss with 4.4°C warming.

7. European wetlands – Mediterranean coast: as above with a linear extrapolation of high range 100% loss with 4.4°C warming.

8. Bangladesh, Sundarbans: based on Qureshi and Hobbie [43] and Smith *et al.* [62] with sea level rise and temperature relationship (for 2100) drawn from Hulme *et al.* [63]. This produces very similar results to an estimate based on 'average' model characteristics. Some models project higher sea level rise and others lower. Assumed relationship is 15% loss for 1.5°C (range 1–1.5°C) and 75% loss for 3.5°C (range 2–3.5°C).

9. Australia, Kakadu region: this estimate is highly uncertain. In the WGII TAR report Gitay *et al.* [42] assert that the wetlands 'could be all but displaced if predicted sea-level rises of 10–30 cm by 2030 occur and

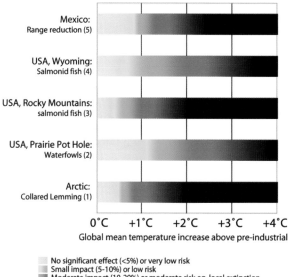

Figure 18.2 Impacts on Animal Species.

are associated with changes in rainfall in the catchment and tidal/storm surges' (p. 308). Here it is assumed that a 30 cm sea level rise displaces 80% of the wetlands and that the sea level rise vs. temperature relationship is drawn from Hulme *et al.* [63] from the HadCM2 and HadCM3. Note that the estimate range from recent models is 1.2–3.1°C for a 30 cm sea level rise.

18.5.2 *Impacts on Animal Species*

Notes on Figure 18.2 and Figure 18.3:

1. Canadian Arctic, collared lemming: based on data in Kerr and Packer [33] with conversion of local temperatures to global mean based on a range of the current AOGCMs; mid-range used. Interpolation is used to estimate range reductions based on data in Kerr and Packer [33].

2. USA, waterfowl population Prairie Pot Hole Region: based on data in Sorenson *et al.* [41] with interpolation of data.

3. USA, reduction of Salmonid fish habitat in Rocky Mountains: based on data in Keleher and Rahel [32] with extrapolations to 5% and 10% reductions. June, July, August temperatures 'upscaled' to global by associating projected JJA temperatures from a range of GCMs for the USA with global mean temperatures using MAGICC/SCENGEN. This is obviously quite uncertain given that temperature changes in the region are likely to be quite different from the USA average, with mountainous regions likely to experience amplification of trends for the continental averages.

4. USA, reduction of Salmonid fish habitat in Wyoming: based on data in Keleher and Rahel [32] with extrapolations to 50% reduction. Upscaling of temperatures as in (3).

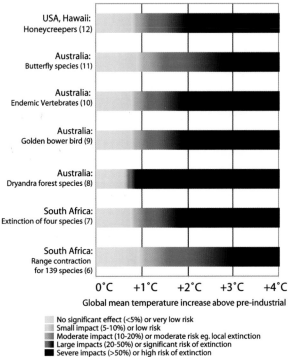

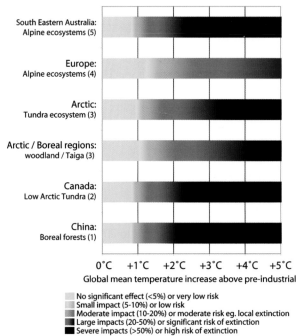

Figure 18.4 Impacts on ecosystems.

Figure 18.3 Impacts on Animal Species (continued).

5. Mexico: highly indicative interpretation of results of Peterson *et al.* [35] for range reductions. The 50% range reduction level is associated with the upper end of their warming scenario, which corresponds to 2.4°C warming above 1861–1890, and this range reduction applies to up to 19% of the entire Mexican fauna. Between present temperatures and 2.4°C a linear scaling is used here. Note that there is projected to be a severe risk of extinction for up to tens of fauna species (0–2.4% of species lose 90% of range for 1.9–2.4°C warming).

6. South Africa, range reductions of large number of animals: highly indicative only, interpretation of results of Erasmus *et al.* [34] for range reductions in the 29 endangered species projected to experience 50% or more range reductions with a warming of 2.4°C (1.9–3.1°C range) (above 1861–1890). The scale assumes that a 50% reduction in the range of these species occurs with 3.1°C. Lower reductions are linearly scaled from 1990 temperatures.

7. South Africa, predicted extinctions: highly indicative only interpretation of results of Erasmus *et al.* [34] for extinctions projected for a 2.4°C increase (1.9–3.1°C range). The scale used assumes that there is a 100% chance of extinction with a 3.1°C increase, zero probability at current temperatures, and the likelihood of extinction increase linearly.

8. Australia, south west Dryandra forest: based on Pouliquen-Young and Newman [20] as cited by Gitay *et al.* [42]. Assumed that 'very large' range reduction meant a 90% reduction, that the loss of

range scale was linear for the present climate to a warming of 1.1°C (above 1861–1890), and that 90% reduction occurs at 1.1°C.

9. Australia, predicted extinction of Golden Bower bird of highland tropical forests, north east Queensland: based on [15] and using range reduction of 90% with a 3°C warming and linear interpolation for range losses between 1990 (0.6°C and 0% range loss) and this level.

10. Australia, 'catastrophic' loss of endemic vertebrates from rainforest in highland tropical rainforests: based on [16] and with similar scaling as above.

11. Australia, large range reduction in range of butterfly species: based on [64] with risk of large range reductions for large numbers of species linearly increasing from zero at 0.6°C to 50% loss for 80% of species at 2.9°C.

12. USA, predicted extinction for honeycreepers in montane forests of Hawaii: based on [65] with risk of extinction to 90% at 3.2°C.

18.5.3 *Impacts on Ecosystems*

Notes on Figure 18.4 and Figure 18.5:

1. Boreal forests, China: based on Ni [66] with linear scaling of loss of boreal forest in China with temperature.

2. Arctic, Canadian Low Arctic Tundra: loss of area is 77% with 3.3°C warming based on [67] and linearly interpolated from zero at 0.6°C.

3. Arctic/Boreal, Boreal woodland/Taiga and Arctic Tundra: loss of ecosystems respectively 44% and

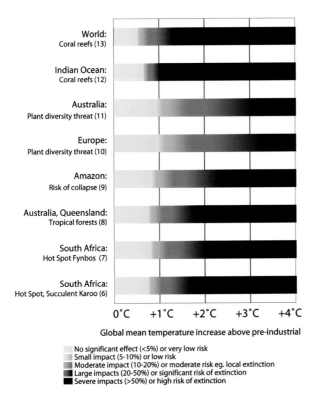

Figure 18.5 Impacts on ecosystems (continued).

57% with 3.8°C warming and scaled linearly from zero at 0.6°C warming. Based on [68].

4. Alpine ecosystems, Europe: highly indicative measure of risk only. Scale is percentage of alpine species losing 90% of their range with linear scaling of the estimated 38% losing this level with a warming of about 4.7°C (range 3.3–4.7°C). This is done only to provide a visual picture of increasing risk with temperature, which is one of the main findings of the literature for this region.

5. Alpine ecosystems, south-eastern Australia: assumes 90% reduction with a warming of 3.8°C (above 1861–1890) with linear scaling of area loss from present climate. Busby [21] found that the alpine zone would be confined to only 6 peaks for a warming of 1.7–3.8°C.

6. Biodiversity Hot Spot, Succulent Karoo, South Africa: based on Midgley and Rutherford at http://www.nbi.ac.za/frames/researchfram.htm. The scale is likelihood of extinction of the 2800 plants endemic to the Succulent Karoo ecosystem, where it is assumed that the systems will no longer exist with 100% certainty with an increase of 2.4°C and that the likelihood of extinction scales linearly upward from zero at current temperatures.

7. Biodiversity Hot Spot, Fynbos, South Africa: based on Midgley *et al.* [69] and linear scaling loss of the area of Fynbos with temperature from zero at present up to 61% loss of area with a 2.4°C increase (above 1861–1890). Ten per cent of endemic

Proteaceae species are projected to suffer complete loss of range, and hence are also very likely to become extinct with a 51–61% area loss in Fynbos.

8. Tropical forests, Highland tropical forests – Australia, Queensland: based on results of Ostendorf *et al.* [70], Hilbert *et al.* [14], Williams *et al.* [16] and Hilbert *et al.* [15], with linear scaling of area losses with local temperature increase. Across results from different assessments this produces fairly consistent estimates.

9. Tropical forests, Amazon: this is speculative drawing on the work of Cowling *et al.* [39] and Cox *et al.* [71] and assuming that there is a 50% risk of collapse with a warming of 2.4°C.

10. Plant diversity threat, Europe: based on Bakkenes *et al.* [38] with scale being fraction of plant species occurring at present within a grid cell in Europe that no longer appear with given level of warming. Assumes linear scaling with temperature increase above the present. As such is indicative only of increasing risk with temperature, the risk being that of extinction or severe range reduction. The absence of plants from a grid cell in 2050 does not imply that the species is globally extinct, only that it is no longer climatically suited to that region. The higher the fraction of species displaced in the model is a measure of the ecological dislocation caused by rapid warming and for some species is indicative of the rising level of extinction risk.

11. Plant diversity threat, Australia: based on Hughes *et al.* [31]. Scaled number of species out of climatic range with temperature above present.

12. Coral reefs – Indian Ocean: based on the work of (Sheppard, 2003 #311) who predicts extinction of reef sites in the southern Indian Ocean for warming in the range 0.9–1.4°C. It is assumed that there is a 90% chance of extinction at a temperature increase of 1.4°C.

13. Coral reefs – global assessment: based on results of Hoegh-Guldberg [12]. For both models used and all reefs studied, annual bleaching occurred by 2040s. Scale is chance of a major bleaching occurring in a decadal period e.g. 10% corresponds to 1 year per decade, 50% to five years out of 10 and 100% to annual bleaching. Scaling is from 0.4°C above 1861–1890 as unusual bleaching began in the 1980s with annual bleaching occurring at 2.3°C above 1861–1890.

Acknowledgements

Malte Meinshausen is thanked for assistance in preparing the graphs presented here, and Claire Stockwell and Kathrin Gutmann are thanked for their assistance in conducting the research used in this paper.

REFERENCES

1. UN: 1992, 'United Nations Framework Convention on Climate Change'. United Nations, New York.

2. Hare, W.L.: 2003, 'Assessment of Knowledge on Impacts of Climate Change – Contribution to the Specification of Art. 2 of the UNFCCC'. Berlin, Externe Expertise für das WBGU-Sondergutachten "Welt im Wandel: Über Kioto hinausdenken. Klimaschutzstrategien für das 21. Jahrhundert", Wissenschaftlicher Beirat ger Bundesregierung Globale Umweltveränderungen, http://www.wbgu.de/wbgu_sn2003_ex01.pdf. http://www.wbgu.de/wbgu_sn2003_ex01.pdf

3. Wigley, T.M.L., Raper, S., Salmon, M., Hulme, M. and McGinnis, S.: 2003, 'MAGICC/SCENGEN 4.1'. Norwich, UK and Boulder, United States, Climate Research Unit, Norwich.

4. Leemans, R. and Eickhout, B.: 2004, 'Another reason for concern: regional and global impacts on ecosystems for different levels of climate change', *Global Environmental Change Part A* **14**, 219–228.

5. Parmesan, C. and Yohe, G.: 2003, 'A globally coherent fingerprint of climate change impacts across natural systems', *Nature* **421**, 37–42.

6. Root, T.L., Price, J.T., Hall, K.R., Schneider, S.H., Rosenzweig, C. and Pounds, J.A.: 2003, 'Fingerprints of global warming on wild animals and plants', *Nature* **421**, 57–60.

7. Leemans, R. and van Vliet, A.: 2004, 'Extreme weather: does nature keep up? Observed responses of species and ecosystems to changes in climate and extreme weather events: many more reasons for concern'. Wageningen, Environmental Systems Analysis Group, Wageningen University: 60.

8. Root, T.L., MacMynowski, D.P., Mastrandrea, M.D. and Schneider, S.H.: 2005, 'From the Cover: Human-modified temperatures induce species changes: Joint attribution', *PNAS* **102**, 7465–7469.

9. Pounds, J.A. and Crump, M.L.: 1994, 'Amphibian Declines and Climate Disturbance – the Case of the Golden Toad and the Harlequin Frog', *Conservation Biology* **8**, 72–85.

10. Pounds, J.A., Fogden, M.P.L. and Campbell, J.H.: 1999, 'Biological response to climate change on a tropical mountain', *Nature* **398**, 611.

11. Williams, C.E.: 2004, 'Status of coral reefs of the world: 2004'. Townsville, Australian Institute of Marine Science: 2 Volumes.

12. Hoegh-Guldberg, O.: 1999, 'Climate change, coral bleaching and the future of the world's coral reefs', *Marine and Freshwater Research* **50**, 839–866.

13. Sheppard, C.R.C.: 2003, 'Predicted recurrences of mass coral mortality in the Indian Ocean', *Nature* **425**, 294–297.

14. Hilbert, D.W., Ostendorf, B. and Hopkins, M.S.: 2001, 'Sensitivity of tropical forests to climate change in the humid tropics of north Queensland', *Austral Ecology* **26**, 590–603.

15. Hilbert, D.W., Bradford, M., Parker, T. and Westcott, D.A.: 2003, 'Golden bowerbird (Prionodura newtonia) habitat in past, present and future climates: predicted extinction of a vertebrate in tropical highlands due to global warming', *Biological Conservation* **In Press, Corrected Proof**.

16. Williams, S.E., Bolitho, E.E. and Fox, S.: 2003, 'Climate change in Australian tropical rainforests: an impending environmental catastrophe', *Proceedings of the Royal Society of London Series B-Biological Sciences* **270**, 1887–1892.

17. Rutherford, M.C., Midgley, G.F., Bond, W.J., Powrie, L.W., Musil, C.F., Roberts, R. and Allsopp, J.: 1999, 'South African Country Study on Climate Change'. Pretoria, South Africa, Terrestrial Plant Diversity Section, Vulnerability and Adaptation, Department of Environmental Affairs and Tourism.

18. Hannah, L., Midgley, G.F., Lovejoy, T., Bond, W.J., Bush, M., Lovett, J.C., Scott, D. and Woodward, F.I.: 2002, 'Conservation of Biodiversity in a Changing Climate', *Conservation Biology* **16**, 264–268.

19. Midgley, G.F., Hannah, L., Millar, D., Rutherford, M.C. and Powrie, L.W.: 2002, 'Assessing the vulnerability of species richness to anthropogenic climate change in a biodiversity hotspot', *Global Ecology and Biogeography* **11**, 445–452.

20. Pouliquen-Young, O. and Newman, P.: 1999, 'The Implications of Climate Change for Land-Based Nature Conservation Strategies.' Perth, Australia, Australian Greenhouse Office, Environment Australia, Canberra, and Institute for Sustainability and Technology Policy, Murdoch University: 91. Final Report 96/1306.

21. Busby, J.R.: 1988, 'Potential implications of climate change on Australia's flora and fauna', in Pearman, G.I. (eds.), *Greenhouse: Planning for Climate Change*, CSIRO Division of Atmospheric Research, Melbourne, pp. 387–388.

22. Bugmann, H.: 1997, 'Sensitivity of forests in the European Alps to future climatic change', *Climate Research* **8**, 35–44.

23. Theurillat, J.P. and Guisan, A.: 2001, 'Potential Impact of Climate Change on Vegetation in the European Alps: A Review', *Climatic Change* **50**, 77–109.

24. Johannessen, O.M., Bengtsson, L., Miles, M.W., Kuzmina, S.I., Semenov, V.A., Genrikh, V.A., Nagurnyi, A.P., Zakharov, V.F., Bobylev, L.P., Pettersson, L.H., Hasselmann, K. and Cattle, H.P.: 2004, 'Arctic climate change: observed and modelled temperature and sea-ice variability', *Tellus A* **56**, 328–341.

25. Derocher, A.E., Lunn, N.J. and Stirling, I.: 2002, 'Polar bears in a changing climate', *Integrative and Comparative Biology* **42**, 1219–1219.

26. Parmesan, C., Root, T.L. and Willig, M.R.: 2000, 'Impacts of extreme weather and climate on terrestrial biota', *Bulletin of the American Meteorological Society* **81**, 443–450.

27. White, T.A., Campbell, B.D., Kemp, P.D. and Hunt, C.L.: 2000, 'Sensitivity of three grassland communities to simulated extreme temperature and rainfall events', *Global Change Biology* **6**, 671–684.

28. Mouillot, F., Rambal, S. and Joffre, R.: 2002, 'Simulating climate change impacts on fire frequency and vegetation dynamics in a Mediterranean-type ecosystem', *Global Change Biology* **8**, 423–437.

29. Walther, G.R., Post, E., Convey, P., Menzel, A., Parmesan, C., Beebee, T.J.C., Fromentin, J.M., Hoegh-Guldberg, O. and Bairlein, F.: 2002, 'Ecological responses to recent climate change', *Nature* **416**, 389–395.

30. Ni, J.: 2001, 'Carbon Storage in Terrestrial Ecosystems of China: Estimates at Different Spatial Resolutions and Their Responses to Climate Change', *Climatic Change* **49**, 339–358.

31. Hughes, L., Cawsey, E.M. and Westoby, M.: 1996, 'Climatic range sizes of Eucalyptus species in relation to future climate change', *Global Ecology and Biogeography Letters* **5**, 23–29.

32. Keleher, C.J. and Rahel, F.J.: 1996, 'Thermal limits to salmonid distributions in the rocky mountain region and potential habitat loss due to global warming: A geographic information system (GIS) approach', *Transactions of the American Fisheries Society* **125**, 1–13.

33. Kerr, J. and Packer, L.: 1998, 'The impact of climate change on mammal diversity in Canada', *Environmental Monitoring and Assessment* **49**, 263–270.

34. Erasmus, B.F.N., Van Jaarsveld, A.S., Chown, S.L., Kshatriya, M. and Wessels, K.J.: 2002, 'Vulnerability of South African animal taxa to climate change', *Global Change Biology* **8**, 679–693.

35. Peterson, A.T., Ortega-Huerta, M.A., Bartley, J., Sanchez-Cordero, V., Soberon, J., Buddemeier, R.H. and Stockwell, D.R.B.: 2002, 'Future projections for Mexican faunas under global climate change scenarios', *Nature* **416**, 626–629.

36. Halloy, S.R.P. and Mark, A.F.: 2003, 'Climate change effects on alpine plant biodiversity: A New Zealand perspective on quantifying the threat', *Arctic Antarctic and Alpine Research* **35**, 248–254.

37. Ni, J.: 2000, 'A simulation of biomes on the Tibetan Plateau and their responses to global climate change', *Mountain Research and Development* **20**, 80–89.

38. Bakkenes, M., Alkemade, J.R.M., Ihle, F., Leemans, R. and Latour, J.B.: 2002, 'Assessing effects of forecasted climate change on the diversity and distribution of European higher plants for 2050', *Global Change Biology* **8**, 390–407.

39. Cowling, S.A., Cox, P.M., Betts, R.A., Ettwein, V.J., Jones, C.D., Maslin, M.A. and Spall, S.A.: 2003, 'Contrasting simulated past and future responses of the Amazon rainforest to atmospheric change', *Philosophical Transactions of the Royal Society of London, in press*.

40. Cox, P.M., Betts, R.A., Collins, M., Harris, P.P., Huntingford, C. and Jones, C.D.: 2004, 'Amazonian forest dieback under climate-carbon cycle projections for the 21st century', *Theoretical and Applied Climatology* **78**, 137–156.

41. Sorenson, L.G., Goldberg, R., Root, T.L. and Anderson, M.G.: 1998, 'Potential effects of global warming on waterfowl populations breeding in the Northern Great Plains', *Climatic Change* **40**, 343–369.

42. Gitay, H., Brown, S., Easterlin, W. and Jallow, B.: 2001, 'Chapter 5: Ecosystems and Their Goods and Services', *Climate Change 2001: Impacts, Adaptation and Vulnerability*, Cambridge University Press, Cambridge, UK, pp. 237–342.

43. Qureshi, A. and Hobbie, D.: 1994, 'Climate change in Asia'. Manila, Asian Development Bank. Cited by World Bank 2000. Chapter 2 Potential Impacts of climate change in Bangladesh.

44. Smith, P., Andren, O., Brussaard, L., Dangerfield, M., Ekschmitt, K., Lavelle, P. and Tate, K.: 1998, 'Soil biota and global change at the ecosystem level: describing soil biota in mathematical models', *Global Change Biology* **4**, 773–784.

45. Nicholls, R.J., Hoozemans, F.M.J. and Marchand, M.: 1999, 'Increasing flood risk and wetland losses due to global sea-level rise: regional and global analyses', *Global Environmental Change-Human and Policy Dimensions* **9**, S69–S87.

46. Najjar, R.G., Walker, H.A., Anderson, P.J., Barron, E.J., Bord, R.J., Gibson, J.R., Kennedy, V.S., Knight, C.G., Megonigal, J.P., O'Connor, R.E., Polsky, C.D., Psuty, N.P., Richards, B.A., Sorenson, L.G., Steele, E.M. and Swanson, R.S.: 2000, 'The potential impacts of climate change on the mid-Atlantic coastal region', *Climate Research* **14**, 219–233.

47. Galbraith, H., Jones, R., Park, R., Clough, J., Herrod-Julius, S., Harrington, B. and Page, G.: 2002, 'Global climate change and sea level rise: Potential losses of intertidal habitat for shorebirds', *Waterbirds* **25**, 173–183.

48. Smith, J.B., Schellnhuber, H-J., Mirza, M.M.Q., Fankhauser, S., Leemans, R., Erda, L., Ogallo, L., Pittock, B., Richels, R. and Rosenzweig, C.: 2001, 'Chapter 19: Vulnerability to climate change and reasons for concern: A synthesis', *Climate Change 2001: Impacts, adaptation and vulnerability*, Cambridge University Press, Cambridge, UK, pp. 915–967.

49. Parry, M., Rosenzweig, C., Iglesias, A., Fischer, G. and Livermore, M.: 1999, 'Climate change and world food security: a new assessment', *Global Environmental Change-Human and Policy Dimensions* **9**, S51–S67.

50. Parry, M., Arnell, N., McMichael, T., Nicholls, R., Martens, P., Kovats, S., Livermore, M., Rosenzweig, C., Iglesias, A. and Fischer, G.: 2001, 'Millions at risk: defining critical climate change threats and targets', *Global Environmental Change* **11**, 181–183.

51. Arnell, N.W., Cannell, M.G.R., Hulme, M., Kovats, R.S., Mitchell, J.F.B., Nicholls, R.J., Parry, M.L., Livermore, M.T.J. and White, A.: 2002, 'The consequences of CO_2 stabilisation for the impacts of climate change', *Climatic Change* **53**, 413–446.

52. Parry, M.L., Rosenzweig, C., Iglesias, A., Livermore, M. and Fischer, G.: 2004, 'Effects of climate change on global food production under SRES emissions and socio-economic scenarios', *Global Environmental Change* **14**, 53–67.

53. Lal, M., Harasawa, H., Murdiyarso, D., Adger, W.N., Adkhikary, S., Ando, M., Anokhin, Y., Cruz, R.V., Ilyas, M. and *et al.*: 2001, 'Chapter 11: Asia', *Climate Change 2001: Impacts, adaptation and vulnerability*, Cambridge University Press, Cambridge, UK, pp. 533–590.

54. Fischer, G., Shah, M., van Velthuizen, H. and Nachtergaele, F.: 2001, 'Global agro-ecological assessment for agriculture in the 21st century'. Laxenburg, Austria, IIASA: 33.

55. Pittock, B., Wratt, D., Basher, R., Bates, B., Finalyson, M., Gitay, H., Woodward, A., Arthington, A., Beets, P. and Biggs, B.: 2001, 'Chapter 12: Australia and New Zealand', *Climate Change 2001: Impacts, adaptation and vulnerability*, Cambridge University Press, Cambridge, UK, pp. 591–639.

56. Martinez-Vilalta, J. and Pinol, J.: 2002, 'Drought-induced mortality and hydraulic architecture in pine populations of the NE Iberian Peninsula', *Forest Ecology and Management* **161**, 247–256.

57. Mendelsohn, R., Morrison, W., Schlesinger, M.E. and Andronova, N.G.: 2000, 'Country-Specific Market Impacts of Climate Change', *Climatic Change* **45**, 553–569.

58. Mendelsohn, R., Schlesinger, M. and Williams, L.: 2000, 'Comparing impacts across climate models', *Integrated Assessment* **1**, 37–48.

59. European Community: 1996, 'Climate Change – Council conclusions 8518/96' (Presse 188-G) 25/26. VI.96.

60. Rijsberman, F.J. and Swart, R.J., eds: 1990, *Targets and Indicators of Climate Change*, Stockholm Environment Institute, 1666.

61. Donnelly, J.P. and Bertness, M.D.: 2001, 'Rapid shoreward encroachment of salt marsh cordgrass in response to accelerated sea-level rise', *Proceedings of the National Academy of Sciences of the United States of America* **98**, 14218–23.

62. Smith, J.B., Rahman, A., Haq, S. and Mirza, M.Q.: 1998, 'Considering Adaptation to Climate change in the sustainable development of Bangladesh. World Bank Report'. Washington, DC, World Bank: 103.

63. Hulme, M., Sheard, N. and Markham, A.: 1999, 'Global Climate Change Scenarios'. Norwich, Climatic Research Unit: 2.

64. Beaumont, L.J. and Hughes, L.: 2002, 'Potential changes in the distributions of latitudinally restricted Australian butterfly species in response to climate change', *Global Change Biology* **8**, 954–971.

65. Benning, T.L., LaPointe, D., Atkinson, C.T. and Vitousek, P.M.: 2002, 'Interactions of climate change with biological invasions and land use in the Hawaiian Islands: Modeling the fate of endemic birds using a geographic information system', *PNAS* **99**, 14246–14249.

66. Charron, D., Thomas M, Waltner-Toews D, Aramini J, Edge T, Kent R, Maarouf A, Wilson J.: 2004, 'Vulnerability of waterborne diseases to climate change in Canada: a review.' *J Toxicol Environ Health*, 1667–77.

67. Malcolm, J.R., Markham, A., Neilson, R.P. and Garaci, M.: 2002, 'Estimated migration rates under scenarios of global climate change', *Journal of Biogeography* **29**, 835–849.

68. Neilson, R.P., Prentice, I.C., Smith, B., Kittel, T. and Viner, D.: 1997, 'Simulated changes in vegetation distribution under global warming', in Waston, R.T., Zinyowera, M.C., Moss, R.H. and Dokken, D.J. (eds), *The Regional Impacions of Climate Change. An Assessment of Vulnerability*, Cambridge University Press, New York, pp. 439–456.

69. Hu, Q. and Feng, S.: 2002, 'Interannual rainfall variations in the North American summer monsoon region: 1900–98', *Journal of Climate* **15**, 1189–1202.

70. Ostendorf, B., Hilbert, D.W. and Hopkins, M.S.: 2001, 'The effect of climate change on tropical rainforest vegetation pattern', *Ecological Modelling* **145**, 211–224.

71. Cox, P.M., Betts, R.A., Collins, M., Harris, P., Huntingford, C. and Jones, C.D.: 2003, 'Amazon dieback under climate-carbon cycle projections for the 21st century'. UK, Hadley Centre. Technical Note 42.

CHAPTER 19

Assessing the Vulnerability of Crop Productivity to Climate Change Thresholds Using an Integrated Crop-Climate Model

A. J. Challinor[1,2], T. R. Wheeler[2], T. M. Osborne[1,2] and J. M. Slingo[1]

[1] *NCAS Centre for Global Atmospheric Modelling, Department of Meteorology, University of Reading, Reading*
[2] *Department of Agriculture, University of Reading, Reading*

ABSTRACT: Extreme climate events and the exceedance of climate thresholds can dramatically reduce crop yields. Such events are likely to become more common under climate change. Hence models used to assess the impacts of climate change on crops need to accurately represent the effects of these events. We present a crop-climate modelling system which is capable of simulating the impact on crop yield of threshold exceedance, changes in the mean and variability of climate, and adaptive measures. The predictive skill of this system is demonstrated for the current climate using both climate-driven simulations and fully coupled crop-climate simulations.

The impacts of climate change on crop productivity are then examined using the A2 emissions scenario. Exceedance of high temperature thresholds at the time of flowering reduces the yield of crops in some areas. The nature of this response can be moderated by the choice of variety, and in some areas this choice makes the difference between an increase and a decrease in yield. Therefore dangerous climate change in this context is related to temperature threshold exceedance and the ability of farming systems to adapt to it. This will vary in a non-linear manner with the climate change scenario used.

19.1 Introduction: Simulation of the Impacts of Climate Change on Crop Productivity

Estimates of the impacts of climate change on crop productivity usually rely on crop simulation models driven by weather data downscaled from General Circulation Models (GCMs). An important consequence of this approach is that differences in the spatial and temporal scales of crop and climate models may introduce uncertainties into assessments of the impacts of climate change (e.g. Mearns et al., 2001). Most crop models are designed to run at the field scale. They can provide good simulations of crop productivity at this scale, but not at the regional scale. However, policy decisions on the stabilisation of greenhouse gases require regional assessments of impacts on food systems. Thus, to provide this information, crop model outputs have to be aggregated to a regional scale. The assumptions implicit in this process are a source of error in regional yield estimation (Hansen and Jones, 2000).

An alternative approach is to design a crop model to operate on spatial and temporal scales close to the scale of the GCM output (Challinor et al., 2003). By using a large area process-based crop model as part of a more integrated modelling approach, errors in the aggregation of yield to the regional scale may be reduced. This paper aims to show how an integrated crop – climate modelling system can be used to assess the impacts of climate variability and change on crop productivity. Such a system can take explicit account of the impact of climate extremes on crop productivity.

19.1.1 *The Importance of Extreme Events and Climate Threshold Exceedance*

Many studies have shown that increases in atmospheric concentrations of CO_2 will benefit the yield of most crops, with the exception of those that have the C4 photosynthetic pathway, such as maize, millet and sugar cane (for example, Kimball, 1983; Idso and Idso, 1994). However, other aspects of climate change are expected to have a negative impact on the yield of annual crops, and these may partly, or entirely, offset the yield gains due to elevated CO_2. For example, warmer mean seasonal temperatures reduce the duration from sowing to harvest of wheat. This results in a reduction in the amount of light captured by the crop leaf canopy, and hence biomass and yield at harvest decline with an increase in temperature (Mitchell et al., 1993; Wheeler et al., 1996a).

Even where the sensitivity of crop yields to the seasonal mean climate is well known, large impacts on crop production can also occur when climate thresholds are transgressed for short periods (Parry et al., 2001). Floods, droughts and high temperature episodes are likely to become more frequent under climate change (IPCC, 2001b) and this will have an impact on crop productivity. Important climate thresholds for food crops include episodes of high temperatures that coincide with critical phases of the crop cycle (Wheeler et al., 2000), as well as changes in the sub-seasonal distribution of rainfall (Wright et al., 1991). Experimental studies have led research in this field and these are beginning to be understood in

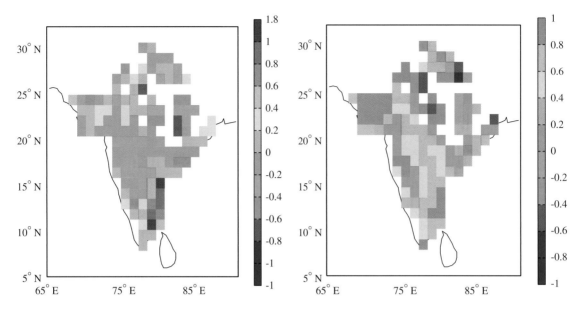

Figure 19.1 Left: mean (24 years) fractional difference between GLAM yield and an empirical fit to GLAM (Yield = b_0 + $b_1D + b_2P + b_3D^2 + b_4P^2 + b_6PD$ where b_i are constants, D is simulated crop duration and P is precipitation during that period). Right: correlation coefficient for the same period between GLAM yields and yields from the empirical fit. GLAM simulations and associated weather data are taken from Challinor et al. (2005a).

terms of simple physiology (Prasad et al., 2000; Ntare et al., 2001).

High temperature events near flowering disrupt pollination and cause yield losses due to reduced numbers of grains or seeds at the harvest. This response has been observed in wheat (Wheeler et al., 1996b), groundnut (Prasad et al., 2000) and soybean (Ferris et al., 1999), amongst others. Such studies have shown that the threshold temperature above which grain-set is reduced is usually between 31 and 37°C, provided that this short term high temperature event coincides with a sensitive stage of the crop such as flowering. The increasing recognition of the importance of weather events and climate thresholds such as these is reflected in crop modelling studies (e.g. Hansen and Jones, 2000; Semenov and Barrow, 1997; Easterling et al., 1996).

19.1.2 *Simulation Methods Used to Date*

In recognition of the socio-economic nature of climate change impacts, integrated assessments of the global impacts of climate change to date often simulate crop yield, land-use change and world food trade (Fischer et al., 2002; Parry et al., 2004). The treatment of crop growth and development in such assessments is frequently based on empirical methods (either parameterisations of crop model functions or direct use of statistical relationships such as those of Doorenbos and Kassam, 1979). This is a pragmatic way forward, but needs to be complemented with more detailed studies of the response of crops to climate. These more detailed studies focus on fundamental processes such as those related to changing CO_2 levels, intra-seasonal weather variability, and climate threshold

exceedance. When these processes begin to impact seriously on yield, statistical relationships developed under the current climate may no longer be valid (Challinor et al., 2005a).

The choice of crop model has been shown to provide a significant source of uncertainty in the simulation of yield under climate change (Mearns et al., 1999). In the present-day climate the use of an empirical regression (also called a yield function) based on crop model output can produce results that differ from direct use of crop model output. The following analysis, based on the use of reanalysis data with the crop model of Challinor et al. (2004), demonstrates this.

An empirical regression of model yields based on simulated crop duration (which is determined by mean temperature) and seasonal rainfall is compared to the model yields in figure 19.1. A good empirical fit to the crop model (right panel) does not necessarily imply that the mean yields simulated by both methods are similar (left panel). For example, in Gujarat (the western-most region shown), where simulated yields correlate significantly (r = 0.4–0.8) with observed yields (Challinor et al., 2005a), the empirical regression provides a good fit (r > 0.8) and yet the difference between the model yields and the regression can be greater than 40%.

Similar issues exist in considering how to use climate information for impacts studies. Different GCMs produce different climates, and any simulated yield changes contingent on those climates may differ in magnitude and sign (e.g. Tubiello et al., 2002). Hence no single simulation can be considered to be a prediction of a future climate. Even if the climate is correctly simulated, the statistics of weather may not be correct. For example, seasonal mean values of rainfall and temperature may be correct, but the daily values may not be realistic. Lack of confidence in daily weather data, coupled with the coarse resolution of GCMs has lead to the use of weather generators to generate downscaled time series for climate change scenarios. The downscaling relationships are based on changes between the current and future climate in the mean and the variability of weather (Semenov and Barrow, 1997). This method has the advantage of not relying on the correct simulation by the GCM of the basic mean state. It has the disadvantage of relying on a set of assumptions, embedded in the weather generator, regarding the relationship between mean climate and weather and between different weather variables. Such weather statistics may not remain constant as climate changes (Jenkins and Lowe, 2003) and correcting for this has inherent uncertainties. The impact of these uncertainties could be significant since the choice of parameters for a weather generator can alter the magnitude and even the sign of the changes in yield associated with climate change (Mavromatis and Jones, 1998).

The variety of methods used to simulate the yields associated with future climate scenarios leads to a large range of predictions and associated uncertainties. Luo and Lin (1999) reviewed estimates of the potential yield impacts of climate change in the Asia-Pacific region. Estimates of yield for future climates using climate models varied in both magnitude and in stated ranges. For example the two estimates of rice yield in Bangladesh (incorporating the CO_2 fertilisation effect) were '-12 to -2%' and -35%. Estimates of yield which did not include the CO_2 effect tended to have larger uncertainties (e.g. -74 to $+32\%$ for spring wheat in Mongolia). When a large range of sites and of GCM scenarios is used, the resulting uncertainty can be very large: Reilly and Schimmelpfennig (1999) projected wheat yield impacts for a doubling of CO_2 of between -100 and $+234\%$ for the USA and Canada. Only by dealing effectively with the disparity in spatial scale between GCMs and crop models can the uncertainty associated with yield estimates be reduced.

19.2 An Integrated Approach to Impacts Prediction

19.2.1 *Scientific Basis*

The scientific basis for a large-area crop model has been established by looking at the relationship between crop yield and weather data on a number of spatial scales (Challinor et al., 2003). Such a large-area model has the advantage of addressing the issues in sections 2.1 and 2.2: use of a process-based model which operates on the spatial scale of the GCM avoids the need for downscaling of weather data whilst maintaining a process-based modelling approach. Also, intra-seasonal variability can be represented and the impact of temperature threshold exceedance can be simulated. Further, full integration of the crop and climate models (see section 3.4) allows the GCM to capture feedbacks between the crop and the

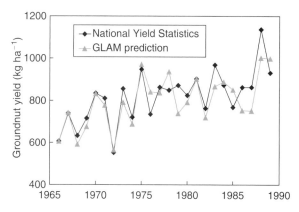

Figure 19.2 All-India groundnut yields simulated using GLAM on a 2.5° by 2.5° grid (Challinor et al., 2004). The time trend in the GLAM yields is taken from the (linear) time trend in observations.

climate and also diurnal temperature variability, which is important in determining the impact of temperature threshold exceedance.

19.2.2 *The General Large-Area Model for Annual Crops*

The General Large-Area Model for Annual Crops (GLAM; Challinor et al., 2004) is a process-based crop model. It has a daily time-step, allowing it to resolve the impacts of sub-seasonal variability in weather. It has a soil water balance with 25 layers which simulates evaporation, transpiration and drainage. Roots grow with a constant extraction-front velocity and a profile linearly related to Leaf Area Index (LAI). LAI evolves using a constant maximum rate of change of LAI modified by a soil water stress factor. Separate simulation of biomass accumulation, by use of transpiration efficiency allows Specific Leaf Area (SLA, the mass of leaf per unit area of leaf) to be used as an internal consistency check: leaf area and leaf mass can be derived independently of each other and used to calculate values of SLA which can be compared to typical observed values. The sowing date is simulated by applying an intelligent planting routine to a given sowing window. The crop is planted when soil moisture exceeds a threshold value. If no such event occurs within the window then crisis planting is simulated on the final day of the sowing window.

19.2.3 *Results for the Current Climate*

The geographical focus of work to date with GLAM is the tropics. Much of the world's food is grown in this region. Also, there is a well-documented dependence on rainfed agriculture across much of the tropics. Farmers rely on monsoon rains to bring sufficient water for crop cultivation. Preliminary work focussed on simulations in the current climate as predictive skill here is seen as a pre-requisite for predictive skill in future climates. Figure 19.2 shows the ability of GLAM to capture interannual

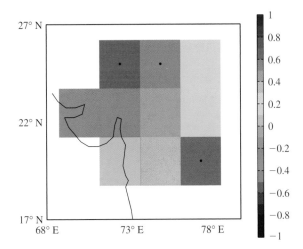

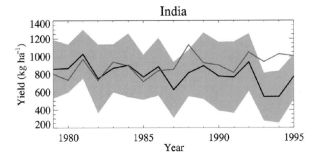

Figure 19.4 Observed FAO groundnut yield statistics (red line) with simulated mean values (black line) and spatial standard deviation (grey shading).

Figure 19.3 Correlation between observed and simulated yields (Challinor et al., 2005b). Dots indicate 95% significance. The simulated yields were formed from an ensemble mean GLAM simulation of crop yield in Gujarat, India. Time series of yield were formed by driving the crop model with each individual ensemble member.

variability in yields when driven with observed weather data. Agreement between simulated and observed yields tends to be greatest in regions where the area under cultivation is greatest, and where there is a strong climate influence on yields. Hence the all-India yields shown mask some regional variability in skill. See Challinor et al. (2004) for a more detailed analysis.

GLAM has also been used with seasonal hindcast ensembles (Challinor et al., 2005b). This study showed that an ensemble of crop yields can contain useful information in both the mean (figure 19.3) and in the spread (not shown). Probabilistic methods of yield estimation are relevant to future as well as current climates, since they provide a tool for the quantification of the uncertainty outlined in section 2.2.

19.2.4 *Fully Coupled Crop-Climate Simulation*

Full integration of crop and climate models is the logical progression of the work described so far. Advantages of a fully coupled crop-climate model include:

- Resolution of the diurnal cycle would enable more accurate simulation of temperature threshold exceedance.
- Feedbacks between the crop and its environment can be simulated. This may have a significant impact on yield for irrigated crops.
- Integration of management decisions such as sowing date allows an assessment of the vulnerability of farming systems to changes in the mean and variability of climate.

Accordingly, the crop growth and development formulations of GLAM have been incorporated into the land

surface scheme of the Hadley Centre atmospheric GCM, HadAM3 (Osborne, 2004). Crop growth is determined according to the GLAM parameterisations in accordance with the simulated weather and climate of HadAM3. Dynamical crop growth within the land surface scheme alters the important surface characteristics for the determination of fluxes to the atmosphere such as leaf area, albedo and roughness length, while the simulated rates of surface evaporation (soil evaporation and/or plant transpiration) will affect the humidity of the crop environment.

Initial evaluation of the coupled crop-climate model has focused on the simulation of groundnut by GLAM throughout the Tropics. Figure 19.4 shows the simulated and observed yields for India. GLAM was not regionally calibrated for these simulations, yet the mean and variability of yields compare well with observations.

The coupled model HadAM3-GLAM was forced with observed interannual variations in sea surface temperatures which play a large role in determining interannual variations in climate; e.g. ENSO variations. Figure 19.5 illustrates the capacity of HadAM3-GLAM to simulate interannual variability of crop growth simulations in response to the simulated variations in climate for two regions in India.

Sowing of the crop is dependent on the onset of the monsoon and exhibits considerable interannual variability at both regions. Subsequent crop biomass production requires the transpiration of considerable amounts of water and is therefore dependent on the amount of water in the soil profile. Consequently, variability in the amount and distribution of the rainfall results in the large range of crop biomass simulated at harvest. For the NW India region, the duration and amount of rainfall is only sufficient to grow one crop. In contrast, the temporal distribution of the rainfall in SE India is more bimodal, allowing a second crop to be sown in 8 out of the 17 years. However, these growing seasons are terminated by the model due to water stress in January or February, indicating a need for supplementary irrigation. These results illustrate the potential of the coupled model to assess the vulnerability of crop production to climate.

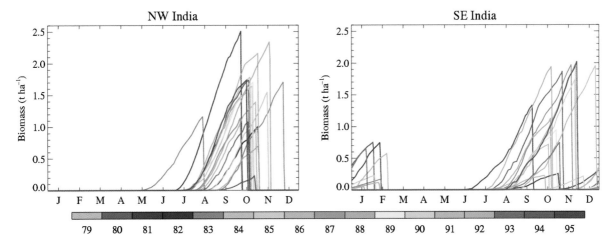

Figure 19.5 Time series of simulated groundnut biomass at two HadAM3 grid points in India. Coloured lines represent growth of the crop during each year from 1979–1995.

19.3 Regional Crop Modelling Study for India 2071–2100 Under the A2 Scenario

19.3.1 *Methods*

Parameterisations of the impacts of high temperature episodes (see section 2.1) have been added to GLAM (Challinor et al., 2005c). These methods are based on the mean 8am–2pm temperatures (T_{AM}) during the flowering stage of the crop. Only flowering that is associated with subsequent pegs and pods (and therefore yield) is considered. Accumulated thermal time is used to determine the start and end dates (t_1 and t_2) of flowering. Daily T_{AM} is examined for the period $t_1 - 6$ to $t_2 + 12$. Temperature threshold exceedance is defined as $T_{AM} > 34°C$ (sensitive variety), $36°C$ (moderately sensitive variety) or $37°C$ (tolerant variety). For each day (i) during the flowering stage, these high temperature events are characterised according to their timing relative to i and their duration in days. Only one of the high temperature events impacts yield. For each event, the following is carried out: (i) two critical temperatures are calculated as a function of the timing and duration of the event; (ii) The fraction of pods setting as a result of the flowers forming on day i (P_i) is reduced linearly from one to zero for values of T_{AM} between these two critical temperatures; (iii) The total fraction of pods setting (P_{tot}) is determined as a sum over all days in the flowering stage, using a prescribed fraction of total flowers forming each day (F_i). The lowest value of P_{tot} is then used to reduce the rate of change of harvest index. Steps (i) and (iii) include parameters which vary according to the crop variety (sensitive, moderately sensitive, or tolerant).

Challinor et al., 2005c did not account for the impact of water stress on pod-set. Hence step (iii) in the description above has been modified accordingly:

$$P_{tot} = \sum_{i=t_1}^{t_2} P_i F_i \min\left(\frac{S_i}{S_{cr}}, 1\right) \tag{1}$$

where S_i is the soil water stress factor (ratio of available water to transpirative demand) and S_{cr} is a threshold value of S_i below which pod-set is affected by water stress. In sensitivity tests, three values of S_{cr} were used (0.2, 0.3 and 0.4) and yields were found to be insensitive to the value chosen. $S_{cr} = 0.2$ was used for all the simulations in this study.

A regional climate simulation from the joint Indo-UK program on climate change was used to drive GLAM for the study presented here. As part of this program the PRECIS regional climate model (http://www.metoffice.com/research/hadleycentre/models/PRECIS.html) was run using boundary conditions derived from global climate models: a coupled general circulation model (HadCM3) was used to simulate changes in climate, and these changes were added to the baseline (current) climate of the atmosphere-only model HadAM3. In order to understand the role of sulphate aerosols, simulations both with and without the sulphur cycle were carried out (see IITM, 2004). Availability of data at the time of the present study limited the scenario used to a 2070–2100 A2 simulation without sulphur. The A2 scenario is one of the most extreme scenarios, with emissions rising monotonically from present-day values (<10 Gt of carbon) to over 25 Gt in 2100 (IPCC, 2001a). Hence the impacts on crop yield presented here are not predictions, but rather a demonstration of both the methods used and of one potentially plausible future scenario.

19.3.2 *Results*

Use of the modified version of GLAM driven by, but not coupled with, regional climate modelling data allows the importance of extremes of temperature and water stress to be assessed. Also, the water-stress parameterisation can be turned off, allowing an assessment of the impact of temperature alone. When used to drive GLAM, the PRECIS simulations of the A2 scenario project an increase in the importance of temperature and water stress near

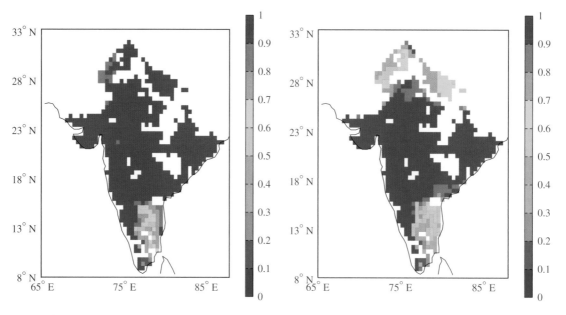

Figure 19.6 Mean fraction of setting pods in groundnut for 1960–1990 (left panel) and 2071–2100 (right panel) as simulated by GLAM, driven by the Hadley centre PRECIS model under the A2 scenario. Both panels show a variety which is moderately sensitive to high temperature stress near flowering.

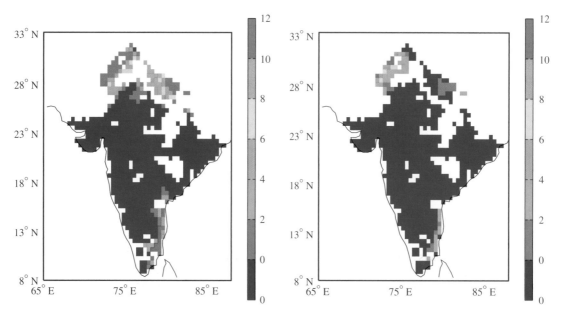

Figure 19.7 Number of years from the period 2071–2100 (Hadley centre PRECIS model under the A2 scenario) when the total fraction of pods setting in groundnut simulated by GLAM is below 50% when temperature stress only is considered. The left panel shows a variety which is sensitive to high temperature stress near flowering and the right panel shows a variety which is tolerant.

flowering (figure 19.6). In particular the north of India shows very little impact on the mean number of pods setting in the current climate, but a significant impact in the 2071–2100 projection.

One possible adaptation to climate change is the cultivation of crops more tolerant to high temperatures. Simulations were performed using two crop types, one

that is sensitive, and one tolerant, to high temperature events. The contrast between these two sets of (figure 19.7) shows the potential importance of crop variety in providing adaptation options for high temperature stress. The choice of variety makes the difference between an increase and a decrease in yields in the north-east of the study region.

19.4 Discussion

19.4.1 *Adaptation to climate change*

The choice of crop variety is only one amongst many possible options for adaptation to high temperature threshold exceedance. Changes in planting date and irrigation levels provide alternative methods of continuing to grow the same crop in a climate with increased incidence of high temperatures. Broader adaptation options include a change to another crop type altogether. Furthermore, adaptation to climate change implies adaptation not only to temperature extremes, but also to other changes, such as those in rainfall, mean temperature and ambient CO_2 levels. Adaptation to these changes may involve the use of a crop with different thermal time and/or water requirements. Adaptation to CO_2 increases may involve changes in applied nutrient and irrigation levels, since the magnitude of the CO_2 fertilisation effect may depend upon these decisions (Tubiello and Ewert, 2002).

It is clear, then, that in determining effective adaptation strategies, it is important to consider all the impacts of CO_2 increases. The range of possible adaptation responses to these impacts depends upon the resources available and upon the uptake time for technological change (see e.g. Easterling et al., 2003); only when these factors are taken into consideration can vulnerability to climate change be assessed (Reilly and Schimmelpfennig, 1999). Ultimately, it is farmers who will have to adapt to climate change, and studies of potential adaptation measures need to be considered within the full socio-economic context of local farming practices (e.g. Easterling et al., 1993). This may mean that adaptation is considered in the context of responses on seasonal timescales (e.g. Gadgil et al., 1999; O'Brien et al., 2000; Kates, 2000).

19.4.2 *Research needs and opportunities*

The choice of crop model, and the way in which climate change simulations are used to drive the crop model, are an important factor in determining the results of an agricultural climate change impacts assessment (section 2.2). Crop models that simulate the impact of key processes, such as high temperature stress, provide an opportunity to quantify the relationship between greenhouse gas emissions and crop productivity (section 4). In particular, off-line studies present a pragmatic way to create the crop yield projections that are associated with climate change projections. Fully interactive crop-climate simulation, whilst being more computationally expensive and less widely tested, provides a tool for the investigation of the impact of coupled vegetation-atmosphere processes and of the diurnal cycle.

Whichever crop modelling methods are chosen, observations of crop yield are critical to the assessment of the accuracy of crop simulations. Many studies use proxies for observed yields, such as yields simulated by a crop model using observed weather (e.g. Hansen and Indeje,

2004). This is clearly problematic if we are to quantify the uncertainty associated with our projections. Ground-truthing of both crop and climate projections for the coming years and decades has an important role in ensuring the reliability of the scenarios that are developed.

19.5 Conclusions

An integrated approach to crop-climate modelling provides tools for the estimation of the vulnerability of food systems to climate variability and change. A number of recent advances have been highlighted: firstly, the simulation of yields under the current climate using the General Large-Area Model for annual crops is presented as a necessary condition for the simulation of the impacts of climate change using GLAM. Secondly, fully coupled GLAM-HadAM3 simulations allow simultaneous estimation of the impact of climate change on farming practices and on yield. Thirdly, off-line studies have shown the importance of crop variety as a means of adaptation to climate threshold exceedance. Fully coupled studies of the impact of climate thresholds would allow the impact of diurnal variability of temperature to be explicitly represented.

The further research needs and opportunities outlined in section 5.2 highlight the potential of both fully coupled and off-line large-area integrated crop-climate modelling. Key processes such as the impact on crop yield of high temperature stress, changes in rainfall and CO_2, and changes in management strategies, can be simulated using such a system. The assessment of the accuracy of yield simulation in current and evolving climates, and the associated data sets of observed yields, have an important role in the development of reliable yield projections with quantified levels of uncertainty.

Acknowledgements

The authors are grateful to the International Crops Research Institute for the Semi-Arid Tropics (ICRISAT) and the Indian Institute for Tropical Meteorology for the crop productivity data and the PRECIS simulations, respectively. The PRECIS simulations were funded by DEFRA through the Indo-UK project. The comments of the reviewer were valuable in producing the manuscript in its final form.

REFERENCES

Challinor, A. J., T. R. Wheeler and J. M. Slingo, 2005c: Simulation of the impact of high temperature stress on the yield of an annual crop. *Agricultural and Forest Meteorology* (accepted).

Challinor, A. J., J. M. Slingo, T. R. Wheeler and F. J. Doblas-Reyes, 2005b: Probabilistic hindcasts of crop yield over western India. *Tellus* 57A (498–512).

Challinor, A. J., T. R. Wheeler, J. M. Slingo, P. Q. Craufurd and D. I. F. Grimes, 2005a: Simulation of crop yields using the ERA40 re-analysis: limits to skill and non-stationarity in weather-yield relationships. *Journal of Applied Meteorology* 44 (4) 516–531.

Challinor, A. J., T. R. Wheeler, J. M. Slingo, P. Q. Craufurd and D. I. F. Grimes, 2004: Design and optimisation of a large-area process-based model for annual crops. *Agricultural and Forest Meteorology* 124 (99–120).

Challinor, A. J., J. M. Slingo, T. R. Wheeler, P. Q. Craufurd, and D. I. F. Grimes (2003, February). Towards a combined seasonal weather and crop productivity forecasting system: Determination of the spatial correlation scale. *J. Appl. Meteorol.* 42, 175–192.

Doorenbos, J. and A. H. Kassam (1979). Yield response to water. FAO Irrigation and Drainage 33, FAO, Viale delle Terme di Caracalla, 00100 Rome, Italy.

Easterling, W. E., N. Chhetri and X. Niu (2003). Improving the realism of modelling agronomic adaptation to climate change: simulating technological substitution. *Climatic Change*, 60, 149–173.

Easterling, W. E., X. F. Chen, C. Hays, J. R. Brandle, and H. H. Zhang (1996). Improving the validation of model-simulated crop yield response to climate change: An application to the EPIC model. *Climate Research* 6 (3), 263–273.

Ferris R., Wheeler T. R., Ellis R. H. and Hadley P. (1999). Seed yield after environmental stress in soybean groan under elevated CO_2. *Crop Science* 39, 710–718.

Fischer, G., M. Shah, and H. van Velthuizen (2002). Climate change and agricultural vulnerability. Technical report, International Institute for Applied Systems Analysis. Available at http://www.iiasa.ac.at/Research/LUC/.

Gadgil, S., P. R. S. Rao, K. N. Rao, and K. Savithiri (1999, September). Farming Strategies for a variable climate. Technical Report 99AS7, CAOS, Indian Institute of Science, Bangalore 560012, India.

Hansen, J. W. and M. Indeje (2004). Linking dynamic seasonal climate forecasts with crop simulation for maize yield prediction in semi-arid Kenya. *Agric. For. Meteorol.* 125, 143–157.

Hansen, J. W. and J. W. Jones (2000). Scaling-up crop models for climatic variability applications. *Agric. Syst.* 65, 43–72.

Idso K. E and Idso S. B. (1994). Plant responses to atmospheric CO_2 enrichment in the face of environmental constraints: a review of the past 10 years' research. *Agricultural and Forest Meteorology* 69, 153–203.

IPCC (2001a). Climate Change 2001: The Scientific Basis. Contribution of Working Group I to the Third Assessment Report of the Intergovernmental Panel on Climate Change. Cambridge University Press. 881 pp.

IPCC (2001b). Climate Change 2001: Impacts, Adaptation, and Vulnerability. Contribution of Working Group II to the Third Assessment Report of the Intergovernmental Panel on Climate Change. Cambridge University Press. 1032 pp.

IITM (2004, December). Indian climate change scenarios for impact assessment. Technical report, Indian Institute of Tropical Meteorology, Homi Bhabha Road, Pune 411 008, India.

Jenkins, G. and J. Lowe (2003). Handling uncertainties in the ukcip02 scenarios of climate change. Technical note 44, Hadley Centre.

Kates, R. W. (2000). Cautionary tales: Adaptation and the global poor. *Clim. Change* 45 (1), 5–17.

Kimbal B. A. (1983). Carbon dioxide and agricultural yield; an assemblage and analysis of 430 prior observations. *Agronomy Journal*, 75, 779–788.

Luo, Q. Y. and E. Lin (1999). Agricultural vulnerability and adaptation in developing countries: The Asia-pacific region. *Clim. Change* 43 (4), 729–743.

Mavromatis, T. and P. D. Jones (1998). Comparison of climate change scenario construction methodologies for impact assessment studies. *Agric. For. Meteorol.* 91 (1–2), 51–67.

Mearns, L. O., W. Easterling, C. Hays, and D. Marx (2001). Comparison of agricultural impacts of climate change calculated from the high and low resolution climate change scenarios: Part I. the uncertainty due to spatial scale. *Clim. Change* 51, 131–172.

Mearns, L. O., T. Mavromatis, and E. Tsvetsinskaya (1999). Comparative response of EPIC and CERES crop models to high and low spatial resolution climate change scenarios. *J. Geophys. Res.* 104 (D6), 6623–6646.

Mitchell, R. A. C., Mitchell, V. J., Driscoll, S. P., Franklin, J. and Lawlor, D.W. (1993). Effects of increased CO_2 concentration and temperature on growth and yield of winter wheat at 2 levels of nitrogen application. *Plant Cell and Environment* 16, 521–529.

Ntare, B. R., J. H. Williams and F. Dougbedji, 2001: Evaluation of groundnut genotypes for heat tolerance under field conditions in a Sahelian environment using a simple physiological model for yield. *J. Ag. Sci.* 136, 81–88.

O'Brien, K., L. Sygna, L. O. Næss, R. Kingamkono, and B. Hochobeb (2000). Is information enough? user responses to seasonal climate forecasts in southern Africa. Technical Report ISSN: 0804–4562, Center for International Climate and Environmental Research – Oslo. Report to the World Bank, AFTE1-ENVGC. Adaptation to climate change and variability in sub-Saharan Africa, Phase II.

Osborne, T. M. 2004: Towards an integrated approach to simulating crop-climate interactions. Phd Thesis, University of Reading. pp. 184.

Parry, M. L., C. Rosenzweig, A. Iglesias, M. Livermore, and G. Fischer (2004). Effects of climate change on global food production under SRES emissions and socio-economic scenarios. *Global Environmental Change – Human and Policy Dimensions* 14 (1), 53–67.

Parry M., Wheeler, T. R. and others (2001). Investigation of thresholds of impact of climate change on agriculture in England and Wales. Final report to the MAFF for Project CC0358.

Reilly, J. M. and D. Schimmelpfennig (1999). Agricultural impact assessment, vulnerability, and the scope for adaptation. *Clim. Change* 43 (4), 745–788.

Prasad, P. V. V., P. Q. Craufurd, R. J. Summerfield, and T. R. Wheeler (2000). Effects of short episodes of heat stress on flower production and fruit-set of groundnut Arachis hypogaea l. *J. Exp. Bot.* 51 (345), 777–784.

Semenov, M. A. and E. M. Barrow (1997). Use of a stochastic weather generator in the development of climate change scenarios. *Climatic Change* 35, 397–414.

Tubiello, F. N. and F. Ewert (2002). Simulating the effects of elevated co_2 on crops: approaches and applications for climate change. *Eur. J. Agron.* 18 (1–2), 57–74.

Tubiello, F. N., C. Rosenzweig, R. A. Goldberg, S. Jagtap, and J. W. Jones (2002). Effects of climate change on us crop production: simulation results using two different GCM scenarios. part I: Wheat, potato, maize, and citrus. *Climate Research* 20 (3), 259–270.

Wheeler T. R., Batts G. R., Ellis R. H., Hadley P. and Morison J. I. L. (1996a). Growth and yield of winter wheat (*Triticum aestivum*) crops in response to CO_2 and temperature. *Journal of Agricultural Science*, 127, 37–48.

Wheeler T. R., Hong T. D. Ellis R. H., Batts G. R., Morison J. I. L. and Hadley P. (1996b). The duration and rate of grain growth, and harvest index, of wheat (*Triticum aestivum* L.) in response to temperature and CO_2. *Journal of Experimental Botany*, 47, 623–630.

Wheeler T. R., Craufurd P. Q., Ellis R. H., Porter J. R. and Vara Prasad P.V. (2000). Temperature variability and the yield of annual crops. *Agriculture, Ecosystems and Environment*, 82, 159–167.

Wright G. C., Hubick K. T. and Farquahar G. D. (1991). Physiological analysis of peanut cultivar response to timing and duration of drought stress. *Australian Journal of Agricultural Research*, 42, 453–470.

CHAPTER 20

Climate Stabilisation and Impacts of Sea-Level Rise

Robert J. Nicholls[1] and Jason A. Lowe[2]

[1] *School of Civil Engineering and the Environment, University of Southampton Highfield, Southampton, United Kingdom*
[2] *Hadley Centre (Reading Unit), Department of Meteorology, The University of Reading, Earley Gate, Reading, United Kingdom*

SUMMARY: Atmospheric temperature stabilisation does not translate into sea-level stabilisation on century timescales: rather it leads to a slower rise in sea level which continues far into the future. This fact reflects the long thermal lags of the ocean system and also the slow response and possible irreversible breakdown of the Greenland and West Antarctic Ice Sheets given global warming. Therefore, while climate stabilisation reduces the coastal impacts of sea-level rise during the 21st century, compared to unmitigated emissions, the largest benefits are in the 22nd century (and beyond). Further, many impacts may only be delayed rather than avoided, depending on the ultimate 'commitment to sea-level rise'. Given these constraints, a realistic stabilisation target for sea level is a maximum rate of rise rather than stabilisation of sea level *per se*. Hence, adaptation and mitigation need to be considered together for coastal areas, as collectively they can provide a more robust response to human-induced climate change than consideration of each policy alone. Mitigation will reduce both the rate of rise and the ultimate commitment to sea level rise, while adaptation is essential to manage the commitment to sea-level rise (the residual rise), most especially for vulnerable coastal lowlands and small islands. However, the timescale of sea-level rise is challenging for policy as long timescales extending beyond 2100 need to be considered to understand the full implications of the different policy choices being made now or in the near future.

20.1 Introduction

A significant global-mean sea-level rise is expected due to human-induced warming during the 21st century. In the Intergovernmental Panel on Climate Change (IPCC) Third Assessment Report (TAR) the projected rise from 1990 to 2100 was 9 to 88 cm with a mid estimate of 48 cm (Church and Gregory, 2001)[1]. Through the 21st century, climate stabilisation would slow but not stop this rise due to the long thermal lags of oceanic response, which is termed the '*commitment to sea-level rise*' (Wigley and Raper, 1993; Nicholls and Lowe, 2004; Meehl et al., 2005). Beyond the 21st century, substantial *additional* rises of sea level appear almost inevitable due to the same effect (Wigley, 2005), with Greenland and Antarctica becoming possible significant additional sources (Nicholls and Lowe, 2004; Lowe et al., 2005; Rapley, 2005; Lenton et al., 2005). While significant contributions to sea level due to the instability of the West Antarctic Ice Shelf (WAIS) are considered unlikely during the 21st century, there are large uncertainties, and it becomes more likely if global warming continues (Vaughan and Spouge, 2002). Similarly, if local temperatures rise above 2.7°C, irreversible melting of Greenland is likely to occur (Gregory et al., 2004). Collectively, these two sources could contribute up to 12 to 13 m of global-mean sea-level rise, although this rise may take a millennium or more to be realised.

Sea-level rise raises significant concern due to the high concentration of natural and socio-economic values in the coastal zone. The coastal zone is a major focus of human habitation and economic activity, as well as being important ecologically and in earth system functioning (Turner et al., 1996; Crossland et al., 2005). In 1990, it was estimated that 1.2 billion (or 23%) of the world's population lived in the near-coastal zone[2], at densities about three times higher than the global mean (Small and Nicholls, 2003). The highest population density occurs below 20-m elevation. Net migration to the coast is also widely reported and under the SRES scenarios[3], the near-coastal population could increase to 2.4 to 5.2 billion people by the 2080s (Nicholls and Lowe, 2004), living in dominantly urban settings (Nicholls, 1995a; Small and Nicholls, 2003).

This paper reviews the implications of climate stabilisation for sea-level rise impacts in coastal areas. The present state of knowledge is synthesised, including an emphasis on the recent paper of Nicholls and Lowe (2004), and the major knowledge gaps are identified.

---

[1] These estimates assume a small contribution from Antarctica but not disintegration.

[2] The area both within 100 km horizontally and 100 m vertically of the coastline.

[3] SRES – Special Report on Emission Scenarios (Nakicenovic and Swart, 2000).

20.2 Sea-level and Climate Change in Coastal Areas

20.2.1 *Sea-level Change*

The local or *relative* change in sea level at any coastal location depends on the following components (Nicholls and Lowe, 2004):

- *Global-mean sea-level rise* due to an increase in the volume of ocean water. During the 20th/21st century this is primarily due to thermal expansion of the ocean as it warms, and the melting of small ice caps due to human-induced global warming (Church and Gregory, 2001). The contribution of ice melt in Greenland is less certain, but most projections for the 21st century indicate that it will be less than small ice caps. Increased precipitation over Antarctica is expected to grow the Antarctic ice sheet, producing a sea-level *fall* that will offset some of the positive contribution from Greenland.
- *Regional oceanic and meteorological change* could cause deviations from the global mean rise due to thermal expansion by up to ±100% over the 21st century (Gregory and Lowe, 2000). These regional variations are caused by the non-uniform patterns of temperature and salinity changes in the ocean, as well as changes in the time mean pattern of atmospheric surface pressure, and changes in the depth mean ocean circulation (Lowe and Gregory, submitted). The detailed patterns resulting from these changes are highly uncertain, differing greatly between models (Gregory et al., 2001).
- *Vertical land movement* (subsidence/uplift) occurs due to a range of geological and human-induced processes such as tectonics, neotectonics, glacial-isostatic adjustment, consolidation and fluid withdrawal (Emery and Aubrey, 1991). During the 21st century, vertical land movement averaged over the entire global coastline is expected to be less than the rise resulting from oceanographic changes. However, at some locations the vertical land movements will still be important, most particularly in areas subject to human-induced subsidence due to ground fluid withdrawals, such as expanding cities in deltaic settings (Nicholls, 1995a; Woodroffe et al., accepted).

In this review, the main focus is global-mean sea-level rise, with some limited consideration of vertical land movement, reflecting the approach of existing analyses.

20.2.2 *Observed Sea-level Trends*

During the 20th century, tide gauge measurements corrected for vertical land movement suggest that global sea levels have risen by 10 to 20 cm (Church and Gregory, 2001). Unambiguous acceleration of the global-mean rise during the 20th century has not been observed, but recent satellite altimetry measurements of the global-mean rise for the last decade of the 20th century show that the rise has been around 3 mm/yr (e.g. Cabanes et al., 2001). However, other periods of above average rise appear to

have occurred during the last 50 years, and it is not yet clear if this is the beginning of accelerated global sea-level rise due to human-induced global warming.

20.2.3 *Future Sea-level Rise Scenarios*

Projections of future sea level need to include both the response to future emissions of greenhouse gases and the ongoing response to past emissions. This latter component is the commitment to sea-level rise and will continue for many centuries into the future, even if the atmospheric radiative forcing were stabilised immediately. The sea-level rise commitment is mainly a result of the long time scales associated with the mixing of heat from the ocean's surface to deep ocean layers (Wigley and Raper, 1993; Church and Gregory, 2001). Estimates using the HadCM3 model suggest that if radiative forcing were (hypothetically) stabilised today, sea level would still eventually rise by more than 1 m due to thermal expansion alone, although this would take more than a 1000 years to occur (Nicholls and Lowe, 2004). Wigley (2005) presents consistent results that suggest a rise of 0 to 90 cm by 2400, with a best estimate of a 30-cm rise at a near-constant rate of 10 cm/century, which will remain sizeable for centuries beyond the modelled period.

A number of scenarios in which greenhouse gas concentrations are eventually stabilised have been produced, notably the IPCC S Scenarios (Enting et al., 1994), the WRE scenarios (Wigley et al., 1996) and the recent post-SRES stabilisation scenarios (Swart et al., 2002). The eventual stabilisation of atmospheric carbon dioxide concentrations at the lower levels being suggested (e.g. 550 ppmv) will require emissions to be reduced substantially below current levels. By reducing emissions enough to achieve a stabilisation of concentration at either 550 ppm or 750 ppm[4] during the 22nd century the rise in global mean sea level may be delayed by up to a few decades during the 21st century (e.g. Figure 20.1). However, unlike atmospheric temperate rise, the unmitigated sea-level rise in the 21st century is only delayed rather than avoided, and by 2250 sea levels under a stabilised climate are still rising with little sign of deceleration (e.g. Mitchell et al., 2000). Figure 20.1 illustrates this important point for a range of climate sensitivities from 1.5°C to 4.5°C.

The combined effect of the commitment to sea-level rise and the relatively slow divergence of different future SRES emissions scenarios over the next few decades means that for a given climate model (and hence climate sensitivity) the future global-mean sea-level rise is almost independent of future emissions to about 2050 (e.g., Figure 20.1). However, after 2050 future emissions become increasingly important in controlling future sea-level rise. Therefore, global-mean sea-level rise appears inevitable during the 21st century and beyond even given substantial

---

[4] These scenarios are henceforth termed S550 and S750, respectively.

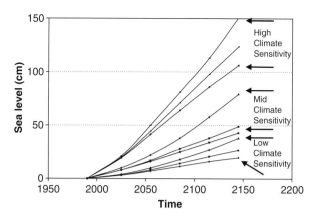

Figure 20.1 Unmitigated and stabilisation scenarios for the global mean rise in sea level (relative to 1990), including the effect of climate sensitivity (spanning the range 1.5 to 4.5°C). In each triplet of lines, unmitigated (IS92a) emissions is the highest line, S750 is the middle line, and S550 is the lowest line. Results are obtained directly from Mitchell et al. (2000) or simulated with a simple climate model (reprinted from Nicholls and Lowe (2004) with permission from Elsevier).

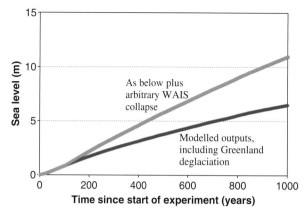

Figure 20.2 Global-mean sea-level rise over the next millennia assuming atmospheric concentrations for CO_2 are increased at 2% per annum and reach four times pre-industrial levels and are then stabilised at year 70. The lower curve is from Lowe et al. (2005) and includes deglaciation of Greenland. The upper curve adds an illustrative WAIS collapse starting at 100 years, and continuing at 0.5 m/century.

mitigation of climate change, but we can influence the future amount and rate of sea-level rise by mitigation, most especially in the 22nd century and beyond.

20.2.4 *Uncertainties in Future Sea-level Rise Scenarios*

The 9 to 88 cm range for global rise in the TAR (Church and Gregory, 2001) does not embrace the full range of uncertainties, including those associated with changes in the major ice sheets, particularly the maritime West Antarctic Ice Sheet (WAIS), which contains enough water to raise global sea levels by up to 6 m (Mercer, 1978). Oppenheimer (1998) considered a number of scenarios in which the WAIS might contribute to future sea-level rise, noting that in the scenario with the greatest likelihood (assuming continued growth in greenhouse gas emissions consistent with the IPCC IS92 emissions scenarios) it could contribute as much as 60 to 120-cm rise per century during a collapse that would take 500 to 700 years. Vaughan and Spouge (2002) concluded from a Delphi analysis that the probability of a WAIS collapse on the time scales suggested by Oppenheimer (1998) was less than 20% and that a sea-level rise contribution from the WAIS of more than 10 mm/yr (or 1 m/century) by the year 2300 was unlikely, at only 5%. Going beyond 2300 the probability of collapse increases under climate change. However, as Vaughan and Spouge (2002) highlight, refining the estimate is difficult because observations cannot yet even identify the potential collapse mechanisms and different types of ice sheet model tend to emphasise different mechanisms or controlling parameters.

The Greenland ice sheet is mostly grounded above sea level and surface melting rather than collapse is how it might contribute to sea-level rise. Recent work suggests

that a local atmospheric temperature rise of 2.7°C could trigger the start of slow melting of the Greenland ice sheet, over a timescale of 1000 years or more (Gregory et al., 2004). This temperature change threshold is breached under most plausible emission scenario pathways, and suggests an ongoing contribution to sea-level rise in addition to the commitment to sea-level rise due to thermal expansion already discussed. Once melting has started and the ice sheet elevation is declining its ultimate fate will depend on the evolving ice sheet topography and the evolving local climate, but there is a concern that a complete loss will occur (Church and Gregory, 2001). If the temperatures were reduced after the ice sheet had been completely lost then it is currently unclear if the ice sheet would reform. Lunt et al. (2004) and Toniazzo et al. (2004) provide conflicting views using different models and experimental set up.

Figure 20.2 illustrates the global-mean sea-level rise under a high (four times pre-industrial CO_2 concentrations) warming scenario including deglaciation of Greenland and thermal expansion, calculated using a complex climate model (HadCM3) coupled to a dynamic ice sheet model (Ridley et al., 2005). The total global rise after 1000 years is about 7 m. This excludes any contribution from WAIS collapse, which could significantly enhance the total rise as also illustrated in Figure 20.2.

Climate mitigation will reduce the risks of the deglaciation of the Greenland ice sheet, most especially if local atmospheric temperature rise can be contained below about 2°C, although to achieve this goal will require stringent mitigation (Lowe et al., 2005). Using a simpler intermediate complexity climate model, simpler treatment of ice sheets, and ignoring Antarctica, Lenton et al. (2005) estimated global sea-level rise to be in the range of 0.5 m to 11.4 m by the year 3000: the smallest rise represents their most extreme mitigation scenario, while the highest rise represents a scenario approaching maximum possible emissions. These simulations suggest that society can significantly influence future sea levels over the millennium timescale via future greenhouse emissions. Further simulations of these long-term changes with more complex climate models and additional scenarios would be useful. Mitigation is also likely to reduce the risk of WAIS collapse, but a lack of understanding of the mechanisms and availability of suitable analytical tools means that this benefit of mitigation is not presently understood in quantitative terms.

20.2.5 *Other Climate Change Impacts in the Coastal Zone*

In addition to rising sea levels, future global climate change is expected to directly alter many other environmental factors relevant to the coast (Table 20.1). Some of the more important issues, especially storm track, frequency and intensity remain quite uncertain.[5] While the effect of mitigation on mean sea-level rise has been investigated (Schimel et al., 1997; Mitchell et al., 2000), the effect of

mitigation on changes in storm characteristics and storm-driven impacts has not been robustly addressed. However, if these and other changes in climate (Table 20.1) are ultimately driven by surface warming then mitigation is likely to have an effect. Furthermore, it may have an effect on shorter time scales than that on sea-level rise.

20.3 Impacts of Sea-level Rise

Sea-level rise has a number of biophysical effects on coastal areas: the most significant are summarised in Table 20.1. Most of these impacts are broadly linear functions of sea-level rise, although some processes such as wetland loss and change show a threshold response and are more strongly related to the rate of sea-level rise, rather than the absolute change. These natural-system effects have a range of potential socio-economic impacts, including the following identified by McLean and Tsyban (2001):

- Increased loss of property and coastal habitats
- Increased flood risk and potential loss of life

Table 20.1 Some climate change and related factors relevant to coasts, their biogeophysical effects and potential benefits of climate mitigation (adapted from Nicholls and Lowe, 2004).

CLIMATE FACTOR	DIRECTION OF CHANGE	BIOGEOPHYSICAL EFFECTS		MITIGATION BENEFITS
Global-mean sea level	+ve	Inundation, flood and storm damage	Surge Backwater effect	Impacts Reduced
		Wetland loss (and change)		
		Erosion		
		Saltwater Intrusion	Surface Waters Groundwater	
		Rising water tables/impeded drainage		
Sea water temperature (of surface waters)	+ve	Increased coral bleaching; Poleward migration of coastal species; Less sea ice at higher latitudes		Impacts Reduced
Precipitation intensity/ Run-off	Often +ve[6]	Changed fluvial sediment supply; Changed flood risk in coastal lowlands;[7]		Impacts Reduced?
Wave climate	Uncertain[8]	Changed patterns of erosion and accretion; Changed storm impacts		Unknown
Storm intensity	Possible +ve for tropical cyclones	Increased storm flooding, wave and wind damage		Impacts Reduced?
Storm track and frequency	Uncertain	Changed occurrence of storm flooding and storm damage		Unknown
Atmospheric CO_2	+ve	Increased productivity in coastal ecosystems; Decreased $CaCO_3$ saturation impacts on coral reefs and acidification of the coastal ocean (Turley et al., 2005)		Impacts Reduced

[5] The first occurrence of a hurricane in the South Atlantic in 2004 (hurricane Catarina) is a noteworthy event, which is certainly of regional significance.

[6] Due to an intensified hydrological cycle.

[7] Changes in catchment management will often be much more important for coastal areas, such as the large Asian deltas in south, south-east and east Asia (e.g., Woodroffe et al., accepted).

[8] Significant temporal and spatial variability expected.

- Damage to coastal protection works and other infra-structure
- Loss of renewable and subsistence resources
- Loss of tourism, recreation, and transportation functions
- Loss of non-monetary cultural resources and values
- Impacts on agriculture and aquaculture through decline in soil and water quality.

The actual impacts of sea-level rise will depend on our ability to adapt to these impacts. Sea-level rise (and other adverse changes) produces *potential impacts*. Successful proactive adaptation can anticipate sea-level rise or other problems and result in reduced *actual impacts*. Successful reactive adaptation in response to these actual impacts can further reduce the level of impacts to *residual impacts*. There is widespread confusion between potential impacts and actual or residual impacts (Nicholls, 2002) and our actual ability to adapt to sea-level rise remains one of the major unknowns in determining actual impacts, with widely divergent views apparent in the literature (Nicholls and Tol, 2005).

The available national-scale assessments generally comprise inventories of the potential impacts to a 1-m rise in sea level, with limited consideration of adaptation (Nicholls and Mimura, 1998). Most impact studies have focused on one or more of the following impact processes: (1) inundation, flood and storm damage, (2) erosion and (3) wetland loss. A range of possible impact indicators have been developed (e.g. IPCC CZMS, 1992; Nicholls, 1995b;

Nicholls and Mimura, 1998; www.survas.mdx.ac.uk). However, as national-scale assessments of the impacts of sea-level rise are based on a range of methodologies and assumptions and are often incomplete in some aspects, these indicators require expert judgement to apply across large samples of coastal nations (e.g. Nicholls, 1995b). The available studies do confirm the importance of the coastal zone and suggest that at least 180 million people (at early 1990s population levels) are exposed to a 1-m rise in sea level (Nicholls, 1995b). As one might expect, low-lying coastal areas are most sensitive to sea-level rise, particularly deltaic and small island settings, as well as coastal ecosystems.

Given the uncertainties in these national studies, larger scale regional and global assessments provide a more consistent basis to assess the impacts of sea-level rise, including under mitigation of climate change. Here results from the Fast Track approaches (Nicholls, 2004; Nicholls and Lowe, 2004) and the FUND model (Tol, 2004) are reviewed.

Using the MAGICC model authored and provided by Wigley (Wigley and Raper, 2001), Nicholls and Lowe (2004) show that while climate mitigation has benefits (i.e. avoids impacts) in all cases, flood impacts still increase with time in nearly all cases, under stabilisation scenarios (Figure 20.3). (These results consider all the sea-level rise scenarios shown in Figure 20.1, and hence cover a wide range of emissions scenarios and model uncertainty, as described by a range of climate sensitivity). Thus, it is

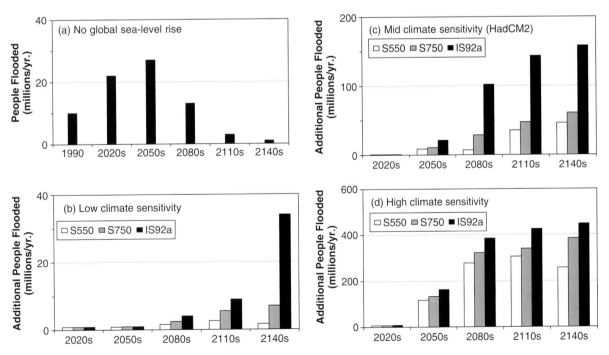

Figure 20.3 Coastal flooding under the IPCC 'S' Stabilisation experiments from 1990 to the 2140s, which compares unmitigated (IS92a) impacts with those under the S750 and S550 stabilisation scenarios. (a) People flooded/year without any global sea-level rise; (b) Additional people flooded/year due to sea-level rise assuming low climate sensitivity; (c) as (b) for mid climate sensitivity (HadCM2); (d) as (b) for high climate sensitivity. Note the varying scale of the y axis. (Reprinted from Nicholls and Lowe (2004) with permission from Elsevier).

not immediately clear what impacts are avoided and what impacts are simply delayed due to mitigation, and some adaptation is required irrespective of future greenhouse emissions. Coastal wetlands are more responsive to mitigation as the impacts are rate dependent, and mitigation is one action that would aid wetland survival (Nicholls and Lowe, 2004).

The effect of stabilisation on people flooded across the different SRES socio-economic scenarios using the method of Swart et al. (2002) is shown in Figure 20.4, and discussed in more detail in Nicholls and Tol (2005). Additional people flooded are shown for three different protection scenarios. (These results can be compared with 'Millions at Risk' (Parry et al., 2001)). In all cases, stabilisation reduces the number of people flooded, but note that the impacts are very sensitive to the protection standard, and increasing protection is much more effective at reducing impacts up to the 2080s than stabilisation. This is not necessarily the case beyond the 2080s. Lastly, impacts are larger in the SRES A2 and B2 worlds than the A1FI world, despite the A1FI world experiencing the largest rise in sea level. The greater impacts reflect different socio-economic conditions and demonstrate that development is an important factor in determining vulnerability to sea-level rise (Nicholls, 2004), with important implications for mitigation and adaptation responses to sea-level rise.

Tol (2004) found that the economic consequences of mitigation reduced its benefits, as economic growth is slowed and less money is available to invest in protection. Nonetheless, the analysis shows net benefits for mitigation in coastal areas through the 21st century, and indicates that future analyses need to consider adaptation and mitigation together to obtain realistic results.

Hence, the commitment to sea-level rise means that an adaptation response to sea-level rise is essential, regardless of mitigation policy. This will need to continue into the 22nd century and beyond and the requirement could be substantial, depending on the magnitude of sea-level rise. Adaptation could comprise protection, accommodation or planned retreat (Klein et al., 2001), or most likely some mixed portfolio of all these options. Nicholls and Tol (2005) based on a cost-benefit analysis approach suggest that a widespread protection response would be an economically-rational response under the SRES scenarios (a rise of up to 38 cm by the 2080s taken from the HadCM3 model (see discussion in Nicholls (2004)), which broadly agrees with several previous analyses. However, other evidence suggests that there are potential limitations to a protection response, particularly as the magnitude of sea-level rise increases (e.g. Nicholls et al., 2005). Protection only manages flood risk and does not remove it – the final response to occasional inevitable disasters remains uncertain and could trigger coastal abandonment. Mitigation provides a mechanism to minimise the occurrence of this situation in the long-term. Therefore, relying solely on adaptation would appear as problematic as depending on mitigation alone.

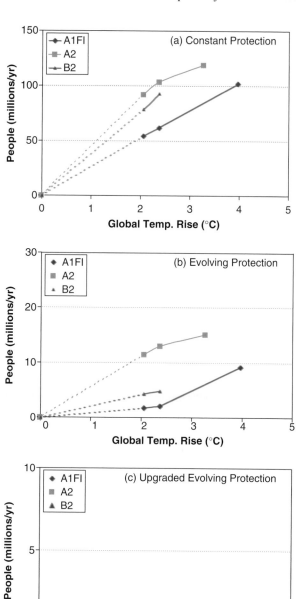

Figure 20.4 The effect of mitigation under the SRES scenarios on additional people flooded in the 2080s, using the runs reported in Nicholls and Tol (2005). Each graph assumes a different protection scenario. (a) Constant Protection assumes constant (1990) protection standards. (b) Evolving Protection assumes dynamic protection upgraded in line with rising GDP/capita, but with no allowance for sea-level rise (i.e. the response is based on present climate variability). (c) Upgrade Evolving Protection includes further upgrade by a factor of 10 (e.g. 1 in 1 defence is upgraded to a 1 in 10 defence, etc., up to a limit of 1 in 1000). Note the difference scales on the y axis.

20.4 Discussion/Conclusions

This paper has briefly explored the potential benefits of mitigation of human-induced climate change in coastal areas. While mitigation could reduce the impacts of several climate change factors as shown in Table 20.1, sea-level rise differs from other climate change factors due to the physical constraint of the ocean's high heat capacity and the long response times of the major ice sheets: there is a 'commitment to sea-level rise', which will continue for hundreds if not thousands of years even given a stable climate. Given these constraints, a realistic stabilisation target for sea level is a target rate of rise rather than stable sea levels (cf. Wigley, 2005). Unmitigated impacts of sea-level rise could be significant and are reduced to varying degrees by mitigation, but never totally avoided. Further, we may only be buying time to adapt, as some impacts are delayed rather than avoided (although more time to adapt is a useful benefit of mitigation). Hence based on the available knowledge, a combination of mitigation and adaptation are required in coastal areas (Nicholls and Lowe, 2004) and both policies need to be assessed in an integrated manner to develop the best response to climate change (Tol, 2004). Given that the commitment to sea-level rise was highlighted in the IPCC First Assessment Report on Climate Science (Warrick and Oerlemans, 1990), it is perhaps surprising that climate policy has not fully recognised the implications of this fact.

The time and space scales of climate change are highly challenging for human institutions and policy development given their global scale and potential impacts generations into the future. This effect is most extreme for coastal areas, where coastal inhabitants are committed to adapt to sea-level rise for hundreds if not thousands of years into the future (Nicholls and Lowe, 2004). More focus on long-term impacts, including methods for meaningful assessments of the risks under different stabilisation pathways would be useful, including analysis beyond 2100 (e.g. Figure 20.3). For instance, collectively the loss of the Greenland Ice Sheet and the collapse of the West Antarctic Ice Shelf could raise global-mean sea levels by up to 10 m or more over the next 1000 years (e.g. Figure 20.2). As suggested by the simulations to the year 3000 by Lenton et al. (2005), mitigation could reduce these risks, and further analyses of sea levels under different emission pathways beyond 2100 are required to inform climate policy about some of these long-term issues. This would need to include the response of the Antarctic and Greenland ice sheets, and should include an analysis of the response of the Greenland Ice Sheet to 'overshoot' (a temporary exceedance of the 2.7°C temperature threshold). A related question which has not been adequately addressed is how far into the future should climate policy (as compared to climate science) consider these issues – decades, centuries or millennia? The authors believe that the scientific results justify a long-term perspective of multiple centuries or more, but it is unclear how widely this view is shared.

Lastly, given the inevitable 'commitment to sea-level rise', the high vulnerability of some coastal settings such as deltas is noteworthy (e.g. McLean and Tsyban, 2001).

Potentially we can adapt, but by what method: protect, accommodate or retreat, or most likely a portfolio of these options (see Evans et al., 2004). Small islands have particular problems as a number of island nations could be completely submerged given a 1 or 2 m rise in sea level (Nurse and Sem, 2001). It is fundamental that the international community invest in the development of adaptive capacity in vulnerable coastal regions in parallel with mitigation efforts. More research on the full range of adaptation options is required so that strategic long-term plans can be made to manage future rises in sea level, including those under stabilisation pathways. The new Dynamic Interactive Vulnerability Assessment (DIVA) tool (www.dinas-coast.net) provides an improved capacity for many of these analyses.

In conclusion, this analysis reinforces the earlier conclusion of Nicholls and Lowe (2004) and Tol (2004) that adaptation and mitigation need to be considered together for coastal areas, as collectively they can provide a more robust response to human-induced climate change than consideration of each policy alone. The research challenge is to better quantify the implications of different stabilisation pathways on sea-level rise, and hence explore the benefits of combined adaptation and mitigation mixtures beyond simple generalisations.

REFERENCES

Cabanes, C., Cazenave, A., and Le Provost, C., 2001. Sea level rise during past 40 years determined from satellite and in situ observations. *Science*, 294(5543): 840–842.

Church, J.A., and Gregory, J.M., 2001. Changes in Sea Level. In: Houghton, J.T., Ding, Y., Griggs, D.J., Noguer, M., van der Linden, P.J., and Xiaosu, D. (eds.) *Climate Change 2001. The Scientific Basis.* Cambridge University Press, Cambridge, pp.639–693.

Crossland, C.J., Kremer, H.H., Lindeboom, H.J., Marshall Crossland, J.I., and Le Tissier, M.D.A. (eds.) 2005. *Coastal Fluxes in the Anthropocene.* The Land-Ocean Interactions in the Coastal Zone Project of the International Geosphere-Biosphere Programme Series: Global Change – The IGBP Series, Springer, Berlin, 232pp.

Emery, K.O., and Aubrey, D.G., 1991. *Sea levels, land levels and tide gauges.* New York: Springer Verlay.

Enting, I.G., Wigley, T.M.L., and Heimann, M., 1994. Future emissions and concentrations of carbon dioxide: key ocean/atmosphere/land analyses, CSIRO Division of Atmospheric Research Technical Paper No. 31.

Evans, E.P., Ashley, R., Hall, J.W., Penning-Rowsell, E.C., Saul, A., Sayers, P.B., Thorne, C.R., and Watkinson, A., 2004. *Foresight Flood and Coastal Defence Project: Scientific Summary: Volume 2, Managing future risks.* Office of Science and Technology, London. 416pp.

Gregory, J.M., Church, J.A., Dixon, K.W., Flato, G.M., Jackett, D.R., Lowe, J.A., Oberhuber, J.M., O'Farrell, S.P., and Stouffer, R.J., 2001. Comparison of results from several AOGCMs on global and regional sea level change 1900–2100. *Climate. Dynamics*, 18(3/4), 225–240.

Gregory, J.M., and Lowe, J.A., 2000. Predictions of global and regional sea level rise using AOGCMs with and without flux adjustment. *Geophysical Research Letters*, 27, 3069–3072.

Gregory, J.M., Huybrechts, P., and Raper, S.C.B., 2004. Threatened Loss of the Greenland Ice-sheet. *Nature*, 428, 616.

IPCC CZMS, 1992. A common methodology for assessing vulnerability to sea-level rise – second revision. In: *Global Climate Change and the Rising Challenge of the Sea.* Report of the Coastal Zone Management Subgroup, Response Strategies Working Group of the Intergovernmental Panel on Climate Change, Ministry of Transport, Public Works and Water Management, The Hague, The Netherlands, Appendix C, 27pp.

Klein, R.J.T., Nicholls, R.J., Ragoonaden, S., Capobianco, M., Aston, J., and Buckley, E.N., 2001. Technological options for adaptation to climate change in coastal zones. *Journal of Coastal Research*, 17(3), 531–543.

Lenton, T.M., 2005. Climate Change to the end of the Millennium, *Climatic Change*, in press.

Lenton, T., Williamson, M., Warren, R., Loutre, M., Goodess, C., Swann, M., Cameron, D., Hankin, R., Marsh, R., and Shepherd, J., 2005. Climate change on the millennial timescale. Technical Report, Tyndall Centre for Climate Change Research, in press.

Lowe, J.A., and Gregory, J.M., submitted. Mechanisms of sea level rise in the Hadley Centre climate model. *Journal of Geophysical Research*.

Lowe, J.A., Gregory, J.M. Ridley, J., Huybrechts, P., Nicholls R.J., and Collins, M., 2005. The Role of Sea Level Rise and the Greenland Ice Sheet in Dangerous Climate Change and Issues of Climate Stabilisation. Proceedings of 'Exeter Meeting on Avoiding Dangerous Climate Change.'

Lunt, D.J., de Noblet-Ducoudré, N., and Charbit S., 2004. Effects of a melted Greenland ice sheet on climate, vegetation, and the cryosphere, *Climate Dynamics*, 23, 679–694.

McLean, R., and Tsyban, A., 2001. Coastal Zone and Marine Ecosystems. In: McCarthy, J.J., Canziani, O.F., Leary, N.A., Dokken, D.J., and White, K.S. (eds.) *Climate Change 2001: Impacts, Adaptation and Vulnerability*, Cambridge University Press, Cambridge, pp. 343–380.

Meehl, G.A., Washington, W.M., Collins, W.D., Arblaster, J.M., Hu, A., Buja, L.E., Strand, W.G., and Teng, H., 2005. How Much More Global Warming and Sea Level Rise? *Science*, 307, 1769–1772.

Mercer, J.H., 1978. West Antartic Ice Sheet and CO_2 Greenhouse Effect: A Threat of Disaster. *Nature*, 271, 321–325.

Mitchell, J.F.B., Johns, T.C., Ingram, W.J., and Lowe, J.A., 2000. The effect of stabilising atmospheric carbon dioxide concentrations on global and regional climate change. *Geophysical Research Letters* 27, 2997–2980.

Nakicenovic, N. and Swart, R. (eds.) 2000. *Special Report on Emissions Scenarios: A Special Report of Working Group III of the Intergovernmental Panel on Climate Change*. Cambridge University Press, Cambridge, 599pp.

Nicholls, R.J., 1995a. Coastal Megacities and Climate Change. *Geojournal*, 37(3), 369–379.

Nicholls, R.J., 1995b. Synthesis of vulnerability analysis studies. *Proceedings of WORLD COAST 1993*, Ministry of Transport, Public Works and Water Management, the Netherlands. pp. 181–216. (downloadable at www.survas.mdx.ac.uk)

Nicholls, R.J., 2002. Rising sea levels: potential impacts and responses. In: Hester, R. and Harrison, R.M. (ed.) *Global Environmental Change*. Issues in Environmental Science and Technology, Number 17, Royal Society of Chemistry, Cambridge, pp.83–107.

Nicholls, R.J., 2004. Coastal Flooding and Wetland Loss in the 21st century: Changes Under The SRES Climate and Socio-Economic Scenarios. *Global Environmental Change*, 14, 69–86.

Nicholls, R.J., and Lowe, J.A., 2004. Benefits of Mitigation of Climate Change for Coastal Areas. *Global Environmental Change*, 14, 229–244.

Nicholls, R.J., and Mimura, N., 1998. Regional issues raised by sea-level rise and their policy implications. *Climate Research*, 11, 5–18.

Nicholls, R.J., and Tol, R.S.J., 2005. Responding to sea-level rise: An analysis of the SRES scenarios. *Philosophical Transactions of the Royal Society, A*, accepted.

Nicholls, R.J., Tol, R.S.J., and Vafeidis, A.T., 2005. Global Estimates Of The Impact Of A Collapse Of The West Antarctic Ice Sheet. *Climatic Change*, In review.

Nurse, L., and Sem, G., 2001. Small Island States. In: McCarthy, J.J., Canziani, O.F., Leary, N.A., Dokken, D.J. and White, K.S. (eds.) *Climate Change 2001: Impacts, Adaptation and Vulnerability*, Cambridge University Press, Cambridge, pp.843–875.

Oppenheimer, M., 1998. Global warming and the stability of the West Antarctic Ice Sheet. *Nature* 393, 325–332 (28 May 1998).

Parry, M., Arnell, N., McMichael, T., Nicholls, R., Martens, P., Kovats, S., Livermore, M., Rosenzweig, C., Iglesias, A., and Fischer, G., 2001. Millions at risk: defining critical climate threats and targets. *Global Environmental Change*, 11(3), 1–3.

Rapley, C., 2005. 'Antarctic Ice Sheets and Sea Level Rise'. Proceedings of Exeter Meeting on 'Avoiding Dangerous Climate Change.'

Ridley, J.K., Huybrechts, P., Gregory, J.M., and Lowe, J.A., 2005. Elimination of the Greenland ice sheet in a high CO_2 environment. *Journal of Climate*. in press.

Schimel, D., Grubb, M., Joos, F., Kufmann, R., Moss, R., Ogana, W., Richels, R., and Wigley, T., 1997. Stabilisation of Atmospheric Greenhouse Gases: Physical, Biological and Socio-economic Implications. [Houghton, J.T., Gylvan Meira Filho, L., Griggs, D.J., and Maskell, K. (eds.)] IPCC Technical Paper III, Intergovernmental Panel on Climate Change.

Small, C., and Nicholls, R.J., 2003. A Global Analysis of Human Settlement in Coastal Zones, *Journal of Coastal Research*, 19(3), 584–599.

Swart, R., Mitchell, J., Morita, T., and Raper, S., 2002. Stabilisation scenarios for climate impact assessment. *Global Environmental Change*, 12, 155–165.

Toniazzo, T., Gregory, J.M., and Huybrechts, P., 2004. Climatic Impact of a Greenland Deglaciation and Its Possible Irreversibility. *Journal of Climate*, 17, 21–33.

Tol, R.S.J., 2004. *The Double Trade-Off Between Adaptation and Mitigation for Sea Level Rise: An Application of FUND*. Research Unit Sustainability and Global Change, Hamburg University and Centre for Marine and Atmospheric Sciences, Hamburg, Germany Institute for Environmental Studies, Vrije Universiteit, Amsterdam, The Netherlands Center for Integrated Study of the Human Dimensions of Global Change, Carnegie Mellon University, Pittsburgh, PA, USA, Working Paper FNU-48.

Turley, C., Blackford, J., Widdicombe, S., Lowe, D., and Nightingale, P., 2005. Reviewing the impact of increased atmospheric CO_2 on oceanic pH and the marine ecosystem. Proceedings of Exeter Meeting on 'Avoiding Dangerous Climate Change.'

Turner, R.K., Subak, S., and Adger, W.N., 1996. Pressures, trends, and impacts in coastal zones: interactions between socio-economic and natural systems. *Environmental Management*, 20(2), 159–173.

Vaughan, D.G., and Spouge, J., 2002. Risk estimation of the collapse the West Antarctic ice sheet, *Climatic Change*, 52(1), 65–91.

Warrick, R.A., and Oerlemans, J., 1990. Sea Level Rise. In: Houghton, J.T., Jenkins, G.J., and Ephraums J.J. (eds.) *Climate Change, The IPCC Scientific Assessment*, Cambridge University Press, Cambridge, pp. 260–281.

Wigley, T.M.L., 2005. The Climate Change Commitment. *Science*, 307, 1766–1769.

Wigley, T.M.L., and Raper, S.C.B., 1993. Future changes in global mean temperature and sea level. In: Warrick, R.A., Barrow, E.M. and Wigley, T.M.L. (eds.) *Climate and Sea Level Change: Observation, Projections and Implications*. Cambridge University Press, Cambridge, pp.111–133.

Wigley, T.M.L., and Raper, S.C.B., 2001. Interpretation of high projections for global-mean warming. *Science* 293: 451–453.

Wigley, T.M.L., Richels R., and Edmonds, J.A., 1996. Economic and environmental choices in the stabilisation of atmospheric CO_2 concentrations. *Nature*, 379, 242–245.

Woodroffe, C.D., Nicholls, R.J., Saito, Y., Chen, Z., and Goodbred, S.L. (accepted). Landscape variability and the response of Asian megadeltas to environmental change, In Harvey, N. (ed) Global Change implications for coasts in the Asia-Pacific region, Springer, Berlin.

SECTION V

Regional Perspectives: Polar Regions, Mid-Latitudes, Tropics and Sub-Tropics

INTRODUCTION

Section V considers, through six papers, climate impacts in five disparate regions: the Arctic, Australia, California, Africa and Asia.

Hassol and Correll summarise the key results from the Arctic Climate Impact Assessment (ACIA). This is a major review, commissioned by the Arctic Council (a forum of eight nations with territory in the Arctic: Canada, Denmark/ Greenland/Faroe Islands, Finland, Iceland, Norway, Russia, Sweden and the USA), carried out over four years and involving some 300 scientists. For the purposes of the ACIA study, the Arctic was defined as the region north of 60°N. There have been substantial changes in climate in this region in recent decades: warming at a rate greater than the global-mean rate; increased precipitation; a reduction in annual sea-ice extent and average ice thickness; and substantial glacier retreat. Although it is not yet possible to attribute these changes to anthropogenic influences with high confidence, the changes observed are all in the directions expected to occur as a result of human interference with the climate system. They therefore provide an early indication of the likely environmental and societal consequences of global warming.

The projected changes are much larger than those that have occurred already. Based on a limited number of emissions scenarios and climate models, a warming of 4–7°C is expected by 2100[1]. Major impacts can be expected on animals (polar bears, seals, caribou/reindeer, etc.), migratory birds, forest fire incidence, both freshwater and marine fish, and roads, buildings, pipelines, etc. as a result of permafrost changes. Most of these present major problems for indigenous peoples who are already having difficulty adapting to ongoing changes. Not surprisingly, most of the consequences are undeniably bad – although a few are positive (such as ship access through Arctic waters, and marine productivity). ACIA made no attempt to define thresholds for dangerous interference, but noted the potentially serious consequences of warming and ice-mass loss in Greenland – discussed in more detail by Lowe et al. in Section I.

Folkestad et al. cover similar ground but take a different approach. They take as a dangerous interference threshold the EU value of 2°C global-mean warming relative to pre-industrial times, and consider what such a change would mean for the Arctic. Using a limited range of emissions scenarios and GCMs they show that the timeframe within which the global temperature might rise to 2°C above the pre-industrial level is between 2026 and 2060[2]. Folkestad et al. claim that a 2°C global-mean warming would be equivalent to 3.2–6.6°C increase in the annual-mean Arctic temperature[3]. In discussing impacts, they give detailed results for changes in vegetation and perennial sea-ice area. Forest extent increases by some 55% while tundra decreases by 42% for 2°C global-mean warming; and, for each 1°C warming, perennial ice in the Arctic Ocean decreases by about 1.5 million km[2].

Steffen and co-authors examine Australian perceptions of dangerous climate change, providing a Southern Hemisphere and arid country perspective that helps complete our global picture. They consider climate impacts on human health, agriculture, water resources, coral reefs, and biodiversity. They show how the distribution of rainfall over Australia has changed markedly over the past 50 years, with both large increases and decreases. A particularly dramatic example comes from southwestern Western Australia, where rainfall underwent a 15% decrease in the mid-1970s. In spite of the overall declining trend, however, crop yields have continued to increase due to changes in practices. At the same time, streamflow in the region decreased substantially; as a consequence Perth is already water-constrained. This case study shows how the same change in climate can have very different consequences for different sectors. Changing SSTs and decreasing ocean pH have a marked effect on coral bleaching events; if these become too frequent coral reefs will be unable to recover – Australia is the 'home' of one of the world's largest reef systems, the Great Barrier Reef. Steffen and co-workers note that there is a lack of knowledge linking the impact of longer-term climate change with the behaviour of major climate features such as ENSO and the ocean circulation. These are important controlling factors for rainfall in Australia, and current research suggests there is a significant risk of further decreases in

---

[1] Editor's note: A wider choice of scenarios and models gives warming as much as 14°C.

[2] Editor's note: Direct use of results from the IPCC TAR, Fig. 9.14, gives 2031 to 2100 as a more comprehensive timeframe.

[3] Editor's note: A more complete assessment of GCMs shows the area-average, annual-mean warming for the Arctic to lie between about 1 and 2.5 times the global-mean warming, so 2°C global-mean warming would lead to Arctic warming of 2–5°C.

rainfall, leading to more severe and more prolonged droughts. They emphasise that the rate of climate change might be more important than the magnitude of the change, and that changes in water availability will be as important as changes in temperature *per se*.

Hayhoe and co-authors present a comprehensive impact assessment for California (one of the world's largest economies) based on two different climate change scenarios (the IPCC SRES B1 and A1FI scenarios) and using results from two different AOGCMs (PCM and HadCM3) downscaled to produce high-resolution information. (Downscaling is particularly important in a region like California where there are marked topographic changes over short distances.) California is a large climate-sensitive region with a wide variety of industries and ecosystems. Results for the different emissions scenarios tend to scale, showing similar patterns but different magnitudes. Temperature increases under the higher emissions A1FI scenario are nearly double those under the lower B1 scenario, leading to proportionally greater impacts on human health, agriculture, water resources, and ecosystems. Results from the two models show larger qualitative differences, highlighting the need to consider a wide range of model results in order to capture model-related uncertainties. Predictions for precipitation changes show important inter-model differences, but some impacts (such as the qualitative effects on snowpack and the seasonal timing of runoff) are robust to these uncertainties. This work highlights the importance of region-specific and spatially-detailed climate impact assessments covering a range of sectors, emissions scenarios and climate models to inform decision-makers of potential 'dangerous' impacts and the outcome of alternative GHG emissions and concentration stabilisation choices.

Nyong and Niang-Diop provide a radically different perspective for what is the poorest and arguably the most vulnerable continent on the planet, Africa. Of the world's least-developed countries, 70% are in Africa. Their high vulnerability is a result not just of climatic factors but also of other stresses such as poverty, disease and conflict. These effects of these vulnerabilities will be felt strongly in the water resources, agriculture, fisheries and health sectors. Under climate change scenarios, this work shows that the majority of crops will decrease in yield, that the area suitable for malaria in southern Africa will double and that coral reefs will be lost due to bleaching. The coral reef issue is a multi-faceted one: not only are reefs important to the tourism industry, but they also play a crucial role in fisheries production and in protecting the coastline from wave action and erosion. Nyong and Niang-Diop place great emphasis on the need for adaptation, noting the important distinction between facilitating adaptation (building on existing capacities, developing new information, raising awareness, removing barriers, making funds and resources available, etc.) and implementing adaptation (making actual changes in operational practices and behaviour, and installing and operating new technologies). An important aspect is the need to integrate indigenous knowledge with modern techniques.

Finally, Harasawa summarises important findings on key vulnerabilities and critical impact levels as reflected in the pertinent literature on East and South-East Asia. The paper makes the general observation that these regions are threatened by a multitude of potential climate change effects, in spite of the considerable adaptive capacities available in the countries considered. In fact, the author presents evidence that the current amount of global warming has already triggered a number of adverse impacts such as the expansion of flood and drought disaster regions in China and the bleaching of coral reefs in the Okinawa islands. Likely future impacts as derived from regional climate change scenarios are then listed, including damages caused by heavy rain episodes, significant changes in Japanese and Korean vegetation cover, adverse consequences of sea-level rise on Asian coastal stability, increasing mortality due to heat strokes and epidemics, and various negative effects on industry, energy production and transportation in the region. The paper concludes by highlighting several future research needs.

The group of papers in this section spans a wide range of regions, climates, impacts and adaptive capacities. The contrast between potential climate impacts and the abilities to cope with these impacts in the developed counties compared with the lesser-developed countries is stark. It highlights the global nature of a problem that respects no geographical boundaries.

CHAPTER 21

Arctic Climate Impact Assessment

Susan Joy Hassol
Lead Author, Impacts of a Warming Arctic, the synthesis report of the Arctic Climate Impact Assessment

Robert W. Corell
Chair, Arctic Climate Impact Assessment, Senior Policy Fellow, American Meteorological Society

ABSTRACT: The Arctic is extremely vulnerable to observed and projected climate change and its impacts. The Arctic is now experiencing some of the most rapid and severe climate change on Earth. Over the next 100 years, climate change is expected to accelerate, contributing to major physical, ecological, social, and economic changes, many of which have already begun. Changes in arctic climate will also affect the rest of the world through increased global warming and rising sea levels. The Arctic Climate Impact Assessment was a four-year effort by an international team of 300 scientists to assess observed and projected climate change and its impacts on the region and the world. It was requested by the Arctic Council, an intergovernmental forum including the eight arctic nations (Canada, Denmark/Greenland/Faroe Islands, Finland, Iceland, Norway, Russia, Sweden, and the United States of America), six Indigenous Peoples organizations, and official observers.

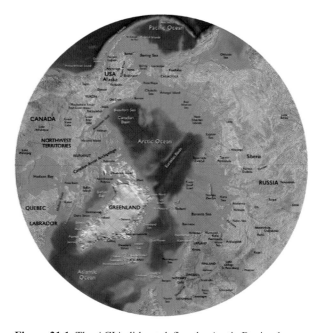

Figure 21.1 The ACIA did not define the Arctic Region by a single boundary line, as the needs of each aspect of the assessment required different definitions to adequately assess the impacts. For example, the boreal forests extend well below the Arctic Circle, while the computer analysis required a simple boundary, and hence, defined the Arctic as north of 60 degrees. These differences are noted in each section of the Assessment.

21.1 Introduction

Climate change is being experienced particularly intensely in the Arctic. Arctic average temperature has risen at almost twice the rate as that of the rest of the world in the past few decades. Widespread melting of glaciers and sea ice and rising permafrost temperatures present additional evidence of strong arctic warming. These changes in the Arctic provide an early indication of the environmental and societal significance of global warming (see Appendix I which lists the Key Findings of the Arctic Climate Impact Assessment).

An acceleration of these climatic trends is projected[1] to occur during this century, due to ongoing increases in concentrations of greenhouse gases in the Earth's atmosphere. While greenhouse gas emissions do not primarily originate in the Arctic, they are projected to bring wide-ranging changes and impacts to the Arctic. These arctic changes will, in turn, impact the planet as a whole. For this reason, people outside the Arctic have a great stake in what is happening there. For example, climatic processes unique to the Arctic have significant effects on global and regional climate. The Arctic also provides important natural resources to the rest of the world (such as oil, gas, and fish) that will be affected by climate change. The melting of arctic glaciers is one of the factors contributing to sea-level rise around the globe.

Climate change is also projected to result in major impacts inside the Arctic, some of which are already underway. Whether a particular impact is perceived as negative or positive often depends on one's interests. For example, the reduction in sea ice is very likely to have devastating consequences for polar bears, ice-dependent seals, and local people for whom these animals are a primary food source. On the other hand, reduced sea ice is likely to increase marine access to the region's resources, expanding opportunities for

---

[1] The findings stated in this chapter are derived from the scientific analysis of over 300 scientists, indigenous peoples, and other experts documented in the eighteen chapters of the ACIA science report, all of which are available on the web at www.acia.uaf.edu or through Cambridge University Press in the Scientific Assessment report, October 2005.

shipping and possibly for offshore oil extraction (although operations could be hampered initially by increasing movement of ice in some areas). Further complicating the issue, possible increases in environmental damage that often accompanies shipping and resource extraction could harm the marine habitat and negatively affect the health and traditional lifestyles of indigenous people.

Climate change is taking place within the context of many other ongoing changes in the Arctic, including the observed increase in chemical contaminants entering the Arctic from other regions, overfishing, land use changes that result in habitat destruction and fragmentation, an increase in ultraviolet radiation reaching the surface, rapid growth in the human population, and cultural, governance, and economic changes. Impacts on the environment and society result not from climate change alone, but from the interplay of all of these changes. The combination of climate change and other stresses presents a range of potential problems for human health and wellbeing as well as risks to other arctic species and ecosystems.

The changes in arctic climate reported in the Arctic Climate Impact Assessment, including shorter warmer winters, increased precipitation, and substantial decreases in snow and ice cover, are projected to persist for centuries. Unexpected and even larger shifts and fluctuations in climate are also possible.

21.1.1 *Why the Arctic Warms Faster than Lower Latitudes*

First, as arctic snow and ice melt, the darker land and ocean surfaces that are revealed absorb more of the sun's energy, increasing arctic warming. Second, in the Arctic, a greater fraction of the extra energy received at the surface due to increasing concentrations of greenhouse gases goes directly into warming the atmosphere, whereas in the tropics, a greater fraction goes into evaporation. Third, the depth of the atmospheric layer that has to warm in order to cause warming of near-surface air is much shallower in the Arctic than in the tropics, resulting in a larger arctic temperature increase. Fourth, as warming reduces the extent of sea ice, solar heat absorbed by the oceans in the summer is more easily transferred to the atmosphere in the winter, making the air temperature warmer than it would be otherwise.

21.2 Observed Changes in Arctic Climate

21.2.1 *Increasing Temperatures and Precipitation*

Records of increasing temperatures, melting glaciers, reductions in extent and thickness of sea ice, thawing permafrost, and rising sea level all provide strong evidence of recent warming in the Arctic. There are regional variations due to atmospheric winds and ocean currents, with some areas showing more warming than others and a few areas even showing a slight cooling; but for the Arctic as a whole, there is a clear warming trend. There are also patterns within this overall trend; for example, in most

places, temperatures in winter are rising more rapidly than in summer. In Alaska and western Canada, winter temperatures have increased as much as 3–4°C in the past 50 years.

Precipitation has increased by roughly 8% across the Arctic over the past 100 years. In addition to the overall increase, changes in the characteristics of precipitation have also been observed. Much of the precipitation increase appears to be coming as rain, mostly in winter, and to a lesser extent in autumn and spring. The increasing winter rains, which fall on top of existing snow, cause faster snowmelt and, when the rainfall is intense, can result in flash flooding in some areas. Rain-on-snow events have increased significantly across much of the Arctic, for example by 50% over the past 50 years in western Russia. Snow cover extent over arctic land areas has declined by about 10% over the past 30 years, with much of the decrease taking place in spring, resulting in a shorter snow cover season.

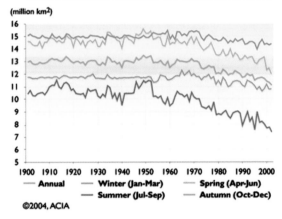

Figure 21.2 Annual average extent of Arctic sea ice from 1900 to 2003. A decline in sea-ice extent began about 50 years ago and this decline sharpened in recent decades, corresponding with the arctic warming trend. The decrease in sea-ice extent during summer is the most dramatic of the trends.

21.2.2 *Declining Sea Ice*

Arctic sea ice is a key indicator and agent of climate change, affecting surface reflectivity, cloudiness, humidity, exchanges of heat and moisture at the ocean surface, and ocean currents. Changes in sea ice also have enormous environmental, economic, and societal implications.

Over the past 30 years, the annual average sea-ice extent has decreased by about 8%, or nearly one million square kilometers, an area larger than all of Norway, Sweden, and Denmark combined, and the melting trend is accelerating. Sea-ice extent in summer has declined more dramatically than the annual average, with a loss of 15–20% of the late-summer ice coverage. There is also significant variability from year to year. September 2005 had the smallest extent of arctic sea-ice cover on record (see Figure 21.3). Sea ice has also become thinner in recent decades, with arctic-wide average thickness reductions estimated at 10–15%,

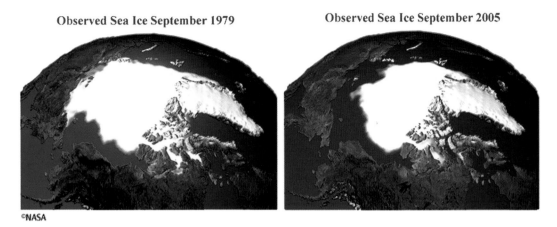

Observed Sea Ice September 1979   **Observed Sea Ice September 2005**

©NASA

Figure 21.3 These two images, constructed from satellite data, compare arctic sea ice concentrations in September of 1979 and 2005. September is the month in which sea ice is at its yearly minimum and 1979 marks the first year that data of this kind became available in meaningful form. The lowest concentration of sea ice on record was in September 2005.

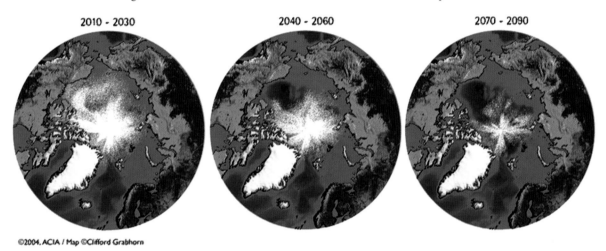

2010 - 2030   2040 - 2060   2070 - 2090

©2004. ACIA / Map ©Clifford Grabhorn

Figure 21.4 Projected Ice Extent. September sea-ice extent, already declining markedly, is projected to decline even more rapidly in the future. The three images above show the average of the projections from five climate models for three future time periods, using the B2 emissions scenario. As the century progresses, sea ice moves further and further from the coasts of arctic land masses, retreating to the central Arctic Ocean. Some models project the nearly complete loss of summer sea ice in this century.

and with particular areas showing reductions of up to 40% between the 1960s and late 1990s.

21.3 Projections of Future Arctic Climate

21.3.1 *Future Changes in Arctic Temperature and Precipitation*

Global climate model projections suggest that the Arctic will warm roughly twice as much as the globe over the course of this century. Under two mid-range IPCC SRES emissions scenarios (B2 and A2) for an average of five climate models, this would result in about 4 to 7°C of warming by 2100 averaged over the Arctic region (the full set of IPCC SRES scenarios suggests a wider range of possible outcomes). Winter temperature increases and increases over the oceans are projected to be substantially greater than the average.

Precipitation is also projected to increase strongly. Over the Arctic as a whole, annual total precipitation is projected to increase by about 20% by the end of the century, with most of the increase coming as rain. The overall increase is projected to be most concentrated over coastal regions and in winter and autumn; increases in these seasons are projected to exceed 30%.

Snow cover extent, which has already declined by 10% over the past 30 years, is projected to decline an additional 10–20% before the end of this century. The decreases in snow-covered area are expected to be greatest in April and May, suggesting a further shortening of the snow season and an earlier pulse of river runoff to the Arctic Ocean and coastal seas. Important snow quality changes are also projected, such as an increase in thawing and freezing in winter that leads to ice formation that in turn restricts the access of land animals to food and nesting sites.

21.3.2 *Future Changes in Arctic Sea Ice*

Sea ice has already declined considerably over the past half century. Additional declines of roughly 10–50% in annual average sea ice extent are projected by 2100. Loss of sea ice in summer is projected to be considerably greater, with a 5-model average projecting more than a 50% decline by the end of this century, and some models showing a near-complete disappearance of summer sea ice (see Figure 21.4).

21.4 Impacts of Arctic Climate Change on the Globe

Because the Arctic plays a special role in global climate, arctic changes have global implications. Here we focus on two of these: increased global warming due to a reduction in arctic surface reflectivity, and increases in global sea level due to melting of land-based ice in the Arctic. Arctic changes will also reverberate to the rest of the planet through potential alterations in ocean circulation patterns, changes in greenhouse gas emissions from arctic ecosystems, and changes in the availability of arctic resources including oil, gas, fish, and habitat for migratory birds.

21.4.1 *Increased Global Warming Due to Surface Reflectivity Changes*

The bright white snow and ice that cover much of the Arctic reflect away most of the solar energy that reaches the surface. As greenhouse gas concentrations rise and warm the lower atmosphere and surface, snow and ice begin to form later in the autumn and melt earlier in the spring. The melting back of the snow and ice reveals the land and water surfaces beneath, which are much darker, and thus absorb more of the sun's energy. This warms the surface further, causing faster melting, which in turn causes more warming, and so on, creating a self-reinforcing cycle by which global warming feeds on itself, amplifying and accelerating the warming trend. This process is already underway in the Arctic with the widespread retreat of glaciers, snow cover, and sea ice. This regional warming accelerates warming at the global scale.

21.4.2 *Sea Level Rise*

Climate change causes sea level to rise by affecting both the density and the amount of water in the oceans. First and most significantly, water expands as it warms, and less-dense water takes up more space. Secondly, warming increases melting of glaciers (land-based ice), adding to the amount of water flowing into the oceans.

The total volume of land-based ice in the Arctic corresponds to a global sea level equivalent of about eight meters. Most arctic glaciers have been in decline since the early 1960s, with this trend speeding up in the 1990s (see Figure 21.5). A small number of glaciers, especially in Scandinavia, have gained mass as increased precipitation outpaced the increase in melting in a few areas.

The Greenland Ice Sheet (a very large collection of glaciers that covers the continent of Greenland) dominates land ice in the Arctic. The area of surface melt on the Ice Sheet increased on average by 16% from 1979 to 2002, an

area roughly the size of Sweden, with considerable variation from year to year. The area of surface melt broke all records in 2002 (see Figure 21.6), with extreme melting occurring at a record high elevation of 2000 meters.

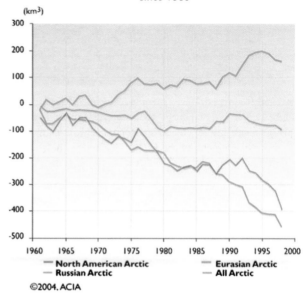

Cumulative Change in Volume of Arctic Glaciers since 1960

Figure 21.5 For the Arctic as a whole, there was a substantial loss in glacial volume shown here from 1961 to 1998. Glaciers in the North American Arctic lost the most mass (about 450 km³), with increased loss since the late 1980s. Glaciers in the Russian Arctic have also had large losses (about 100 km³). Glaciers in the European Arctic show an increase in volume because increased precipitation in Scandinavia and Iceland added more to glacial mass than melting removed over that period.

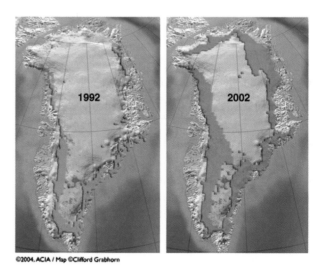

Figure 21.6 Seasonal surface melt extent on the Greenland Ice Sheet has been observed by satellite since 1979 and shows an increasing trend. The melt zone shown here for 1992 and 2002, where summer warmth turns snow and ice around the edges of the ice sheet into slush and ponds of melt-water, has been expanding inland and to record high elevations in recent years.

Satellite data show an increasing trend in melt extent since 1979. This trend was interrupted in 1992, following the eruption of Mt. Pinatubo, which created a short-term global cooling as particles spewed from the volcano reduced the amount of sunlight that reached the Earth.

Recent studies of glaciers in Alaska indicate an accelerated rate of melting. The associated sea-level rise is nearly double the estimated contribution from the Greenland Ice Sheet during the past 15 years. This rapid retreat of Alaska's glaciers represents about half of the estimated loss of mass by glaciers worldwide, and the largest contribution by glacial melt to rising sea level yet measured.

Projections from global climate models suggest that the contribution of arctic glaciers to global sea-level rise will accelerate over the next 100 years, amounting to roughly four to six centimeters by 2100. Recent research suggests that this estimate should be higher due to the increase in arctic glacial melt during the past two decades.

Over the longer term, the arctic contribution to global sea-level rise is projected to be much greater as ice sheets continue to respond to climate change and to contribute to sea-level rise for thousands of years. Climate models indicate that the local warming over Greenland is likely to be up to two to three times the global average. Ice sheet models project that sustained local warming of that magnitude would eventually lead to a virtually complete melting of the Greenland Ice Sheet (over a period of perhaps 1000 years or more), with a resulting sea-level rise of about seven meters.

Sea-level rise is projected to have serious implications for coastal communities and industries, islands, river deltas, harbors, and the large fraction of humanity living in coastal areas worldwide.

21.5 Impacts of Climate Change in the Arctic

21.5.1 *Shifting Vegetation Zones*

Climate-induced changes in arctic landscapes are important to local people and animals in terms of food, fuel, culture, and habitat. These changes also have the potential for global impacts because many processes related to arctic landscapes affect global climate and resources. Some changes in arctic landscapes are already underway and future changes are projected to be considerably greater.

The major arctic vegetation zones include the polar deserts, tundra, and the northern part of the boreal forest. Climate change is projected to cause vegetation shifts because rising temperatures favor taller, denser vegetation, and will thus promote the expansion of forests into the arctic tundra, and tundra into the polar deserts. The timeframe of these shifts will vary around the Arctic. Where suitable soils and other conditions exist, changes are likely to be apparent in this century. Where they do not, the changes can be expected to take longer. These vegetation changes, along with rising sea levels, are projected to

shrink tundra area to its lowest extent in at least the past 21,000 years, greatly reducing the breeding area for many birds and the grazing areas for land animals that depend on the open landscape of tundra and polar desert habitats. Not only are some threatened species very likely to become extinct, some currently widespread species are projected to decline sharply.

Many animal species from around the world depend on summer breeding and feeding grounds in the Arctic, and climate change will alter some of these habitats significantly. For example, several hundred million migratory birds migrate to the Arctic each summer and their success in the Arctic determines their populations elsewhere. Important breeding and nesting areas are projected to decrease sharply as treeline advances northward, encroaching on tundra, and because the timing of bird arrival in the Arctic might no longer coincide with the availability of their insect food sources. At the same time, sea-level rise will erode the tundra extent from the north in many areas, further shrinking important habitat. A number of bird species, including several globally endangered seabird species, are projected to lose more than 50% of their breeding area during this century.

21.5.2 *Forest Disturbances: Insects and Fires*

Increased insect outbreaks due to climate warming are already occurring and are almost certain to continue. Increasing climate-related outbreaks of spruce bark beetles and spruce budworms in the North American Arctic provide two important examples. Over the past decade, areas of Alaska and Canada have experienced the largest and most intense outbreaks of spruce bark beetles on record. There has also been an upsurge in spruce budworm outbreaks, and the entire range of white spruce forests in North America is considered vulnerable to such outbreaks under projected warming. Large areas of forest disturbance create new opportunities for invasive species from warmer climates and/or non-native species to become established.

Fire is another major disturbance factor in the boreal forest and it exerts pervasive ecological effects. The area burned in boreal western North America has doubled over the past thirty years, and it is forecast to increase by as much as 80% over the next 100 years under projected warming. The area of boreal forest burned annually in Russia averaged four million hectares over the last three decades, and more than doubled in the 1990s. As climate continues to warm, the forest fire season will begin earlier and last longer. Models of forest fire in parts of Siberia suggest that a summer temperature increase of 5.5°C would double the number of years in which there are severe fires, and increase the area of forest burned annually by nearly 150%.

21.5.3 *Animal Species*

Many animal species will be affected by increasing arctic temperatures. In the marine environment, the sharp decline in sea ice is likely to have devastating impacts on polar bears. The earliest impacts of warming would be expected to occur at the southern limits of the polar bears' distribution, such as James and Hudson Bays in Canada, and such impacts have already been documented. The condition of adult polar bears has declined during the last two decades in the Hudson Bay area, as have the number of live births and the proportion of first-year cubs in the population. Polar bears in that region suffered 15% declines in both average weight and number of cubs born between 1981 and 1998. Other ice-dependent species such as ringed seals, walrus, and some species of marine birds are also likely to be negatively impacted.

Terrestrial animal species also face threats due to warming. Caribou and reindeer herds depend on the availability of abundant tundra vegetation and good foraging conditions, especially during the calving season. Climate-induced changes are projected to reduce the area of tundra and the traditional forage (such as mosses and lichens) for these herds. Freeze–thaw cycles and freezing rain are also projected to increase, reducing the ability of caribou and reindeer populations to access food and raise calves. Future climate change could thus mean a potential decline in caribou and reindeer populations, threatening human nutrition and a whole way of life for some indigenous arctic communities.

Freshwater species face climate-related changes to their environments that include increasing water temperatures, thawing permafrost, and reduced ice cover on rivers and lakes. Southernmost species are projected to shift northward, competing with northern species for resources. The broad whitefish, Arctic char, and Arctic cisco are particularly vulnerable to displacement as they are fundamentally northern in their distribution. As water temperatures rise, spawning grounds for cold-water species will shift northward and are likely to be diminished. As southerly fish species move northward, they may introduce new parasites and diseases to which arctic fish are not adapted, increasing the risk of death for arctic species. The implications of these changes for both commercial and subsistence fishing in far northern areas are potentially devastating as the most vulnerable species are often the only fishable species present.

Marine fisheries are largely controlled by factors such as local weather conditions, ecosystem dynamics, and management decision; projecting impacts of climate change on marine fish stocks is thus highly problematic. There is some chance that climate change will induce major ecosystem shifts in some areas that would result in radical changes in species composition with unknown consequences. Barring such shifts, moderate warming is likely to improve conditions for some important fish stocks such as cod and herring, as higher temperatures and reduced ice cover could possibly increase productivity of their prey and provide more extensive habitat.

21.5.4 *Coastal Erosion*

The effects of rising temperatures are altering the arctic coastline and much larger changes are projected to occur during this century as a result of reduced sea ice, thawing permafrost, and sea-level rise. Thinner, less extensive sea ice creates more open water, allowing stronger wave generation by winds, thus increasing wave-induced erosion along arctic shores. Sea-level rise, increasing storm surge heights, and thawing of coastal permafrost exacerbate this problem. Dozens of arctic communities are already threatened by these changes, and some are already planning to relocate. Hundreds more could be at risk in the future.

21.5.5 *Thawing Permafrost*

Transportation and industry on land, including oil and gas extraction and forestry, will increasingly be disrupted by the shortening of the periods during which ice roads and tundra are frozen sufficiently to permit travel. For example, warming has caused the number of days per year in which travel on the tundra is allowed under Alaska Department of Natural Resources standards to drop from over 200 to about 100 in the past 30 years, resulting in a 50% reduction in days that oil and gas exploration and extraction equipment can be used. In addition, as frozen ground thaws, many existing buildings, roads, and pipelines are likely to be destabilized, requiring costly repair and replacement. Projected warming and its effects will need to be taken into account in the design of all new construction, requiring measures that will increase costs.

21.5.6 *Marine Access*

Observed and projected reductions in sea ice suggest that the Arctic Ocean will have longer seasons of less sea-ice cover of reduced thickness, implying improved ship accessibility around the margins of the Arctic Basin (although this will not be uniformly distributed). Increased access to arctic resources will also raise new issues relating to sovereignty, security, and safety. In some areas, such as the Canadian Arctic, increased ice movement due to warming under the complex sea ice conditions of the archipelago could actually make shipping more difficult, particularly in the first few decades of this century. The risk of oil spills and other industrial accidents in the challenging arctic environment raises additional concerns.

21.5.7 *Indigenous Communities*

Across the Arctic, indigenous people are already reporting the effects of climate change. Local landscapes, seascapes, and icescapes are becoming unfamiliar. Climate change is occurring faster than indigenous knowledge can adapt and is strongly affecting people in many communities. Unpredictable weather, snow, and ice conditions make travel hazardous, endangering lives. Impacts of climate change on wildlife, from caribou on land, to fish in the rivers, to seals and polar bears on the sea ice, are having

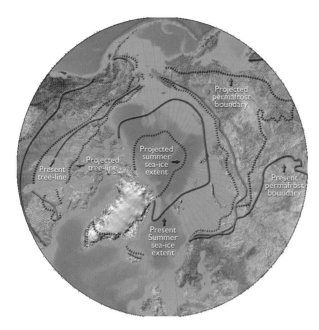

Figure 21.7 Changes in summer sea-ice extent and treeline are projected to occur by the end of this century. The change in the permafrost boundary assumes that present areas of discontinuous permafrost will be free of any permafrost in the future, and this is likely to occur beyond the 21st century.

enormous effects, not only for the diets of Indigenous Peoples, but also for their cultures and their very identities.

21.6 Concluding Thoughts

Despite the fact that a relatively small percentage of the world's greenhouse gas emissions originate in the Arctic, human-induced changes in arctic climate are among the largest on Earth. As a consequence, the changes already underway in arctic landscapes, communities, and unique features provide an early indication for the rest of the world of the environmental and societal significance of global climate change. As this assessment illustrated, changes in climate and their impacts in the Arctic are already being widely noticed and felt, and are projected to become much greater. These changes will also reach far beyond the Arctic, affecting global climate, sea level, biodiversity, and many aspects of human social and economic systems.

APPENDIX

The Key Findings of the Arctic Climate Impact Assessment
(All Reports of the ACIA are available at www.acia.uaf.edu)
1. **Arctic climate is now warming rapidly and much larger changes are projected.**
   - Annual average arctic temperature has increased at almost twice the rate as that of the rest of the world

over the past few decades, with some variations across the region.
- Additional evidence of arctic warming comes from widespread melting of glaciers and sea ice, and a shortening of the snow season.
- Increasing global concentrations of carbon dioxide and other greenhouse gases due to human activities, primarily fossil fuel burning, are projected to contribute to additional arctic warming of about 4–7°C over the next 100 years.
- Increasing precipitation, shorter and warmer winters, and substantial decreases in snow cover and ice cover are among the projected changes that are very likely to persist for centuries.
- Unexpected and even larger shifts and fluctuations in climate are also possible.

2. **Arctic warming and its consequences have worldwide implications.**
   - Melting of highly reflective arctic snow and ice reveals darker land and ocean surfaces, increasing absorption of the sun's heat and further warming the planet.
   - Increases in glacial melt and river runoff add more freshwater to the ocean, raising global sea level and possibly slowing the ocean circulation that brings heat from the tropics to the poles, affecting global and regional climate.
   - Warming is very likely to alter the release and uptake of greenhouse gases from soils, vegetation, and coastal oceans.
   - Impacts of arctic climate change will have implications for biodiversity around the world because migratory species depend on breeding and feeding grounds in the Arctic.

3. **Arctic vegetation zones are very likely to shift, causing wide-ranging impacts.**
   - Treeline is expected to move northward and to higher elevations, with forests replacing a significant fraction of existing tundra, and tundra vegetation moving into polar deserts.
   - More productive vegetation is likely to increase carbon uptake, although reduced reflectivity of the land surface is likely to outweigh this, causing further warming.
   - Disturbances such as insect outbreaks and forest fires are very likely to increase in frequency, severity, and duration, facilitating invasions by non-native species.
   - Where suitable soils are present, agriculture will have the potential to expand northward due to a longer and warmer growing season.

4. **Animal species' diversity, ranges, and distribution will change.**
   - Reductions in sea ice will drastically shrink marine habitat for polar bears, ice-inhabiting seals, and some seabirds, pushing some species toward extinction.
   - Caribou/reindeer and other land animals are likely to be increasingly stressed as climate change alters

their access to food sources, breeding grounds, and historic migration routes.

- Species ranges are projected to shift northward on both land and sea, bringing new species into the Arctic while severely limiting some species currently present.
- As new species move in, animal diseases that can be transmitted to humans, such as West Nile virus, are likely to pose increasing health risks.
- Some arctic marine fisheries, which are of global importance as well as providing major contributions to the region's economy, are likely to become more productive. Northern freshwater fisheries that are mainstays of local diets are likely to suffer.

5. **Many coastal communities and facilities face increasing exposure to storms.**
   - Severe coastal erosion will be a growing problem as rising sea level and a reduction in sea ice allow higher waves and storm surges to reach the shore.
   - Along some arctic coastlines, thawing permafrost weakens coastal lands, adding to their vulnerability.
   - The risk of flooding in coastal wetlands is projected to increase, with impacts on society and natural ecosystems.
   - In some cases, communities and industrial facilities in coastal zones are already threatened or being forced to relocate, while others face increasing risks and costs.

6. **Reduced sea ice is very likely to increase marine transport and access to resources.**
   - The continuing reduction of sea ice is very likely to lengthen the navigation season and increase marine access to the Arctic's natural resources.
   - Seasonal opening of the Northern Sea Route is likely to make trans-arctic shipping during summer feasible within several decades. Increasing ice movement in some channels of the Northwest Passage could initially make shipping more difficult.
   - Reduced sea ice is likely to allow increased offshore extraction of oil and gas, although increasing ice movement could hinder some operations.
   - Sovereignty, security, and safety issues, as well as social, cultural, and environmental concerns are likely to arise as marine access increases.

7. **Thawing ground will disrupt transportation, buildings, and other infrastructure.**
   - Transportation and industry on land, including oil and gas extraction and forestry, will increasingly be disrupted by the shortening of the periods during which ice roads and tundra are frozen sufficiently to permit travel.
   - As frozen ground thaws, many existing buildings, roads, pipelines, airports, and industrial facilities are likely to be destabilized, requiring substantial rebuilding, maintenance, and investment.

- Future development will require new design elements to account for ongoing warming that will add to construction and maintenance costs.
- Permafrost degradation will also impact natural ecosystems through collapsing of the ground surface, draining of lakes, wetland development, and toppling of trees in susceptible areas.

8. **Indigenous communities are facing major economic and cultural impacts.**
   - Many Indigenous Peoples depend on hunting polar bear, walrus, seals, and caribou, herding reindeer, fishing, and gathering, not only for food and to support the local economy, but also as the basis for cultural and social identity.
   - Changes in species' ranges and availability, access to these species, a perceived reduction in weather predictability, and travel safety in changing ice and weather conditions present serious challenges to human health and food security, and possibly even the survival of some cultures.
   - Indigenous knowledge and observations provide an important source of information about climate change. This knowledge, consistent with complementary information from scientific research, indicates that substantial changes have already occurred.

9. **Elevated ultraviolet radiation levels will affect people, plants, and animals.**
   - The stratospheric ozone layer over the Arctic is not expected to improve significantly for at least a few decades, largely due to the effect of greenhouse gases on stratospheric temperatures. Ultraviolet radiation (UV) in the Arctic is thus projected to remain elevated in the coming decades.
   - As a result, the current generation of arctic young people is likely to receive a lifetime dose of UV that is about 30% higher than any prior generation. Increased UV is known to cause skin cancer, cataracts, and immune system disorders in humans.
   - Elevated UV can disrupt photosynthesis in plants and have detrimental effects on the early life stages of fish and amphibians.
   - Risks to some arctic ecosystems are likely as the largest increases in UV occur in spring, when sensitive species are most vulnerable, and warming-related declines in snow and ice cover increase exposure for living things normally protected by such cover.

10. **Multiple influences interact to cause impacts to people and ecosystems.**
    - Changes in climate are occurring in the context of many other stresses including chemical pollution, over-fishing, land use changes, habitat fragmentation, human population increases, and cultural and economic changes.
    - These multiple stresses can combine to amplify impacts on human and ecosystem health and wellbeing. In many cases, the total impact is greater than the sum of its parts, such as the combined

impacts of contaminants, excess ultraviolet radiation, and climatic warming.

- Unique circumstances in arctic sub-regions determine which are the most important stresses and how they interact.

REFERENCES

ACIA, *Impacts of a Warming Arctic*, Arctic Climate Impact Assessment, Cambridge University Press, 2004, 140 pages. Available in pdf at www.acia.uaf.edu

Arctic Climate Impact Assessment, Cambridge University Press, 2005, 1000 + pages. Available in pdf at www.acia.uaf.edu. The Chapters are:

Chapter 1: An Introduction to the Arctic Climate Impact Assessment (Lead Authors: H. Huntington and G. Weller)

Chapter 2: Arctic Climate – Past and Present (Lead Author: G. McBean)

Chapter 3: The Changing Arctic: Indigenous Perspectives (Lead Authors: H. Huntington and Shari Fox)

Chapter 4: Future Climate Change: Modeling and Scenarios for the Arctic Region (Lead Authors: V. M. Kattsov and E. Källén)

Chapter 5: Ozone and Ultraviolet Radiation (Lead Authors: B. Weatherhead, A. Tanskanen, and A. Stevermer)

Chapter 6: Cryosphere and Hydrology (Lead Author: J. E. Walsh)

Chapter 7: Arctic Tundra and Polar Desert Ecosystems (Lead Author: T. V. Callaghan)

Chapter 8: Freshwater Ecosystems and Fisheries (Lead Authors: F. J. Wrona, T. D. Prowse, and J. D. Reist)

Chapter 9: Marine Systems (Lead Author: H. Loeng)

Chapter 10: Principles of Conserving the Arctic's Biodiversity (Lead Author: M. B. Usher)

Chapter 11: Management and Conservation of Wildlife in a Changing Arctic Environment (Lead Author: D. R. Klein)

Chapter 12: Hunting, Herding, Fishing and Gathering: Indigenous Peoples and Renewable Resource Use in the Arctic (Lead Author: M. Nuttall)

Chapter 13: Fisheries and Aquaculture (Lead Authors: H. Vilhjálmsson and A. Håkon Hoel)

Chapter 14: Forests, Land Management and Agriculture (Lead Author: G. P. Juday)

Chapter 15: Human Health (Lead Authors: J. Berner and C. Furgal)

Chapter 16: Infrastructure: Buildings, Support Systems, and Industrial Facilities (Lead Author: A. Instanes)

Chapter 17: Climate Change in the Context of Multiple Stressors and Resilience (Lead Authors: J. J. McCarthy and M. L. Martello)

Chapter 18: Summary and Synthesis of the ACIA (Lead Author: G. Weller)

CHAPTER 22

Evidence and Implications of Dangerous Climate Change in the Arctic

Tonje Folkestad[1], Mark New[2], Jed O. Kaplan[3], Josefino C. Comiso[4], Sheila Watt-Cloutier[5], Terry Fenge[5], Paul Crowley[5] and Lynn D. Rosentrater[6]

[1] *WWF International Arctic Programme, Oslo, Norway*
[2] *Centre for the Environment, Oxford University, Oxford, England*
[3] *European Commission Joint Research Centre, Ispra, Italy*
[4] *NASA Goddard Space Flight Center, USA*
[5] *Inuit Circumpolar Conference, Ontario, Canada*
[6] *LDR Consulting, Oslo, Norway*

ABSTRACT: In the Arctic even a slight shift in temperature, raising averages to above freezing, can bring about rapid and dramatic changes in an ecosystem that is defined by being frozen. Various threshold levels of global warming (e.g. 1.5, 2, 3, 4°C) have been used to examine what constitutes dangerous climate change. And based on the resulting impacts literature some governments and non-governmental organizations have stated their clear political support for keeping the global-mean temperature increase to less than 2°C above pre-industrial levels. In this paper we have examined the bio-physical changes in the Arctic associated with a global temperature increase of 2°C over pre-industrial levels in order to understand some of the regional implications of dangerous climate change.

The UN Framework Convention on Climate Change, penned at the Rio Earth Summit in 1992 and signed by nearly 200 nations (including the United States), sets the policy framework for international efforts to tackle the climate problem. Its guiding principle is to avoid 'dangerous anthropogenic interference with the climate system'. The scientific community has pursued this goal by examining 'dangerous climate change' from the perspective of catastrophic events [1], national sovereignty [2], and changes to ecosystems [3]. It remains, however, a crucial task for policymakers to agree on the level of warming that can be called dangerous.

22.1 Evidence and Implications of Dangerous Climate Change

Climate models used to predict the consequences of increased greenhouse gas concentrations in the atmosphere all exhibit a warming over the Arctic that is larger than the global mean warming. This Arctic 'amplification' was first noted by Budyko and colleagues [4, page 167]. Therefore, an inter-comparison of global climate models (GCMs) was made with two objectives: (1) to provide an estimate of the time-range within which global mean temperature might increase to 2°C above its pre-industrial level, and (2) to describe the possible changes in arctic climate that will accompany such an increase. Results from six coupled ocean-atmosphere GCMs, each driven by four separate forcing scenarios (Table 22.1), indicate that the Earth will have warmed by 2°C, relative to their 'pre-industrial' control climate[1] temperatures, by between 2026 and 2060. The geography of the Arctic, with its land-sea distribution and snow/ice albedo feedbacks, along with

more minor changes in cloud and heat transport, produces an amplified regional warming for latitudes greater than 60°N with ranges of 3.2–6.6°C overall, 4–10°C in winter and 1–3.5°C in summer. In each of the GCMs that were evaluated the amplification is similar for fast- and slow-warming scenarios. In other words, changes in the Arctic will be similar regardless of when a global change of +2°C occurs. However, a faster global warming will necessarily produce more rapid warming in the Arctic. This amplification of arctic temperature changes means that the rates of increase in mean annual temperature are likely to be between 0.45 and 0.75°C per decade, but possibly even as large as 1.55°C per decade. In line with the absolute changes in temperature, rates of warming are largest in winter (0.8–3.0°C and 0.5–2.0°C per decade over the Arctic Ocean and land areas respectively) and lowest in summer (0.1–0.55°C and 0.2–1.2°C per decade over the Arctic Ocean and land areas respectively).

Precipitation is predicted to increase by 5–25% over the region as whole, with the largest changes in winter (5–30%) and smaller changes in summer (3–18%). Although information on the proportion of precipitation that falls as snow was not available for these GCMs, the increased warming implies that a higher fraction of precipitation will fall as rain. Currently most summer precipitation, except in the central Arctic Ocean, falls as

---

[1]Control simulations in GCMs are long integrations with constant external radiative forcing (CO_2, solar radiation, etc.). For the models used here, 'pre-industrial' temperatures are defined using control simulations with CO_2 levels corresponding to those before the period of rapid industrialisation (before about 1850).

Table 22.1 GCM-scenario combinations used in this study, and the Transient Climate Response (TCR) of each model. TCR is a measure of a GCM's sensitivity to CO_2 forcing (and by inference, total GHG forcing) and is defined by the IPCC [5, Figure 9.1] as the temperature change at the year of CO_2 doubling, when the climate model is forced by a 1% annual compound increase in CO_2 from pre-industrial concentrations. Estimates of future CO_2 concentrations and total radiative forcing arising from these emissions scenarios can be found in IPCC [5, Appendix II].

| Model | TCR (°C) | Scenarios | | | |
		IS92aGG	IS92aGS	SRES A2	SRES B2
HadCM3	2.0	✓	✓	✓	✓
ECHAM4	1.4	✓	✓	✓	✓
CCSRNIES	1.8/3.1 g	✓	✓	✓	✓
CGCM1	1.96	✓	✓	—	—
CGCM2	No data	—	—	✓	✓
GFDLR30	1.96	—	—	✓	✓
CSIROMk2	2.0	✓	✓	✓	✓

Table 22.2 Changes in Arctic biome area under 2°C warming scenarios.

| | Forest | | Tundra | | Other | |
	$km^2 \times 1000$	% change	$km^2 \times 1000$	% change	$km^2 \times 1000$	% change
Present	5591.6		7366.2		136.5	
10th percentile 'cool'	6314.5	12.9	6659.2	−9.6	120.6	−11.6
Robust mean	8710.3	55.8	4275.0	−42.0	109.0	−20.1
Mean	8839.2	58.1	4148.1	−43.7	107.0	−21.6
90th percentile 'warm'	10485.7	87.5	2455.9	−66.7	152.7	11.9

rain. For each season, we therefore expect the proportion of the Arctic that receives wet precipitation to increase, but for snow mass to increase where it remains cold enough for snow.

Four scenarios representing 80% of the range in the amplitude of local temperature and precipitation anomalies in the Arctic at the time of a 2°C global warming were derived from the ensemble of seven GCMs forced by the series of different emissions scenarios. These monthly climate change patterns were then used to simulate future vegetation in the Arctic using the biogeochemistry-biogeography model BIOME4. The effect of a 2°C global warming suggests a potential for greater changes in terrestrial arctic ecosystems during the 21st century than have occurred since the end of the last major glacial epoch. Forest extent increases in the Arctic on the order of $3 \times 10^6 km^2$ or 55%, with a corresponding reduction of 42% in tundra area. Tundra types generally shift north with the largest reductions in the prostrate dwarf-shrub tundra, where nearly 60% of habitat is lost. Modelled shifts in the northern limit of trees reach up to 400 km from the present tree line, but may be limited by dispersion rates.

Surface temperature has been shown to be highly correlated with sea ice concentrations in the seasonal ice regions, so historical satellite records of surface temperatures from the late 1970s to 2003 were analysed to assess the magnitude of recent warming. This record shows that the Arctic has been warming at a rate of 0.46°C per decade. Since the figure falls within the range of projected changes from our GCM exercise we took it as a reasonable conservative estimate of future warming in order to investigate associated changes in sea ice cover.

Regression analysis of the satellite record indicates that for every 1°C increase in annual temperature in the Arctic region, the perennial ice in the Arctic Ocean decreases by about $1.48 \times 10^6 km^2$, with the correlation coefficient being significant but only −0.57. We used this trend to project how the perennial ice cover may look in the years 2025, 2035 and 2060 when temperatures are expected to reach the 2°C global increase. Maps indicate considerable decline in the perennial ice cover with changes mainly around the peripheral seas as the ice edge moves progressively to the north with time. While our assumption of a linear trend is likely invalid, a similar technique accurately predicted the perennial ice cover during the last three years.

Retreat of the arctic sea ice will have deleterious effects on ice-living seals, polar bears, and walrus, thus leading to profound cultural and economic impacts for Inuit and other northern indigenous peoples. Observations by Inuit lend credence to the designation 'dangerous climate change': environmental indicators for when and where to go hunting, and when and when not to travel are no longer reliable; melting permafrost has altered the landscape, led to

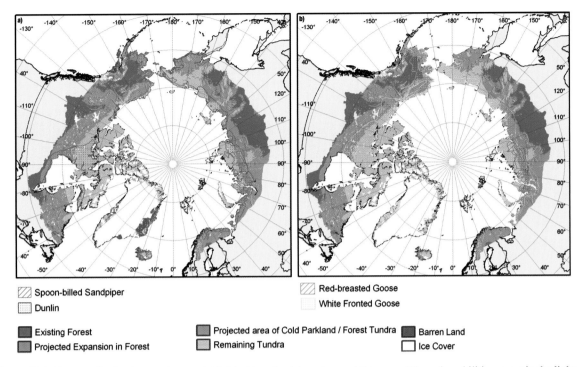

Spoon-billed Sandpiper
Dunlin

Red-breasted Goose
White Fronted Goose

Existing Forest
Projected Expansion in Forest

Projected area of Cold Parkland / Forest Tundra
Remaining Tundra

Barren Land
Ice Cover

Figure 22.1 Current distributions and potential habitat loss for (a) waders and (b) geese. The vulnerabilities occur in the light green areas, which illustrate the expansion of forests into taiga, and in the pink areas showing the disappearance of tundra. Analysis courtesy of Christoph Zöckler, UNEP/WCMC.

increased rates of erosion and displaced coastal communities; and in some regions traditional ice cellars used to store country food have lost their preservative value. In short, climate change is already threatening traditional ways of life among arctic peoples, leaving very little time for Inuit and other northern indigenous people to adapt.

22.2 Local Dangers Have Global Consequences

It is important to realize that changes in the Arctic will not only affect people and species locally: they have global consequences as well. For example, biodiversity on a global scale may be threatened as local habitats for migratory species disappear. Arctic tundra is the main breeding habitat for more than 20 million individual geese and waders that over-winter in the mid-latitudes of Europe, Asia, and North America. Many of these species will be severely impacted by the loss of tundra ecosystems projected for a rise in temperatures of 2°C. Figure 22.1 shows the current distributions and potential habitat loss for (a) waders and (b) geese. Species like the dunlin (*Calidris alpina*) and spoon-billed sandpiper (*Eurynorhynchus pygmeus*) may lose up to 45% of their breeding habitat if global temperature increases by 2°C; red-breasted goose (*Branta ruficollis*) and white-fronted goose (*Anser albifrons*) could lose up to 50%.

Changes in the Arctic can also intensify the warming effect across the planet and will contribute significantly

to global sea level rise. Sea ice keeps the planet relatively cool by reflecting solar radiation back into space. Since seawater absorbs more heat from the sun than ice does, once the permanent sea ice begins melting the warming effect increases globally. On land, warming over Greenland will lead to substantial melting of the Greenland Ice Sheet, contributing to increases in sea levels around the world. The tens of millions of people living in low-lying areas, such as Bangladesh, Bangkok, Calcutta, Dhaka, Manila, and the US states of Florida and Louisiana, are particularly susceptible to rising sea levels. Over the long term (on the scale of centuries) Greenland contains enough melt water to raise global sea level by about seven meters.

In the autumn of 2004, the eight countries with arctic territories – Canada, Denmark, Finland, Iceland, Norway, the Russian Federation, Sweden and the United States – released the most comprehensive study of regional climate to date: the Arctic Climate Impact Assessment. This four-year study, conducted by more than 250 scientists and members of indigenous organizations from throughout the region, noted that Arctic average temperature has risen at nearly twice the rate as that of the rest of the world in recent decades, contributing to profound environmental changes [6]. Such changes are in accord with those expected to occur as a result of increased emissions of carbon dioxide and other greenhouse gases. This could be an indication that we are already moving into the era of dangerous climate change.

REFERENCES

1. Hansen, J., 2004. Defusing the global warming time bomb. Scientific American **March**:68–77.
2. Barnett, J., and Adger, W. N., 2003. Climate dangers and atoll countries. Climatic Change **61**:321–337.
3. O'Neill, B. C., and Oppenheimer, M., 2002. Dangerous Climate Impacts and the Kyoto Protocol. *Science* **296**:1971–1972.
4. MacCracken, M. C. et al. (eds), 1990. Prospects for future climate. Lewis Publishers.
5. IPCC, 2001. Climate change 2001: the scientific basis. Contribution of Working Group I to the third assessment report of the Intergovernmental Panel on Climate Change. Cambridge University Press.
6. ACIA 2004. Impacts of a Warming Arctic: Arctic Climate Impact Assessment. Cambridge University Press.

CHAPTER 23

Approaches to Defining Dangerous Climate Change: An Australian Perspective

Will Steffen[1], Geoff Love[2] and Penny Whetton[3]
[1]*Australian Greenhouse Office, Canberra, Australia*
[2]*Bureau of Meteorology, Melbourne, Australia*
[3]*CSIRO, Division of Atmospheric Research, Melbourne, Australia*

ABSTRACT: This paper presents some Australian perspectives on the issue of what constitutes 'dangerous climate change'. The approach is based on a sectoral analysis of (i) the degree of risk to climate change based on sensitivity to past or projected increases in temperature and changes in precipitation, and (ii) the limits of the sectors ability to adapt to climate change. The sectors/systems included in the analysis are human health, agriculture, water resources, coral reefs and biodiversity. A synthesis of the sectoral analyses gives some insights into important factors for defining dangerous climate change from an Australian perspective: the importance of frequency and intensity of extreme events; rates of change, in addition to the magnitude of change; the overriding importance of water availability; interactions of climate impacts with other aspects of global change, such as the direct effects of increasing atmospheric CO_2; and the need to consider the resilience of the impacted system and the capacity to increase it. Some emerging issues, such as possible shifts in the behaviour of important modes of climate variability like ENSO and the stability and resilience of the Earth System as a whole under anthropogenic forcing, are briefly discussed.

23.1 General Perspectives

Perceptions of dangerous climate change by decision-makers and the community are influenced both gradually as the symptoms of climate change impact on daily lives, and more rapidly as major events are shown to be anthropogenic climate change related (see Wigley, 2004, for a discussion of the complexities in defining 'dangerous' climate change). Because of this duality the science community must analyse the impacts on, and adaptability of, local and regional scale biological and sociological systems to ongoing climate change, as well as assess the likelihood of larger scale extreme events. The role of the science community is to establish what (and if) relationships exist between specific events and changing greenhouse gas concentrations in the atmosphere, and it is the communitys role to decide what is acceptable and what poses unacceptable dangers on the basis of the science advice. This paper will present some scientific perspectives on the issue of dangerous climate change from a southern hemisphere and Australian viewpoint.

The southern hemisphere, which is largely oceanic, differs in many ways from the northern hemisphere but is no less important for understanding global climate. For example, the Southern Ocean accounts for some 40% of global oceanic carbon sink (Takahashi et al., 1997), and the Antarctic circumpolar current and associated deep water formation at the continental margins around Antarctica are important components of the global oceanic circulation and the climate system (Murray, 2000). However, much less is known about the behaviour of the Southern Ocean overturning circulation than about its North Atlantic counterpart. Changes in the thermohaline circulation have been linked to abrupt climate change (Clark et al., 2002), and they are thus an important factor in defining dangerous climate change.

There is a risk, however, in limiting approaches to defining dangerous climate change to only one perspective. With regard to resource management, we propose that dangerous climate change can best be defined from the perspective of the various systems or sectors that are impacted by a changing climate. The critical issues to be examined are the level of vulnerability of systems or sectors to climate now and what measures they can take to adapt to a changing climate in the future. In many instances, planned adaptation can increase the coping range of a sector to a changing climate (Figure 23.1, from Jones and Mearns, 2004), thus modifying the perception within the sector of dangerous climate change. A particularly important aspect of the analysis is to examine the limits to this adaptation – at what point can a system no longer adapt to a changing climate and significant damage or disruption to the system occurs. Limits to adaptability are often related to nonlinear changes in the impacted system; such changes can often be triggered by gradual change in climate.

23.2 Sectoral Vulnerabilities

Vulnerable Australian sectors that the community perceives to be impacted by climate change provide examples to explore the magnitudes and/or rates of climate change that may cause unacceptable levels of damage or risk. A sensitivity approach to past or projected increases in temperature

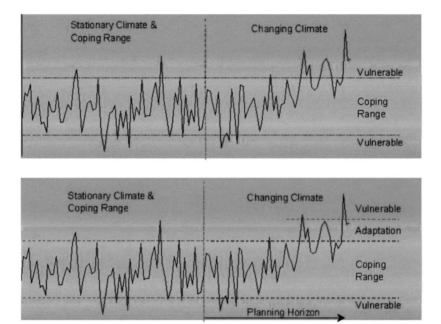

Figure 23.1 The role of planned adaptation in expanding the coping range and reducing vulnerability in response to a changing climate (from Jones and Mearns, 2004).

and changes in precipitation can provide a first-order analysis of the degree of risk of various impacted systems.

23.2.1 *Human Health*

Increases in heat-related deaths due to temperature extremes certainly constitute one direct way to assess dangerous climate change. In February 2004, Australia experienced a record heatwave, about six months after the European heatwave of August 2003. In Australia, mean maximum temperatures for the 1–22 February 2004 period were 5–6°C above average throughout large areas of eastern Australia and reached 7°C above average in parts of New South Wales (National Climate Centre, 2004). Sydney experienced 10 successive nights with minimum temperatures over 22°C (previous record of six) and Adelaide had 17 successive days over 30°C (previous record of 14). About two-thirds of continental Australia recorded maximum temperatures over 39°C in the 1–22 February 2004 period.

The heatwave led to a range of impacts on human health, with severe heat stress and collapses due to heat stress reported in Adelaide and Sydney. In Brisbane, where the temperature peaked at 41.7°C on the weekend of 21–22 February, the Queensland ambulance service recorded a 53% increase in ambulance call-outs. The commissioner of the ambulance service described it as '…the most significant medical emergency in the south-east corner (of Queensland) on record'. Currently, about 1100 heat-related deaths occur annually in Australias temperate cities, with the projected rise in temperature due to anthropogenic climate change leading to a substantial increase in heat-related deaths in all Australian cities by 2050 (McMichael et al., 2003).

The 2004 heatwave occurred against a background of a long-term increase in the frequency of hot days and nights in Australia (Figure 23.2, from Collins et al., 2000), superimposed on considerable year-to-year variability. Although the attribution of a single extreme event, such as the 2004 Australian heatwave, to climate change might never be possible, the risk of such extreme events occurring may be increased by human influences on climate. For example, an analysis of European mean summer temperatures, based on model simulations with and without anthropogenic greenhouse gas forcing, indicated that human influences could have more than doubled the risk of a summer heatwave of the intensity of that of 2003 (Stott et al., 2004).

23.2.2 *Agriculture*

Australian agriculture has adapted to one of the most variable climates on the planet, and there are undoubtedly lessons learnt relevant to agricultural management in the context of climate change elsewhere. Nevertheless, major droughts in Australia typically cause declines in Gross Domestic Product (GDP) of around 1% (A\$6.6 billion), with much larger regional impacts in affected areas. The 2002–2003 drought is estimated to have cost 1.6% of GDP (A\$10 billion) and about 70,000 jobs. There are concerns about the sectors ability to adapt to a potentially drier climate and to more/hotter droughts.

The abrupt change in rainfall in the southwestern corner of Western Australia, in which average winter rainfall decreased by 10–20% around the mid-1970s (Figure 23.3a), provides an example of impacts and adaptation in Australian agriculture as much of the Western Australian wheat production zone lies in the region affected by the rainfall anomaly. Figure 23.3b shows the change in wheat

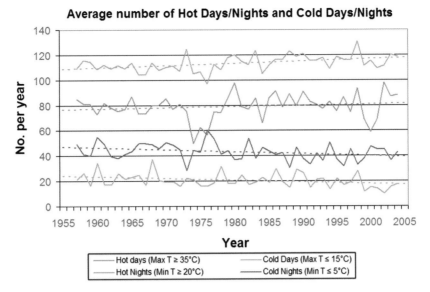

Figure 23.2 Australian average number of hot days (daily maximum temperature ≥35°C), cold days (daily maximum temperature ≤15°C), hot nights (daily minimum temperature ≥20°C) and cold nights (daily minimum temperature ≤5°C) per year. Note that annual averages of extreme events are based only on observation sites that have recorded at least one extreme event (hot day/night or cold day/night) per year for more than 80% of their years of record. Dashed lines represent linear lines of best fit.

yields over the same period. Contrary to expectations, wheat yield has actually increased over the period, owing to a range of changes in technology and management that were adopted over the period for other reasons (sometimes called autonomous adaptation) (Wigley and Tu Qipu, 1983). An example of such management change was the widespread adoption of no-till agriculture, which conserves soil moisture and allows the same or greater yield to be obtained with less rainfall. In addition, the drier climate has led to less water-logging of crops in some areas of southwestern Western Australia.

A multi-industry research project on climate adaptation in Australias rural industries has revealed some insights into the communitys perception of dangerous climate change. First, sensitivity to climate is highly industry specific with opposing effects in some cases; what might be considered as dangerous climate change by one industry could be perceived as beneficial by another industry in the same region! In general, Australian producers view abrupt changes as more dangerous than changes in means, even when the changes in extreme events associated with slow changes in underlying means are considered. Given the experience of rural Australia in dealing with a highly variable climate now, there is a level of confidence in most rural industries that their coping capacity is already high and can be raised even further in response to climate change (cf. Figure 23.1). A common perception is that the adaptive capacity of agricultural industries is high and that climate change will become a problem only if it is sufficiently abrupt. A further important point is that Australian agriculture is adapted to deal with extreme events (e.g. droughts), so long as they are infrequent enough to allow a sufficient number of productive years to ensure long-term profitability. A significant shift in the

patterns of extreme events could pose a major threat to long-term viability.

23.2.3 *Water Resources*

The demands of irrigated agriculture, biodiversity protection and urban supply are placing Australias scarce water resources under increasing pressure. Higher temperatures in the future and possible rainfall decreases are likely to increase water demand and reduce supply, further increasing the pressure on this key resource. Increases in the intensity of daily rainfall are likely to place increased pressure on urban drainage capacity and catchment management. A recent survey of the potential impacts of climate change in Australia concluded that, despite the large uncertainties in climate change projections for this century, there is a significant risk that climate change will exacerbate the pressure on Australian water supplies in the coming decades (Pittock, 2003; Jones et al., 2002).

Changes in the water supply of the city of Perth over the past several decades exemplify the risks facing major Australian cities from a changing climate. Although it is not yet known with a high degree of certainty whether anthropogenic climate change was the primary driver or a major contributing factor for the abrupt drop in rainfall in Western Australia in the mid-1970s, the impact on Perth has nevertheless been profound (for further information see the Indian Ocean Climate Initiative website at www.ioci.org.au). The drop in rainfall was amplified in terms of runoff so that the 10–20% decrease in rainfall (Figure 23.3a) translated into a ca. 50% decrease of streamflow into Perth's water supply dams (Figure 23.4). The city's initial response was to increase the extraction of groundwater, but this has now been augmented by the construction of a desalination plant

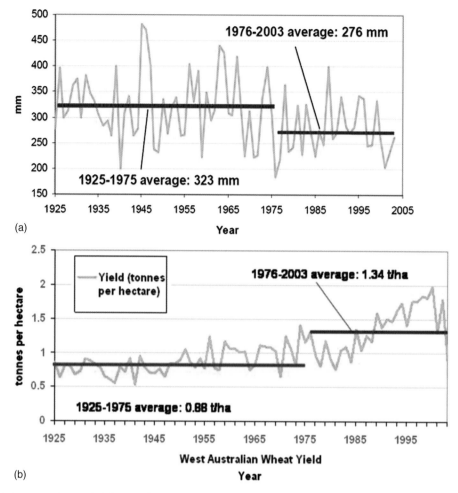

Figure 23.3 (a) Average winter rainfall (mm) for southwestern Western Australia for the period 1925–2003 (Australian Bureau of Meteorology); (b) Annual wheat yield (tonnes per hectare) for Western Australia for the period 1925–2003 (Australian Bureau of Statistics). Most of Western Australia's wheat production occurs in the southwestern corner of the state and is grown during the winter season.

to treat seawater for domestic uses. Both of these are short-term measures and the longer-term security of Perth's water supply remains a critical issue for Western Australia. In fact, the water supply issue was a major factor in a recent state election campaign.

The examples of Perth's water supply and that of the Western Australian wheat growers demonstrate how differently the same change in climate can affect different sectors. Because the adaptive capacity of the wheat growers appears to be much greater than that of urban water users in Perth, the farmers were able to maintain and even enhance profitability while rainfall dropped. Thus, they may not judge the decrease in rainfall as dangerous climate change. The city of Perth, on the other hand, may come to a very different conclusion regarding the definition of dangerous climate change.

23.2.4 *Coral Reefs*

Reefs are under pressure from increasing acidity in the surface waters (cf. paper by Turley et al., this volume),

increasing sea surface temperature and other human-induced stresses. Globally, there has been a sharp increase in mass bleaching events over the past century, with unusually widespread bleaching events in the 1997–98 period, in association with a major El Niño event (Figure 23.5, from Lough, 2000). Australia's Great Barrier Reef, a global biodiversity hotspot and one of the best managed reefs in the world in terms of minimising stresses from local sources, suffered major bleaching events during the 1997–98 period and may be significantly affected by climate change in the future under even moderate emission scenarios.

A case study of the potential impacts of rising water temperature on the Great Barrier Reef (Done et al., 2003) demonstrates how a modelling approach, based on our best understanding of coral ecology and its response to rising water temperature, can be employed to generate useful information for defining dangerous climate change. Figure 23.6 shows the projected response of a coastal reef to increasing water temperature due to anthropogenic climate change. The approach has a number of valuable

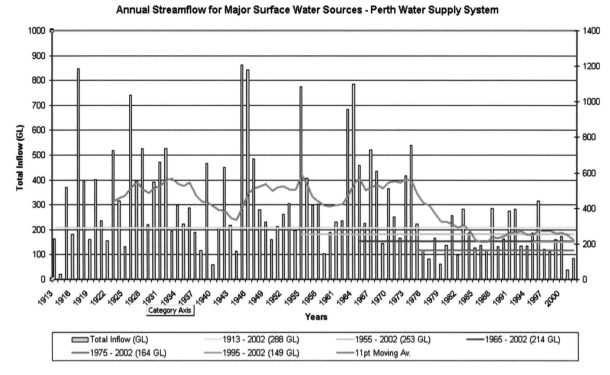

Annual Streamflow for Major Surface Water Sources - Perth Water Supply System

Figure 23.4 Annual streamflow into Perth's water supply dams, with averages before and after the rainfall decrease of about 10–20% (depending on location) that occurred in 1974–75 (Western Australia Water Corporation, via Ian Foster; see Foster, 2002).

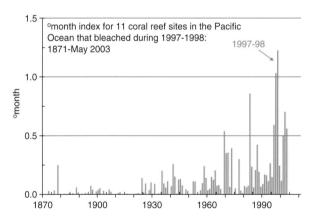

Figure 23.5 Coral bleaching records showing the large number of events recorded in 1998 (from Lough, 2000).

features for addressing the issue of dangerous climate change:

- The degree of damage is related to the number of days in which a temperature threshold is exceeded, which in turn is related to a judgement about the level of impact (from 'low' to 'catastrophic').
- Projections of future change are given as probability distributions rather than as a single, or just a few, projections.
- Two scenarios are used to show that the probability of high or catastrophic impacts can change significantly with the CO_2 emission scenario.

The strength of this approach is that it lends itself to a risk analysis, in which stakeholders can evaluate a wide range

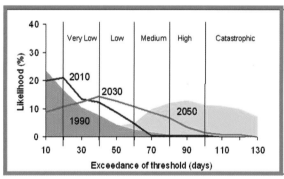

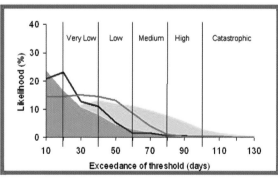

Figure 23.6 Predicted coral bleaching days per annum for a coastal reef in Australia's Great Barrier Reef region in 1990, 2010, 2030 and 2050, plotted as probability distribution functions and indicating impact levels from 'very low' to 'catastrophic', as defined in Done et al. (2003). The top panel is based on a fossil fuel-intensive energy scenario with high CO_2 emissions (IPCC SRES A1FI). The bottom panel is based on a scenario with a transition to energy systems with significantly lower CO_2 emissions (IPCC SRES A1T) (from Done et al., 2003).

of potential futures based on sensitivities to critical parameters and judge for themselves the level of risk they are willing to take with an impacted system. Thus, the approach is policy-relevant and not policy-prescriptive, through identification of the issues for which value judgements are needed.

23.2.5 *Biodiversity*

Australia has the highest fraction of endemic (found only in a particular area or region) biotic diversity of any continent but also has the highest current extinction rate of any continent for vertebrates, largely due to land-use change and introduction of exotic species. The ability of species to adapt to future changes in climate by migrating with climatic zones will be limited by habitat fragmentation resulting primarily from land-use change. Climate change may thus exacerbate loss of species.

In an Australian context, some of the most vulnerable ecosystems to climate change are those near their upper temperature limit already. For example, the Australian Alps are an important habitat for many plant and animal species yet are a low mountain range (ca. 2000 m in altitude) and cover a small area. There is little scope for these natural ecosystems to adapt to a warmer climate as there is no scope for them to move upwards. For example, the distribution of high mountain vegetation is related primarily to summer temperature, with increasing summer temperature likely to lead to the expansion of woody vegetation into areas now occupied by herbaceous species (Hughes, 2003). For species dependent on reliable snow cover for survival (e.g. the pygmy possum), the future also looks problematic. The IPCC SRES range of scenarios lead to projections of an 18–66% reduction in area of snow cover by 2030 and a 39–96% reduction by 2070 (Whetton, 1998).

Such effects are not confined to the alpine regions. In north Queensland a modest 1°C increase in mean annual temperature above the 1961–1990 average is estimated to lead to a 50% decrease in the area of highland rainforest environments (Hilbert et al., 2001). In addition, given their small area now, these environments will become increasingly fragmented as suitable habitat becomes confined to separate islands of higher elevation (refugia). Modelling studies suggest that many of the regions endemic vertebrates would lose much of their climatically suitable habitats with relatively small shifts in climate. For example, the golden bowerbird would lose 63% of its current habitat with only 1°C of warming and 98% of its habitat with a 3°C increase in mean annual temperature (Hilbert et al., 2003).

As with other analyses of the impacts of climate change on biodiversity (e.g. papers by Leemans and Hare, this volume), in general there is relatively little adaptive capacity in many of Australias natural ecosystems, especially when the compounding effects of habitat fragmentation due to land-use change and introduction of exotic species are taken into account. Thus, managed systems such as agroecosystems have a much greater adaptive capacity than natural ecosystems; the latter will generally experience deleterious effects at less severe levels of climate change. Examination of these and other sectoral analyses gives some insights into guidelines for defining dangerous climate change from an Australian perspective:

- Changes in frequency and intensity of extreme events will be more important than changes in underlying mean values.
- Rates of change may be as, or more, important than the magnitude of mean changes.
- Changes in water availability will be as important as changes in temperature or rainfall.
- Interactions with other aspects of global change (e.g., the direct effects of increasing atmospheric CO_2 on ocean acidity and on terrestrial ecosystems) must be considered.
- Resilience of the impacted system, and the capacity to increase it, is critical to defining dangerous climate change.

The sectoral studies also suggest a more specific approach that can be used to integrate both scientific information and societal value judgements in assessing what constitutes dangerous climate change. The first step is to define a damage scale as quantitatively as possible for the vulnerable system. The second step, which involves a value judgment and must be based on a broad consultation with society in general, is to define what is acceptable damage and what is not. The third step is to relate the damage scale to climatic or atmospheric change parameters; that is, to generate a damage function related to the stressor. Finally, an analysis can be undertaken to determine what level of change in the climate-related parameters leads to unacceptable damage in the impacted system. It should be emphasized that the application of this approach must be tailored to the specific characteristics of the impacted system under study. The definition of what is dangerous climate change can thus be quite different across different systems or sectors.

23.3 Emerging Issues

Large-scale discontinuities in the climate system, often called *abrupt changes*, provide another approach to defining dangerous climate change. The issue here is the degree or rate of climate change at which the risk of such abrupt changes becomes unacceptably high. The most well-known example of a large-scale discontinuity is the potential shift in the thermohaline circulation in the north Atlantic Ocean.

From an Australian perspective, abrupt shifts in the behaviour of known modes of climate variability such as ENSO or the Asian Monsoon system would have potentially very significant consequences. The ENSO phenomenon exerts a strong influence on eastern Australia, and a strong El Niño event can cause a 1% decrease in Australia's GDP. Long-term records of ENSO behaviour show high variability in both intensity and periodicity (Moy et al., 2002). More recent records based on sea surface temperature estimates from coral in the equatorial Pacific Ocean show that the El Niño–La Niña oscillation pattern shifted around 1900 from a 10-year period to a 4-year period (Urban et al., 2000). Any significant change towards more El Niño-like mean conditions in the future would create significant problems for contemporary Australian society and would almost surely be considered as dangerous climate change by Australians.

Finally, the context of anthropogenic climate change, coupled with other changes in the global environment due to human activities, must be considered in the broadest definition of dangerous change. The ice core records from the Vostok and Dome C sites show that for the last 730,000 years at least, the Earth has cycled between two fundamental states – glacial and interglacial. The current concentration of CO_2 in the atmosphere (ca. 380 ppm) is already about 100 ppm higher than the level associated with interglacial states and thus represents a doubling of the operating range between glacial (ca. 180 ppm) and interglacial states (ca. 280 ppm). This increase has occurred at a rate at least 10 times and possibly 100 times faster than increases of CO_2 concentration at any other time during the past 420,000 years (Falkowski et al., 2000).

Such rapid changes in atmospheric composition raise questions about the stability and resilience of the Earth System as a whole (Steffen et al., 2004). The possible weakening and collapse later this century of the terrestrial carbon sinks that are currently absorbing a significant fraction of anthropogenic CO_2 emissions (Cox, this volume) suggests that anthropogenic forcing could trigger a shift in the Earth System to a new state with persistently higher greenhouse gas concentrations and a higher mean temperature than the interglacial state. Such a potential state change in the Earth System could constitute the ultimate definition of dangerous climate change.

REFERENCES

Clark P.U., Pisias N.G., Stocker T.F., Weaver A.J., 2002. The role of thermohaline circulation in abrupt climate change. Nature 415:863–869.

Collins D.A., Della-Marta P.M., Plummer N., Trewin B.C., 2000. Trends in annual frequencies of extreme temperature events in Australia. Aust. Met. Mag. 49:277–292.

Done T.J., Whetton P., Jones R., Berkelmans R., Lough J., Skirving W., Wooldridge S., 2003. Global climate change and coral bleaching on the Great Barrier Reef. Australian Institute of Marine Science. Final Report to the State of Queensland Greenhouse Task Force through the Department of Natural Resources and Mines.

Falkowski P., Scholes R.J., Boyle E., Canadell J., Canfield D., Elser J., Gruber N., Hibbard K., Hogberg P., Linder S., Mackenzie F.T., Moore B. III, Pedersen T., Rosenthal Y., Seitzinger S., Smetacek V., Steffen W., 2000. The global carbon cycle: A test of knowledge of Earth as a system. Science 290:291–296.

Foster I., 2002. Climate change position paper. Western Australia Department of Agriculture, Perth, 31 May 2002

Hilbert D.W., Bradford M., Parker T., Westcott D.A., 2004. Golden bowerbird (Prionodura newtonia) habitat in past, present and future climates: Predicted extinction of a vertebrate in tropical highlands due to global warming. Biological Conservation 116:367–377.

Hilbert D.W., Ostendorf B., Hopkins M., 2001. Sensitivity of tropical forests to climate change in the humid tropics of North Queensland. Austral Ecology 26:590–603.

Hughes L., 2003. Climate change and Australia: Trends, projections and research directions. Austral Ecology 28:423–443.

Jones R.N., Mearns L.O., 2004. Assessing Future Climate Risks. In: Lim B., Spanger-Siegfried E., Burton I., Malone E., Huq S. (eds), *Adaptation Policy Frameworks for Climate Change: Developing Strategies, Policies and Measures*, Cambridge University Press, Cambridge and New York, 119–144.

Jones R.N., Whetton P.H., Walsh K.J.E., Page C.M., 2002. Future impacts of climate variability, climate change and land use change on water resources in the Murray Darling Basin: Overview and draft program of research [electronic publication]. Murray-Darling Basin Commission, Canberra, A.C.T, 27 pp. Available from: http://publications.mdbc.gov.au/

Lough J.M., 2000. Unprecedented thermal stress to coral reefs? Geophysical Research Letters 27:3901–3904.

McMichael A., Woodruff R., Whetton P., Hennessy K., Nicholls N., Hales S., Woodward A., Kjellstrom T., 2003. Human health and climate change in Oceania: A risk assessment. Commonwealth of Australia, 126 pp. (www.health.gov.au/puhlth/strateg/envhlth/climate/).

Moy C.M., Seltzer G.O., Rodbell D.T., Anderson D.M., 2002. Variability of El Niño/Southern Oscillation activity at millennial timescales during the Holocene epoch. Nature 420:162–165.

Murray J.W., 2000. The oceans. In: Jacobson M.C., Charlson R.J., Rodhe H., Orians G.H. (eds), Earth System Science, Academic Press, pp. 230–278.

National Climate Centre, 2004. Eastern Australia experiences record February heatwave. Bulletin of the Australian Meteorological and Oceanographic Society 17:27–29.

Pittock B. (ed.), 2003. Climate change: An Australian guide to the science and potential impacts. Australian Greenhouse Office, Canberra, 239 pp. www.greenhouse.gov.au/science/guide/pubs/science-guide.pdf

Steffen W.L., Sanderson A., Tyson P.D., Jaeger J., Matson P.A., Moore B. III, Oldfield F., Richardson K., Schellnhuber H.J., Turner B.L. II, Wasson R.J., 2004. Global change and the Earth System: A planet under pressure. Springer-Verlag, 336 pp.

Stott P.A., Stone D.A., Allen M.R., 2004. Human contribution to the European heatwave of 2003. Nature 432:610–613.

Takahashi T., Feely R.A., Weiss R.F., Wanninkhof R.H., Chipman D.W., Sutherland S.C., Takahashi T.T., 1997. Global air-sea flux of CO_2: An estimate based on measurements of sea-air pCO_2 difference. Proc. Nat. Acad. Sci. USA 94:8292–8299.

Urban F.E., Cole J.E., Overpeck J.T., 2000. Influence of mean climate change on climate variability from a 155-year tropical Pacific coral record. Nature 407:989–993.

Whetton P.H., 1998. Climate change impacts on the spatial extent of snow-cover in the Australian Alps. In: Green K. (ed.) Snow: A Natural History; An Uncertain Future, Australian Alps Liaison Committee, Canberra, Australia, pp. 195–206.

Wigley T.M.L., 2004. Choosing a stabilization target for CO_2. Climatic Change 67:1–11.

Wigley T.M.L., Tu Qipu, 1983. Crop-climate modelling using spatial patterns of yield and climate: Part 1, Background and an example from Australia. Journal of Climate and Applied Meteorology 22: 1831–1841.

CHAPTER 24

Regional Assessment of Climate Impacts on California under Alternative Emission Scenarios – Key Findings and Implications for Stabilisation

Katharine Hayhoe[1,2], Peter Frumhoff[3], Stephen Schneider[4], Amy Luers[3] and Christopher Field[5]

[1] *ATMOS Research & Consulting*
[2] *Department of Geosciences, Texas Tech University*
[3] *Union of Concerned Scientists*
[4] *Department of Biological Sciences and Institute for International Studies, Stanford University*
[5] *Department of Global Ecology, Carnegie Institution of Washington*

ABSTRACT: Regional estimates of climate change impacts and their relationship to temperature and other indicators of change provide essential input to decisions regarding 'acceptable' levels of global mean temperature change, long-term stabilisation levels, and emissions pathways. Here, we present an assessment of regional impacts for California based on projections by two climate models (HadCM3 and PCM) for the SRES higher and lower emissions scenarios (A1FI and B1). Temperature increases under the higher scenario are nearly double those of the lower scenario by 2100, with proportionally greater impacts on human health, ecosystems, water resources, and agriculture. This analysis provides the basis for our discussion of perspectives on the role of impact assessments in increasing both our confidence in projections and our understanding of potential impacts under alternative scenarios.

24.1 Introduction

Evaluating the potential impacts of climate change at the local to regional scale is essential to assessing regional vulnerabilities. Comparison of potential impacts under a range of future greenhouse gas (GHG) concentrations provides a means of quantifying the degree to which alternative emission scenarios could mitigate impacts.

Based on a comprehensive analysis of projected impacts for California under the SRES A1FI (higher) and B1 (lower) emission scenarios [1,2], we argue that the magnitude of future climate change depends substantially upon the GHG emission scenarios we choose. Although not directly linked to climate policy-based stabilisation levels, the scenarios examined bracket a large part of the range of IPCC non-intervention emissions futures with atmospheric concentrations of CO_2 reaching 550 ppm (B1) and 970 ppm (A1FI) by 2100.

The majority of impacts seen under the A1FI scenario are substantially greater than those under the B1 scenario, suggesting that climate change and many of its impacts scale with the quantity and timing of GHG emissions. However, even under the lower B1 scenario, many ecological, water, agricultural and health-related impacts may exceed socially or politically acceptable thresholds. The more than 190 signatories to the UNFCCC agreed to stabilise GHG concentrations at levels that will prevent 'dangerous anthropogenic interference with the climate system' [3]. This suggests that some of the results reported in [2] or here for California might well be interpreted by decision-makers and stakeholders as requiring a response under the UNFCCC. In addition, the enormous uncertainties in impacts and the potential for severe risk in several key areas (e.g. wildfire, El Niño-driven winter storms) is sufficiently large so as to compound the societal challenge of specifying a level of change that should be avoided. Whether to postpone decisions in favour of waiting for further research findings or to implement precautionary steps via policy decisions is the value choice faced by decision-makers aware of the potentially large impacts of climate change such as those presented here.

Disaggregation of impacts by sector allows for a broad picture of the degree to which increases in temperature and other indicators of climate change may affect California and other world regions. Here, we first summarise the projected impacts for California under a higher vs. a lower emissions scenario as presented by [2]. This is the first study to use projections from multiple climate models and the higher and lower SRES scenarios to examine climate change impacts at the regional level. For this reason, we next examine the implications of our findings for evaluating a given temperature or stabilisation target aimed at preventing long-term and serious impacts at the regional and global level. Finally, we build on the lessons learned during this experience to highlight some of the key scientific, methods- and resource-related issues that may have contributed to the policy relevance of this study.

24.2 Climate Change Impacts on California

California is one of the largest economies in the world, and is characterized by its diverse range of climate zones, limited water supply, and the economic importance of

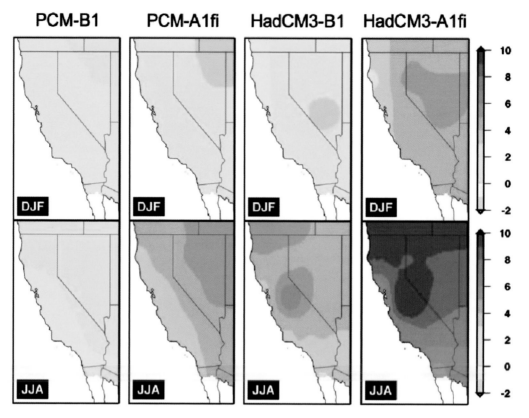

Figure 24.1 Downscaled projected winter (DJF) and summer (JJA) temperature change (°C) for 2070–2099 relative to 1961–1990. Projections shown are for the HadCM3 and PCM models for the SRES A1FI (higher) and B1 (lower) scenarios. By the end of the century, consistently greater temperature anomalies emerge for the higher emissions scenarios, with larger increases and therefore heightened inter-scenario differences particularly in summer. Statewide, SRES B1 to A1FI winter temperature projections for the end of the century are 2.2–3°C and 2.3–4°C for PCM and HadCM3 winter, and 2.2–4°C and 4.6–8.3°C for PCM and HadCM3 summer.
Source: Hayhoe et al., 2004a [2].

climate-sensitive industries such as agriculture [6]. As such, it provides a challenging test case to evaluate impacts of regional-scale climate change under alternative emissions scenarios. Using a lower- and a moderate-sensitivity[1] climate model from the IPCC AR4 database (PCM and HadCM3 respectively), the higher and lower SRES emissions scenarios (A1FI and B1), and two well-documented statistical downscaling methods [4,5], the potential impact of climate change on temperature-sensitive sectors in California was evaluated, with initial results presented in [2].

Annual temperature increases over California nearly double from the lower B1 to the higher A1FI scenario by 2100. Statewide, the range in projected average temperature increases is higher than previously reported [6,7],

particularly for summer temperature increases, which equal or exceed increases in winter temperatures in all cases (Figure 24.1). These results are consistent with observed state-wide temperature trends over the last 50 years [8] and are further supported by recent findings that development of irrigated land over the last century may have acted to damp historical summer temperature increases relative to those of winter in the Central and Imperial Valleys [9]. Downscaled monthly mean temperature projections show consistent spatial patterns across California, with lesser warming along the southwest coast (likely due to the moderating influence of the ocean[2]) and increasing warming to the north and northeast (Figure 24.1).

Heatwaves and associated impacts on a range of temperature-sensitive sectors are substantially greater under the higher scenario, with some inter-scenario differences apparent by mid-century. For many of California's urban areas, the duration, intensity and number of heatwave events are projected to increase linearly with average

---

[1] Global climate sensitivities for 13 of the latest-generation AR4 AOGCMs under a doubling of CO_2 range from 2.1 to 4.4°C. PCM is at the lowest end of the range (2.1°C), while HadCM3 lies near the middle (3.3°C). For the area over California, projected end-of-century temperature change from the same models under the SRES B1 scenario (a transient scenario under which CO_2 doubles by 2100) relative to current-day temperatures ranges from 0.6–2.8°C, with PCM again falling near the lower end of the range (at 0.9°C) and HadCM3 being nearer to the top of the range (at 2.4°C).

[2] Summer temperature increases in central and northern coastal areas may be additionally moderated if the coastal upwelling continues to increase over this next century as it has in the past [10,11,12].

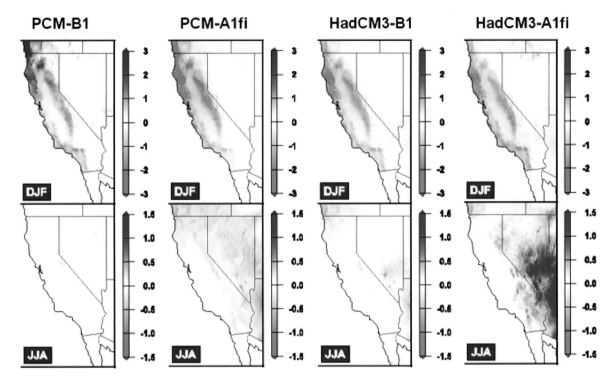

Figure 24.2 Downscaled projected winter (DJF) and summer (JJA) precipitation change (mm/day; note scale difference between winter and summer plots) for 2070–2099 relative to 1961–1990. Projections shown are for the HadCM3 and PCM models for the SRES A1FI (higher) and B1 (lower) scenarios. The majority of scenarios suggest a net decrease by the end of the century. Geographical patterns of precipitation change are consistent across models and scenarios, with the greatest decreases (increases for the PCM B1 scenario) occurring on the northwest coast and along the eastern Central Valley and western Sierras. *Source:* Hayhoe et al., 2004a [2].

summer temperatures, particularly in inland areas. By 2070–2099, extreme heatwaves that occurred once or twice per decade during the historical period are projected to make up 1/4 to 1/2 of annual heatwaves for inland cities such as Sacramento, Fresno, and the inland suburbs of Los Angeles under B1 and 1/2 to 3/4 of the heatwaves each year under A1FI [13]. Under these scenarios, heat-related mortality is also projected to increase, with the largest increases in acclimatized mortality rates (100% to 1000%) expected for coastal cities such as San Francisco and Los Angeles that are relatively unaccustomed to extreme heat. When the effects of ozone-related health impacts are included (difficult to estimate quantitatively, as they depend strongly on assumed future emissions of air pollutants), the net projected impact of increasing heat on human health is even greater.

Rising temperatures over California, exacerbated in most cases by decreasing winter precipitation[3] (Figure 24.2), produce substantial reductions in snowpack in the Sierra Nevada Mountains (Figure 24.3) with cascading impacts on California streamflow, water storage and supply.

Under even the lowest emissions scenario, 1 April snowpack is projected to decrease 25–40% before mid-century and 30–70% before 2100, with peak runoff shifting 1–3 weeks earlier. Under the higher scenario, 1 April snowpack decreases before 2100 are on the order of 70–90%. The response of snowpack to temperature varies by elevation, with lower levels being more sensitive to local temperature changes on the order of 2.5°C or less (displaying a logarithmic relationship between temperature increases and snowpack amount), while snowpack at higher elevations decreases most sharply for temperature changes greater than 2.5°C (displaying an exponential dependence on temperature). In addition to snowpack losses, the proportion of years projected to be dry or critical increases from 32% in the historical period to 50–64% by the end of the century. Under both scenarios, these changes have the potential to disrupt the current highly-regulated California water system by reducing the value of rights to mid- and late-season natural streamflow while boosting the value of rights to stored water.

The combination of decreased water availability and rising temperatures also threatens some of California's $30 billion agricultural industry. Perennial crops such as oranges, grapes and other fruit may be more vulnerable to climate change because there are few options for short-term adaptation, and long-term adaptation such as switching to new cultivars or shifting the location of orchards is

---

[3]The four model/scenario combinations used in this study cover the standard deviation of projected end-of-century precipitation for California (~ +10 to −15%) based on a 14-AOGCM intercomparison (this work) as well as from 84 simulations created by probabilistic sampling from 12 state-of-the-art models and 3 socioeconomic scenarios [14].

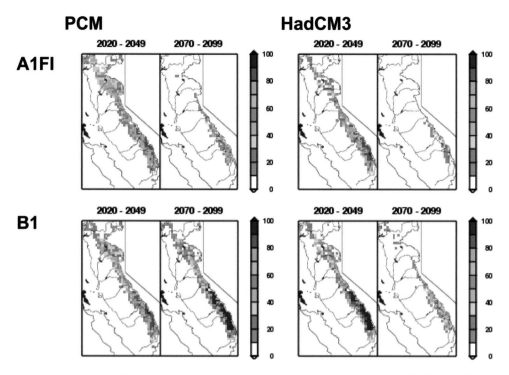

Figure 24.3 Projected percentage of snowpack remaining in the Sierra Nevada region for the HadCM3 and PCM models under the A1FI and B1 scenarios for 2020–2049 and 2070–2099 relative to 1961–1990 average. Total snow water equivalent losses by the end of the century range from 29–72% for the B1 scenario to 73–89% for the A1FI scenario. Losses are greatest at elevations below 3,000 m, ranging from 37–79% for B1 to 81–94% for A1FI by the end of the century. Figures courtesy of E. Maurer.

expensive. Citrus is sensitive to high temperatures during particular growth phases, and high temperatures can also affect wine grape quality. According to both PCM and HadCM3 projections, Central Valley regions may already experience impaired grape quality before mid-century, while California's major grape-growing regions (with the sole exception of the Cool Coastal areas) are projected to be either marginal or impaired before 2100 under both the lower B1 and higher A1FI scenario. Response to temperature increases varies by region, with the quality of wine grapes from present-day grape-growing regions currently at the lower end of the temperature range improving before declining, while other regions (Napa, Sonoma and the Central Valley) display a logarithmic dependency and thus higher sensitivity to smaller-scale temperature changes on the order of 2°C or less (Figure 24.4). Once more, this suggests that a B1-like scenario may be insufficient to prevent some serious repercussions on California's agricultural industry[4].

The distribution of California's diverse vegetation types is also projected to change substantially. Reductions in the extent of alpine/subalpine forest and the displacement of evergreen conifer forest by mixed evergreen forest are driven primarily by temperature, becoming pronounced under A1FI by the end of the century. Other

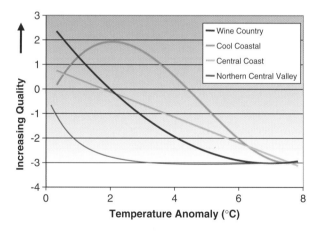

Figure 24.4 Projected impact of temperature increases on wine grape quality as measured on a sliding scale from 'Impaired' (−3) through 'Marginal' (−1), 'High/Low Optimal' (+1) and 'Mid-Optimal' (+3) for four of the major grape-growing regions in California. This result illustrates the potential for non-linear response from temperature-dependent impacts to increasing temperature over time or due to a higher emissions scenario.

vegetation more sensitive to precipitation and fire frequency (grass, shrubland) does not exhibit consistent inter-scenario differences. There is a clear link between climate and fire in California, where fires occur during the dry summer season and area burned is generally higher following a wet El Niño winter [15]. Little projected

---

[4]Note: This analysis does not consider the potentially beneficial effects of CO_2 fertilization on agricultural crops.

change in summer rainfall (Figure 24.2), coupled with warmer temperatures, an expanding population and increasing human settlement in previously wild areas make it likely that the economic costs and damages due to wildfire will increase in the future. Numerous factors interact to determine the net effect of climate on fire – climate change may cause fire hazard conditions to increase in one area but decrease in another, while changes in the number of rare but extreme fires that cause the most damage [16] could be different again. Hence, in this study it was not possible to quantify the response of fire risk to temperature increases over California, although based on the above reasoning we believe it is likely that fire risk will increase.

Finally, California is also highly vulnerable to El Niño events, as demonstrated by the severe winter storms and floods that have devastated the California coast in recent years. During El Niño events, some of the most significant increases in heavy rainfall events have been observed, as well as several of the largest floods on record [17]. Several studies have found that storm characteristics have been changing over the last century, creating a greater flood risk during the last half of the 21st Century and particularly during the 1990s [18,19]. It is uncertain whether these trends are likely to continue, although some observational evidence suggests that the shift in increasing El Niño event frequencies may be linked to increasing global temperatures [20,21]. However, it is clear that even without a change in the frequency or intensity of winter storms, long-term sea level rise due to climate change would expose the coast to severe flooding and erosion, damage to coastal structures and real estate, and salinity intrusion into vulnerable coastal aquifers.

Overall, we found that extreme heat and related impacts on a range of temperature-sensitive sectors in California were substantially greater under the higher-emissions scenario, with some inter-scenario differences apparent by mid-century. Inter-scenario differences for other sectors more sensitive to non-temperature-related drivers were not as distinct, obscuring the response of these sectors to a given temperature or CO_2 stabilisation target. The extent to which these results are model-dependent will be better understood as more such analyses are conducted for California as well as other world regions.

24.3 Relationship to GHG Stabilisation Levels and Temperature Change

Estimates of the relationship between impacts and temperature or other drivers provide critical information for deliberations on global mean temperature change, long-term stabilisation levels, and candidate emissions pathways for preventing 'dangerous anthropogenic interference' with the climate system. Notably, since their publication the results of this study of climate change impacts on California have been highlighted in several climate-related

initiatives and policy discussions at the state and regional level [22,23].

In June 2005, California Governor Schwarzenegger signed an Executive Order setting a goal for California of reducing its greenhouse gas emissions to 2000 levels by 2010; to 1990 levels by 2020; and to 80% below 1990 levels by 2050 [24]. The UK has proposed a similar target of a 60% reduction to 1990 levels by 2050, but has presented this within the global context of CO_2 stabilisation at 550 ppm [25], while the European Union has endorsed a global mean temperature change target of 2°C relative to pre-industrial times [26].

It is not yet clear whether these targets are sufficient to prevent all potential instances of 'dangerous anthropogenic interference'. Mastrandrea & Schneider [27], for example, estimate a 20% risk of exceeding dangerous thresholds of change (as represented by IPCC Working Group 2) for a 2°C warming, a risk that increases steadily with temperature. The actual warming if the world follows the SRES B1 scenario is uncertain, but CO_2 levels reach (although do not yet stabilise at) 550 ppmv by 2100, making this scenario a reasonable starting point for a discussion of potential impacts on California under a target selected to prevent 'dangerous interference'.

The risk of significant impacts is already relatively high under the B1 scenario (Table 24.1). Regional temperature changes of 2–3.5°C may represent unacceptable stress on water supply (through reduction of Sierra snowpack by 2070–2099 and shifts in peak streamflow to earlier in the year) and human health (through heat-related mortality), as well as on some perennial crops in regions currently at their upper temperature limits. These impacts intensify for the larger increases in temperature associated with the A1FI scenario. A1FI impacts include near-complete (70–89%) loss of Sierra Nevada snowpack (Figure 24.3) and severely impaired grape growing conditions for all but the cool coastal regions (Figure 24.4). If future analyses for other regions bear out these findings, consideration of emission scenarios leading to a CO_2 (or CO_2-equivalent [28]) stabilisation level lower than 550 ppm may be warranted.

It is also essential to note that many impacts are moderated or even controlled by climate influences other than temperature. In California, for example, this includes impacts on ecosystem and vegetation shifts, coastal impacts from sea level rise and El Niño-driven winter storms, wildfire extent and damages, and the net impact of climate change on the water supply system. In cases such as these, temperature change may act as an exacerbating factor rather than the primary driver. Precipitation and changes in atmospheric circulation patterns, socioeconomic change, human choices and behaviour all play roles of equal or greater importance. Hence, it is difficult to determine with high confidence how the impacts on these systems would scale with temperature. Resolving emission targets for impacts based on their relationship to temperature tells only part of the story. Instead, a

Table 24.1 Comparison of projected end-of-century (2070–2099) temperature-dependent impacts relative to 1961–1990 under the SRES B1 (lower) vs. the A1FI (higher) emissions scenarios for the HadCM3 and PCM models. Impacts for precipitation-related impacts are not shown as they were not found to exhibit a significant dependence on GHG emissions and scenario.

	B1		A1FI	
	PCM	HadCM3	PCM	HadCM3
Increase in number of heatwave days per year (days)				
Los Angeles	+27	+44	+63	+100
San Francisco	+24	+53	+80	+121
San Bernardino	+30	+50	+63	+93
Sacramento	+30	+50	+71	+100
Fresno	+29	+46	+68	+94
Increase in acclimatized heat-related mortality (%)				
Los Angeles	102	260	400	670
San Francisco	280	300	1200	610
San Bernardino	200	190	270	290
Sacramento	20	110	270	260
Fresno	−30	−60	70	30
Change in April 1 snowpack				
1000–2000 m	−65	−95	−87	−97
2000–3000 m	−22	−73	−75	−93
3000–4000 m	+15	−33	−48	−68
All elevations	−29	−73	−72	−89
Change in wine grape quality				
Wine country	Impaired	Marginal	Impaired	Impaired
Currently optimal (mid)				
Cool Coastal	Optimal	Optimal	Optimal	Impaired
Currently optimal (low)	(mid-high)	(mid-high)	(high)	
Central Coast	Marginal	Marginal	Marginal	Impaired
Currently optimal (mid/high)				
San Joaquin Valley	Impaired	Impaired	Impaired	Impaired
Currently marginal				
Central Valley	Impaired	Impaired	Impaired	Impaired
Currently marginal/impaired				

Source: Hayhoe et al., 2004a,b [2,13].

multi-dimensional threshold incorporating a number of key drivers (e.g. temperature, precipitation, population, settlement patterns, water demand) would be more suitable, as simultaneous changes in multiple factors could push a system over the limits of acceptable change.

24.4 Key Considerations for Regional Assessments

A number of factors contribute to the policy relevance of any given impact assessment. In this section, we review key scientific, methodological, resource, and communication-related issues that played a role in facilitating the use of this study's findings by local, state and regional decision-makers to justify and build support for climate initiatives.

The primary scientific issues considered during this study consisted of:

- Consideration of multiple and consistent socio-economic futures in order to accurately capture a plausible range of future projections.

- Use of multiple, reasonably credible and tested climate models with varying regional responses to climate change to look for robust conclusions despite model dependence.

- Resolution of appropriate spatial and temporal scales of climate forcings and change to capture regional to local-scale influences on climate.

- Consistent representation of the uncertainty involved in socioeconomic and climate change projections, in a manner that can be easily understood and assimilated into estimates of impacts and their relationship to temperature or other indices of change.

- Recognition of the multiple types, interactions, and need for consistency between drivers of change, including climate, socio-economic development, human decisions, and policy.

The success of this assessment depended on the relevant scientific expertise of a multi-disciplinary team of nineteen scientists, most based in California. Fields of study included climate modelling and downscaling; regional hydrology and extremes; ecology and agriculture; dynamic vegetation modeling; the economics of water management, supply and demand; human health and welfare; and the social dimensions of climate change. Uniformity of underlying assumptions was ensured by providing team members with consistent socio-economic projections and climate simulations in the format required for analysis in each sector. The benefits of an integrated team with regional expertise allowed a credible search for robust analyses based on previously-evaluated methods that could then be re-applied to a consistent set of socio-economic assumptions and resulting climate projections. Other essential resources include up-to-date climate model projections for a range of socio-economic and stabilisation scenarios, downscaling methods, region-specific impact models, and adequate computing facilities. Although the question of both physical and intellectual resources often poses a challenge for scientists from developing nations (see [29]), many resources already exist and are freely available[5].

Regional assessments should not be merely a collection of independently-conducted studies or a means for scientists to inform regional specialists and stakeholders [30]. Information can flow in both directions and there are many scientific as well as societal advantages to engaging those whose welfare and livelihoods are threatened by climate change. Decision-making at the local to state level based on the latest information has a greater likelihood of preparing for possible impacts, thereby increasing the potential for successful adaptation and/or mitigation. Such communication requires awareness of the needs and cultures of different stakeholders, decision-makers, and the media as dissemination of results does not always occur via the standard formats with which most scientists are familiar. In California, outreach efforts based on these findings have resulted in significant dialogues among the author team, representatives of non-profit organizations, and key stakeholder groups including state legislatures, private industry, labor unions, and public utilities. Roundtable discussions allowed stakeholders to ask scientists questions about climate change, and scientists to explore values and perspectives on related issues. These dialogues have helped shape further research plans on more targeted questions about potential impacts and adaptive capacity of specific sectors and populations, as well as creating opportunities for a large body of informed California scientists and stakeholders to speak to the implications of climate change for their region.

These scientific and technical considerations highlight the philosophical underpinnings of the California analysis in order to make our underlying assumptions transparent to others pursuing the same course. The greater the number of replicates and similar evaluations, the higher our collective ability to assess the robustness of the projected impacts on both a regional and global level for a given emissions scenario or stabilisation target.

24.5 Conclusions

Assessing the potential impacts of climate change on the local to regional-scale level is essential for identifying the vulnerabilities these systems may display under future change. Through considering a range of potential socio-economic futures and climate projections, we can obtain estimates of the responses of key sectors to a given level of change. Value judgments regarding the tolerability of these impacts can then be made, enabling the establishment, evaluation and revision of emission targets aimed at avoiding unacceptable levels of change.

In a recent study for California [2], we estimated the degree to which key sectors' vulnerabilities and their relationship to temperature change can be quantified. Large and consistent increases in temperature and extreme heat drove significant impacts on temperature-sensitive sectors under both lower and higher emissions scenarios, with the most severe impacts under the higher A1FI scenario (Table 24.1). These findings support the conclusion that climate change and many of its impacts scale with the quantity and timing of emissions. As such, they represent a point of departure for assessing the outcome of changes in emission trajectories driven by climate-specific policies, and the extent to which lower emissions can reduce the likelihood and thus the risks of "dangerous anthropogenic interference with the climate system" as the UNFCCC states.

Acknowledgements

The analysis of climate change impacts for California presented here includes the work of colleagues Dan Cayan, Ed Maurer, Norman Miller, Suzanne Moser, Kimberly Nicholas Cahill, Elsa Cleland, Larry Dale, Ray Drapek, Michael Hanemann, Larry Kalkstein, James Lenihan, Claire Lunch, Ron Neilson, Scott Sheridan and Julia Verville. This manuscript benefited greatly from the editorial comments of Tom Wigley as well as two anonymous reviewers.

---

[5]The latest IPCC AR4 Working Group 1 simulations from 20+ AOGCMs (including stabilisation scenarios, commitment scenarios, fixed percentage CO_2 increases, and various SRES scenarios) in addition to a large library of PCM and CCSM simulations and scientific visualization tools are available by registration via the Earth System Grid portal (www.earthsystemgrid.org).

REFERENCES

1. Nakićenović, N. and Swart R., 2000. Special Report on Emissions Scenarios. Cambridge University Press, Cambridge, UK.
2. Hayhoe, K., Cayan D., Field C., Frumhoff P., Maurer E., Miller N., Kalkstein L., Lenihan J., Lunch C., Neilson R., Sheridan S. & Verville J., 2004a. Emission Scenarios, Climate Change and Impacts on California. Proc. Natl. Acad. Sci., 101, 12422–12427.

3. United Nations (1990) United Nations Framework Convention on Climate Change.

4. Wood, A., Maurer E., Kumar A. & Lettenmaier D., 2002. Long-range experimental hydrologic forecasting for the eastern United States, J. Geophys. Res., 107, art. no. 4429.

5. Dettinger, M., Cayan D., Meyer M. & Jeton A., 2004. Simulated hydrologic responses to climate variations and change in the Merced, Carson, and American River basins, Sierra Nevada, California, 1900–2099. Clim. Change 62:283–317.

6. Wilkinson, R., Clarke K., Goodchild M., Reichman J. & Dozier J., 2002. The Potential Consequences of Climate Variability and Change for California: The California Regional Assessment, United States GCRP. Available online at: www.usgcrp.gov/usgcrp/nacc/california.htm

7. Wilson, T., Williams L., Smith J. & Mendelsohn. R., 2003. Global Climate Change and California: Potential Implications for Ecosystems, Health, and the Economy. Prepared for California Energy Commission by Electric Power Research Institute. Available online at: www.energy.ca.gov/pier/reports/500-03-058cf.html

8. National Climatic Data Center, 2005. Climate at a Glance – California State Summary. Available online at: http://www.ncdc.noaa.gov/oa/climate/research/cag3/ca.html

9. Sloan, L. et al., 2005. Have urban and agricultural land use changes affected California's climate? Presented at the Second Annual Climate Change Research Conference, Sacramento, Sept. 2005. Available online at: www.climatechange.ca.gov/events/2005_conference/presentations/2005-09-14/2005-09-14_SLOAN.PDF

10. Bakun, A., 1990. Global Climate Change and Intensification of Coastal Ocean Upwelling. Science 247, 198–201.

11. Diffenbaugh, N.S., Snyder M.A. & Sloan L.C., 2004. Could CO_2-induced land-cover feedbacks alter near-shore upwelling regimes? PNAS 101, 27–32.

12. Snyder, M.A., Sloan L.C., Diffenbaugh N.S. & Bell J.L., 2003. Future climate change and upwelling in the California Current. Geophysical Research Letters 30, 1823, doi: 10.1029/2003GL017647 (2003).

13. Hayhoe, K., Kalkstein L., Moser S. & Miller N., 2004b. Rising Heat and Risks to Human Health: Technical Appendix. A report of the Union of Concerned Scientists, Available online at: www.climatechoices.org

14. Michael D. Dettinger, 2005. From climate-change spaghetti to climate-change distributions for 21st Century California. San Francisco Estuary and Watershed Science. Vol. 3, Issue 1 (March 2005), Article 4. Available online at: http://repositories.cdlib.org/jmie/sfews/vol3/iss1/art4

15. Westerling, A.L. and Swetnam T.W., 2003. Interannual to Decadal Drought and Wildfire in the Western United States. EOS, 84, 545, 554–555.

16. Fried, J.S., Torn M.S. & Mills E., 2004. The impact of climate change on wildfire severity: A regional forecast for Northern California. Climatic Change, 64, 169–191.

17. Cayan, D., Barnett T. & Riddle L., 1997. The 1997–1998 El Niño – Increased potential for winter storms, flooding and coastal impacts in California. Available online at: meteora.ucsd.edu/elnino/ENSO.html

18. Westerling, A., 2001. Climate variability and large storm surge on the Pacific coast of the United States. Presented at the 81st Annual Meeting of the American Meteorological Society, January, 2001.

19. Bromirski, P.D., Flick R.E. & Cayan D.R., 2003. Storminess Variability along the California Coast. J. Clim. 16, 982–993.

20. Trenberth, K. & Hoar T., 1997. El Niño and climate change. Geophys. Res. Lett. 24, 3057–3060.

21. Tsonis, A., Hunt A. & Elsner J., 2003. On the relation between ENSO and global climate change, Met. & Atmos. Phys., 84, 229–242.

22. West Coast Governors' Global Warming Initiative, 2004. Staff Recommendations to the Governors. Available online at: http://www.climatechange.ca.gov/westcoast/documents/

23. Baker, A., 2005. Public Briefing of the Executive Order S-3-05. Available online at: http://www.climatechange.ca.gov/climate_action_team/meetings/2005-07-28_meeting/index.html

24. The Governor of the State of California, 2005. Executive Order S-3-05. Available online at: http://www.dot.ca.gov/hq/energy/ExecOrderS-3-05.htm

25. Defra, 2003. The Energy White Paper. Available online at: www.dti.gov.uk/energy/whitepaper/index.shtml

26. European Environment Agency, 2004. Impacts of Europe's Changing Climate. Available online at: reports.eea.eu.int/climate_report_2_2004/en

27. Mastrandrea, M. & Schneider S., 2001. Integrated Assessment of Abrupt Climatic Changes, Clim. Policy 1, 433–449.

28. Wigley, T.M.L., 2004: Choosing a stabilization target for CO_2. Climatic Change 67, 1–11.

29. Jones, P., Amador J., Campos M., Hayhoe K., Marin M., Romero J. & Fischlin A., 2004. Generating climate change scenarios at high resolution for impact studies and adaptation: Focus on developing countries, Proceedings of the International Workshop on Adaptation to Climate Change, Sustainable Livelihoods and Biological Diversity, in press.

30. Cash, D. & Moser S., 2000. Linking global and local scales: designing dynamic assessment and management processes. Global Env. Change, 10, 109–120.

CHAPTER 25

Impacts of Climate Change in the Tropics: The African Experience

Anthony Nyong
Department of Geography and Planning, Faculty of Environmental Sciences, University of Jos, Nigeria

Isabelle Niang-Diop
ENDA, Senegal

ABSTRACT: Global warming is already happening as shown from instrumental records, with its impact being felt most by the world's poorest people, particularly those in Africa. Food production, water supplies, public health, and people's livelihoods are being threatened. If current climate trends continue, climate models predict that by 2050 Sub-Saharan Africa will be warmer by 0.5°C–2.0°C and drier with 10% less rainfall in the interior and with water loss exacerbated by higher evaporation rates. An increase in the frequency of extreme events such as drought and floods is also predicted, along with a shift in their seasonal patterns. It is important that efforts be made to stabilize greenhouse gas in the atmosphere at such a level that would reduce the current trend in global warming and its negative effects. This requires that the industrialized nations cut their emissions of greenhouse gases considerably by the middle of this century. Even if this cut is achievable, sea level rise and global warming would continue to increase over centuries because of the inertia of the earth systems. Adaptation is therefore necessary to complement mitigation efforts, particularly for the developing countries. It is imperative that Africa's adaptive capacity be enhanced to handle climate-related risks, building on the continent's rich indigenous knowledge systems and experiences. Particular attention should be paid to the development of appropriate technology for adaptation, while removing the bottlenecks that hinder the operation of the various adaptation funds that have been created under the UNFCCC.

25.1 Introduction

Article 2 of the United Nations Framework Convention on Climate Change (UNFCCC) defines the ultimate objective of UNFCCC as 'the stabilization of greenhouse gas concentrations in the atmosphere at a level that would prevent dangerous anthropogenic interference with the climate system. Such a level should be achieved within a time-frame sufficient to allow ecosystems to adapt naturally to climate change, to ensure that food production is not threatened and to enable economic development to proceed in a sustainable manner' (UNFCCC, 1992). Issues that are still to be defined and agreed upon, based on the contents of Article 2 are:

- What stabilized *concentration level* should/may be adopted?
- What is *dangerous* anthropogenic interference?
- What *time frame* is necessary to achieve the objective?

To stabilize the atmospheric CO_2 concentration at 550 ppm or lower as proposed by the European Union and other Parties to the UNFCCC requires significant cuts in emissions of greenhouse gases by industrialised countries (Hare, 2003). Even if this stabilization is achieved, sea level rise and global warming would continue to increase over centuries because of the inertia of the earth systems. The

implication is that global warming will continue to be a burden, particularly to Africa, which is considered the most vulnerable region to climate change, due to the extreme poverty of many Africans, frequent natural disasters such as droughts and floods, and its over-dependence on agriculture (IPCC, 2001). Some of these stresses are to a large extent exacerbated by climate change.

Much of Africa's physical environment is very harsh, consisting of either very arid regions or very wet coastal areas. Droughts and floods will be the two most important avenues through which the impacts of global warming will be felt in Africa. Droughts have mainly affected the Sahel, the Horn of Africa and Southern Africa where about 40% of Africa's population lives. Floods are recurrent in some countries and even countries located in dry areas (Tunisia, Egypt, and Somalia) have not been flood-safe (Kabat *et al.*, 2002). Some countries even experience both floods and droughts in the same year.

Besides the physical environment, much of the continent is grappling with other crises arising from structural adjustment policies, trade liberalization, globalization, conflicts, poor governance, malnutrition, poverty and a high disease burden, particularly from malaria and HIV/AIDS. Africa has one of the highest population growth rates in the world, with a current rate of about 2.4% per year (UNPOP, 2004). An unmanaged population growth resulting in a sustained pressure on natural resources will act as a catalyst for the propagation of the adverse impacts of climate change in Africa. Recent estimates of poverty and human development show that almost all the African countries are classified as being in the low human development category. Of the 49 countries that make up the Least Developed Countries (LDCs), 33 (70%) are in Africa, while the rest are in Asia. Compared with least developed countries in Asia, poverty in African LDCs is widening and deepening. The number of Africans living in extreme poverty in these countries rose dramatically from 89.6 million to 233.5 million between 1960 and 1990 (Ben-Ari, 2002). Exposures to multiple risks and stresses, such as those outlined above, increase the vulnerability of Africa and hamper its socio-economic development, even without the addition of climate change. Africa has experienced more human conflicts in recent years than any other continent. Human conflicts for instance, in semi-arid parts of Africa (covering a third of the continent) are largely driven by water sharing between pastoralists and agriculturalists during periods of drought. (For details on climate-induced conflicts in the West African Sahel, see Nyong and Fiki, 2005, Fiki and Lee, 2004.)

25.2 Africa's Changing Climate

Observational records show that the continent of Africa has been warming through the 20th century at the rate of about 0.5°C per century with slightly larger warming in the June–November seasons than in December–May (Hulme *et al.*, 2001). The warming trend observed is consistent with changes in the global climate and is likely to be a signal of the anthropogenic greenhouse effect. Since the mid-1970s, precipitation has declined by about −2.4 ± 1.3% per decade in tropical rainforests in Africa, this rate being stronger in West Africa (−4.2 ± 1.2% per decade) and in north Congo (−3.2 ± 2.2% per decade). Overall, in the West Africa/north Congo tropical rainforest belt rainfall levels were 10% lower in the period 1968–1997 than in the period 1931–1960 (Nicholson *et al.*, 2001).

Results from GCMs based on SRES show that by the 2050s and 2080s, northern and southern Africa could experience a warming of up to 7° and 5°C respectively (Hulme *et al.*, 2001). Modelled results of the effects of CO_2 stabilization at 550 ppm (by 2150) and 750 ppm (by 2250) for the Sahel region show a reduction in warming of 2.9° and 2.1°C respectively, and stabilization at 750 ppm could delay warming by around 40 years across Africa (Arnell *et al.*, 2002). Climate models predict a decline in summer precipitation in southern and northern Africa. In West and East Africa, a slight increase is predicted in precipitation up to 15° Latitude in the dry season. In Africa, climate change is not only about global warming; it is also associated with changes in climate variability and changes in the frequency and magnitude of extreme events, such as more droughts and floods.

Several models project not just higher average temperatures and lower rainfall in semi arid regions of Africa, but an increasing probability of El Niño – Southern Oscillation events (ENSO), which have become more frequent, persistent and intense since the mid-1970s (IPCC, 2001; Devereux and Edwards, 2004). The two regions in Africa with the most dominant ENSO influences are in eastern equatorial Africa during the short October–November rainy season and in southeastern Africa during the main November–February wet season. The recurrent drying of the Sahel region since the 1970s seems to be linked with a positive trend in the equatorial Indian Ocean sea surface temperatures (Giannini *et al.*, 2003).

25.3 Vulnerability to Climate Variability and Change

Africa is highly vulnerable to climate change and several of the studies on which this conclusion is based have largely looked at the impacts from abrupt climatic events. Little attention has been paid to the possibility of imperceptible changes in climate accumulating until thresholds are crossed that could cause entire systems to collapse. This risk is greatest in Africa where much of the livelihood and socioeconomic systems depend on rainfall-sustained agro-ecology. In the following sections, we review the vulnerability to and impact of current and projected future climate variability and change on Africa in major sectors: water resources, agriculture, coastal zones, health and ecosystems and livelihoods.

25.3.1 *Water Resources*

Rainfall is highly variable in the Sahel region, Sudan, Burkina Faso, Cape Verde, Chad, Gambia, Guinea-Bissau, Mali, Mauritania, Niger, Senegal and parts of the Greater Horn of Africa and most stable in East and West Gulf of Guinea: Cameroon, Central African Republic, Equatorial Guinea, Gabon, Guinea, Liberia, Nigeria, and Sierra Leone. Several African countries have experienced droughts since the 1970s which have led to a general decrease in river discharges, and a consequent reduction in lake areas. Some regions in Africa are much more generally water-stressed than others. Even without climate change, and based on SRES population projections, countries in eastern and northern Africa are more water-stressed, particularly under the A2 scenario. Table 25.1 shows the numbers of people in each of the regions in Africa living in watersheds with less than 1000 m^3/capita/year, both in millions and as a percentage of total regional population, in the absence of climate change. Introducing the effects of climate change, models predict a reduction in the global total number of people living in water-stressed watersheds. According to Arnell (2004) more people in western Africa might experience a reduction in water stress. In northern, central and southern

Table 25.1 People (millions) living in water-stressed watersheds by regions in Africa without Climate Change.

Region	1995	%	2025 A1/B1	2025 A2	2025 B2	2055 A1/B1	2055 A2	2055 B2	2085 A1/B1	2085 A2	2085 B2
Northern Africa	124.4	94	209.9	239.9	201.4	269	403.1	264.6	302.2	603.0	310.7
Western Africa	0.1	0	34.3	35.8	30.8	89.8	118.4	113.3	70.1	217.6	245.7
Central Africa	0	0	25.7	26.8	2.4	35	41.5	38.0	33.5	45.2	46.0
Eastern Africa	6.5	5	34.2	40.6	27.0	52.3	186.7	255.4	74.2	283.7	316.1
Southern Africa	3.1	2	35.6	37.6	32.9	50.3	60.8	126.7	48.2	100.5	187.6

Source: Arnell, N.W. (2004): 'Climate change and global water resources: SRES emissions and socio-economic scenarios', *Global Environmental Change*, 14:31–52.

Africa, considerably more people will be adversely affected by climate change, experiencing an increase in water stress. In southern Africa, the area having water shortages will have increased by 29% by 2050, the countries most affected being Mozambique, Tanzania and South Africa (Schultz *et al.*, 2001).

In the Nile region, most scenarios of water availability estimate a decrease in river flow up to more than 75% by the year 2100. Reductions in annual Nile flows in excess of 20% will induce an interruption of normal irrigation practices and reduce agricultural productivity (Dixon *et al.*, 2003). The likelihood of this situation is greater than 50% by the year 2020. Besides reducing agricultural productivity, it could also lead to conflicts as the current allocation of water that was negotiated during periods of high flow would become untenable and need further negotiations with neighbouring countries.

25.3.2 Agriculture

Agriculture is a very important economic sector in Africa and accounts for more than 40% of total export earnings, one third of the national income, employs between 70% and 90% of the total labor force and supplies up to 50% of household food requirements and up to 50% of household incomes (Desanker, 2001. The overall economic growth and development in Africa depend primarily on the performance of agriculture in driving incomes and employment. Unfortunately, agriculture and agro-ecological systems are most vulnerable to climate change, especially in Africa. It is estimated that by the 2080s, the net balance of changes in cereal-production potential for sub-Saharan Africa will very likely be negative, with net losses of up to 12% of the region's current production (Gitay *et al.*, 2001). It is also estimated that up to 40% of sub-Saharan countries will lose a rather substantial share of their agricultural resources (implying a loss at 1990 prices of US$10–60 billion). The distribution of these losses is not uniform as certain countries will be affected more than others. Even with CO_2 stabilization, it is estimated that cereal crop yields in Africa will still decrease by 2.5–5% by the years 2080s (Arnell *et al.*, 2002).

Livestock is closely linked to rainfall and changes in annual precipitation, therefore changes in rain-fed livestock numbers in Africa will be directly proportional to changes in annual precipitation. Given that several GCMs predict a decrease in precipitation in the order of 10–20% in the main semi-arid zones of Africa where most livestock is held, there is the possibility that climate change will have a negative impact on pastoral livelihoods through a reduction in water availability and biomass (IPCC, 2001). Although it is generally believed that this reduction will be potentially balanced by positive effects of CO_2 enrichment, substituting grassland area by trees (with enrichment of CO_2) may place additional stress on livelihoods derived from rangelands (Desanker *et al.*, 2001).

Significant portions of people in African countries depend on fish for protein, thus near-term impacts on the fishery sector could affect human nutrition and health. Climate change is projected to alter freshwater temperatures, water chemistry and circulation, which could affect fisheries production (Costa, 1990). Fisheries will also be affected by sea level rise through either coastal erosion or inundation, which could destroy fisheries infrastructures and fishing villages and could also affect important ecosystems involved in reproduction and larval growth of fishes. In Congo, it is estimated that more than 50% of the fish coming from the Conkouati lagoon could disappear due to an increased penetration of sea water in the lagoon (République du Congo, 2001). In Cameroon, with a potential inundation of low lying areas in the estuary (for a 1.0 m sea level rise), about 53% of all the fishing villages could be displaced inducing the migration of 6,000 fishermen (Republic of Cameroon/UNEP, 1998). In Kenya, a decrease in fish catch (between 10–43%) was observed following the 1998 coral mortality induced by a strong ENSO event (McClanahan *et al.*, 2002).

The food security threat posed by climate change is greatest for Africa, where agricultural yields and per capita food production have been steadily declining, and where population growth will double the demand for food, water and forage in the next 30 years (Davidson *et al.*, 2003). It is estimated that climate change will place an additional 80–125 million people (±10 million) at risk of hunger by the 2080s, 70–80% of whom will be in Africa (Parry *et al.*, 1999). It has been found that ENSO events have been closely correlated with all weather-related famines in the Horn of Africa for at least the past 200 years (Davis, 2001).

25.3.3 *Health*

In recent years, it has become clear that climate change and variability will have direct and indirect impacts on diseases that are endemic in Africa. Vector-borne diseases such as malaria and Rift Valley Fever increase dramatically during periods of above normal temperature and rainfall (Githeko and Ndegwa, 2001). Cholera, a water- and food-borne disease, has been known to cause large-scale severe epidemics during periods of strong El Niño. For instance, following the 1997/98 El Niño event, malaria, Rift Valley Fever and cholera outbreaks were recorded in many countries in East Africa (WHO, 1998). Meningitis, a disease associated with low humidity, causes epidemics before the rains in West Africa, the Sahel and recently in eastern Africa (Desanker *et al.*, 2001). Under climate change, the meningitis belt in the drier parts of West and Central Africa will expand to the eastern region of the continent. Africa accounts for about 85% of all deaths and diseases associated with malaria (Van Lieshout *et al.*, 2004). Flooding could facilitate breeding of malaria vectors and consequently malaria transmission in arid areas. In South Africa, it is estimated that the area suitable for malaria will double by 2100 and that 7.8 million people will be at risk (5.2 million being people who have never experienced malaria) (Republic of South Africa, 2000). Climate change could place an additional population of 21–67 million people in Africa at risk of malaria epidemics by the 2080s, the greatest population at risk being located in eastern and southern Africa, particularly in the highlands (Van Lieshout *et al.*, 2004).

25.3.4 *Ecosystems*

In Africa, coastal zones are characterized by the presence of high productivity ecosystems (mangroves, estuaries, deltas, coral reefs), which constitute the basis for important economic activities like tourism and fisheries. Africa's coastal zone is also characterized by a concentration of populations and industries. For example, 40% of the population of West Africa lives in coastal cities and it is expected that the 500 km coast between Accra (Ghana) and the Niger delta (Nigeria) will become a continuous urban megalopolis with more than 50 million people by 2020. With this concentration of human populations come problems of pollution, pressure on infrastructure, resource and land use conflicts, and overexploitation of ecosystems and species. In contrast to other mega cities in the developed world that would be more resilient to climate change, the African mega cities, due to the concentration of poor populations in potentially-hazardous areas, will be less resilient to climate change (Klein *et al.*, 2002). In countries where important agricultural products come from

the coastal zones, potential losses in crop revenues are another concern since they could be at risk of inundation and salinization of soils. Economic values at risk within these coastal zones represent a high percentage of the GDP (between 5.8 and 54.2%).

Ecosystems are not only the foundation of the economy of most African countries, but also contain a number of plants and animals which constitute about 20% of all known species (Biggs *et al.*, 2004). With climate change, most of these species are threatened. About one-fifth of southern African bird species migrate on a seasonal basis within Africa and a further one-tenth migrate annually between Africa and the rest of the world (Hockey, 2000). If climatic conditions or very specific habitat conditions at either terminus of these migratory routes change beyond the tolerance of the species involved, significant losses of biodiversity could result. In South Africa, isolated plant communities, particularly at high altitudes will be affected by temperature rise. Changes in the seasonal distribution of rainfall could affect fire regimes and plant phenological cues, especially in the southern Cape (Tyson *et al.*, 2002). Models indicate that in South Africa, the savanna and the Nama-Karoo biomes will advance at the expense of the grasslands. In Malawi, climate change could induce a decline of nyala (*Tragelaphu*s) and zebra (*Equiferus*) in the Lengwe and Nyika national parks because these species may be unable to adapt to climate-induced habitat changes (Dixon *et al.*, 2003).

Coral reefs play a crucial role in fisheries production and in protecting the coastline from wave action and erosion. In Africa, coral reefs are mainly present along the Indian coasts (East Africa and Indian Ocean islands) and also in the Cape Verde and Sao Tome and Principe islands. Of the 18 richest endemic coral reef centers, six are located in Africa and will need specific protection and conservation measures (Roberts *et al.*, 2002). Major coral bleaching events have occurred in recent times. The last major event on the eastern coast of Africa was in 1998, which resulted in an average of 30% mortality of corals in the western Indian Ocean region, and for Mombasa and Zanzibar decreases in tourism value of coral reefs were estimated to be about US$ 12–18 million (Payet and Obura, 2004). Other potential consequences of coral bleaching could be an increase in the number of people affected by toxins due to the consumption of contaminated marine animals.

25.4 Adapting to Climate Change

Responding to climate change, as contained in Article 2 of the UNFCCC encompasses two strategies: (i) mitigation: controlling greenhouse gases to stabilize climate change at an acceptable limit, and (ii) adaptation: adjustments to the impact of climate change given existing levels of greenhouse gasses in the atmosphere. While the continent has signed up to international agreements to reduce the emission of greenhouse gas, the most viable option for dealing with the impacts of climate change is adaptation. Successful adaptation depends upon technological advances, institutional arrangements, availability of financing, and information exchange.

African governments at the country and regional levels have embarked on programmes of actions that aim at reducing the impacts of and adapting to climate change in Africa. Some of these programmes have been externally funded or assisted by international organizations. Others have been funded through national government budgets. Regional and Africa-wide collaborations have also been established to deal with the issues of climate change. Despite these efforts, virtually all assessments conclude that Africa has the lowest adaptive capacity, hence its high vulnerability to climate change impacts.

The process of adaptation comprises a number of different activities, carried out by different public and private agents. Most simply put, one can distinguish between facilitating adaptation and implementing adaptation. Facilitating adaptation includes developing information and raising awareness, removing barriers to adaptation, making available financial and other resources for adaptation and otherwise enhancing adaptive capacity. Implementing adaptation includes making the actual changes in operational practices and behaviour, and installing and operating new technologies. While adaptation can be facilitated by external agencies, adaptation should be implemented by the local people, from the scale of the household to the national/regional levels. At the national level, this can be done through mainstreaming adaptation into national developmental policies and programmes, as financial constraints may not allow for the creation of specific climate change projects. Recognizing the local and national domains of adaptation activities is a first step towards a successful and sustainable adaptation.

25.5 The Way Forward

Capacity Building: It is important to identify and strengthen the capacity of the various regional institutions whose mandates include climate and other environmental issues, so as to carry out those mandates. Just as capacity is not static but requires continuous renewal, so is capacity building a continual process of improvement within an individual, organization or institution; not a one-time event. It is essentially an internal process, which only may be enhanced or accelerated by outside assistance, for instance by donors. Capacity building emphasizes the need to build on what exists, to utilize and strengthen existing capacities, rather than arbitrarily thinking of starting from scratch. Building Africa's capacity to adapt to climate change is about complex processes of changing people's mindsets and behaviour and introducing more efficient technologies and systems. This has two important implications, as emphasized widely in the literature.

First, capacity building takes a long time and requires a long-term commitment from all involved. Second, success of capacity building efforts should not be measured in terms of disbursements or outputs with little attention to sustainability, but in terms of deliverable targets in set timelines.

Use of Appropriate Technology: As far as possible low cost and appropriate technology options must be favoured. In Nigeria, for example, the practice of deepening existing wells in a river floodplain, and adopting simple rainwater harvesting technologies showed good results, indicating that good scientific information and low cost technology can provide appropriate adaptation that can also increase the level of resources management (Tarhule and Woo, 2002). Better than investing in completely new technologies, there are opportunities to enhance existing traditional technologies. The development and testing of these varieties should not be restricted to controlled experimental farms, but should be tested in real-world situations with local farmers. It is also recognized that investments in education, road infrastructures, and agricultural research, which need a constant involvement of the public sector (through policy actions and investments) could increase agricultural productivity while reducing poverty in rural areas. In fact, new techniques can be adopted if they can be shown to increase not just economic, but social benefits to the farmers.

Integration of Indigenous Knowledge: Indigenous knowledge has provided local communities in Africa with the capability of dealing with past and present vulnerabilities to climatic extremes and other stresses. There are documented successful traditional farming techniques to conserve biodiversity while managing soils so that the soil-plant relationship is maintained. Indigenous knowledge should be integrated into formal climate change mitigation and adaptation strategies. In as much as we acknowledge the importance of indigenous practices in climate change mitigation and adaptation, they should not be developed as substitutes of modern techniques. It is important that the two are complements and learn from each other in order to produce 'best practices' for mitigation and adaptation (Nyong *et al.*, 2005). A 'best practice' is the result of articulating indigenous knowledge with modern techniques – a mix that proves more valuable than either one on its own. The interaction between the two different systems of knowledge can also create a mechanism of dialogue between local populations and climate change professionals, which can be meaningful for the design of projects that reflect people's real aspirations and actively involve the affected communities.

Funding Adaptation: While several adaptation funds have been set up to assist developing countries to adapt to climate change, the mechanism for drawing from these funds are still vague. It is therefore important that such funds be made operational and easily accessible to Africa.

25.6 Conclusion

Africa is highly vulnerable to climate change and its vulnerability results largely from the continent's dependence on agriculture. Africa's high vulnerability to climate change is exacerbated by other stresses such as poverty, wars and conflicts, limited technological development, a high disease burden and a rapid population growth rate. The impacts of climate change in Africa, interacting with these other stresses, are capable of hindering Africa's development.

There must be substantial and genuine reductions in greenhouse gas emissions by the principal emitters, complemented by effective and sustainable adaptation. This implies that Africa's capacity to adapt to climate change must be strengthened through capacity building and the facilitation of adaptation through a fusion of top-down and bottom-up approaches. Local communities in Africa have lived with large-scale variations in climate change and have developed indigenous knowledge systems that have enabled them to cope with their impacts. One major factor that has served as an impediment to successful indigenous adaptation is poverty. Implementing successful and sustainable adaptation would require among other things, the integration of indigenous knowledge systems into Western adaptation science.

The bottleneck that surrounds the operation of the various adaptation funds needs to be removed so that poor countries, including those in Africa, can begin to draw upon such funds to finance adaptation. Experience has shown that the cost of adaptation is usually much lower than the losses from a climate-induced disaster. Planning for adaptation is a proactive way of dealing with disasters, as it shifts emphasis from disaster management, which is very expensive, to disaster reduction.

REFERENCES

Arnell, N.W. (2004): 'Climate change and global water resources: SRES emissions and socio-economic scenarios', *Global Environmental Change*, 14:31–52.

Arnell, N.W., Cannell, M.G.R., Hulme, M., Kovats, R.S., Michell, J.F.B., Nicholls, R.J., Parry, M., Livermore, M.J., and A. White (2002): 'The consequences of CO_2 stabilisation for the impacts of climate change', *Climatic Change*, 53(4):413–446.

Ben-Ari, N. (2002): 'Poverty is worsening in African LDCs', *Africa Recovery, United Nations*, 16(2–3):9.

Biggs, R., Bohensky, E., Desanker, P., Fabricius, C., Lynam, T., Misselhorn, A.A., Musvoto, C., Mutale, M., Reyers, B., Scholes, R.J., Shikongo, S., and A.S. Van Jaarsveld (2004): Nature Supporting People: *The Southern African Millennium Ecosystem Assessment. Integrated Report*. Council for Scientific and Industrial Research, Pretoria, South Africa.

Costa, M.J. (1990): 'Expected effects of temperature changes on estuarine fish populations'. In: J.J. Beukema et al. (eds.) *Expected Effects of Climatic Change on Marine Ecosystems*, Kluwer Academic Publishers, pp 99–103.

Davis, M. (2001): *Late Victorian Holocausts: El Niño, Famines and the Making of the Third World*, New York, Verso.

Desanker, P., Magadza, C., Allali, A., Basalirwa, C., Boko, M., Dieudonne, G., Downing, T., Dube, P.O., Githeko, A., Gihendu, M.,

Gonzalez, P., Gwary, D., Jallow, B., Nwafor, J., Scholes, R. (2001). Africa. In: McCarty, J.J., Canziani, O.F., Leary, N.A., Dokken, D.J., and K.S. White (eds). *Climate change 2001: Impacts, adaptation, and vulnerability*. Cambridge University Press, 487–531.

Devereux, S. and J. Edwards (2004): 'Climate Change and Food Security', *Climate Change and Development*, 35(3): 22–30.

Dixon, R.K., Smith, J., and S. Guill (2003): 'Life on the edge: vulnerability and adaptation of African ecosystems to global climate change'. *Mitigation and Adaptation Strategies for Global Change*, 8(2):93–113.

Fiki, C. and B. Lee (2004): Conflict generation, conflict management and self-organizing capabilities in drought-prone rural communities in north-eastern Nigeria: A case study. Journal of Social Development In Africa, 19(2):25–48.

Giannini, A., R. Saravanan, and P. Chang, (2003), Oceanic forcing of Sahel rainfall on interannual to interdecadal time scales. *Science*, 302, 1027–1030.

Gitay, H., Brown, S., Easterling, W., Jallow, B., Antle, J., Apps, M., Beamish, R., Chapin, T., Cramer, W., Frangi, J., Laine, J., Erda, L., Magnuson, J., Noble, I., Price, J., Prowse, T., Root, T., Schulze, E., Sirotenko, O., Sohngen, B., Soussana, J. (2001): 'Ecosystems and their goods and services'. In: McCarty, J.J., Canziani, O.F., Leary, N.A., Dokken, D.J., and K.S. White, (eds). *Climate change 2001: Impacts, adaptation, and vulnerability*. Cambridge University Press, 235–342.

Githeko, A.K., and W. Ndegwa (2001): 'Predicting malaria epidemics using climate data in Kenyan highlands: a tool for decision makers'. *Global Change and Human Health*, 2:54–63.

Hare, W. (2003): *Assessment of Knowledge on Impacts of Climate Change – Contribution to the Specification of Art. 2 of the UNFCCC*. WBGU, Potsdam, Berlin.

Hockey, P.A.R. (2000): 'Patterns and correlates of bird migrations in sub-Saharan Africa'. *Journal of Birds*, 100:401–417.

Hulme, M., Doherty, R., Ngara, T., New, M. and D. Lister (2001): 'African Climate Change: 1900–2100'. *Climate Research*, 17:145–168.

IPCC (2001): *Climate change 2001: Synthesis report. A contribution of Working Groups I, II and III to the Third Assessment Report of the Intergovernmental Panel on Climate Change*. Cambridge University Press, Cambridge, 398 pp.

Kabat, P., Schulze, R.E., Hellmuth, M.E., and J.A. Veraart (eds) (2002): 'Coping with impacts of climate variability and climate change in water management: A scoping paper'. *Dialogue on Water and Climate*, Wageningen, DWC-Report no. DWCSSO-01 (2002). 114 pp.

Klein, R.J.T., Nicholls, R.J., and F. Thomalla (2002): 'The resilience of coastal megacities to weather-related hazards: a review'. In: Kreimer, A., Arnold, M., Carlin, A. (eds) *Proceedings of the future of disaster risk: Building safer cities*. World Bank, Washington D.C., 111–137.

McClanahan, T., Maina, J., and L. Pet-Soede (2002): 'Effects of the 1998 Coral Mortality Event on Kenyan Coral Reefs and Fisheries'. *Ambio*, 31(7–8):543–550.

Nicholson, S.E. (2001): 'Climatic and environmental change in Africa during the last two centuries'. *Climate Research*, 17(2):123–144.

Nyong, A.O. and C. Fiki (2005): *Drought-Related Conflict Generation, Resolution and management in Sahelian Northern Nigeria*. Paper presented at the International Workshop on Climate Change and Human Security, Oslo, Norway, 21–23 June, 2005.

Nyong, A.O., Adesina, F. and B. Osman (2005): 'The Value of Indigenous Knowledge in Climate Change Mitigation and Adaptation Strategies in the African Sahel'. *Mitigation and Adaptation Strategies for Global Change* (In Press).

Nyong, A.O. and P.S. Kanaroglou (1999): 'Modelling Seasonal Variations in Domestic Water Demand in Rural Northern Nigeria: Implications for Policy'. *Environment and Planning A*. 34(4): 145–158.

Parry, M.L., Rosenzweig, C., Iglesias, A., Livermore, M., and G. Fischer (2004): 'Effects of climate change on global food production under SRES emissions and socio-economic scenarios'. *Global Environmental Change*, 14(1):53–67.

Payet, R., and D. Obura (2004): 'The negative impacts of human activities in the Eastern African region: An International waters perspective'. *Ambio*, 33(1–2):24–33.

Ramankutty, N., Foley, J.A., and J. Olejniczak (2002): 'People on the land: Changes in global population and croplands during the 20th century'. *Ambio*, 31(3):251–257.

Republic Of Cameroon/UNEP (1998). *Country Case study on climate change impacts and adaptation assessments.* Volume 2: Cameroon. UNEP, Nairobi.

Roberts, C.M., McLean, C.J., Veron, J.E.N., Hawkins, J.P., Allen, G.R., McAllister, D.E., Mittermeier, J.P., Schueler, F.W., Spalding, M., Wells, F., Vynne, C., and T.B. Werner (2002): 'Marine biodiversity hotspots and conservation priorities for tropical reefs'. *Science*, 295(5558):1280–1284.

Republique Du Congo (2001): *Communication Nationale Initiale.* Ministère de l'Industrie Minière et de l'Environnement. Brazzaville, 56 pp.

Republic Of South Africa (2000): *South Africa Initial National Communication to the United Nations Framework Convention on Climate Change.* Pretoria.

Schulze, R., Meigh, J. and M. Horan (2001): 'Present and potential future vulnerability of eastern and southern Africa's hydrology and water resources', *South African Journal of Science*, 97:150–160.

Shiklomanov, I.A. (ed.) (1998). *Comprehensive Assessment of the Freshwater Resources of the World.* WMO, Geneva, 88 pp.

Tarhule, A. and M.K. Woo (2002). 'Adaptations to the dynamics of rural water supply from natural resources: a village example in semi-arid Nigeria'. *Mitigation and Adaptation Strategies for Global Change*, 7(3):215–237.

Tyson, P., Odada, E., Schulze, R., Vogel, C. (2002). 'Regional-Global change linkages: Southern Africa'. In: Tyson, P., Fuchs, R., Fu, C., Lebel, L., Mitra, A.P., Odada, E., Perry, J., Steffen, W., Vriji, H. (eds). *Global-Regional Linkages in the Earth System.* Springer, Berlin, The IGBP Series, 3–73.

UNFCCC, (1992). *United Nations Framework Convention on Climate Change*, Geneva, Switzerland, United Nations.

United Nations Population Division (2004). WWW.un.org/esa/population/unpop.htm.

Van Lieshout, M., Kovats, R.S., Livermore, M.T.J., and P. Martens, (2004). 'Climate change and malaria: analysis of the SRES climate and socio-economic scenarios'. *Global Environmental Change*, 14(1): 87–99.

WHO (1998): *World Health Forum: The State of the World Health*, 1997 Report, 18, 248–260.

CHAPTER 26

Key Vulnerabilities and Critical Levels of Impacts in East and Southeast Asia

Hideo Harasawa
National Institute for Environmental Studies, Japan

26.1 Introduction

The Asian region, especially East and Southeast Asia, is very vulnerable to climate change. In the Third Assessment Report (TAR) of the Intergovernmental Panel on Climate Change (IPCC), the Asian region was divided into four sub-regions based on broad climatic and geographical features: boreal Asia, arid and semi-arid Asia, temperate Asia, and tropical Asia. The potential impacts and risks due to climate changes for these regions according to TAR are listed in Table 26.1 [1].

26.2 Observed Climate Change Impacts

26.2.1 *Climate Change Observed in Japan*

The IPCC TAR shows that the change in the global surface temperature for the last 100 years was 0.6°C.

Throughout the 20th century the Japan Meteorological Agency monitored the annual mean surface temperature at 17 observation sites where human responsibility for temperature change due to urbanization could be considered minimal. The temperature remained at lower levels until 1940, to turn sharply upwards in the 1960s and 1990s. The temperature increase in the 20th century was 1.0°C, which is above the mean global value of 0.6°C. Temperatures began to rise in the mid-1980s, and temperatures in the 1990s were clearly higher than before. The rise in temperature in urban areas over the past 100 years has been more than 2°C, with the rise in Tokyo reaching 3°C. This large rise in the urban areas is caused both by global warming and the heat island phenomenon [2].

Recent research shows an increasing trend in the number of days with a maximum temperature higher than 35°C, while during the 1990s there was a decreasing trend in

Table 26.1 Potential Impacts for Sectors in Asia [1].

Agriculture	Potential Impacts
Boreal Asia **Arid and semi-arid Asia** **Temperate Asia** **Tropical Asia**	• Crop production and aquaculture would be threatened by a combination of thermal and water stresses, sea-level rise, increased flooding, and strong winds associated with intense tropical cyclones.
Water resources	• Freshwater availability is expected to be highly vulnerable to anticipated climate change.
Boreal Asia	• Surface runoff would increase during spring and summer periods and would be pronounced in boreal Asia.
Arid and semi-arid Asia	• There could be drier conditions in arid and semi-arid Asia during summer, which could lead to more severe droughts.
Temperate Asia **Tropical Asia**	• Increased precipitation intensity, particularly during the summer monsoon, could increase flood-prone areas in temperate and tropical Asia.
Ecosystem and biodiversity	• The dangerous processes of permafrost degradation resulting from global warming would increase the vulnerability of many climate-dependent sectors affecting the economy in boreal Asia. • The frequency of forest fires is expected to increase in boreal Asia. • Climate change would exacerbate threats to biodiversity resulting from land-use/cover-change and population pressure in Asia.
Coastal Resources **Temperate Asia**	• The large deltas and low-lying coastal areas of Asia could be inundated by rises in sea-levels.
Tropical Asia	• Tropical cyclones could become more intense. Combined with rises in sea levels, this would enhance the risk of loss of life and of properties in low-lying coastal areas of cyclone-prone countries of Asia.
Human health **Temperate Asia** **Tropical Asia**	• Warmer and wetter conditions may increase the incidence of heat-related and infectious diseases in tropical and temperate Asia.

incidences of extremely low temperatures. In July 2004 many places in Japan experienced exceptionally high, record-breaking temperatures, and more than 600 heat-stroke cases were referred to hospitals in the Tokyo metropolitan area. Extremely high temperatures and heatwaves, a major factor in global warming or urban heat island effects, are now a matter of public concern but have not yet been examined using observed data. In July 2004, some locations in Japan experienced very heavy precipitation caused by the Baiu-front and typhoons, which caused much damage to society and to human activities. The general public and the insurance companies are now concerned that there may be a relationship between current global warming and extreme events.

26.2.2 Observed Impacts

The impacts of current global warming have been identified in the Asian region. The major impacts of climate change observed in China are as follows:

- Rise in sea level in coastal areas from 1–3 mm per year;
- Glacier area in Northwest China now reduced by 21%;
- Spring flowering of plants has advanced by 2–4 days;
- Area of drought disaster in the North China has expanded, resulting in severe agricultural losses; flood disaster area in South China also has expanded causing severe human and economic losses;
- Coral reef in maritime areas in Guangxi Province and Hainan Province has shown signs of albinism.

In Japan many impacts of global warming have been identified [3]. Living organisms and ecosystems detect warming and respond in various ways. Among the phenological observations conducted nationwide by the Meteorological Agency since 1953, the changes in the flowering date of the Japanese cherry (*Prunus yedoensis*) are particularly striking. These trees now flower five days earlier on average than they did 50 years ago. There are a number of other examples of warming.

- Rise in sea level of 2 mm per year observed over the past 30 years.
- Reduction in the thickness of snow in the ravine at Tsurugidake, Toyoma Prefecture.
- *Omiwatari* ('the divinity's pathway' observed at Lake Suwa in winter) has been seen only infrequently in recent years because of a series of warm winters.
- Decreased alpine flora in Hokkaido, the northernmost island in Japan, and other high mountains.
- Expanded distribution of southern broad-leaved evergreen trees such as the Chinese evergreen oak.
- The Nagasakiageha butterfly (*Papilio memnon thunbergii*), whose northern border was Kyushu and Shikoku Islands, appeared in Mie Prefecture in the 1990s and in the Tokyo area in the early 2000s.
- The southern tent spider, seen only in western Japan during the 1970s, appears in the Kanto Region in the 1980s.

- Expansion of the wintering grounds of the white-fronted goose to Hokkaido.
- Bleaching of the coral reef in the Okinawa islands.
- Ermine and grouse, with habitat on mountains such as Hakusan and Tateyama, shift to higher elevations.

These and other indications of diverse changes have been observed and demonstrate that the impacts of warming have begun to appear in the snow, ice, organisms, and ecosystems of Japan.

26.3 Future Impacts

Impact research has progressed since TAR. Some cases are introduced in this section.

26.3.1 Development of Regional Climate Scenario

Most impact research has used regional climate scenarios derived from a general circulation model (GCM) output, making it difficult to predict regional impacts for small countries in Asia. Fortunately, thanks to the development of GCMs and regional climate models (RCMs), impact research has been able to use regional climate model outputs in recent years. For example, the Global Warming Research Initiative launched by the Council for Science and Technology Policy of Japan makes close linkages between climate model research and impact/risk research. For impact studies the Japan Meteorological Research Institute has developed a regional climate model (RCM20) with a spatial resolution of about 20 km. Figure 26.1 shows current and future temperatures in Kanto District, including Tokyo, as predicted by RCM20 [4].

26.3.2 Future Heavy Rain Damage

Figure 26.2 shows the probability of heavy rain over the next 100 years using the RCM20. Using these simulated results, Wada [5] predicted the future probability of heavy rain using the statistical Gumbel method. According to

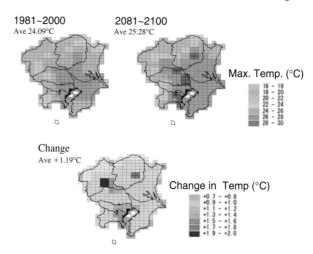

Figure 26.1 Change in maximum temperature in August in Kanto District using RCM20.

this figure, some areas will suffer from rainfall heavy enough to cause damage to human lives and assets.

From the recent GCM results developed by the National Institute for Environmental Studies (NIES), the Center for Climate System Research (CCSR), and the Japan Marine Science and Technology Center (JAMSTEC) using the Earth Simulator, Emori et al. [6] predicted that the amount of heavy rain (>100 mm/day) would double or triple by 2100.

26.3.3 *Impacts on Vegetation*

1. **Predicted impacts on vegetation in Japan**
Figure 26.3 shows the current and predicted distribution of beech forests, which are typical of the cool temperate zone and distributed widely in Japan. At the southern limit of their distribution, global warming will cause these forests to develop into evergreen forests. Matsui et al. [7] predict that about 90% of beech forests will disappear when the annual average temperature increases by about 4°C in 2090.

2. **Impact on Korean forest**
Figure 26.4 shows the extent of the damage to forest vegetation caused by climate change on the Korean peninsula by 2100. Whether current forest vegetation will be damaged or not can be ascertained by comparing the potential velocity of forest moving (VFM) with the velocity of vegetation zone shift that is estimated in the light of the climate change scenario. The VFM is assumed to be in the 0.25–2.0 km/year range. In the SRES A2 scenario, where the temperature increase is higher than in the other SRES scenarios, the extinction area will be 2.08% of the Korean peninsula if VFM is assumed to be 0.25 km/year.

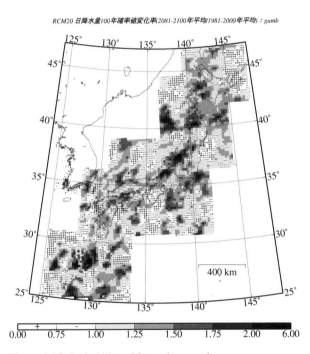

Figure 26.2 Probability of future heavy rain.

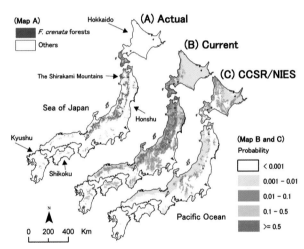

Figure 26.3 (A) Actual distribution and probability distributions of Japanese beech (*Fagus crenata*) (B) current climatic conditions and (C) the CCSR/NIES climate change scenario in the 2090s (modified from Matsui et al., 2004, by Dr. N. Tanaka (Forestry and Forest Products Research Institute).

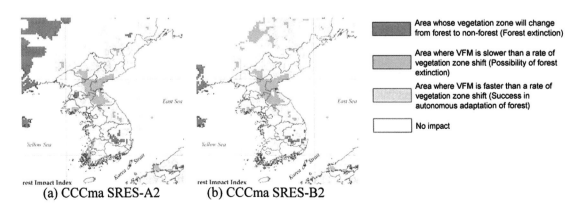

(a) CCCma SRES-A2   (b) CCCma SRES-B2

Figure 26.4 Impact of climate change on natural vegetation in Korea in 2100.

3. Vegetation change in Southeast Asia

Lasco et al. [8] predicted the potential impacts of climate change on Philippine forests quantitatively, using GIS and the Holdridge Life Zones. Three synthetic scenarios, each of precipitation (50%, 100%, and 200% increases) and air temperature (1°C, 1.5°C, and 2°C increases) were used. The research showed that dry forests (more than 1 million ha) are the most vulnerable to climate change. However, the wet and rain forest life zones will significantly expand as dry and moist forests become wetter.

26.3.4 *Impacts Due to Rise in Sea Levels [2]*

Coastal zones contain the habitats of organisms that are extremely vulnerable to climate change, one of these being coral reefs. Coral reefs grow upwards at a rate of about 40 cm per 100 years [3]. Thus, if the future rise in sea level exceeds that rate, reefs will not be able to keep pace. Even more serious is the rising sea-water temperature. The optimum water temperature for coral reef growth is 18–28°C. If a high water temperature of more than 30°C

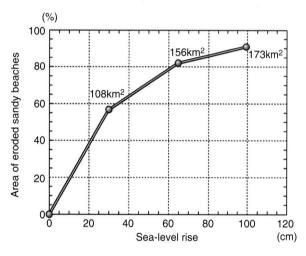

Figure 26.5 SLR and sandy beach [3].

continues, the algae coexisting symbiotically with the coral will separate from the reef, and the coral will discolor and die (coral bleaching). Coral bleaching occurred on a large scale in all parts of the world after the El Niño/La Niña of 1997–1998 [3]. If such phenomena occur more frequently in the future, they will likely cause serious damage to precious coral reef ecosystems.

Another major problem is the erosion of sandy coastlines. While the main causes of erosion are a decreasing sediment supply and the blocking of longshore sand transport, a rise in sea level will accelerate beach erosion [3]. If the sea level rises by 30 cm, it is estimated that at least 57% of the sandy beaches in Japan will be eroded (Figure 26.5). If the sea level rises by 65 cm to 1 m, sandy beach erosion will reach as much as 82%–90%.

Tidelands, which support rich ecological communities, are no exception. As tidelands are cut off from the hinterland by dikes or other structures, they cannot recede inland even if sea levels rise, and they are eroded. Thus tidelands, which have an extremely gentle mean slope of 1/300, will lose an area 150 m wide if there is a rise in sea level of 50 cm. If tidelands continue to disappear in this way, there is likely to be a huge impact on migratory birds such as snipes and plovers.

Combined impacts of high tide, storm surge, and sea level rise were predicted by Mimura et al. [9]. Figure 26.6 shows an example of their research. Some 15% of land in Bangladesh will disappear with a 1 m rise in sea level, and if a cyclone makes landfall, about half of the land will be inundated.

26.3.5 *Heightened Health Risks*

Rising temperatures will have a direct impact on human health, increasing the overall death rate from heat stroke and other disorders. The elderly and people with underlying medical conditions will be at greatest risk. Figure 26.7 shows the maximum daily temperature and the number of heatstroke patients transported to hospital in 2004 [10]. Worsening atmospheric pollution and epidemics of

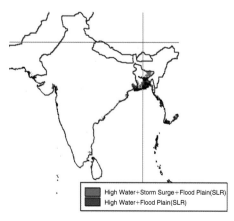

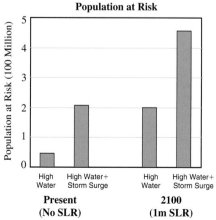

Figure 26.6 High tide, storm surge, and sea level rise (SLR).

vector-borne infectious diseases such as malaria and dengue are also of concern [3].

26.3.6 *Impacts on Industry, Energy, and Transportation*

Fukushima et al. [11] estimated future climate change impacts on the skiing industry in Japan. Using a model and the relationship between daily snow depth and the number of skiers, changes in the number of skiers in the seven ski areas were predicted for several scenarios with respect to air temperature changes; for example, a drop of more than 30% in visiting skiers was forecast for almost all skiing areas in Japan except the northern region (Hokkaido) and/or high altitude regions (in the center of the main island) should the air temperature increase by 3°C.

The direct impact of global warming on industry, and energy at the currently projected level resulting from climate fluctuations and changes (significance, speed, time period), is expected to be small because there is spare capacity to meet changing demands and long-term facility renewal is under way to deal with changing supply needs. Table 26.2 shows examples of the direct effects of changes

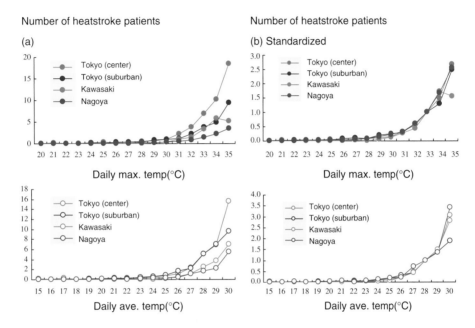

Figure 26.7 Heat wave and its impacts on Tokyo (Summer in 2004).

Table 26.2 Sensitivity of industries to climate change [12].

Element	Industry	Energy (electricity)
Change in amount and pattern of rainfall	Water demand (industrial/municipal water use)	Hydraulic power generation
	Water deficit/drought and food/product manufacturing	Management and control of dam facilities Impoundment of cooling water
Temperature increase	Cooling/warming apparatus Insulated houses/buildings Industries sensitive to seasonal change Winter: clothing, air conditioning Summer: summer products, beverages	Control of snowmelt water by damming and storage
Water temperature increase (sea water, fresh water)	Demand for natural gas to regulate aquaculture water	Decline in turbine efficiency (electricity generation) Increased adhesion of living creatures to turbines
Sea level rise	Location of industry for insurance Embankment building	Inundation of coastal facilities/equipment
Moisture		Demand for air conditioners/coolers
Typhoon	Factory/facility, Transportation/communication, Telecommunication industry	Typhoon-proof design; measures to address salt water intrusion, thunderstorm, snowfall Natural energy generation (wind energy)

due to global warming on the industrial and energy (electric power) sectors.

The relationship between the summer temperature and the consumption of summer products was analyzed (Harasawa and Nishioka, 2001). The authors indicate that if the mean June–August temperature rises by 1°C, consumption of summer products such as air conditioners, beer, soft drinks, clothing, and electricity will increase by about 5%. Table 26.3 shows a summary of the impacts of global warming on industries and energy sectors in Japan.

26.4 Future Research Needs

Research continues to clarify the impacts of global warming in an extremely broad range of areas. Table 26.4 lists the matrix of pertinent research activities to date in Japan. Numerous results have been obtained for Japan for terrestrial ecosystems; agriculture, forestry, and fisheries; and coastal zones.

In impact and risk studies a wide range of research is needed, including detection of emerging impacts,

Table 26.3 Impacts of climate change on industries and energy sectors [12].

Changes in Climate Parameters	Impacts
1°C temperature increase from June to August	About 5% increase in consumption of summer products
Extension of high temperature period	Increase in use of air conditioners and of consumption of beer, soft drinks, ice cream
Increase in thunder storms	Damage to information devices and facilities
1°C temperature increase in summer	Increase in electricity demand by about 5 million kW
	Increase in electricity demand in factories to enhance production
Increase in annual average temperature	Increase in household electricity consumption in southern Japan Decrease in total energy consumption for cooling, warming in northern Japan
Change in amount and pattern of rainfall	Hydro electric power generation, management and implementation of dams, cooling water management
1°C increase in cooling water	0.2–0.4% reduction in electricity generation electricity at thermal power plants 1%–2% reduction in nuclear power plant production

Table 26.4 Impacts research matrix[3].

	Water resources Water environment	Terrestrial ecosystem	Agriculture, forestry and fishery	Ocean environment	Coastal zones	Land preservation, disaster prevention, and human settlement	Industry Energy	Human health
Impact detection		OOO		OO	O			O
Element studies on assessment methodology etc.	OO	OOO	OOO	O	OOO	OO	O	OOO
National assessment Impact map	O	OOO	OOO		OOO	O		O
Threshold of impacts Vulnerable sectors and areas	O	OO	OO	OO	OO		O	OO
Economic assessment								
Adaptation	O		OO	O	O	O	O	O
Impacts on the Asia and Pacific region	O	OO	OO		OO			O

OOO: Results obtained in most areas. O: Studies in limited areas.
OO: Results obtained in some areas. Blank: No studies or unapparent situation.

impacts on individual sectors, nationwide assessments, identification of impacts thresholds and vulnerable regions, and adaptation strategies and measures. Many of the studies to date have focused on elementary aspects, such as methods of predicting impacts. However, to link these with counter-measures against global warming, we need clear answers to the following question:

- How greatly will these impacts affect the countries in question? (e.g. number of people at risk and financial losses)
- Which sectors in which regions will experience the severest impacts?
- Threshold of impacts – by how many degrees can the surface temperature heat up and by how many centimeters can sea levels rise for the impact on Asian countries to become intolerable?
- When will these intolerable impacts occur?

Measures against warming can be classified as either measures to mitigate global warming or those to adapt to a warmer world. Major efforts are clearly needed to prevent warming; however, we must also investigate adaptive measures to eliminate the harmful effects of warming, as we cannot completely prevent warming through the institutional and technical countermeasures currently in existence. While improving the accuracy of impact forecasts, we must also investigate adaptive measures for the severe impacts that will appear.

REFERENCES

1. IPCC, 2001: Climate Change 2001 Impacts, Adaptation, and Vulnerability, Cambridge University Press, 1032pp.
2. Japan Meteorological Agency, 2002: Climate of Japan in 20th Century, 116pp. (In Japanese)
3. Ichikawa, A. ed., 2004: Global Warming – The Research Challenges A Report of Japan's Global Warming Initiative, Springer, 160pp.
4. Kurihara, K., 2004: Development of regional climate model for impact studies, Proceedings of Global Warming Research Initiative Symposium, 17–18. (In Japanese)
5. Wada, K., 2004: Assessment of natural disaster risk due to global warming, Proceedings of Global Warming Research Initiative Symposium, 21–22. (In Japanese)
6. Emori, S. et al., 2005: Validation, parameterization dependence, and future projection of daily precipitation simulated with a high-resolution atomospheric GCM, Geophysical Research Letters, 32, L06708, doi:10.1029/2004GL022306.
7. Matsui, T. et al., 2004: Probability distributions, Fagus crenata forests following vulnerability and predicted climate sensitivity in changes in Japan. Journal of Vegetation Science, 15, 605–614.
8. Lasco, R.D. et al., 2005: Assessment of climate change impacts on and vulnerability of forests ecosystems in the Philippines using GIS and the Holdrige Life Zones, forthcoming.
9. Mimura, N. et al., 2004: Personal Communication.
10. National Institute for Environmental Studies, 2004: Global Warming and Human Health Web, (in Japanese) http://www.nies.go.jp/impact/index.html
11. Fukushima, T., et al., 2002: Influences of air temperature change on leisure industries: case study on ski activities. Mitigation and Adaptation Strategies for Global Change, 7, 173–189.
12. Harasawa, H. and S. Nishioka, eds., 2001: Global Warming Impacts on Japan, Kokon Pub. (in Japanese)

SECTION VI

Emission Pathways

INTRODUCTION

The papers in this section are diverse, but they all focus attention on the probabilities of exceeding different concentration or temperature thresholds along alternative pathways.

Two papers, one authored by Mastrandrea and Schneider and the other by Meinshausen, explore how probabilistic approaches to climate sensitivity might inform policy makers of the risks of exceeding levels of Dangerous Anthropogenic Interference (DAI) for various stabilisation levels. Both make it clear that there is no one-to-one association between concentration targets and temperature targets.

This complication has been compounded by recent research which indicates that the range of climate sensitivity values that are possibly consistent with the historical record is much wider than suggested by the IPCC, in particular the probability of high values. While the most likely level of temperature change for an effective doubling of greenhouse gas concentrations above pre-industrial levels lies between 2–4°C, more extreme change cannot be ruled out. Stainforth, Allen, Frame and Piani, in fact, report on the size of uncertainty inherent in GCM modelling using distributed computational power; they showed that the response to even a relatively low stabilisation level (doubled concentrations) could be as much as 11°C.

Stainforth et al. join Allen, Andronova, Booth, Dessai, Frame, Forest, Gregory, Hegeri, Knutti, Fiani, Secton and Stainforth in highlighting problems that are inherent in constraining climate sensitivity on the basis of observational data. To them, it is important to note that the various density functions that have been authored recently are products not only of the underlying data, but also of subjective expert judgements employed in their construction. Analyses of transient responses seem to indicate that the uncertainty in the climate sensitivity (CS) is less important on the shorter timeframe of stabilisation than equilibrium responses. Allen et al. also suggest an alternative approach that would consider the more well-constrained relationship between cumulative emissions and maximum temperature change.

Several authors explore the transient implications of multi-gas emission trajectories. Den Elzen and Meinshausen noted the importance of considering the non-CO_2 gases and aerosols when analysing pathways.

Tol and Yohe argue that rapid reductions of global greenhouse gas emissions could lead to an initial increase in temperature due to aerosol reductions, although they make no precise reference to the fact that reduced aerosol emissions will likely materialise with or without reduced CO_2 emissions. Jones, Cox and Huntingford show that carbon cycle feedbacks currently underestimated will have significant impacts on future pathways by requiring 20–30% greater emissions cuts to meet 550 ppmv CO_2 stabilisation goals, and changes in optimal trajectories. Climate-carbon cycle feedbacks are also critical in determining the feasibility of 'overshoot' scenarios which may rely on natural carbon sinks to reduce CO_2 levels to the stabilisation target.

Den Elzen and Meinshausen expand the exploration of transient emissions trajectories by considering the likelihood of overshooting a temperature target, like the EU target of 2°C above the pre-industrial level. They conclude that there would be a relatively high risk of exceeding such the target if concentrations were stabilised at 550 ppm CO_2 equivalent. Some risk would even persist with stabilisation at 400 ppm. They did note, though, that the risk of overshooting a temperature target can be reduced by letting concentrations peak (and then decrease) before they are stabilised. For example, meeting the EU temperature target could perhaps be achieved if emissions peaked around 2015, with subsequent decreases by 2050 dependent on the eventual stabilisation level (−10% for stabilising at 550 ppmv CO_2 eq. and about 15% more for each 50 ppmv lower stabilisation target). In the case of low stabilisation targets (400–500 ppmv CO_2 equivalent), concentrations temporarily exceed the target levels before they return to their ultimate stabilisation targets by 2150.

They, and others, highlight the consequences of delaying action on climate change. Meinshausen shows that delays are possible, but at the cost of requiring more rapid emissions reduction later. Kallbekken and Rive in fact show results in which a 20-year delay of action could result in required rates of emission reduction 3–9 times greater than that required for a more immediate response to meet the same temperature target.

Tol and Yohe also offer a word of caution by demonstrating the possibility of a dangerous climate policy mitigation that would slow economic growth to such an extent that vulnerability to climate change might actually be higher (particularly in developing countries).

CHAPTER 27

Probabilistic Assessment of 'Dangerous' Climate Change and Emissions Scenarios: Stakeholder Metrics and Overshoot Pathways

Michael D. Mastrandrea[1] and Stephen H. Schneider[1,2]

[1] *Center for Environmental Science and Policy, Stanford University, Stanford, CA*
[2] *Department of Biological Sciences, Stanford University, Stanford, CA*

ABSTRACT: Climate policy decisions driving future greenhouse gas mitigation efforts will strongly influence the success of compliance with Article 2 of the United Nations Framework Convention on Climate Change, avoiding 'dangerous anthropogenic interference with the climate system' (DAI). However, success will be measured in very different ways by different stakeholders, suggesting a spectrum of possible definitions for DAI. The likelihood of avoiding a given threshold for DAI is dependent in part upon uncertainties in the climate system – notably, the range of uncertainty in climate sensitivity. We combine a set of probabilistic global average temperature metrics for DAI with probability distributions of future climate change produced from a combination of several published climate sensitivity distributions, and a range of proposed concentration stabilization profiles differing in both stabilization level and approach trajectory – including overshoot profiles. These analyses present a 'likelihood framework' to differentiate future emissions pathways with regard to their potential for preventing DAI. Our analysis of overshoot profiles in comparison with non-overshoot profiles demonstrates that overshoot of a given stabilization target can significantly increase the likelihood of exceeding 'dangerous' climate impact thresholds, even though equilibrium warming in our model is identical for non-overshoot concentration stabilization profiles having the same target.

27.1 Introduction

27.1.1 *Article 2 and Climate Policy*

Article 2 of the United Nations Framework Convention on Climate Change (UNFCCC) states its ultimate objective as: 'Stabilization of greenhouse gas concentrations in the atmosphere at a level that would prevent dangerous anthropogenic interference with the climate system.' This level should be achieved within a timeframe sufficient to allow ecosystems to adapt naturally to climate change, to ensure that food production is not threatened, and to enable economic development to proceed in a sustainable manner [1]. Thus, 'dangerous anthropogenic interference' (DAI) may be characterized in terms of the consequences (or impacts) of climate change [2]. While the evaluation of DAI can be informed by scientific evidence and analysis, it is ultimately a normative decision, influenced by value judgments, socio-political processes, and factors such as development, equity, sustainability, uncertainty, and risk. The perception of DAI will likely be different depending on geographical location, socio-economic standing, and ethical value system. However, plausible uncertainty ranges for some DAI thresholds can be quantified from current scientific knowledge [3], which can inform the development of policies to avoid potentially 'dangerous' outcomes. More than 180 signatories to the UNFCCC have committed to prevention of DAI, and we argue that this long-term policy goal can lend important guidance for near-term climate change policy decisions. We believe that climate policy should be conceptualized and policy options compared in terms of preventing or reducing the probability of 'dangerous' climate impacts. Such a risk-management framework is familiar to policymakers and appropriate for climate policy decisions, which by necessity require decision-making under uncertainty [4,5,6].

Due to the complexity of the climate change issue and its relevance to international policymaking, careful consideration and presentation of uncertainty is essential when communicating scientific results [7,8,9,10]. As expressed in the Intergovernmental Panel on Climate Change (IPCC) Third Assessment Report (TAR), the scientific community can provide essential information underpinning decisions on what constitutes DAI [11]. For instance, scientific research can provide information on the intensity and spatial scale of climate impacts associated with future climate change. Further, the scientific community can provide specific probabilistic guidance on the implications of different policy choices and their respective likelihood of avoiding 'dangerous' climate impacts. We present a probabilistic framework for differentiating climate policy options by assessing their likelihood of avoiding thresholds for DAI. We apply this framework to a range of emissions pathways resulting in stabilization of atmospheric greenhouse gas concentrations, with and without overshoot of the stabilization concentration. These emissions pathways imply different development scenarios and magnitudes and timing of climate mitigation efforts.

27.2 DAI Metrics

27.2.1 *Aggregate Metric*

In Mastrandrea and Schneider [12], we presented a cumulative density function (CDF) of the threshold for DAI, based on the IPCC 'Reasons for Concern about Climate Change' [3; here, Figure 27.1]. Each category represents a semi-independent 'consensus estimate' of a metric for measuring 'concern' about the climate system. One interpretation of these metrics is as indicators of the level of global mean temperature change associated with DAI in the categories presented. Specifically, we view the increasing scale and intensity of impacts represented by the color gradient in each category as an estimate not only of physical climate impacts, but also of societal perceptions

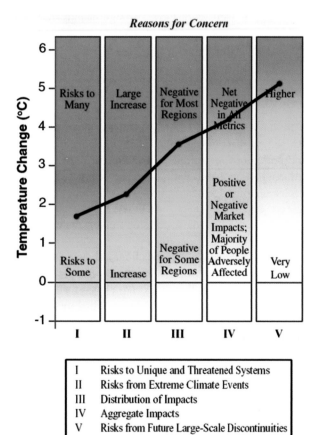

Figure 27.1 An adaptation of the IPCC Reasons for Concern figure from [12], with the thresholds used to generate our CDF for DAI-Ø. The IPCC figure conceptualizes five reasons for concern, mapped against global temperature increase. As temperature increases, colors become redder, indicating increasingly widespread and/or more severe negative impacts. We use the transition-to-red thresholds for each reason for concern to construct a CDF for DAI-Ø.

of 'danger' from those impacts. Interpreted in this way, increasing temperatures will progressively exceed thresholds in each metric and cumulatively contribute to the likelihood that the climate change occurring will be perceived to be 'dangerous' by an ever-widening group of stakeholders and decision makers. In other words, as warming intensifies, more and more stakeholders will perceive that DAI thresholds are being exceeded (based on their own value-driven assessments of what constitutes DAI in various metrics), cumulatively adding to the global perception of 'danger' from climate change.

In [12], we represented this accumulation of 'danger' by assigning data points at the threshold temperature above which each metric becomes red (solid black line in Figure 27.1), and assuming that crossing each threshold cumulatively adds an equal probability (20%) of reaching a global 'aggregate' DAI threshold (see [12] for a more complete description of methods). This aggregation method acts to average the thresholds of each impact metric, producing a median, 50th percentile threshold of 2.85°C above current temperatures, and a 90% confidence interval of [1.45°C, 4.65°C] [12]. Following [13], from which much of this analysis is drawn, we label this CDF DAI-Ø, since it is an average of impacts accumulating across all five metrics in Figure 27.1. We presented a 'traceable account' [14] of our assumptions in producing this aggregation, and we believe a similar account should be made each time such a definition is created by any analyst, as others, such as Wigley [15], have done.

27.2.2 *Aggregation Methods*

This aggregation method used to produce DAI-Ø is not intended to represent our (or other analysts') preferred assessment of DAI, though it might represent some stakeholder assessments. Rather, it represents the simplest first order summation of these impact metrics. To clarify the thinking behind our use of this method, a brief discussion follows. Maintaining a traceable account of any aggregation first requires disaggregation into individual 'stakeholder metrics', followed by transparent re-aggregation, explicitly choosing weights on each metric to represent different assessments of impacts and value positions associated with these assessments. As we discussed in [12], there are many ways that DAI could be interpreted from Figure 27.1, or from other sources (e.g., [16]), which suggests measuring vulnerabilities in terms of the 'Five Numeraires': market losses, human 'excess deaths', species extinctions, increasing inequity, or loss of quality of life. Some stakeholders may value impacts in one category above all others, or may factor information from several impact categories into their evaluations of DAI. An individual who, to some extent, values multiple impact categories, but who is not convinced that crossing the lowest threshold under consideration will constitute 'dangerous' change in his estimation, may choose 'weights' for each individual threshold and derive an averaged threshold somewhere between the individual thresholds in a manner similar to DAI-Ø. Another individual who recognizes multiple climate impacts may respond to the existence of multiple, additive risks from climate impacts by increasing his risk aversion and choosing an even lower threshold for DAI than that suggested by the impact category with the lowest threshold. However, we use the DAI-Ø metric in this work not to represent an individual evaluation of DAI, but as a demonstration that at some stage there must be, implicitly or explicitly, an aggregation of stakeholder values in any internationally-negotiated climate policy target based on preventing DAI. The history of the international climate negotiations (Alliance of Small Island States (AOSIS)-proposed targets versus negotiated Kyoto targets for example) has shown that some policymakers are willing to delay action until 'enough' impacts have accumulated, or set as a target a level of climate change which may cross thresholds for DAI for some particularly vulnerable populations. During the negotiations leading to the creation of the Kyoto Protocol, AOSIS submitted a draft protocol requiring 20% cuts in emissions by 2005 for industrialized nations. Clearly, the Kyoto targets are not as stringent as this target proposed by one stakeholder group. Given these circumstances, we believe the averaging method we present is an appropriate aggregation method to demonstrate our probabilistic framework.

As discussed previously, the stakeholder assessments of DAI which underlie such a global aggregation can vary widely. We present below, again, as a framework for methods to analyze DAI, an initial step in disaggregating our DAI-Ø metric by interpreting each 'reasons for concern' category as representing a limited number of stakeholder prime interests, and to show how that can lead to very different DAI thresholds.

27.2.3 *Stakeholder Metrics*

To represent possible stakeholder DAI metrics and the diversity of possible evaluations of DAI, we produce two new CDFs, (DAI-I, DAI-V) based on individual 'reasons for concern' categories (Columns I and V in Figure 27.1), which reflect possible policy perspectives presented by generalized stakeholders in the international climate debate. An AOSIS member, a conservation biologist seeking to preserve biodiversity, or others sympathetic to these values may focus their evaluation of DAI on Column I, risks to unique and threatened systems, which represents temperature change associated with risks to unique human settlements such as low-lying small island nations, vulnerable coastal states like Bangladesh, or Arctic indigenous cultures dependent on sea-ice, and/or to unique or vulnerable ecosystems, like mountaintop species communities. A mid-latitude nation, or a nation with high adaptive capacity and little concern for impacts elsewhere in the world, might ignore considerable impacts to other regions of the world and be most concerned with abrupt nonlinear global climate changes, basing their evaluation

of DAI on Column V, risks from future large-scale discontinuities such as ocean circulation alterations or deglaciation of Greenland. We construct these stakeholder CDFs based on the increasing climate risk in each of these categories indicated by the 'reasons for concern' color scale. We define our DAI-I and DAI-V CDFs by constructing normal distributions with median equal to the transition from yellow to orange in each category (orange signifying a medium level of impacts), and two-standard deviation (2σ) length equal to the distance from this transition to the beginning of the color scale (the transition from white to yellow). This analysis yields a median threshold for DAI-I of 1.2°C above current temperatures with a 90% confidence interval of [.3°C, 2.1°C], and a median threshold for DAI-V of 4.15°C with a 90% confidence interval of [3°C, 5.3°C]. Research published after the TAR has indicated that some abrupt nonlinear global changes, such as breakdown of the Greenland or Western Antarctic ice sheets, may be triggered by lower temperature thresholds than those currently indicated in Column V [17, e.g.]. Therefore, if this recent information were taken into account, as it is likely to be in the next IPCC assessment in 2007, a stakeholder basing their evaluation of DAI on Column V would likely produce a distribution for DAI thresholds of lower temperature increases than the one reported here, which is based on information now five years old.

We do not use the same 'transition to red' thresholds used in the aggregate DAI-Ø metric to define our stakeholder median thresholds. In the aggregate metric, the 'transition to red' in each impact category is used as a threshold for 'danger' in that category for aggregation across impact categories. However, nontrivial impacts occur in each category at temperatures below the 'transition to red', indicated by the yellow and orange regions. To account for these impacts, we use the 'transition to orange' threshold as median for our stakeholder metrics. As noted earlier, any particular association of a color transition on Figure 27.1 with DAI is arbitrary, as was our 'transition to red' choice for DAI-Ø in Figure 27.1. Thus, our choice for a probabilistic DAI metric based on an individual category is likewise arbitrary – but reasonable, we believe, to demonstrate the framework. As with all our metric definitions, there are other comparably plausible methods by which to construct such distributions. Our purpose, as mentioned earlier, is to demonstrate a quantitative framework for analysis and policy debate, not to offer our model-dependent numerical results as 'answers' or recommendations.

Finally, an important geopolitical stakeholder in the climate policy debate is the European Union (EU). With the withdrawal of the United States from the Kyoto Protocol, the EU now ranks as the largest current and historical emitter of greenhouse gases involved in negotiating binding international mitigation policies. The European Council, in keeping with a precautionary approach and perhaps influenced by the recognition of multiple, additive climate risks as discussed above, has adopted a long-term policy goal of limiting global average temperature increase to 2°C above *pre-industrial* temperatures. In a growing number of studies, this threshold is also designated as a temperature limit above which dangerous climate impacts may occur [18,19]. Clearly, a policy target such as this cannot capture the probabilistic nature of any assessment of future climate impact thresholds, nor is it intended to do so. Temperature increases below 2°C may still induce dangerous changes in some sectors or regions – and increases above 2°C may not – all subject to the variety of definitions and metrics of 'danger' discussed above. The EU threshold has both political and analytical history, so we adopt the EU policy goal as a third 'stakeholder threshold', DAI-EU.

27.3 Emissions Pathways and DAI

27.3.1 *Stabilization Profiles*

The UNFCCC [1] has called for parties to consider 'stabilization of greenhouse gas concentrations', and much recent international debate has centered on the desirable level of atmospheric greenhouse gas concentrations at which to stabilize, or the level in keeping with avoiding DAI. Many research efforts have produced concentration stabilization profiles to drive models investigating the climate implications of various emissions pathways. (In this paper, we make an effort, as in [20], to differentiate between 'emissions scenarios', which represent descriptions of possible future states of the world and the characteristics relevant for emissions, 'emissions pathways', which represent time-evolving paths for global emissions of greenhouse gases and aerosols, and 'concentration profiles', which represent time-evolving trajectories for atmospheric concentrations of greenhouse gases and aerosols). O'Neill and Oppenheimer [21], for example, produce stabilization profiles which reach a range of CO_2-equivalent stabilization levels through three approach categories: slow approach (labeled SC by [21]), rapid approach (RC), and overshoot approach (OS). These differ from many stabilization profiles in that they consider emissions of aerosols and all known significant radiatively active gases beyond CO_2, and a wider range of approach pathways to final stabilization levels. Though these profiles are only a subset of plausible profiles, they are representative of the middle of the range of published profiles – yet are different enough to allow us to clearly demonstrate the probabilistic frameworks we are offering in this analysis, especially for overshoot profiles that many analysts have suggested will be the most likely characteristic future path for greenhouse gas concentrations over the next centuries.

27.3.2 *Risk Assessment*

The temperature profile associated with an emissions pathway is dependent upon uncertainty in our understanding

of the climate system and our capacity to model it. Thus, any stabilization level for greenhouse gases can produce a distribution of possible temperature increases, some of which may exceed a given threshold for DAI, some of which may not. Analysis of stabilization profiles and their likelihood of success in achieving the goal of avoiding DAI requires explicit treatment of the uncertainty in the climate system. Our approach is quantitative: probabilistic analyses of temperature distributions associated with each profile. To quantify this probabilistic framework, we apply our DAI metrics to representative concentration stabilization profiles generated by emissions pathways from the three approach categories in [21], for 500 ppm and 600 ppm CO_2-equivalent (CO_2e) stabilization levels. All profiles reach their target stabilization level by the year 2200, with the OS profiles peaking in 2100 at a level 100 ppm CO_2e above the final target. These stabilization profiles, well within the published range, are primarily offered to demonstrate the framework, not to bound future outcomes, as higher and lower emissions pathways are still quite plausible.

27.4 Temperature Projections

27.4.1 *Climate Sensitivity Uncertainty*

Much of the uncertainty in current understanding of the climate system is captured by the so-called climate sensitivity, defined as the equilibrium global mean surface temperature increase from a doubling of atmospheric CO_2. Using general circulation models (GCMs), the IPCC has long estimated the climate sensitivity to lie somewhere between 1.5°C and 4.5°C [22], without indicating the relative probability of values within – let alone outside – this range. Recent studies produce distributions wider than the IPCC range, with significant probability of climate sensitivity above 4.5°C [23,24,25, e.g.]. The likelihood of avoiding any given temperature threshold for DAI is extremely sensitive to the uncertainty associated with climate sensitivity, as demonstrated by Mastrandrea and Schneider [12], and Hare and Meinshausen [26]. Specifically, this likelihood is very dependent on the upper bound of the climate sensitivity distribution, as larger climate sensitivities will contribute most to the likelihood of DAI threshold exceedence.

For any stabilization profile, differences in climate sensitivity will lead to very different projected temperature profiles. The IPCC TAR presents future global average temperature profiles by forcing a simple climate model tuned to several complex Atmosphere-Ocean GCMs (AOGCMs), driven by the 6 SRES illustrative scenarios [22]. Each 'tuning' employs a different climate sensitivity, resulting in the temperature ranges presented in the TAR for each illustrative scenario. O'Neill and Oppenheimer also use a simple climate model of the type used in the

TAR to produce temperature profiles based on their stabilization profiles, and choose three climate sensitivities within the IPCC range to produce a sensitivity analysis. Neither analysis, nor other recent research [27], relates emissions pathways, which implicitly or explicitly require climate mitigation policy decisions, to probability functions for avoiding DAI. O'Neill and Oppenheimer do compare the future temperature profiles generated by their emissions pathways to thresholds for individual climate impacts that may be considered 'dangerous', and consider the sensitivity of their results to three values for climate sensitivity, but they do not produce probability distributions for their results. A probabilistic linkage, one we also make in this paper, has become a focus in recent literature [20,26,28].

In our analysis, we use three probability distributions from two published sources [23,24]: the expert and uniform prior distributions from Forest *et al.*, and the probability distribution from Andronova and Schlesinger including solar forcing (labeled T2 by [23]). Use of these probability distributions allows us to sample a range of uncertainty in climate sensitivity representative of the range reported in current publications.

27.4.2 *Climate Modeling*

To generate consistent temperature time series for application of our DAI metrics, and to explore the probabilistic range for future temperature change implied by uncertainty in climate sensitivity, we use the radiative forcing time series for the SC, RC and OS profiles for 500 and 600 ppm CO_2e stabilization levels to force a simple two-box climate model. This model was originally developed by Schneider and Thompson [29], and modified by Nordhaus [30] for use in the DICE model. We modify the Nordhaus version to reduce the timestep from ten years to one (See Appendix A for model details).

Owing to the many model-dependent assumptions inherent in the use of such highly simplified models, we emphasize that our quantitative results using this simple model are not intended to be taken literally, but we do suggest that the probabilistic framework and methods be taken seriously: they produce relative trends and general conclusions that better represent a risk-management approach than estimates made without probabilistic representation of outcomes. The demonstrated application of threshold metrics for DAI to emissions pathways extends the risk-management framework presented in [12], introducing a method for assessing the probability of DAI for future climate profiles produced by other climate models and stakeholder metrics.

27.4.3 *Probabilistic Temperature Time Series*

We generate temperature time series by running this simple climate model [29] using a range of climate sensitivities sampled from our climate sensitivity distributions. For a given emissions pathway, running the model under each different climate sensitivity will produce a different temperature time series. Thus, by sampling many times from a climate sensitivity distribution, running the model with each value for climate sensitivity, and recording the temperatures produced, we can generate probability distributions for future temperature change, for each emissions pathway in any given year, based on the uncertainty in climate sensitivity. All temperatures are expressed as temperature increase above the year 2000.

In this paper, we present distributions based on an aggregation of the separate results using each of the three climate sensitivity distributions listed in Section 4.1, despite recognizing that, when using distributions produced with different methodologies, it is better methodological practice to present results separately, as we did in [12, Figure 27.2]. As our primary purpose here, however, is to demonstrate a framework for probabilistic analysis, and since the choice of any of these three climate sensitivity distributions does not change the qualitative properties of our analysis, we choose to present aggregate results for the sake of clarity of framework and to reduce the sheer number of similar figures. Our presentation of results is intended to demonstrate our probabilistic framework, and presenting separately the results using each climate sensitivity distribution requires, for each analysis step, either one very busy figure or three separate figures displaying essentially the same information. We believe such complexity would obscure the demonstration of our analysis methods while adding little intellectual value. Given that there is no assessment of the differential confidence that could be assigned to each published distribution nor a basis for choosing one distribution as most likely, we choose to aggregate over these three distributions, presenting results approximating the range of uncertainty among published climate sensitivity distributions. Use of other published climate sensitivity distributions would shift up or down the probabilities of exceedence of DAI thresholds reported in the next section, depending on whether the new distributions have a greater or lesser probability of high values for climate sensitivity than those currently used.

We also generate probability distributions for the equilibrium temperature predicted by each combination of climate sensitivity and radiative forcing stabilization level. The equilibrium temperature increase (which will lag the radiative forcing by many decades owing to the thermal inertia of the climate system) for a given combination of climate sensitivity and radiative forcing can be calculated as

$$\Delta T_{EQ} = (\Delta F / \Delta F_{2x}) \times \Delta T_{2x} \qquad (27.1)$$

where ΔT_{EQ} is the equilibrium temperature increase above pre-industrial levels, ΔF is the radiative forcing in W/m^2 for a particular stabilization level, ΔF_{2x} is the radiative forcing estimate for a doubling of atmospheric CO_2, and ΔT_{2x} is the climate sensitivity. For these calculations, we set $\Delta F_{2x} = 3.71 \ W/m^2$, as suggested by [31].

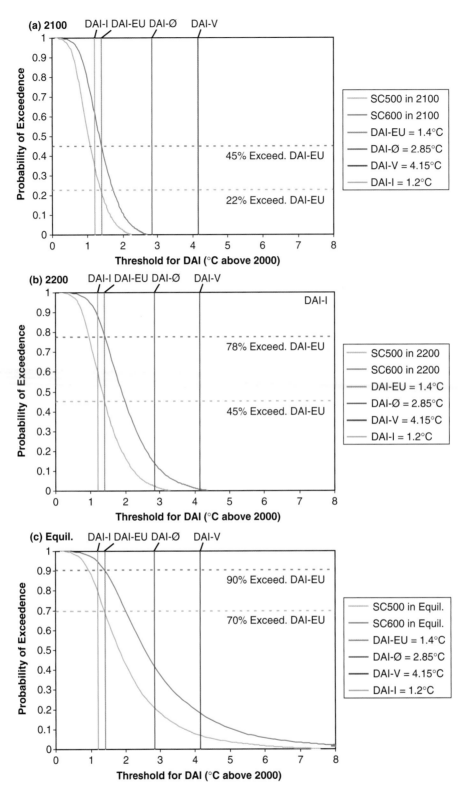

Figure 27.2 Comparison of the probability of exceedence of the indicated thresholds for DAI for two concentration profiles, one stabilizing at 500 ppm CO_2e (SC500), one stabilizing at 600 ppm CO_2e (SC600). Panels a) and b) display probabilities of exceedence for transient temperature increase above 2000 in 2100 and 2200, respectively, and panel c) displays probabilities of exceedence in equilibrium. The curves are generated from probability distributions for future temperature increase generated by running the simple climate model described in Appendix A for each concentration profile, sampling from the probability distributions for climate sensitivity employed in this paper. As reported in the text, probabilities of exceedence of the DAI-EU threshold are indicated in each panel. Adapted from [13].

27.5 DAI Analysis

27.5.1 *Probabilistic DAI Analysis*

We map our aggregate metric (DAI-Ø) and stakeholder metrics (DAI-I, DAI-V, DAI-EU) for DAI onto probability distributions for transient global average temperature increase at 2100 and 2200 for each emissions pathway's concentration profile, and for each equilibrium temperature increase and radiative forcing stabilization level (500 ppm CO_2e, 600 ppm CO_2e). This analysis allows us to characterize and compare emissions pathways in terms of their likelihood of exceedence of DAI thresholds. Figure 27.2 compares results for two emissions pathways in 2100 (a), in 2200 (b), and at equilibrium (c), a 'slow change' (SC) profile stabilizing at 500 ppm CO_2e, and an SC profile stabilizing at 600 ppm CO_2e, both of which approach their stabilization level more slowly than the other (RC, OS) profile types. The figure displays the relationship for each profile between the temperature increase above 2000 levels chosen as the median threshold for DAI in each metric and the probability of exceedence of that threshold. The lower the threshold for DAI, of course, the higher will be the probability of exceedence. These curves are calculated for each concentration profile from the probability distributions (PDFs) for temperature increase in a given year described above (Section 4.3). These PDFs can be used to construct cumulative density functions (CDFs) by integrating the PDFs. Any point on one of the curves in Figure 27.2 is equal to one minus the corresponding point on a CDF constructed from the PDF for temperature increase above 2000 at the indicated time (2100, 2200, in equilibrium). In other words, any point on a CDF for temperature increase at a given time indicates the probability (between zero and one) that the temperature increase at that time is equal to or below the increase at that level. If a threshold for DAI were set at that level, this probability would represent the probability of compliance with that threshold. One minus the probability of compliance represents the probability of exceedence (see the examples below and in Figure 27.2). Panels a) and b) indicate transient temperature increase, while temperatures approach the equilibrium distributions displayed in panel c), provided atmospheric concentrations are stabilized indefinitely.

To evaluate the probability of exceedence of our metrics for DAI, we indicate, in Figure 27.2, the median DAI threshold of our aggregate DAI metric, DAI-Ø, the median thresholds of our DAI-I and DAI-V metrics, and the DAI-EU threshold. (The DAI-EU threshold is defined as 2°C above pre-industrial temperatures, while we present temperature distributions of temperature increase above 2000. Therefore, we express the DAI-EU threshold as 1.4°C, based on the central estimate of 0.6°C warming over the 20th century in the IPCC TAR [22]). For example, the SC500 emissions pathway we use has a probability of exceedence of the DAI-EU threshold of 21% in 2100, 45% in 2200, and 70% in equilibrium, under the

assumption set used in this analysis, while the SC600 pathway we use has exceedence probabilities of 45%, 78%, and 90%, respectively.

As mentioned previously, the quantitative results from our simple model should not be viewed as high-confidence indicators of future outcomes. The specific probabilities of exceedence of DAI thresholds presented in Figure 27.2 are highly dependent on the model formulation (see Section 5.3 and Appendix A) and probability distribution for climate sensitivity we use. However, the qualitative features of the trends we present are likely to reflect similar features which could be obtained using more complex models, and we present these results to demonstrate the probabilistic characterization of emissions pathways as a framework we believe can be informative to policy makers when evaluating climate policy options. In addition to the profiles presented in Figure 27.2 (SC500, SC600), we examine similar results for stabilization profiles from the other approach categories for stabilization presented in [21] (RC500, RC600, OS500, OS600), and will highlight three properties of these results, beyond reinforcing the well-established finding that stabilization at a higher level of atmospheric greenhouse gas concentration increases the probability of exceeding most thresholds for DAI:

1. The pathway to stabilization has significant impact on the probability of exceeding thresholds for DAI.

   This similarly unsurprising result is still quite relevant to ongoing international and national climate policy-making. Comparing trajectories which stabilize at the same level (SC500, RC500, and OS500, e.g.) indicates that faster accumulation of greenhouse gases makes exceedence more likely to occur and to occur earlier (see Figure 27.3, e.g.). Additionally, overshoot of the stabilization level leads to significantly greater likelihood of exceedence. We explore these characteristics in more detail in Section 5.2.

2. Stabilization at 500 ppm CO_2e can still impose a significant probability of DAI for some stakeholders (e.g., those adhering to DAI-I), while stabilization at a level as high as 600 ppm CO_2e may produce a relatively low probability of exceeding some DAI thresholds (e.g., those focusing on DAI-V) before 2200.

   With the exception of the SC500 pathway (37%), all six pathways imply a greater than one in two chance by the year 2100 that temperature increases will induce impacts perceived to be 'dangerous' under the DAI-I median threshold (1.2°C above 2000 temperatures). However, under the median threshold for the DAI-V metric (4.15°C), none of the pathways imply a greater than one in twenty chance through the year 2200. This contrast reinforces the point made earlier and in [12] that what is perceived as 'dangerous' is a value judgment, represented in this analysis by thresholds based on two 'reasons for concern' (Section 2.3 above). Further, given historical and projected greenhouse gas emissions and the inertia in the

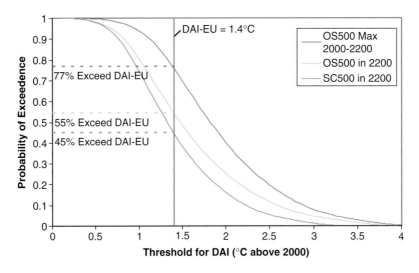

Figure 27.3 Comparison of the probability of exceedence of the DAI-EU threshold for overshoot (OS500) and non-overshoot (SC500) concentration profiles stabilizing at 500 ppm CO_2e. The green and yellow curves display probabilities of exceedence for transient temperature increase above 2000 in 2200, while the red curve displays probabilities of exceedence for the maximum temperature reached sometime between 2000 and 2200 for the overshoot (OS500) concentration profile. While there is only a modest increase in the probability of exceedence of the DAI-EU threshold in 2200 under the two profiles, there is a significant increase in the probability of at least a temporary exceedence of the DAI-EU threshold prior to 2200. Adapted from [13].

climate system, avoiding thresholds such as the DAI-I median threshold with a high degree of reliability (e.g., >90%) may well prove infeasible, except through a temperature profile which temporarily overshoots this threshold and may trigger impacts 'dangerous' to certain stakeholders during such an overshoot (see section 5.2 below).

3. Equilibrium temperature increases far exceed transient temperature changes of the next two centuries.

For all pathways, equilibrium temperature increases (and probabilities of exceedence of DAI thresholds) are greater than those seen through 2200 (see Figure 27.2). How much greater is dependent on the ocean model formulation used (see Section 5.3 and Appendix A), which has a major influence on the transient approach to equilibrium temperature increase. Our results suggest that even a stabilization level of 500 ppm CO_2e can imply significant probabilities of DAI threshold exceedence for at least some stakeholders, in the long-term. How policy makers might value very long-term risks is a function of their concern for sustainability and their method of discounting.

27.5.2 *Irreversibility and Path-Dependence*

The complex response of the climate system to external forcings may include abrupt nonlinear climate changes and other impacts essentially irreversible on time scales relevant to policymaking [e.g., 32,33]. Paleoclimatic data and scientific understanding of current components of the climate system indicate that such changes are possible in the future due to anthropogenic forcings [e.g.,

34,35]. We examine stabilization profiles that differ in their transition to the final stabilization level. While overshoot profiles may reduce the mitigation costs of reaching a given stabilization level [36,37,38], they may also increase the climate impacts associated with a given stabilization goal [21,39]. Further, the additional transient warming induced by overshoot stabilization profiles may exceed temperature thresholds for irreversible, abrupt nonlinear climate changes or impacts (like species extinctions), which will persist long after the temporary threshold exceedence. The higher and more rapid rise of temperatures will increase the probability of DAI exceedence for overshoot profiles, compared to monotonically increasing profiles reaching the same stabilization level.

As an illustration, Figure 27.3 displays curves similar to Figure 27.2 for the overshoot OS500 profile in 2200 and the maximum temperature increase reached using that profile between 2000 and 2200. For comparison, the relationship between probability of exceedence and threshold for DAI under the SC500 profile in 2200 (identical to that in Figure 27.2.b) is also displayed. The concentration overshoot increases the probability of exceeding any given threshold for DAI relative to non-overshoot profiles with the same stabilization level. As stated previously, the SC500 profile has a 45% chance of exceeding the DAI-EU threshold (vertical line in Figure 27.3) in 2200. The OS500 profile has a 55% chance of exceedence in 2200, and a 77% chance of exceedence for the maximum temperature increase. In this case, maximum temperatures under the OS500 profile are reached in or prior to 2200. While the OS500 profile only modestly increases the probability of exceedence of the DAI-EU threshold in 2200 compared to the SC500 profile (55%–45%), the

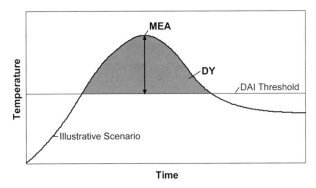

Figure 27.4 Visual representation of the Maximum Exceedence Amplitude (MEA) and Degree Year (DY) metrics. We introduce these tools to differentiate emissions pathways by the degree to which they exceed thresholds for DAI. For the illustrative temperature profile displayed here, MEA is measured as the maximum temperature increase reached above the indicated threshold for DAI (horizontal gray line), and DY is measured as the cumulative exceedence of that threshold by the profile (gray shading). Adapted from [13].

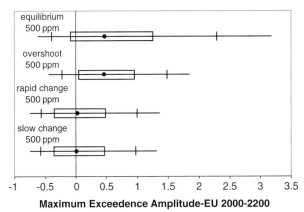

Figure 27.5 Box-and-whisker diagrams for the Maximum Exceedence Amplitude (MEA) above the DAI-EU threshold, MEA-EU. The diagrams indicate the 95% confidence interval (full horizontal line), 90% confidence interval (vertical tick marks), 50% confidence interval (box), and median value (dot). The lower three diagrams display the distribution of MEA-EU between 2000 and 2200 for the three concentration profiles stabilizing at 500 ppm CO_2e. For comparison, the top box-and-whisker diagram displays the MEA-EU distribution in equilibrium for stabilization of 500 ppm CO_2e. The overshoot concentration profile increases the median and overall range for MEA-EU, compared to the non-overshoot profiles. Adapted from [13].

overshoot significantly increases the probability (77%) of temporary overshoot of the threshold prior to 2200.

We present two metrics for further evaluating the increase in risk of DAI associated with an overshoot profile, pictured in Figure 27.4 for an illustrative profile. First, we define the Maximum Exceedence Amplitude (MEA) above a given DAI threshold as the maximum difference between the DAI threshold and a calculated temperature profile through a given time period. If a temperature profile never exceeds the DAI threshold, the MEA will be negative, representing the closest approach to the threshold during that period. Figure 27.5 shows box and whisker diagrams for MEA-EU (MEA for the DAI-EU threshold), for 2000–2250, for the three profiles stabilizing at 500 ppm CO_2e (SC500, RC500, OS500), and the equilibrium temperature distribution for stabilization at 500 ppm CO_2e. Each box and whisker indicates the 95% confidence interval (horizontal line), 90% confidence interval (vertical lines), 50% confidence interval (box), and median value (dot), and represents the distribution of MEA-EU values from calculations using our aggregate climate sensitivity distribution (see Section 4.3). The non-overshoot profiles both have a 50% chance of exceeding the DAI-EU threshold by the year 2250. The overshoot profile has more than a 75% of at least temporary exceedence within that timeframe, with a median MEA-EU comparable to that in equilibrium.

The MEA metric provides information about the maximum temperature reached in a given timeframe, and the magnitude of exceedence of a given threshold for DAI. It does not provide information about the duration of exceedence. A prolonged period of temperatures above a threshold for DAI is likely to induce more severe impacts than a short exceedence of equal magnitude. Degree days are a commonly used metric to measure cumulative departure

from a given temperature level. For example, heating and cooling degree days are used to measure energy demand, and growing degree days are used in agriculture and pest control. We adopt a similar metric, degree years, as a measure of both the length and magnitude of exceedence of a given threshold for DAI. We define degree years (DY), for a given time period n, temperature profile T, and threshold DAI, as:

$$DY = \sum_{t=1}^{n} \begin{cases} T(t) - DAI, & T(t) > DAI \\ 0, & T(t) < DAI \end{cases} \quad (27.2)$$

where T(t) is the temperature increase above 2000 in a given year, and only positive values contribute to the sum. Degree years are the sum of the magnitudes of threshold exceedence in each year of a given time period – in other words, the area under the temperature profile curve but above the DAI threshold for that period (gray shading in Figure 27.4). Figure 27.6 shows box and whisker diagrams for DY-EU (DY for the DAI-EU threshold), for the period 2000–2200, calculated for the three profiles stabilizing at 500 ppm CO_2e, as in Figure 27.5. Each box and whisker again represents the distribution of DY-EU values from calculations using our aggregate climate sensitivity distribution, and indicates the same confidence intervals. In parallel with Figure 27.5, the overshoot profile greatly increases the degree years accumulated above the DAI-EU threshold compared to both non-overshoot profiles, with the median DY value nearly tripling between

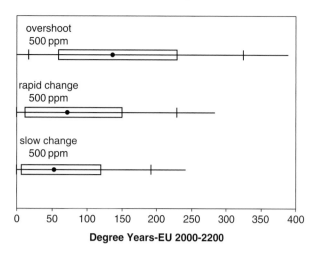

Figure 27.6 Box-and-whisker diagrams for Degree Years (DY) above the DAI-EU threshold, DY-EU. The diagrams indicate the 95% confidence interval (full horizontal line), 90% confidence interval (vertical tick marks), 50% confidence interval (box), and median value (dot). The diagrams display the distribution of DY-EU between 2000 and 2200 for the three concentration profiles stabilizing at 500 ppm CO_2e. The overshoot concentration profile increases both the median and variance of the DY-EU distribution, compared to the non-overshoot profiles. Adapted from [13].

the slow change profile (SC500) and the overshoot profile (OS500).

Together, the MEA and DY metrics characterize in two measures the implications of a climate change profile with respect to a given threshold for DAI. MEA cannot identify the length of exceedence of a threshold, and DY cannot distinguish between a short but large exceedence and a long but low-magnitude exceedence. The metrics provide complementary information.

27.5.3 *Model Uncertainty*

The temperature distributions and threshold exceedence probabilities in this paper are highly sensitive to the choice of climate sensitivity distribution and ocean model formulation. As stated previously, [12] and [26] both address the sensitivity of transient and equilibrium temperature distributions to the choice of climate sensitivity distribution. The ocean model formulation (box-diffusion vs. box-advection-diffusion, e.g.) will not affect the equilibrium temperature distribution, but it can significantly affect the evolution of the transient approach to equilibrium. That is, models with deep oceans and slower response times will cross thresholds for DAI more slowly than models with faster response times. This is strictly true when using a simple climate model with a single equilibrium warming level for a given radiative forcing. Some nonlinear processes not included in our simple model can create multiple equilibria and path dependence [e.g., 40]. In such models, overshoot scenarios could imply

lower thresholds for DAI than those we report here with this linear model. We have not explicitly quantified the uncertainty created by this additional uncertain component, but we qualitatively explore some of the implications of ocean model formulation in Appendix A.

27.6 Conclusions

Identification of the severity, spatial extent, and salience of impacts determines the level that can be labeled as 'dangerous anthropogenic interference with the climate system.' Determining this level is ultimately a value-laden process, one that will undoubtedly lead to different levels for different stakeholders in different regions of the world, applying different perceptions and values to the question. Despite the layered complexity of determining DAI, we believe the probabilistic risk management framework we present here and in [12] is an effective method for informing the policy process and evaluating the implications of alternative policy choices with respect to DAI. In this paper, we demonstrate a probabilistic framework for evaluating and comparing emissions pathways with respect to their potential for compliance with thresholds for DAI, applying three possible probabilistic metrics for DAI, and the DAI-EU threshold, to several stabilization profiles.

For this demonstration, we use individual thresholds from these metrics, in order to demonstrate that an evaluation of the effectiveness of a given emissions pathway or set of policy choices in meeting a policy goal, such as avoiding exceedence of the EU threshold of 2°C above pre-industrial temperatures or avoiding exceedence of the median threshold from DAI-Ø, must explicitly consider the uncertainties inherent in the linkage between atmospheric greenhouse gas concentrations and temperature increase, represented here by uncertainty in climate sensitivity. However, our probabilistic DAI metrics represent a further level of uncertainty regarding the range of possible definitions of 'danger' across or within impact categories. Even with perfect knowledge about future emissions and the response of the climate system to those emissions, the assessment of DAI would require treatment of uncertainty about the level of impacts associated with DAI. We have not yet conducted a thorough probabilistic analysis of all of these sources of uncertainty.

For our probabilistic metrics, we provide results at different levels of aggregation. The DAI-Ø metric represents a global aggregation of one interpretation of the thresholds for DAI presented in the IPCC Reasons for Concern, a simple representation of an aggregation of differentiated stakeholder values. Such aggregation is inherent in any international climate policy negotiation. Our 'stakeholder metrics', DAI-I and DAI-V, represent assessments of DAI at a more disaggregated level. Such assessments are also likely to be an inherent component of international climate policy negotiations. Further disaggregation ultimately leads to small-scale definitions of vulnerability or 'danger' for certain locations, impact sectors, or populations. We propose that the probabilistic framework we demonstrate here can be applied at any level along this continuum. The most difficult part of any such assessment will be to represent quantitatively the stakeholder metrics of various groups and regions, and then to perform an analysis that, via a traceable account of alternative aggregation weighting schemes, helps to guide decision-makers at all scales to determine within this risk management framework how much risk of exceedence of DAI thresholds they are willing to accept.

Acknowledgements

The authors would like to thank B. O'Neill, M. Oppenheimer, P. Baer, N. Andronova, and C. Forest for data provided for this analysis, and M. Meinshausen for extremely helpful comments on a previous draft. S.H. Schneider acknowledges partial support from the Winslow Foundation. M.D. Mastrandrea acknowledges support from the Center for Environmental Science and Policy, Stanford Institute for International Studies. This research was made possible in part through support from the Climate Decision Making Center. This Center has been created through a cooperative agreement between the National Science Foundation (SES-034578) and Carnegie Mellon University. This chapter is adapted from [13].

APPENDIX A

This Appendix explores an additional uncertain factor of the modeled atmosphere-ocean system beyond climate sensitivity – the formulation of the ocean model component, which will have a significant influence on the modeled transient climate response to anthropogenic radiative forcing. In turn, the pathway of approach to equilibrium will greatly affect the likelihood of transient temperature increase exceeding DAI thresholds within a given timeframe (before 2100, e.g.).

The two-box climate model we use is a simple example of a linearized, one-dimensional box-diffusion (BD) model. The two-box model is of the form:

$$T(t) = T(t - 1) + \sigma_1\{F(t) - \lambda T(t - 1) \\ - \sigma_2[T(t - 1) - T_{LO}(t - 1)]\}$$

$$T_{LO}(t) = T_{LO}(t - 1) + \sigma_3[T(t - 1) - T_{LO}(t - 1)]$$

where $T(t)$ is the temperature in the upper box in year t, $T_{LO}(t)$ is the temperature in the lower box in year t, $F(t)$ is the radiative forcing above pre-industrial levels of the upper box in year t, and λ, σ_1, σ_2, and σ_3 are constants as defined in [30]. We adjust σ_1 and σ_3 to use a one-year timestep by dividing the DICE values of σ_1 and σ_3 by ten, making the values 0.0226 and 0.002, respectively.

An alternative approach is to use a box-advection-diffusion (BAD) model. The rate of mixing between upper

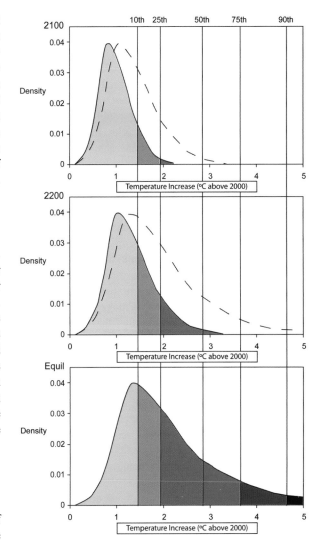

Figure A1 Probability density functions (PDFs) corresponding to the CDFs (orange curves) shown in Figure 27.2 for the SC500 stabilization scenario. The dashed PDFs in the top two panels (2100, 2200) use a climate model with a faster response time. In this model temperature increase approaches the equilibrium distribution far more rapidly – the distribution for 2200 is already very similar to that in equilibrium (bottom panel). The vertical lines and color shading in each panel represent percentile thresholds from the DAI-Σ metric and the outcomes which exceed each threshold. A faster response time shifts forward in time exceedence of these DAI thresholds, though equilibrium exceedence is identical.

and lower oceanic regions determines the thermal inertia of the system, and thus the nature of the transient response. A BD model's response time is typically proportional to the inverse square of the radiation damping term (equal to the radiative forcing for a doubling of atmospheric CO_2 divided by the climate sensitivity), whereas a BAD model has a response time inversely proportional to the radiation damping term raised to a power between 1 and 2-empirically, closer to 1. Harvey [41] describes the transient response of these two model formulations to a step function doubling of atmospheric CO_2. He demonstrates that the BAD model exhibits significantly lower e-folding response time compared to the BD model – in other words, the approach to equilibrium is much more rapid using the BAD model.

In Figure 27.2 of this paper, we show transient temperature CDFs for 2100 and 2200 and the equilibrium temperature CDFs for two stabilization scenarios. In Figure A1 we display a simple representation of alternate transient characteristics and their implications for the likelihood of DAI threshold exceedence. Figure A1 presents the probability density functions (PDFs) corresponding to the CDFs (orange curves) shown in Figure 27.2 for the SC500 stabilization scenario. The dashed PDFs in the top two panels (2100, 2200) represent the response of the climate model used in this paper when heat transfer to the deep ocean box is shut off, significantly lowering the effective heat capacity of the modeled climate system. The probability distribution for temperature increase approaches the equilibrium distribution (bottom panel) far more rapidly – the distribution for 2200 is already very similar to that in equilibrium. The vertical lines in each panel represent percentile thresholds from the DAI-Ø metric, and demonstrate that a more rapid approach to equilibrium advances in time the increasing likelihood of exceedence of the DAI thresholds reported in the figure. Larger areas under the curves of the dashed PDFs fall above each threshold for DAI than under the solid curves (the colored areas).

In summary, both the climate sensitivity and the ocean model formulation will affect the magnitude and rate of approach to equilibrium, and thus have major influence on the probability of exceedence of any specified DAI threshold.

REFERENCES

1. *United Nations Framework Convention on Climate Change* (UNFCCC, 1992; http://www.unfccc.de).
2. Patwardhan, A., Schneider, S.H., Semenov, S.M. (2003) IPCC Concept Paper, http://www.ipcc.ch/activity/cct3.pdf
3. Smith, J.B., Schellnhuber, H.-J., Mirza, M.M.Q., Fankhauser, S., Leemans, R., Erda, L., Ogallo, L., Pittock, B., Richels, R., Rosenzweig, C. *et al.* (2001) in *Climate Change 2001: Impacts, Adaptation, and Vulnerability – Contribution of Working Group II to the Third Assessment Report of the Intergovernmental Panel on Climate Change*, eds. McCarthy, J.J., Canziani, O.F., Leary, N.A., Dokken, D.J., White, K.S. (Cambridge University Press, Cambridge, UK), pp. 913–967.
4. Manne, A., Richels, R.G. (1995) *Energy J.* **16**, 1–37.
5. Hammitt, J.K., Lempert, R.J., Schlesinger, M.E. (1992) *Nature* **357**, 315–318.
6. Webster, M.D. (2002) *Energy J.* **23**, 97–119.
7. Schneider, S.H. (1997) *Environ. Model. Assess.* **2**, 229–248.
8. Morgan, M.G., Henrion, M., eds. (1990) *Uncertainty: a Guide to Dealing with Uncertainty in Quantitative Risk and Policy Analysis* (Cambridge University Press, New York, NY).
9. Pittock, R.N., Jones, R.N., Mitchell, C.D. (2001) *Nature* **413**, 249.
10. Schneider, S.H., Turner, B.L., Morehouse Garriga, H. (1998) *J. Risk Res.* **1**, 165–185.
11. Watson, R.T., Albritton, D.L., Barker, T., Bashmakov, I.A., Canziani, O., Christ, R., Cubasch, U., Davidson, O., Gitay, H., Griggs, D. *et al.* (2001) in *Climate Change 2001: Synthesis Report, Summary for Policy Makers – Contribution of Working Groups I, II, and III to the IPCC Third Assessment Report*, ed. Watson, R.T. (Cambridge University Press, Cambridge, UK) pp. 37–43.
12. Mastrandrea, M.D., Schneider, S.H. (2004) *Science* **304**, 571–575.
13. Schneider, S.H., Mastrandrea, M.D. (2005) *Proc. Natl. Acad. Sci.* **102**, 15728–15735.
14. Moss, R.H., Schneider, S.H. (2000) in *Guidance Papers on the Cross Cutting Issues of the Third Assessment Report of the IPCC.* Eds. Pachauri, R., Taniguchi, T., Tanaka, K. (World Meteorological Organization, Geneva) pp. 33–51.
15. Wigley, T. (2004) *Clim. Change* **67**, 1–11.
16. Schneider, S.H., Kuntz-Duriseti, K., Azar, C. (2000) *Pac. and Asian J. Energy* **10**, 81–106.
17. Oppenheimer, M., Alley, R.B. (2004) *Clim. Change* **64**, 1–10.
18. Grassl, H., Kokott, J., Kulessa, M., Luther, J., Nuscheler, F., Sauerborn, R., Schellnhuber, H.-J., Schubert, R., Schulze, E.-D. (2003) German Advisory Council on Global Change (WBGU) Special Report, http://www.wbgu.de/wbgu_sn2003_engl.pdf
19. Hare, W. (2003) German Advisory Council on Global Change (WBGU), http://www.wbgu.de/wbgu_sn2003_ex01.pdf
20. Meinshausen, M., Hare, W., Wigley, T.M.L., van Vuuren, D.P., den Elzen, M.G.J., Swart, R. *Clim. Change* (in press).
21. O'Neill, B.C., Oppenheimer, M. (2004) *Proc. Natl. Acad. Sci.* **101**, 16411–16416.
22. Cubasch, U., Meehl, G.A., Boer, G.J., Stouffer, R.J., Dix, M., Noda, A., Senior, C.A., Raper, S., Yap, K.S., Abe-Ouchi, A. *et al.* (2001) in *Climate Change 2001: The Scientific Basis – Contribution of Working Group I to the IPCC Third Assessment Report*, eds. Houghton, J.T., Ding, Y., Griggs, D.J., Noguer, M., van der Linden, P.J., Xiaosu, D. (Cambridge University Press, Cambridge, UK) pp. 525–582.
23. Andronova, N.G., Schlesinger, M.E. (2001) *J. Geophys. Res.* **106**, 22605–22612.
24. Forest, C.E., Stone, P.H., Sokolov, A.P., Allen, M.R., Webster, M.D. (2001) *Science* **295**, 113–117.
25. Stainforth, D.A., Aina, T., Christensen, C., Collins, M., Faull, N., Frame, D.J., Kettleborough, J.A., Knight, S., Martin, A., Murphy, J.M. *et al.* (2005) *Nature* **433**, 403–406.
26. Hare, W., Meinshausen, M. (2005) *Clim. Change* (in press).
27. Andreae, M.O., Jones, C.D., Cox, P.M. (2005) *Nature* **435**, 1187–1190.
28. Knutti, R., Joos, F., Mueller, S.A., Plattner, G.K., Stocker, T.F. *Geophys. Res. Let.* (in press).
29. Schneider, S.H., Thompson, S.L. (1981) *J. Geophys. Res.* **86**, 3135–3147.
30. Nordhaus, W.D. (1994) *Managing the Global Commons: the Economics of Climate Change* (MIT Press, Cambridge, MA).
31. Myhre, G., Highwood, E.J., Shine, K.P., Stordal, F. (1998) *Geophys. Res. Let.* **25**, 2715–2718.
32. Schneider, S.H. (2004) *Global Environ. Change* 14, 245–258.
33. Oppenheimer, M. (1998) *Nature* **393**, 325–332.
34. Gregory, J.M., Huybrechts, P., Raper, S.C. (2004) *Nature* **428**, 616.
35. Stocker, T. F., Schmittner., A. (1997) *Nature* **388**, 862–865.
36. Wigley, T.M.L., Richels, R., Edmonds, J.A. (1996) *Nature* **379**, 240–243.
37. Hammitt, J.K. (1999) *Clim. Change* **41**, 447–468.
38. Swart, R., Mitchell, J., Morita, T., Raper, S. (2002) *Global Environ. Change* **12**, 155–165.
39. Wigley, T.M.L. (2004) *Benefits of Climate Policies: Improving Information for Policymakers* (Organization for Economic Cooperation and Development, Paris).
40. Mastrandrea, M.D., Schneider, S.H. (2001) *Clim. Pol.* **1**, 433–449.
41. L.D.D. Harvey, *J. Geophys. Res.* **91**, 2709 (1986).

CHAPTER 28

What Does a 2°C Target Mean for Greenhouse Gas Concentrations? A Brief Analysis Based on Multi-Gas Emission Pathways and Several Climate Sensitivity Uncertainty Estimates

Malte Meinshausen
Swiss Federal Institute of Technology (ETH Zurich), Environmental Physics, Department of Environmental Sciences, Zurich, Switzerland

ABSTRACT: Assuming a policy target of how much climate change is tolerable – e.g. in terms of global mean temperature rise – how low do greenhouse gas concentrations need to peak or stabilize to prevent this target being exceeded? Drawing from a set of 11 published climate sensitivity uncertainty estimates, the probability of exceeding 2°C equilibrium warming (e.g. European Union target) is found to lie between 63% and 99% for stabilization at 550 ppm CO_2 equivalence. Only at levels around 400 ppm CO_2 equivalence or below, could the probability of staying below 2°C in equilibrium be termed 'likely' for most of the climate sensitivity PDFs. The probability of exceeding 2°C ranges from 8% to 57% in this case. Going beyond the analysis of equilibrium warming levels, the transient probabilities of exceeding 2°C by 2100 are estimated. The analysis is based on 'Equal Quantile Walk' multi-gas emission pathways that stabilize at 550, 475 and 400 ppm CO_2eq. Given that stabilization at 400 ppm became infeasible to reach without overshooting, the latter pathway overshoots its ultimate 400 ppm stabilization level by peaking at 475 ppm. The results suggest that such an overshooting in terms of concentrations would increase the probability of overshooting the 2°C target. Hence, the maximal greenhouse gas concentration might be more relevant for achieving a 2°C target than the ultimate stabilization level. Thus, future research and policy might want to focus rather on the peaking concentrations over the 21st century instead of the ultimate stabilization level.

28.1 Introduction

Reviews of the scientific literature on climate impacts often conclude that a temperature increase of 2°C above pre-industrial levels cannot be assumed to be free of (potentially large-scale) adverse impacts. For example, the loss of the Greenland ice-sheet may be triggered by a local temperature increase of approximately 2.7°C (Huybrechts *et al.*, 1991; Gregory *et al.*, 2004), which could correspond to a global mean temperature increase of less than 2°C. This loss is likely to cause a sea level rise of seven meters over the next 1000 years or more (Gregory *et al.*, 2004). Hurricanes might increase in their intensity (Knutson and Tuleya, 2004; Emanuel, 2005), leading to an increase in the number of category-5 hurricanes, such as 'Katrina' in August 2005. Similarly, unique ecosystems such as coral reefs, the Arctic, and alpine regions are increasingly under pressure and may be severely damaged by global mean temperature increases of 2°C or below (Smith *et al.*, 2001; Hare, 2003; ACIA, 2004). For rising temperatures beyond 2°C, increasing risks of extreme events, distribution of climate impacts or aggregated effects on markets are becoming a growing reason for concern (Smith *et al.*, 2001). In addition, strong positive carbon cycle feedbacks are increasingly likely (Friedlingstein *et al.*, 2003; Jones *et al.*, 2003a; Jones *et al.*, 2003b), which would lead to even more climate change beyond the direct effect of anthropogenic emissions. Similarly, potentially large, but still very uncertain, methane releases might occur from thawing permafrost or ocean methane hydrates (Archer *et al.*, 2004; Buffet and Archer, 2004).

Despite the increasing knowledge on climate impacts, science will never be able to suggest a single threshold of what constitutes 'dangerous' climate change, as this is a value judgment, a political decision to take. In 1996, the European Council adopted a climate target that reads 'the Council believes that global average temperatures should not exceed 2 degrees above pre-industrial level'. This target has since been reaffirmed by the EU on a number of occasions, such as March 2005[1].

Starting from such a policy target, this analysis attempts to estimate the probability that certain concentration levels are consistent with such a 2°C threshold (and other temperature thresholds), both in equilibrium and over the medium-term until 2100. Thus, this analysis is deliberately focused narrowly on one jigsaw piece, the link between concentrations and temperatures, and our current understanding of the main uncertainty in regard to this link.

This uncertain link between greenhouse gas concentrations and temperatures is an important one in the vast body of literature that addresses more comprehensive

---

[1] EU Presidency Council conclusions, 23 March 2005, Brussels, available at http://europa.eu.int/rapid/pressReleasesAction.do?reference=DOC/05/1&fo

research questions, such as finding so-called 'optimal' mitigation action in a cost-benefit analysis framework (e.g. Nordhaus, 1992), probabilistic climate change forecasts including emission uncertainty (see e.g. Webster *et al.*, 2003; Richels *et al.*, 2004), optimal hedging strategies (e.g. Yohe *et al.*, 2004) and tolerable emissions corridors based on physical climate change thresholds (e.g. Toth *et al.*, 2003b; 2003a).

The response of the climate system to increased concentrations of greenhouse gases is often expressed in terms of 'climate sensitivity', defined as the warming which would ultimately occur for doubled CO_2 concentrations. Regrettably from a policy perspective, there is a relatively large uncertainty as to what the climate sensitivity actually is. A growing number of studies appeared over recent years that attempt to constrain this climate sensitivity uncertainty by observations in order to provide probability density functions (PDFs). Eleven such PDFs are used in the present study. The climate sensitivity PDFs can be used to derive the likelihood that an equilibrium temperature is exceeded for a given greenhouse gas stabilization level. The simple calculus is provided (see Section 28.2.2).

It is less straightforward though to estimate the medium-term probabilistic temperature implications of a given concentration path. One reason is that it is not clear how fast the climate system reaches equilibrium. In other words, the climate inertia, largely determined in simple climate models by the ocean mixing factors, is uncertain. Moreover, there are still significant uncertainties in regard to the net direct and indirect aerosol cooling effect (see e.g. Anderson *et al.*, 2003), which masks some of the positive greenhouse gas forcing. These uncertainties of climate sensitivity, inertia and aerosol forcing are somewhat interrelated, as e.g. the historic temperature observations could hardly be reproduced by models with a high climate sensitivity, low inertia and low aerosol forcing. High climate sensitivity values seem more likely in conjunction with high inertia and/or high aerosol forcing. This study takes account of this interdependency, as the maximum likelihood estimators of ocean diffusivity and aerosol forcing are applied to each climate sensitivity value. Applied methods are briefly highlighted in Section 28.2 and the Appendix.

The results of this analysis are provided in two main sections: Section 28.4 is about equilibrium considerations and section 28.5 on the medium-term temperature implications up to 2100, which might be more policy-relevant. More specifically, a continuum of stabilization levels between 350 ppm and 650 ppm CO_2 equivalence concentrations is analyzed in terms of their consistency with different temperature-based climate targets in equilibrium (Section 28.4). The subsequent section focuses on the medium-term temperature evolution before the climate system reaches equilibrium. Estimates of the probabilistic, transient temperature implications are provided for three multi-gas emissions paths (Section 28.5). Caveats are briefly discussed in section 28.6. Section 28.7 concludes.

28.2 Method

In addition to a brief description of the used set of probability density functions of climate sensitivity, this section describes the underlying methods for the presented equilibrium (Section 4) and medium-term (transient) results (Section 5). The probability of exceeding certain warming levels in the medium-term by 2100 critically depends on the concentration profile and consequently on the assumed emissions path. For example, whether or not a stabilization path is first overshooting the ultimate stabilization level critically influences the probability of staying below 2°C. Three 'Equal Quantile Walk' multi-gas emission pathways are briefly presented.

28.2.1 *Climate Sensitivity PDFs*

Climate sensitivity is the expected equilibrium warming for doubled CO_2 concentrations. Therefore, if climate sensitivity were 3°C, the equilibrium warming resulting from a stabilization at 550 ppm CO_2 (or the equivalent thereof) would be approximately 3°C above pre-industrial levels, given that pre-industrial CO_2 concentrations were 278 ppm ($2 \times 278 = \sim 556$ ppm). The conventional uncertainty range stated in the past three IPCC Assessment Reports is 1.5°C to 4.5°C for climate sensitivity (Houghton *et al.*, 1990; Houghton *et al.*, 1996; Houghton *et al.*, 2001). This conventional uncertainty range has been translated by Wigley and Raper (2001) into a lognormal PDF for climate sensitivity, which is used here alongside the observationally constrained PDFs. Since the IPCC Third Assessment Report, some studies (Andronova and Schlesinger, 2001; Forest *et al.*, 2002; Gregory *et al.*, 2002; Knutti *et al.*, 2003; Murphy *et al.*, 2004; Frame *et al.*, 2005; Knutti and Meehl, submitted; Piani *et al.*, submitted)[2] attempted to constrain the climate sensitivity by using recent observations. The applied methods, prior assumptions and used observational data vary among those studies. This analysis merely uses a large set of the probability density functions (PDFs) as published (see Figure 28.1). No post-processing of the climate sensitivity PDFs has been applied, i.e. the climate sensitivities were not re-weighted to account for different prior assumptions (see Frame *et al.*, 2005). However, where necessary, climate sensitivity PDFs have been truncated at 10°C to ensure better comparability as some studies did not explore higher sensitivities. This truncation, whether done for this analysis or implicitly applied by the authors of above publications when making prior assumptions, does reduce the upper end of projected equilibrium warming a bit and is important to keep in mind when reviewing the results.

---

[2] Note that the climate sensitivity PDF by Andronova and Schlesinger used here is the composite one that includes solar and aerosol forcing.

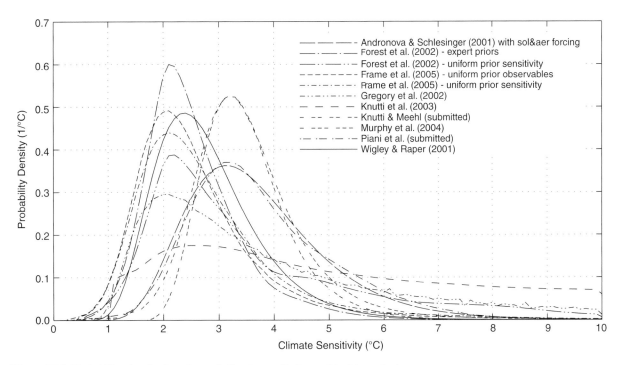

Figure 28.1 Probability density functions of climate sensitivity used in this analysis.

28.2.2 *Equilibrium Calculations*

Using a standard formula for the radiative forcing ΔQ caused by increased CO_2 concentrations C above pre-industrial levels C_o ($\Delta Q = 5.35*\ln(C/C_o)$ with $C_o = 278$ ppm) (Ramaswamy *et al.*, 2001), one can derive equilibrium warming levels ΔT_{eq} for any CO_2 (equivalent) concentration and climate sensitivity ΔT_{2x} by $\Delta T_{eq} = \Delta T_{2x}(\Delta Q/5.35 * \ln(2))$. Thus, the likelihood $P(\Delta T_{eq} \geq \Delta T_{crit})$ of exceeding a certain warming threshold ΔT_{crit} in equilibrium when stabilizing CO_2 (equivalent) concentrations at level C can be calculated as the integral

$$P(\Delta T_{eq} \geq \Delta T_{crit}) = \int_{\Delta T_{crit}}^{\infty} P\left[\Delta T_{2x} = x \frac{\ln(2)}{\ln\left[\dfrac{C}{C_o}\right]}\right]dx,$$

where $P(\Delta T_{2x} = \Delta t)$ is the assumed probability density function (PDF) for climate sensitivity ΔT_{2x}, or alternatively as

$$P(\Delta T_{eq} \geq \Delta T_{crit}) = 1 - P\left[\Delta T_{2x} \leq \Delta T_{crit} \frac{\ln(2)}{\ln\left[\dfrac{C}{C_o}\right]}\right],$$

where $P(\Delta T_{2x} \leq \Delta t)$ is the corresponding cumulative distribution function (CDF) for climate sensitivity ΔT_{2x}. The presented equilibrium results assume natural forcing to be the same in equilibrium as in pre-industrial times. As for the transient calculations, the CO_2 equivalent (CO_2eq) concentrations are here defined as the CO_2 concentrations that

would correspond to the same radiative forcing as caused by all human-induced increases in concentrations of greenhouse gases, tropospheric ozone and sulphur aerosols.

28.2.3 *Transient Calculations*

In contrast to the parameterized equilibrium calculations, transient probabilistic temperature evolutions were computed for this study with a simple climate model, namely the upwelling diffusion energy balance model MAGICC 4.1 by Wigley, Raper, Hulme *et al.*, (Wigley, 2003a). The probabilistic treatment of uncertainties has been confined to climate sensitivity on the basis of the above-cited PDF estimates (see Figure 28.1). Other key parameters, namely ocean diffusivity and sulphate forcing, are adapted depending on the climate sensitivity using maximum likelihood estimates (MLE) derived from a simple historical constraint test based on 20th century global mean temperatures (Folland *et al.*, 2001; Jones *et al.*, 2001; Jones and Moberg, 2003) and 1957–1994 ocean heat uptakes (Levitus *et al.*, 2000; Levitus *et al.*, 2005). This observational constraint is similar to the ones used in some earlier studies to derive a PDF for climate sensitivity (cf. e.g. Knutti *et al.*, 2003). However, as the focus of this analysis is to compare the implications of different climate sensitivity PDFs, rather than deriving a joint PDF for higher dimensional parameter spaces, the observational constraints are merely used to derive maximum likelihood estimates for other key climate system properties for a given climate sensitivity. This procedure supports the calculation of quantiles for the transient temperature implications based on the quantiles of published climate sensitivity CDFs, without being inconsistent with

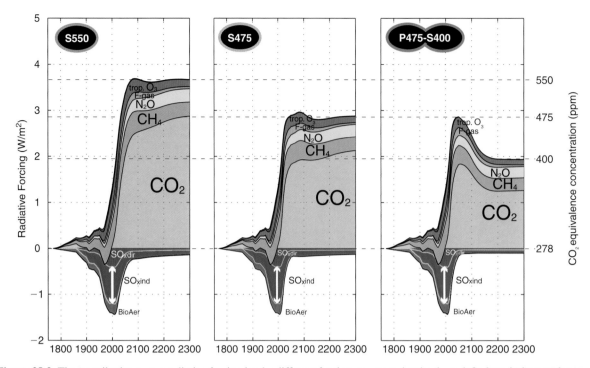

Figure 28.2 The contribution to net radiative forcing by the different forcing agents under the three default emissions pathways for stabilization at (a) 550, (b) 475 and (c) 400 ppm CO_2 equivalent concentration with the latter pathway (c) peaking at 475 ppm. The upper line of the stacked area graph represents net human-induced radiative forcing. The net cooling due to the direct and indirect effect of SO_x aerosols and aerosols from biomass burning is depicted by the lower negative boundary, on top of which the positive forcing contributions are stacked (from bottom to top) by CO_2, CH_4, N_2O, fluorinated gases, tropospheric ozone and the combined effect of fossil organic & black carbon. Note that a significant reduction of SO_x aerosol emissions (and consequently radiative forcing) for the near future is implied by the pathways (cf. Figure 28.4). The vertical arrows qualitatively indicate the large uncertainty of SO_x aerosol forcing (see methods section and Appendix).

historic temperature (and heat uptake) observations (see as well method description in Appendix).

Given that the heat uptake estimates by Levitus *et al.* (Levitus *et al.*, 2000; Levitus *et al.*, 2005) are dependent to a large degree on the infilling method for non-observed data points, which adds considerable uncertainty, two simple constraining methods were applied. Method A uses only the global mean temperature as a constraint to derive maximum likelihood estimates for a given climate sensitivity, while method B uses in addition the constraint provided by ocean heat content change (see Appendix).

Specifically, the simple climate model is run subsequently with different climate sensitivity quantiles, e.g. 5%, 10% 15%... of a specific CDF, while adapting each time the ocean mixing and sulphate forcing parameters to their respective maximum likelihood estimates. The resulting global mean temperatures profiles then correspond to the quantiles of the underlying climate sensitivity. Note that this procedure takes account of the interdependency between climate sensitivity, ocean mixing and sulphate cooling, but not of the uncertainty distributions in the latter two parameters. For other climate model parameters, this analysis assumes IPCC TAR 'best estimate' parameters, such as those related to carbon cycle feedbacks. Solar forcing is assumed according to

Lean *et al.*, (1995) and volcanic forcing is assumed according to Ammann *et al.*, (2003).

28.3 Emission Pathways

In order to assess probabilistic temperature evolutions over time, three multi-gas emission pathways have been designed which stabilize at CO_2 equivalence levels of 550 ppm ($3.65 W/m^2$), 475 ppm ($2.86 W/m^2$) and 400 ppm ($1.95 W/m^2$). The latter pathway is assumed to peak at 475 ppm before returning to its ultimate stabilization level around the year 2150 (see Figure 28.2). This overshooting of the ultimate stabilization level is partially justified by the already substantial present net forcing levels and the attempt to avoid sudden drastic reductions in the presented emission pathways. All pathways are within the range of the lower mitigation scenarios in the literature (cf. Hare and Meinshausen, 2004).

The presented multi-gas emission pathways were derived by the 'Equal Quantile Walk' (EQW) method (Meinshausen *et al.*, in press) on the basis of 54 existing IPCC SRES and Post-SRES scenarios (Nakicenovic and Swart, 2000; Swart *et al.*, 2002). The emissions that have been adapted to meet the pre-defined stabilization targets

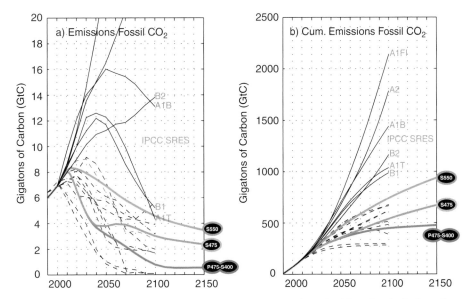

Figure 28.3 Global fossil CO_2 emissions (a) and cumulative fossil CO_2 emissions (b). Note that the indicated cumulative emissions may be up to 100–150 GtC lower, if landuse CO_2 emissions do not decline as steeply as depicted in Figure 28.4a. Emissions of the illustrative IPCC SRES non-mitigation scenarios (Nakicenovic and Swart, 2000) are depicted for comparative purposes (thin lines) together with some of the lower mitigation scenarios available in the literature (Nakicenovic and Riahi, 2003; Azar *et al.*, submitted) (thin dotted lines).

of 550 ppm, 475 ppm and 400 ppm CO_2 equivalence include those of all major greenhouse gases (fossil CO_2, land use CO_2, CH_4, N_2O, HFCs, PFCs, SF_6), ozone precursors (VOC, CO, NO_x) and sulphur aerosols (SO_x). The basic idea behind the 'Equal Quantile Walk' method is that emissions for all gases of the new emission pathway are in the same quantile of the existing distribution of IPCC SRES and post-SRES scenarios. In other words, if fossil CO_2 emissions are assumed in the lower 10% region of the existing SRES and Post-SRES scenario pool, then methane, N_2O and all other emissions are designed to also be in the pool's respective lower 10% region. More details, a comparison with other mitigation pathways, such as the CO_2 WRE profiles, and caveats in regard to the EQW method are described in Meinshausen *et al.*, (in press)[3].

In the following, the actual emissions of the presented EQW pathways are briefly described. Fossil CO_2 emissions increase from 6 GtC/yr in 1990 to their peak value of approximately 8 GtC/yr around 2010–2015 in all three pathways. Thereafter, fossil emissions are decreasing, down to 7.5 GtC/yr and 4 GtC/yr in 2035 to limit maximal CO_2eq concentrations at 550 ppm and 475 ppm, respectively. Cumulative fossil CO_2 emissions for the 550 ppm, 475 ppm and 400 ppm stabilization until 2150 are 940 GtC, 670 GtC and 470 GtC for the three pathways (see Figure 28.3b). However, these cumulative fossil CO_2 emissions would need to be about 100 to 150 GtC lower to reach the same CO_2 concentrations, if landuse CO_2 emissions were kept at present levels of approximately 1 GtC/yr, instead of being reduced as shown in Figure 28.4a.

The global greenhouse gas (GHG) emissions of the presented EQW emission pathways can be summarized by their GWP-weighted sum for illustrative purposes[4] (Figure 28.4f). Under the default scenario for stabilization at 550 ppm CO_2eq, GHG emissions (CO_2, CH_4, N_2O, HFCs, PFCs and SF_6) are approximately 10% below their 1990 levels by 2050. For stabilization at 475 ppm CO_2eq, global GHG emissions are about 45% lower by 2050 compared to 1990 levels. For stabilization at 400 ppm with an initial peaking at 475 ppm CO_2eq, global emissions are approximately 55% lower by the year 2050 compared to 1990. Interestingly, emissions up to 2040 are the same, whether CO_2 equivalence concentrations will stabilize at 475 ppm or whether the world aims for a lower stabilization level with a temporary overshoot up to 475 ppm. The presented EQW emission paths and their gas-by-gas emissions are available for download at www.simcap.org.

28.4 Equilibrium Results

28.4.1 *The Probability of Exceeding 2°C*

At 550 ppm CO_2 equivalence (corresponding approximately to a stabilization at 475 ppm CO_2 only), the likelihood of exceeding 2°C is very high, ranging between 63% and 99% for the different climate sensitivity PDFs

---

[3] Data and software available at www.simcap.org

[4] Note that the Global Warming Potentials (GWPs) were not applied in any of the underlying calculations for deriving CO_2 equivalence concentrations (which is a different concept than CO_2 equivalent emissions). The GWPs, specifically the 100 year GWPs (IPCC 1996), were simply used here to present the different greenhouse gas emissions in a manner consistent with the current practice in policy documents, such as the Kyoto Protocol.

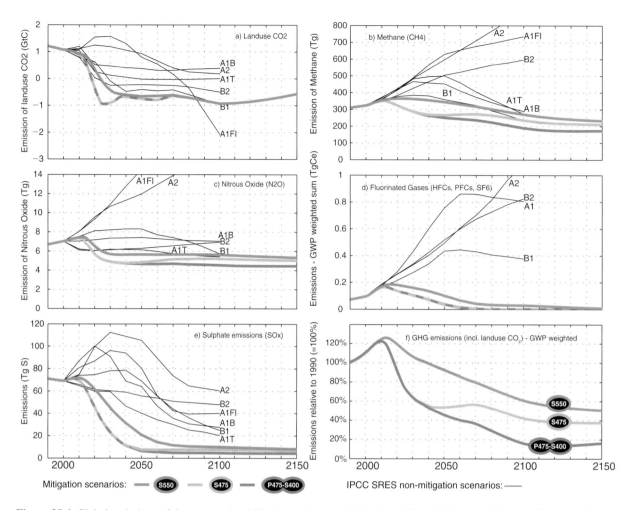

Figure 28.4 Global emissions of the presented stabilization pathways: (a) Landuse CO_2, (b) Methane CH_4, (c) Nitrous Oxide N_2O, (d) the GWP weighted sum of CF_4, C_2F_6, HFC125, HFC134a, HFC143a, HFC227ea, HFC245ca, and SF_6, (e) sulphate aerosol emissions SO_x and the GWP weighted sum of all greenhouse gases ((a) to (d) and Figure 28.3a) All gases are treated separately in the emission pathways and the climate model despite being aggregated here for illustrative purposes. Emissions of the illustrative IPCC SRES non-mitigation scenarios are depicted in thin lines.

Table 28.1 The probability of exceeding 2°C warming above pre-industrial levels in equilibrium for different CO_2 equivalence stabilization levels. Upper bound, mean and lower bound are given for the set of eleven analyzed climate sensitivity PDFs (see text and cf. Figure 28.5).

CO_2eq stabilization level (ppm)	350	400	450	475	500	550	600	650	700	750
Upper Bound	31%	57%	78%	90%	96%	99%	100%	100%	100%	100%
Mean	7%	28%	54%	64%	71%	82%	88%	92%	94%	96%
Lower Bound	0%	8%	26%	38%	48%	63%	74%	82%	87%	90%

with a mean of 82% (see Table 28.1 and Figure 28.5). In other words, an equilibrium warming below 2°C could be categorized as 'unlikely' using the IPCC WGI terminology[5]. If greenhouse gas concentrations were to be stabilized at 475 ppm CO_2eq then the likelihood of exceeding 2°C would be lower, in the range of 38% to 90% (mean 64%), but still significant. In other words, all 11 analyzed climate sensitivity PDFs suggest that there is either a 'medium likelihood' or an 'unlikely' chance to stay below 2°C in equilibrium for stabilization at 475 ppm CO_2 equivalence. Only for a stabilization level of 400 ppm CO_2eq and below can warming below 2°C be roughly classified as 'likely' (probability of exceeding 2°C between 8% and 57% with mean 28%). The likelihood of exceeding 2°C at equilibrium is further reduced,

[5] See IPCC TAR Working Group I Summary for Policymakers: Virtually certain (>99%), very likely (90%–99%), likely (66%–90%), medium likelihood (33%–66%), unlikely (10%–33%), very unlikely (1%–10%), exceptionally unlikely (<1%).

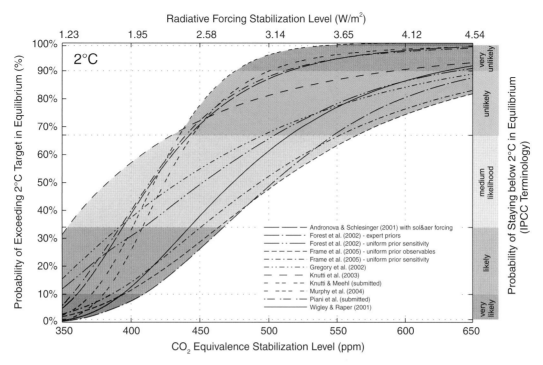

Figure 28.5 The probability of exceeding 2°C global mean equilibrium warming for different CO_2 equivalent stabilization levels.

0% to 31% (mean 7%), if greenhouse gases were stabilized at 350 ppm CO_2eq (see Table 28.1 and Figure 28.5).

28.4.2 *The Probability of Exceeding Other Temperature Levels*

For comprehensive climate impact assessments across different stabilization levels, it is warranted to also include the lower risk/higher magnitude adverse climate impacts that can be expected at higher temperature levels. Clearly, this analysis does not provide any impact assessment, but provides estimates of likelihoods that certain temperature levels might be exceeded for different stabilization levels. Given that climate sensitivity PDFs largely differ on the probability of very high climate sensitivities (>4.5°C), it is not surprising that a rising spread of probability is obtained for higher warming thresholds. For stabilization at 550 ppm CO_2eq the probability of exceeding global mean temperatures of 3°C is still substantial, ranging from 21% to 69%. Furthermore, 7 out of the 11 analyzed climate sensitivity PDFs suggest that the probability of exceeding 4°C equilibrium warming is between 10% and 33% for stabilization at 550 ppm CO_2 equivalence. Given that a 4°C global mean temperature rise is expected to cause rather disastrous climate impacts on multiple scales (Smith *et al.*, 2001), such a probability between 10% to 33% seems non-negligible (see Figure 28.6).

28.5 Transient Results Until 2100

The significant influence that climate inertia can have on 2100 warming becomes apparent from the transient

temperature results obtained by different methods to set other climate model parameters depending on the climate sensitivity (Figure 28.7). As described in the Appendix, two different methods have been applied to account for the dependency between climate sensitivity, ocean diffusivity and sulphate forcing (see Figure 28.9 in Appendix). Compared to method B ('Combined constraint'), method A ('Temperature Constraint') suggests higher ocean diffusivity values and hence higher climate inertia, especially for high climate sensitivities (see Figure 28.10 in Appendix). Thus, transient temperatures at the upper end of the projected temperature distribution are lower for the high inertia. The lower end of the projected temperature distributions differs less between the methods, since both methods suggest similar maximum likelihood estimators (MLE) of ocean diffusivity and of aerosol forcing for low climate sensitivities (see Figure 28.10 in Appendix).

The effect of higher climate inertia for stabilization pathways, such as the presented 550 ppm and 475 ppm, is simply that the equilibrium warming is more slowly approached (compare upper end of temperate distributions in panel *a* and *d* or *b* and *e* of Figure 28.7). Specifically, in the case of the 550 ppm stabilization path, the transient temperature by 2100 is estimated to lie between 1.4°C and 2.9°C for the 'high-inertia' method A and between 1.4°C and 3.6°C for the 'low-inertia' method B (cf. 5% to 95% confidence levels, marked by dashed lines in Figure 28.7a and Figure 28.7d).

For pathways with overshooting concentrations, the effect of climate inertia is that the overshooting of temperature is dampened or even completely 'shaved off'. Thus – due to the inertia of the climate system – the peak

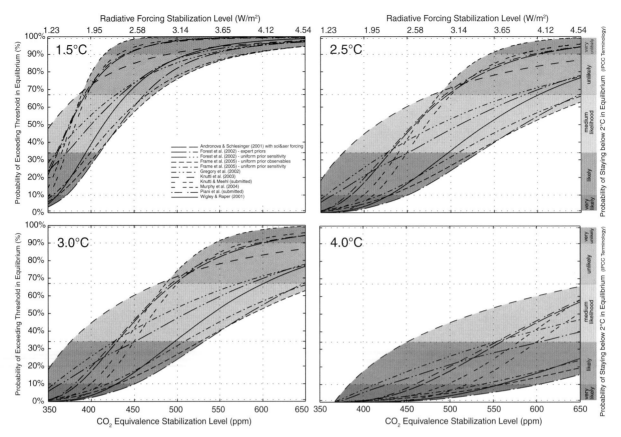

Figure 28.6 The probability of exceeding (a) 1.5°C, (b) 2.5°C, (c) 3°C and (d) 4°C global mean equilibrium warming for different CO_2 equivalent stabilization levels.

at 475 ppm CO_2eq before the stabilization at 400 ppm CO_2eq does not translate into a comparable peak in global mean temperatures. Nevertheless, the initial peak at 475 ppm CO_2eq – rather than the 400 ppm CO_2eq stabilization level – seems to be decisive when addressing the question of whether a 2°C temperature threshold will be crossed (see Figure 28.7). For this particular emission path, both methods A and B suggest that the probability of staying below 2°C is lower during the medium term up to 2100 – compared to equilibrium. Specifically, the probability of staying below 2°C is decreased between 5% and 35% due to the overshooting up to 475 ppm – depending on the climate sensitivity PDF and applied method to derive other key parameters (see Figure 28.8). In other words, the factor that determines whether a 2°C target will be achieved or not, is the overshooting level of 475 ppm rather than the stabilization level of 400 ppm.

28.6 Caveats

This section discusses some of the limitations and caveats of the present analysis. Obviously, this analysis is not an integrated assessment of tolerable warming levels, or an elaboration of so-called 'economically optimal' emission reductions by weighing avoided climate impacts against net mitigation costs. Starting from one example of a

policy target, namely 2°C above pre-industrial levels, the focus of this study is merely on one jigsaw piece, the uncertain link between concentrations and temperatures – illustrated with a set of prescribed emission pathways for different concentration stabilization levels.

One obvious limitation is that this analysis presents equilibrium and transient probabilities of exceeding certain warming levels merely based on climate sensitivity PDFs. Thus, no probabilistic treatment of forcing, carbon cycle feedback and ocean mixing uncertainties has been undertaken. This does only affect the presented transient results, though. The projected transient temperature ranges would widen, if those uncertainties were treated similarly to the climate sensitivity. The reason why this analysis does limit the probabilistic treatment to climate sensitivity is twofold. Firstly, the present uncertainty in climate sensitivity is likely to contribute the most to the overall medium and long-term temperature response uncertainty for a given emission path. Secondly, this analysis attempts to derive results that are representing the current uncertainty range published in the literature. Since joint PDFs for the key climate system parameters were in most studies either not computed or published, the presented analysis takes a second-best approach. Furthermore, even if joint PDFs were provided in the cited studies, ocean mixing parameters are not transferable between models in most cases, which would considerably complicate the computation of

Method: Temperature Constraint

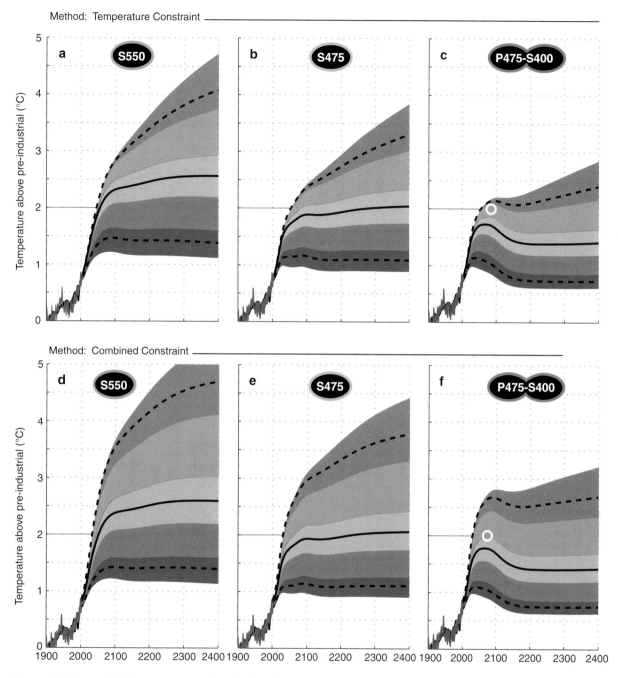

Method: Combined Constraint

Figure 28.7 The probabilistic temperature implications for pathways that stabilize at (a,d) 550 ppm, (b,e) 475 ppm, and (c,f) 400 ppm CO_2 equivalent concentrations, based on the climate sensitivity PDF that is derived from the conventional IPCC uncertainty range 1.5°C to 4.°C (Wigley and Raper, 2001). The upper (lower) panels depict results when applying the maximum likelihood estimates for ocean mixing and sulphate aerosol forcing as derived by Method A (B). Shown are the median (solid lines), and 90% confidence interval boundaries (dashed lines), as well as the 1%, 10%, 33%, 66%, 90%, and 99% percentiles (borders of shaded areas). The historic temperature record and its uncertainty are shown from 1900 to 2003.

probabilistic temperatures based on different studies' joint PDFs. Alternatively, joint PDFs could have been derived by a historical constraining method, e.g. such as the one applied here for finding maximum likelihood parameters. However, this would change the character of this analysis, which attempts to provide results based on the range reported in the literature spanning different models and constraining methods.

There is a potential limitation in regard to the climate sensitivity PDFs drawn from the literature. Given that all these PDFs are influenced to some degree by rather arbitrary assumptions in regard to the prior distributions, one could have attempted to normalize all PDFs, e.g. by re-weighting them to adjust the PDFs as if they all had been derived by a common set of prior assumptions. See Frame *et al.*, (2005) for a detailed discussion on this

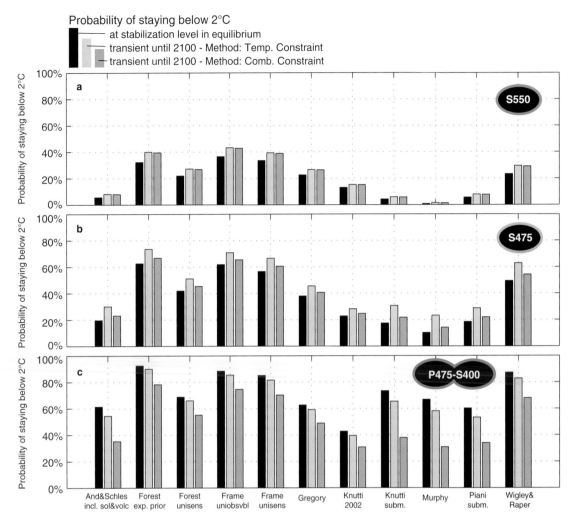

Figure 28.8 The probabilities of staying below 2°C for the three emission pathways that stabilize at 550 ppm CO$_2$ equivalence concentrations (a), 475 ppm (b) and 400 ppm after overshooting stabilization concentrations up to 475 ppm(c). Results are given for a set of 11 different probability density functions (PDFs) for climate sensitivity (see text). The equilibrium probabilities of staying below 2°C (black bars) are a little lower than the transient probabilities up to 2100 (light and dark blue bars) for the 550 ppm and 475 ppm stabilization pathways that do not overshoot concentrations (a,b). In contrast, the probabilities of staying below 2°C are lower during the transient period up to 2100 than for equilibrium for the latter emission pathway that overshoots its ultimate stabilization level of 400 ppm (see text for discussion).

issue. However, given that the prior assumptions are to some degree arbitrary, the different non-post-processed PDFs reflect this component of uncertainty.

In addition to the above, there are some caveats and limitations that apply specifically to the presented transient results. First of all, a simple energy balance upwelling diffusion model is used for all transient calculations, instead of an intermediate complexity model as used in Knutti *et al.* (in press), or coupled general circulation models (AOGCMs). However, the used simple climate model (MAGICC 4.1 by Wigley, Raper *et al.*) has been calibrated against AOGCMs, and used extensively over the past decade – as, for example, for the global mean temperature projections in the IPCC Third Assessment Report (see e.g. Wigley and Raper, 2001).

A further potential limitation is the case-study type of approach based on three mitigation pathways. Many

different emission pathways can lead to the same CO$_2$ (equivalence) stabilization level (see e.g. Swart *et al.*, 2002; Wigley, 2003b; Knutti *et al.*, in press). In general, though, the cumulative emissions are a relatively robust pathway characteristic across emissions pathways that lead to the same concentration stabilization level. Thus, pathways with higher emissions in earlier decades generally imply lower emissions towards the middle and end of the 21st century. Furthermore, pathways with higher emissions in the earlier decades – compared to those presented – lead to (higher) overshooting of the ultimate CO$_2$ equivalence concentrations. As a consequence of (higher) overshooting, the probability of exceeding any given temperature threshold is likely to be increased temporarily compared to the results shown below (see Figure 28.8), although this depends on the rate at which CO$_2$ equivalence concentrations are brought down again after peaking.

28.7 Discussion and Conclusion

The presented results attempt to sketch what the probabilities are of exceeding a given temperature threshold that one must be willing to accept when embarking on one emission pathway or another. The main conclusions that can be drawn from this analysis:

First, the results indicate that a 550 ppm CO_2 equivalent stabilization scenario is clearly not in line with a climate target of limiting global mean temperature rise to 2°C above pre-industrial levels. Even for the most 'optimistic' estimate of a climate sensitivity PDF, the risk of overshooting 2°C is 63% in equilibrium (cf. Figure 28.5).

Second, there is also a substantial risk of overshooting high temperature levels for stabilization at 550 ppm CO_2eq. Assuming a climate sensitivity PDF, which is consistent with the conventional IPCC 1.5°C to 4.5°C range (Wigley and Raper, 2001), the risk of overshooting 4°C as a global mean temperature rise is still 9%. Assuming the recently published climate sensitivity PDF by Murphy *et al.* (Murphy *et al.*, 2004), the risk of overshooting 4°C is as high as 25% (cf. Figure 28.6).

Third, the probability of overshooting 2°C can be substantially reduced for lower stabilization levels. For stabilization at 400 ppm CO_2 equivalence, seven out of eleven climate sensitivity PDFs suggest that staying below 2°C warming in equilibrium is 'likely' or 'very likely' based on the IPCC terminology for probabilities. For stabilization at 475 ppm CO_2 equivalence, the chance of staying below 2°C in equilibrium is still rather limited given that all analyzed PDFs suggest it to be a 'medium likelihood' or 'unlikely' event (cf. Figure 28.5).

Fourth, the presented results suggest that if CO_2 equivalence concentrations temporarily overshoot a 400 ppm stabilization level by 75 ppm, the probability of staying below 2°C is decreased (see Figure 28.8). However, the overshooting of such relatively low stabilizations levels might be a necessity, not an option, given that a peaking at 475 ppm CO_2 equivalence is already asking for substantial emission reductions in the coming two to three decades. It follows that future research and policy might want to focus on the peaking concentrations over the coming century rather than ultimate stabilization levels over the coming century if it is deemed likely that a 2°C temperature target is required to prevent 'dangerous anthropogenic interference with the climate system' (Art. 2 UNFCCC).

Acknowledgements

The author is most grateful to discussions with, tips from and editing support by Nicolai Meinshausen, Bill Hare, Stefan Rahmstorf, Myles Allen, Reto Knutti, Claire Stockwell, Fiona Koza, Paul Baer and Michèle Bättig. I thank as well Gary Yohe and an anonymous reviewer for the critical review, which helped to substantially improve the analysis. Dieter Imboden is warmly thanked for his support. Finally, the author would like to thank Tom Wigley for providing vital assistance and the MAGICC 4.1 climate model. Clearly, all errors are due to the author, not the colleagues.

APPENDIX

A Simple Constraining Method to Limit Inconsistencies Between Model Runs and Observations

As briefly outlined in the method section, a simple historical constraint has been applied in order to take into account the dependency between climate sensitivity, ocean mixing and sulphate aerosol forcing. The constraining datasets of global mean temperature and ocean heat uptake were used in many earlier studies (e.g. Knutti *et al.*, 2002; Knutti *et al.*, 2003; Frame *et al.*, 2005). The dependency arises from the fact that some combinations of these three parameters seem to clearly contradict observations of global mean temperature over the past century.

Method A, the 'Temperature Constraint' method, simply finds the maximum likelihood estimator for both ocean diffusivity and (indirect) aerosol forcing based on global mean temperature observations over the last century. Given that the indirect aerosol forcing scales linearly with aerosol emissions in the same way that the direct aerosol forcing does, it is only necessary to constrain one of these aerosol forcing parameters. For a given climate sensitivity, the likelihood of ocean diffusivity and aerosol forcing is determined as the sum of least squares residuals, weighted by the time-dependent measurement uncertainty of global temperature observations (Folland *et al.*, 2001; Jones *et al.*, 2001; Jones and Moberg, 2003) and internal variability ($\sigma_{ctlr,T}$) estimated from a HadCM3 control run ($\sigma = 0.119°C$), as provided in Harvey and Wigley, (2003). Specifically, the likelihood L_i for parameter combination 'i' is estimated by $\exp(-0.5 * SSR_{A,i})$, where the weighted sum of squared residuals $SSR_{A,i}$ is calculated as:

$$SSR_{A,i} = \sum_{t=1900}^{2000} \frac{(T_t^{obs} - T_{t,i}^{mod})}{2(\sigma_{meas,T,t}^2 + \sigma_{ctlr,T}^2)}$$

with $\sigma_{meas,T,t}$ being the time-dependent measurement uncertainty, T_t^{obs} the global mean temperature observations and $T_{t,i}^{mod}$ being the global mean temperature results of parameter combination i.

The maximum likelihood estimators are then calculated by the best fit of 961 model setups sampled from 31 ocean diffusivity and 31 aerosol forcing parameters for each climate sensitivity quantile (see Figure 28.9 and 28.10).

Method B, the 'Combined Constraint' method, uses in addition to the global mean temperature constraint, the ocean heat content data and its measurement uncertainty as provided by Levitus *et al.* (2000). The likelihood $\exp(-0.5 * SSR_{B,i})$ of each parameter combination *i* is

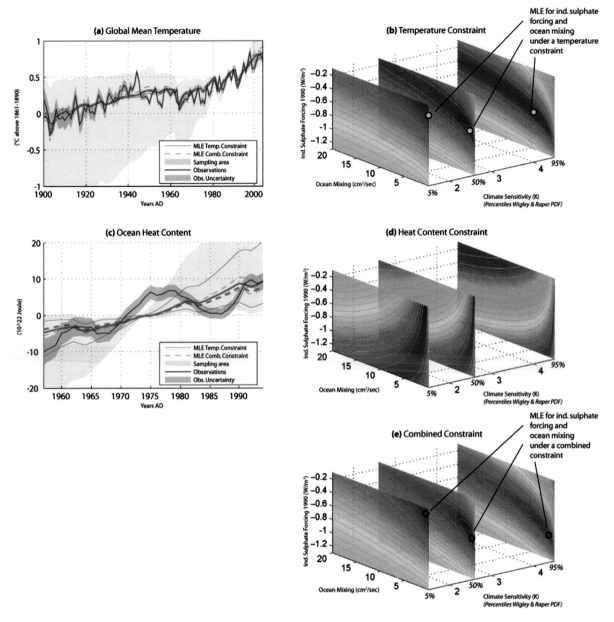

Figure 28.9 Overview of applied methods to derive maximum likelihood estimators for ocean diffusivity and sulphate aerosol forcing given a certain climate sensitivity. 961 combinations of ocean diffusivity and aerosol forcing were run with the simple climate model for each analyzed quantile of climate sensitivity (here shown for illustrative 5%, 50% and 95% quantiles of the IPCC lognormal PDF by Wigley and Raper (2001) – see panels b, d and e). The effective sampling area in the observable space of global mean temperature and ocean heat content is depicted by the light grey shaded areas in panels a and c, respectively. The weighted 'sum of squared residuals' likelihood of parameter combinations based on a comparison with observations of temperature (Folland *et al.*, 2001; Jones *et al.*, 2001; Jones and Moberg, 2003) and heat content (Levitus *et al.*, 2000) is illustrated by the contour lines and coloring of the slices in panels b and d. The combination of both constraints is shown in panel e. Maximum likeliood estimates are shown as bright and dark blue circles for method A ('temperature constraint') in panel b and method B ('combined constraint') on panel e, respectively.

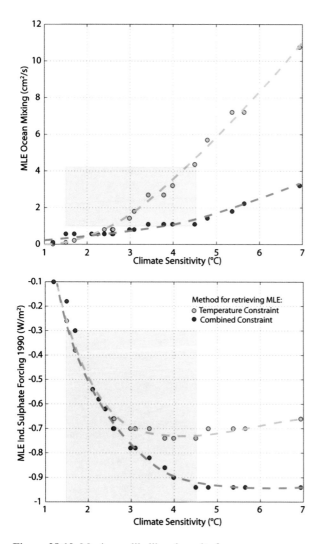

Figure 28.10 Maximum likelihood results for ocean mixing/ocean diffusivity (panel a) and indirect sulphate aerosol forcing in 1990 (panel b). Shown are results for the two different methods, which maximize the model match of global mean temperatures (method A 'temperature constraint') and both temperatures and heat uptake (method B 'combined constraint'). For comparison, the grey shaded areas denote the uncertainty ranges as used for global-mean warming projections in the IPCC Third Assessment Report for climate sensitivity (1.5°C to 4.5°C), ocean diffusivity (1.3 cm/s to 4.1 cm/s) and the total (indirect and direct) aerosol forcing characterized by the forcing in 1990 ($-0.7\,\text{W/m}^2$ to $-1.7\,\text{W/m}^2$ – here plotted as $-0.3\,\text{W/m}^2$ to $-1.3\,\text{W/m}^2$ range as the model uses a standard value for direct aerosol forcing of $0.4\,\text{W/m}^2$) (see Cubasch *et al.*, 2001; Wigley and Raper, 2001; Wigley 2005).

thus estimated by:

$$SSR_{B,i} = \sum_{t=1900}^{2000} \frac{(T_t^{obs} - T_{t,i}^{mod})}{2(\sigma_{meas,T,t}^2 + \sigma_{ctlr,T}^2)}$$

$$+a \sum_{t=1957}^{1994} \frac{(H_t^{obs} - H_{t,i}^{mod})}{2(\sigma_{meas,H,t}^2 + \sigma_{ctlr,H}^2)}$$

where $\sigma_{meas,T,t}$ is the time-dependent measurement uncertainty of ocean heat content, as provided by (Levitus *et al.*, 2000), H_t^{obs} the ocean heat content observations and $H_{t,i}^{mod}$ being the ocean heat content model results of parameter combination i. The internal variability ($\sigma_{ctlr,H}$) is estimated from the detrended observational time series.

Note that both methods A and B use very simplified measurements of fit, as they don't take into account the autocorrelation of the time series. Method A is likely to be relatively robust on whether the autocorrelation is taken into account or not. However, method B is sensitive to the autocorrelation, as the autocorrelation structure of global-mean temperatures is roughly four times shorter compared to the ocean heat uptake observations. Furthermore, the ocean heat content observations are likely to be subject to larger uncertainties than global mean temperatures. The availability of (deep ocean) observations is even sparser compared to surface temperature series. Therefore, the chosen in-filling method to complement actual observations with data for the non-observed ocean bins has a large influence. In addition, even complex models can currently only reproduce the trend, not the variability of the heat content observations provided by Levitus *et al.* (2000; 2005) – see e.g. Reichert *et al.* (2002) and Barnett *et al.*, (2001). Here, a pragmatic approach has been chosen by applying a weight factor 'a' to the ocean heat uptake residuals, assumed as 1/20. Clearly, method B should therefore only be regarded as a sensitivity result. Applying a weight factor 'a' of 1/4 or larger, the relationship between climate sensitivity and ocean diffusivity would be less pronounced (ocean diffusivity would be roughly constant and rather low – not shown).

The results obtained under both methods seem to be generally in line with what earlier studies suggest, namely that the climate inertia is likely to be high, if the real climate sensitivity is high and vice versa (Hansen *et al.*, 1985; Raper *et al.*, 2002). The results of method A, however, seem to better match the uncertainty range for ocean diffusivity as used for the global-mean temperature projections in the IPCC TAR – at least over the climate sensitivity range from 1.5°C to 4.5°C (cf. Figure 28.10). Method B does seem to deliver parameter settings that fit the observed ocean heat content changes better. However, given the high uncertainty and short observational period of the heat content data, it is not clear whether Method B is actually delivering more reliable results. One could argue on the contrary that the insensitivity of ocean mixing towards climate sensitivity under method B seems unrealistic, although this is speculative at the moment. It seems thus an open question as to which of these two or other constraining methods are best suited for this task, which needs to be addressed in future studies. Future studies will benefit from reproducing the original constraining methods (Andronova and Schlesinger, 2001; Forest *et al.*, 2002; Gregory *et al.*, 2002; Knutti *et al.*,

2003; Murphy *et al.*, 2004; Frame *et al.*, 2005; Knutti and Meehl, submitted; Piani *et al.*, submitted), in order to directly derive transient probabilities of overshooting different temperature thresholds for different mitigation pathways based on joint PDFs.

REFERENCES

ACIA: 2004, *Impacts of a Warming Arctic – Arctic Climate Impact Assessment*, Cambridge University Press, Cambridge, UK.

Ammann, C.M., Meehl, G.A., Washington, W.M. and Zender, C.S.: 2003, 'A monthly and latitudinally varying volcanic forcing dataset in simulations of 20th century climate', *Geophysical Research Letters* **30**.

Anderson, T.L., Charlson, R.J., Schwartz, S.E., Knutti, R., Boucher, O., Rodhe, H. and Heintzenberg, J.: 2003, 'Climate forcing by aerosols – a hazy picture', *Science* **300**, 1103–1104.

Andronova, N.G. and Schlesinger, M.E.: 2001, 'Objective estimation of the probability density function for climate sensitivity', *Journal of Geophysical Research-Atmospheres* **106**, 22605–22611.

Archer, D., Martin, P., Buffett, B., Brovkin, V., Rahmstorf, S. and Ganopolski, A.: 2004, 'The importance of ocean temperature to global biogeochemistry', *Earth and Planetary Science Letters* **222**, 333–348.

Azar, C., Lindgrena, K., Larsonb, E., Möllersten, K. and Yand, J.: submitted, 'Carbon capture and storage from fossil fuels and biomass – Costs and potential role in stabilizing the atmosphere', *Climatic Change*.

Barnett, T.P., Pierce, D.W. and Schnur, R.: 2001, 'Detection of anthropogenic climate change in the world's oceans', *Science* **292**, 270–274.

Buffet, B. and Archer, D.: 2004, 'Global Inventory of Methane Clathrate: Sensitivity to Changes in the Deep Ocean', *Earth and Planetary Science Letters*, 185–199.

Cubasch, U., Meehl, G.A., Boer, G.J., Stouffer, R.J., Dix, M., Noda, A., Senior, C.A., Raper, S. and Yap, K.S.: 2001, 'Projections of Future Climate Change', in Houghton, J.T., Ding, Y., Griggs, D.J., Noguer, M., van der Linden, P.J., Dai, X., Maskell, K. and Johnson, C.A. (eds.), *Climate Change 2001: The Scientific Basis*, Cambridge University Press, Cambridge, UK, pp. 892.

Emanuel, K.: 2005, 'Increasing destructiveness of tropical cyclones over the past 30 years', *Nature* **436**, 686–688.

Folland, C.K., Rayner, N.A., Brown, S.J., Smith, T.M., Shen, S.S.P., Parker, D.E., Macadam, I., Jones, P.D., Jones, R.N., Nicholls, N. and Sexton, D.M.H.: 2001, 'Global temperature change and its uncertainties since 1861', *Geophysical Research Letters* **28**, 2621–2624.

Forest, C.E., Stone, P.H., Sokolov, A., Allen, M.R. and Webster, M.D.: 2002, 'Quantifying Uncertainties in Climate System Properties with the Use of Recent Climate Observations', *Science* **295**, 113–117.

Frame, D.J., Booth, B.B.B., Kettleborough, J.A., Stainforth, D.A., Gregory, J.M., Collins, M. and Allen, M.R.: 2005, 'Constraining climate forecasts: The role of prior assumptions', *Geophysical Research Letters* **32**.

Friedlingstein, P., Dufresne, J.L., Cox, P.M. and Rayner, P.: 2003, 'How positive is the feedback between climate change and the carbon cycle?' *Tellus Series B-Chemical and Physical Meteorology* **55**, 692–700.

Gregory, J.M., Huybrechts, P. and Raper, S.C.B.: 2004, 'Climatology: Threatened loss of the Greenland ice-sheet', *Nature* **428**, 616.

Gregory, J.M., Stouffer, R.J., Raper, S.C.B., Stott, P.A. and Rayner, N.A.: 2002, 'An observationally based estimate of the climate sensitivity', *Journal of Climate* **15**, 3117–3121.

Hansen, J., Russell, G., Lacis, A., Fung, I., Rind, D. and Stone, P.: 1985, 'Climate Response-Times – Dependence on Climate Sensitivity and Ocean Mixing', *Science* **229**, 857–859.

Hare, B. and Meinshausen, M.: 2004, 'How much warming are we committed to and how much can be avoided?' PIK Report.

Potsdam, Potsdam Institute for Climate Impact Research: 49. No. 93 http://www.pik-potsdam.de/publications/pik_reports

Hare, W.: 2003, 'Assessment of Knowledge on Impacts of Climate Change – Contribution to the Specification of Art. 2 of the UNFCCC'. Potsdam, Berlin, WBGU – German Advisory Council on Global Change http://www.wbgu.de/wbgu_sn2003_ex01.pdf

Harvey, L.D.D. and Wigley, T.M.L.: 2003, 'Characterizing and comparing control-run variability of eight coupled AOGCMs and of observations. Part 1: temperature', *Climate Dynamics* **21**, 619–646.

Houghton, J., Ding, Y., Griggs, D.J., Noguer, M., van der Linden, P.J. and Xiaosu, D., eds: 2001, *Climate Change 2001: The Scientific Basis; Contribution of Working Group I to the Third Assessment Report of the Intergovernmental Panel on Climate Change (IPCC)ge*, Cambridge University Press, Cambridge, UK, 944.

Houghton, J., Meiro Filho, L.G., Callender, B.A., Harris, N., Kattenberg, A. and Maskell, K.: 1996, *Climate Change 1995: the Science of Climate Change*, Cambridge University Press, Cambridge, 572.

Houghton, J.T., Jenkins, G.J. and Ephraums, J.J.: 1990, *IPCC First Assessment Report: Scientific Assessment of Climate Change – Report of Working Group I*, Cambridge University Press, Cambridge, 365.

Huybrechts, P., Letreguilly, A. and Reeh, N.: 1991, 'The Greenland Ice-Sheet and Greenhouse Warming', *Global and Planetary Change* **89**, 399–412.

Jones, C., Cox, P.M., Essery, R.L.H., Roberts, D.L. and Woodage, M.J.: 2003a, 'Strong carbon cycle feedbacks in a climate model with interactive CO_2 and sulphate aerosols', *Geophysical Research Letters* **30**, 1479–1483.

Jones, C.D., Cox, P. and Huntingford, C.: 2003b, 'Uncertainty in climate-carbon-cycle projections associated with the sensitivity of soil respiration to temperature', *Tellus Series B-Chemical and Physical Meteorology* **55**, 642–648.

Jones, P.D. and Moberg, A.: 2003, 'Hemispheric and large-scale surface air temperature variations: An extensive revision and an update to 2001', *Journal of Climate* **16**, 206–223.

Jones, P.D., Osborn, T.J., Briffa, K.R., Folland, C.K., Horton, E.B., Alexander, L.V., Parker, D.E. and Rayner, N.A.: 2001, 'Adjusting for sampling density in grid box land and ocean surface temperature time series', *Journal of Geophysical Research-Atmospheres* **106**, 3371–3380.

Knutson, T.R. and Tuleya, R.E.: 2004, 'Impact of CO_2-induced warming on simulated hurricane intensity and precipitation: Sensitivity to the choice of climate model and convective parameterization', *Journal of Climate* **17**, 3477–3495.

Knutti, R., Joos, F., Mueller, S.A., Plattner, G.K. and Stocker, T.F.: in press, 'Probabilistic climate change projections for stabilization profiles', *Geophysical Research Letters*.

Knutti, R. and Meehl, G.A.: submitted, 'Constraining climate sensitivity from the seasonal cycle in surface temperature.' *Journal of Climate*.

Knutti, R., Stocker, T.F., Joos, F. and Plattner, G.-K.: 2002, 'Constraints on radiative forcing and future climate change from observations and climate model ensembles', *Nature* **416**, 719–723.

Knutti, R., Stocker, T.F., Joos, F. and Plattner, G.K.: 2003, 'Probabilistic climate change projections using neural networks', *Climate Dynamics* **21**, 257–272.

Lean, J., Beer, J. and Bradley, R.S.: 1995, 'Reconstruction of solar irradiance since 1610: Implications for climate change', *Geophysical Research Letters* **22**, 3195–3198.

Levitus, S., Antonov, J. and Boyer, T.: 2005, 'Warming of the world ocean, 1955–2003', *Geophysical Research Letters* **32**.

Levitus, S., Antonov, J.I., Boyer, T.P. and Stephens, C.: 2000, 'Warming of the world ocean', *Science* **287**, 2225–2229.

Meinshausen, M., Hare, B., Wigley, T.M.L., van Vuuren, D., den Elzen, M.G.J. and Swart, R.: in press, 'Multi-gas emission pathways to meet climate targets', *Climatic Change*, 50.

Murphy, J.M., Sexton, D.M.H., Barnett, D.N., Jones, G.S., Webb, M.J., Collins, M. and Stainforth, D.A.: 2004, 'Quantification of modelling uncertainties in a large ensemble of climate change simulations', *Nature* **430,** 768–772.

Nakicenovic, N. and Riahi, K.: 2003, 'Model runs with MESSAGE in the Context of the Further Development of the Kyoto-Protocol'. Berlin, WBGU – German Advisory Council on Global Change**:** 54. Report-No.: WBGU II/2003 available at http://www.wbgu.de/wbgu_sn2003_ex03.pdf

Nakicenovic, N. and Swart, R., eds: 2000, *IPCC Special Report on Emissions Scenarios*, Cambridge University Press, Cambridge, UK, 612.

Nordhaus, W.: 1992, 'An optimal transition path for controlling greenhouse gases', *Science* **258,** 1315–1319.

Piani, C., Frame, D.J., Stainforth, D.A. and Allen, M.R.: submitted, 'Constraints on climate change from a multi-thousand member ensemble of simulations.' *Geophysical Research Letters*.

Ramaswamy, V., Boucher, O., Haigh, J., Hauglustaine, D., Haywood, J., Myhre, G., Nakajiama, T., Shi, G.Y. and Solomon, S.: 2001, 'Radiative Forcing of Climate Change', in Houghton, J.T., Ding, Y., Griggs, D.J., Noguer, M., van der Linden, P.J., Dai, X., Maskell, K. and Johnson, C.A. (eds.), *Climate Change 2001: The Scientific Basis*, Cambridge University Press, Cambridge, UK, pp. 892.

Raper, S.C.B., Gregory, J.M. and Stouffer, R.J.: 2002, 'The Role of Climate Sensitivity and Ocean Heat Uptake on AOGCM Transient Temperature Response', *Journal of Climate* **15,** 124–130.

Reichert, B.K., Schnur, R. and Bengtsson, L.: 2002, 'Global ocean warming tied to anthropogenic forcing', *Geophysical Research Letters* **29.**

Richels, R., Manne, A. and Wigley, T.M.L.: 2004, 'Moving beyond concentrations: the challenge of limiting temperature change', AEI-Brookings Joint Center for Regulatory Studies. Working-Paper 04–11.

Smith, J.B., Schellnhuber, H.J. and Mirza, M.Q.M.: 2001, 'Vulnerability to Climate Change and Reasons for Concern: A Synthesis', in McCarthy, J.J., Canziani, O.F., Leary, N.A., Dokken, D.J. and White, K.S. (eds.), *Climate Change 2001: Impacts, Adaptation, and Vulnerability*, Cambridge University Press, Cambridge, UK, pp. 1042.

Swart, R., Mitchell, J., Morita, T. and Raper, S.: 2002, 'Stabilisation scenarios for climate impact assessment', *Global Environmental Change* **12,** 155–165.

Toth, F.L., Bruckner, T., Füssel, H.-M., Leimbach, M. and Petschel-Held, G.: 2003a, 'Integrated Assessment of Long-Term Climate Policies: Part 1 – Model Presentation', *Climatic Change* **56,** 37–56.

Toth, F.L., Bruckner, T., Füssel, H.-M., Leimbach, M. and Petschel-Held, G.: 2003b, 'Integrated Assessment of Long-Term Climate Policies: Part 2 – Model Results and Uncertainty Analysis', *Climatic Change* **56,** 57–72.

Webster, M.D., Forest, C.E., J, R., M.H, B., Kicklighter, D., M, M., Prinn, R., Sarofim, M., Sokolov, A., Stone, P. and Wang, C.: 2003, 'Uncertainty Analysis of Climate Change and Policy Response', *Climatic Change* **61,** 295–320.

Wigley, T.M.L.: 2003a, 'MAGICC/SCENGEN 4.1: Technical Manual'. Boulder, Colorado, UCAR – Climate and Global Dynamics Division available at http://www.cgd.ucar.edu/cas/wigley/magicc/index.html

Wigley, T.M.L.: 2003b, 'Modelling climate change under no-policy and policy emissions pathways'. OECD Workshop on the Benefits of Climate Policy: Improving Information for Policy Makers. Paris, France, OECD: 32. OECD: ENV/EPOC/GSP(2003)7/FINAL

Wigley, T.M.L.: 2005, 'The climate change commitment', *Science* **307,** 1766–1769.

Wigley, T.M.L. and Raper, S.C.B.: 2001, 'Interpretation of high projections for global-mean warming', *Science* **293,** 451–454.

Yohe, G., Andronova, N. and Schlesinger, M.: 2004, 'Climate – To hedge or not against an uncertain climate', *Science* **306,** 416–417.

CHAPTER 29

Observational Constraints on Climate Sensitivity

Myles Allen[1], Natalia Andronova[2], Ben Booth[1,3], Suraje Dessai[4], David Frame[1], Chris Forest[5], Jonathan Gregory[3], Gabi Hegerl[6], Reto Knutti[7], Claudio Piani[1], David Sexton[3] and David Stainforth[1]

[1] *Dept. of Physics, University of Oxford, Parks Road, Oxford, UK*
[2] *Dept. of Atmospheric, Oceanic and Space Sciences, University of Michigan, USA*
[3] *The Met Office, Fitzroy Road, Exeter, UK*
[4] *Tyndall Centre and School of Environmental Sciences, University of East Anglia, Norwich, UK*
[5] *Earth, Atmos. and Planetary Sciences, MIT, Cambridge, USA*
[6] *Nicholas School, Duke University, Levine SRC, Durham, USA*
[7] *National Centre for Atmospheric Research, Boulder, Colorado, USA*

ABSTRACT: Climate sensitivity, or the equilibrium warming resulting from a doubling of carbon dioxide levels, cannot be measured directly, since the real climate system will never be subjected to a carbon dioxide doubling and then allowed to come into equilibrium. Because we can neither observe sensitivity directly nor find observable quantities that are directly proportional to it over the full range of values that are consistent with current observations, any estimate of the probability that a given greenhouse gas stabilisation level might result in a 'dangerous' equilibrium warming turns out to be dependent on subjective prior assumptions of the investigators and not purely on constraints provided by actual climate observations. In contrast, we can observe the strength of atmospheric feedbacks, or the change in top-of-atmosphere energy flux in response to a surface temperature change, much more directly than climate sensitivity itself. The net strength of these feedbacks is directly related to the inverse of the climate sensitivity, or the range of stabilisation concentrations consistent with a target temperature rise. Hence, policies that focus on a maximum temperature rise, accepting uncertainty in the stabilisation concentration that may be required to achieve it, are better informed by climate observations than policies that focus on a target stabilisation concentration, accepting uncertainty in the resulting long-term equilibrium warming.

29.1 Introduction

The communiqué issued by the meeting of G8 leaders in Gleneagles in July 2005 contained the statement: 'We reaffirm our commitment to the United Nations Framework Convention on Climate Change and to its ultimate objective to stabilise greenhouse gas concentrations in the atmosphere at a level that prevents dangerous anthropogenic interference with the climate system.' But what is that level? Other papers in this volume have discussed the possible impacts of various degrees of warming, and a common assumption seems to be emerging that a warming greater than 2°C above pre-industrial temperatures should be avoided. There is clearly room for debate regarding the likely scale and acceptability of the impacts of a 2°C warming, but in this paper we will focus on a more pragmatic problem. Assuming we decide we want to avoid a 2°C warming (or some other target temperature), what target stabilisation concentrations of greenhouse gases will deliver that objective?

Our incomplete knowledge of the properties of the climate system means that we cannot guarantee, given the information available today, that any particular stabilisation level will definitely avoid a 2°C warming. For example, if the equilibrium warming resulting from a doubling of carbon dioxide levels, (referred to herein as 'climate sensitivity' or S [1]), turns out to be 4.5°C, the upper end of the 'traditional' range of 1.5–4.5°C used by the Intergovernmental Panel on Climate Change (IPCC) since the 1980s, then the long-term equilibrium warming resulting from present-day concentrations of all well-mixed greenhouse gases (i.e. assuming anthropogenic aerosol cooling is eventually eliminated) would be around 3°C, with over 1.8°C of this resulting from present-day carbon dioxide levels alone. Many policy studies (e.g. Edmonds and Smith, this volume) assume, for illustration, values of S around half this amount, or 2–2.5°C, but current evidence suggests that it is unlikely that the climate sensitivity will turn out to be that low. Hence, basing policy on the assumption that $S \leqslant 2$°C represents a considerable gamble.

The fact that we can only quantify the probability of a given level of warming resulting from a given stabilisation level has long been recognised by the scientific community. This paper will go further: there are some crucial risks associated with any given stabilisation level that we cannot even quantify objectively. We will argue that estimates of risks that depend on the shape of the upper tail of the distribution for climate sensitivity are inherently subjective, or dependent on prior assumptions of climate researchers and not on actual climate observations.

The reasons for the difficulty of placing an objective upper bound on climate sensitivity are quite fundamental, and are therefore unlikely to be overcome by any 'magic bullet' observation in the foreseeable future. Moreover, the problem becomes worse the more extreme the outcome considered: better estimates of, say, aerosol forcing of climate might improve (make less subjective) our estimate of the probability of $S > 4.5°C$ without having much impact on our estimates of the probability of $S > 7°C$. Of course, we know that the odds of $S > 7°C$ will always be less than the odds of $S > 4.5°C$, but if the consequences of a 7°C warming are completely catastrophic, while the consequences of a 4.5°C warming are merely very bad, we still need to know how much less likely they are. Indeed, since risk is conventionally defined as probability times impact, estimates of the aggregate risk associated with a given stabilisation concentration may well be dominated by low probability, very high impact outcomes, such as S turning out to be $>7°C$.

29.2 The Problem, and an Analogy

The properties of the climate system that we can observe provide progressively less information on the relative likelihood of different values of S the higher the values of S we consider, so any estimates of risk that depend on the shape of the upper tail of the distribution of values for climate sensitivity that are consistent with current observations are inherently more subjective than estimates of risk that depend on the lower tail of the distribution. Since this sounds like a rather technical point, and it has some very profound implications for the scientific justification of different policy targets, let us clarify it with an analogy. Consider the problem of measuring the speed of a car with a pair of synchronised clocks, with one observer recording the start time over a 10 m interval, and another observer recording the finish time. Suppose, in comparing the clocks afterwards, the start and finish times differ by 0.5 s: the most likely value of the speed is 20 m/s, or 72 km per hour. We know, however, that the observers' reaction times are only accurate to $±0.2$ s (1 standard error). Assuming they are Gaussian and independent, the 5–95% confidence interval on the transit time is 0.03–0.97 s. These observations place a lower bound on the speed of the car (there is a less than 5% chance it is travelling at less than 10 m/s, or 36 km/hour) but they do not give us an upper bound: the difference in transit time of a car travelling at 500 km/hour versus 1,000 km/hour is less than 0.04 s, well under the precision of the observation.

We might use prior knowledge (of the law-abiding nature of the driver, for example) to rule out higher speeds, but they are not ruled out by these observations: indeed, if we begin by assuming all speeds are equally likely, so a speed between 1 and 2 km/hour is as likely as a speed between 100 and 101 km/hour before the observation is made, then we cannot formally place an upper bound on the speed at all, because an infinite range of high speeds and only a finite range of low speeds are equally consistent with our observations. Of course, one could argue that assuming all speeds are equally likely is a silly prior assumption to make, because it is harder to maintain a speed between 100 and 101 km/hour than it is to maintain a speed between 1 and 2 km/hour. This argument might suggest a 'prior distribution' in which equal percentage increases in speed are deemed equally likely, making a speed between 1 and 2 km/hour as likely as one between 100 and 200 km/hour: even with this 'logarithmic' prior, it is straightforward to show that the range of speeds consistent with these observations is still unbounded.

The problem arises, of course, because our observations are proportional to a quantity (transit time) that is inversely proportional to the quantity we are trying to measure (speed). Precisely the same problem arises with climate sensitivity. We show below that the results of numerous studies, supported by basic physical principles, indicate that observable properties of the climate system, such as the warming attributable to greenhouse gases to date, tend, in the limit of high S, to be proportional to $1/S$ rather than S, so if any of the uncertainties in these observations follow a normal Gaussian distribution, they provide no formal upper bound on S unless this upper bound is assumed *a priori*. Thus, the probability of an extreme equilibrium warming for a given greenhouse gas stabilisation concentration cannot be inferred solely and directly from observations.

The definition of 'high S', or the point of this transition to observable quantities scaling with $1/S$ rather than S, depends on the observable quantity in question. For studies based on the response to short-term forcing such as volcanic eruptions or studies in which uncertainty in forcing dominates uncertainty in the response, relevant observables may scale with $1/S$ over most of the range of interest. For the longest-term change for which we have accurate forcing estimates, which is the transient response to greenhouse forcing to date, the transition occurs for values of S that are greater than 6–7°C. Hence this is not an issue for 'best-guess' estimates of sensitivity in the traditional 1.5–4.5°C range, but for pinning down the 'worst case', most dangerous, scenarios.

29.3 Why this Matters for Stabilisation Targets

While we can all agree that a warming $>7°C$ in response to a stabilisation concentration of 550 ppm carbon dioxide equivalent is improbable, before we settle on 550 ppm as a stabilisation target, we need to know just how improbable it is. This is the problem: we still cannot quantify this probability objectively, and may not be able to do so for the indefinite future. We might argue it is under 30%, because we can quantify the probability of a $>4.5°C$ warming in response to 550 ppm stabilisation and we know the probability of a $>7°C$ warming must be less than that. Since, however, a $>7°C$ warming could well be catastrophic, many stakeholders would doubtless like to know before signing up to a 550 ppm stabilisation scenario whether it allows a 1% chance of this outcome (which many might find unacceptable) or a 0.001% chance (which would put it into the category, along with the Earth being hit by a comet in the next century, of exciting outcomes that are probably not worth worrying about). If there are consequences of indefinite stabilisation that we definitely do not like and whose probability we cannot quantify, the question must be asked whether attempting to specify a target stabilisation concentration is an appropriate policy objective at all.

An interesting corollary is that other policy objectives might be much better constrained by current observations. To return to the car-speed analogy: our observations imply that the car will arrive at a point 10 km away in 0.5–16 minutes. While this is clearly a broad range, it follows directly from the observations without any obscure reliance on prior assumptions because the quantity we are trying to estimate (time of arrival) is directly proportional to the quantity we can measure (transit time). This time-of-arrival is directly analogous to the 'radiative forcing', F, due to the increase in carbon dioxide that would yield a particular long-term equilibrium warming, such as 2°C. For the equilibrium response, radiative forcing (proportional to the logarithm of the carbon dioxide increase) is equal to the long-term temperature change divided by the sensitivity $F = \Delta T/S$. Hence the F required to deliver a particular ΔT is proportional to $1/S$, and hence directly proportional to observable properties of the climate system in the limit of high S. If the ultimate policy target is defined in terms of a maximum allowable warming, and we simply decide we will do what it takes to get us there, then the probability we are interested in becomes the chance that this might require (expensive) stabilisation below 350 ppm-equivalent, for example. This probability is objectively determined by available climate observations, while the probability that a 550 ppm-equivalent stabilisation might yield a (catastrophic) >7°C warming is not. Hence, to be scientifically justifiable, it is essential that the ultimate policy target remains defined in terms of a temperature change to be avoided, and that we ensure this is not replaced by a 'dangerous greenhouse gas concentration', which is impossible to define objectively. We now proceed to explain in more detail how these issues arise in the context of a number of recent studies.

29.4 Bayesian Formulation of Estimates of Climate Sensitivity

Any attempt to estimate climate sensitivity using observations requires a model or set of models that predict both S and some observable quantity(-ies) (which are, it is hoped, related to sensitivity) given a range of values of unknown climate system properties represented by choices of parameters, subsystems or even entire models [1]. This model could be just a simple energy conservation equation,

through to a full general circulation climate model. Even studies that are not explicitly Bayesian in their approach can be recast in these terms, to separate or clarify the respective roles of data and model formulation. The probability distribution function (PDF) for S given a set of observations, *data*, can then be expressed in terms of Bayes' Theorem:

$$P(S \,|\, data) = \frac{P(data \,|\, S)P(S)}{P(data)},$$

where $P(data \,|\, S)$ is proportional to the 'likelihood' that these observations would be simulated by a model whose sensitivity lies within the range S to $S + dS$, where dS is a small increment. In studies where a subset of otherwise equally plausible models all have the same sensitivity, $P(data \,|\, S)$ is simply the average likelihood of *data* taken across this subset. Given a 'prior' sampling strategy for models or model parameters, $P(S)$ is proportional the implied probability that the sensitivity is between S and $S + dS$ before these data are considered, known as the 'prior predictive distribution', or simply the 'prior'. $P(data)$ is a constant required to ensure all probabilities sum to 100%.

29.5 The Role of the Prior or 'Prior Predictive Distribution'

The appropriate prior $P(S)$ is ambiguous, primarily because the observations used in *data* must play no role in the prior for Bayes' Theorem to apply, and it is impossible to separate prior beliefs about climate system properties from knowledge of many relevant climate observations. Many of the studies cited, for example, use observed temperature changes over the past century as an observational constraint, but these will be well known to anyone working in the field and hence difficult to disentangle from prior assumptions. This complicates the application of conventional methods of determining $P(S)$ such as expert elicitation [10]. Only the very simplest models can be set up without direct reference to climate data, and in these simple models the implications of any prior for *data* will be sufficiently transparent to be impossible for the expert to ignore. Climate observations are used extensively in the formulation, evaluation and refinement of more complex models, in ways that are poorly documented and may be obscure even to the model developers themselves.

29.6 Focussing on Likelihood

Given this problem, it is essential to distinguish the role of observational constraints from prior assumptions in policy guidance. Here we will focus on estimated likelihoods, $P(data \,|\, S)$, from a range of studies. These are equivalent to estimated distributions for S if and only if

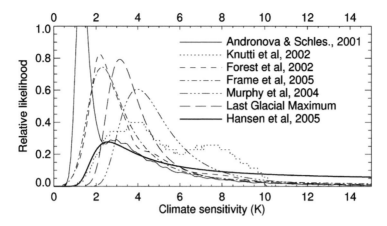

Figure 29.1 Likelihood functions, $P(data \mid S)$, for a range of studies and data sources plotted against S. Thin solid, dotted and short dashed lines: simple and intermediate-complexity models driven with a range of possible forcing profiles and compared to observed transient climate change over the 20th century [2,4,9]. Dash-dot line: simple model driven with greenhouse forcing alone and compared to 20th century attributable greenhouse warming estimated from a pattern-based detection and attribution analysis [3]. Dash-treble-dot line: mean likelihood as a function of sensitivity of members of a perturbed physics ensemble generated from a general circulation model and compared with observations of present-day climatology [11]. Long-dashed line: likelihood of Last Glacial Maximum paleo-climate data as a function of sensitivity assuming forcing of $6.6 \pm 1.5 \, \text{W/m}^2$ and temperature change of $5.5 \pm 0.5°\text{C}$ ($1-\sigma$ ranges) [12]. Thick solid line: likelihood of 2003 energy budget estimates assuming external forcing of $1.8 \pm 0.42 \, \text{W/m}^2$, global energy imbalance of $0.85 \pm 0.08 \, \text{W/m}^2$ and surface temperature change relative to pre-industrial of $0.65 \pm 0.025°\text{C}$ ($1-\sigma$ ranges) [6]. These distributions are approximately equivalent to the distribution of possible equilibrium warming on a 550 ppm CO_2 stabilisation scenario if all values for this warming are assumed equally likely before the constraint of the data is applied.

the prior $P(S)$ is constant for all values of S over which the likelihood is non-zero. Such a 'uniform' prior for S is used in many studies because it makes clear the information provided by specific observations. This focus is also consistent with conventional climate change detection studies, which typically assume a uniform prior in warming attributable to greenhouse gases, even though physical reasoning would assign, for example, a low probability to negative values (greenhouse-induced cooling).

Evaluating $P(data \mid S)$ requires a representation of the expected discrepancy between actual observations and those simulated by the model due, for example, to observational uncertainty or internal climate variability. The magnitude of this 'noise' term plays a crucial role in how fast $P(data \mid S)$ declines as the model-data fit deteriorates away from the 'best-fit' combination of parameters. Noise properties also cannot be observed directly, and must be based either on model-simulated variability (possibly augmented by information about the errors in observations) or estimated from residual model-data discrepancies. If modelled noise omits or underestimates sources of model-data discrepancy in the real world, uncertainties in S will be underestimated.

Figure 29.1 shows likelihood functions $P(data \mid S)$ from a selection of studies [2,3,4,5,6,9,11] based on a wide range of data sources. Other studies have also attempted to estimate sensitivity [8,14] but stopped short of expressing their results in terms of an explicit probability density function, making it difficult to include them on this Figure. Both of these studies [8,14] find a most likely values of

S in the vicinity of 2–3°C, but do not systematically explore the likelihood of higher values: it would be interesting to do so. Ref. [8] notes that higher values of sensitivity might be allowed if they assume a reduced volcanic forcing due to the Krakatoa eruption in the 19th century. They interpret this as highlighting the role of uncertainty in past forcing, but it also highlights the importance of the assumption that the same climate system properties (sensitivity, the parameterisation of heat uptake by the ocean and so forth) apply to the short-term response to volcanoes as apply to the long-term response to greenhouse forcing. This may be the case [14], but since the dynamical response to volcanic and greenhouse forcing could be very different, in ways that would be hard to represent in an energy balance model, inferring sensitivity to greenhouse warming from the response to volcanoes will always be problematic [3].

The data sources behind Figure 29.1 include: various aspects of present-day climatology (time mean quantities including the seasonal cycle), the relationship between temperature changes and energy fluxes into and out of the atmosphere-ocean system, and the transient response to external forcing over the 20th century. Studies [5] and [6] obtain similar results from an analysis of the global energy budget although only [5] express their results in terms of a PDF. We show the numbers given in ref. [6] expressed as a PDF to illustrate the link between it and these other studies.

All studies show a highly asymmetric distribution, with low simulated values of S being inconsistent with these

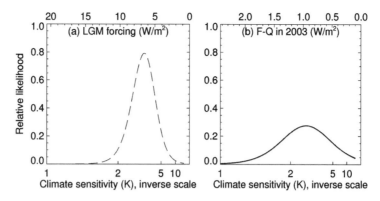

Figure 29.2 Likelihood functions for (a) LGM forcing and (b) net forcing minus global energy imbalance in 2003 based on paleo-climate evidence and combined TOA flux and ocean heat content change observations as a function of climate sensitivity. Note that, because forcing is inversely proportional to sensitivity if the temperature change is relatively well known (which is the case in both situations), a Gaussian likelihood function in forcing translates into a Gaussian in inverse sensitivity, or feedback parameter.

data sources, but only a relatively gentle decline of likelihood $P(data \mid S)$ towards high values of S. The reason for this is that, at least for high sensitivities, the constraints provided by the observations in all these studies actually correspond to constraints on the climate feedback parameter, λ, or the additional energy radiated to space per degree of surface warming, which is inversely proportional to climate sensitivity. If, as all these studies assume, sensitivity is constant over time, then there is a one-to-one correspondence between λ and S and we can unambiguously equate $P(data \mid \lambda)$ with $P(data \mid S)$ if $S = F_{2 \times CO_2}/\lambda$, where $F_{2 \times CO_2}$ is the radiative forcing due to doubling carbon dioxide.

To see why, consider the global energy conservation equation $F - Q = \lambda \Delta T$, where F is external forcing, Q the global energy imbalance and ΔT the perturbation temperature. In all of these studies, the dominant uncertainties are on the left hand side of this equation, in the forcing and energy imbalance terms. If the uncertainties in $F - Q$ are Gaussian, this yields a Gaussian uncertainty range in λ. Even if they are not Gaussian over the full range, we only need for some aspect of uncertainty in $F - Q$ to behave in a Gaussian manner in the limit of small $F - Q$ for the problem to arise.

This is illustrated in Figure 29.2 for two of the cases shown in Figure 29.1 plotted against λ, or an inverse scale in S. Figure 29.2a shows $P(data \mid S)$ for temperature and forcing at the Last Glacial Maximum (LGM), assuming an effective forcing of $6.6 \pm 1.5\,\mathrm{W/m^2}$ and temperature change of $5.5 \pm 0.5°C$ ($1-\sigma$ ranges, with uncertainties modelled as t-distributions with – rather generously given the paucity of independent LGM proxy observations – 10 degrees of freedom) [12]. The upper axis shows the LGM forcing corresponding to each value of S or λ assuming the best-guess values of ΔT (Q can be considered zero in this case because the timescales considered are long enough that the system will be close to equilibrium, at least for the atmospheric feedbacks relevant to present-day sensitivity). The likelihood distribution is essentially symmetric in the forcing because fractional uncertainty

in ΔT is much smaller than fractional uncertainty in forcing, so the latter makes only a small contribution to the overall uncertainty.

The equilibrium response to changes in forcing over the Holocene or Glacial-Interglacial cycles would provide a constraint giving a linear relationship between *data* and S only if the relevant radiative forcing were known and the dominant uncertainty were in the paleo-temperature response. Unfortunately, the reverse is true: paleo-temperatures are better known than paleo-forcing, or at least the component of paleo-forcing that is relevant to sensitivity today, so such studies also provide a constraint on λ rather than S and so are relatively ineffective at ruling out high sensitivities unless these are excluded *a priori*. Any constraint based on Holocene or Last Glacial Maximum climate must assume that the sensitivity of the present-day climate to a forcing dominated by carbon dioxide change is the same as, or can be inferred directly from, the sensitivity to a change dominated by solar forcing and a combination of solar forcing and strong cryospheric feedbacks, which is problematic.

Figure 29.2b shows $P(data \mid S)$ for present-day global energy budget observations, assuming external forcing F of $1.8 \pm 0.42\,\mathrm{W/m^2}$, global energy imbalance Q of $0.85 \pm 0.08\,\mathrm{W/m^2}$ (these error ranges may well be optimistic, but we are following reference [6], and they serve to illustrate our point) and surface temperature change relative to pre-industrial times, ΔT, of $0.65 \pm 0.025°C$ ($1-\sigma$ ranges). Again, since uncertainty in $F - Q$ is dominant, the distribution corresponds closely to the distribution of uncertainty in $F - Q$ given the best-guess values of ΔT, shown by the top axis. Ref. [6] notes that they do not rule out higher values of climate sensitivity, while asserting that their results 'favour an intermediate value' (the peak of the likelihood function occurs slightly over 2°C). The problem is that they do not favour it very much, since likelihood declines only very slowly towards high values. A very similar distribution is obtained by ref. [5], who also focus on the global energy budget, and note explicitly that their range is unbounded at the higher end because of the uncertainty in forcing.

A naïve method of generating a distribution for climate sensitivity from, for example, energy budget observations, would be to generate a distribution of possible values of $F - Q$, compute $F_{2 \times CO_2} \Delta T / (F - Q)$ for each member of the distribution, and plot a histogram of the result. Given that the uncertainty in ΔT is relatively small, this is equivalent to assuming all values of $F - Q$, and hence all values of S^{-1}, are equally likely before the constraint of observations on $F - Q$ is applied. This is a very strong prior assumption: it implies that a value of the climate sensitivity between 0.91 and 1.0°C is as likely as a value between 5 and 10°C before the study is performed (in both cases S^{-1} differs by 0.1). If such a prior is to be imposed, it is essential that policy-makers are told about it, and comparison of results between studies is not meaningful unless similar priors are imposed. Failure to do so means that we could appear to rule out high values of climate sensitivity simply by choosing to constrain it using an observable variable that is strongly non-linear in sensitivity, with the gradient $\partial (data)/\partial S$ tending towards zero as S increases, which is clearly absurd.

In studies focussing on the transient response [2,3,4,9,15], a simple Taylor expansion of the transient temperature response to any external forcing F given a constant effective heat capacity c and feedback parameter λ

$$c \frac{d\Delta T}{dt} = F - \lambda \Delta T$$

shows that the first sensitivity-dependent term to emerge in the limit of either high sensitivity, short timescale or high heat capacity (or any combination thereof) is proportional to λ, not S. For example, suppose F increases linearly from equilibrium starting conditions. The warming at the time the forcing corresponds to a CO_2 doubling (defined as the standard 'transient climate response' if F is due to a 1%/year compound CO_2 increase) is given by the simple formula (see the discussions in refs. [7,16]):

$$\Delta T = S - \frac{S^2}{S_0} \left(1 - \exp \left(-\frac{S_0}{S} \right) \right)$$

where S_0 is a constant that depends on c and the rate of forcing increase. If $S \ll S_0$ then $\Delta T \propto S$, while if $S \gg S_0$ then $\Delta T \propto S^{-1}$. Hence, again, observations provide a constraint on λ, not S, in the high-S limit. For typical values of climate system properties, S_0 is in the region of 6–7°C (lower for more rapid changes in external forcing), but the transition to $\Delta T \propto S^{-1}$ occurs rapidly for values of S higher than this. Hence, even if the forcing is known precisely and the only uncertainty is in the response, this problem of ΔT ceasing to scale with S becomes an issue for values of S greater than around 6–7°C.

There is no such simple explanation why observable aspects of climatology should scale with λ rather than S,

as is assumed by ref. [11], but this is intuitively plausible since the same processes control climatology as control the top of atmosphere energy budget, particularly if the amplitude of the seasonal cycle is included in the definition of climatology. The method of ref. [11] is equivalent to assuming that all perturbations to λ are equally likely before the constraints of their perturbed-physics ensemble and comparison with observations are made. The resulting likelihood distribution is close to Gaussian in λ and hence, when plotted against sensitivity, strongly asymmetric.

Note that the dash-double-dot curve shown in Figure 29.1 is not the same as the likelihood-weighted distribution shown in ref. [11], because the latter contains two additional weighting steps: first, regions of parameter space were given more weight if a large nominal change in a parameter gave a small change in λ. With the benefit of hindsight, this could be difficult to justify given that it makes results strongly dependent on the functional form of the parameters, which is often an accident of model development history. It is supported by expert opinion, but opinions on the relative likelihood of different parameter intervals were only elicited after their impact on the sensitivity was known, making it difficult to disentangle expert opinion on parameters from expert opinion on sensitivity. Second, ref. [11] weights all values of sensitivity by S^{-2} before plotting distributions or computing confidence intervals. This is justified if all values of λ really are deemed equally likely before the study is performed, but corresponds to a highly non-uniform prior distribution for sensitivity.

29.7 Relative Likelihood of High Versus Very High Values of Sensitivity

One of the problems with interpreting Figure 29.1 is that all these distributions have been normalised to have equal area, which makes differences between distributions at the lower end of the range impact on estimated likelihoods at the upper end. Since different aspects of the observations determine lower and upper bounds, this has some rather bizarre consequences: for example, if two studies were to rely on the same data to give the upper bound on sensitivity, but one of them had a very weak constraint on the lower bound, that study would, simply by conservation of probability, imply a lower upper bound, as probability 'leaks out' to low values. This is misleading: failure to place a lower bound on sensitivity should not increase our confidence in the upper bound.

This paradox can be resolved if we recall that what the data really provide is estimates of relative likelihood, so there is no reason why the normalisation should be applied over the full range of sensitivities. To examine how likelihood decreases as we increase sensitivity above what is conventionally deemed a 'high but feasible' value such as 4°C, we simply re-normalise the distributions shown in

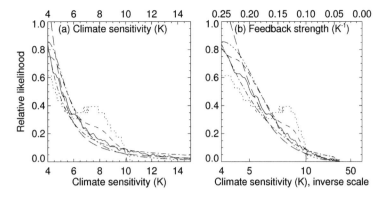

Figure 29.3 (a) Likelihood functions from Figure 29.1 normalised to give equal likelihood of $S > 4°C$.
(b) Likelihood functions $P(data | \lambda)$ plotted against λ, expressed as $S^{-1} = \lambda / F_{2 \times CO_2}$, which corresponds to the fractional increase in CO_2 concentration per degree of warming. These distributions are equivalent to the distribution of possible concentration targets consistent with a given temperature target if all concentration targets are assumed equally likely before the constraint of the data is applied. Note that while the distributions in (a) show 'fat' upper tails, corresponding to a weak constraint, the distributions in (b) are much closer to Gaussian, indicating a direct link to the observable quantities used to constrain them.

Figure 29.1 to have equal area above 4°C. This allows us to address questions like 'if sensitivity is likely (meaning, in the IPCC jargon, >66% probability) to be less than 4°C, what do these studies tell us about how likely is it to be less than 7°C?'

Results are shown in Figure 29.3a, which reproduces the curves in Figure 29.1 normalised to have equal likelihood of S greater than 4°C. The similarity between the likelihood functions is particularly striking given the wide disparity in methods and data sources used, and suggests some underlying explanation that goes beyond simple coincidence. Refs. [4] and [9] show a cut-off at 10°C, but this was imposed in the way these studies were set up.

The reason for this similarity is indicated by Figure 29.3b, which shows exactly the same likelihood functions plotted against the climate feedback parameter, expressed in terms of inverse sensitivity. The distributions in Figure 29.3b are much closer to Gaussian as we approach low values of λ, corresponding to high sensitivities, suggesting that, for all these data sources, observable properties of the climate system tend to vary linearly with λ rather than S, at least in the limit of high S. This result is confirmed by quantitative analysis of the properties of the upper tails of these distributions.

Crucially, if the relationship between *data* and λ is linear, then the relationship between *data* and S is non-linear, with the rate of change $\partial(data)/\partial S$ tending towards zero as S increases. In practical terms, this means that a change in climate system properties that takes a 5°C to a 10°C sensitivity has less impact on any of these observable properties of the climate system than one that takes a 3°C to a 5°C sensitivity. In the majority of studies quoted, values of S in double figures would not be excluded at the 5% level if we begin by assuming all values of sensitivity are equally likely over the range zero to 15°C. If, drawing all these sources of information together, we conclude that the sensitivity is 'likely' (>66% chance) <4°C, we can only conclude it is 'very likely' (>90% chance) less than ~7°C,

and only 'virtually certain' (>99% chance) less than some indeterminate number in the double figures. Hence, at present, the only way of ruling out these high values of equilibrium warming is *a priori*: they are not and, for many data sources, cannot be excluded by the comparison of models with observations.

29.8 Non-Linear Climate Change and Time-Dependent Sensitivity

The entire discussion so far has assumed that the climate system has a single value of the climate sensitivity, and that this value does not change over time or in response to climate change itself. Numerous studies have shown that this assumption may well prove incorrect: the sensitivity of the system to short-term forcing may prove different from its sensitivity to long-term forcing, and even a gradual global warming (through its impact on sea-ice feedbacks, for example) may change the climate sensitivity itself [13]. This possibility further re-inforces the fundamental message of this paper, which is that we cannot rely on estimates of the likelihood of high long-term equilibrium warming that are based on observations available today.

For example, we might, in analysing feedback processes that are active today, conclude that a >7°C warming was substantially less likely than a >6°C warming for a given stabilisation concentration. Further research might, however, reveal that once the system had warmed up by 6°C, a threshold is crossed that releases entirely new feedback processes, making a further degree of warming inevitable. Hence, once we account for this additional feedback, the probability of a >7°C warming is actually the same as that of a >6°C warming. We do not, of course, have positive evidence that such feedbacks exist, or they would be incorporated into our current estimates, but the larger the temperature change, the more likely we are to see a qualitative change in system behaviour. Hence, the possibility of such non-linearities and threshold behaviour further reinforces the tentative nature of estimates of risk that depend on the shape of the upper tail of the distribution for climate sensitivity.

29.9 Implications for the Definition of 'Dangerous Climate Change'

The fact that we cannot, on physical grounds, place a firm upper limit on climate sensitivity has important practical implications. Any estimate of the probability that a given greenhouse gas stabilisation level might result in a 'dangerous' equilibrium warming is critically dependent on subjective prior assumptions of the investigators, encapsulated in $P(S)$, and not on constraints provided by actual climate observations. Hence, it is premature to suggest that we can provide an objective assessment of the risks associated with different stabilisation levels given the information provided by current climate observations. Note that this does not preclude an objective assessment of the risks of future transient climate change associated with specific concentration pathways: the problem of the non-linear relationship between observable quantities and forecast response applies specifically to stabilisation scenarios.

The linear relationship between *data* and λ that applies over a wide range of data sources means that the distribution of atmospheric CO_2 concentrations consistent with a given temperature stabilisation target is much easier to constrain with presently-available observations than the distribution of equilibrium warming consistent with a given CO_2 concentration. The reason is that the first depends on λ, which is constrained by data, while the second depends on S, which is not. This has a very specific practical implication. It is much easier to quantify the risks associated with a strategy aiming at a given temperature target (for example, the risk of needing steeper, and hence more expensive, cuts in emissions in the future if the climate sensitivity turns out to be higher than expected), than it is to quantify the risks associated with a strategy aiming at a given concentration stabilisation target (for example, the risk of that stabilisation target giving a higher-than-expected, and hence more dangerous, warming).

Devising policies to avoid a given level of warming still requires us to consider different future concentrations of greenhouse gases, so from a purely scientific perspective, the point made in this paper may seem rather moot (although it does help clarify why different studies have reported such a wide range of upper bounds on climate sensitivity). The main message is for policy definition: it is essential that we do not fall into the trap of assuming that stabilisation below 450 ppm is 'safe' just because our best estimate of climate sensitivity is in the region of 3°C, when the observations allow a substantial, and impossible to quantify, chance of a much higher climate sensitivity. Concentration targets will have to be

revised as the true climate sensitivity emerges: it is essential, therefore, that 'keeping concentrations below X ppm CO_2-equivalent' does not become the ultimate objective of policy.

To avoid dangerous climate change, we will have to do what it takes to avoid a dangerous level of warming, and we will ultimately only find out what it takes by doing it, as atmospheric greenhouse gas levels stabilise and we observe how the system responds. The uncomfortable conclusion for policy makers must be that the commitment to avoid dangerous climate change must be made before we know what that commitment will require in terms of long-term greenhouse gas stabilisation concentrations.

Acknowledgements

We are grateful to Tom Wigley and the reviewers for many excellent suggestions that substantially improved the readability and clarity of this manuscript, and look forward to continuing to explore these issues in the future. Work at Oxford and the Hadley Centre was supported by Defra under contract PECD 7/12/37 and sub-contract PB/B3891, with additional support from the UK Natural Environment Research Council and the US NOAA/DoE through the International Detection and Attribution Group. NA was supported by NSF Award No ATM-0084270, SD by FCT Portugal Grant No SFRH/BD/4901/2001) and GH by NSF Award No ATM-0002206.

REFERENCES

1. Andronova, N.G., Schlesinger, M.E., Dessai, S., Hulme, M. and Li, B. The concept of climate sensitivity, history and development, in 'Human-Induced Climate Change: An Interdisciplinary Assessment', M. E. Schlesinger et al. (Eds.), Cambridge University Press (2005)
2. Andronova, N.G. and Schlesinger, M.E. Objective estimation of the probability density function for climate sensitivity, *Journal of Geophysical Research*, **106**, 22,605–22,612 (2001)
3. Frame, D.J., Booth, B.B.B., Kettleborough, J.A., Stainforth, D.A., Gregory, J.M., Collins, M. and Allen, M.R. Constraining climate forecasts: The role of prior assumptions, *Geophysical Research Letters*, **32**, L09702, doi: 10.1029/2004GL022241 (2005)
4. Forest, C.E., Stone, P.H., Sokolov, A.P., Allen, M.R. and Webster. M.D. Quantifying uncertainties in climate system properties with the use of recent climate observations, *Science*, **295**, 113–117 (2002)
5. Gregory, J.M., Stouffer, R.J., Raper, S.C.B., Rayner, N. and Stott, P.A. An observationally-based estimate of the climate sensitivity, *Journal of Climate*, **15**, 3117–3121 (2002)
6. Hansen, J., Nazarenko, L., Ruedy, R., Sato, M., Willis, J., Del Genio, A., Koch, D., Lacis, A., Lo, K., Menon, S., Tovakov, T., Perlwitz, J., Russell, G., Schmidt, G.A. and Tausnev, N. Earth's energy imbalance: Confirmation and implications, *Science*, **308**, 1431–1435 (2005)
7. Hansen, J., Russell, G., Lacis, A., Fung, I., Rind, D. and Stone, P. Climate response times: Dependence on climate sensitivity and ocean mixing, *Science*, **229**, 857–859 (1985)
8. Harvey, L. D. D. and Kaufmann, R. K. Simultaneously Constraining Climate Sensitivity and Aerosol Radiative Forcing, *Journal of Climate*, **15**, 2837–2861 (2002)
9. Knutti, R., Stocker, T.F., Joos, F. and Plattner, G. K. Constraints on radiative forcing and future climate change from observations and climate model ensembles, *Nature*, **416**, 719–723 (2002)

10. Morgan, M. G. and Keith, D. W., Subjective judgements by climate experts, *Environmental Policy Analysis*, **29**, 468–476 (1995)

11. Murphy, J., Sexton, D., Barnett, D., Jones, G., Webb, M., Collins, M. and Stainforth, D. Quantification of modelling uncertainties in a large ensemble of climate change simulations, *Nature,* **430**, 768–772 (2004)

12. Schneider von Deimling, T., Held, H., Ganopolski, A. and Rahmsdorf, S. Climate sensitivity estimated from ensemble simulations of glacial climate, submitted, (2005)

13. Senior, C.A. and Mitchell, J.F.B. The time dependence of climate sensitivity, *Geophysical Research Letters*, **27**, 2685–2688 (2000)

14. Wigley, T.M.L., Ammann, C.M., Santer, B.D. and Raper, S.C.B. The effect of climate sensitivity on the response to volcanic forcing, *Journal of Geophysical Research*, **110**, D09107, doi:10.1020/2004JD005557 (2005)

15. Wigley, T.M.L., Jones, P.D. and Raper, S.C.B. The observed global warming record: What does it tell us? *Proceedings of the National Academy of Sciences*, **94**, 8314–8320 (1997)

16. Wigley, T.M.L. and Schlesinger, M.E. Analytical solution for the effect of increasing CO_2 on global mean temperature, *Nature*, **315**, 649–652 (1985)

CHAPTER 30

Of Dangerous Climate Change and Dangerous Emission Reduction

Richard S.J. Tol[a,b,c,d] and Gary W. Yohe[e]

[a] *Research Unit Sustainability and Global Change, Hamburg University and Centre for Marine and Atmospheric Science, Hamburg, Germany*
[b] *Institute for Environmental Studies, Vrije Universiteit, Amsterdam, The Netherlands*
[c] *Engineering and Public Policy, Carnegie Mellon University, Pittsburgh, PA, USA*
[d] *Currently visiting: Princeton Environmental Institute and Department of Economics, Princeton University, Princeton, NJ, USA*
[e] *Economics Department, Wesleyan University, Middletown, CT, USA*

ABSTRACT: At an individual level, danger is seen as a non-negligible chance of a serious loss of welfare. At a global level, danger is also seen in welfare terms, but it is impossible to define a global welfare function. Alternative *hypothetical* global welfare functions, informed by different interpretations of the "burning embers" diagram (Figure 19.1 of the IPCC Third Assessment Report) for example, give conflicting advice on appropriate stabilisation targets. Using Monte Carlo analyses of the FUND model, we explore the emission reduction implications of alternative definitions of *dangerous climate change* to show the possibility of *dangerous emission reductions*; i.e., shooting at stabilization targets that are so low that reduced economic growth actually increases vulnerability to climate change.

30.1 Introduction

The climate change community has been preoccupied with danger since the advent of the United Nations Framework Convention on Climate Change (the UNFCCC). This is part of a wider trend in which societies, as they become more prosperous and satisfy more of their needs, can begin to address involuntary risks that loom further in the distance. In the climate arena, of course, this trend was amplified in Rio when the world decided to avoid 'dangerous interference with the climate system.'

This has led to a series of studies in which researchers try to define and measure danger so as to provide guidance for long-term emission reduction policies. Some studies set out to define danger in an objective manner (Azar and Rodhe, 1997; Dessai *et al.*, 2004; Grübler and Nakicenovic, 2001; Gupta and van Asselt, 2004; Hare, 2003; Parry *et al.*, 1996, 2001; Swart *et al.*, 1989; WBGU, 1995). A few other studies, heeding the advice of Funtowicz and Ravetz (1994) that values cannot be avoided in such an endeavour, try to define danger in consultation with a typically unrepresentative set of stakeholders (Berk *et al.*, 2002; ECF and PIK, 2004). Curiously, few of the studies stopped to wonder whether such an exercise would be fruitful. Therefore, in Section 2, we present some thoughts on the impossibility of defining danger in a way that would support or even represent a consensus.

Climate change is seen as dangerous because of its negative impacts. Climate policy is therefore about avoiding such impacts. In Section 3, we present estimates of the impacts that could be avoided by emission abatement. Despite some 15 years of intense discussion and study of climate policy, there are only a few attempts

to estimate avoided damage (Nicholls and Lowe, 2004; Tol, 2005a).

Section 3 ignores uncertainty and vulnerability. In Section 4, we correct this omission. Uncertainty would increase the need for emission reduction, in most circumstances at least, but competition between adaptation and mitigation for scarce resources may act as a brake on overly stringent emission abatement.

Sections 3 and 4 are based on results from an integrated assessment model called the Climate Framework for Uncertainty, Negotiation and Distribution (FUND). There are a large number of assumptions in FUND; a single book chapter cannot present and discuss all of these. A list of publications, which together document the model, and the source code can be found at http://www.uni-hamburg.de/Wiss/FB/15/Sustainability/fund.htm. Section 5 offers some concluding remarks.

30.2 Dangerous Definitions

The ultimate objective of the United Nations Framework Convention on Climate Change is to 'stabilise atmospheric concentrations of greenhouse gases' in order to avoid 'dangerous interference with the climate system' which is in turn usually understood as causing unacceptable impacts. According to Webster's (1996), 'danger' is 'liability or exposure to harm or injury'. The first part of this definition refers to probability, the second part to consequence. For an action to be called dangerous, the probability of harm should not be too large; otherwise it would simply be deemed silly, stupid or irresponsible. The probability should also not be so small that it is inconsequential – the lower threshold probability depends on the harm potentially caused. An individual would thus call an action dangerous if it entails of chance of harm, so that probability times impact is considerable. Avoiding danger would imply reducing either probability or harm.

For an individual, harm or injury is easily defined, and an economist would speak of a substantial loss of utility. Article 2 of the UNFCCC is not about individual danger, however. It is, instead, about collective danger. Individual and collective notions of probability are readily reconciled in many cases,[1] but individual and collective notions of harm are not in nearly every case.[2] The notion of harm combines fact – what would be the impacts of climate change? – and value – how bad are these impacts? The definition of collective harm requires agreement on uncertain and unknown facts, as well as the reconciliation of widely divergent values. This is quite unlikely. Attempts by a few to dictate the rest are unlikely to succeed and less likely to last if they do. That is why societies spend so much effort constructing social and political institutions that are designed to reflect the 'public interest'.

It follows that any attempt to define scientifically what constitutes 'dangerous interference with the climate system' is bound to fail: The needed value judgements have no role in science, and values cannot be objectively aggregated. It is therefore not surprising that 15 years of trying to define danger has brought so little progress.[3] It is true that the attempts to define 'dangerous interference' have brought useful discussion about the seriousness of climate change and the need to reduce greenhouse gas emissions. Moreover, they have focused attention on extreme events and the possibility of abrupt change – possible climate impacts that most could eventually agree would be uncomfortable if not intolerable. However, such discussion is perhaps more fruitful if it could be held without implications under international law.

It may be impossible to agree on dangerous interference, but perhaps it is possible to agree on climate policy.

A look at the 'burning embers' diagram of the IPCC TAR shows that this exercise can be futile, as well. This diagram stems from the reasons for concern of the synthesis chapter of WG2 (Smith *et al.*, 2001). Although there are five more or less burning embers, there are only four reasons to be concerned about climate change: Unique systems, justice and equity, aggregate impacts, and large-scale discontinuities. Those concerned about the damage done to unique systems – such as coral reefs, small island communities, and butterfly species – would argue that the present climate change is already intolerable. Those concerned about large-scale discontinuities – such as a shutdown of the thermohaline circulation or the collapse of the West Antarctic Ice Sheet – would argue that we can tolerate a few more degrees of warming. In short, people with different notions of dangerous interference would advocate different intensities of emission reduction. One group would call for the immediate implementation of stringent emissions controls; the other would argue for waiting or, if willing to approach the question from a risk management perspective, a modest hedge against future problems.

30.3 Avoided Impacts

Since the definition of 'dangerous' will be decided ultimately, if at all, by negotiation at a global scale, scientific analysis can only estimate the impacts of climate change that would be avoided through greenhouse gas emission reduction. Although the UNFCCC dates back to 1992, only few quantitative analyses of avoided impacts of climate change were published.

Avoided impacts are an essential part of climate change policy analysis, unless one accepts an external ultimate target and seeks to reach this at minimum cost (as in, e.g. Manne and Richels, 1999). The method of tolerable windows is one option for policy analysis, but most of its implementations have focused on greenhouse gas concentrations and temperatures rather than on impacts (e.g. Petschel-Held *et al.*, 1999).[4] Cost-benefit analysis of greenhouse gas emission reduction does estimate avoided impacts. However, in a cost-benefit analysis, the marginal costs of emission reduction are balanced with the marginal costs of climate change (e.g. Nordhaus, 1993; Tol, 1999). Cost-benefit studies therefore report the marginal damage costs of climate change, rather than the total avoided damage. See Tol (2005b) for a review of this literature.

Corfee-Morlot and Agrawala (2004) recently edited a book titled 'The benefits of climate change policies'. Although this volume contains many useful insights into climate change impacts, it does not provide an estimate of the damages avoided by emission reduction. The recent collection of climate change impact papers edited by Parry (2004) is also silent on avoided impacts. Nicholls

---

[1] There are some issues with subjective probabilities, particularly those subjective probabilities that do not respond to evidence.

[2] Indeed, Arrow's (1951) impossibility theorem prevents this. Unless society consists of less than three individuals or the economy of less than three goods, unless all preferences are equal, or unless someone dictates either tastes or outcomes, individual utility (and hence harm) cannot be consistently aggregated to social welfare (and hence harm). Note that, although couched in economic terms, Arrow's impossibility theorem is mathematical. As such, it holds as much for economic issues as it does for other issues. The impossibility theorem says that it is impossible to consistently aggregate individual preferences to a collective preference, regardless of the aggregation method and the nature of the preferences. Arrow's impossibility is old and awkward, not just for attempts to define 'dangerous interference', but for economics and public policy as a whole. There have been many attempts to overcome it, but each has its own problems; and none can be smoothly combined with the realities of climate change (Tol, 2001; Tol and Verheyen, 2004).

[3] A referee argued that Article 2 of the UNFCCC gives a legal imperative to avoid and hence to define dangerous interference, but surely the application of agreements can be tempered by the complexity of reality.

[4] See Toth *et al.* (2000) to see one attempt to extend the tolerable windows approach to climate change impacts.

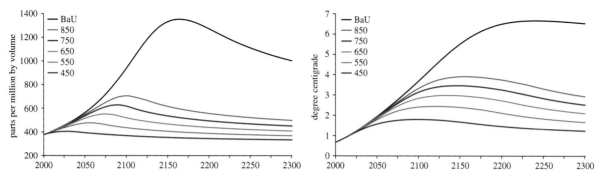

Figure 30.1 The atmospheric concentration of carbon dioxide (left panel) and the global mean temperature (right panel) according to the no control scenario (BaU) and the five stabilisation scenarios, characterised by their peak concentrations of 850, 750, 650, 550 and 450 ppm.
Source: Tol (2005a).

and Lowe (2004) do estimate avoided impacts of sea level rise. They find that mitigation can substantially reduce flood impacts, but also point to the slow response of the sea level to global warming and, hence, mitigation. Tol (forthcoming, b) also estimates avoided sea level rise impacts, including, in contrast to Nicholls and Lowe (2004), the costs of emission reduction. Tol (forthcoming, b) shows that the effect of the costs of emission reduction on vulnerability to sea level rise is minor. In contrast, Tol and Dowlatabadi (2001) show that this effect is large for infectious diseases.

Tol (2005a) is the first to provide a comprehensive estimate of the climate change impacts avoided by greenhouse gas emission reduction. The model used has several advantages. It includes many climate change impacts: agriculture, forestry, water resources, sea level rise, energy consumption, human health, and ecosystems. The sectoral impacts are modelled in an internally consistent way. Vulnerability varies with development. Impacts are expressed in many indicators, as well as in a single, consistent superindicator (welfare-equivalent income loss).[5] Model results are smooth but non-monotonic. Valuation is based on willingness-to-pay, that is, the analysis informs our decision to buy a better climate for our grandchildren. Impacts feedback on the assumed development path, but this feedback is small (Fankhauser and Tol, 2005). The costs of emission reduction include both exogenous and endogenous technological change. The model has also disadvantages. It relies on reduced-form impact models. Valuation relies on direct costs and benefit transfer. Climate scenarios are crude. Interactions between impacts are not well-captured. Climate change is assumed to be smooth, and low probability yet high impact scenarios are ignored. The analysis is based on a single no-policy scenario, and uncertainties about climate change, its impacts, and the valuation of impacts are ignored as well. Therefore, the

results should be interpreted with caution. The results are illustrative only.

Five alternative stabilisation scenarios are run. The scenarios are characterised by their *peak* concentration of carbon dioxide equivalent, which is varied from 450 ppm to 850 ppm[6] in steps of 100 ppm. *Stabilisation levels* are lower than *peak* levels; stabilisation requires that carbon dioxide emissions are driven to zero; the mixed lifetime of atmospheric carbon dioxide implies that overshoot is difficult to avoid. Carbon dioxide emissions are reduced such that marginal emission reduction costs are equal for all regions; and such that marginal emission reduction costs increase with the discount rate. The marginal costs of methane and nitrous oxide emission reduction equal the marginal costs of carbon dioxide emission reduction, corrected for the global warming potential. This implementation is approximately cost-effective.

Table 30.1 shows the maximum market impacts for the 16 regions for the no control and the stabilisation scenarios.[7] Table 30.2 shows the same information for non-market impacts. Emission reduction clearly reduces peak impacts. Three things are noteworthy. Firstly, the largest gain in avoided impacts is from moving from the no control scenario to the peak at 850 ppm scenario. Deeper emission cuts avoid more damage, but the additionally avoided damage gets smaller and smaller. This is as would be expected given that most 'optimal policy' experiments call for modest (but persistent) emission reductions. Secondly, in some cases, the maximum impact is insensitive to emission abatement. This is particularly true for non-market impacts in poor regions. Infectious diseases explain this. The main impacts would occur in the first decades of the 21st century during which climate is hardly influenced by any emissions abatement while people are still poor

---

[5] Note that this superindicator tacitly assumes a social welfare function, and even a global welfare function (e.g. Fankhauser *et al.*, 1997). The welfare function is the standard Negishi one of 'one dollar, one vote', which best describes current policy.

---

[6] Note that the business as usual scenario has relatively high emissions. Note also that the analysis extends to 2300. In alternative business as usual scenarios, carbon dioxide concentrations may not reach 750 ppm by 2100.

[7] Maximum or peak impacts are probably the best way to represent 'dangerous interference'.

Table 30.1 Maximum climate change damage (%GDP) on market sectors for the no control scenario and five stabilisation scenarios for the 16 FUND regions.

Name	CO₂[a]	GMT[b]	USA	CAN	WEU	JPK	ANZ	EEU	FSU	MDE	CAM	SAM	SAS	SEA	CHI	NAF	SSA	SIS
BaU	1352	6.7	0.94	0.65	1.89	0.49	1.33	2.65	9.80	2.42	2.40	0.66	2.34	2.24	9.61	6.66	3.08	8.38
850	704	3.9	0.14	−0.20	0.34	−0.18	−0.59	0.72	3.82	0.26	0.26	0.15	0.21	0.34	0.16	2.77	1.11	0.72
750	629	3.5	0.07	−0.25	0.20	−0.20	−0.70	0.53	3.24	0.02	0.06	0.08	0.03	0.15	−0.58	2.30	0.88	−0.06
650	554	3.0	0.02	−0.23	0.06	−0.22	−0.77	0.36	2.66	−0.19	−0.10	0.04	−0.12	0.06	−1.14	1.78	0.63	−0.71
550	479	2.5	−0.01	−0.21	−0.08	−0.19	−0.72	0.20	2.09	−0.35	−0.24	0.02	−0.24	−0.03	−1.54	1.15	0.34	−1.22
450	406	1.8	−0.02	−0.17	−0.11	−0.15	−0.62	0.05	1.46	−0.46	−0.30	0.01	−0.35	−0.10	−1.82	0.42	0.07	−1.72

[a] Maximum atmospheric concentration of carbon dioxide, in parts per million by volume.
[b] Maximum increase global mean surface air temperature since pre-industrial times, in degrees centigrade.
Source: Tol (2005a).

Table 30.2 Maximum climate change damage (%GDP) on non-market sectors for the no control scenario and five stabilisation scenarios for the 16 FUND regions.

Name	CO₂[a]	GMT[b]	USA	CAN	WEU	JPK	ANZ	EEU	FSU	MDE	CAM	SAM	SAS	SEA	CHI	NAF	SSA	SIS
BaU	1352	6.7	0.31	0.23	0.26	0.26	0.23	0.18	1.22	0.18	0.22	0.23	0.09	0.19	0.21	0.53	4.66	0.15
850	704	3.9	0.25	0.18	0.21	0.22	0.18	0.13	1.22	0.11	0.14	0.15	0.07	0.11	0.14	0.53	4.66	0.15
750	629	3.5	0.24	0.17	0.20	0.21	0.17	0.12	1.22	0.09	0.13	0.14	0.07	0.10	0.13	0.53	4.66	0.15
650	554	3.0	0.23	0.16	0.19	0.19	0.15	0.10	1.22	0.09	0.12	0.13	0.07	0.08	0.10	0.53	4.66	0.15
550	479	2.5	0.21	0.14	0.17	0.18	0.14	0.08	1.22	0.07	0.12	0.13	0.07	0.07	0.09	0.53	4.67	0.15
450	406	1.8	0.18	0.12	0.15	0.16	0.11	0.06	1.73	0.07	0.13	0.33	0.57	0.04	0.06	0.56	4.92	0.17

[a] Maximum atmospheric concentration of carbon dioxide, in parts per million by volume.
[b] Maximum increase global mean surface air temperature since pre-industrial times, in degrees centigrade.
Source: Tol (2005a).

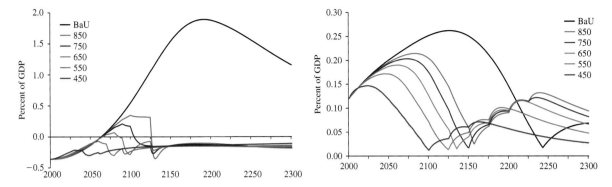

Figure 30.2 The monetised damages of climate change in Western Europe for the no control and the five stabilisation scenarios. Market impacts are displayed in the left panel, non-market damages in the right panel. *Source*: Tol (2005a).

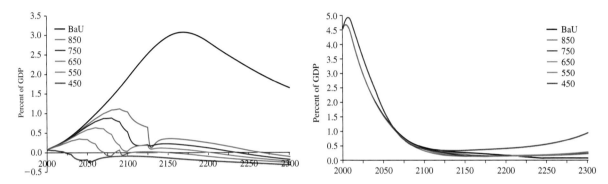

Figure 30.3 The monetised damages of climate change in Sub-Saharan Africa for the no control and the five stabilisation scenarios. Market impacts are displayed in the left panel, non-market damages in the right panel. *Source*: Tol (2005a).

enough to be vulnerable to these afflictions. Thirdly, damages *increase* for some regions between peak concentrations of 550 ppm and 450 ppm. One reason is that emission reduction becomes so costly that economic growth is slowed down, and vulnerability to climate change increases. Tol and Dowlatabadi (2001) first pointed out this possibility. Tol (forthcoming, a) presents an extensive sensitivity analysis, showing that this result only holds for vector-borne diseases and the poorest countries. Another reason is that abatement is so stringent with a 450 ppm target that sulphur emissions fall substantially as well. Abatement thereby removes the sulphur veil and accelerates regional warming.

Figure 30.2 shows the market and non-market impacts for Western Europe for the no control and the stabilisation scenarios. In the long run, market impacts are more or less equal for the stabilisation scenarios, because the global mean temperatures converge (cf. Figure 30.1). All stabilisation scenarios show considerably lower market impacts (roughly 1.5% of GDP) than the no control scenario. In the medium run, each stabilisation scenario avoids the peak in market damages seen in the no control scenario (almost 2.0% of GDP). The stabilisation scenarios vary by about 0.5% of GDP. In the short run, the stabilisation scenarios

show slightly higher market impacts than the no control scenario because of the reduction in SO_2 emissions.

Different results emerge for the non-market impacts. Unlike market impacts, non-market impacts are always negative, but they are also smaller.[8] In the no control scenario, impacts go up first, then down, then up again. This is because non-market impacts are largely driven by the absolute value of the rate of warming (i.e., cooling causes damage as much as warming). In the stabilisation scenarios, the maximum rate of warming is lower and earlier; the switch from cooling to warming is earlier as well, and cooling is faster for higher peak concentrations. In the stabilisation scenarios, the graphs are less smooth. This is because the rate of warming is partly driven by methane emission control; in the later years, emission control is constant for periods of 25 years, while methane has a lifetime of some 10 years only.

Figure 30.3 shows the market and non-market impacts for Sub-Saharan Africa for the no control scenario and the stabilisation scenarios. The market impacts look similar

---

[8] This result is particular to the *FUND* model. It derives from the fact that health impacts are small – as avoided cold deaths offset increasing heat deaths, and as vector-borne diseases disappear with economic growth.

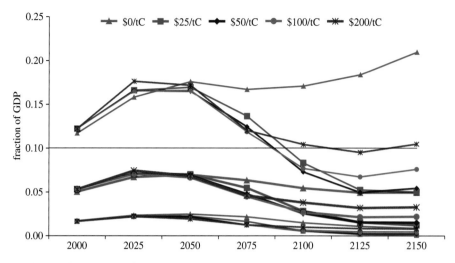

Figure 30.4 The 10, 50 and 90 percentile of climate change impacts in Sub-Saharan Africa as a function of time and intensity of emission reduction policy.

to those in Western Europe, but impacts are more negative; the stabilisation scenarios differ more in the long run; and the differences in the short-run are less pronounced as Africa has less sulphur to remove. The non-market impacts look very different. The main effect is not climate but development driven. Impacts fall from about 4.5% of GDP to a fraction of that, as Africans are assumed to rapidly grow rich enough to control diarrhoea and malaria. In fact, highest impacts are seen in the scenario that keeps CO_2 concentrations below 450 ppm, as mitigation crowds out public health care (cf. Tol, forthcoming, a).

30.4 Uncertainty and the Dangers of Mitigation

The analysis above can be rightly criticised for ignoring uncertainty. Tol (2005b) estimates the uncertainty about the marginal damage costs of carbon dioxide emissions. For a consumption discount rate of 5%, a typical value for governments who are members of the Organization of Economic Cooperation and Development (the OECD), the median estimate is 7$/tC. This justifies only modest emission reduction. However, the 90 (95) percentile is 35$/tC (62$/tC). For a 3% discount rate, these numbers are 33$/tC, 125$/tC and 165$/tC, respectively. With uncertainties so large, it is foolish to ignore them.

Suppose, for the sake of argument, that we were to adopt the following definition of dangerous interference: a 10% probability that 10% of the people would suffer harm equivalent to a 10% loss of income. This definition is as arbitrary as any. A 10% loss of income would make most people we know pretty miserable. A fraction of 10% of the population would imply that, if randomly picked, everyone would at least know someone very well who would suffer this fate. And a 10% probability is simply 'not insignificant'.

Unfortunately, current impact models are not able to explore the above definition. The main issue is the fraction

of population. Current models work with large regions, assuming that the people within those regions are all the same. Therefore, Figure 30.4 looks at a 90-percentile of monetised impacts in Sub-Saharan Africa, which is the region most vulnerable to climate change, and which has about 10% of the world's population. The 90-percentile already exceeds the threshold 10% income-equivalent loss, mostly due to climate-change-induced malaria and diarrhoea deaths. Starting at a damage of about 12% of GDP in 2000, the 90-percentile increases to some 22% of GDP in 2150 without emission abatement.

Figure 30.4 also shows impacts if carbon dioxide emissions are taxed at 25/50/100/200$/tC in 2000, rising by 5% a year. Moderate emission abatement reduces the probability of damage exceeding the 10% of GDP threshold, and a 50$/tC tax is more effective than a 25$/tC tax. However, if emissions are taxed at 200$/tC, impacts increase again. The reason is that emission reduction is so expensive (recall that 200$/tC is the start tax) that economic growth slows and basic health care improves more slowly. As a result, vulnerability to diseases such as malaria and diarrhoea decreases at a slower rate, and impacts go up. Tol and Dowlatabadi (2001) further discuss this mechanism. In this case, abatement policies do not avoid dangerous interference, but exacerbate it.

30.5 Conclusion

This paper makes three points. Firstly, we think that it will be impossible to come up with an objective definition of dangerous interference (or even a definition of 'dangerous' upon which everyone can agree). Attempts to do so should therefore be replaced by focused efforts to calibrate the costs and benefits of climate policy (adaptation and mitigation) in many metrics, to understand the feasibilities of these policies, to quantify the relative likelihoods of potentially large-scale climate impacts, to design strategies for broadening the appeal of hedging and risk management approaches to the climate issue, and to explore the efficiency and equity implications of alternative policy instruments.

Secondly, we continue to see that the most serious impacts of climate change can be avoided by relatively modest emission reduction, not only because incrementally avoided impacts become smaller as abatement becomes more stringent, but also because a risk management approach to minimizing the potential cost of avoiding an abrupt change in the future suggests a similar response. We do not appeal to a full-blown cost-benefit analysis to make this point; nor is it dependent on the specification of a particular target being deemed optimal or void of danger.

Thirdly, and perhaps most importantly, we see the potential for overly ambitious mitigation to increase vulnerability to climate change by slowing economic growth in parts of the developing world.

These findings are of course subject to a range of caveats. This is less true for the first result – that 'danger' has no meaningful definition at social level. This claim is based on a mathematical truism and supported by common sense. Qualitatively, the second and third results cannot be contested either, though their quantitative illustrations must be viewed with the usual amount of scepticism. The second result – that the first steps of emission reduction avoid more damage than the subsequent steps – depends solely on the result that more and faster climate change is worse than less and slower climate change. The third result – that too much mitigation may increase impacts – hangs on the assumption that emission abatement has opportunity costs, especially with regard to adaptation to climate change. The question is therefore not about the sign of the results, but about their magnitude.

The findings presented here rely on a single model, which is a medium-sized integrated assessment model with thousands of lines of code and hundreds of assumptions on parameters. These findings also rely on the controversial method of monetising environmental and health impacts; even if one accepts the principles of monetisation, application is difficult and uncertain.

Nonetheless, international climate policy will have to proceed without the clear guidance of an objective assessment of what to achieve. In this sense, climate policy is not unique. And, as in other arenas, a moderate start seems more advisable than rushing into deep emission cuts.

Acknowledgements

Paul Baer and an anonymous referee had constructive and interesting comments on an earlier version. Financial support by the Michael Otto Foundation for Environmental Protection and the Princeton Environmental Institute is gratefully acknowledged. GY acknowledges the contribution of B. Belle in helping him focus on what is important.

REFERENCES

Arrow, K.J. (1951), *Social Choice and Individual Values*, Wiley and Sons, New York.

Azar, C. and Rodhe, H. (1997), 'Targets for Stabilization of Atmospheric CO_2', *Science*, **276**, 1818–1819.

Berk, M.M., van Minnen, J.G., Metz, B., Moomaw, W., den Elzen, M.G.J., van Vuuren, D.P. and Gupta, J. (2002), *Climate Options for the Long Term (COOL) Global Dialogue – Synthesis Report*, Report 490200003, RIVM, Bilthoven.

Corfee-Morlot, J. and Agrawala, S. (2004), *The Benefits of Climate Change Policies*, OECD, Paris.

Dessai, S., Adger, W.N., Hulme, M., Turnpenny, J., Köhler, J. and Warren, R. (2004), 'Defining and Experiencing Dangerous Climate Change', *Climatic Change*, **64**, 11–25.

ECF and PIK (2004), *What is Dangerous Climate Change? Initial Results of a Symposium on Key Vulnerable Regions, Climate Change and Article 2 of the UNFCCC*, European Climate Forum and Potsdam Institute for Climate Impact Research, Potsdam.

Fankhauser, S. and Tol, R.S.J. (2005), 'On Climate Change and Economic Growth', *Resource and Energy Economics*, **27**, 1–17.

Fankhauser, S., Tol, R.S.J. and Pearce, D.W. (1997), 'The Aggregation of Climate Change Damages: A Welfare Theoretic Approach', *Environmental and Resource Economics*, **10**, 249–266.

Funtowicz, S.O. and Ravetz, J.R. (1994), 'Uncertainty, Complexity and Post-Normal Science', *Environmental Toxicology and Chemistry*, **13** (12), 1881–1885.

Grübler, A. and Nakicenovic, N. (2001), 'Identifying dangers in an uncertain climate', *Nature*, **412**, 15.

Gupta, J. and van Asselt, H. (eds.) (2004), *Re-evaluation of the Netherlands Long-Term Climate Targets*, Institute for Environmental Studies, Vrije Universiteit, Amsterdam.

Hare, W. (2003), *Assessment of Knowledge on Impacts of Climate Change – Contribution to the Specification of Art. 2 of the UNFCCC*, Wissenschaftlicher Beirat der Bundesregierung Globale Umweltveränderungen, Berlin (http://www.wbgu.de/wbgu_sn2003_ex01.pdf)

Keller, K., Bolker, B.M., and Bradford, D.F. (2004), 'Uncertain climate thresholds and optimal economic growth', *Journal of Environmental Economics and Management*, **48**, 723–741.

Manne, A.S. and Richels, R.G. (1999), 'The Kyoto Protocol: A Cost-Effective Strategy for Meeting Environmental Objectives?', *Energy Journal Special Issue on the Costs of the Kyoto Protocol: A Multi-Model Evaluation*, 1–24.

Nicholls, R.J. and Lowe, J.A. (2004), 'Benefits of mitigation of climate change for coastal areas', *Global Environmental Change*, **14**, 229–244.

Nordhaus, W.D. (1993), 'Rolling the 'DICE': An Optimal Transition Path for Controlling Greenhouse Gases', *Resource and Energy Economics*, **15**, 27–50.

Parry, M.L. (ed.) (2004), 'Climate Change', *Global Environmental Change*, **14** (1), 1–99.

Parry, M., Arnell, N.W., McMichael, T., Nicholls, R., Martens, W.J.M., Kovats, S., Livermore, M., Rosenzweig, C., Iglesias, A. and Fischer, G. (2001), 'Millions at risk: defining critical climate change threats and targets', *Global Environmental Change*, **11**, 181–183.

Parry, M.L., Carter, T.R. and Hulme, M. (1996), 'What Is a Dangerous Climate Change?', *Global Environmental Change*, **6** (1), 1–6.

Petschel-Held, G., Schellnhuber, H.-J., Bruckner, T., Toth, F. L. and Hasselmann, K. (1999), 'The Tolerable Windows Approach: Theoretical and Methodological Foundations', *Climatic Change*, **41** (3–4), 303–331.

Smith, J.B., Schellnhuber, H.-J., Mirza, M.M.Q., Fankhauser, S., Leemans, R., Lin, E., Ogallo, L., Pittock, B., Richels, R.G., Rosenzweig, C., Tol, R.S.J., Weyant, J.P. and Yohe, G.W. (2001), 'Vulnerability to Climate Change and Reasons for Concern: A Synthesis', Chapter 19, pp. 913–967, in McCarthy, J.J., Canziani, O.F., Leary, N.A., Dokken, D.J. and White, K.S. (eds.), *Climate Change*

2001: Impacts, Adaptation, and Vulnerability, Cambridge University Press, Cambridge.

Swart, R.J., de Boois, H. and Rotmans, J. (1989), 'Targeting Climate Change', *International Environmental Affairs*, **1**, 222–234.

Tol, R.S.J. (1999), 'Spatial and Temporal Efficiency in Climate Change: Applications of *FUND*', *Environmental and Resource Economics*, **14** (1), 33–49.

Tol, R.S.J. (2001), 'Equitable Cost-Benefit Analysis of Climate Change', *Ecological Economics*, **36** (1), 71–85.

Tol, R.S.J. (2005a), *The Benefits of Greenhouse Gas Emission Reduction: An Application of* FUND, Research Unit Sustainability and Global Change **FNU-64**, Hamburg University and Centre for Marine and Atmospheric Science, Hamburg.

Tol, R.S.J. (2005b), 'The Marginal Damage Costs of Carbon Dioxide Emissions: An Assessment of the Uncertainties', *Energy Policy*, **33** (16), 2064–2074.

Tol, R.S.J. (forthcoming, a), 'Emission Abatement versus Development as Strategies to Reduce Vulnerability to Climate Change: An Application of *FUND*', *Environment and Development Economics*.

Tol, R.S.J. (forthcoming, b), 'The Double Trade-Off between Adaptation and Mitigation for Sea Level Rise: An Application of *FUND*', *Mitigation and Adaptation Strategies for Global Change*.

Tol, R.S.J. and Dowlatabadi, H. (2001), 'Vector-borne diseases, development & climate change', *Integrated Assessment*, **2**, 173–181.

Tol, R.S.J. and Verheyen, R. (2004), 'State Responsibility and Compensation for Climate Change Damages – A Legal and Economic Assessment', *Energy Policy*, **32** (9), 1109–1130.

Toth, F.L., Cramer, W.P. and Hizsnyik, E. (2000), 'Climate Impact Response Functions: An Introduction', *Climatic Change*, **46** (3), 225–246.

WBGU (1995), *Scenario zur Ableitung CO2-Reduktionsziele und Umsetzungsstrategien – Stellungnahme zur ersten Vertragsstaatenkonferenz der Klimarahmenkonvention in Berlin* (Scenario to Deduct CO_2 Reduction Targets and Implementation Strategies – Position for the First Conference of the Parties of the Framework Convention on Climate in Berlin), Wissenschaftlicher Beirat der Bundesregierung Globale Umweltveränderungen, Dortmund.

Webster's (1996), *New Universal Unabridged Dictionary*, Barnes and Noble Books, New York.

CHAPTER 31

Multi-Gas Emission Pathways for Meeting the EU 2°C Climate Target

Michel den Elzen[1] and Malte Meinshausen[2]

[1] *Netherlands Environmental Assessment Agency (MNP), The Netherlands*
[2] *Climate and Global Dynamics Division, National Center for Atmospheric Research (NCAR), USA*

ABSTRACT: This study presents a set of multi-gas emission pathways for CO_2-equivalent concentration stabilization levels at 400, 450, 500 and 550 ppm, along with an analysis of their global and regional reduction implications and implied probability of achieving the EU climate target of 2°C. The effect of different assumptions made for baselines, and technological improvement rates on the resulting emission pathways is also analysed. For achieving the 2°C target with a probability of more than 60%, greenhouse gas concentrations need to be stabilized at 450 ppm CO_2-equivalent or below; if the 90% uncertainty range for climate sensitivity is believed to be 1.5 to 4.5°C. A stabilisation at 450 (400) ppm CO_2-equivalent requires global emissions to peak around 2015, followed by substantial overall reductions in the order of 30%–40% (50%–55%) compared to 1990 levels in 2050 (including land use CO_2). In 2020, Annex I emissions (excl. land use CO_2) need to be approximately 15%–20% (25%) below 1990 levels, and non-Annex I emissions also need to be reduced compared to their baseline emissions.

31.1 Introduction

The aim of this study is to develop multi-gas abatement pathways for the set of the six greenhouse gases covered under the Kyoto Protocol that are compatible with the long-term EU climate target of limiting the global mean temperature increase to 2°C above pre-industrial levels (1861–1890), as adopted in 1996 and recently reconfirmed by the European Council in March 2005. We also analysed the associated probability of these pathways overshooting the EU climate target of 2°C in equilibrium (Hare and Meinshausen, 2004). See 'What does a 2°C target mean for greenhouse gas concentrations?' by Meinshausen in this volume. Earlier analysis of emission pathways leading to climate stabilization focuses mainly on CO_2 only (e.g. Enting et al., 1994; Wigley et al., 1996; Swart et al., 1998; Hourcade and Shukla, 2001). Consistent information on the reduction potential for the non-CO_2 gases has been lacking for a long time, which is why most studies on the implications of a multi-gas reduction strategy are more recent (e.g. Reilly et al., 1999; Eickhout et al., 2003). So far, there are roughly five ways of accounting for non-CO_2 emissions: (i) simple scenario assumptions independent of the CO_2 emission level, for example, the common non-intervention scenario (SRES A1B) for non-CO_2 emissions in the IPCC Third Assessment Report (Cubasch et al. 2001); (ii) 'scaling', concentrations or radiative forcing, which are proportionally scaled with CO_2: e.g. 23% of CO_2 forcing (see Raper and Cubasch, 1996); (iii) accounting for source-specific reduction potentials for all gases, as in the post-SRES scenarios (Morita et al., 2000; Swart et al., 2002); (iv) different approaches assuming cost-optimal implementation of available reduction options over the greenhouse gases, sources and regions (van Vuuren et al., 2005) and/or over time (Manne and Richels, 2001); and (v) meta-approaches that make use of the multi-gas characteristics in existing scenarios derived by any of the previous approaches (Meinshausen et al. in press).

Here we focus on a cost-optimisation variant (iv), which closely reflects the political reality of pre-set caps on aggregated emissions and individual cost-optimising actors. Specifically, the actors are assumed to choose a cost-minimizing mix of reductions across the different greenhouse gases to achieve the preset global emission level for each five-year period. We will focus on the development of multi-gas pathways for the greenhouse gas concentration stabilization targets 400, 450, 500 and 550 ppm CO_2-equivalent[1], so as to achieve more certainty in reaching the EU 2°C target (e.g. Hare and Meinshausen, 2004). See 'What does a 2°C target mean for greenhouse gas concentrations?' by Meinshausen in this volume. For these targets, we assume a certain overshooting (or peaking), i.e. concentrations may first increase to an 'overshooting' concentration level, then decrease before stabilizing.

This study also explores the step that succeeds the development of global emission pathways: i.e. the issue of differentiating post-2012 commitments, in other words, how to allocate the global emission reductions on a regional level. A detailed analysis of abatement costs is outside the scope of this study. The underlying global abatement

---

[1] 'CO_2-equivalence' summarises the climate effect ('radiative forcing') of all human-induced greenhouse gases, tropospheric ozone and aerosols.

cost estimates for the presented pathways are only briefly discussed in the Appendix and den Elzen and Meinshausen (2005). Regional costs compatible with earlier pathways to 550 and 650 ppm CO_2-eq. (Eickhout et al., 2003) are presented in den Elzen et al. (2005a).

The next section presents the overall method used for this analysis of linking global emission pathways with climate targets. Section 31.3 presents the global-mean temperature implications of the presented pathways. Global emission implications of the multi-gas pathways are presented in section 31.4. Section 31.5 analyses the regional emission implications. The final section 31.6 concludes.

31.2 Method for Developing Emission Pathways with Cost-Effective Multi-Gas Mixes

The applied method focuses on a cost-effective split among different greenhouse gas reductions for given emission limitations on Global Warming Potential (GWP)-weighted and aggregated emissions. Thus, the method reflects the existing policy framework with preset caps on GWP-weighted overall emissions under the assumption of cost-minimizing national strategies.[2] The emissions are iteratively adapted to meet the pre-defined climate targets and include those of all major greenhouse gases (fossil CO_2, CH_4, N_2O, HFCs, PFCs and SF_6, i.e. the so-called six Kyoto greenhouse gases), ozone precursors (VOC, CO and NO_x) and sulphur aerosols (SO_x). For our method we used two models in combination: FAIR 2.0 and SiMCaP.

The FAIR (Framework to Assess International Regimes for the differentiation of commitments) 2.0 model developed at the MNP in the Netherlands (www.mnp.nl/fair) is a policy decision-support tool, which aims to assess the environmental and abatement costs implications of climate regimes for differentiation of post-2012 commitments (den Elzen and Lucas, 2005; den Elzen et al., 2005b). For the calculation of the emission pathways, only the (multi-gas) abatement costs model of FAIR is used. This model distributes the difference between a baseline and a global emission pathway over the different regions, gases and sources following a least-cost approach, taking full advantage of the flexible Kyoto Mechanisms (emissions trading) (see den Elzen et al., 2005a). For this purpose, it makes use of (time-dependent) Marginal Abatement Cost (MAC) curves for the different regions, greenhouse gases and sources as described below. The FAIR model also uses baseline scenarios, i.e. potential greenhouse gas emissions in the absence of climate policies, from the integrated

climate assessment model IMAGE (IMAGE-team, 2001) and the energy model, TIMER (van Vuuren et al., 2004).

The SiMCaP model ('Simple Model for Climate Policy Assessment') was developed at the Swiss Federal Institute of Technology (ETH) in Zurich, Switzerland (www.simcap.org). The SiMCaP pathfinder module makes use of an iterative procedure to find emission paths that correspond to a predefined arbitrary climate target. The global climate calculations make use of the simple climate model, MAGICC 4.1 (Wigley and Raper, 2001; 2002; Wigley, 2003).

The integration of both models, the 'FAIR–SiMCaP' 1.0 model, combines their respective strengths: (i) to calculate the cost-optimal mixes of greenhouse gas reductions for a global emissions profile under a least costs approach (FAIR) and (ii) to find the global emissions profile that is compatible with any arbitrary climate target (SiMCaP).

For a brief discussion of the underlying global cost estimates for the presented emission pathways, see the Appendix. It should be noted that the applied approach does not derive cost-effective pathways over the whole scenario period *per se*, but focuses on a cost-effective split among different greenhouse gas reductions for given emission limitations on GWP-weighted and aggregated emissions. In this way, we are also not analysing the effect of postponing abatement actions (benefits from technology development in time and from discounting costs further in the future) versus early action strategies (accounting for the inertia, uncertainty etc.). Therefore the pathways in our approach do not depend on the discount rate.

More specifically, the FAIR–SiMCaP calculations consist of four steps (Figure 31.1):

1. Using the SiMCaP model to construct a parameterised global CO_2-equivalent emission pathway, which is here defined by sections of linear decreasing or increasing emission reduction rates (see also Appendix A in den Elzen and Meinshausen, 2005). This CO_2-equivalent emission pathway includes the anthropogenic emissions of six Kyoto greenhouse gases (using the 100-year GWPs (IPCC, 2001)). One exception is formed by the land use and land use change-related (hereafter simply land use) CO_2 emissions; this is because no MAC curves are available for these, although the option of sink-related uptakes is parameterised in FAIR as one mitigation option. The land use CO_2 emissions are described by the baseline scenario. Up to 2012, the pathway incorporates the implementation of the Kyoto Protocol targets for the Annex I regions excluding Australia and the USA. The USA follows the proposed greenhouse-gas intensity target (White-House, 2002), which is close to a number of business as usual projections.

2. The abatement costs model of FAIR is used to allocate the global emissions reduction objective (except land

---

[2] Note though that as a consequence of using GWPs, neither the applied approach nor the international framework necessarily find cost-effective pathways over time for meeting an absolute temperature target, like 2°C, because a too large emphasis might be put on short-lived gases in the near-term.

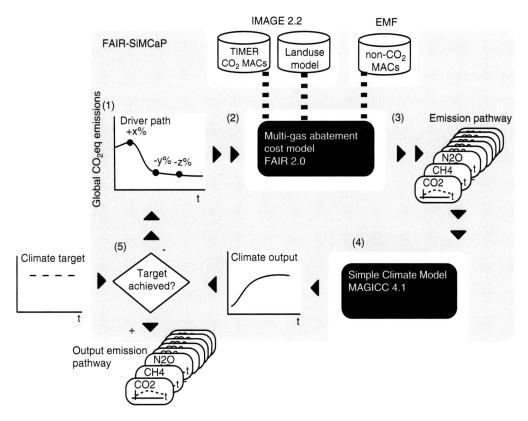

Figure 31.1 The FAIR–SiMCaP model. The calculated global emission pathways were developed by using an iterative procedure as implemented in SiMCaP's 'pathfinder' module, using MAGICC to calculate the global climate indicators, the multi-gas abatement costs and the FAIR model to allocate the emissions of the individual greenhouse gases and the IMAGE 2.2 and TIMER model for the baseline emissions scenarios along with the MAC curves.

use CO_2 emissions): i.e. the difference between the baseline emissions and the global emission pathway (see Figure 31.2) of step 1. Here a least-cost approach (cost-optimal allocation of reduction measures) is used for five year intervals over the 2000–2100[3] period for the six Kyoto greenhouse gases; 100-year GWP indices, different number of sources (e.g. for CO_2: 12; CH_4: 9; N_2O: 7) and seventeen world regions are employed, taking full advantage of the flexible Kyoto Mechanisms. Figure 31.2 shows the contribution of the different greenhouse gases in the global emissions reduction to, in this case, reach the 450 ppm CO_2-equivalent concentration level for two baseline scenarios (as briefly described later in this section). The figure clearly shows that up to 2025, there are potentially large incentives for sinks and non-CO_2 abatement options (cheap options), so that the non-CO_2 reductions and sinks form a relatively large share in the total reductions. Later in the scenario period, the focus is more on the CO_2 reductions, and the contribution of

most gases becomes more proportional to their share in baseline emissions.

Different sets of baseline- and time-dependent MAC curves for different emission sources are used here. Response curves from the TIMER energy model are used for the energy CO_2 emissions (van Vuuren et al., 2004), including technological developments, learning effects and system inertia. For CO_2 sinks the MAC curves of the IMAGE model are used (van Vuuren et al., 2005). For non-CO_2, exogenously determined MAC curves from EMF-21 (DeAngelo et al., 2004; Delhotal et al., 2004; Schaefer et al., 2004) are used; these are based on detailed abatement options. As these curves were constructed for 2010 only, increases in the abatement potentials due to technology process and removal of implementation barriers are assumed. Here, a relatively conservative value of an increasing potential (at constant costs) for all other non-CO_2 MAC curves of 0.4% per year is assumed. There are still some remaining agricultural emission sources of CH_4 and N_2O, where no MAC curves were available (e.g. for N_2O agricultural waste burning, indirect fertilizer, animal waste and domestic sewage). As it is unlikely that these sources remain unabated under ambitious climate targets, we assumed a linear reduction towards a maximum

---

[3] After 2100, there are no MAC estimates, and here the CO_2-equivalent emission reductions rates are assumed to apply to each individual gas, except where non-reducible fractions (0.65) have been defined (N_2O, CH_4).

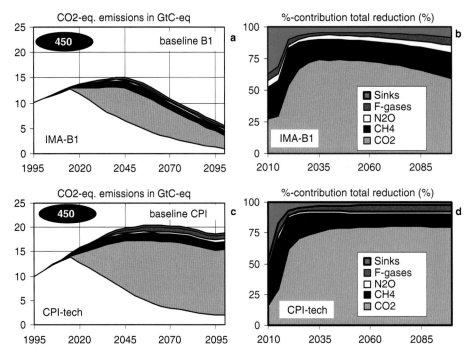

Figure 31.2 Contribution of greenhouse gases in total emission reductions under the emission pathways for a stabilization at 450 ppm CO_2-equivalent concentration of the IMA-B1 (a,b) and CPI+tech scenario (c,d).

of 35% reduction compared to the baseline levels within a period of 30 years (2040). For a detailed description of the MAC curves we refer to van Vuuren et al. (2005) and den Elzen and Meinshausen (2005).

3. The greenhouse gas concentrations, and global temperature and sea level rise are calculated using the simple climate model MAGICC 4.1.

4. Within the iterative procedure of the SiMCaP model, the parameterisations of the CO_2-equivalent emission pathway (step 1) are optimised (repeat step 1, 2 and 3) until the climate output and the prescribed target show sufficient matches.

These emission pathways have been developed for three underlying baseline scenarios:

1. CPI: the Common POLES IMAGE (CPI) baseline (van Vuuren et al., 2005) scenario with the land use CO_2 emissions of this scenario and with the default MAC curves. The CPI scenario assumes a continued process of globalisation, medium technology development and a strong dependence on fossil fuels. This corresponds to a medium-level emissions scenario when compared to the IPCC SRES emissions scenarios (Figure 31.2c).

2. CPI + tech: the CPI baseline scenario with the land use CO_2 emissions of the IMA-B1 scenario (less deforestation) and with MAC curves assuming additional technological improvements: (1) for the MAC curves of energy CO_2, an additional technological improvement factor of 0.2%/year; (2) for the MAC curves of

the non-CO_2 gases, a technological improvement rate of 1%/year instead of 0.2%/year and (3) for the sources of non-CO_2 gases, where no MAC curves were available, a maximum reduction of 80% instead of 35% in 2040.

3. IMA-B1: the IMAGE IPCC SRES B1 baseline (IMAGE-team, 2001) scenario with the land use CO_2 emissions of this scenario and the default MAC curves. This scenario assumes continuing globalisation and economic growth, and a focus on the social and environmental aspects of life (Figure 31.2a).

31.3 Emission Pathways and their Temperature Implications

This section presents various global multi-gas emission pathways for stabilization at CO_2-equivalence levels of 550 ppm (3.65 W/m²), 500 ppm (3.14 W/m²), 450 ppm (2.58 W/m²) and 400 ppm (1.95 W/m²). The latter three pathways are assumed to peak at 525 ppm (3.40 W/m²), 500 ppm (3.14 W/m²) and 480 ppm (2.92 W/m²) before they return to their ultimate stabilization levels around 2150 (Figures 31.3 and 31.4). This peaking is partially reasoned by the already substantial present net forcing levels (Hare and Meinshausen, 2004) and the attempt to avoid drastic sudden reductions in the emission pathways presented. These lower two stabilization pathways are within the range of the lower mitigation scenarios in the literature (Swart et al., 2002; Nakicenovic and Riahi,

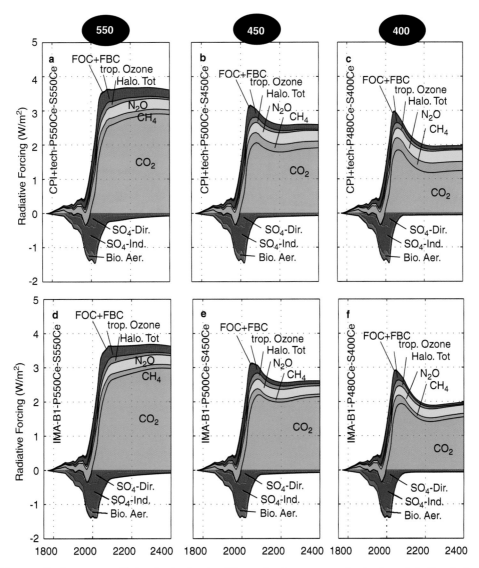

Figure 31.3 The contribution to net radiative forcing by the different forcing agents under the three default emission pathways for a stabilization at (a,d) 550, (b,e) 450 and (c,f) 400 ppm CO$_2$-equivalent concentration after peaking at (b,e) 500 and (c,f) 475 ppm, respectively for the (a–c) CPI+tech and (d–f) IMA B1 baseline scenarios. The upper line of the stacked area graph represents net human-induced radiative forcing. The net cooling due to the direct and indirect effect of SOx aerosols and aerosols from biomass burning is depicted by the lower negative boundary, on top of which the positive forcing contributions are stacked (from bottom to top) by CO$_2$, CH$_4$, N$_2$O, fluorinated gases (including the cooling effect due to stratospheric ozone depletion), tropospheric ozone and the combined effect of fossil organic & black carbon.

2003; Azar et al., 2004) (Figure 31.4). Note that actual CO$_2$ concentrations might differ between scenarios for the same CO$_2$-equivalent concentration due to different non-CO$_2$ reductions. For example, 550 CO$_2$-eq. corresponds approximately with 475–500 ppm CO$_2$, and 400 ppm CO$_2$-eq. corresponds with 350–375 ppm CO$_2$ only (Figure 31.3). No emission pathways for 450 and 400 ppm CO$_2$-eq. level were derived for the CPI baseline, given the standard MAC curves.

Figure 31.5 shows the probabilistic temperature implications of the overshoot concentration profiles based on the climate sensitivity (IPCC lognormal) PDF of Wigley

and Raper (2001), assuming the conventional 1.5 to 4.5°C climate sensitivity uncertainty range at a 90% confidence interval, for the emission pathways under the B1 scenario. In these transient calculations, we included the natural forcings (i.e. solar and volcanic forcings) (as in Hare and Meinshausen, 2004). The results under the other scenarios are similar.

An important caveat is that these transient calculations only take account of the uncertainty in climate sensitivity, but assume other parameters (as e.g. ocean diffusivity, sulphate aerosol forcing etc.) according to the IPCC Third Assessment Report 'best guess' values. This is

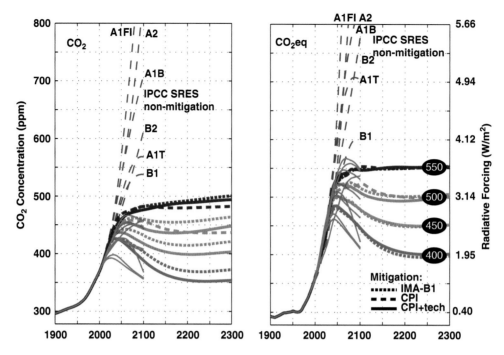

Figure 31.4 The CO_2 (a) and CO_2-equivalent (b) concentrations for the stabilization pathways at 550, 500, 450 and 400 ppm CO_2-equivalent concentrations for the three baseline scenarios (CPI, CPI+tech and IMA-B1). For comparison, the concentration implications of the IPCC-SRES non-mitigation scenarios (grey dotted lines) and the lower range of published mitigation scenarios (Swart et al. 2002; Nakicenovic and Riahi, 2003; Azar et al. 2004) (grey solid lines) are also plotted (see details in Hare and Meinshausen, 2004).

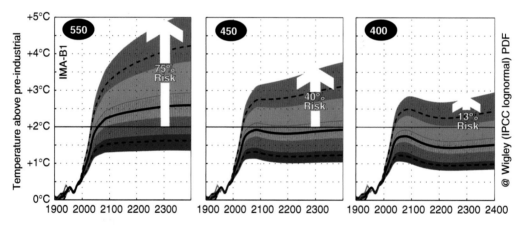

Figure 31.5 The probabilistic temperature implications for the stabilization pathways at 550 ppm, 450 ppm and 400 ppm CO_2-equivalent concentrations for the B1 baseline scenario based on the climate sensitivity PDF by Wigley and Raper (2001) (IPCC lognormal). Shown are the median (solid lines) and 90% confidence interval boundaries (dashed lines), as well as the 1%, 10%, 33%, 66%, 90%, and 99% percentiles (borders of shaded areas). The historical temperature record and its uncertainty from 1900 to 2001 is shown (grey shaded band) (Folland et al. 2001).

clearly a simplification, given that neither the dependency between climate sensitivity and other parameters nor their uncertainty is reflected in the transient temperature results. However, the presented transient results compare well with studies that take account of the dependency. See 'What does a 2°C target mean for greenhouse gas concentrations?' by Meinshausen in this volume. Furthermore, the uncertainty in long-term (equilibrium)

global mean temperature levels resulting from different concentration stabilization levels are clearly dominated by the uncertainty in climate sensitivity.

Due to the inertia of the climate system, the peak of radiative forcing (3.14 W/m^2) before stabilization at 450 ppm CO_2-eq. (2.58 W/m^2) does not translate into a comparable peak in global mean temperatures. However, for the 400 ppm CO_2-eq. stabilization pathway presented,

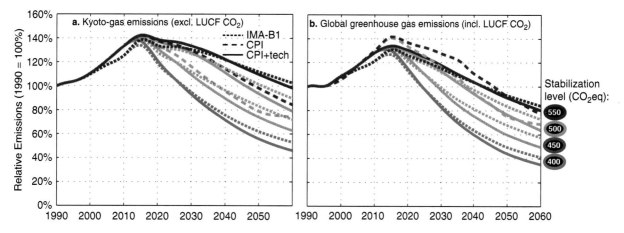

Figure 31.6 Global emissions relative to 1990 excluding (a) and including (b) land use CO_2 emissions for the stabilization pathways at 550, 500, 450 and 400 ppm CO_2-equivalent concentrations for the three scenarios (CPI, CPI+tech and IMA-B1).

the initial peak at 480 ppm CO_2-eq. seems to be decisive with regard to the question of whether the 2°C or any other temperature threshold will be crossed (Figure 31.5). Figure 31.5 shows that for a stabilization at 550 ppm CO_2-eq. (corresponding approximately to a 475 ppm CO_2 only stabilization), the risk of overshooting 3°C is still about 33%. There is even a risk of about 10% that 4°C is exceeded in the long term. The probability that warming exceeds 2°C is very high, approximately 75%. For the long-term stabilization at 500 ppm CO_2-eq. (approximate 450 ppm CO_2 stabilization) too, the probability of exceeding 2°C is likely, about 60% (not shown). Only for a stabilization at 400 ppm CO_2-eq. (approximately 350–375 ppm CO_2 stabilization) and, to a lesser extent, at 450 ppm CO_2-eq. (about 400 ppm CO_2 only stabilization), is the possibility of equilibrium warming exceeding 2°C strongly reduced, to less than about 13% and 40% respectively.

31.4 The Global Emission Implications

The emissions of the pathways for stabilization at 550, 450 and 400 ppm CO_2-eq. concentrations can be summarized in their GWP-weighted sum of six Kyoto gases emissions, as illustrated in Figure 31.6. Clearly, there are different pathways that can lead to the ultimate stabilization level. Here, we assume that the global emission reduction rates should not exceed an annual reduction of 2.5%/year for all default pathways (at least not over longer time periods). The reason is that a faster reduction might be difficult to achieve given the inertia in the energy production system: electrical power plants, for instance, have a technical lifetime of 30 years or more. Fast reduction rates would require early replacement of existing fossil-fuel-based capital stock, which may be associated with large costs. A maximum rate of 2%/year is hardly exceeded for

the majority of the post-SRES mitigation scenarios, apart from some lower stabilization scenarios. As a result of the assumed onset of reductions from the baseline emissions, reduction takes place relatively early, and global emissions peak around 2015–2020. For all stabilization pathways, the global reduction rates remain below 2.5%/year for the whole scenario period, except for the pathways at 400 ppm CO_2-eq., with maximum reduction rates of 2.5–3%/year over 20 years.[4]

Greenhouse gas emission reductions *excluding* and *including* land use CO_2 emissions are analysed here. Given the assumption of these static land use scenarios with decreasing emissions, the quantified reduction requirements obviously differ, depending on whether the reduction requirements refer to all greenhouse gas emissions including land use CO_2 or Kyoto gas emissions (excl. land use CO_2). In general, emission pathways for the CPI + tech and B1 baselines have slightly higher greenhouse gas emissions (excl. land use CO_2) compared to the pathways under the CPI baseline for the same concentration target, because the land use CO_2 emissions for the CPI + tech and B1 scenario are assumed to be lower.

By 2050, global greenhouse gas emissions (excl. land use CO_2) will have to be near 40–45% below 1990 levels for stabilization at 400 ppm CO_2-eq. For higher stabilization levels, e.g. 450 ppm CO_2-eq. stabilization, greenhouse gas emissions (excl. land use CO_2) may be higher, namely 15–25% below 1990 levels. However, if land use CO_2 emissions do not decrease as rapidly as assumed here, but continue at presently high levels, an additional reduction of Kyoto-gas emissions (excl. land use CO_2) by around 10% are required up to 2050.

---

[4] A further delay in peaking of global emissions in 10 years doubles maximum reduction rates to about 5%/year, and very likely leads to high costs (den Elzen and Meinshausen, 2005).

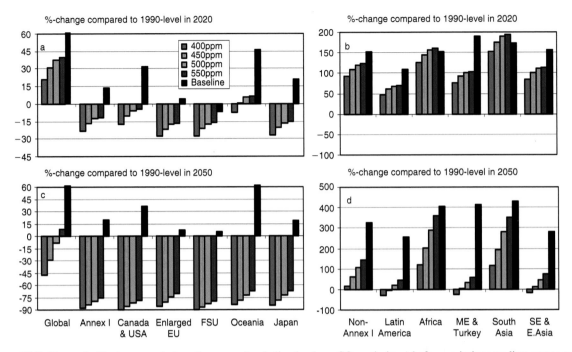

Figure 31.7 Change in Kyoto-gas emission allowances (excluding land use CO_2 emissions) before emissions trading compared to 1990 levels in 2020 (upper) and 2050 (lower) for the Annex I regions (a,c) and non-Annex I regions (b,d) for the Contraction & Convergence approach for the stabilization pathways at 550, 500, 450 and 400 ppm CO_2-equivalent concentrations under the CPI+tech scenario.

Global greenhouse gas emissions (incl. land use CO_2) will have to decrease to 5–10% below 1990 levels by 2050 for stabilization at 550 ppm CO_2-eq. For stabilization at 500 ppm CO_2-eq., global greenhouse gas emissions would need to be 15–25% below 1990 levels in 2050. The reduction requirements now become as high as 50–55% and 30–40% below 1990 levels in 2050 to reach the 400 ppm and 450 ppm CO_2-eq. target, respectively (instead of 40–45% and 15–25%, respectively) (see Figure 31.6b). These reductions are about 10–15% higher than the reductions of the Kyoto gas emissions excluding land use CO_2.

In general, when we compare the reductions for the different concentration levels, we find that about 15–20% additional reductions by 2050 are needed for every 50 ppm lower stabilization level. We also see that higher near-term emissions need to be compensated by lower future emissions (compare CPI with B1 of the 500 ppm level, for example).

31.5 The Regional Emission Implications

This section presents regional emission allowances that follow from the global emission pathways. We chose one out of many possible options for the international regime of differentiating future (post-2012) commitments: the Contraction & Convergence approach. This approach is selected here, as it is a widely known and transparent

approach despite concerns in regard to its political feasibility. The approach defines emission allowances on the basis of convergence of per capita emission allowances (starting after 2012) of all countries (including the USA)[5] in 2050 under a contracting global emissions pathway (Meyer, 2000). Figure 31.7 gives the change in the regional emission allowances of the six Kyoto gases (excluding land use CO_2) compared to the 1990 levels for 2020 and 2050 for the CPI + tech scenario.

This analysis suggests that Annex I commitments need to be strongly intensified after 2012, if global emissions should follow any of the presented pathways. In 2020, Annex I Kyoto-gas emissions (excluding land use CO_2) need to be reduced by about 25% in comparison with 1990 levels for 400 ppm, and about 15–20% for 450 ppm stabilization. The reductions compared to the baseline are about 10–15% higher. In 2050, the reductions below 1990 levels stand at about 90% (400 ppm) and 80% (450 ppm), respectively (see Figure 31.7).

---

[5] There are a number of reasons to assume that the US might join a post-2012 regime, whatever it may be called. Avoiding future disasters like the aftermath of Hurricane Katrina may play a part in this, as well as high oil prices and the motivation of becoming less dependent on foreign fossil fuel reserves. Obviously, there is no certainty that this will happen. However, it is hard to conceive of any global climate regime that is compatible with stabilising GHG concentrations at 550 ppm equivalent or lower if the USA decides against joining the international effort to reduce emissions after 2012 or further delays its involvement, as analysed in den Elzen and Meinshausen (2005).

Most non-Annex I regions will need to reduce their emissions by 2020 compared to baseline levels, but emissions may increase compared to 1990. For the low-income regions (for example, Southern Asia (India)) emission allowances may even exceed baseline emissions under 500 and 550 ppm CO_2-eq. For the middle and high income non-Annex I regions, the reductions compared to the baseline emissions are below the reductions for the Annex I regions, about 25% and 35% by 2020, but increase to about 70% and 80% by 2050 for 450 and 400 ppm, respectively. These non-Annex I reductions in 2050 are less than the Annex I reductions compared to their baseline emissions.

31.6 Conclusions

This study describes a method to derive multi-gas pathways that closely reflect the existing international framework of pre-set caps on aggregated emissions and individual cost-optimising actors. Thus, cost-optimal mixes of greenhouse gases reductions are derived for a given global emission pathway. The presented emission pathways stabilize CO_2-equivalent concentration at 550, 500, 450 and 400 ppm. The presented lower pathways allow overshooting, i.e. concentrations peak before stabilizing at lower levels, e.g. going up to 480–500 ppm CO_2-equivalent before going down to levels such as 400 or 450 ppm CO_2-equivalent later on.

The emission pathways leading to a 550 ppm CO_2-equivalent stabilization are unlikely to meet the EU 2°C climate target. In order to achieve such a 2°C target with a probability of more than 85% (60%) (assuming the probabilistic density function for climate sensitivity of Wigley and Raper, 2001), greenhouse gas concentrations need to be stabilized at 400 (450) ppm CO_2-equivalent or lower (see 'What does a 2°C target mean for greenhouse gas concentrations?' by Meinshausen in this volume). This, in turn, requires global emissions to peak around 2015 in order to avoid global reduction rates exceeding more than 2.5%/year, followed by substantial overall reductions by as much as 40–45% (15–25%) in 2050 compared to 1990 levels, excluding land use CO_2 emissions. The reduction requirements become as high as 50–55% (30–40%) below 1990 levels in 2050 for all greenhouse gas emissions, including land use CO_2.

Finally, the analysis of the post-2012 regime for future commitments, Contraction & Convergence, shows that Annex I emissions in 2020 will need to be reduced by about 15–25% below 1990 levels for 400–450 ppm CO_2-eq. Non-Annex I emissions may increase compared to their 1990 levels, but need to deviate from their baseline emissions as soon as possible. For the advanced developing countries, this could be as early as 2015. In general, reaching lower levels of greenhouse gas concentrations requires earlier reductions and faster participation of the non-Annex I countries compared to higher levels of greenhouse gases.

APPENDIX

Global Emission Abatement Costs

In its Third Assessment Report (TAR), the IPCC presents estimates for macro-economic costs (i.e. loss in GDP growth) of stabilization of the CO_2 concentration. For stabilization of the CO_2 concentration at 450 ppm (comparable to 500–525 ppm CO_2-eq.), GDP reductions for 2050 have to be 1.0–4.0% (see Figure 8.18 in Hourcade and Shukla (2001).The range is primarily derived from the assumption of different baseline scenarios (B1 to A1FI, respectively). These are global estimates, with some sectors and also regions (e.g. the oil-exporting regions) being likely to be more severely affected (e.g. van Vuuren et al. 2003).

These GDP costs have to be seen in perspective though. On the one hand, such long-term GDP abatement costs are approximately equivalent to a delay of only a couple of years with respect to a point in time, while the world might experience a twenty-fold increase in its GDP around 2100 compared to present levels (Azar and Schneider, 2002; 2003). Furthermore, the climate damage avoided and ancillary benefits are not included in such cost estimates, although they might be comparable in scale, if not much greater.

Here, we present some results of the global abatement costs as a percentage of world GDP for the different CO_2-equivalent concentration levels. Before presenting the costs, it should be noted that these costs only represent the direct-cost effects based on MAC curves but not the various linkages and rebound effects via the economy or impacts of carbon leakage. In other words, there is no direct link with macro-economic indicators such as GDP losses or other measures of income of utility loss. The cost figures are also very dependent on our assumptions about abatement potentials and reduction costs for all greenhouse gases. For a further discussion on the limitations, but also the strengths of this cost methodology we refer to den Elzen et al. (2005b).

Global costs increase for lower stabilization levels. The emission pathways show an increase of the costs up to 2050, and then a general decrease as GDP growth outstrips the growth in calculated abatement costs for most of the pathways (Figure 31.8).

The Figure also shows that the global abatement costs are even more influenced by the baseline emissions and the assumed improvements in technical change of the abatement potentials and costs, than the final concentration stabilization level, as was also concluded by the IPCC. More specifically, the baseline emissions directly determine the reductions that are required to reach the emission profile for stabilization. The economic assumptions also obviously influence the relative cost measures such as GDP losses or abatement costs such as percentage of GDP.

Another crucial uncertainty is the rate at which the abatement costs for CO_2 and non-CO_2 emission reductions

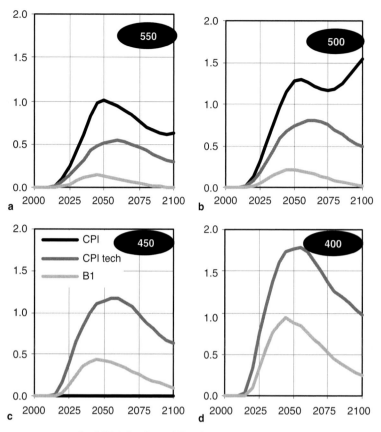

Figure 31.8 Global abatement costs as % of GDP for the stabilization pathways at (a) 550 ppm, (b) 500 ppm, (c) 450 ppm and (d) 400 ppm CO$_2$-equivalent concentrations for the three baseline scenarios (CPI, CPI+tech and IMA-B1).

develop in time (compare the CPI and CPI + tech baseline scenario). Given these uncertainties and limitations (mainly that ancillary benefits are not included and climate damage avoided), the results should be taken as qualitatively indicative, but not as quantitatively robust.

REFERENCES

Azar, C., Linddgren, K., Larson, E. and Möllersten, K.: 2004, 'Carbon capture and storage from fossil fuels and biomass – Costs and potential role in stabilizing the atmosphere', *Climatic Change* **(in press)**.

Azar, C. and Schneider, S.H., 2002. Are the economic costs of stabilising the atmosphere prohibitive? Ecological Economics, **42** (1–2): 73–80.

Azar, C. and Schneider, S.H., 2003. Are the economic costs of (non-) stabilizing the atmosphere prohibitive? A response to Gerlagh and Papyrakis. Ecological Economics, **46** (3): 329–332.

Cubasch, U., Meehl, G.A., Boer, G.J., Stouffer, R.J., Dix, M., Noda, A., Senior, C.A., Raper, S. and Yap, K.S.: 2001, 'Projections of Future Climate Change', in Houghton, J.T., Ding, Y., Griggs, D.J., Noguer, M., van der Linden, P.J., Dai, X., Maskell, K. and Johnson, C.A. (eds.), *Climate Change 2001: The Scientific Basis*, Cambridge University Press, Cambridge, UK.

DeAngelo, B.J., DelaChesnaye, F.C., Beach, R.H., Sommer, A. and Murray, B.C.: 2004, 'Methane and nitrous oxide mitigation in agriculture', *Energy Journal* **(in press)**.

Delhotal, K.C., DelaChesnaye, F.C., Gardiner, A., Bates, J. and Sankovski, A.: 2004, 'Mitigation of methane and nitrous oxide emissions from waste, energy and industry', *Energy Journal* **(in press)**.

den Elzen, M.G.J. and Lucas, P.: 2005, 'The FAIR model: a tool to analyse environmental and costs implications of climate regimes', *Environmental Modeling and Assessment* **10**, 115–134.

den Elzen, M.G.J., Lucas, P. and van Vuuren, D.P.: 2005a, 'Abatement costs of post-Kyoto climate regimes', *Energy Policy* **33**, pp. 2138–2151.

den Elzen, M.G.J. and Meinshausen, M.: 2005, 'Meeting the EU 2°C climate target: global and regional emission implications', Netherlands Environmental Assessment Agency (MNP), Bilthoven, the Netherlands. MNP-report 728001031 (www.mnp.nl/en)

den Elzen, M.G.J., Schaeffer, M. and Lucas, P.: 2005b, 'Differentiating future commitments on the basis of countries' relative historical responsibility for climate change: uncertainties in the 'Brazilian Proposal' in the context of a policy implementation', *Climate Change* **71**, 277–301.

Eickhout, B., den Elzen, M.G.J. and van Vuuren, D.P.: 2003, 'Multi-gas emission profiles for stabilising greenhouse gas concentrations'. Bilthoven, the Netherlands, Netherlands Environmental Assessment Agency (MNP). MNP-report 728001026 www.mnp.nl\en

Enting, I.G., Wigley, T.M.L. and Heimann, M.: 1994, 'Future emissions and concentrations of carbon dioxide'. Mordialloc, Australia: 120

European Council: 1996, 'Communication on Community Strategy on Climate Change, Council Conclusions', European Council, Brussels.

European Council: 2005, 'Presidency conclusions', European Council, Brussels.

Folland, C.K., Rayner, N.A., Brown, S.J., Smith, T.M., Shen, S.S.P., Parker, D.E., Macadam, I., Jones, P.D., Jones, R.N., Nicholls, N. and Sexton, D.M.H.: 2001, 'Global temperature change and its uncertainties since 1861', *Geophysical Research Letters* **28**, 2621–2624.

Hare, W.L. and Meinshausen, M.: 2004, 'How much warming are we committed to and how much can be avoided?' Potsdam, Germany, Potsdam Institute for Climate Impact Research (PIK). PIK report No. 93, www.pik-potsdam.de/publications/pik_reports

Hourcade, J.-C. and Shukla, P.R.: 2001, 'Global, regional and national costs and ancillary benefits of mitigation', in Metz, B., Davidson, O., Swart, R., Pan, J. (eds.), *Climate Change 2001: Mitigation; Contribution of Working Group III to the Third Assessment Report of the IPCC*, Cambridge University Press, Cambridge, UK.

IMAGE-team: 2001, 'The IMAGE 2.2 implementation of the SRES scenarios. A comprehensive analysis of emissions, climate change and impacts in the 21st century'. Bilthoven, the Netherlands. CD-ROM publication 481508018

IPCC: 2001, *Climate Change 2001: Mitigation*, Cambridge University Press, Cambridge, UK.

Manne, A.S. and Richels, R.G.: 2001, 'An alternative approach to establishing trade-offs among greenhouse gases', *Nature* **410,** 675–677.

Meinshausen, M., Hare, B., Wigley, T.M.L., van Vuuren, D., den Elzen, M.G.J. and Swart, R.: in press, 'Multi-gas emission pathways to meet climate targets', *Climatic Change*, (**in press**).

Meyer, A.: 2000, *Contraction & Convergence. The global solution to climate change*, Green Books, Bristol, UK.

Morita, T., Nakicenovic, N. and Robinson, J.: 2000, 'Overview of mitigation scenarios for global climate stabilization based on new IPCC emission scenarios (SRES)', *Environmental Economics and Policy Studies* **3,** 65–88.

Nakicenovic, N. and Riahi, K.: 2003, 'Model runs with MESSAGE in the Context of the Further Development of the Kyoto-Protocol'. Berlin, WBGU – German Advisory Council on Global Change: 54

Raper, S.C.B. and Cubasch, U.: 1996, 'Emulation of the results from a coupled general circulation model using a simple climate model', *Geophysical Research Letters* **23,** 1107–1110.

Reilly, J., Prinn, R.G., Harnisch, J., Fitzmaurice, J., Jacoby, H., Kicklighter, D., Stone, P., Sokolov, A. and C., W.: 1999, 'Multi-gas Assessment of the Kyoto Protocol', *Nature* **401,** 549–555.

Schaefer, D.O., Godwin, D. and Harnisch, J.: 2004, 'Estimating future emissions and potential reductions of HFCs, PFCs and SF6', *Energy Journal* (**in press**).

Swart, R., Berk, M., Janssen, M., Kreileman, E. and Leemans, R.: 1998, 'The safe landing analysis: risks and trade-offs in climate change', in Alcamo, J., Leemans, R. and Kreileman, E. (eds.), *Global change scenarios of the 21st century. Results from the IMAGE 2.1 model*, Elseviers Science, London, pp. 193–218.

Swart, R., Mitchell, J., Morita, T. and Raper, S.: 2002, 'Stabilisation scenarios for climate impact assessment', *Global Environmental Change* **12,** 155–165.

van Vuuren, D.P., de Vries, H.J.M., Eickhout, B. and Kram, T.: 2004, 'Responses to Technology and Taxes in a Simulated World', *Energy Economics* **26,** pp. 579–601.

van Vuuren, D.P., Eickhout, B., Lucas, P.L. and den Elzen, M.G.J.: 2005, 'Long-term multi-gas scenarios to stabilise radiative forcing', *Energy Journal, forthcoming*.

White House: 2002, 'Executive Summary of Bush Climate Change Initiative'.

Wigley, T.M.L.: 2003, 'MAGICC/SCENGEN 4.1: Technical Manual'. Boulder, CO, UCAR – Climate and Global Dynamics Division http://www.cgd.ucar.edu/cas/wigley/magicc/index.html

Wigley, T.M.L. and Raper, S.C.B.: 2001, 'Interpretation of high projections for global-mean warming', *Science* **293,** 451–454.

Wigley, T.M.L. and Raper, S.C.B.: 2002, 'Reasons for larger warming projections in the IPCC Third Assessment Report', *Journal of Climate* **15,** 2945–2952.

Wigley, T.M.L., Richels, R. and Edmonds, J.A.: 1996, 'Economic and environmental choices in the stabilisation of CO_2 concentrations: choosing the "right" emissions pathway', *Nature* **379,** 240–243.

CHAPTER 32

Why Delaying Emission Cuts is a Gamble

Steffen Kallbekken and Nathan Rive
CICERO Center for International Climate and Environmental Research, Oslo, Norway

ABSTRACT: In the debate on avoiding dangerous climate change the aspect of feasibility of the mitigation options has often been missing. We introduce this aspect and show through an illustrative modeling exercise that, if we decide to delay emissions reductions, and the overall environmental effectiveness of global mitigation efforts is to remain the same in terms of medium to long-term temperature change, we must be willing and able to undertake much more substantial emission reductions than with earlier emission reductions. In our illustration, a 20-year delay in emission reductions means that we must reduce emissions at an annual rate that is 3–9 times greater than with immediate emission reductions. If we are not able to achieve such higher rates, delaying emission cuts will inevitably result in higher temperatures in the short to medium term. While the inertia of the climate system creates this result, the inertia of the socio-economic system gives us reason to be concerned about it. Unless we are willing to accept higher temperatures, choosing to delay emission cuts is a gamble that feasibility will increase over time as a result of the delay. That is, the act of delaying must somehow be correlated with improved feasibility of global emissions cuts.

32.1 Introduction

While there is general agreement that avoiding 'dangerous anthropogenic interference with the climate system' (United Nations, 1992) will require long-term reductions in greenhouse gas (GHG) emissions, there has been much debate concerning the timing and magnitude of these reductions. The choice of an appropriate emissions reduction pathway has received significant attention in the literature. Wigley *et al.* (1996) argued that significant emissions reductions should be postponed, in light of slow capital turnover, and technological improvement with time. Grubb (1997) counter-argued for near-term abatement by highlighting the abatement technologies available today, the importance of induced technological change and learning-by-doing, and the risks of significant long-term climate impacts.

In the literature on long-term climate agreements, it is typically assumed that the approach to climate policy will be consistently both (economically) rational and knowledgeable. This is reflected in recommendations to frame our actions with a long-term climate target that avoids 'dangerous' climate change (e.g. O'Neill and Oppenheimer, 2002; Corfee-Morlot and Höhne, 2003) or to develop optimal long-term climate policies that account for both the costs and benefits of mitigation policy (e.g. Nordhaus and Boyer, 2001).

Our approach is to revisit the timing debate without assuming consistent and rational long-term climate policies. This is based on the argument that it is more realistic to assume that future agreements (whether one global or a more fragmented regime) will be based on what is feasible *at each point in time* – rather than on some optimized or cost-effective long-term mitigation scenario. There is little reason to believe binding targets will be set for the next 50 or 100 years, or that future generations would feel obliged to stick to them. This feasibility is determined by such constraints as the trade-off between the economic, environmental, social, and political costs and benefits of mitigation – particularly for the most influential political actors – as well as concerns such as enforcement, public pressure, fairness and burden-sharing.

Instead of making uncertain assumptions about how feasibility might improve through for example technological innovation, we ask how much feasibility must increase if environmental integrity is to be maintained when emission cuts are delayed. We then discuss whether such an increase in feasibility seems reasonable in light of the arguments previously put forward in the debate.

32.2 An Illustrative Modeling Exercise

In this section, we generate a number of illustrative 'what if' scenarios to highlight the differences between 'early' and 'delayed' emission reduction scenarios. We are particularly interested in the relationship between the environmental outcomes of early and delayed emission reduction scenarios, and annual rates of global emissions reductions.

We use the DEEP economic model (Kallbekken, 2004) to generate three initial emission scenarios for a 100-year period beyond the first phase of the Kyoto Protocol. The first is an *Early action* scenario, where we assume that when the commitments under the Kyoto Protocol expire in 2012, new agreements will be in place such that, on aggregate, global GHG emissions continue to be reduced at an annual rate of 0.3% (discussed below).

In the second and third scenarios, we assume that no emission cuts take place during the 20-year period following the Kyoto Protocol (2013–2033). This, of course, does not exclude the possibility of other climate action, such as technological development taking place during this period. After 20 years, emission reductions are agreed upon: as with the *Early action* case, at an annual rate of 0.3%. As the emissions growth during the intervening years (2013–2033) is uncertain, we run both a high and a low emissions growth scenario using economic growth rates from the SRES A1B and B2 scenarios respectively (Nakicenovic and Swart, 2000). We call these two scenarios *Delayed high* and *Delayed low*. These emission scenarios are fed into a simple climate model (Fuglestvedt *et al.*, 2000) to obtain projections of temperature change.

In terms of climate policy, we define 'feasibility' as the maximum annual rate of emission reductions that can be agreed upon. Thus, when all the various factors that determine feasibility have played their role, we end up with one number (that is, the overall annual global emissions reductions) that can used to express the overall trade-off between these factors. There is, however, little empirical evidence on which to base any estimate of what this maximum rate is today, and will be in the future. Our only point of reference is the Kyoto Protocol. The Protocol requires industrialized countries to keep their CO$_2$-equivalent 2008–2012 emissions at (on average) 5% below their 1990 emissions, which corresponds to an annual emissions reduction rate of about 0.3% (from the previous year). The commitments under the Kyoto Protocol can of course be interpreted in many different ways. However, as a starting point for our illustration, we will assume that global emission reductions at a constant annual rate of 0.3% are feasible today, and that this rate will remain constant. Compared to the reduction rates assumed to be feasible in other studies, this is a conservative assumption (e.g. Yohe *et al.*, 2004).

Our assumptions, of course, are simplified for the purpose of the exercise, and it is stressed that the important assumption is not the exact level of this rate, but rather that there is some limit to what emission reductions are feasible in any given year. This simplification excludes two important possibilities. Firstly, the feasible annual rate of global emissions reductions could increase with time. Secondly, there might be a 'phase-in' to global emissions reductions, such as the 9-year delay between signing the Kyoto Protocol and implementing the emission cuts. We exclude these possibilities because we are primarily interested in the differences arising from the *timing* of the start of emission reductions, not what happens once the reductions are underway. In effect, we are assuming that the progression of these events will be identical in both cases, and simply cancel them out (see Figures 32.1 and 32.2).

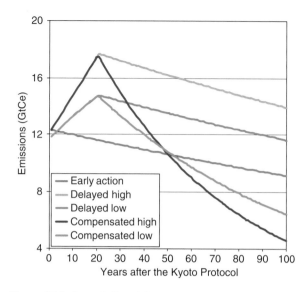

Figure 32.1 Annual Global GHG emission profiles.

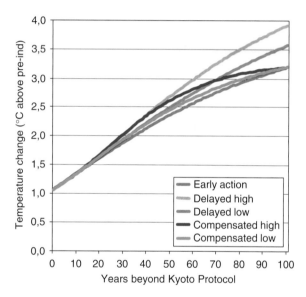

Figure 32.2 Global mean temperature change profiles.

Table 32.1 shows the projected temperature change under our three emission scenarios.[1] As we have assumed that it is not feasible to reduce global emissions at a rate greater than 0.3% per year, the temperature projections for each scenario show how much the global mean temperature would increase over a 100-year period from the end of the first phase of the Kyoto Protocol. We find that if the feasibility constraint does not change, then the global mean temperature will be 0.4–0.7°C higher on a 100-year time horizon if we delay action by 20 years than if we take immediate action after 2012.

---

[1]We have used the 50% probability interval for climate sensitivity as reported in Murphy *et al.*, 2004 (3.5°C for a doubling of CO$_2$-concentrations). While climate sensitivity does, of course, have an impact on projected temperature change, it does not have an impact on the comparison of early and delay emission reductions.

Table 32.1 Global mean temperature increase (°C above pre-industrial levels) 100 years after Kyoto Protocol under Early and Delayed scenarios. Results from the simple climate model with SRES emissions are shown for comparison.

Early action	Delayed high	Delayed low	BAU A1B (in 2100)	BAU B2 (in 2100)
3.2	3.9	3.6	4.0	3.6

Table 32.2 Annual emission reductions that achieve same maximum temperature increase 75, 100 and 125 years beyond the Kyoto Protocol as with Early action.

Time horizon (years beyond KP)	Early Action ΔT (°C above pre-ind)	Compensated high (% per year)	Compensated low (% per year)
75	2.8	2.6%	1.4%
100	3.2	1.7%	1.1%
125	3.5	1.3%	0.9%

Many of the arguments used in favor of delaying emission cuts relate to the benefits of technological improvements over time. One claim is that investing in technological improvements will bring down the cost of emission reductions, and make greater emission reductions feasible. To this end, a climate change technology agreement between the United States, Japan, Australia, China, India and South Korea was signed in July 2005 (Shanahan, 2005). Another important argument is that overly expensive emission reductions could divert resources from technological development, and thus keep further emission reductions expensive, and reduce the feasibility of greater cuts in the future.

In light of this, we generate two further scenarios to consider the case where feasibility does improve with a delay in emissions cuts, and ask what annual emission reductions are required if we are to achieve the same environmental effectiveness with delayed action as with early action. In other words, we ask how much other mitigation efforts during the period of no emission cuts must increase feasibility to compensate for the advantages of early action if the delayed policy option is to perform at least as well as early action. These two scenarios are called *Compensated high* and *Compensated low*. The high and low refers to high and low economic growth rates borrowed from SRES (see Figure 32.1).

We use the global mean temperature change in a given year as the indicator of environmental effectiveness of our emission reduction scenarios. Using such an indicator is to some extent at odds with humanity's general concern for *all* future climate change. However, because of discounting, there is a higher concern for the near future than the distant future. We thus present our results for three different indicator years: 75, 100, and 125 years beyond the first phase of the Kyoto Protocol. Additionally, because we allow for temperature overshoot, temperatures continue to rise well into the next centuries and these indicator years provide the maximum temperature change for their respective horizons.[2]

To find the required annual emission reductions we will use an iterative algorithm that runs both the economic and climate models. Table 32.2 shows what annual emission reductions are required in the Compensated scenarios in order to achieve the same temperature change after 75,

100 and 125 years as the early action scenario does with 0.3% annual reductions. In this illustration, if we are to reach the same temperature with delayed emission cuts as with immediate emission cuts and annual reductions of 0.3%, we must be able to reduce emissions at an annual rate that is 3–9 times greater (Figure 32.1). The results show that the longer the time horizon, the smaller the 'penalty' on delaying emission reductions.

There are two reasons why annual emission reductions must be steeper in the Compensated scenarios: (1) starting from greater global emissions it takes more time to reduce emissions to the same level, and (2) the effect of reduced emissions on temperature is postponed due to the long response time for CO_2 and the inertia of the climate system, requiring emission cuts below the level of the early action scenario in the later years.

One additional insight is that *any* delay in emission cuts will inevitably produce higher temperatures over some time horizon. The temperature change in our *Delayed* scenarios will always be higher than in the *Early action* scenario. The temperature change in the *Compensated* scenarios will be higher than the *Early action* scenario in every year leading up to the indicator year (up to 0.25°C higher) as shown in Figure 32.2.

In addition to the time horizon, one important and non-obvious parameter choice in our modeling exercise is what global emission reductions we assume to be feasible initially. If we assume that the feasible annual reductions under *Early action* are much smaller (0.1%), then *Compensated* action would require emission cuts of 0.7 and 1.3% per year for low and high growth respectively. If instead we assume that the feasible rate is three times as high (i.e. 0.9%), the compensated reduction rates must be 2.1% and 2.9% for low and high growth respectively. This shows that *Compensated* action requires more stringent annual emission cuts regardless of our assumptions about initial feasibility.

32.3 Discussion

Our results show that to uphold a given level of environmental effectiveness, we must be willing and able to undertake much more rapid annual emission reductions if we decide to delay. What is remarkable in our results is

---

[2]In terms of the magnitude of the overshoot, if we extend the Early Action scenario until 2500 temperatures start to stabilize at around 4.8°C above pre-industrial levels by the middle of the 25th century.

just how much greater (in the range of 3–9 times) the annual emissions reductions must be as a result of a 20 year delay.

Of course, the results are illustrative, and like any experiment are dependent on our assumptions. However, they provide a rough guide to what we can expect by delaying emissions cuts by 20 years. Furthermore, if we had not exempted the possibility of accelerating the rate of emission reductions over time, and of a 'phase-in' period, this would only have emphasized our results.

In terms of achieving any given long-term climate target, this means that if we want to wait before taking on binding emission reductions, instead of undertaking relatively modest action today (that is, follow up the Kyoto Protocol with further emissions cuts), we must be certain that our technological capacity and political willingness for undertaking mitigation will improve substantially with the delay of emissions cuts. That is, the act of delaying must itself be correlated with an improvement in the feasibility of global emissions cuts. Any feasibility improvements that occur once reductions have started (i.e. learning by doing) will be featured in the early case as well and thus provides no comparative advantage.

There are at least two key issues that need to be considered: technological development and the inertia of the socio-economic system. Technological change can have a significant impact on the costs of abatement, and on the feasibility of emissions reductions. Thus, there are obvious implications for the choice of timing of GHG abatement. Delaying action would allow time for research and development (R&D), and the unrestricted economic growth during the interval of no climate policy may increase our capacity for technological progress. However, if no action is taken in the near-term, the technological limits of our mitigation capabilities are unlikely to improve substantially. For example, efficiency improvements have not been achieved without external price shocks (Schneider and Azar, 2001) and government R&D has fallen in recent years (Margolis and Kammen, 1999).

The term 'inertia of the socio-economic system' refers to the fact that there are many obstacles to undertaking any major change in the socio-economic system. One advantage of postponing significant action in the near-term is that it avoids 'premature retirement of existing capital stocks and takes advantage of the natural rate of capital stock turnover' and allows more 'time to retrain the workforce and for structural shifts in the labor market and education' (IPCC, 2001; section 10.4.3).

However, comparing the *Early* and *Compensated* scenarios, we find that inertia provides significant arguments for early emission cuts. Firstly, the gradual emissions reductions afforded under the *Early* action scenario will be considerably less painful to economy than those in the *Compensated* cases, as they will require less rapid changes to the socio-economic system.

Secondly, we should also be concerned with the capital investments made in the coming decades. For instance, a large number of power plants will be built in the near-term, and will have an economic lifespan of 30–40 years. Under the *Compensated* cases, we can expect that a greater number of these power plants will be fossil fuel fired compared to the *Early* action scenario due to the higher economic growth, and absence of climate regulations in the non-Annex B regions. This will contribute further to the lock-in into existing technologies. Overcoming the 'lock-in' of fossil fuel power plants, as well as other parts of our carbon economy does not become easier by extending the life of our existing energy infrastructure.

Our model runs showed that if feasibility remains unchanged, the temperature change over the 75, 100, and 125-year horizons could be around 0.4–0.7°C higher than under *Early action* (Table 32.2). This implies that we could further exceed levels of temperature change that could be 'dangerous'.

If, on the other hand, feasibility does increase as a result of postponing emission cuts, then it may still be possible to limit temperature increases to the same levels as with early action. These improvements must in some way result from the period of no emission cuts, either through technological development or public pressure, and they must subsequently be sustained throughout the mitigation period. There are several reasons why this is unlikely to happen.

Thus, those who believe our greenhouse gas emissions can produce dangerous climate change, but wish to postpone emission reductions, are taking a gamble that delaying emissions cuts will dramatically improve the feasibility of future emission cuts.

REFERENCES

Corfee-Morlot, J. and Höhne, N. 2003, 'Climate change: long-term targets and short-term commitments', Glob. Env. Change 13, 277–293.

Grubb, M. 1997, 'Technologies, energy systems, and the timing of CO_2 emission abatement – An overview of economic issues', Energy Policy 25(2), 159–172.

Fuglestvedt, J.S., Berntsen, T., Godal, O. and Skodvin, T. 2000, 'Climate implications of GWP-based reductions in greenhouse gas emissions', Geophys. Res. Lett. 27 (3), 409–412.

IPCC: 2001, 'Climate Change 2001: Mitigation', Intergovernmental Panel on Climate Change, Cambridge University Press, Cambridge, UK.

Kallbekken, S. 2004, 'A description of the Dynamic analysis of Economics of Environmental Policy (DEEP) model', CICERO Report 2004:01, CICERO, Oslo.

Margolis, R.M. and Kammen, D.M. 1999, 'Under-investment: The Energy Technology and R&D Policy Challenge', Science 285, 690–691.

Murphy, J.M., Sexton, D.M.H., Barnett, D.N., Jones, G.S., Webb, M.J., Collins, M. and Stainforth, D.A. 2004, 'Quantification of modelling uncertainties in a large ensemble of climate change simulations', Nature 430, 768–772.

Nakicenovic, N. and Swart, R., (eds): 2000, 'Special Report on Emission Scenarios', Cambridge University Press, Cambridge, UK.

Nordhaus, W.D., and Boyer, J. 2001, 'Warming the World', The MIT Press, Boston, USA.

O'Neill, B.C. and Oppenheimer, M. 2002, 'Dangerous Climate Impacts and the Kyoto Protocol', Science 296, 1971–1972.

Schneider, S.H., and Azar, C. 2001, Are Uncertainties in Climate and Energy Systems a Justification for Stronger Near-term Mitigation Policies? Report prepared for Pew Center on Global Climate Change, Arlington.

Shanahan, D. 2005, 'New Asia-Pacific climate plan' *The Australian*, July 27, 2005. Available online: http://www.theaustralian.news.com.au/printpage/0,5942,16060815,00.html

United Nations: 1992, 'United Nations framework convention on climate change', UNFCCC. Available online: http://www.unfccc.int/

Wigley, T.M.L., Richels, R. and Edmonds, J.A. 1996, 'Economic and Environmental Choices in the Stabilisation of Atmospheric CO_2 Concentrations', Nature 379, 242–245.

Yohe, G., Andronova, N. and Schlesinger, M. 2004, 'To Hedge or Not Against an Uncertain Climate Future?', Science 306, 416–417.

CHAPTER 33

Risks Associated with Stabilisation Scenarios and Uncertainty in Regional and Global Climate Change Impacts

David Stainforth, Myles Allen, David Frame and Claudio Piani
Department of Physics, University of Oxford, Oxford, UK

ABSTRACT: Any stabilisation level for greenhouse gas concentrations implies an acceptance of a certain degree of climate change. The choice of a stabilisation level as a political or societal goal can therefore only be made in the context of the predicted effects of different choices. However, the science of how the earth's climate responds to changing concentrations of greenhouse gases, and particularly the probabilistic analysis of such responses, is still in its infancy. The climateprediction.net project has found that the response to even a relatively low stabilisation level could be substantial (greater than $11°C$ for a doubling of CO_2). This is consistent with previous work using simpler models but by using complex models we are able to extract ranges of response for multiple variables, on both a global and regional level. Such results are of profound significance in terms of the risks associated with political decisions and the methodology of impact assessments.

33.1 Introduction

Planning for climate change mitigation and adaptation needs to consider a range of possible futures. Even if anthropogenic greenhouse gas (GHG) emissions exactly follow some emission scenario there are significant uncertainties in how the climate might respond. Intrinsic uncertainties result from the chaotic nature of the climate system and further uncertainties result from our lack of scientific understanding. Over recent years there have been a number of attempts to quantify these uncertainties and thus produce probabilistic statements regarding the effects of climate change. Most of these have been at a global level.

Using a complex climate model we have undertaken a grand ensemble of climate simulations. As described below, this approach was supported by the e-science, distributed computing methodology of climate*prediction*.net. In this way we have found model versions which are as realistic as other state-of-the-art climate models but with climate sensitivities (the equilibrium global mean temperature change with doubling levels of carbon dioxide) ranging from less than $2°C$ to more than $11°C$. This has significant implications for any choice of stabilisation level much above pre-industrial values (e.g. present-day levels), because such a choice implies acceptance of a risk of extreme climate change.

We present here the method and analysis which leads to this result as well as the associated ranges for precipitation in northern and southern Europe. Such regional information is critical for mitigation and adaptation planning. We also discuss procedures for extracting ranges/distributions for climate variables at a regional level. The development of analysis methods to assess the **probability** of such responses is extremely problematic and cannot be simply inferred from ensemble distributions [1]. Research is ongoing in this subject in a number of academic disciplines. Nevertheless it is already possible to carry out risk analyses for a variety of societal vulnerabilities from these types of results. The existence of the climate*prediction*.net dataset and the regional information it contains suggests a possibility for new procedures for impact analyses. Rather than such assessments being based on generalised, average information from the modellers, it may be more appropriate for impact assessments to search the dataset to find the range of combined precipitation and temperature responses. For instance, in assessing the flood risks or agricultural impacts it would be possible to search the dataset and use the range of combined precipitation and temperature behaviour.

33.2 Background

The IPCC Third Assessment Report [2] provided uncertainty estimates based on the range of behaviour found in general circulation models (GCMs) and concluded that the climate sensitivity was likely to be in the range of 1.5 to $4.5°C$. There are fundamental problems with interpreting this range as an objective probabilistic statement. First only order ten GCMs were available so it is statistically inappropriate to identify any behaviour which only has, say, a 5% probability of occurring. And yet such possibilities could be crucial in the decision making process. Second, the climate modelling community worldwide is not large so it is not surprising that modellers share methodological approaches. Consequently the models are not independent, a fundamental barrier to deductions of objective probabilities.

There have been several more recent studies using observations of past climate to constrain the future climate response [3,4,5]. These studies allow for the possibility of high sensitivities (>6°C) although the probability assigned to them varies substantially.

33.3 Grand Ensembles

The science of climate change is still a young discipline. An enormous amount has been achieved in a very short period of time and it is clear, as concluded by the IPCC [2], that the earth is warming and that "most of the warming observed [since 1950] is attributable to human activities" [2]. Consequently the scientific basis for societal action on a global level is clear. Nevertheless, it is important not to overstate what science can tell us in this field. In particular, there has been relatively little attention paid to uncertainty analyses so it is not surprising that the IPCC identified as a high priority for action the need to "improve methods to quantify uncertainties of climate projections and scenarios, including long-term ensemble simulations using complex models".

There are three sources of uncertainty in climate change predictions. They are:

33.3.1 *Response Uncertainty*

This reflects our incomplete understanding of the climate system. It is not possible to carry out experiments on the real climate so we use climate models, but there are large uncertainties in how such models are constructed. To evaluate such uncertainties we have used a perturbed-physics ensemble (PPE). Such ensembles consist of large numbers of simulations which are identical except for the values given to certain parameters; the different parameter combinations produce different "model versions". The parameters are perturbed from their standard values within a range considered plausible by experts in the relevant parameterisation schemes. There are hundreds of uncertain parameters in a GCM and parameter perturbations combine non-linearly [1,3] (i.e. it is not possible to predict the effect of changing multiple parameters simultaneously, by changing one at a time) so it is necessary to carry out a sampling of parameter space [1,6], requiring tens of thousands of simulations. Since this is beyond the capacity of conventional super-computing facilities we have used a distributed computing approach in which more than 100,000 people from 150 countries have volunteered the unused computing capacity of their personal computers [6].

In the future it will be necessary to take this concept of model perturbations further by changing entire parameterisation schemes and repeating such experiments using different GCMs.

33.3.2 *Natural Variability*

This is a consequence of the chaotic nature of the climate system such that very small changes at one point in time can lead to completely different states at some future time. It is addressed in models using initial condition (IC) ensembles in which small changes are made to the starting conditions of the simulation.

33.3.3 *Forcing Uncertainty*

This represents the familiar uncertainty in future factors which influence climate, including anthropogenic GHG emissions, natural GHG emissions (e.g. volcanic activity) and external natural forcing (e.g. changes in solar radiation). In climate models they are represented using scenarios of different future forcings. Typically an ensemble of simulations is carried out, each representing a different scenario of possible future forcings.

Since these three sources of uncertainty interact non-linearly it is necessary to investigate them using one "grand ensemble" (i.e. ensemble of ensembles) as illustrated in Figure 33.1. The first climate*prediction*.net experiment comprises a grand ensemble exploring model response and natural variability uncertainty using the Hadley Centre GCM HadSM3 at standard climate resolution. Within each ensemble member (simulation) the response

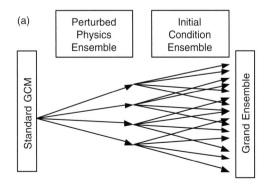

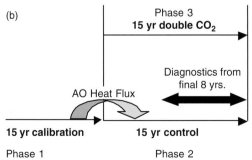

Figure 33.1 Schematic of the experimental design. A grand ensemble is an ensemble of ensembles designed to explore uncertainty resulting from model construction, initial conditions and forcing. (a) The standard GCM has parameters perturbed to create a large PPE and for each member of this ensemble an IC ensemble is created, producing a grand ensemble of simulations. (b) For each member of the grand ensemble 45 years of simulation are undertaken, including 15 exploring the response to doubling the concentrations of CO_2 in the atmospheric component of the model.

to changing forcing is explored using a double CO_2 scenario.

33.4 Uncertainty in Global Temperature Change

Over 80,000 simulations have been completed to date (April 2005) and an analysis of the uncertainty in climate sensitivity from an initial subset of simulations (2578) is contained in reference [1]. Model versions have been found with climate sensitivities ranging from less than 2°C to more than 11°C. In reference [1] the problems associated with interpreting the distribution as a probability distribution are discussed and comparisons are made with other state-of-the-art climate models (the models from the second Coupled Model InterComparison Project {CMIP II}) demonstrating that it is not possible to rule out high sensitivities on the basis of the ability of model versions to simulate observations.

The existence of such model versions enables the hitherto impossible study of a wide range of sensitivities, with GCMs. We hope that such studies will reveal constraints on the possible range of future behaviour but such constraints have not yet been identified.

33.5 Uncertainty in Regional Changes

33.5.1 *Results from the Grand Ensemble*

A significant benefit from a grand ensemble of GCM simulations is that regional information can be extracted. Figure 33.2 shows the distribution of mean precipitation

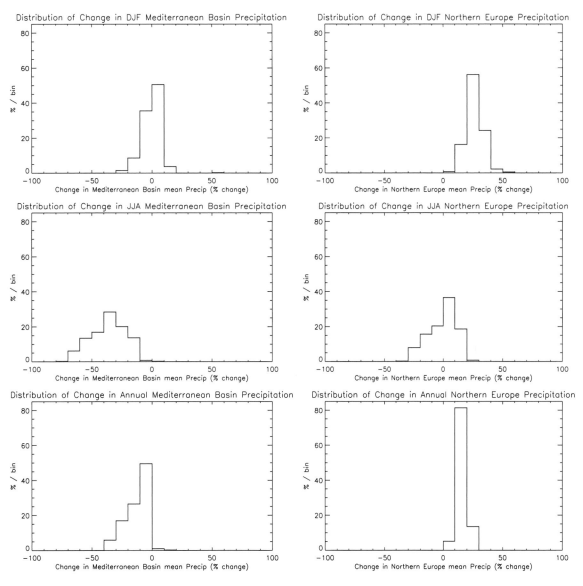

Figure 33.2 The distribution of changes in precipitation in response to doubling levels of CO_2 for northern hemisphere winter (DJF = December/ January/ February) and northern hemisphere summer (JJA = June/ July/August) for the Mediterranean basin [−10:40 longitude, 30:50 latitude] and northern Europe (−10:40 longitude, 40:75 latitude). The changes are calculated as the difference between the control and double CO_2 phases, of the mean precipitation in the region averaged over years eight to fifteen after the start of the phases (see Figure 1b).

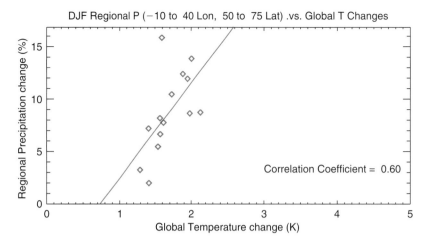

Figure 33.3 Northern European winter precipitation against annual mean global mean temperature change in the CMIP II transient simulations.

change 8–15 years after CO_2 concentrations are doubled, for northern European and Mediterranean winter, summer and annual mean. These results are based on the same dataset and quality control procedures used in reference [1]. All model versions show an increase in northern European annual mean precipitation (0–30%) and a decrease in the Mediterranean basin annual mean precipitation (0–40%). However, this annual mean behaviour masks more extreme behaviour on a seasonal basis. Virtually all model versions show an increase in northern European winter precipitation of more than 10%, and up to 50%. In the Mediterranean basin the summer decrease is between 10% and 70%. By contrast there is no clear indication of even the sign of the change in Mediterranean basin winter and northern European summer precipitation. Such information is available for other regions and variables and should provide valuable inputs to mitigation and adaptation planning. Furthermore it will be possible to extract distributions based on the combined behaviour of a number of variables e.g. temperature, cloud cover, surface pressure etc. which should enable improved uncertainty analyses in impact studies.

As highlighted in reference [1] it is important not to interpret these distributions as probability density functions because they are highly dependent on the choice of perturbations explored in the perturbed-physics ensemble. Furthermore, these regional distributions are not the equilibrium response. Low sensitivity model versions have reached equilibrium after eight years but high sensitivity model versions are still adjusting after fifteen. These ranges are therefore likely to represent a lower bound on the range of potential climatic response in these variables. For global mean temperature it is possible to use fitting techniques to extract the equilibrium response [1]. A combination of re-running simulations in-house and scaling pattern analyses may help provide ranges for equilibrium regional responses.

33.5.2 *Alternative Methods*

Beyond looking simply at the regional/seasonal distributions from the grand ensemble it is possible to extract such distributions using other methods. For instance examination of the data can reveal correlations between predicted variables of interest and better constrained or observable quantities. For example, Figure 33.3 shows northern European winter precipitation against global annual mean temperature from the transient simulations of the CMIP II ensemble. The correlation appears to be good suggesting that in this case we can convert a distribution in global temperature to one in northern European seasonal precipitation. Examination of the grand ensemble supports this result and provides the opportunity to look for such constraints in a statistically robust way. Analysis of the grand ensemble is also revealing **patterns** of observations which may help constrain predicted quantities; not surprisingly such patterns vary according to the predicted quantity of interest.

33.6 **Conclusions**

33.6.1 *Implications for Stabilisation Levels*

The disturbing conclusion of this work is that currently we can provide neither an upper bound on climate sensitivity nor an objective probability distribution for this quantity [1]. This has profound implications for any choice of stabilisation level. In our experiment a stabilisation scenario of twice pre-industrial CO_2 levels has been studied; ~550 ppm. The results suggest that in such a scenario the response in global mean temperature could range from less than 2°C to more than 11°C. The associated increase in northern European winter precipitation could be at least 10% to 50% and the decrease in Mediterranean basin summer precipitation could be at least 10% to 70%. While the lower ends of these ranges could provide tolerable targets it would be unwise to dismiss the possibility that the response could be extreme.

33.6.2 *Risk Analyses*

Until recently climate science has been restricted in its ability to even attempt objective probabilistic statements of potential climate response, as reflected in the conclusion of the IPCC Third Assessment Report [2]. It is not surprising therefore that the first probabilistic analyses are broadening the range of responses which should be considered. We cannot yet provide objective probabilities on climate change forecasts but requiring model behaviour to be consistent with our knowledge of past climate provides a range of possible responses. This allows for the development of risk analyses which will be hugely beneficial in mitigation and adaptation planning, particularly as further regional and seasonal information becomes available. In many circumstances it is not important to have a probability distribution but rather to be able to rule out possible futures. For instance in planning a specific flood protection scheme the design and costs may vary little for certain levels of climate change, but change dramatically after a certain point e.g. 100% (or 10%) seasonal precipitation increase. In that case, if studies of the type presented herein, suggest that that point will not be met (or is very likely to be met) then it provides information of significant value for the organisations concerned.

33.6.3 *Impact Assessments*

Impact assessments involve a further range of assumptions and therefore uncertainties in the analysis process and are typically based on mean predictions from the climate science community. The availability of grand ensemble data provides the possibility for impact assessments to integrate climate uncertainty into their analysis procedures. They should be able to interrogate the dataset and extract information on the combined distributions of the variables relevant for their particular impact or vulnerability study.

Furthermore, the methodology of climate*prediction*.net provides the opportunity for a range of additional experiments and for actively involving the general public in attempts to understand the possible consequences of climate change.

REFERENCES

1. Stainforth, D.A. et al. Uncertainty in Predictions of the Climate Response to Rising Levels of Greenhouse Gases. Nature, 433, 403–406 (2005).
2. Houghton, J.T., Ding, Y., Griggs, D.J., Noguer, M., van der Linden, P.J., Dai, X., Maskell, K. and Johnson, C.A. (Eds) (2001). Climate Change 2001: The Science of Climate Change, Cambridge University Press.
3. Murphy, J.M. et al. Quantifying uncertainties in climate change from a large ensemble of general circulation model predictions. Nature, 430, 769–772 (2004).
4. Knutti, R. et al. Constraints on Radiative Forcing and Future Climate Change from Observations and Climate Model Ensembles, Nature, 416,719–723 (2002).
5. Forest, C.E. et al. Quantifying Uncertainties in Climate System Properties with the use of Recent Climate Observations. Science, 295,113–117 (2002).
6. Stainforth, D. et al. Environmental Online Communication Ch. 12. (ed. Arno Scharl). Springer-Verlag London Ltd., ISBN:1-85233-783-4, 2004.

CHAPTER 34

Impact of Climate-Carbon Cycle Feedbacks on Emissions Scenarios to Achieve Stabilisation

Chris D. Jones[1], Peter M. Cox[2] and Chris Huntingford[3]

[1] Hadley Centre, Met Office, Exeter, UK
[2] Centre for Ecology and Hydrology, Winfrith Technology Centre, Dorchester, Dorset, UK
[3] Centre for Ecology and Hydrology, Wallingford, Oxon, UK

ABSTRACT: As atmospheric concentrations of CO_2 increase due to burning of fossil fuels, stabilisation scenarios are receiving increasing amounts of interest both politically and scientifically, leading to the question, 'what emissions pathway is required to lead us to a given climate/CO_2 state?' At present, about half of anthropogenic CO_2 emissions are absorbed naturally, but there is growing consensus that this fraction will reduce due to the action of climate change on the natural carbon cycle. Such climate-carbon cycle feedbacks will therefore influence the amount of carbon emissions required to stabilise atmospheric CO_2 levels.

Here we quantify the impact that climate change will have on the world's natural carbon cycle and how this will affect the amount of CO_2 emissions which are permissible to achieve a stabilised climate in the future. Our simulated feedbacks between the climate and the carbon cycle imply a reduction of 21–33% in the integrated emissions (between 2000 and 2300) for stabilisation, with higher fractional reductions necessary for higher stabilisation concentrations. Any mitigation or stabilisation policy which aims to stabilise atmospheric CO_2 levels must take into account climate-carbon cycle feedbacks or risk significant underestimate of the action required to achieve stabilisation.

34.1 Introduction

As atmospheric concentrations of greenhouse gases (and most notably carbon dioxide, CO_2) increase due to burning of fossil fuels, there is growing recognition that this will cause major changes in climate. For some regions of the world, this may lead, ultimately, to 'dangerous climate change'. For this reason stabilisation scenarios are receiving increasing amounts of interest both politically and scientifically. Instead of asking where a 'business-as-usual' increase in CO_2 emissions will take us, the question becomes 'what emissions pathway is required to lead us to a given climate/CO_2 state?', thereby ensuring a stable climate into the future.

At present, approximately half of anthropogenic CO_2 emitted is absorbed naturally, by the land surface and the oceans (Schimel et al., 1996; Jones and Cox, 2005). Without this, atmospheric CO_2 concentration would be far higher than the current value of approximately 375 parts per million (ppm). Projections of future rises in CO_2 generally assume that this natural mitigation will continue, and this is included in calculations of emission pathways required to achieve stabilisation. However, the behaviour of the natural carbon cycle is dependent on climate change itself. That is, as the climate responds to increased atmospheric greenhouse gas concentrations, these climate changes act to reduce the uptake of CO_2 by the terrestrial and ocean biospheres, and lead to higher CO_2 levels than would otherwise be the case. We will refer to these throughout this work as 'climate-carbon cycle feedbacks'.

There is recent modelling evidence that feedbacks between the climate system and carbon cycle will have a significant impact on future relationships between emissions and atmospheric CO_2 concentrations. For a prescribed scenario of CO_2 emissions, Cox et al. (2000) found that the natural components of the carbon cycle have, in response to future evolving climate, a reduced capability to mitigate CO_2 emissions and thus provide a positive feedback; in fact the land surface eventually turns into a major natural source. Friedlingstein et al. (2001) found a weaker positive feedback with the terrestrial carbon sink reduced but not becoming a source of carbon. In fact a comparison of ten coupled climate-carbon cycle models found overwhelming evidence that the feedbacks are positive: although the strength is uncertain, the impact of climate change will be to reduce both the terrestrial and oceanic carbon cycles' ability to take up anthropogenic CO_2 (Friedlingstein et al., 2005). In the same way, therefore, it is likely that these positive feedbacks will reduce the magnitude of CO_2 emissions that lead to stabilised CO_2 levels.

The IPCC Third Assessment Report (Prentice et al., 2001) briefly alludes to the impact of the carbon cycle on stabilisation emissions but does not quantify the associated magnitude of emissions reductions. Joos et al. (1999) used a 'low order' model to quantify the ocean carbon cycle impact but did not discuss the terrestrial behaviour. Friedlingstein et al. (2001) show how their positive feedbacks reduce the emissions required to stabilise atmospheric CO_2 in an idealised $4 \times CO_2$ experiment.

Here we explicitly quantify the impact of climate-carbon cycle feedbacks on realistic stabilisation emissions scenarios. We do not attempt to define 'dangerous' in the context of dangerous climate change – this remains a political question and is discussed further throughout this book – but we do address the question of what emission profile achieves stabilisation at a particular CO_2 level. In particular we re-examine the stabilisation scenarios proposed by Wigley et al. (1996) (hereafter referred to as the 'WRE' scenarios), which lead to stabilised atmospheric CO_2 levels, but do not account for feedbacks in the global carbon cycle. The important conclusion of this study is that compared to the previous projections of Wigley et al. (1996), climate-carbon cycle feedbacks significantly reduce the total 'permissible emissions' to achieve any given stabilisation level.

34.2 Emission Profiles to Achieve Stabilisation

34.2.1 *General Circulation Model Simulations*

State-of-the-art coupled atmosphere ocean general circulation models (AOGCMs), such as HadCM3 (Gordon et al., 2000), are the best tool for making predictions of future climate change over the coming centuries because of the detail with which they are able to represent the processes involved. We use a version of HadCM3 with a fully interactive carbon cycle (HadCM3LC, Cox et al., 2001). Land-atmosphere and ocean-atmosphere fluxes of carbon are modelled explicitly. Thus to make simulations of an evolving climate in response to anthropogenic emissions, we prescribe specific emission scenarios of CO_2 and the model simulates the resulting atmospheric concentrations of CO_2.

In this study, however, we calculate the emission profiles required to achieve a stabilised level of CO_2. To accomplish this, we perform simulations with prescribed profiles of atmospheric CO_2 (and non-CO_2 greenhouse gases). Throughout the simulation, the resulting climate and CO_2 state determines the carbon fluxes into and out of the natural terrestrial and oceanic carbon cycle. The 'permissible emissions' are therefore the difference between the rate of change of atmospheric CO_2 and the modelled natural carbon fluxes. Two GCM simulations were performed, corresponding to profiles WRE450 and WRE550 (stabilisation at 450 ppm and 550 ppm respectively; Wigley et al., 1996). The results of these predictions of emissions for the period 1860 to 2300 are presented in Figure 34.1. It is apparent that the permissible emissions calculated with HadCM3LC (red lines) are significantly reduced compared to the previous estimates of Wigley et al. (1996) (black lines). For WRE450, HadCM3LC predicts that permissible emissions are reduced by about 2–3 GtC yr^{-1} through much of the 21st century and are still almost 1 GtC yr^{-1} lower by 2300. For WRE550, permissible emissions are reduced by up to 5 GtC yr^{-1} by the latter half of the 21st century and are

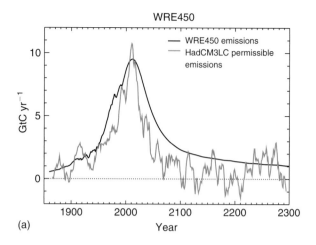

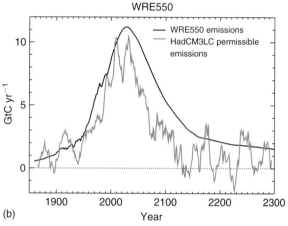

Figure 34.1 Stabilisation emissions for (a) WRE450 scenario and (b) WRE550 scenario, both *with* (red lines, as simulated by HadCM3LC) and *without* (black lines, as in Wigley et al., 1996) climate-carbon cycle feedbacks. The prominent short-term variability in the GCM results has no impact on the results and should be discounted – it is a result of the interannual variability in the natural fluxes (which are as large as 2 GtC yr^{-1}, Jones et al., 2001) which is assigned to the emissions term because it is not present in the smooth, prescribed d(CO_2)/dt term.

still 1 GtC yr^{-1} lower by 2300. This represents a significant reduction in fossil fuels that may be burnt whilst achieving climate stabilisation at CO_2 concentrations of either 450 or 550 ppm.

This comparison of HadCM3LC against the model of Wigley et al. (1996) is justified by noting that Figure (3a) of Cox et al. (2000) shows how the coupled GCM behaves very similarly to their model in the absence of climate change. Hence we assume that the differences shown here are predominantly due to the climate feedbacks rather than the use of a different carbon cycle model.

Figure 34.2 shows how the effect of climate carbon cycle feedbacks on total cumulative emissions from 1860 to 2300 is to reduce them from 1260 to 800 GtC in the WRE450 case and from 1810 to 1130 GtC in the WRE550 case. The estimates of Wigley et al. (1996) are given as

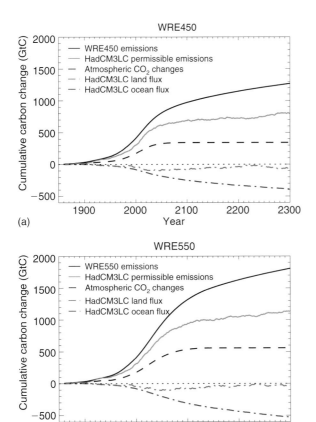

(a)

(b)

Figure 34.2 Cumulative changes in carbon stores for (a) WRE450 scenario and (b) WRE550 scenarios. Atmospheric carbon (dashed black line), terrestrial carbon (green line), ocean carbon (blue line). Anthropogenic stabilisation emissions both *with* (red line, as simulated by HadCM3LC) and *without* (solid black line, as in Wigley et al., 1996) climate-carbon cycle feedbacks.

thick solid black lines, and simulations by HadCM3LC are solid red lines. The dashed black line represents the amount of carbon as CO_2 in the atmosphere (expressed in units of GtC change in atmospheric carbon). The blue and green lines show the accumulated change by the ocean and terrestrial biosphere respectively (negative values imply an accumulated uptake of carbon drawn down from the atmosphere) and hence emissions are given by the atmospheric concentrations minus ocean/terrestrial uptake. In other words, the permissible emissions (red line) equals the prescribed atmospheric carbon change (dashed black line) minus oceanic/terrestrial uptake (i.e. *subtract* the green and blue dashed lines).

In both scenarios, the terrestrial biosphere initially takes up carbon, but later carbon is released back into the atmosphere: there is predicted to be a sink-to-source transition whereby climate change is reducing the permissible emissions to follow the given scenario of CO_2 concentration. This transition occurs around the middle of the 21st century in each case, and carbon release occurs earlier in the WRE550 scenario due to the stronger climate change

associated with the higher CO_2 levels. As in the 'business as usual' climate carbon cycle simulation of Cox et al. (2000), the terrestrial carbon loss is primarily a result of increased soil respiration across the globe driven by higher temperatures, and in some regions there may also be direct vegetation loss (e.g. in Amazonia, Betts et al., 2004; Cox et al., 2004). Such climate feedbacks on the terrestrial ecosystem are not included in the original work of Wigley et al. (1996), and hence our projections which include terrestrial ecosystem functioning stress the importance of including feedbacks in simulations of future CO_2 behaviour.

Ocean carbon storage increases monotonically in both experiments, although at steadily decreasing rates. The ocean uptake is primarily driven by the difference between atmospheric and oceanic CO_2 concentration and hence as the ocean takes up more carbon (and the atmosphere is stabilised at a constant level) this difference decreases, and so does the rate of carbon uptake. The impact of climate change is again to reduce this rate through warming-induced stratification of surface waters and reduced overturning circulation (Friedlingstein et al., 2005).

34.2.2 *Further Stabilisation Simulations with a 'Simple' Model*

Unfortunately, the computational cost of such GCM simulations with currently available computer power greatly restricts the possible number of simulations. Hence to examine a wide range of scenarios (from stabilisation at 450 ppm to 1000 ppm) we have extended the results from the two GCM experiments by using a 'simple model'. This simple model has been calibrated to reproduce the results of the GCM for the original 'business as usual' experiment of Cox et al. (2000) and has been tested to ensure that it reproduces the results of the two GCM stabilisation experiments presented above. Description, formulation and details of the calibration of the simple model are given in Jones et al. (2003a). However, it is noted here that the simple model can capture the features of the carbon cycle as depicted in the WRE450 and WRE550 HadCM3LC simulations described above. A caveat to this is that inherent lags in the full GCM simulation are not captured as the simple model here does not simulate the rate of ocean heat uptake but rather changes instantaneously to follow the radiative forcing of the CO_2 changes. Hence there is a tendency for the simple model to slightly overestimate the strength of the terrestrial carbon sink in the early 21st century and underestimate it towards the end of the simulation. Figure 34.3 shows the success of the simple model in recreating the GCM results. For simplicity, the simple model experiments neglect all climate forcing other than from CO_2. The opposing effects of non-CO_2 greenhouse gases and sulphate aerosols are assumed to approximately cancel during the historical period, and their future impacts are not the focus of this study.

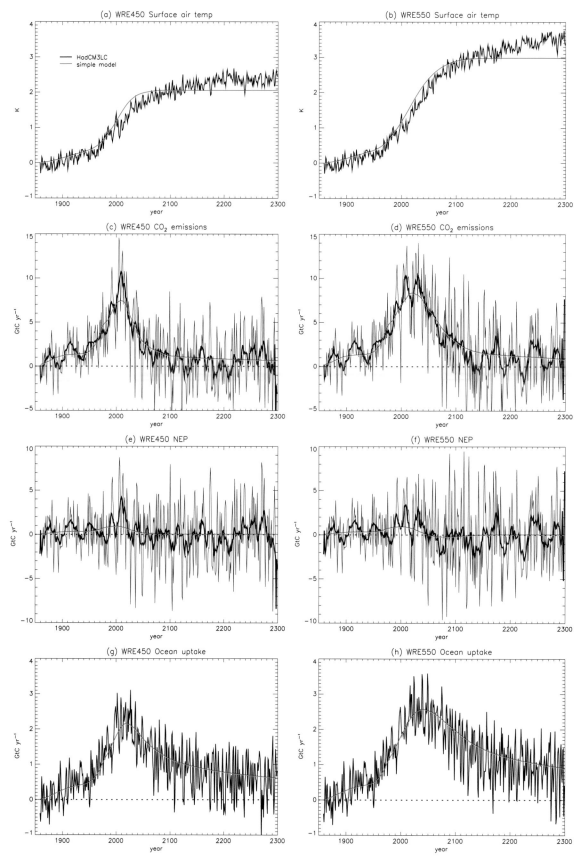

Figure 34.3 Comparison of simple model and HadCM3LC results for WRE450 (left hand column) and WRE550 (right hand column) stabilisation experiments. Simple model results (red lines) and HadCM3LC results (black lines) for global mean temperature change (top row), 'permissible' anthropogenic emissions (second row: annual GCM results in thin line, 10-year smoothed data in thick line), NEP (third row: annual GCM results in thin line, 10-year smoothed data in thick line) and ocean carbon uptake (bottom row).

Table 34.1 Cumulative emissions totals (GtC) from 2000 to 2300 for the 5 WRE stabilisation scenarios with and without carbon cycle feedbacks.

	2000–2300 Cumulative Emissions (GtC)				
Stabilisation level (ppm)	450	550	650	750	1000
With feedbacks	598	995	1321	1600	2162
Without feedbacks	757	1351	1859	2308	3219
% reduction due to feedbacks	21	26	29	31	33

The simple model also allows a decoupling of the climate and carbon-cycle response to CO_2 allowing simulations where the climate feedback on the carbon cycle is 'switched off'. Simulations are performed for the other stabilisation profiles considered by Wigley et al. (1996) (i.e. leading to stable atmospheric CO_2 values of 650 ppm, 750 ppm and 1000 ppm). For each of these five scenarios, experiments are performed with and without climate-carbon cycle feedbacks. The total emissions in each case are summarised in Table 34.1. Figure 34.4 shows the simulated profiles of permissible emissions for these simulations, both with climate-carbon cycle feedbacks (red lines) and without (black lines).

The results for WRE450 and WRE550 are similar to those shown previously for HadCM3LC, but differ slightly due to the CO_2-only forcing used in the simple model, which enables better simulation of the observed emissions during the historical period. The results from all the scenarios are qualitatively very similar. Each WRE scenario already requires an eventual decrease in anthropogenic emissions below present-day levels in order to stabilise CO_2 levels. But the impact of climate-carbon cycle feedbacks is to reduce the permissible emissions further. In each case the peak emissions permissible for each scenario, the level of emissions by 2300 and the total (cumulative) emissions over the period are all reduced as a result of the climate-carbon cycle feedbacks. The total cumulative emissions are reduced by 21% in the WRE450 case and 33% in the WRE1000 case, as summarised in Table 34.1 and shown in Figure 34.5. The higher the level of stabilisation, the greater the level of reduction required in the total emissions compared with the case of no climate feedbacks. This is due to the greater amount of climate change associated with the higher stabilisation levels and hence the greater reduction in the strength of the natural carbon sink. The percentage reduction appears to level off, however, asymptoting to around 34% for CO_2 levels greater than 1000 ppm.

Figure 34.5 also shows the cumulative emissions from 2000 up to 2100 and up to 2200 (the subdivisions within the bars). For stabilisation at low levels it is clear that the majority of permissible emissions are 'used up' during the 21st century: emissions after 2100 are a small fraction of the total. For stabilisation at higher levels, a greater proportion of permissible emissions are available after 2100. In other words, for the WRE profiles of CO_2 concentration, cumulative emissions up to 2100 show less variation across different stabilisation levels than do

cumulative emissions after 2100, although this feature is clearly dependent on the rate at which the profiles reach the stabilisation level.

34.3 Discussion

The Coupled Climate-Carbon Cycle Model Intercomparison Project (C4MIP, Friedlingstein et al., 2005) has studied and compared the behaviour of the climate-carbon cycle feedback between ten coupled climate carbon cycle models. It found significant uncertainty in the strength of the feedback, but all models agreed that the feedbacks are positive and therefore in the context of CO_2 stabilisation would result in a reduction in permissible emissions.

The C4MIP analysis shows that the uncertainty is not confined to any single process, but contains significant contributions from all of these: climate sensitivity to a doubling of CO_2 (Andreae et al., 2005), sensitivity of respiration to temperature (Jones et al., 2003*a*), CO_2 fertilisation (Cramer et al., 2001; Adams et al., 2004), vegetation productivity sensitivity to climate (Cramer et al., 2001; Adams et al., 2004; Matthews et al., 2005) and oceanic uptake sensitivity to raised CO_2 and changed climate (Sarmiento et al., 1998; Doney et al., 2004). All of these sensitivities feature in the feedback analysis of Friedlingstein et al. (2003) and Friedlingstein et al. (2005).

HadCM3LC has the strongest feedback of the C4MIP models with a gain, roughly twice that of the mean of the ten models. Despite some of the C4MIP models clustering about a feedback strength of about half that of HadCM3LC, there is still no consensus on the magnitude of the components of the feedback, with different models producing similar feedback strengths by very different mechanisms. HadCM3LC has had aspects of the carbon cycle extensively validated against observations. It captures the large-scale terrestrial and oceanic patterns of fluxes measured by the TransCom 3 inversion study (Gurney et al., 2002), especially when all relevant climate forcings of 20th century climate are included (Jones et al., 2003*b*) which correct much of the overestimate of present day warming and CO_2 increase seen in Cox et al. (2000). It is also able to capture the carbon cycle sensitivity to climate variability and short-term transient changes such as those caused by ENSO (Jones et al., 2001) and the Mt. Pinatubo eruption (Jones and Cox, 2001). The atmospheric and terrestrial components of the model have

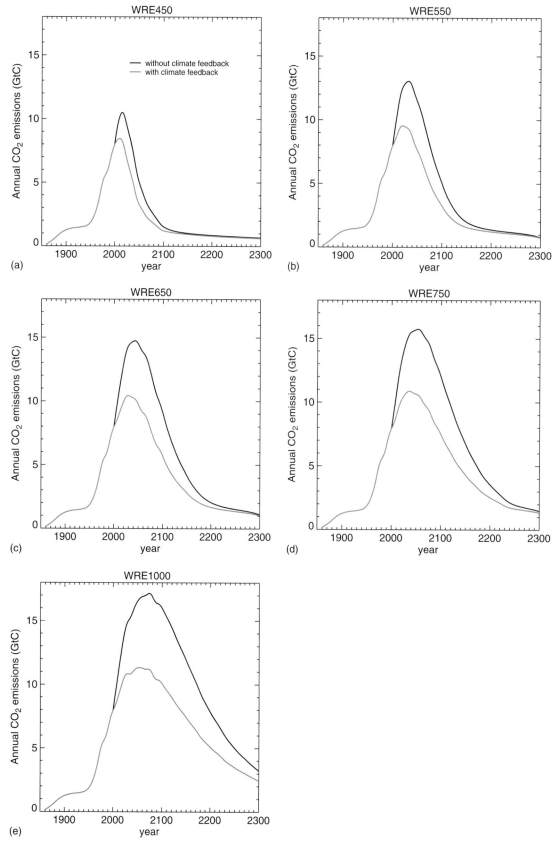

Figure 34.4 Stabilisation emissions for (a) WRE450, (b) WRE550, (c) WRE650, (d) WRE750 and (e) WRE1000 scenarios, both *with* (red lines) and *without* (black lines) climate-carbon cycle feedbacks as simulated by the simple model.

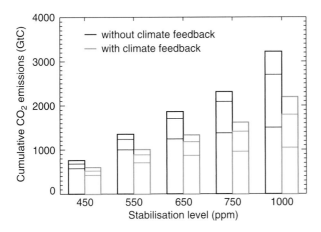

Figure 34.5 Cumulative emissions totals from 2000 to 2300 for the 5 WRE stabilisation scenarios, both *with* (red lines) and *without* (black lines) climate-carbon cycle feedbacks as simulated by the simple model. The lower and upper subdivisions within the bars show cumulative emissions up to 2100 and up to 2200 respectively.

additionally been validated over the historical period against site-specific flux tower measurements and finer scale inversion estimates (Jones and Warnier, 2004). Hence, while uncertainties remain, the ability to recreate present day behaviour increases confidence in the predictive capability for future change. The strength of feedback presented here cannot be ruled out by observations and a simple analytical model suggests that terrestrial sink-to-source transition may be inevitable beyond some critical CO_2 level (Cox et al., 2005).

The historical record of temperature and CO_2 offers little constraint on climate sensitivity due to the uncertainty in the climate forcing (in particular of aerosols) (Gregory et al., 2002; Forest et al., 2002; Andreae et al., 2005).

It should also be noted that future carbon cycle behaviour, and hence implied permissible emissions, would be affected by other processes not yet included in our modelling. Limitation of plant growth by nitrogen or other nutrients, natural fire activity and impacts on the terrestrial carbon cycle from anthropogenic land use change are not included here.

Further uncertainty arises because future anthropogenic emissions will come from a combination of fossil fuel burning and land use change. In deriving our permissible emissions consistent with the modelled carbon fluxes we do not differentiate between their possible sources. However, fossil fuel burning is associated with SO_2 release and other particulate pollution, which may exert a negative radiative forcing, although this is expected to reduce in future as a result of clean-air technology. Land use change exerts its own biogeophysical forcing of climate through changes to albedo, surface roughness and hydrology (Betts, 2000; Betts et al., 2004; Sitch et al., 2005). Although it has yet to be resolved whether this biogeophysical forcing is sufficient to counter the biogeochemical

forcing from CO_2 release (Brovkin et al., 1999; Matthews et al., 2003) it is likely to be substantial.

The non-uniqueness of stabilisation pathways will be considered in a future study, as there are many CO_2 profiles, and associated emissions, to stabilise at a given level. However, initial analysis indicates that the cumulative emissions to stabilise by different pathways were relatively insensitive to the chosen pathway. Cumulative emissions are the balance between accumulated CO_2 in the atmosphere and the change in the terrestrial and oceanic carbon storage. Generally, in the long-term these are more dependent on the final state than the pathway to achieve it, although this may not be strictly true in extreme cases if rapid rates of climate change, or 'overshoot' and subsequent recovery, caused the climate system to cross some irreversible threshold such as Amazon dieback (Cox et al., 2000, 2004) or a sudden drop in ocean carbon uptake due to THC collapse (Joos et al., 1999). The economic implications, however, of different routes to stabilisation may be important. Small reductions in the short term may increase the need for more rapid, and potentially much more expensive, reductions in the future (Meinshausen, 2005).

The very long-term limit which permissible emissions approach is determined by the persistent natural sinks (Prentice et al., 2001) such as transport of anthropogenic carbon to the deep ocean. Over periods much longer than our simulations even this will diminish, leaving only much smaller sink terms such as accumulation in peatlands or carbonate compensation in the ocean. Hence we would expect the lines in figure 34.1 to decrease to just a couple of tenths of GtC yr^{-1} over millennial timescales.

34.4 Conclusions

In this study we have attempted to quantify the impact climate change will have on the world's natural carbon cycle and how this will affect the amount of CO_2 emissions which are permissible to achieve a stabilised climate in the future. We use a climate model able to explicitly simulate interactions between the climate and carbon cycle and find that climate-carbon cycle feedbacks significantly reduce the permissible emissions for stabilisation of atmospheric CO_2 concentration. Feedbacks consistent with the Hadley Centre climate-carbon cycle GCM imply a reduction of 21–33% in the integrated emissions (between 2000 and 2300) for stabilisation, with higher fractional reductions necessary for higher stabilisation concentrations.

We recognise that uncertainties in climate model formulations (including their climate sensitivity and ecosystem response to climate change) mean that there are significant uncertainties in any projection of emissions profile required to achieve stabilisation. Further, we note that HadCM3LC has the largest climate-carbon cycle feedback strength of all of the C4MIP models although it validates well against available observations. However all such models exhibit some degree of positive feedback, and hence in the context of stabilisation scenarios all would imply some further reduction in permissible emissions. The uncertainty in these results is thus in the amount of reduction of permissible emissions rather than in the fact that some reduction will be required as a result of climate change.

We conclude, therefore, that any mitigation or stabilisation policy which aims to prevent 'dangerous' climate change through stabilisation of atmospheric CO_2 levels must take into account climate-carbon cycle feedbacks and their associated uncertainty. Failure to do so may lead to a significant underestimate of the action required to achieve stabilisation.

Acknowledgements

This work was supported by the UK DEFRA Climate Prediction Programme under contract PECD 7/12/37, and the UK Natural Environment Research Council.

REFERENCES

Adams, B., A. White, and T. M. Lenton, An analysis of some diverse approaches to modelling terrestrial net primary productivity, *Ecological Modelling*, **177**, 353–391, 2004.

Andreae, M. O., C. D. Jones, and P. M. Cox, Strong present-day aerosol cooling implies a hot future, *Nature*, **435**, doi:10.1038/nature03671, 2005.

Betts, R. A., P. M. Cox, M. Collins, P. P. Harris, C. Huntingford, and C. D. Jones, The role of ecosystem-atmosphere interactions in simulated Amazonian precipitation decrease and forest dieback under global climate warming, *Theor. Appl. Climatol.*, **78**, 157–175, 2004.

Betts, R. A. Offset of the potential carbon sink from boreal forestation by decreases in surface albedo. *Nature*, **408**, 187–190, 2000.

Brovkin, V., A. Ganapolski, M. Claussen, C. Kubatzki, and V. Petoukhov. Modelling climate response to historical land cover change. *Global Ecol. Biogeogr.*, **8**, 509–517, 1999.

Cox, P. M., R. A. Betts, M. Collins, P. P. Harris, C. Huntingford, and C. D. Jones, Amazonian forest dieback under climate-carbon cycle projections for the 21st Century, *Theor. Appl. Climatol.*, **78**, 137–156, 2004.

Cox, P. M., R. A. Betts, C. D. Jones, S. A. Spall, and I. J. Totterdell, Acceleration of global warming due to carbon-cycle feedbacks in a coupled climate model, *Nature*, **408**, 184–187, 2000.

Cox, P. M., R. A. Betts, C. D. Jones, S. A. Spall, and I. J. Totterdell, Modelling vegetation and the carbon cycle as interactive elements of the climate system, in Meteorology at the Millennium, edited by R. Pearce, pp. 259–279, Academic Press, 2001.

Cox, P. M., C. Huntingford, and C. D. Jones, Conditions for Sink-to-Source transitions and runaway feedbacks from the land carbon-cycle, 2005, this Volume.

Cramer, W., A. Bondeau, F. Woodward, I. Prentice, R. Betts, V. Brovkin, P. Cox, V. Fisher, J. Foley, A. Friend, C. Kucharik, M. Lomas, N. Ramankutty, S. Stitch, B. Smith, A. White, and C. Young-Molling, Global response of terrestrial ecosystem structure and function to CO_2 and climate change: Results from six dynamic global vegetation models, *Global Change Biol.*, **7**, 357–374, 2001.

Doney, S. C., K. Lindsay, K. Caldeira, J. M. Campin, H. Drange, J. C. Dutay, M. Follows, Y. Gao, A. Gnanade-sikan, N. Gruber, A. Ishida, F. Joos, G. Madec, E. Maier-Reimer, J. C. Marshall, R. J. Matear, P. Monfray, A. Mouchet, R. Najjar, J. C. Orr, G. K.

Plattner, J. Sarmiento, R. Schlitzer, R. Slater, I. J. Totterdell, M. F. Weirig, Y. Yamanaka, and A. Yool, Evaluating global ocean carbon models: the importance of realistic physics, *Global Biogeochem. Cycles*, doi: 10.1029/2003GB002150, 2004.

Forest, C. E., P. H. Stone, A. P. Sokolov, M. R. Allen, and M. D. Webster, Quantifying uncertainties in climate system properties with the use of recent climate observations, *Science*, **295**, 113–117, 2002.

Friedlingstein, P., L. Bopp, P. Ciais, J. Dufresne, L. Fairhead, H. LeTreut, P. Monfray, and J. Orr, Positive feedback between future climate change and the carbon cycle, *Geophys. Res. Let.*, **28**, 1543–1546, 2001.

Friedlingstein, P., P. M. Cox, R. A. Betts, V. Brovkin, I. Fung, B. Govindasamy, C. D. Jones, M. Kawamiya, K. Lindsay, D. Matthews, T. Raddatz, P. Rayner, E. Roeckner, S. Thompson, and N. Zeng, Climate-carbon cycle feedback analysis, results from the C4MIP model intercomparison, 2005 (J. Clim, accepted).

Friedlingstein, P., J. L. Dufresne, P. M. Cox, and P. Rayner, How positive is the feedback between climate change and the carbon cycle?, *Tellus. B*, **55B**, 692–700, 2003.

Gregory, J. M., R. J. Stouffer, S. C. B. Raper, P. A. Stott, and N. A. Rayner, An observationally based estimate of the climate sensitivity, *J. Climate*, **15**, 3117–3121, 2002.

Gurney, K. R., R. M. Law, A. S. Denning, P. J. Rayner, D. Baker, P. Bousquet, L. Bruhwiler, Y. H. Chen, P. Ciais, S. Fan, I. Y. Fung, M. Gloor, M. Heimann, K. Higuchi, J. John, T. Maki, S. Maksyutov, K. Masarie, P. Peylin, M. Prather, B. C. Pak, J. Randerson, J. Sarmiento, S. Taguchi, T. Takahashi, and C. W. Yuen, Towards robust regional estimates of CO_2 sources and sinks using atmospheric transport models, *Nature*, **415**, 626–630, 2002.

Jones, C. D., M. Collins, P. M. Cox, and S. A. Spall, The carbon cycle response to ENSO: A coupled climate-carbon cycle model study, *J. Climate*, **14**, 4113–4129, 2001.

Jones, C. D. and P. M. Cox, Modelling the volcanic signal in the atmospheric CO_2 record, *Global Biogeochem. Cycles*, **15**, 453–466, 2001.

Jones, C., P. Cox, and C. Huntingford, Uncertainty in climate-carbon cycle projections associated with the sensitivity of soil respiration to temperature, *Tellus. B*, **55B**, 642–648, 2003a.

Jones, C. D., P. M. Cox, R. L. H. Essery, D. L. Roberts, and M. J. Woodage, Strong carbon cycle feedbacks in a climate model with interactive CO_2 and sulphate aerosols, *Geophys. Res. Let.*, **30**, doi:10.1029/2003GL016867, 2003b.

Jones, C. D. and M. Warnier, Climate-land carbon cycle simulation of the 20th century: assessment of HadCM3LC C4MIP phase 1 experiment, Hadley Centre Technical Note 59, Hadley Centre, Met Office, Met Office, Exeter, EX1 3PB, UK, 2004.

Jones, C. D. and P. M. Cox, On the significance of atmospheric CO_2 growth-rate anomalies in 2002–03, *Geophys. Res. Let.*, **32**, doi10.1029/2005GL023027, 2005.

Joos, F., G. K. Plattner, T. F. Stocker, O. Marchal, and A. Schmittner, Global warming and marine carbon cycle feedbacks on future atmospheric CO_2. *Science*, **284**, 464–467, 1999.

Matthews, H. D., A. J. Weaver, M. Eby, and K. J. Meissner, Radiative forcing of climate by historical land cover change. *Geophys. Res. Lett.*, **30**, doi:10.1029/2002GL016098, 2003.

Matthews, H. D., M. Eby, A. J. Weaver, and B. J. Hawkins, Primary productivity control of the simulated climate-carbon cycle feedback, *Geophys. Res. Let.*, **32**, doi:10.1029/2005GL022941, 2005.

Meinshausen, M. What does a 2°C Target Mean for Greenhouse Gas Concentrations? A Brief Analysis Based on Multi-Gas Emission Pathways and Several Climate Sensitivity Uncertainty Estimates, this volume.

Prentice, I. C., G. D. Farquhar, M. J. R. Fasham, M. L. Goulden, M. Heimann, V. J. Jaramillo, H. S. Kheshgi, C. Le Quere, R. J. Scholes, and D. W. R. Wallace. The carbon cycle and atmospheric carbon dioxide. In J. T. Houghton, Y. Ding, D. J. Griggs, M. Noguer, P. van der Linden, X. Dai, K. Maskell, and C. I. Johnson, editors, Climate Change 2001: The scientific basis. Contribution of Working

Group I to the Third Assessment Report of the Intergovernmental Panel on Climate Change, Chapter 3, pages 183–237. Cambridge University Press, 2001.

Sarmiento, J., T. Hughes, R. Stouffer, and S. Manabe, Simulated response of the ocean carbon cycle to anthropogenic climate warming, *Nature*, **393**, 245–249, 1998.

Schimel, D., D. Alves, I. Enting, M. Heimann, F. Joos, D. Raynaud, T. Wigley, M. Prather, R. Derwent, D. Enhalt, P. Fraser, E. Sanhueza, X. Zhou, P. Jonas, R. Charlson, H. Rodhe, S. Sadasivan, K. P. Shine, Y. Fouquart, V. Ramaswamy, S. Solomon, J. Srinivasan, D. Albritton, R. Derwent, I. Isaksen, M. Lal, and D. Wuebbles, Radiative forcing of climate change, in Climate Change 1995. The Science of Climate Change, edited by J. T. Houghton, L. G. M. Filho, B. A. Callander, N. Harris, A. Kattenberg, and K. Maskell, Chapter 2, pp. 65–131, Cambridge University Press, Cambridge, 1996.

Sitch, S., V. Brovkin, W. von Bloh, D. van Vuuren, B. Eickhout, and A. Ganopolski. Impacts of future land cover changes on atmospheric CO_2 and climate. *Global Biogeochem. Cycles*, **19**, doi: 10.1029/ 2004GB002311, 2005.

Wigley, T. M. L., R. Richels, and J. A. Edmonds, Economic and environmental choices in the stabilisation of atmospheric CO_2 concentrations, *Nature*, **379**, 242–245, 1996.

SECTION VII

Technological Options

INTRODUCTION

Technology is both a cause and a potential solution to the challenge of the anthropogenic climate change. During the last two centuries, since the onset of the industrial revolution, technological change has been instrumental in providing ever-increasing affluence and human wellbeing in the world. In doing so it has contributed toward a whole host of emerging environmental problems, ranging from indoor air pollution to global climate change. The ever-larger access to and use of fossil energy sources, starting with coal, has increased exponentially carbon dioxide emissions during the last two centuries. The unintended consequences are rising atmospheric carbon dioxide concentrations well above the range observed during the last million years. At the same time, technology holds the promise to help radically reduce greenhouse gas emissions and other adverse impacts of human activates on a wide range of planetary processes. Nobel Laureate Paul Crutzen has suggested that the present era should be called the Anthropocene, to symbolise this unprecedented influence of one single species on the planet Earth. Technology is no doubt an essential part of this equation, both in the sense of improving human wellbeing and in endangering it at the same time. Thus, technology can amplify as well as alleviate adverse impacts of human activities.

The role of technology was largely ignored in the first round of global modelling efforts and scenarios in the early 1970s, but it has recently moved to the forefront of both science and policy in addressing climate change. Today, technological change is an integral component of most emissions-reduction studies and assessments. Diffusion of zero or low-emitting technologies is considered not only to make deep emissions reductions possible during this century but, with the advent of new and advanced technologies, also substantially more affordable than today.

Technological change is associated with long timespans. It takes decades to a century to diffuse new technologies and achieve the full replacement of the capital stock. Timespans associated with anthropogenic climate change are at least as long. Both technology and climate change are very slow to change, due to the inherent inertia of their underlying systems. Both are inherently cumulative in nature, meaning that their consequences are large and emerging changes in systems they affect are fundamental. Technology is highly malleable in the long run. Mitigation

of climate change is also a long-term challenge as it cannot be resolved in the near future. This means that in the long run, technology needs to be central element of response strategies to climate change.

This section includes seven papers that consider the role of technology in climate change from multiple perspectives. Most of the papers assess contribution of different technologies to reducing future carbon dioxide emissions (often against some baseline without explicit measures and policies directed at curbing emissions) either by assuming or by invoking measures and policies directed at achieving their widespread diffusion.

Three of the seven papers in this section use modelling approaches to analyse the technology needs for achieving deep emissions reductions. The papers differ both with respect to their methodological approach and temporal and spatial resolution. The paper by Jae Edmonds and Steve Smith deploys an integrated assessment model MiniCam to investigate technology needs for limiting future temperature increase to two degrees Celsius. The two stabilisation goals of stabilizing the concentrations and stabilising the temperature are related but are associated with substantial uncertainties. Especially uncertain is the climate sensitivity to future increase in atmospheric concentrations. Despite these deep uncertainties, it is certain that concentrations need to be stabilised in order to halt further temperature increases. Keigo Akimoto and Toshimasa Tomoda deploy an energy-systems engineering model NE21+ to investigate technology needs for achieving atmospheric stabilisation of carbon dioxide concentrations at different future levels. The third modelling approach by Terry Barker et al. is based on initial results achieved by a new economic model E3MG with induced technological change. Policies are introduced that lead to emissions stabilization by inducing diffusion of carbon-saving technologies with a co-benefit of promoting economic growth and development. This is in a sharp contrast to the first two modelling approaches that generally associate loss of economic output in the range of a few per cent with the investment in mitigation measures and policies.

Jae Edmonds and Steve Smith examine energy technology implications of limiting the change in mean global surface temperature to 2°C relative to pre-industrial temperatures. As mentioned above, this entails deep uncertainties particularly concerning the climate sensitivity to doubling of greenhouse gas concentrations. For example, the (median) sensitivity of 2.5°C implies stabilisation of carbon dioxide concentrations at less than 500 ppmv, with a global emissions peak occurring within the next two decades, followed by a decline in emissions to half the current levels by the end of the century. The authors use an integrated assessment model MiniCam to assess the implications of limiting global temperature increase. A portfolio of technological options is required to meet the assumed 2°C limit including multigas mitigation measures with a substantial share of carbon dioxide capture and storage. An important finding of the analysis is that the value of technology improvements, beyond already significant technological change in the reference case, is found to be exceptionally high, denominated in trillions of 1990 US$. These results are very sensitive with respect to the assumed value of the climate sensitivity. For values of 3.5°C or greater, authors conclude that it may be impossible to limit temperature change below 2°C, while for values of 1.5°C and less it may be a trivial matter requiring little deviation from a reference IPCC SRES emissions path until after the middle of the century.

Keigo Akimoto and Toshimasa Tomoda evaluate a large portfolio of technological options and their costs for 77 world regions within the framework of an energy-systems model DNE21+. The analysis starts with two IPCC SRES baseline scenarios and considers three alternative carbon dioxide stabilisation levels beyond the end of the century, 650, 550 and 450 ppmv. The results of the analysis include the marginal costs of carbon dioxide emissions of about $100, 120 and 290 per ton of (elemental) carbon by 2100 for 650, 550 and 450 ppmv, respectively. Consequently, the stabilisation also leads to GDP loss compared to the baselines. However, the stabilisation costs are more sensitive to the choice of the baseline than the choice of a stabilisation target. This is an important result, indicating that the nature of the development path itself is crucial for determining the costs of mitigation. Across all of these different stabilisation scenarios, capture and storage is an important technology option for reducing carbon emissions particularly in the developed countries, while energy saving is particularly important in the developing countries. This is, to an extent, a surprising result given that energy requirements are much larger in the developed countries. The authors conclude that these differences in reduction options between developed and developing countries would be more beneficial to the latter by making them more economically competitive under these emission reduction schemes.

The paper by Terry Barker et al. assesses the costs of stabilising atmospheric concentrations of carbon dioxide at three levels 450, 500 and 550 ppmv compared to a baseline that derives from one of the IPCC SRES scenarios. The authors use an integrated assessment model E3MG to introduce two policy instruments at increasing rates for achieving the three stabilisation targets. The policies include emission trading permits for the energy industries and carbon taxes for the rest of the economy, with tax revenues being recycled to maintain fiscal neutrality.

These assumptions lead to ever-higher real cost of fossil fuels and prompt a shift toward low-carbon technologies. An interesting result is that the ensuing world-wide wave of extra investment over the century raises the rate of economic growth, which is endogenous in the model. In contrast to other modelling approaches in this chapter that indicate some loss of economic activities associated with stabilisation, Barker et al. obtain a purely economic benefit to result from stabilisation. This finding complements previous results in the induced technological change literature, showing reductions in the mitigation costs due to the cumulative nature of technological learning. The approach is important as it provides a different perspective on stabilisation costs compared to other approaches that indicate additional costs of mitigation compared to the baseline that lead to a GDP loss and not a gain as shown by Barker et al.

The other four papers in this section assess technology potentials, deployment and diffusion of mitigation technologies and the associated mitigation costs. One of them, by Bert Metz and Detlef Van Wuuren, reviews technology portfolios and costs in different integrated assessment models required to achieve low greenhouse gas stabilisation levels. The paper by Rob Socolow also assesses the potential mitigation contributions of a whole portfolio of technologies whereby each class of technologies contributes one 'wedge' toward the overall reduction with respect to a baseline emissions trajectory. The other two papers pursue feasibility and potentials of two broad classes of mitigation technologies – Peter Read's paper considers a broad portfolio of biomass options and the paper by Jon Gibbins et al. assesses carbon removal with subsequent carbon disposal and storage. The first three papers offer global perspectives while the fourth one focuses more on the options for the UK. All four consider assess potentials of mitigation options well beyond the Kyoto commitments through the middle of the century.

The paper by Metz and Van Wuuren assess technology and policy strategies that lead to stabilisation of greenhouse gases at 550 ppmv (carbon dioxide equivalent concentrations) and whether stabilisation at such a low level is possible based on available technologies and affordable costs. They conclude that the combined technical potential of different options as reported in the literature to be, in principle, sufficient to achieve such stabilisation levels. However, they indicate that the deployment and application of these technologies is more uncertain as it requires further development, technology transfer, and widespread diffusion. In general, effective and efficient stabilisation strategies need to rely on a portfolio of options (changing over time) to achieve least costs. Exclusion of mitigation options would increase costs. Multi-gas mitigation strategies, emission trading, optimal timing and vigorous technology development, deployment and diffusion are all required to keep costs of stabilisation relatively low. For low stabilisation levels, marginal costs will increase steadily as more and more expensive measures are required. If the most efficient implementation is chosen, current studies estimate global costs in the order of a maximum of a few per cent GDP loss by the year 2050. They argue that the sustainable development strategies and corresponding behavioural attitudes would make the achievement of low-level stabilisation easier and help to benefit from additional co-benefits, such as increased energy security and environmental protection.

Rob Socolow also assesses the potential mitigation contributions of a whole portfolio of technologies for achieving atmospheric carbon dioxide, whereby each class of technologies contributes one 'wedge' towards the overall reduction with respect to a baseline emissions trajectory. A wedge is one billion tons of (elemental) carbon per year of emissions savings by the middle of the century. Individual wedges include energy efficiency, carbon capture and storage, nuclear and renewable energy. The author argues that mitigation policies for implementing seven stabilisation wedges should place humanity on a path toward stabilising the climate at a carbon dioxide concentration of some 500 ppmv (carbon dioxide only). Further, he assumes that these wedges would not be deployed without deliberate climate mitigation measures and policies. The concept of stabilisation wedges is further explained by introducing 'virtual' wedges that are achieved as a result of the continued development of the global economy even in the absence of carbon policy. These wedges are already embedded in almost all 'baseline' scenarios in the literature. Thus, the stabilisation wedges must be achieved over and above the structural shifts, energy efficiency gains, and energy system decarbonisation that are likely to occur in the next 50 years even without carbon policy. The framework could contribute new elements to global carbon policy, promoting internationally co-ordinated commercialisation of low-carbon technology.

Peter Read reviews the potential of bioenergy in conjunction with carbon capture and storage as a technology strategy for achieving deep emissions reductions. He argues for development and early deployment of these technologies as a precautionary strategy for avoiding abrupt climate change. The basic idea is that bioenergy is carbon-emissions neutral in the sense that emitted carbon dioxide is reabsorbed by the biomass regrowth. This all assumes a sustainable bioenergy production. In addition, the author argues that the biomass carbon could be captured and stored. This is essentially the same as proposed from carbon capture and storage technologies for reducing emissions from fossil energy sources (see paper by Jon Gibbins et al. below). The difference is that, in the case of biofuels this technology has negative emissions, namely it would lead to a net removal of carbon from the atmosphere and thus a reduction of atmospheric carbon dioxide concentrations. In other words, bioenergy in conjunction with carbon capture and storage is equivalent to direct carbon removal from the atmosphere. Its deployment could become interesting as one of the few technologies available for reducing atmospheric carbon dioxide concentrations.

The author recognises that the sheer scale of required operations – including massive changes in land-use patterns, potential conflicts with food production and huge carbon storage capacities – are all daunting from the current perspective.

Jon Gibbins et al review the scope for future carbon capture and storage technologies for achieving CO_2 emissions reductions from electricity generation in the United Kingdom. Among the conclusions, the research team suggests that large (approx. 45%) reductions in CO_2 emissions from UK electricity generation could be achieved by as early as 2020 by including CCS in the mitigation strategy. The team also conclude that CCS technologies have considerable potential for future emissions reductions globally, and that making new power plants at least 'capture ready', if not actually built to capture CO_2 from the outset, is particularly important in economies where large numbers of new power plants are being built. Two policy instruments are used to achieve these targets: emission trading permits for the energy industries and carbon taxes for the rest of the economy, with the revenues recycled to maintain fiscal neutrality. These are applied at escalating rates 2011–2050 to allow for early action under the UNFCCC. Extra investment is induced by the permit schemes and taxes since they lead to substantial increases in the real cost of burning fossil fuels according to their carbon content. This prompts a switch to low-carbon technologies. The ensuing world-wide wave of extra investment over the century to 2100 raises the rate of economic growth, which is endogenous in the model. There is a purely economic benefit in stabilisation, although small, which increases with more demanding targets.

All seven papers imply that a wide portfolio of technologies would have to be deployed and adopted throughout the world to achieve the emissions cuts required to stabilise carbon dioxide concentrations and temperature change. In the first approximation, concentrations and temperature change are a function of cumulative emissions. This implies that future global emissions trajectories have to curve through a maximum some time this century (during the next decade or two for stabilization at relatively low levels of, say, 400–500 ppmv and a few decades later for high levels), and proceed to decline well below current levels towards the end of the century. This is a tall order from the current perspective, as global emissions have been increasing unabated at close to 2% per year for the, last two centuries. This explains the relative consensus, that a comprehensive technology portfolio is required, among the seven papers reflecting a much broader literature.

What is more controversial, however, is whether now-known technologies can achieve this momentous global undertaking or whether fundamentally new options, such as fusion, that are still technically not feasible, might be required. Be that as it may, the need for vigorous technological development, deployment and diffusion is in indicated in all seven papers that will not occur under the 'business as usual' conditions. In different ways all seven papers call for new policies and institutions that could rise to the challenge of mobilising the appropriate environment for widespread adoption of a wide range of mitigation technologies. Biomass and carbon capture and storage are almost universally present in mitigation and stabilisation scenarios in the literature. This is why they were afforded a special attention in the two of the seven papers. Even these two important groups of options, however, are not foreseen to take the full burden of mitigation efforts even under the most of the optimistic assumptions, strengthening rather than weakening the portfolio argument.

This diversity in modelling approaches and mitigation technology perspectives is important as it highlights some of the deep uncertainties in our understanding of socioeconomic driving forces of climate change and relationship between technological change and society. It mirrors some of the essential controversies associated with the economic and technological dimensions of climate change. It perhaps explains somewhat why some of the countries are committed to proactive measures directed at mitigating climate change, while others rely more on voluntary policies. One robust finding of all seven contributions to this chapter is that fundamental technological and associated institutional changes are needed to stabilise atmospheric concentrations of greenhouse gases, despite the deep uncertainties that surround the science and politics of climate change.

Our sincere thanks go to the authors and the anonymous reviewers. The contributors to this volume and the reviewers have been particularly generous with their time and efforts, not only in preparing draft manuscripts but also in revising them substantially in light of many fruitful exchanges and the customary peer review process.

CHAPTER 35

How, and at What Costs, can Low-Level Stabilization be Achieved? – An Overview

Bert Metz and Detlef van Vuuren
Netherlands Environmental Assessment Agency, MNP/RIVM, The Netherlands

ABSTRACT: In order to prevent 'dangerous anthropogenic interference with the climate system', stabilization of greenhouse gases at low levels (at 550 ppmv CO_2 equivalent or below) might be needed. This paper discusses some of the current literature on whether stabilization at such a low level is possible, based on available technologies and on stabilization scenarios (including their costs). The combined technical potential of different options as reported in the literature seems, in principle, to be sufficient to achieve low-level stabilization. Application of these technologies, however, is more uncertain as it requires further development, technology transfer and widespread diffusion. Effective and efficient stabilization strategies use a portfolio of options (changing over time) to achieve a least cost approach. For low-level stabilization marginal costs will increase steadily as more and more expensive measures are required. Costs in terms of welfare loss, compared to a situation without climate policy measures, depend highly on the underlying socioeconomic development. If the most efficient implementation (including multi-gas strategies, maximum participation of countries in a global emission trading regime, optimal timing of reduction actions) is chosen, current studies estimate global costs in the order of a maximum of a few per cent GDP loss by the year 2050. Costs for individual countries may differ greatly. By looking at mitigation/stabilization costs in a wider context we can help to identify co-benefits of policies for achieving sustainable development, energy security or environmental goals, and so reduce costs.

35.1 Introduction

Article 2 of the UNFCCC calls for stabilization of greenhouse gas concentrations at such a level to avoid 'dangerous interference with the climate system' (UNFCCC, 1992). Uncertainties in the climate system inhibit the ability to unambiguously determine 'safe' concentration levels below which this condition can be considered fulfilled, and in fact, such a step necessarily involves all kinds of value judgements (Rayner and Malone, 1998). Nevertheless, in recent years, literature has been published that suggest that a framework to make Article 2 operational can be developed (Mastandrea and Schneider, 2004) and that limiting global mean temperature change to around 2°C above pre-industrial levels could be interpreted as a reasonable level to avoid some of the most dangerous risks of climate change (Corfee Morlot et al., 2005; ECF and PIK, 2004; IPCC, 2001a; Leemans and Eickhout, 2004; O'Neill and Oppenheimer, 2002; Schneider and Lane, 2005). The EU and several EU member states, in fact, have decided to interpret Article 2 in terms of a maximum temperature increase target of 2°C (EU, 1996; 2004; 2005).

The step from limits on global mean temperature to a concentration stabilization target is again beset with uncertainty, of which the most dominant is the uncertainty in climate sensitivity. Recent literature handles these uncertainties by calculating the probability to which various greenhouse gas concentration levels are able to reach this temperature target (Meinshausen, 2005; Richels et al., 2004). While most previous literature concentrated on stabilization of the CO_2 concentration at 450 ppmv CO_2 in order to comply with the 2° target, more recent literature points out that stabilization at much lower levels might be necessary to make it 'likely' to stay below the 2° target. Specifically, attention has shifted from 450 ppmv CO_2 – i.e. 550 ppmv CO_2-eq. if all main greenhouse gases are included[1] – to more ambitious targets in the order of 400 ppmv CO_2/450 ppmv CO_2-eq or even lower (Den Elzen and Meinshausen, 2005; Meinshausen, 2005). In the context of these targets, this paper reviews existing literature on technological options and emission scenarios to identify what kind of measures could achieve such low levels of stabilization of greenhouse gases in the atmosphere.

The literature on stabilization scenarios was summarized in the IPCC Third Assessment report (TAR) (Morita et al., 2001). Most mitigation studies at the time were looking at CO_2-only policies, and the lowest stabilization level was 450 ppmv CO_2. In fact, the number of studies looking at this lower range level was considerably less than

---

[1] The equivalent CO_2 concentration expresses the additional radiative forcing of all greenhouse gases as the equivalent CO_2 concentration that would result in that same level of forcing. This metric will be used throughout this paper, in addition to CO_2 concentration levels, as a metric to capture all major greenhouse gases.

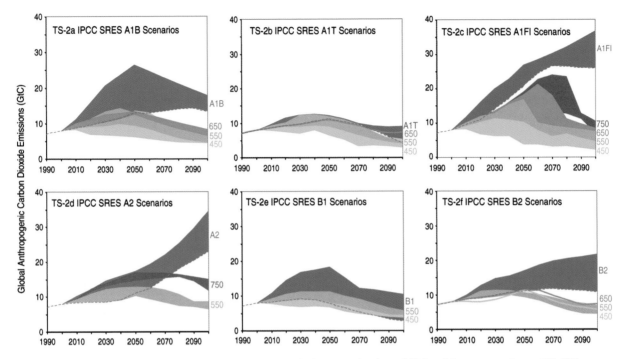

Figure 35.1 Comparison of the IPCC SRES scenarios and emission scenarios for stabilising CO_2 concentration at 450–750 ppmv. *Source*: IPCC, 2001b.

the number of studies looking at higher stabilization levels (see also Swart et al., 2002).

Figure 35.1 shows the gap between 450 ppmv CO_2 stabilization scenarios and the various SRES baseline scenarios, based on studies assessed in the IPCC TAR. An important conclusion from this assessment was the significant impact of the socioeconomic development of the various world regions and the underlying preferences on the magnitude of the mitigation challenge. Two low emission baselines from the IPCC set are A1T and B1; in A1T this is a result of very optimistic assumptions on technology development for carbon-free technologies and in B1 this is a result of the assumptions made on environmentally-friendly lifestyles and large-scale practice of energy efficiency. Looking at the A1T and B1 reference scenarios, the gap between baseline emissions and the emissions required for stabilization at 450 ppmv is relatively small. Looking at the high baselines in the SRES set, A2 and A1FI, this gap is large. The A2 scenarios represent a world of low economic growth, slow technology development and high coal use, hence high emissions. A1FI represents a high-growth world, with technology development mostly focussed on fossil fuel use. In terms of cumulative emissions for reaching 450 ppmv CO_2 to be avoided between 2000 and 2100, the mitigation challenge (i.e. the total amount of CO_2 emissions to be avoided in the period 2000–2100) differs from 1500 $GtCO_2$ (or about 40% of cumulative emissions) from the lower set of baselines, up to 6000 $GtCO_2$ for the higher set (around 70% of cumulative emissions or more). While acknowledging this broad range of uncertainty, it might be

helpful to focus on an average value within this range: i.e. 3000 $GtCO_2$ for CO_2 alone and 4000 $GtCO_2$-eq for all Kyoto gases. This is in the order of a 60% reduction of cumulative emissions (by 2100 this means a reduction of annual global emissions of about 80% below current levels). For stabilization at 450 CO_2-eq, comparable figures would be a reduction of 4500 $GtCO_2$-eq, or a 70% reduction of cumulative emissions (implying a 90% reduction in annual global emissions by 2100) (Den Elzen and Meinshausen, 2005).

Is such a reduction feasible? What kind of technologies would be required? Again, uncertainties play an important role. First of all, even if we focus on the technologies that are already available today, their potential and costs in the next 100 years are very uncertain. One element of this is the development of cost. The effectiveness of these technologies also depends on widespread deployment and diffusion – on a scale at which none of these technologies have been previously applied. Social acceptance will play a role as well. Will societies accept the use of nuclear power as a means to reduce emissions? Will they accept large-scale storage of CO_2? In other words, what current technologies can deliver over the next 100 years depends greatly on the socioeconomic circumstances that will be prevalent. While realising this, we will nevertheless make an attempt to explore the question of the technical and economic feasibility of stabilization at 550 ppmv CO_2-equivalent without assuming the appearance of radically new technologies beyond the scope of those that are already applied at demonstration scale today.

Table 35.1 Estimated technical potential of renewable energy sources.

Energy source	Potential by 2020–2025	Long-term technical potential (EJ/yr)
Hydro	35–55[1]	>130[1], 30–50[2]
Geothermal	4[1]	>20[1]
Wind	7–10[1]	>130[1], 230–640[2], 340[3]
Ocean thermal	2[1]	>20[1]
Solar	16–22[1], 2–230[2]	>2600[1], 160–5000[2], 1330[3]
Biomass	72–137[1]	>1300[1], 280–450[2], 275 to 1105[3]
Total renewables	130–230[1]	640[1]–>7000

Estimates as reported in [1](Nakicenovic et al., 2000), [2](WEA, 2000), [3](Hoogwijk, 2004).

35.2 Technology Options

A range of technology options exists for reducing greenhouse gas emissions and enhancing sinks and reservoirs (Moomaw et al., 2001):

- Energy efficiency improvement.
- Decarbonization of the energy system
  - increasing the use of low or zero carbon energy sources (gas, nuclear, biomass, wind, solar)
  - applying CO_2 capture and storage.
- Biological carbon sequestration and/or reducing deforestation emissions.
- Reducing other greenhouse gases from industry, agriculture, waste.

What is the potential of the various technology options for delivering the emission reductions indicated above? The answer to this question depends on the constraints applied. In the literature some guidance is given by differentiating between technical, economic and market (or implementation) potentials of mitigation measures. The technical potential normally excludes cost considerations. The economic potential looks at what is economically feasible under certain conditions, and the market potential is what is actually realized. The latter thus accounts for implementation barriers such as lack of information, limitation in technology diffusion or limited acceptability of certain options. In reality, however, these definitions are not strictly applied and all estimates feature scenario-dependent aspects.

Here, we focus mostly on the technical potential, but we also provide an indication of the costs associated with these potentials. A crucial scenario-dependent element is technology development; more specifically, the assumptions made on cost reductions and efficiency improvements. Different assumptions about technology development create a wide range of estimates for the potential of the options discussed. With these considerations in mind, we will briefly discuss the technical potential of the different options, their limitations and include a rough indication of the costs involved.

35.2.1 *Energy Efficiency Improvement*

Estimates for the potential of energy efficiency improvement for avoiding GHG emissions vary widely (Moomaw

et al., 2001). Technical studies generally focus on possible efficiency improvement over the next few decades, finding a large potential reduction of over 50–70% of current emissions. Obviously, part of that potential is already captured in the baseline while, in some cases, lifestyle changes may also result in higher emissions. The largest potential is typically found in the building and transport sectors. In the industrial sector, current potential could be somewhat smaller, but history has shown that policies are more effective in promoting efficiency in this sector. For the 2000–2020 period, the potential from efficiency improvement has been estimated at about 200 $GtCO_2$ – or 25% of emissions (Moomaw et al., 2001). On the basis of this, a conservative estimate of the potential over the 2000–2100 period would be in the order of 1000–1500 $GtCO_2$ (assuming a constant potential of 25%). The costs of these options differ widely but typically range from negative costs up to 50 US$/$tCO_2$.

35.2.2 *Renewable Energy*

Several estimates have been published for the technical potential of renewable energy (see Berndes et al., 2003; Hoogwijk, 2004; Nakicenovic et al., 2000; WEA, 2000). In the most recent estimate for biofuels, Hoogwijk (2004) estimates the potential for primary biofuels to range from 165 to 655 EJ in 2050 and 275 to 1105 EJ in 2100, which is large compared to the total energy consumption of 400 EJ today. An important factor underlying these wide ranges are the different assumptions on land availability. Wide ranges in the order of a few hundred up to a few thousand EJ (see Table 35.1) are also reported for wind-based electricity and solar-based electricity. It is important to note, however, that even the lowest estimates are equal to several times the present world electricity consumption. For PV and wind, the problem of integrating each into the existing energy system may play an important role in determin-ing the market potential. Renewable energy options are currently among the more expensive options (about 40–60US$/$tCO_2$), but costs are likely to go down substantially over time.

35.2.3 *Nuclear Power*

The potential for nuclear electricity is similarly difficult to quantify. If restricted to current technologies and uranium resources, the potential might be in the order of 300–400 $GtCO_2$. However, new discoveries of uranium resources, use of thorium, more efficient technologies (including breeders) and production of uranium from sea water could, at least in theory, imply that this option is almost without technical limits. Costs in the next decades could be in the order of 15–120 US\$/tCO_2 when replacing natural gas power plants (Sims et al., 2003). A crucial uncertainty for nuclear energy is the social acceptability of large-scale use of nuclear power.

35.2.4 *CO_2 Capture and Storage*

The potential for CO_2 capture and storage is also significant. Based on recent assessments (IEA, 2004; IPCC, 2005), total cumulative capacity in geological storage sites (enhanced oil recovery fields, depleted oil and gas fields, unminable coal seams and unused saline formations) can be conservatively estimated to be at least two thousand $GtCO_2$-eq. Possible additional storage in the deep oceans is not included in these estimates. Costs for this option consist of the capture, transport and storage costs of CO_2, but also the reduced efficiency of plants using fossil fuels. Despite the fact that this option has only been implemented in a few industrial-scale projects so far, studies are optimistic about the technical feasibility. Overall costs are estimated to be in the order of 20–90 US\$/tCO_2, with possibilities for significant cost decreases in the coming decades (IEA, 2004; IPCC, 2005; Sims et al., 2003). The recent IPCC (2005) assessment indicates that energy and economic models suggest that CO_2 capture and storage will begin to deploy at a significant level in the electricity sector when CO_2 prices begin to reach approximately 25–30 US\$/t CO_2.

35.2.5 *Biological Sequestration*

Enhanced biological sequestration of CO_2 in forests and soils by specific management measures could add another 350 $GtCO_2$, cumulatively up to 2050 (Kauppi et al., 2001). Several uncertainties need to be considered. First of all, land availability is important for sequestering carbon by forestation. For other forms of biological sequestration, the temporary nature of the sequestration and need for land for agriculture may restrict potential. In any case, the costs of biological sequestration are assessed to be relatively low (in the order of 10–50 US\$/tCO_2).

35.2.6 *Non-CO_2 Gases*

The total estimated emissions of non-CO_2 gases covered by the Kyoto Protocol (methane, nitrous oxide, hydrofluorocarbons, perfluorocarbons and sulphur hexafluoride) during the 21st century come to about 1200–1800 $GtCO_2$-eq. across a range of different models and scenarios (van Vuuren et al., 2005; Weyant et al., 2005). Most of these emissions originate from agricultural activities (fertilizer use, animal husbandry, rice production). Recently, attempts

Table 35.2 Cumulative technical potential for avoiding GHG emissions in the 2000–2100 period (in $GtCO_2$ equivalent).

Technology option	Cumulative technical potential 2000–2100 ($GtCO_2$ equiv.)
Energy efficiency improvement	>1000
Renewables	>3000
Nuclear	>300
CO_2 capture and storage	>2000
Biological sequestration	>350
Non-CO_2 GHG reduction	>500

have been made to quantify the potential of reduction measures for these gases (Delhotal et al., 2006; Schaefer et al., 2006). For most energy-related and industrial sources of non-CO_2 gases it was possible to identify reduction measures and possible routes of implementation. For agricultural sources, techniques currently identified as being applicable cover a much smaller percentage of total emissions. The total technical potential identified currently covers about 500 $GtCO_2$ cumulative over 2000–2100. Costs of these measures are generally low, and typically in the order of 0–50 US\$/tCO_2 (Delhotal et al., 2006; Schaefer et al., 2006).

Table 35.2 shows the overall technical potential to be much larger than the high end of the range needed for stabilising CO_2 concentrations between 350 and 450 ppmv. On this basis of similar calculations, the IPCC Third Assessment report concluded earlier that, 'most model results indicate that known technological options could achieve a broad range of atmospheric CO_2 stabilization levels, such as 550 ppmv, 450 ppmv or below, over the next 100 years or more, but implementation would require associated socioeconomic and institutional changes.' (IPCC, 2001c). The caveat contained in this statement refers to the necessary conditions for implementation, including the diffusion and transfer of mitigation technologies to developing countries, as well as the willingness to accept the cost of such stabilization strategies. These aspects will be explored in subsequent sections.

35.3 Stabilization Scenarios

At the time of the Third Assessment Report, the lowest stabilization scenarios found in the literature looked typically at stabilization at 450 ppmv CO_2, more-or-less congruent with 550 ppmv CO_2-eq. In more recent years, a few studies have become available in which more ambitious targets are investigated.

35.3.1 *Option Portfolios*

The available literature on strategies leading to stabilization at 550 ppmv CO_2-eq, shows that in order to minimize costs

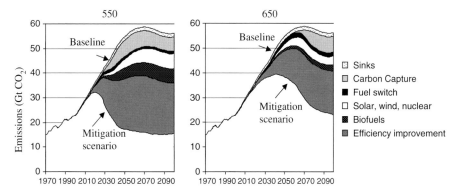

Figure 35.2a and b Emission reduction through mitigation measures for stabilising GHG concentrations at 550 (left) and 650 (right) ppmv CO_2-equivalent (PCC2050 case).
Source: van Vuuren et al., 2003.

and risks, robust mitigation and stabilization strategies will make use of a portfolio of options varying from country to country according to specific national circumstances (Bashmakov et al., 2001). Modelling studies of least-cost approaches to stabilization at various levels show large differences in the optimal portfolio composition as a result of differences in assumptions and coverage of options (Morita et al., 2001).

Figure 35.2 gives a typical example of the outcome of such studies for a multi-gas stabilization at 550 and 650 ppmv CO_2-eq (van Vuuren et al., 2003). The contribution of non-CO_2 emission reductions in the portfolios is modest, particularly at lower stabilization levels. However, it shifts the mitigation action somewhat from the energy sector to the agriculture and industry sector and avoids the most expensive options to reduce CO_2 emissions. In this study, energy efficiency is the dominant option for avoiding CO_2 emissions until around 2030, while renewables, nuclear energy and CO_2 capture and storage contribute more to the portfolio later on, with efficiency improvement continuing to be very important. Other studies generally show a similar trend over time, but the contribution of supply-side versus demand-side options may differ. The lower the stabilization level, the earlier the more expensive options will have to be applied.

Looking into stabilization levels below 550 ppmv CO_2-eq, we see that recent studies (Azar et al., 2005; Nakicenovic and Riahi, 2003) show a much larger contribution from CO_2 capture and storage, and non-fossil energy supply options. This may involve very large penetration levels of renewables, possibly in combination with hydrogen in combination with CO_2 capture and storage. The latter, together with the use of biofuels, allows use of zero carbon fuels in the transport sector. Azar et al. (2005) show that an attractive option for very low stabilization levels is the use of biomass-fuelled power plants in combination with CO_2 capture and storage. Biofuels take up CO_2 during their growth phase; normally this CO_2 is emitted again when the biofuels are burned, creating a CO_2-neutral option. If these CO_2 emissions are prevented by capturing CO_2 gases from flue gases of electric

power plants, followed by secure storage, a system would be created that absorbs CO_2 from the atmosphere on a net basis. The costs are equal to the additional costs of biofuels (compared to fossil fuels) and capture and storage, but for more ambitious stabilization strategies this option could be used very effectively within a larger portfolio of options. Its feasibility depends on costs and the potential for biofuel production. It is important to note that the studies quoted have, in fact, studied mitigation on the basis of already low emission baselines. It is obvious that the very low stabilization levels may be much more difficult to reach when starting from socio-economic conditions that are unfavourable for strong mitigation action, as reflected by the A2 and A1FI reference scenarios.

The assumption in the modelling studies of least-cost implementation of mitigation options reflects a system of full emissions trading between all countries, so that measures are taken where costs are lowest. In reality this may not be the case, leading to different outcomes of the global portfolio and a higher cost level than reported in these studies. National circumstances will lead to specific portfolios for individual countries.

As indicated above, most scenario studies show a portfolio of options as part of their 'optimal' stabilization strategy instead of just one or two big options. There are various reasons for this, including the limited potential of various options compared to the overall mitigation objective, increasing costs with large-scale application, differences between regions and sectors, and timing and implementation issues. A portfolio approach reduces also the risk of choosing a few 'winner' technologies that could turn out to be failures. On the other hand, there are risks involved in a portfolio approach in particular, due to spreading R&D budgets, reduced economies of scale and reduced learning-by-doing. Nevertheless, most studies still show a larger portfolio to be more attractive.

In looking at ambitious stabilization scenarios, an important conclusion is that the inertia present in the energy system calls for a smooth and thus early transition. The extended infrastructure that will be installed and/or replaced in the coming decades implies that decision-makers need to

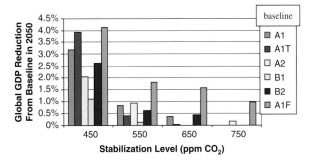

Figure 35.3 Costs for different stabilization scenarios. GDP losses in 2050 for stabilizing CO_2-concentration from various SRES baselines (CO_2-only).
Source: Hourcade et al., 2001.

plan now for such long-term reduction strategies, since low stabilization targets will otherwise be impossible to reach without premature retirement o`f capital investments. For higher stabilization targets, in contrast, there is much more flexibility in the timing of reduction.

35.3.2 Costs

Cost estimates are uncertain. This uncertainty is a consequence of uncertainty in baseline trends, effectiveness of policies, flexibility of economies to adjust to higher energy prices, technology development and assumed international policies. Costs of stabilization at levels from 450 ppmv CO_2 upward were assessed in the IPCC Third Assessment Report (Hourcade et al., 2001) for various reference scenarios. While costs only increase moderately, going down from 750 ppmv to 550 ppmv targets, much sharper increases were observed for targets in the range of 450 ppmv, unless baseline emissions are very low.

The global average costs for 450 ppmv CO_2 stabilization (for a global least cost approach) are reported by IPCC to be in the order of a 1–4% lower GDP by 2050 (Hourcade et al., 2001), compared to a situation without stabilization (see Figure 35.3). These estimates assume cost-optimal implementation of options without transaction costs, but on the other hand these costs do not reflect the benefits of avoided climate change damages or co-benefits (see further). Azar and Schneider (2002) point out that this translates into a very small reduction of annual global average economic growth rates. However, GDP effects could be higher during parts of a long-term period and for specific countries and regions. van Vuuren et al. (2003) show that different schemes to differentiate reduction commitments among regions may result in large costs differences between regions and across different schemes. In their results, regions with high per capita emissions and income (OECD) are confronted with medium costs, while regions with high per capita emissions, but a medium income (CIS, Middle East & Turkey, possibly Latin America) are confronted with relatively high costs, and regions with low per capita emissions and low income (Africa and developing Asia) are

confronted with the lowest costs and can even gain from emissions trading. Similar results were also found in other models (Criqui et al., 2003).

A recent study by Bollen et al. (2004) that looked at reductions in 2030 consistent with a 450 CO_2 stabilization trajectory, shows changes in Gross National Income (which includes expenditures or incomes from traded emission allowances) of -0.6% to -1.8% for the EU-25, -1.4 to -1.8% for Russia, -1.3 to $+5.8\%$ for the Middle East and $+0.8$ to $+0.2\%$ for developing countries in Asia and Africa, the first number being for global participation in emission trading and the second number for emission trading without developing countries in Asia and Africa. Although we cite just one study here, the results are consistent with the larger range of global results reported by IPCC (Hourcade et al., 2001).

Obviously, cost estimates highly depend on the options that are considered. Including non-CO_2 emission reduction in a stabilization strategy lowers the costs (van Vuuren et al., 2005), although at low stabilization levels the emphasis over time shifts to CO_2. On the whole, multi-gas stabilization strategies might be about 30–40% cheaper than CO_2-only strategies (van Vuuren et al., 2005). Excluding options from a stabilization portfolio or limiting the application of one or more options (such as nuclear energy or CO_2 capture and storage) would still allow us to achieve low stabilization levels, but at a higher cost (Azar et al., 2005; IPCC, 2005; Nakicenovic and Riahi, 2003). IPCC (2005) in its recent assessment of CO_2 capture and storage indicates that costs of stabilization, for a range of stabilization levels, can be reduced with 30% or more when including CO_2 capture and storage in a portfolio of measures. Azar et al. (2005) indicate cost decreases of 30–50%, when including CO_2 capture and storage in a 450 ppmv CO_2 stabilization case, and 50–80% for a 350 ppmv CO_2 stabilization case.

35.3.3 Implementation Issues

Implementing stabilization strategies, particularly for low atmospheric concentration stabilization levels, requires a broad and extensive array of policies and measures sustained over a long period of time. Experience so far with modest action, as a result of national programmes in the light of Kyoto Protocol commitments, shows that such policies and measures are hard to introduce and implement. Several studies analyse the many barriers to introducing such policies and spreading the use of emission abatement technologies within countries and between industrialized and developing countries (Sathaye et al., 2001). There is a multitude of potential obstacles, ranging from lack of awareness, vested interests, prices not reflecting environmental impacts (externalities), cultural and behavioural barriers to change and (in the case of spreading technologies to developing countries) the lack of an effective enabling environment for new investments. Above all, stabilization at low levels will require an awareness of the importance of climate change policies in different parts of

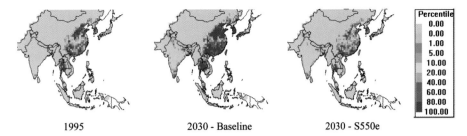

Figure 35.4 Risks of acidification in Asia: (a) in 1995, (b) under the baseline scenario and (c) in the case of stabilization at 550 ppmv (assuming the same air pollution control policies as under (b).
Source: van Vuuren et al., 2003.

the world to enhance the socioeconomic and political circumstances to implement the type of measures discussed above.

35.3.4 Co-benefits

Climate change mitigation policies cannot be seen in isolation from other policies to achieve development goals, energy security goals or other environmental objectives. There are sometimes trade-offs because of conflicting interests (strong climate change mitigation may interfere with maintaining a strong export position on coal or oil), but there are also many potential synergies that can make implementation of mitigation policies easier and cheaper (IPCC, 2001b).

Several studies have looked at the synergies between climate change mitigation and air pollution control. Many air pollutants and greenhouse gases have common sources, implying that controlling one of them can also reduce emissions for the other. The most typical response for controlling air pollution (caused by such substances as sulphur dioxide and nitrogen oxides) is to add end-of-pipe emission control technologies. These technologies generally do not have synergistic value. For control of greenhouse gas emissions, in contrast, systemic changes in the energy system are required that also reduce emission of air pollutants. Synergies between climate change and air pollution control can become apparent in terms of additional emission reductions for air pollutants and/or reduction of air pollution control costs when implementing climate policies.

In low-income countries, taking care of the potential synergies of climate change policies and air pollution policies could be even more important than in high-income countries. An example of this is given by van Vuuren et al. (2003). In this study, reduced acidification risks in Asia as a consequence of climate policies were identified by coupling the global energy model TIMER to the RAINS-Asia model. In 1995, 4% of the ecosystems in the total RAINS Asia region experienced high acidification risks due to deposition of sulphur dioxide above the critical loads. The risks are unevenly distributed across the region, and are especially high in East China with areas where up to 100% of the ecosystems are threatened. Under the baseline assumptions,

the number of ecosystems with a sulphur deposition above the critical loads will increase substantially (see Figure 35.4). The largest increase occurs in China (from almost 6% to almost 10% of total ecosystem). In addition, a large share of ecosystems will receive deposition above critical loads in Thailand, Malaysia, Indonesia, the Korean peninsula, Japan and India in 2030. By introducing climate policies, the situation improves considerably, as shown for the 550 ppmv CO_2-eq stabilization case. As a result of climate policies for Asia as a whole, the exceedance of critical loads is reduced to approximately the 1995 level (or about a 50% reduction compared to baseline). While the exact results reported in this study are obviously uncertain, the fact that the changes in the energy system induced by climate policies will create co-benefits for other environmental problems is relatively uncontroversial.

Climate policy measures also interact with energy security objectives. There could be both co-benefits and trade-offs. Long-term energy security is, amongst other factors, influenced by the diversity of energy sources used, the remaining energy resources and their quality, the share of energy imported and the political stability of exporting regions (Jansen et al., 2004). Climate policy measures that reduce coal use could therefore lead to a decrease of long-term energy security. Introducing CO_2 capture and storage could counteract this effect. Renewable energy use and energy efficiency too could lead to an increase in energy security. The number of studies on climate policies reporting quantitatively on the relationship with energy security is still limited, but there seems to be a growing interest in this area.

35.4 Conclusions

The discussion of the technical potential of different options to reduce greenhouse gas emissions suggests that stabilization of atmospheric concentrations at low levels, i.e. 550 ppmv CO_2-eq or below, is technically feasible with the technologies known today. However, it requires a very broad portfolio of policies and appropriate socioeconomic and political circumstances. Without these circumstances, the implementation and diffusion prospects of these technologies are highly uncertain. More sustainable development and a high level of innovation and international co-operation (as included in the IPCC B1 (sustainable development) or A1T (high-tech) SRES scenarios – see Figure 35. 1) make it easier to reach low-level stabilization, while a very fossil-fuel intensive economy or lack of international cooperation (as included in the A1FI (fossil-fuel intensive) or A2 (regional focus; low-tech) IPCC SRES scenario – see Figure 35.1) would make it very difficult. For all stabilization strategies, the biggest problem does not seem to be the technologies or the costs, but overcoming the many political, social and behavioural barriers to implementing mitigation options.

Excluding options from a portfolio of actions does increase the cost of stabilization. Current studies on mitigation strategies indicate that multi-gas strategies, emission trading, optimal timing and strong technology development, diffusion and transfer are essential to keep costs of low-level stabilization relatively low. Although considerable uncertainties exist, least-cost strategies are estimated to reduce global average GDP by 2050 by 1–4% for 550 ppmv CO_2-eq stabilization. For specific regions, countries or sectors costs could be higher. In considering costs it is important to take into account potential co-benefits or interactions of climate policies in terms of other environmental objectives (air pollution, acid deposition), energy security or other development goals.

REFERENCES

Azar, C., Lindgren, K., Larson, E. and Möllersten, K., 2005. Carbon capture and storage from fossil fuels and biomass – Costs and potential role in stabilizing the atmosphere. Climatic Change, (in press).

Azar, C. and Schneider, S.H., 2002. Are the economic costs of stabilizing the atmosphere prohibitive? Ecological Economics, 42: 73–80.

Bashmakov, I., Jepma, C., Bohm, P., Gupta, S., Haites, E., Heller, T., Montero, J.P., Pasco-Font, A., Stavins, R., Turkson, J., Xu, H. and Yamaguchi, M., 2001. Policies, Measures, and Instruments. In: B. Metz, O. Davidson, R. Swart and J. Pan (Editors), Climate Change 2001 – Mitigation; Contribution of Working Group III to the Third Assessment Report of the IPCC. Cambridge University Press, Cambridge.

Berndes, G., Hoogwijk, M. and van den Broek, R., 2003. The contribution of biomass in the future global energy supply: a review of 17 studies. Biomass and Bioenergy, 25(1): 1–27.

Bollen, J.C., Manders, A.J.G. and Veenendaal, P.J.J., 2004. How much does a 30% emission reduction cost? Macroeconomic effects of post-Kyoto climate policy in 2020. CPB Document no 64, Netherlands Bureau for Economic Policy Analysis, The Hague.

Corfee Morlot , J., Smith, J., Agrawala, S. and Franck, T., 2005. Article 2, Long-Term Goals and Post-2012 Commitments: Where Do We Go From Here with Climate Policy? Climate Policy, 5: 251–272.

Criqui, P., Kitous, A., Berk, M.M., Den Elzen, M.G.J., Eickhout, B., Lucas, P., van Vuuren, D.P., Kouvaritakis, N. and Vanregemorter, D., 2003. Greenhouse gas reduction pathways in the UNFCCC Process upto 2025 – Technical Report. CNRS-IEPE, Grenoble, France.

Delhotal, K.C., DelaChesnaye, F.C., Gardiner, A., Bates, J. and Sankovski, A., 2006. Mitigation of methane and nitrous oxide emissions from waste, energy and industry. Energy Policy, forthcoming.

Den Elzen, M.G.J. and Meinshausen, M., 2005. Multigas emission pathways for meeting the EU 2 degrees climate target, this volume.

ECF and PIK, 2004. What is dangerous climate change? Initial Results of a Symposium on Key Vulnerable Regions, Climate Change and

Article 2 of the UNFCCC. European Climate Forum and Postdam Institute for Climate Impact Research, http://www.european-climate-forum.net/pdf/ECF_beijing_results.pdf

EU, 1996. Communication of the Community Strategy on Climate Change. Council conclusions. European Council. Brussels, Council of the EU.

EU, 2004. Climate Change: Medium and longer term emission reduction strategies, including targets. Council (Environment) conclusions. Doc no: 16298/04/ENV 711/ENER 274/FISC 262/ONU 120, Council of the EU. Information Note from General Secretariat to Delegations, 22 December 2004, Brussels. http://eu.eu.int/ueDocs/cms_Data/docs/pressData/en/envir/83237.pdf

EU, 2005. Council of the European Union, Presidency conclusions, March 22–23. http://ue.eu.int/ueDocs/cms_Data/docs/pressData/en/ec/84335.pdf

Hoogwijk, M., 2004. On the global and regional potential of renewable energy sources, Utrecht University, http://www.library.uu.nl/digiarchief/dip/diss/2004-0309-123617/full.pdf

Hourcade, J.C., Shukla, P., Cifuentes, L., Davis, D., Edmonds, J., Fisher, B., Fortin, E., Golub, A., Hohmeijer, O., Krupnick, A., Kverndokk, S., Loulou, R., Richels, R., Segenovics, H. and Yamati, K., 2001. Global, regional and National Costs and Ancillary Benefits of Mitigation. In: B. Metz, O. Davidson, R. Swart and J. Pan (Editors), Climate Change 2001: Mitigation; Contribution of Working Group III to the Third Assessment Report of the IPCC. Cambridge University Press, Cambridge.

IEA, 2004. CO_2 capture and storage. International Energy Agency, Paris.

IPCC, 2001a. Climate Change 2001, Impacts, adaptation and vulnerability, Contribution of Working Group II to the Third Assessment Report of the Intergovernmental Panel on Climate Change, J. McCarthy, C. OF, N. Leary, D. Dokken and K. White (Editors), Cambridge University Press, Cambridge.

IPCC, 2001b. Climate Change 2001: Mitigation, Contribution of Working Group III to the Third Assessment Report of the Intergovernmental Panel on Climate Change, B. Metz, O. Davidson, R. Swart and J. Pan (Editors), Cambridge University Press, Cambridge.

IPCC, 2001c. Climate Change 2001: Synthesis Report, R.T. Watson (Editor), Cambridge University Press, Cambridge.

IPCC, 2005. Special Report on CO_2 capture and storage, B. Metz, O.R. Davidson, H. de Coninck and L.M. Meyer (Editors), Cambridge University Press, Cambridge.

Jansen, J.C., van Arkel, W.G. and Boots, M.G., 2004. Designing indicators of long-term energy supply security. ECN-C—04-007, ECN, Petten, The Netherlands.

Kauppi, P., Sedjo, R., Apps, M., Cerri, C., Fujimori, T., Janzen, H., Krankina, O., Makundi, W., Marland, G., Masera, M., Nabuurs, G.J., Razali, W. and Ravindranath, N.H., 2001. Technological and Economic Potential of Options to Enhance, Maintain, and Manage Biological Carbon Reservoirs and Geo-engineering. In: B. Metz, O. Davidson, R. Swart and J. Pan (Editors), Climate Change 2001: Mitigation; Contribution of Working Group III to the Third Assessment Report of the IPCC. Cambridge University Press, Cambridge.

Leemans, R. and Eickhout, B.E., 2004. Another reason for concern: regional and global impacts on ecosystems for different levels of climate change. Global Environmental Change, 14: 219–228.

Mastandrea, M.D. and Schneider, S.H., 2004. Probabilistic Integrated Assessment of dangerous climate change. Science, 304: 571–574.

Meinshausen, M., 2005. On the risk of overshooting 2°C, Avoiding Dangerous Climate Change, Exeter.

Moomaw, W.R., Moreira, J.R., Blok, K., Greene, D.L., Gregory, K., Jaszay, T., Kashiwagi, T., Levine, M., McFarland, M., Siva Prasad, N., Price, L., Rogner, H.H., Sims, R., Zhou, F. and Zhou, P., 2001. Technological and Economic Potential of Greenhouse Gas Emissions Reduction. In: B. Metz, O. Davidson, R. Swart and J. Pan (Editors), Climate Change 2001: Mitigation; Contribution of Working Group III to the Third Assessment Report of the IPCC. Cambridge University Press, Cambridge.

Morita, T., Robinson, J., Adegbulugbe, A., Alcamo, J., Herbert, D., Lebre la Rovere, E., Nakicenivic, N., Pitcher, H., Raskin, P., Riahi, K., Sankovski, A., Solkolov, V., Vries, B.d. and Zhou, D., 2001. Greenhouse gas emission mitigation scenarios and implications. In: B. Metz, O. Davidson, R. Swart and J. Pan (Editors), Climate Change 2001: Mitigation; Contribution of Working Group III to the Third Assessment Report of the IPCC. Cambridge University Press, Cambridge.

Nakicenovic, N., Alcamo, J., Davis, G., De Vries, B., Fenhamm, J., Gaffin, S., Gregory, K., Gruebler, A., Jung, T.Y., Kram, T., Lebre La Rovere, E., Michaelis, L., Mori, S., Morita, T., Pepper, W., Pitcher, H., Price, L., Riahi, K., Roehrl, A., Rogner, H.H., Sankovski, A., Schlesinger, M., Shukla, P., Smith, S., Swart, R., Van Rooyen, S., Victor, N. and Zhou, D., 2000. IPCC Special Report on Emissions Scenarios. Cambridge University Press, Cambridge.

Nakicenovic, N. and Riahi, K., 2003. Model runs with MESSAGE in the context of the further development of the Kyoto protocol. International Institute for Applied Systems Analysis.

O'Neill, B.C. and Oppenheimer, M., 2002. Climate Change: Dangerous Climate Impacts and the Kyoto Protocol. Science, 296: 1971–1972.

Rayner, S. and Malone, E., 1998. Human Choice & Climate Change. Battelle Press, Columbus Ohio.

Richels, R.G., Manne, A.S. and Wigley, T.M.L., 2004. Moving Beyond Concentrations: The Challenge of Limiting Temperature Change. Working Paper No. 04-11, AEI-Brookings Joint Center, Washington D.C.

Sathaye, J., Bouille, D., Biswas, D., Crabbe, P., Geng, L., Hall, D., Imura, H., A., J., L., M., Peszko, G., Verbruggen, A., Worrell, E. and Yamba, F., 2001. Barriers, Opportunities, and Market Potential of Technologies and Practices. In: B. Metz, O. Davidson, R. Swart and J. Pan (Editors), Climate Change 2001: Mitigation; Contribution of Working Group III to the Third Assessment Report of the IPCC. Cambridge University Press, Cambridge.

Schaefer, D.O., Godwin, D. and Harnisch, J., 2006. Estimating future emissions and potential reductions of HFCs, PFCs and SF6. Energy Policy, forthcoming.

Schneider, S.H. and Lane, J., 2005. An overview of dangerous climate change, this volume.

Sims, R.E.H., Rogner, H.H. and Gregory, K., 2003. Carbon emission and mitigation costs comparison between fossil fuel, nuclear and renewable energy resources for electricity generation. Energy Policy, 31: 1315–1326.

Swart, R., Mitchell, J., Morita, T. and Raper, S., 2002. Stabilisation scenarios for climate impact assessment. Global Environmental Change, 12(3): 155–165.

UNFCCC, 1992. United Nations Framework Convention on Climate Change, http://www.unfccc.int/resources

van Vuuren, D.P., Den Elzen, M.G., Berk, M.M., Lucas, P.L., Eickhout, B., Eerens, H. and Oostenrijk, R., 2003. Regional costs and benefits of alternative post-Kyoto climate regimes. Report 728001025/2003, National Institute for Public Health and the Environment, Bilthoven, The Netherlands. http://www.mnp.nl/en/publications/2003

van Vuuren, D.P., Weyant, J. and De la Chesnaye, F., 2005. Multigas scenarios to stabilise radiative forcing. In: M. Hoogwijk (Editor), IPCC expert meeting on emission scenarios, Washington D.C. www.ipcc.ch/meet/washington.pdf

WEA, 2000. World Energy Assessment – Energy and the challenge of sustainability. UNDP – United Nations Development Programme, New York.

Weyant, J.P., De la Chesnaye, F.C. and Blanford, G., 2005. Overview of EMF – 21: Multigas Mitigation and Climate Change. Energy Journal, (in press).

CHAPTER 36

Stabilization Wedges: An Elaboration of the Concept

Robert Socolow
Princeton University, USA

ABSTRACT: We have earlier introduced the stabilization wedge as a useful unit for discussing climate stabilization. A wedge is 1 GtC/yr of emissions savings in 2055, achieved by a single strategy that will not occur without deliberate attention to global carbon. Implementing seven wedges should place humanity, approximately, on a path to stabilizing the climate at a concentration less than double the pre-industrial concentration, leaving those at the helm in the following 50 years in a position to drive CO_2 emissions to net zero emissions; arguably, the tasks of the two half-centuries are comparably difficult. We elaborate on the concept of the stabilization wedge that is achieved through carbon policy by introducing the 'virtual' wedge that is achieved as a result of the continued decarbonization of the global economy even in the absence of carbon policy. Virtual wedges are already embedded in almost all 'baseline' scenarios, because the decarbonization of the global economy is a robust historical trend. Thus, the stabilization wedges must be achieved over and above the structural shifts, energy efficiency gains, and energy system decarbonization that are likely to occur in the next 50 years even without carbon policy.

We discuss stabilization wedges of energy efficiency, calling attention to the importance of avoiding investments in durable capital facilities, like power plants and apartment buildings, that are energy-inefficient or carbon-wasteful. We briefly explore wedges of capture and storage, nuclear energy, and renewable energy. The wedges framework highlights the importance of early involvement of the developing countries in mitigation activity. The wedges framework, therefore, may be able to contribute new elements to global carbon policy, by aligning the concept of differentiated responsibilities across countries with a global commitment to the internationally coordinated commercialization of low-carbon technology.

36.1 The Stabilization Wedges

Stephen Pacala and I recently presented an extremely simple way of visualizing the mitigation required in the coming half century to set the world onto a path toward stabilization of the climate at a concentration less than double the pre-industrial concentration (Pacala and Socolow, 2004). Our approach has four features:

1. *We focus on the next 50 years.* Interim targets at mid-century help divide the work among generations. A 50-year perspective is long enough to allow major changes in infrastructure and consumption patterns, but it is also short enough to be heavily influenced by decisions made today. It is a time frame, looking forward, with which many businesses are comfortable, and a time frame, looking backward, that is contained in a single human memory. It is the time frame of a scientific career.

2. *We approximate stabilization below doubling by a 'flat trajectory': zero emissions growth (ZEG) for the next 50 years.* Achieving ZEG (a global CO_2-equivalent emissions rate in 2055 no larger than today's) delivers a far more tractable climate problem to later generations than if we postpone action for 50 years. The emission rate must fall in the second half of this century, descending to net zero emissions (emissions balanced by sinks) near the end of the century.

3. *We approximate the baseline, or Business As Usual (a world that pays no deliberate attention to global carbon), by a 'ramp trajectory': linear growth leading to a doubling of global CO_2-equivalent emissions by mid-century.* This approximation is at the center of many clouds of estimates. Thus, achieving stabilization below doubling requires, approximately, halving the anticipated mid-centuries emissions. Restricting attention to fossil-fuel carbon, emissions today are 7 GtC/yr and are heading for 14 GtC/yr by mid-century. Between them, the flat trajectory and the ramp trajectory form the 'stabilization triangle', as seen in Figure 36.1. The interim ZEG target requires removing 7 GtC/yr of emissions in 2055 by actions generated by deliberate attention to global carbon.

4. *We introduce the 'wedge', as a useful unit for quantifying actions that reduce global carbon emissions.* A wedge is 1 GtC/yr of emissions savings in 2055, achieved by a single strategy (Y displaces X) that will not occur without deliberate attention to global carbon. Assuming linear growth in emissions avoided, a wedge reduces emissions by 25 GtC over the next half-century. Achieving ZEG requires creating, roughly, seven wedges.

There are, of course, major simplifications here. Scientific uncertainty shrouds our current understanding of carbon

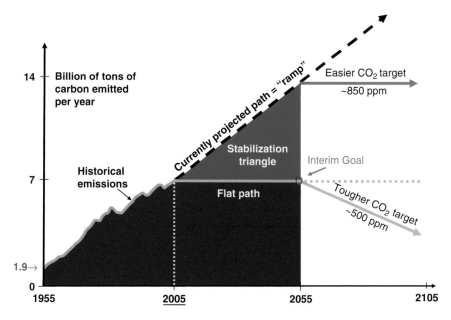

Figure 36.1 The 'stabilization triangle' is an idealization of the first 50 years of action required to achieve stabilization of the atmospheric CO_2 concentration below double the pre-industrial concentration. The triangle is bounded by (1) the Year 2055; (2) a 'flat trajectory' of constant global carbon emissions at the current rate of 7 GtC/yr, intended to approximate the first 50 years of a 500 ppm stabilization trajectory; and (3) a 'ramp trajectory', where emissions climb linearly to twice current rates, intended to approximate Business As Usual, i.e., a world inattentive to global carbon. The stabilization triangle is divided into seven 'wedges' of avoided emissions, each of which grows linearly from zero today to 1 GtC/yr in 2055.

sinks; if carbon fertilization is a powerful effect, the land sink in 2055 could be about 3 GtC/yr larger than if it is absent. Political uncertainty shrouds the choice of stabilization target: the mid-century emissions target changes by about 2 Gt/yr when the stabilization target changes by 50 ppm. As will be discussed further below, economic uncertainty shrouds the size of the future global economy and the extent to which specific wedge technologies will be adopted even in a world that has no focus on carbon.

36.2 Dividing Responsibility Between the Next Two Half-Centuries

The consistency of flat emissions for 50 years with stabilization below doubling can be understood with the help of the lower emissions trajectory in Figure 36.2 (Socolow et al., 2004a). With a particular model of the ocean and land sinks, stabilization at 500 ppm can be achieved provided that emissions in 2055–2105 descend linearly to net-zero emissions and after 2105 slowly decline to match the declining ocean sink so as to remain at net zero emissions. By net-zero emissions, we mean, of course, emissions that are balanced by removals by the land and ocean sinks. In our particular model, net-zero emissions in 2105 for the lower trajectory are about 3 GtC/yr.

In Figure 36.2 the upper emissions trajectory is an extrapolation of the ramp trajectory that forms the upper edge of the stabilization triangle: from 2055 to 2105 emissions

are constant at 14 GtC/yr, and from 2105 to 2155 they descend linearly to net-zero emissions. The result is stabilization at 850 ppm, triple the pre-industrial CO_2 concentration. In the particular sense of Figure 36.2, the choice represented by the Stabilization Triangle is a choice between 'beating doubling' and 'accepting tripling'. Accept tripling, and significant action can be delayed for most of the next half century. Insist on beating doubling, and work needs to begin now.

In Figure 36.2, the four 50-year segments of the upper and lower trajectories can be thought of as the four legs in two 200-year relay races, leading to stabilization at 500 ppm and 850 ppm, respectively. The baton is passed from the first runner to the second in 2055, from the second to the third in 2105, and from the third to the fourth in 2155, with the finish in 2205. The assignments of the second, third, and fourth runners for 850 ppm stabilization imitate those of the first, second, and third runners for 500 ppm stabilization.

If the first and second runners in the relay race have roughly equally difficult tasks, there will be intergenerational equity, at least for the next 100 years. I offer my intuition that the lower trajectory in Figure 36.2 is an approximation to such equity. It could be comparably hard to achieve a global emissions rate in 2055 no higher than today's and to cut the 2055 emissions rate in half between 2055 and 2105.

If the first runner delivers a world in 2055 with a slowly falling population, institutions that promote efficient energy

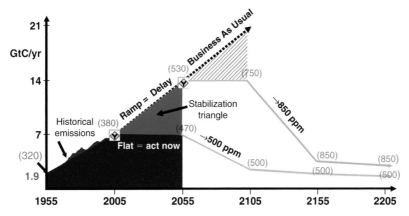

Figure 36.2 The two trajectories forming the Stabilization Triangle lead to stabilization at less than double the pre-industrial CO_2 concentration (500 ppm) and at triple that concentration (850 ppm), assuming the emissions shown here. The 500 ppm trajectory requires a linear reduction of emissions from 2055 to 2105 leading to net zero emissions (emissions equal to land plus ocean sinks) in 2105. The 850 ppm trajectory assumes the same sequence of actions as the 500-ppm trajectory, but delayed by 50 years: a flat trajectory from 2055 to 2105, followed by a linear descent to net zero emissions in 2155. Atmospheric CO_2 concentrations, in ppm, are in parentheses. The future net land sink (uptake minus deforestation) is arbitrarily assumed constant at 0.5 GtC/yr; this sink could be either stronger from carbon fertilization or weaker (even, a net source) from positive biological feedback effects, like peat decomposition. The ocean sink, modeled with HILDA [Siegenthaler and Joos, 1992], is 2.2 GtC/yr today. It is 2.8 GtC/yr, 1.9 GtC/yr, 1.5 GtC/yr, and 1.3 GtC/yr in 2054, 2104, 2154, and 2204, respectively, for 500 ppm stabilization. For 850 ppm stabilization, the four corresponding values are 4.1 GtC/yr, 4.4 GtC/yr, 3.2 GtC/yr, and 2.5 GtC/yr.
Source: Socolow et al., 2004a.

use in both the industrialized and the still developing world, well-established permitting processes for carbon capture and storage, photovoltaic electricity already commercialized and expanding rapidly, and a nuclear fuel cycle under international management, then, in the second half-century, substantial further progress in these areas can be expected. And if research and development has been vigorous, in the second half of the century some revolutionary technologies, such as air scrubbing technology, carbon storage in minerals, nuclear fusion, nuclear thermal hydrogen, and artificial photosynthesis, should be ready for scaling up to provide 'second-period wedges'. The first runner will have left the second runner with a job that, arguably, can be accomplished with an effort that is roughly equally heroic.

Critically important during the next half century is significant growth in the level of research and development focused on climate change mitigation (Hoffert et al., 2002). As Pacala and Socolow, 2004 assert: 'fundamental research is vital to develop the revolutionary mitigation strategies needed in the second half of this century and beyond.' However, it is important for the research community to acknowledge that the tools are at hand to get started. To speak exclusively of revolutionary technology available in 50 years or so is almost guaranteed to generate a kind of paralysis. The public and decision makers may well conclude that the best course is to wait for the revolutionary technology and meanwhile to avoid taking action. Such an embrace of delay may be accompanied by a reluctance to make the very commitments to expanded research and development that could lead to the revolutionary technologies. A societal commitment to the necessary research

and development is more likely in a world that is already engaged in implementing climate change mitigation strategies.

36.3 Achieving Specific Stabilization Wedges

The world will base its choice between beating doubling and accepting tripling on two assessments. It will assess the benefits from less damage suffered, and it will assess the costs of achieving the necessary mitigation strategies. The benefits from reduced damage will be uncertain, but already, especially in Europe, political leaders are judging the incremental damage to be unacceptable. They are asking for guidance, therefore, regarding the availability of mitigation strategies. In the wedges framework mitigation is viewed as the implementation of parallel campaigns to scale up several already commercialized technologies so as to fill the stabilization triangle. There is no silver bullet. No single campaign can accomplish even half the job. A portfolio of strategies is required. The list of candidates for the portfolio is sufficiently long, however, that not every strategy is needed. Three reasons make me optimistic that such a program of parallel campaigns to fill the stabilization triangle can succeed: (1) the world has a notoriously inefficient energy system, (2) the world is just beginning to put a price on carbon; (3) most of the 2055 physical plant is not yet built.

Accepting the analysis above as at least a useful point of departure, the challenge of below-doubling stabilization reduces to an evaluation of potential wedges. Table 36.1 reproduces the table of candidate wedges from Pacala and Socolow, 2004[1]; the table provides estimates of the size of a wedge for 15 separate strategies[2]. A wedge is two million one-megawatt windmills displacing coal power. A wedge is two billion personal vehicles achieving 60 miles per U.S. gallon (mpg) on the road instead of 30 mpg. A wedge is capturing and storing the carbon produced in 800 large (1000-megawatt, i.e., 1 GW) modern coal plants.

The wedge is a useful unit of action, because it permits quantitative discussion of cost, pace, risk, and trade-off. The wedges listed in Table 36.1 involve technologies already deployed somewhere in the world at commercial scale. No fundamental breakthroughs are needed. However, every wedge is hard to accomplish, because huge scale-up is required, and scale-up introduces environmental and social problems not present at limited scale. (See the right hand column of Table 36.1.) For many of the wedges, Table 36.1 shows the extent of scale-up required to go from today's level of deployment to a full wedge. Since there are already the equivalent of 40 thousand one-megawatt windmills deployed globally, for example, a wedge from wind displacing coal for power requires a factor-of-50 increase.

36.4 Virtual Wedges

In Pacala and Socolow, 2004, a single baseline scenario was chosen (the ramp trajectory of Figure 36.1), and it was specified as little as possible. This was deliberate, because we wished to keep the focus on the distinction between a world oblivious to carbon management (the baseline) and a world investing heavily in carbon management, and not to be distracted by the abundance of baselines discussed in the scenarios literature (Nakicenovic et al., 2000; Riahi and Roehrl, 2000a; Riahi and Roehrl, 2000b; IPCC, 2001). However, the wedges analysis can be made richer by consideration of alternative baselines and, in turn, the wedges methodology can provide a precise language for discussing any specific baseline.

Embedded in any baseline scenario will be many activities like those in Table 36.1. A 50-fold expansion in wind power might occur without deliberate attention to carbon, for example. In that case, the expansion is already

embedded in the baseline and cannot contribute to the stabilization triangle. A stabilization wedge of wind would require the construction of a *second* two million one-megawatt windmills.

It is useful to introduce some vocabulary to describe the total amount of carbon-saving activity already embedded in the baseline and the constituents of that total. Consider Figure 36.3. Sitting on top of the stabilization triangle shown in Figure 36.1 is a 'virtual triangle', whose lower boundary is the baseline scenario and whose upper boundary is a 'virtual reference scenario', where carbon emissions grow in exact proportion to economic activity. The virtual triangle embeds all of the activity in the baseline scenario that causes carbon emissions to grow more slowly than the economy. Again, the time frame is the next 50 years.

Most baseline scenarios embed two trends that make carbon emissions grow more slowly than the economy: a falling energy intensity of the global economy (primary energy production growing more slowly than the economy) and a falling carbon intensity of primary energy (carbon emissions growing more slowly than primary energy production, sometimes called the 'decarbonization' of primary energy). The first trend reflects structural shifts (such as the shrinking role of energy-intensive industries) and increasingly efficient energy technology; nearly all scenarios assume that this trend continues. The second trend reflects a long period of decline in the share of coal and increase in the share of nuclear energy, renewable energy, and natural gas. Some scenarios show the trend toward decarbonization reversing, and 'recarbonization' of primary energy production emerging, as the use of coal for power and synthetic fuels expands.

Virtual wedges fill the virtual triangle just as stabilization wedges fill the stabilization triangle. Virtual wedges have the same units as real wedges and measure the extent to which some specific carbon emission reduction activity is present, relative to the virtual reference scenario. If past trends continue over the next 50 years, there will be many virtual wedges. Figure 36.3 shows a virtual triangle whose vertical side at 2055 runs from 14 GtC/yr to 25 GtC/yr, so it contains 11 virtual wedges[3]. In principle, every wedge listed in Table 36.1 can be a virtual wedge except the sixth, seventh, and eight wedge, all associated with CO_2 capture and storage. There are no virtual wedges of CO_2 capture

---

[1] We have changed 2004 to 2005 and 2054 to 2055 in Table 36.1.

[2] The list of 15 wedges in Pacala and Socolow, 2004 was not intended to be complete. Industrial energy efficiency could have been added to the list, as could the substitution of low-carbon electricity for 700 GW of coal in at least three forms: as 700 GW of geothermal electricity, 700 GW of hydropower, or 2000 GW (peak capacity) of high-temperature solar thermal electricity (via parabolic trough or dish concentrators). Pacala and Socolow, 2004 discussed (in the supporting online material) but did not estimate the size of a wedge that would be achieved by substituting decarbonized electricity for carbon fuel used directly, for example, in heat pumps for space heating and plug-in hybrid vehicles. Also Pacala and Socolow, 2004 did not estimate the wedges that might be achieved by reducing emissions of non-CO_2 greenhouse gases like methane. As explained in Pacala and Socolow, 2004: "because the same [baseline] carbon emissions cannot be displaced twice," wedge technologies often compete, e.g., to reduce carbon emissions in transport.

---

[3] The choice of 11 virtual wedges, while only illustrative, conforms to an observation in Pacala and Socolow, 2004 that if, for 50 years, the Gross World Product were to grow at 3% per year while global CO_2 emissions grew at 1.5% per year, 11 virtual wedges would result (though the phrase 'virtual wedges' is not used in this reference). A further distinction is necessary, however, once exponential growth rather than linear growth is introduced. In Pacala and Socolow, 2004, a 'wedge' is defined in two ways: either as the prevention of 25 GtC from entering the atmosphere over 50 years (an 'area wedge') or as the reduction of the carbon emissions rate by 1 GtC/yr 50 years from now (here, a '2055 wedge'). An area wedge and a 2055 wedge are identical for linear growth, but not for exponential growth. The calculation in Pacala and Socolow, 2004 results in 11 *area* wedges.

Table 36.1 Potential wedges: Strategies available to reduce the carbon emission rate in 2055 by 1 GtC/year, or to reduce carbon emissions from 2005 to 2055 by 25 GtC.

	Option	Effort by 2055 for one wedge, relative to 14 GtC/year BAU	Comments, issues
Energy Efficiency and Conservation	Economy-wide carbon-intensity reduction (emissions/$GDP)	Increase reduction by additional 0.15% per year (e.g., increase U.S. goal of reduction of 1.96% per year to 2.11% per year)	Can be tuned by carbon policy
	1. Efficient vehicles	Increase fuel economy for 2 billion cars from 30 to 60 mpg	Car size, power
	2. Reduced use of vehicles	Decrease car travel for 2 billion 30-mpg cars from 10,000 to 5,000 miles per year	Urban design, mass transit, telecommunicating
	3. Efficient buildings	Cut carbon emissions by one-fourth in buildings and appliances projected for 2055	Weak incentives
	4. Efficient baseload coal plants	Produce twice today's coal power output at 60% instead of 40% efficiency (compared with 32% today)	Advanced high-temperature materials
Fuel shift	5. Gas baseload power for coal baseload power	Replace 1400 GW 50%-efficient coal plants with gas plants (4 times the current production of gas-based power)	Competing demands for natural gas
CO_2 Capture and Storage (CCS)	6. Capture CO_2 at baseload power plant	Introduce CCS at 800 GW coal or 1600 GW natural gas (compared with 1060 GW coal in 1999)	Technology already in use for H_2 production
	7. Capture CO_2 at H_2 plant	Introduce CCS at plants producing 250 MtH_2/year from coal or 500 MtH_2/year from natural gas (compared with 40 MtH_2/year today from all sources)	H_2 safety, infrastructure
	8. Capture CO_2 at coal-to-synfuels plant	Introduce CCS at synfuels plants producing 30 million barrels per day from coal (200 times Sasol), if half of feedstock carbon is available for capture	Increased CO_2 emissions, if synfuels are produced *without* CCS
	Geological storage	Create 3500 Sleipners	Durable storage, successful permitting
Nuclear Fission	9. Nuclear power for coal power	Add 700 GW (twice the current capacity)	Nuclear proliferation, terrorism, waste
Renewable Electricity and Fuels	10. Wind power for coal power	Add 2 million 1-MW-peak windmills (50 times the current capacity) 'occupying' 30×10^6 ha, on land or off shore	Multiple uses of land because windmills are widely spaced
	11. PV power for coal power	Add 2000 GW-peak PV (700 times the current capacity) on 2×10^6 ha	PV production cost
	12. Wind H_2 in fuel-cell car for gasoline in hybrid	Add 4 million 1-MW-peak windmills (100 times the current capacity)	H_2 safety, infrastructure
	13. Biomass fuel for fossil fuel	Add 100 times the current Brazil or U.S. ethanol production, with the use of 250×10^6 ha (1/6 of world cropland)	Biodiversity, competing land use
Forests and Agricultural Soils	14. Reduced deforestation, plus reforestation, afforestation and new plantations.	Decrease tropical deforestation to zero instead of 0.5 GtC/year, and establish 300 Mha of new tree plantations (twice the current rate)	Land demands of agriculture, benefits to biodiversity from reduced deforestation
	15. Conservation tillage	Apply to all cropland (10 times the current usage)	Reversibility, verification

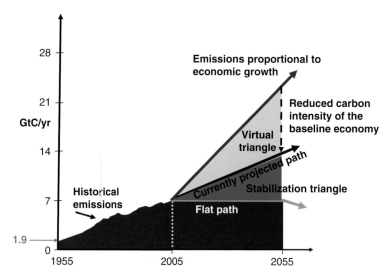

Figure 36.3 Stabilization triangle with seven wedges and virtual triangle with 11 wedges. The virtual triangle represents the CO_2 emissions reductions embedded in nearly all 'baseline' scenarios. These reductions arise from the persistence of the robust trend of decarbonization of the global economy.

and storage in baseline scenarios, because it is generally assumed that, in the absence of carbon policy, it is always cheaper to vent CO_2 than to capture and store it.

Do wedges get used up? Are the first two million one-megawatt windmills more expensive or cheaper than the second two million one-megawatt windmills? The first two million will be at more favorable sites, but the second two million will benefit from the learning acquired building the first two million. The question generalizes to almost all the wedges. Geological storage of CO_2, storage of carbon in soils, uranium fuel, natural gas fuel, semiconductor materials for photovoltaic collectors, land for biomass, river valleys for hydropower – all present the same question: will saturation or learning dominate availability? Where saturation dominates, a wedge strategy gets used up. Where learning dominates, the number of available wedges expands. The debate here closely parallels the debate about whether the world will 'run out' of oil.

From the perspective of wedges at risk of being used up, two questions are nearly identical: 1) Are two stabilization wedges of strategy A available, or only one? 2) Does a virtual wedge of strategy A pre-empt the use of strategy A for stabilization?

36.5 A Carbon-Efficient Global Economy

Achieving a carbon-efficient world, one with half the expected carbon emissions at mid-century, requires attention to carbon flows throughout the global economy. One must confront both end-use energy consumption and energy production, all economic sectors (buildings, vehicles, factories, and farms), and economies at all levels of development (Brown et al., 1998; IPCC, 2001).

Of particular concern is the turnover of physical capital. Although many of today's additions to the world's current

carbon-consuming physical capital, like new vehicle engines, have a lifetime of a decade or two, many other additions lock in carbon demand half a century from now. The new capital that will function for a half-century or more comes in all sizes: from the boiler and steam turbine in a power plant, to the window and roof in an apartment building or private home. Retro-fitting such physical capital after construction is usually far more costly than opting for energy efficiency in the first place[4]. To achieve dramatic reductions in carbon emissions over the next half-century, one must be vigilant about today's long-lived capital investments.

This argument for vigilance regarding new capital investments is not well appreciated. Among carbon policy analysts, there are more frequent arguments for delay than for prompt action. Arguments for delay are based on an understandable concern for avoiding the costs of premature retirement of existing capital stock. But these arguments are not adequately tempered by concern for the creation of carbon-inappropriate new stock.

One sign of this unbalanced attention to the demography of the capital stock is inadequate attention to the carbon consequences of capital formation in developing countries. At present, much of the world's addition to its capital stock (its new power plants, steel mills, and apartment buildings, for example) is taking place in the developing countries, and this is expected to remain the case throughout the next 50 years[5]. Accordingly, it makes little sense to divide the world into 1) Annex I countries whose assignment is to mitigate, and 2) non-Annex I countries whose relationship to the carbon problem is to

---

[4] This sentence is found nearly verbatim on p. 11 of (Socolow, R., et al., 2004b).

[5] This and the preceding sentence are found nearly verbatim on p. 11 of (Socolow, R., et al., 2004b).

suffer impacts and be compensated for them. Both the industrialized and the developing countries need to see mitigation in developing countries as very much in their own self-interest.

From such a perspective, the concept of 'leapfrogging' rises to prominence. Leapfrogging describes the introduction of advanced technology in developing countries ahead of its introduction in industrialized countries. Today, little is done to encourage a developing country to introduce low-carbon technology, like advanced coal technology, before it is extensively tested in industrialized countries. Yet, by going first, the developing country will build fewer facilities that become a liability when a price is later put on CO_2 emissions. Leapfrogging is a path to globally coordinated learning about the potential of new carbon-responsive technology.

36.6 Carbon Capture and Storage

In Figure 36.1, carbon emissions from fossil fuel use are as large in 2055 as today. Moreover, Figure 36.1 is consistent with the extraction from the earth of even greater amounts of carbon in 2055 than today. The rate of extraction of carbon in fossil fuels from the earth may grow, even though CO_2 emissions to the atmosphere stay constant, if some of the CO_2 released while energy is produced is prevented from reaching the atmosphere.

Interfering with CO_2 emission in this way requires a two-step process known as 'carbon capture and storage'. The first step, carbon capture, typically creates a pure, concentrated stream of CO_2, separated from the other products of combustion. The second step, carbon storage, sends the concentrated CO_2 to a destination other than the atmosphere[6].

Opportunities for CO_2 capture are abundant. The natural gas industry routinely generates capturable streams of CO_2 when natural gas, after coming out of the ground, is scrubbed of CO_2 before shipment by pipeline or tanker. Refineries making hydrogen for internal use are generating, as a byproduct, capturable streams of CO_2. Capturable streams of CO_2 will also be generated where technologies are deployed to convert coal or natural gas into hydrogen or synthetic hydrocarbon fuels. In a world focused on CO_2, all of these streams are candidates for capture, instead of venting to the atmosphere.

The most promising storage idea is 'geological storage', where the CO_2 is placed in deep sedimentary formations. (Alternate carbon storage ideas include storage of CO_2 deep in the ocean and storage of carbon in solid form as carbonates). CO_2 capture and storage has the potential to be implemented wherever there are large point sources of CO_2, such as at power plants and refineries. The storage space available below ground is probably large enough to make CO_2 capture and storage a compelling carbon mitigation option.

In all situations where CO_2 capture and storage is under consideration, there may be opportunities to 'co-capture and co-store' other pollutants, like sulfur, with the CO_2. With co-capture, the costs of above-ground pollution control will be reduced, and perhaps total pollution control costs and total environmental emissions as well.

36.7 Non-Carbon Energy Supply

Non-carbon energy supply comes in two principal varieties: nuclear energy and renewable energy. Both, in principle, can produce wedges of electricity by backing out coal electricity. Both can also produce wedges of fuel by backing out hydrocarbon fuels used directly. An example of the latter is the production of electrolytic hydrogen and its use in vehicles instead of gasoline or diesel fuel (National Research Council, 2004). However, it turns out that non-carbon electricity can save about twice as much carbon when used to displace coal-based electricity as when used to produce hydrogen that displaces gasoline[7]. A wedge of wind power can arise from either two million one-megawatt windmills backing out efficient coal power plants or four million one-megawatt windmills making hydrogen for cars and backing out efficient gasoline cars. Thus, from a climate perspective, in most parts of the world, the optimal use of nuclear energy, hydro-energy (falling water), wind or wave energy, solar thermal energy, geothermal energy, and photovoltaic energy, will be to provide electricity, as long as coal power (without CO_2 capture and storage) is still around. There will, of course, be special situations, such as Iceland, where the case for electrolytic hydrogen as a carbon emission reduction strategy may be compelling.

A wedge of nuclear power is its displacement of 700 modern 1000-megawatt coal plants. Today's stock of nuclear power plants is about half this large. Thus, if, over the next 50 years, today's global fleet of nuclear power plants were to be phased out in favor of modern coal plants, about half a wedge of additional CO_2 emissions reductions would be required to compensate. This half-wedge would not be required if current nuclear reactors were replaced with new ones, one-for-one[8].

36.8 Policy

Do the wedges have policy relevance? Can ideas based on wedges supplement the targets, trading, international assistance mechanisms, and other already identified elements of global carbon management? As already noted, the

---

[6] This and the following three paragraphs are found nearly verbatim on pp. 16–17 of (Socolow, R., et al., 2004b). See also (Socolow, 2005).

[7] The factor of two here, of course, depends on several assumptions. See the supporting online material in Pacala and Socolow, 2004.
[8] This paragraph is found nearly verbatim on p. 15 of (Socolow, R., et al., 2004b).

wedges framework encourages the planning of multiple parallel campaigns. Perhaps the Framework Convention's call for differentiated responsibilities across countries can be met in part by differentiated assignments for the commercialization of wedge technologies. In particular, wedge-based global carbon policy could reapportion some of the initiative in global carbon agreements in favor of a greater early role for developing countries. For example, a developing country already investing heavily in new capital stock could assume responsibility for commercializing the first stages of certain specific wedges. Compensation for first movers would be a collective responsibility.

Although champions of particular wedges are often unenthusiastic about other wedges, there is actually much common ground. Advocates of particular wedges, for example, might all agree on the following six principles:

1. It is already time to act.
2. It is too soon to pick 'winners'.
3. Subsidy of early stages is often desirable.
4. At later stages, markets help to determine the best wedges.
5. The best wedges for one country may not be the best for another.
6. The environmental and social costs of scale-up need attention.

Each specific wedge has benefits beyond its effect on climate. Rural development is positively affected by harnessing renewable energy, for example. Co-benefits may be crucial in eliciting the collaborations and coalitions necessary to achieve agreement on early action.

36.9 A World Transformed by Deliberate Attention to Carbon

If those alive today bring about the dramatic reductions in CO_2 emissions that appear to be our assignment for the next 50 years, the world will be so transformed that the options for the following 50 years will be myriad. Institutions for carbon management that reliably communicate the price of carbon will have become well entrenched. If wedges of nuclear power are achieved, strong international enforcement mechanisms to control nuclear proliferation will have emerged. If wedges of carbon capture and storage are achieved, a well-accepted permitting regime will have been created, governing the conversion of coal, oil sands, and perhaps methane clathrates to electricity and fuels. If wedges of renewable energy and terrestrial carbon sink management are achieved, land reclamation will have prospered. If hydrogen is widely used at small scale, in buildings and vehicles, ways to handle hydrogen safely will have been devised and the chicken-and-egg-like problems of establishing a hydrogen infrastructure will have been solved. If energy efficiency gains are large, urban space will have been used in new ways, and advanced technologies for buildings and vehicles will have

been widely deployed. A planetary consciousness will have become much more widespread.

Not an unhappy prospect!

Acknowledgements

The concepts of the stabilization triangle and the stabilization wedge emerged from my collaboration with Steve Pacala. These ideas were further developed with the help of colleagues at Princeton, including David Denkenberger, Jeff Greenblatt, Roberta Hotinski, Tom Kreutz, Harvey Lam, Michael Oppenheimer, Jorge Sarmiento, Bob Williams, and the late David Bradford. I am also indebted to Jeff Greenblatt for running the model that led to Figure 36.2. The current paper benefited from critical readings of our earlier writing on wedges by Marty Hoffert, Klaus Keller, Klaus Lackner, Chris Mottershead and Keywan Riahi and from the reading of a previous draft of this paper by Nebojsa Nakicenovic, Marty Hoffert, Tom Wigley, and an anonymous reviewer.

REFERENCES

Brown, M. A., M. D. Levine, J. P. Romm, A. H. Rosenfeld, and J. G. Koomey, 1998. 'Engineering-economic studies of energy technologies to reduce greenhouse gas emissions: Opportunities and challenges.' *Annual Review of Energy and the Environment, 1998.* Vol. 23, pp. 287–385.

Hoffert, M. I., et al., 2002. Advanced technology paths to global climate stability: energy for a greenhouse planet. *Science* **298**, 981–987.

IPCC, 2001. *Climate Change 2001: Mitigation. Contribution of Working Group III to the Third Assessment Report of the Intergovernmental Panel on Climate Change*, Cambridge University Press, Cambridge, UK.

Nakicenovic, N., et al., 2000: *Special Report on Emissions Scenarios: A Special Report of Working Group III of the Intergovernmental Panel on Climate Change*, Cambridge University Press, Cambridge, UK.

National Research Council, 2004. *The Hydrogen Economy: Opportunities, Costs, Barriers, and R&D Needs.* Washington, D.C., National Academy Press. http://www.nap.edu/books/0309091632/html/

Pacala, S., and R. Socolow, 2004. 'Stabilization wedges: Solving the climate problem for the next 50 years with current technologies.' *Science* 305: 968–972. The article and its detailed supporting online material are available at the website of Princeton University's Carbon Mitigation Initiative: http://www.princeton.edu/~cmi/

Riahi, K., and R.A. Roehrl, 2000a. Greenhouse gas emissions in dynamics-as-usual scenarios of economic and energy development, *Technological Forecasting and Social Change*, Vol. **63**, pp 175–206.

Riahi, K., and R.A. Roehrl, 2000b. Energy technology strategies for carbon dioxide mitigation and sustainable development. *Environmental Economics and Policy Studies* 3(2): 89–123.

Siegenthaler, U., and F. Joos, 1992. 'Use of a simple model for studying oceanic tracer distributions and the global carbon cycle', *Tellus*, 44B(3), 186–207.

Socolow, R., Pacala, S. and Greenblatt, J. 2004a. 'Wedges: Early Mitigation with Familiar Technology,' *Proceedings of the 7th International Conference on Greenhouse Gas Control Technologies*, (GHGT-7), September 5–9, 2004, Vancouver, BC, Canada.

Socolow, R., Hotinski, R., Greenblatt, J. and Pacala, S., 2004b. 'Solving the Climate Problem: Technologies Available to Curb CO_2 Emissions', *Environment*, Vol. 46, No. 10, pp. 8–19. December 2004.

Socolow, R. 2005. 'Can We Bury Global Carbon?' *Scientific American*, July 2005, pp. 49–55.

CHAPTER 37

Costs and Technology Role for Different Levels of CO_2 Concentration Stabilization

Keigo Akimoto and Toshimasa Tomoda
Research Institute of Innovative Technology for the Earth (RITE)

ABSTRACT: In order to evaluate the costs and technological options by region for various targets of CO_2 emission reduction, we have developed a world energy systems model that spans a century and comprises 77 regional divisions. We conducted several case studies for this evaluation using IPCC SRES A1 and B2 scenarios as baselines; and three CO_2 emission stabilization targets of 650, 550 and 450 ppmv for each baseline. Further, in order to evaluate regional differences, an additional case study was conducted, reflecting the current world situation of reduction in emissions. With regard to the 550 ppmv stabilization case under the SRES B2 scenario, we assumed that all Annex I countries except the US comply with the Kyoto target, the US achieves its target of emissions proportionate to GDP in 2010, and all the Annex I countries achieve the UK-proposal target after 2010, i.e. 61% and 77% reduction in 2050 and 2100, respectively. The analysis results also give the marginal cost of CO_2 emission reduction in 2100 under the SRES B2 scenario with the emission trading at about 100, 120, and 290 \$/tC for 650, 550, and 450 ppmv, respectively. However, the stabilization costs are more sensitive to the baseline; and CO_2 capture and storage is important to reduce the stabilization cost; further, cost-effective technological options differ between developed and developing countries.

37.1 Introduction

The official international discussion on the post-Kyoto regimes was beginning in 2005. In this discussion, the reduction target should be carefully examined from the long-term viewpoint of stabilizing atmospheric CO_2 concentration 'at a level that would prevent dangerous anthropogenic interference with the climate system'. In addition, it is now widely acknowledged that even a modest level of stabilization cannot be attained without emission reduction in developing countries. However, information on reduction costs and technological options for emission reduction is indispensable to determine the reduction targets for both developed and developing countries. This paper aims to contribute to this discussion by investigating cost-effective technological options for CO_2 concentration stabilization by using a world energy systems model of high regional resolution. High regional resolution, in general, is desirable for the global analysis because the cost of energy transportation is relatively high and dependent on the distance between the regions and because there exist large regional differences in energy demand growth, energy resources, energy technology level etc.

37.2 Energy Systems Model

This section briefly describes the model to be used in this study. The model, which we call DNE21+ [1], is a linear programming energy systems model of dynamic optimization type. Its timespan ranges up to the end of the 21st century with representative time points at 2000, 2005, 2010, 2015, 2020, 2025, 2030, 2040, 2050, 2075 and 2100.

To account for existing regional differences and evaluate the regional effects, this model divides the world into 77 regions: countries of interest are treated as independent regions, and large-area countries such as the US, Canada, Australia, China, India, Brazil and Russia are further disaggregated into 3–8 regions to consider transportation costs of energy and CO_2 in more detail. The total world cost of energy systems is minimized over the time period from 2000 to 2100.

The energy supply sectors are modelled from the bottom up (technology specific) and the end-use energy sectors from the top down (technology aggregated). Primary energy sources of eight types are explicitly modelled: natural gas, oil, coal, biomass, hydro and geothermal, photovoltaics, wind and nuclear. As technological options, various types of energy conversion technologies are explicitly modelled besides electricity generation. These include oil refinery, natural gas liquefaction, coal gasification, water electrolysis, methanol synthesis etc. The age of energy-conversion plants is taken into account. Five types of CCS (Carbon capture and storage) technologies are also considered: 1) injection into oil wells for EOR operation, 2) storage in depleted natural gas wells, 3) injection into coal-beds for ECBM operation, 4) sequestration in aquifers and 5) sequestration in ocean.

The end-use sector of the model is disaggregated into four types of secondary energy carriers: 1) solid fuel, 2) liquid fuel, 3) gaseous fuel, and 4) electricity. Electricity demand is expressed by the load duration curves characterized by four types of time periods, and the relationship between electricity supply and demand is formulated for each of the four periods. Future energy demand, in the absence of a climate policy, is exogenously provided by the

Table 37.1 Assumed fossil fuel potentials in the world.

	Anthracite and bituminous	Sub-bituminous		Lignite
Coal	424	208		253

		Conventional		Unconventional
	Remaining Reserves	Undiscovered (Onshore)	Undiscovered (Offshore)	
Oil	137	60	44	2,342
Natural gas	132	59	52	19,594

Unit: Gtoe (gigatons of oil equivalent).
Source: [2,3,4].

Table 37.2 Assumed facility costs and energy required for CO_2 capture, and potentials and costs of CO_2 storage.

	Facility cost (US$/(tC/day))[†]	Energy requirement (MWh/tC)[†]
CO_2 chemical recovery from coal fueled power	59,100–52,000	0.792–0.350
CO_2 chemical recovery from gas fueled power	112,500–100,000	0.927–0.719
CO_2 physical recovery from gasification plants	14,500	0.902–0.496

	Facility cost (US$/kW)[†]	Generation efficiency (% LHV)[†]
IGCC with CO_2 capture (physical recovery)	1,700–1,470	34.0–49.0

	Sequestration potential (GtC)	Sequestration cost ($/tC)[‡]
Oil well (EOR)	30.7	81–118[††]
Depleted gas well	40.2–241.5	34–215
Coal-bed (ECBM)	40.4	113–447[††]
Aquifer	856.4	18–143
Ocean	–	36[‡‡]

[†] Cost reduction and energy efficiency improvement are assumed to proceed with time; [‡] cost of CO_2 capture excluded; [††] the proceeds from recovered oil or gas excluded; [‡‡] the cost includes that of CO_2 liquefaction.
Source: e.g. [1,7,8].

energy type, region and time point as the reference scenario, and the model explores the energy supply system of the least cost. In cases of climate policy, energy savings take place and the final energy demands are reduced. The reductions in final energy demands are determined through the long-term price elasticity: we adopted an elasticity of -0.3 for electricity and -0.4 for the three non-electricity energies. The model again explores the energy supply system of the least cost, maintaining the relationship of long-term price elasticity between energy prices and demands.

The 77 regions are linked to each other by the trading of eight items: coal, crude oil, synthetic oil, methane, methanol, hydrogen, electricity, and CO_2. In addition, CO_2 emission permits are also modelled as an inter-regional trading item.

This model treats only CO_2 emission from fossil fuel combustions and not CO_2 emission reductions through biological sequestration.

37.3 Input Data Assumptions

Most of the assumed potentials of primary energy and CO_2 storage are based on GIS data, which are easily processed

to generate the corresponding potential for any one of the regions. Table 37.1 summarizes the world fossil fuel potentials assumed in the model. The world potentials of hydropower, wind power, and photovoltaics are assumed to be 14,400 [2], 12,000 and 1,271,000 TWh/yr, respectively. These potentials correspond to 94%, 78% and 8,300% of the world total electricity generation in the year 2000, respectively. The cost of hydropower is in the range of 20–180 $/MW based on the cost category of generation, as well as economically- and technically-exploitable capacities; and the cost of wind power and photovoltaics is in the 56–118 and 209–720 $/MWh range in the year 2000. Estimates of the potential and cost of wind power and photovoltaics are based on GIS data of wind speeds and solar radiation. The potential of biomass energy is estimated from the area of forest [5] by country and the accumulation rate by climate zone. The assumed world potential is about 3,970 Mtoe/yr and the supply costs range between 171 and 1,000 $/toe [6] depending on ease of access. Table 37.2 shows the assumed facility costs and the required energy of CO_2 capture technologies; it also summarizes the assumptions of the potentials and costs of CO_2 sequestration. Advances in technology are assumed exogenously. The cost reduction for all types of wind power and photovoltaics is 1.0% and 3.4% per year [9] up to the year 2050, respectively. Facility costs of fossil fuel power plants are derived from a report of NEA/IEA [10], and the assumed efficiencies of coal, oil and natural gas fuel power plants are 22–52, 20–60 and 24–62%-LHV. These ranges include regional differences and technology improvements up to the year 2050 when the technology improvement is assumed to be saturated. For example, the assumed efficiencies of coal fuel power plants in 2000 are 22–42% depending on the regions, and those in 2050 are 27–52%. Advances in CO_2 capture technologies are shown also in Table 37.2.

Future scenarios of population, reference GDP and reference final energy demands are derived from the A1 and B2 baseline scenarios of IPCC SRES [11,12]. Energy savings in end-use sectors are modelled in the top-down fashion using the long-term price elasticity. The elasticity of electricity and non-electricity energy savings is assumed to be −0.3 and −0.4, respectively.

37.4 Analysis Results and Discussion

To evaluate the costs and technological options for CO_2 concentration stabilizations, we adopted the three stabilization targets of 650 ppmv (S650), 550 ppmv (S550), and 450 ppmv (S450) by IPCC WG I [13] under the two baselines of SRES A1 and B2. The range of stabilization evaluated in this study corresponds to the range evaluated in the Chapter 2 of IPCC WGIII TAR [14].

We conducted an additional case study to consider regional differences under S550 on the SRES B2-base. For 2010, we assumed the Kyoto target for all the Annex I countries except the US – for which the target reduction in CO_2 intensity until 2010 is 2%/yr. post-2010, we assumed the UK's proposed target [15] for all the Annex I countries, i.e. a reduction in CO_2 emissions to 39% and 23% in 2050 and 2100 respectively, relative to that of 1990. The CO_2 emissions of the non-Annex I countries are constrained not to exceed the difference between the S550 emissions and the allowable maximum emissions of the Annex I countries. Further, the target for each developing country is determined such that the allocation among the Non-Annex I countries is proportional to their 1990 emissions.

37.4.1 *Costs for Different CO₂ Concentration Stabilization Levels and Different Baselines*

Figure 37.1 shows the marginal costs (CO_2 shadow prices) for atmospheric CO_2 concentration stabilization at 650, 550 and 450 ppmv with CO_2 emission trading under the SRES A1-/B2-base baseline. The marginal cost for the 650, 550 and 450 ppmv stabilization under the SRES B2-base is around 80, 90, and 150 \$/tC in 2050 and 100, 120 and 290 \$/tC in 2100, respectively. According to the analysis, the cost would be more sensitive to the baseline population, GDP and final energy demands — than to the concentration target. For example, the marginal cost even

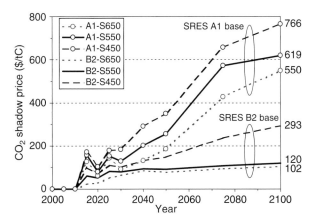

Figure 37.1 CO_2 shadow prices for CO_2 concentration stabilization at 650, 550, and 450 ppmv with CO_2 emission trading under SRESA1-/B2-base baseline.

for the 650 ppmv under SRES A1 is around 550 \$/tC in 2100, which is much higher than that for 450 ppmv under the SRES B2 baseline. Since the baseline is largely influenced by the assumed energy efficiency of the future, we should not forget, in the discussion of mitigation costs, the importance of future energy savings which are pursued regardless of climate change.

37.4.2 *Technology Role for Stabilization*

Figure 37.2 shows world primary energy production targets for achieving the 550 and 450 ppmv stabilizations with CO_2 emission trading, and Figure 37.3 shows world CO_2 emissions and the effects of sequestration. The selection of technologies is crucial to least-cost emission reduction. The importance of low-carbon fossil fuels, nuclear power and renewables increases with the lowering of stabilization levels. Energy savings are also important. The cumulative amount of CO_2 sequestration from 2000–2050 is 48 and 75 GtC for 550 and 450 ppmv, respectively; the amount from 2000–2100 is 270 and 360 GtC for 550 and 450 ppmv, respectively. CCS is one of the key technologies for the least-cost stabilization. RD&D of CCS is underway widely in the world and further efforts, especially on risk assessment, will help bring about wide deployments as the analysis results show.

37.4.3 *Regional Differences in the Kyoto Protocol plus UK-proposed Case*

In this section, regional differences in the costs and role of technology are discussed in the case of the Kyoto Protocol plus the UK proposal, under the stabilization of 550 ppmv with the SRES B2-base. Figure 37.4 shows the regional costs up to the year 2050. The cost varies widely across regions and the costs for EU, Japan, Canada, and New Zealand in 2010 are very high, primarily due to the effects of ageing infrastructure. The lifetime of the majority of average power plants is assumed to be 30 years and technical options for emission reduction are restricted in

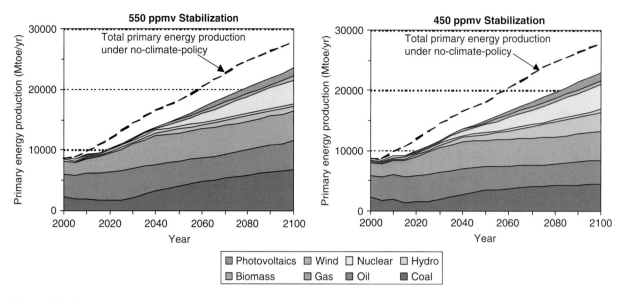

Figure 37.2 World primary energy production for CO_2 concentration stabilization at 550 and 450 ppmv with CO_2 emission trading.

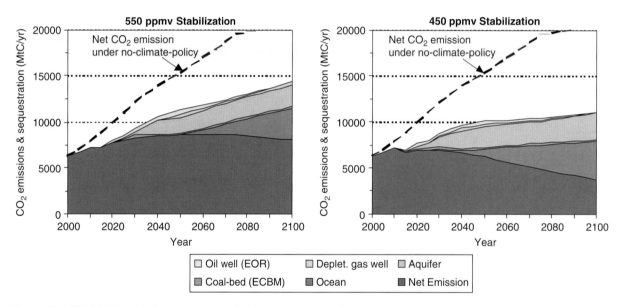

Figure 37.3 World CO_2 emission and sequestration for CO_2 concentration stabilization at 550 and 450 ppmv with CO_2 emission trading.

2010 because of the remaining lifetime of existing plants. The low marginal cost for the US in 2010 is due to its less stringent GDP-dependent carbon intensity target. Also, it is noteworthy that even a 550 ppmv stabilization cannot be achieved without emissions reduction in developing countries.

Figure 37.5 shows the CO_2 emission reduction effects of different technological options. The method of estimating the reduction effect due to each technological option is provided in [10]. Early application of CCS will be more cost-effective for stabilization without emission trading than with trading. No trading requires larger CO_2 emission reductions by developed countries, where energy-saving and renewable energy opportunities are generally limited. The emission-reduction effect of energy saving,

fuel switching among fossil fuels, nuclear power, renewables, and CO_2 geological and ocean sequestration in 2020 and 2050 is around 0.7, 0.2, 0.6, 0.4 and 1.0 GtC/yr and 1.8, 0.4, 1.1, 1.4 and 3.0 GtC/yr, respectively. The best portfolio contains a variety of technologies. This resulting technology mix is due to assumed cost supply curves that are modelled by multiple-step functions of different shape for each region, e.g. seven steps of production costs for conventional oil and gas, five steps of supply costs for hydro and geothermal, wind, and photovoltaics, considering the ease of access, though the DNE21+ is a linear programming model. Further, the explicit treatment of the load duration for electricity in the model also helps bring about the technology mix. For this reason, the portfolio even for one region contains a variety of technologies. In

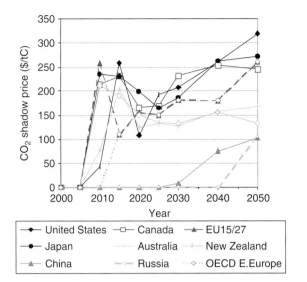

Figure 37.4 CO$_2$ shadow prices for CO$_2$ concentration stabilization at 550 ppmv with the Kyoto Protocol and UK-proposed target.

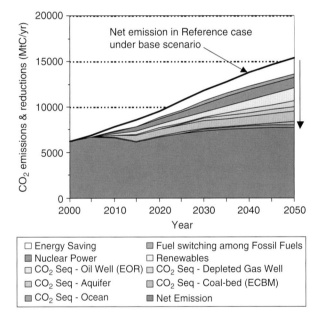

Figure 37.5 CO$_2$ emission reduction effects of technological options for the 550 ppmv stabilization with the Kyoto Protocol and UK-proposed target.

addition, the model has many regions and the cost-supply curve of a certain technology is different among the regions, which makes the world best portfolio more complicated.

In order to identify regional differences in the emission reduction options, energy intensity (primary energy consumption per GDP) and carbon intensity (CO$_2$ emission per primary energy consumption) are shown in Tables 37.3 and 37.4, respectively, for I) no-climate policy and II) 550 ppmv stabilization without emission trading. Energy intensity, particularly in developing countries, should

decrease from the viewpoint of achieving cost-effectiveness even in the no-climate-policy case. A moderate acceleration of energy intensity improvement would be required for most developing *and* developed countries in the stabilization case, as compared with the no-climate-policy case. On the other hand, naturally, while CO$_2$ intensity reduction is not necessitated for the no-climate-policy case, it is required more in developed countries than in developing countries for 550 ppmv stabilization. While CO$_2$ intensity reduction would bring fewer economic benefits, energy efficiency improvement will bring more. Economic competitiveness – particularly that of energy-intensive sectors – in developing countries relative to developed countries would increase more under this reduction regime than under the no-climate-policy regime.

37.5 Conclusion

An advantage of the model approach is that a variety of analysis results are self-consistent. In addition, the results of this study are consistent for the 77 regions because most of the assumed data are derived from the same databases. We conducted a number of case studies; one group is the concentration stabilization at different levels with the emission trading and the other one is a rather complicated emission constraint reflecting the current world situations with regard to the Kyoto Protocol and the UK proposal without emission trading.

The analysis results of the first group show that the marginal cost of CO$_2$ emission reduction under the IPCC SRES B2 baseline is 100, 120 and 290 \$/tC in 2100 for 650, 550 and 450 ppmv stabilizations, respectively; this increases with the lowering of the level of stabilization. The marginal cost, however, is more sensitive to the baseline; the cost even for 650 ppmv under SRES A1 is much higher than that for 450 ppmv under SRES B2. To achieve these stabilizations, a variety of technological options including CCS are required from the viewpoint of cost effectiveness.

The analysis results of the other case without emission trading show that the US enjoys a low marginal cost of emission reduction in 2010 when it achieves its target of carbon intensity of GDP, as compared with the other developed countries of Annex I, which comply with the Kyoto Protocol. The results also indicate that the reduction of energy intensity, i.e. energy saving, is required particularly for developing countries, while the reduction of carbon intensity, e.g. CO$_2$ capture and storage, is required particularly for developed countries. It can be stated that differences in reduction options between developed and developing countries would be more beneficial to the latter by making them more economically competitive under these emission reduction schemes.

These findings are valuable for exploring the effective measures of global warming mitigation after the Kyoto Protocol.

Table 37.3 Energy intensity for no-climate-policy and the stabilization at 550 ppmv with the Kyoto Protocol and UK-proposed target.

	Historical data			No-climate-policy			Stab. at 550 ppmv w.o. ET		
	1980	1990	2000	2010	2030	2050	2010	2030	2050
United Kingdom	1.41	1.14	1.00	0.84	0.81	0.77	0.79	0.79	0.76
United States	1.48	1.16	1.00	0.84	0.82	0.75	0.84	0.73	0.64
Canada	1.33	1.10	1.00	0.83	0.83	0.80	0.78	0.83	0.80
Japan	1.14	0.96	1.00	0.85	0.83	0.72	0.83	0.79	0.71
Russia			1.00	0.55	0.28	0.16	0.51	0.26	0.15
Brazil	0.93	0.95	1.00	0.99	0.75	0.60	0.93	0.73	0.52
Saudi Arabia	0.38	0.72	1.00	0.99	0.76	0.53	0.97	0.60	0.48
China	3.31	2.00	1.00	0.58	0.33	0.24	0.49	0.28	0.19
India	1.44	1.22	1.00	0.68	0.40	0.28	0.60	0.38	0.31
Indonesia	1.15	0.96	1.00	0.75	0.46	0.35	0.69	0.42	0.33

Unit: Y2000 = 1.00.

Table 37.4 Carbon intensity for no-climate-policy and the stabilization at 550 ppmv with the Kyoto Protocol and UK-proposed target.

	Historical data			No-climate-policy			Stab. at 550 ppmv w.o. ET		
	1980	1990	2000	2010	2030	2050	2010	2030	2050
United Kingdom	1.25	1.15	1.00	1.11	1.25	1.27	0.90	0.64	0.50
United States	1.05	1.02	1.00	1.15	1.24	1.16	1.00	0.44	0.26
Canada	1.06	0.98	1.00	1.10	1.26	1.22	0.77	0.38	0.22
Japan	1.14	1.05	1.00	1.14	1.40	1.48	0.82	0.53	0.33
Russia			1.00	0.92	0.99	0.91	0.94	1.01	0.68
Brazil	0.96	0.88	1.00	1.01	1.33	1.43	1.07	1.05	0.77
Saudi Arabia	1.19	1.12	1.00	1.09	1.08	1.05	1.09	0.86	0.57
China	0.89	0.99	1.00	1.06	0.98	0.91	1.04	0.93	0.73
India	0.67	0.87	1.00	1.10	1.05	0.99	1.07	1.02	0.98
Indonesia	0.65	0.78	1.00	1.03	0.98	0.88	1.03	0.97	0.93

Unit: Y2000 = 1.00.

Acknowledgements

This study is in part supported by New Energy and Industry Technology Development Organization (NEDO), Japan.

REFERENCES

1. Akimoto, K., et al., 2004. Role of CO₂ Sequestration by Country for Global Warming Mitigation after 2013. In: E.S. Rubin, D.W. Keith and C.F. Gilboy (Eds.), Proceedings of 7th International Conference on Greenhouse Gas Control Technologies. Volume 1: Peer-reviewed Papers and Plenary Presentations, IEA Greenhouse Gas Programme, Cheltenham, UK.
2. WEC, 2001. Survey of Energy Resources 2001 (CD-ROM). London: World Energy Council.
3. USGS, 2000. U.S. Geological Survey World Petroleum Assessment 2000–Description and Results. http://greenwood.cr.usgs.gov/energy/WorldEnergy/DDS-60/
4. Rogner, H-H., 1997. An assessment of world hydrocarbon resources. Annual Review of Energy and Environment, Vol. 22: 217–262.
5. FAO, 2002. Year Book of Forest Products 2000. Rome: FAO.
6. Akimoto, K., Tomoda, T., Fujii, Y., Yamaji, K., 2004. Assessment of Global Warming Mitigation Options with Integrated Assessment Model DNE21. Energy Economics, Vol. 26: 635–653.
7. David J. and Herzog, H., 2000. The Cost of Carbon Capture. In: Proceeding of 5th Conference of Greenhouse Gas Control Technologies. VIC: CISRO PUBLISHING. p. 985–990.
8. Hendriks, C.A. et al., 2001. Costs of Carbon Dioxide Removal by Underground Storage, In: Proceedings of 5th International Conference on Greenhouse Gas Control Technologies. VIC: CISRO PUBLISHING. p. 967–972.
9. EPRI/DOE, 1997. Renewable Energy Technology Characterizations. EPRI Topical Report TR-109496. CA: EPRI.
10. NEA/IEA, 1998. Projected Costs of Generating Electricity – update 1998, Paris: OECD.
11. Nakicenovic, N. et al. (Eds), 2000. Special Report on Emissions Scenarios. Cambridge: Cambridge University Press.
12. Task Group on Scenarios for Climate Impact Assessment (TGCIA), 2002. Socioeconomic Data for TGCIA. http://sres.ciesin.columbia.edu/tgcia/hm.html.
13. IPCC WGI, 1996. Climate Change 1995: The Science of Climate Change. Cambridge: Cambridge University Press.
14. IPCC WGIII, 2001. Climate Change 2001: Mitigation. Cambridge: Cambridge University Press.
15. Defra, 2003. The scientific case for setting a long-term emission reduction target. http://www.defra.gov.uk/

CHAPTER 38

Avoiding Dangerous Climate Change by Inducing Technological Progress: Scenarios Using a Large-Scale Econometric Model

Terry Barker[1], Haoran Pan[1], Jonathan Köhler[1,2], Rachel Warren[2] and Sarah Winne[2]

[1] *Faculty of Economics, University of Cambridge*
[2] *Tyndall Centre, University of East Anglia*

ABSTRACT: This paper addresses the question of the costs of stabilising atmospheric concentrations of carbon dioxide at three levels 550 ppm, 500 ppm and 450 ppm, with the emissions modelled to 2100. Two policy instruments are used to achieve these targets: emission trading permits for the energy industries and carbon taxes for the rest of the economy, with the revenues recycled to maintain fiscal neutrality. These are applied at escalating rates in 2011–2050 to allow for early action under the UNFCCC. Extra investment is induced by the permit schemes and taxes since they lead to substantial increases in the real cost of burning fossil fuels according to their carbon content. This prompts a switch to low-carbon technologies. The ensuing world-wide wave of extra investment over the century to 2100 raises the rate of economic growth, which is endogenous in the model. There is a purely economic benefit in stabilisation, although small, which increases with more demanding targets.

This finding complements the literature showing reductions in the modelled costs of achieving stabilisation when induced technological change (ITC) is taken into account, but which generally assume that GDP is largely exogenous. The approach is novel in the treatment of technological change within long-term economic models since it is based on the theory of demand-led growth and a panel-data analysis of the global energy system 1970–2001 using formal econometric techniques and thus provides a different perspective on stabilisation costs. In particular, a sectoral and regionally specific analysis is presented using the model E3MG (energy-environment-economy model of the globe), coupled to the simple climate model MAGICC (Wigley and Raper, 1997), which are both components of the Community Integrated Assessment System (CIAS) of the UK Tyndall Centre.

38.1 Introduction

As part of the research programme on Integrated Assessment at the Tyndall Centre, a world macroeconomic model (E3MG environment-energy-economy global) is being developed to investigate policies for climate change, as a module of an Integrated Assessment Modelling system or IAM. In coupling economic models with meteorological and atmospheric chemistry models of climate change, long timescales are necessary because changes in CO_2 concentrations enhance the greenhouse effect over time periods of 50–100 years and more.

In projecting the future, the approach of this paper is first to consider the past. Looking back over the last 200 years, the socio-economic system seems to be characterized by ongoing fundamental change, rather than convergence to any equilibrium state. Maddison (2001) takes a long view of global economic growth over the last millennium. He finds growth rates to be very different across countries and over time, and ascribes the comparatively high rates of growth to technological progress and diffusion. He argues that the increase in growth rates that emerged in Europe since 1500, and that became endemic from 1820, were founded on innovations in banking and accounting, transport and military equipment, scientific thinking and engineering. He also finds that inequalities between nations in per capita GDP have increased (in particular since World War 2) and not diminished over time. These three features of growth (technological progress, diversity across nations and time periods, and increasing inequalities) are evident in the solutions underlying the scenarios reported below.

These ideas are supported by quantitative studies identifying the causes of economic growth. Technological progress associated with investment is intimately related to Denison's (1967, 1985) causal factors[1] (capital, economies of scale and knowledge) accounting for 57% of growth. More recently, Wolff (1994a, 1994b) has found strong correlations between investment embodying technological change and growth in OECD economies.

---

[1] Denison's study of US growth 1929–1982 attributes the average long-run rate of about 3% pa to six factors: about 25% to labour at constant quality, about 16% to improvements in labour quality as from education, 12% to capital, 11% to improved allocation of resources, e.g. labour moving from traditional agriculture to urban manufacturing, 11% to economies of scale and 34% to growth of knowledge.

Technology is important for climate change analysis for two reasons: first, technological progress is implicated in anthropogenic climate change; and second, a change to a low-carbon society will require widespread development and mass deployment of new, low-carbon technologies. Such large-scale changes have been a feature of 'advanced' society in the last 200 years. The industrial revolution of the first part of the 19th century was founded on burning coal as never before. The use of motor vehicles and aircraft, powered by oil products, is still diffusing through the world, providing the most serious challenge for policy to reduce the rate of climate change. Both coal and oil were essential to the transformations of economies and societies.

In modelling this process, we have combined an econometric, long-run model of demand-led growth with an energy technology model to derive the costs of moving from a baseline to stabilisation targets of 550, 500 and 450 ppmv CO_2, using the instruments of emission permit trading and carbon taxes, with and without incorporation of endogenous technical change in the model. At this stage only preliminary estimates of the stabilisation costs for particular concentration targets are provided, since we have yet to carry out an uncertainty analysis of both the economic and climate models which is required to provide estimates of the potential ranges of mitigation benefits and costs for each level of stabilisation. For example, the use of different parameters in the representation of the carbon cycle in the MAGICC climate model would strongly affect whether or not the actual scenario runs used here would deliver stabilisation of CO_2 concentrations in 2100. Furthermore, a wider range of policies than carbon taxes and permit trading might be considered useful. The purpose of the paper is thus not to provide specific estimates of stabilisation costs and outcomes, although we do so, but to use a novel treatment of the economy and innovation processes to illustrate the influence of endogenous and induced technical change on estimates of stabilisation costs and benefits.

38.2 Modelling Economic Growth, Technological Change and the Costs of Stabilisation

In modelling long-run economic growth and technological change, we have followed the "history" approach[2] of cumulative causation (Kaldor, 1957, 1972, 1985; Setterfield, 1997, 2002), which focuses on gross investment (Scott, 1989) and trade (McCombie and Thirlwall, 1994, 2004), in which technological progress is embodied in gross investment. Long-run growth and structural change through socio-technical systems, called 'Kondratiev

waves', are described by Freeman and Louçã (2001) and Geels (2002) and modelled by Köhler (2005). Kondratiev waves characterise long-term economic development, and embody changes in economic structures which have major impacts on the forms of energy use. The invention of the steam engine and the subsequent industrial revolution, and of the internal combustion engine and subsequent diffusion of motor cars, are two obvious examples. Growth in this approach is thus dependent on waves of investment in new technologies.

Grubb, Köhler and Anderson (2002) explain that many energy-environment-economy (E3) models do not incorporate induced technical change[3] (ITC), but instead use the older concept of technology as exogenous 'manna from heaven'. A meta-analysis of costs of mitigation (Barker et al., 2002) also found that technological change in the post-SRES models (Morita et al., 2001) is treated largely as exogenous to the system. Hourcade and Shukla (2001) review modelling studies of costs of stabilisation in post-SRES mitigation scenarios from top-down general economic models[4] and report the results of a model comparison study (pp. 548–9). They identify widely-differing costs of stabilisation at 550 ppmv by 2050 of 0.2–1.75% GDP, mainly influenced by the size of the emissions in the baseline. Hourcade and Shukla (2001, pp. 550–552) explain that a critical factor affecting the timing and cost of cost-effective emission abatement in the model results is the treatment of technological change. The studies incorporating ITC suggest that it could reduce stabilisation costs substantially: ITC greatly broadens the scope of technology-related policies and usually increases the benefits of early action, which accelerates development of cheaper technologies. This is the opposite of the result from models with exogenous technical change, which can imply waiting for better technologies to arrive.

More recent work seems to confirm these findings. For example, Manne and Richels (2004) and Goulder (2004) also found that ITC lowers mitigation costs and that more extensive reductions in GHGs are justified than with exogenous technical change. Nakicenovic and Riahi (2003) noted how the assumption about the availability of future technologies was a strong driver of stabilisation costs. Edmonds et al. (2004) studied stabilisation at 550 ppmv CO_2 in the SRES B2 world using the MiniCAM model and showed a

---

[2] This is contrast to the mainstream equilibrium approach adopted in most economic models of the costs of climate stabilisation. See (DeCanio, 2003) for a critique and (Weyant, 2004) for a discussion of technological change in this approach. Barker (2004) compares the equilibrium with a 'space-time' economics approach to modelling mitigation.

[3] In the models, exogenous or autonomous technological change is that which is imposed from outside the model, usually in the form of a time trend affecting energy demand or the growth of world output. If, however, the choice of technologies is included within the models and affects energy demand and/or economic growth, then the model includes endogenous technological change (ETC). With ETC, further changes can generally be induced by economic policies, hence the term induced technological change (ITC); therefore ITC implies ETC, as assumed throughout the rest of this paper.

[4] Bottom-up models use systems engineering and technology-specific data to represent emissions in different sectors of the economy, whilst top-down models calculate macroeconomic quantities typically represent the different technologies in use through relatively aggregated production functions for each sector of the economy.

reduction in costs of a factor of 2.5 in 2100 using a baseline incorporating technical change. Edmonds considers that advanced technology development to be far more important as a driver of emission reductions than carbon taxes. Van Vuuren et al. (2004) also concluded that technology development is a key in achieving emission reductions as a result of carbon taxes: omitting technology development reduced the efficacy of the carbon tax by 50% in their model. Weyant (2004) concludes that stabilisation will require development on a large scale of new energy technologies and that costs would be reduced if many technologies are developed in parallel and there is early adoption of policies to encourage technology development.

The results from the bottom-up energy-engineering literature give a different perspective. Following the work in particular of IIASA (e.g. Grübler et al., 1999), models investigating induced technical change emerged during the mid- and late 1990s. These models show that ITC can alter results in many ways. Nakicenovic and Riahi (2003) also note the great significance of the choice of baseline scenario in driving stabilisation costs. However, this influence is itself largely due to the different assumptions made about technological change in the baseline scenarios. In an intriguing and path-breaking finding, Gritsevskyi and Nakicenovic (2000) using the MESSAGE model were able to identify some 53 clusters of least-cost technologies allowing for endogenous technological learning and projecting global CO_2 emissions to 2100. The outcomes modelled under uncertainty were strongly grouped into two sets of high and low emission scenarios (p. 909) 'demonstrating a kind of implicit bifurcation across the range of possible emissions' (see Figure 38.1). Since the scenarios are all similar in cost, this suggests that a decarbonised economy may not cost any more than a carbonised one if technology learning curves are taken into account – a general finding that is supported by the results presented

below. Other key findings are that there is a large diversity across alternative energy technology strategies, a finding that also emerges below, and that it is not possible to choose an 'optimal' direction of energy-system development (p. 920). The IPCC Third Assessment Report on such modelling suggests (Watson et al., 2001 p. 109) that up to 5 GtC a year reduction by 2020 (some 50% of baseline projections) might be achieved by current technologies, half of the reduction at no direct cost, the other half at direct costs of less than \$100/tC-equivalent. This does not include new technologies and there is no reason not to expect that the savings would continue as real costs of carbon rise. Pacala and Socolow (2004) argue that the portfolio of available technologies is large enough to solve the climate problem by 2050 without revolutionary new technology, although they do not put a cost on action or explain what incentives are necessary.

38.3 The Approach to Modelling the Economy and Technological Change

The contribution that this paper makes is to introduce a novel approach to the modelling of technological change in the literature on the costs of climate stabilisation and embed this in a macro-econometric model. The theoretical basis of the approach is that economic growth is demand-led and supply-constrained[5]. The direction of causation is as follows: (1) Climate policies lead to higher productive investment; then (2) the higher investment leads to higher output and growth in the short term; and (3) the higher actual short-term growth then leads to higher long-term growth by raising the productive potential of the global economy. The theory requires that the extra investment is an outcome of market forces, not imposed by government, i.e. it has to be induced, profitable and integrated within the system so that it is part of the widening and deepening of markets that is intrinsic to economic growth. The steps in the argument can be elaborated as follows.

1. *Climate policies lead to higher productive investment.* In the long run all physical capital wears out, becomes obsolete or too expensive to maintain, so it has to be replaced. The broad issue for climate policy concerns the scale and nature of replacement low-carbon capital compared to replacement traditional carbon capital in the long run, with the mix between the two determined in the business investment decision by the real price of carbon, and this price in turn being affected

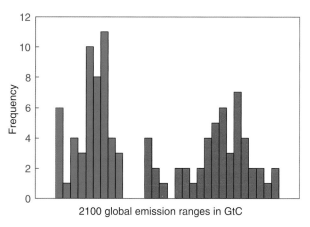

Figure 38.1 Future technological clusters: global emissions 2100.
Source: Gritsevskyi, A. and Nakicenovic, N. (2000), p. 909.
Note: Each bar represents the percentage frequency in ranges covering over 13,000 'optimal' scenarios from 53 different technological dynamics, in yielding the 2100 outcome for emissions. All the bars add to 100%.

[5] See (Setterfield, 1997, 2002) for reviews of post-Keynesian theory of growth; see (Palley, 2003) for a discussion of how long-run supply is affected by actual growth. The modern theory of demand-led growth begins with Harrod's knife-edge model (1939, 1948) and was developed by Kaldor (1957, 1972, 1985), Scott (1989) and McCombie and Thirlwall (1994) among others. The supply-side modern theory of economic growth (see Aghion and Howitt, 1998) goes back to Solow (1956, 1957), with endogenous growth theory developed by Romer (1986, 1990).

by policy. Obviously, fossil-fuel stations with carbon capture require more investment in relation to output than the conventional stations they replace; and hybrid or fuel-cell vehicles require more capital than conventional vehicles. However, the fuels and other inputs used also require investment, so the outcome could go either way. In the event, we found that the higher the carbon price, the higher the global investment, i.e. low-carbon production of energy in the global system is more capital-intensive than high-carbon production. The potential for learning-by-doing is also higher for low-carbon capital, and this affects the rate of adoption of these technologies and the potential for faster economic growth (see below). The literature (e.g. McDonald and Schrattenholzer, 2001) suggests that currently the fossil-fuel technologies are further down their learning curves, and so have fewer economies of specialisation and scale to be reaped in the future than low-carbon technologies. In addition, fossil-fuel inputs may face more severe decreasing returns to scale as the most profitable fuel reserves get used first, compared to those for low-carbon technologies many of which are just beginning to utilise their potential reserves (e.g. utilisation of wind, tides and waves, and of waste).

2. *The higher investment leads to higher output and growth in the short-term.*

   This is partly the usual Keynesian multiplier effect, but at a global scale. The world economy being closed, the leakage into imports from extra domestic investment leads to an increase in world exports. The exports in turn lead to more output, investment, and consumption in the exporting countries and demand continues to grow.

3. *The higher actual short-term growth then leads to higher long-term growth by raising the productive potential of the global economy.*

   The short-term growth becomes long-term because the higher output of the engineering industries (those most immediately affected, but the extra demand is soon diffused across all industries) brings about higher supply in at least four ways.

   i.   Technological progress is accelerated. The climate change policies lead to a transformation of two systems: electricity production and transportation. Both systems have much greater interactions and effects on the modern economy than their fossil-fuel use alone would suggest, so it is not unreasonable to argue that changing their technologies would have accelerating effects on technological change in general. This is the technological cluster argument – advances in one area, e.g. fuel cells, would spill over into many other areas than the originating purpose, e.g. transportation.

   ii. There is ample evidence of increasing returns to scale and the associated specialization as the market expands (this does not have to involve technological

progress, although the two strongly interact). This phenomenon has long been recognized in economics as an explanation of growth in supply from Adam Smith to modern growth theory.

   iii. The paper introduces an additional effect on exports, that of R&D and investment. This is a supply-side effect on export demand, which eases the balance of payment constraint on growth. The literature on the effect is clear on the direction of causation: R&D and R&D-inspired investment in the exporting country leads to higher exports[6], with other effects on exports, such as activity in the importing country and the relative prices and costs of the exports taken into account.

   iv. Finally, demand-led growth increases the quantity and quality of the labour supply and encourages its re-allocation to more productive sectors. Participation rates of labour in the working-age population increase, more labour moves from traditional, usually rural, sectors into modern sectors of the economy, especially in developing countries. It is clear from the data that these are long-term processes.

This approach involves the use of econometric estimation to identify the effects of technological change on exports and energy demand and embed these in a large post-Keynesian non-linear simulation model. The modelling has evolved from the work of the Cambridge Growth Project, when it developed a dynamic version of its UK model (Barker and Peterson, 1987). The effects of induced technological change modelled this way turn out to be sufficiently large in a closed global model to increase slightly the long-run growth of the system.

The effect of investment and R&D on export performance, which drives our long-run results on endogenous technological change, goes back to Posner's technological gap theory (1961). Since the 1990s it has been the topic of substantial empirical research, especially for UK and German trade, and has been found for different countries and regions at the individual plant,

---

[6]There is a two-way causation between exports and investment, but the estimations of E3MG export equations and those in the literature (e.g. León-Ledesma, 2000) allow for both spurious correlations (by using cointegration techniques) and the independence of the explanatory variables (by using instrumental variables or some other means). Causation is very difficult to prove in macroeconomic behaviour, but there are convincing results at the micro level. Two independent, explicit and thorough studies of the direction of causation from innovation to export performance using German microeconomic data come to the unambiguous conclusion that the direction of causation is from R&D innovation to export volumes (Ebling and Janz, 1999; Lachenmaier and Wößmann, 2005). However these studies are concerned with R&D expenditure and other measures of innovation rather than R&D-enhanced gross accumulated investment, the indicator of technological progress adopted in E3MG. There is a close relationship between market R&D expenditures and gross investment so it is very difficult to distinguish separate effects in empirical work.

industrial sector and macro economy levels.[7] The underlying hypothesis is that higher investment and/or R&D is associated with higher quality and innovatory products and therefore exports, and that this leads to higher demand for the exports. (For the demand to be effective in the long run, there must also be an increase in supply, which is realised by economies of specialisation and scale in production and higher employment and labour productivity.)

The approach has been developed to include a regionalised version of the bottom-up technology ETM model within the top-down macroeconomic model, E3MG. Thus, like the WGBU study (Nakicenovic and Riahi, 2003) and that of (McFarland et al., 2004) which are also based on the linkage of top-down and bottom-up models, our modelling approach avoids the typical optimistic bias often attributed to a bottom-up engineering approach, and unduly pessimistic bias of typical macroeconomic approaches. The advantages of using this combined approach have recently been reviewed (Grubb, Köhler and Anderson, 2002).

This modelling explains how low-carbon technologies are adopted as the real cost of carbon rises in the system, with learning-by-doing[8] reducing the unit costs of the technologies as the scale of adoption increases. A rise in the costs of fossil fuels resulting from increases in CO_2 permit prices and carbon taxes thus induces extra investment in low-carbon technologies, and this is shown to be larger and earlier than the investment in fossil technologies in the baseline. The carbon tax revenues and 50% of the permit revenues are assumed to be recycled in the form of reductions in other indirect taxes on consumers. The outcome is that the extra investment and implied accelerated technological change in the stabilisation scenarios leads to extra exports and investment more generally, and higher economic growth. The literature on the economics of stabilisation has been dominated by issues of efficient allocation of resources, rather than sources of growth, and has focussed on economic costs rather than benefits, and, as Azar and Schneider (2002) point out, occasionally exaggerating them. If the economic issue becomes whether mitigation policies might lead to higher growth, then mitigation policies may be seen to provide net economic benefits, so that investment-led climate

policies enhance economic development, albeit with only small but positive effects on economic growth.

The approach requires a set of assumptions to reduce the complexity of the problem. The main ones adopted for the modelling and for the results reported below are as follows:

1. Population growth and migration is exogenous at CPI baseline levels (see below), and the assumption is adopted of sufficient labour being available from productivity growth or structural change to meet the demand for products.
2. Monetary and fiscal policy. Independent central banks are assumed to hold the rate of consumer price inflation constant.
3. The econometric equations in the model are reduced to two sets: energy and export demand. Except for investment by the electricity and vehicles industries, other behavioural equations are treated as being in fixed proportions to their main determinants.
4. The emission permit scheme and the carbon taxes have their effects in raising prices of energy products in proportion to their carbon content where ever they are imposed, and revenues are recycled as reductions in indirect taxes to maintain inflation neutrality. The high rates required, especially for 450 ppm, may prove impractical, if not politically impossible. Thus the scenarios show how high the emission prices and tax rates have to rise to achieve the targets.

The top-down model, E3MG, is a 20-region, structural, annual, dynamic, econometric simulation model based on data covering the period from 1970–2001, and projected forward to 2100. The database contains information about the historic changes by region and sector in emissions, energy use, energy prices and taxes, input-output coefficients, and industries' output, trade, investment and employment. This is supplemented by data on macroeconomic behaviour and bi-lateral trade. These data are used to estimate a set of econometric equations using cointegration techniques proposed by Engel and Granger (1987) and developed by Abadir (2004) for neo-Keynesian theories of markets, which do not clear through prices alone and with long-run solutions which are not necessarily in equilibrium. E3MG requires as inputs dynamic profiles of population, energy supplies, baseline GDP, government expenditures, tax and interest and exchange rates; and it derives outputs of carbon dioxide and other greenhouse gas emissions, SO_2 emissions, energy use and GDP and its expenditure and industrial components. The model covers 12 energy carriers, 19 energy users, 28 energy technologies and 42 industrial/product sectors.

The emphasis in the modelling in this paper is on two sets of estimated equations[9] included in the model: aggregate

---

[7] For the plant level results, see (Roper and Love, 2001); for industries, see (Greenhalgh, 1990; Greenhalgh et al., 1994; and Wakelin, 1998); for the macro-economy, see (Magnier and Toujas-Bernate, 1994; Fagerberg, 1988 and León-Ledesma, 2000).

[8] The modelling of technological change through learning-by-doing or R&D/accumulating knowledge or both is reviewed by Scott (1989) and Ruttan (2002). The model assumes that learning-by-doing is prevalent, and that all R&D must be associated with gross investment for it to be effective, i.e. R&D is not an independent source of technological change (see Schmookler, 1966 and Scherer, 1982). If R&D were independent, then a policy of funding research directly, e.g. into carbon capture or nuclear fusion, may be preferable to inducing technological progress through raising the real price of carbon through climate policy. Since we are uncertain about which is effective, there is a case for both, but if real carbon prices remain low there will not be the incentive to introduce low-carbon technologies even if they are discovered.

[9] In technical terms, these sets of equations have been estimated by instrumental variables in a cointegrating general-to-specific framework (see Barker and DeRamon, 2005, for details), assuming a long-run relationship that can be projected over the next 100 years.

energy demand by 19 fuel users and 20 regions and exports of goods and services by 41 industries and 20 regions. Each sector in each region is assumed to follow a different pattern of behaviour within an overall theoretical structure, implying that the representative agent assumption (that all the observations can be assumed to come from the same underlying distribution) is invalid. This means that we are assuming that the behaviour of each sector-region is not assumed to be the same as that of the average of the group. In order to represent long-term structural change in economies, the model also includes assumptions about how the input-output coefficients might change in the long term. These changes were assessed by considering the role of new technologies over the next 50 years (Dewick et al., forthcoming), although this feedback is not included in the scenarios reported here and the coefficients are left at baseline levels.

The bottom-up model is an annual, dynamic technology model, referred to here as the ETM model (Anderson and Winne, 2004). It is based on the concept of a price effect on the elasticity of substitution between competing technologies. Existing economic models usually assume constant elasticities of substitution between competing technologies. Although the original ETM is not specifically regional and is not estimated by formal econometric techniques, it does model, in a simplified way, the switch from carbon energy sources to non-carbon energy sources over time. It is designed to account for the fact that a large array of non-carbon options is emerging, though their costs are generally high relative to those of fossil fuels. However, costs are declining relatively with innovation, investment and learning-by-doing. The process of substitution is also argued to be highly non-linear, involving threshold effects. The ETM models the process of substitution, allowing for non-carbon energy sources to meet a larger part of global energy demand as the price of these sources decrease with investment, learning-by-doing, and innovation.

One component of the ETM is the learning curve (McDonald and Schrattenholzer, 2001). The importance of including a learning curve in the model cannot be underestimated, as the technology costs do not simply decline as a function of time, but decrease as experience is gained by using a particular technology. As investment is made in 'new' technologies, learning takes place and the cost of the new technology lowers so that it becomes competitive with the 'old' technologies. For each type of energy demanded there is usually a technology or fuel of choice – what might be termed a 'marker' technology – against which the alternatives will have to compete. In the ETM, the total capital and operating costs of using this fuel per unit output are used as a basis or marker for expressing the relative costs of the alternatives. Even though the marker technology may comprise the majority of the market, there are always so-called niche markets and opportunities where the non-carbon technology is cheaper than then marker. ETM provides a simple model of the process of switching from a marker technology to the possible

substitutes. This substitution process may be accelerated if a carbon tax is implemented.

The econometric approach used here also is not limited by the assumptions necessary in those macroeconomic models which handle inter-temporal optimisation. In order to do so, such models have to assume, *inter alia*, that the social planner has perfect foresight with no uncertainty, and that perfectly functioning markets exist.

38.4 Derivation of Pathways and Scenarios to 2100

The Common POLES-IMAGE (CPI) baseline has been taken as a starting point (Van Wuuren et al., 2004). This baseline itself derives from the IMAGE IPCC SRES A1B and B2 baselines. CPI assumes continued globalisation, medium technology, continued development, and strong dependence on fossil fuels. Population follows the UN medium projections for 2030, and the UN long-term medium projection between 2030 and 2100. Further details may be found in (van Vuuren et al., 2004). This baseline is used for the population assumptions of E3MG and projections made for government expenditures and per capita household consumption have been made for each region assuming the average growth rate will slow after 2050. With other components of GDP endogenous in the model, GDP (in $ at year 2000 prices and exchange rates) is calculated. Economic growth is near the historic average at 2.3% p.a. in 2000–2100, with higher rates to 2050 and lower rates thereafter. The solution includes for each region, sectoral output, employment, energy use and prices and emissions. It is used to provide two baseline sets of carbon dioxide emissions, one in which E3MG and ETM allow technological change as a projection of the estimated effects and through learning by doing, and another in which they do not. In both, no new permit schemes or carbon taxes are applied.

Three stabilisation scenarios are used in this study, selected to span the range adopted for international model comparison studies, in which carbon dioxide concentrations stabilise at 450, 500 and 550 ppmv by 2100. Cumulative emissions of CO_2 to 2100 are derived from the MAGICC model (Wigley and Raper, 1997) as used by the IPCC (Watson et al., 2001). The E3MG model is then used to derive a cost-effective emission pathway which keeps cumulative emissions within these limits prescribed by the MAGICC model. Costs of stabilisation are then calculated relative to the baseline. Many other studies of stabilisation costs (e.g. Nakicenovic and Riahi (2003), Van Vuuren et al., 2004) also use the MAGICC climate model (Wigley and Raper, 1997) to represent the relationship between emissions and concentrations. It is a set of linked reduced form models emulating the behaviour of a GCM. It consists of coupled gas-cycle, radiative forcing, climate and ice-melt models integrated into a single package. It calculates the annual-mean global surface air temperature and sea-level implications of emission scenarios for greenhouse gases and sulphur dioxide. Although MAGICC and E3MG both model emissions scenarios detailing non-CO_2 greenhouse gases, we do not consider the costs of reducing these gases and their effects in this first analysis[10].

These emission scenarios are also subject to exogenously defined dates at which countries join in permit and carbon tax schemes. By default the permit scheme covers the energy industries only (electricity supply, the fossil fuel and energy-intensive sectors covering metals, chemicals, mineral products and ore extraction) and starts at small values from 2011, which are assumed to escalate in real terms[11] until 2050, then stay constant again in real terms until 2100. This stylised profile has been designed to illustrate the implications of emissions targets to 2050 that have been adopted by several governments (e.g. the UK's 60% target); it divides the projection period into 2010–2050, when policy leads to changes in real carbon prices, and 2050–2100, when the system responds to these changes. 50% of the permits are allocated freely to the energy users on the basis of their past emissions (grandfathering) and the rest are auctioned (this rule is adopted to prevent excessive profits in the energy sectors from the sale of permits under conditions in which these industries have market power because they have a large share of regional electricity generation[12]). The CO_2 emissions from the rest of the economy are assumed to be covered by a carbon tax at the same rate as the permit prices. The revenues are assumed to be recycled in each region independently. The auction revenues are used along with the revenues from carbon taxes to reduce indirect taxes in general (such as the USA's sales taxes or the EU's Value Added Tax) as the instrument to help maintain general price stability.

38.5 Results for Alternative Mitigation Policies

These assumptions are essentially profiles for the rates required to achieve the stabilisation targets. The profiles

Table 38.1 CO_2 Emission permit prices and carbon tax rates $(2000)/tC.

Scenario	2020	2030	2040	2050–2100
	No ITC			
550 ppmv	34	68	102	136
500 ppmv	55	110	165	220
450 ppmv	170	340	510	680
	ITC			
550 ppmv	15	30	45	60
500 ppmv	24.9	49.8	74.7	99.6
450 ppmv	100	200	300	400

Source: E3MG2.1sp2.

Table 38.2 CO_2 and GHG emissions 2000–2100.

	CO_2 Gt-C per year			
Scenario	2000	2020	2050	2100
Baseline	7.7	11.5	13.2	13.0
550 ppmv	7.7	10.8	10.5	4.7
500 ppmv	7.7	10.4	9.0	2.2
450 ppmv	7.7	9.0	5.7	0.9
	GHG Gt-C-eq			
450 ppmv	11.1	11.5	8.4	3.6

Source: E3MG2.1sp2.

are raised or lowered in proportion to reduce CO_2 emissions sufficiently to achieve the targets. The rates assuming induced technological change are shown to 2100 in Table 38.1 together with the outcomes for CO_2 in Table 38.2 and GDP in Table 38.3. The results are tabulated for the baseline and three stabilisation scenarios, with the results also shown with and without induced technological change, using the permit and carbon tax rates to achieve stabilisation[13].

There are three features worth mentioning about the rates. First, the inclusion of endogenous technological change reduces the rate substantially, almost by half, in all the scenarios. Second, there are the modest levels required for the 550 ppmv target with ITC, with permit prices starting at $1.5/tC in 2011 and rising to $15/tC by 2020. These rates are sufficient to increase energy efficiency appreciably and shift the electricity system to a mixture of low-carbon options including renewables, coal and gas with sequestration, and nuclear depending on region and local conditions. Third, the rates for the 500 ppmv target are only slightly above those for the 550 ppmv target. The reason is that the small increase is a sufficient incentive to cause the conversion from gasoline to electric vehicles

---

[10] If the CO_2 emission pathway does not result in stabilisation in the full integrated analysis, policy parameters are adjusted in E3MG until a consistent solution is achieved. We judged (see Figure 38.3 below) that the concentrations projected by MAGICC were sufficiently close to the targets given the uncertainties for the conclusions of the paper to hold.

[11] E3MG does not optimise global welfare, or the time profile of emission allowance prices, so a choice of a time profile of rates by assumption is necessary. Rates escalate because there is a presumption that announcement effects reduce costs (Watson et al., 2001; Goulder, 2004).

[12] Barker and Rosendahl (2000) in a study of ancillary benefits of GHG mitigation in the EU find that free allocation of permits leads to large profits in the energy industries, compared to the baseline, but profits can be maintained in the long run if only 50% of the permits are allocated freely and 50% are auctioned (p. 21). Goulder has also addressed this issue, using a general equilibrium approach. A recent paper concludes: "Under a wide range of parameter values, profits can be maintained in both "upstream" (fossil-fuel-supplying) and "downstream" (fossil-fuel-using) industries by freely allocating less (and sometimes considerably less) than 50 percent of pollution permits." (Goulder et al., forthcoming, p. 4)

[13] These results are uncertain. A sensitivity analysis of the results to the choice and estimation of parameters in E3MG in principle requires repeated re-estimation of parameters under different assumptions, with associated projections. This is a major exercise planned to assess the uncertainties in the projections, but was not possible for this paper.

Table 38.3 GDP 2000–2100.

	$(2000) trillion			%pa		
Scenario	2000	2050	2100	2000–2050	2050–2100	2000–2100
Baseline						
No ITC	30.7	125.5	275.2	2.86	1.58	2.22
ITC	30.7	133.4	314.1	2.98	1.73	2.35
550 ppmv						
No ITC	30.7	126.6	278.2	2.88	1.59	2.23
ITC	30.7	134.7	317.7	3.00	1.73	2.37
500 ppmv						
No ITC	30.7	127.6	280.3	2.89	1.59	2.24
ITC	30.7	135.3	319.6	3.01	1.73	2.37
450 ppmv						
No ITC	30.7	129.5	283.7	2.92	1.58	2.25
ITC	30.7	138.2	324.7	3.06	1.72	2.39

Source: E3MG2.1sp2.

largely over the years to 2050. The modelling of the conversion is highly non-linear, since it requires a system change, and the permit/tax rates required are very uncertain. As the transport sector decarbonises, it requires more electricity, and this further accelerates the move to low-carbon technologies in the electricity sector. The 450 ppmv target becomes even more difficult to achieve. Permit prices with ITC start at \$10/tC in 2011 and rise to \$100/tC by 2020 and \$400/tC by 2050. The easier, lower-cost options for reducing emissions have been exhausted, and the extra growth stimulated by the higher investment is also encouraging the demand for energy in general.

Table 38.2 and Figure 38.2 show the emission pathways and Figure 38.3 the CO_2 concentrations, both with and without induced technological change. The effect of introducing endogenous technological change into the baseline is to reduce emissions substantially. Taking the system as a whole, the effects of technological change as modelled are simultaneously to reduce energy demand directly through improvements in efficiency but increase economic growth and so increase energy demand indirectly, offsetting some of the effects of the improvements in energy efficiency. This relates to the 'rebound effect' found in studies of energy efficiency (Herring, 2004; Frondel, 2004) in which the expected reductions in energy use do not occur because the extra real income provided by the improvement in efficiency leads to more energy use. At the global, long-run, scale, technology drives energy efficiency, but it also, more significantly, drives economic growth and the outcome is lower energy use and CO_2 emissions.

Table 38.3 and Figure 38.4 show the outcomes for GDP, also with and without induced technological change. In all the stabilisation scenarios, GDP in 2100 is higher than the corresponding baseline with or without ITC. The extra investment in the stabilisation scenarios leads to faster growth. The growth rates 2050–2100 are almost unchanged. The faster growth takes place during the

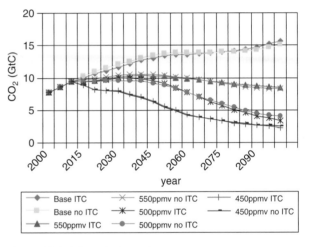

Figure 38.2 CO_2 emission pathways.
Source: UK Tyndall Centre and Cambridge Econometrics, E3MG2.1 SP2 model solutions annual to 2020, every 10 years to 2100.

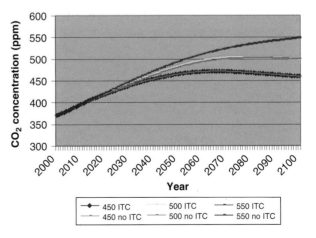

Figure 38.3 Illustrative CO_2 concentrations.
Source: UK Tyndall Centre and Cambridge Econometrics, E3MG2.1 SP2 model solutions annual to 2020, every 10 years to 2100.

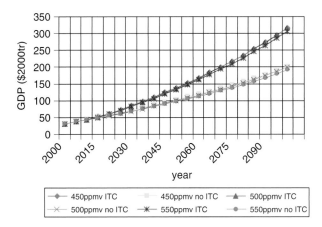

Figure 38.4 The effects of endogenous technological change on GDP.
Source: UK Tyndall Centre and Cambridge Econometrics, E3MG2.1 SP2 model solutions annual to 2020, every 10 years to 2100.

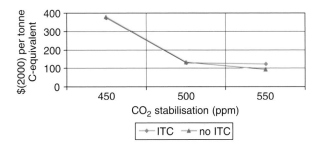

Figure 38.5 The global price of CO_2 permits in E3MG solutions for 2050–2100.
Source: UK Tyndall Centre and Cambridge Econometrics, E3MG2.1 SP2 model solutions.

period of extra investment to 2050, when structural change in developing countries is accelerated. The substantial effects of including endogenous technological change are apparent in the baseline projections, with very small effects of decarbonisation on global economic growth. This is not a surprise. Energy demand and supply is very small in relation to the rest of the economy, around 3–4% of value added, and technological change is led by improvements in the use of machinery and information technology and communications. These improvements allow long-run growth to proceed by saving on scarce resources such as labour and energy. The growth itself ultimately comes from the demand by consumers for goods and services, promoted by technological and marketing innovations. Table 38.3 also shows the extent to which higher growth is induced by the extra investment as a result of the increases in real carbon prices. At 550 ppmv with ITC the overall effects are very small, about 1 percentage point above the baseline with ITC by 2100. When the stabilisation targets are more demanding, the extra investment required leads to a small increase in the growth rate, and GDP is about 3–5% higher by 2100.

Figure 38.5 shows the global permit price and carbon tax rate[14] calculated to achieve the three stabilization targets by 2100, expressed in year 2000 \$s and exchange rates. The social costs of GHG mitigation to achieve stabilisation are substantially reduced when technological change is endogenous.

38.6 Conclusions

These are the early results of a substantial data collection, estimation and modelling project, adopting an econometric and technological approach to the estimation of the costs of stabilisation. We have made economic growth endogenous in an econometric model by allowing exports and energy demand to respond to technological progress. The main conclusion is that general technological change alone seems unlikely to lead to decarbonisation. Improvements in energy efficiency are partly offset in their effects on CO_2 emissions by the effects of higher growth in exports, incomes and therefore the demand for energy. This phenomenon is a global, macroeconomic counterpart to the rebound effect found in microeconomic studies of energy policies.

We conclude that the applications of current cost-effective technologies can decarbonise the world economy, supporting the conclusions of Pacala and Socolow (2004), provided they are specifically driven by increases in the real prices of carbon arising from emission permit schemes and taxes. Our conclusion is conditional on model uncertainties and assumptions, and on specific fiscal polices, with half the permits being freely allocated and the other half auctioned and all government revenues from the permits and taxes recycled back to consumers. If policies are successful in raising real carbon prices, under conditions of macroeconomic stability (so that inflation is unaffected and governmental fiscal rules are followed) then the extra investment is expected to lead to slightly higher global growth and incomes, even for almost complete global decarbonisation.

Acknowledgements

E3MG has been developed by the UK Tyndall Centre for Climate Change Research, funded by the UK Research Councils. The large-scale version 2 used in this paper could not have been developed without substantial contributions from Cambridge Econometrics Ltd in the forms of some data provision and of procedures for data

---

[14] The social cost of carbon can be calculated by expressing all the climate change damages in money terms and discounting to obtain present values. However, the uncertainties are such that we have chosen not to follow this approach and instead, we have adopted stabilisation targets assumed to be chosen by governments acting under the UNFCCC and calculated the required extra cost of achieving them in terms of CO_2 permit prices and tax rates.

management and estimation, which were undertaken by teams headed by Rachel Beaven and Sebastian De-Ramon, and including Dijon Antony, Ole Lofnaes, Michele Pacillo and Hector Pollitt. The authors are very grateful for these contributions.

REFERENCES

Abadir, Karim Maher (2004) 'Cointegration theory, equilibrium and disequilibrium economics', *The Manchester School* 72 (1), 60–71.

Aghion, P. and Howitt, P. (1998) *Endogenous Growth Theory.* Cambridge, Mass.; London : MIT Press.

Anderson, D. and Winne, S. (2004) Modelling innovation and threshold effects in climate change mitigation Tyndall Working Paper 59, www.tyndall.ac.uk

Azar, C. and Schneider, S. (2002) 'Are the economic costs of stabilizing the atmosphere prohibitive?, Ecological Economics, 42: 73–80.

Barker, Terry (2004) 'Economic theory and the transition to sustainability: a comparison of general-equilibrium and space-time-economics approaches', Tyndall Working Paper No. 62, Tyndall Centre, University of East Anglia, November.

Barker Terry, William Peterson (1987) (eds) *The Cambridge Multisectoral Dynamic Model of the British Economy*, Cambridge University Press.

Barker, T., Köhler, J. and Villena, M. (2002) The Costs of Greenhouse Gas Abatement: A Meta-Analysis of Post-SRES Mitigation Scenarios. *Environmental Economics and Policy Studies* 5(2), 135–166.

Barker, T. and Knut Einer Rosendahl (2000) 'Ancillary Benefits of GHG Mitigation in Europe: SO_2, NO_x and PM_{10} reductions from policies to meet Kyoto targets using the E3ME model and EXTERNE valuations', *Ancillary Benefits and Costs of Greenhouse Gas Mitigation*, Proceedings of an IPCC Co-Sponsored Workshop, March, 2000, OECD, Paris.

Barker, T. and De-Ramon, S. (2005) 'Testing the representative agent assumption: the distribution of parameters in a large-scale model of the EU 1972–1998', *Applied Economic Letters*, forthcoming.

DeCanio, S. (2003) *Economic Models of Climate Change A Critique*, Palgrave-Macmillan, New York.

Denison, E.F. (1967) *Why Growth Rates Differ.* Washington D.C.: The Brookings Institution.

Denison, E.F. (1985) *Trends in American Economic Growth, 1929–1982*, Brookings Institution.

Dewick, P., Miozzo, M. and Green, K. (eds) *Technology, Knowledge and the Firm: Implications for Strategy and Industrial Change*, Edward Elgar, Cheltenham, UK.

Dowlatabadi, H. (1998) 'Sensitivity of climate change mitigation estimates to assumptions about technical change', *Energy Economics* 20, 473–493.

Ebling, G. and Janz, N. (1999) 'Export and innovation activities in the German service sector: empirical evidence at the firm level', *ZEW Discussion Paper* 99–53, Mannheim.

Edmonds, J., Clarke, C., Dooley, D., Kim S.H. and Smith, S.J. (2004) Stabilisation in a B2 world: insights on the roles of carbon capture and disposal, hydrogen, and transportation technologies. *Energy Economics* 26, 517–537.

Engle, R.F., Granger, C.W.J. (1987) 'Cointegration and Error Correction: Representation, Estimation and Testing', *Econometrica*, 55.

Fagerberg, J. (1988), 'International Competitiveness.' *Economic Journal*, 98, pp. 355–374.

Freeman, C. and Louçã, F. (2001) *As Time Goes By*, OSO Monographs, Oxford.

Frondel, M. (2004) 'Energy conservation, the rebound effect, and future energy and transport technologies: an introduction to energy conservation and the rebound effect', *International Journal of Energy Technology and Policy*, Vol. 2, No. 3, 2004.

Geels, F. 2002 *Understanding the Dynamics of Technological Transitions*, Twente University Press, Enschede, NL.

Goulder Lawrence, H. (2004) 'Induced technological change and climate policy', Pew Center on Global Climate Change, http://www.pewclimate.org.

Goulder, L., Lans Bovenberg, A. and Derek, J. Gurney (forthcoming) 'Efficiency costs of meeting industry-distributional constraints under environmental permits and taxes', *Rand Journal of Economics*.

Greenhalgh, C. Taylor, P. and Wilson, R. (1994) 'Innovation and Export Volumes and Prices: A Disaggregated Study', *Oxford Economic Papers*, 46, 102–134.

Greenhalgh, Christine (1990) 'Innovation and Trade Performance in the UK', *Economic Journal*, 100, 105–118.

Gritsevskyi, A. and Nakicenovic, N. (2000) Modelling uncertainty of induced technological change, *Energy Policy* 28, 907–921.

Grubb, M., Kohler, J. and Anderson, D. (2002) Induced Technical Change in Energy and Environmental Modeling: Analytical Approaches and Policy Implications. *Ann. Ren. Energy Environ.* 27, 271–308.

Grübler A., Nakicenovic, N. and Victor, D.G. (1999) 'Modeling technological change: implications for the global environment', *Annu. Rev. Energy Environ.* 24:545–569.

Harrod, R. (1939), 'An essay in dynamic theory', *The Economic Journal*, Vol. 49, No. 193, pp. 14–33.

Harrod, R. F. (1948) *Towards a Dynamic Economics*, Macmillan, London.

Herring, H. (2004) 'The rebound effect and energy conservation', in Cleveland, C. (Ed.): *Encyclopeadia of Energy*, Academic Press/Elsevier Science, pp. 237–244.

Hourcade, J-C and Shukla P. (2001) 'Global, regional, and national costs and ancillary benefits of mitigation', Chapter 8 in Third Assessment Report WG III of the IPCC, 2001. Climate Change 2001: Mitigation, Cambridge University Press.

Kaldor, N. (1957) 'A model of economic growth', *Economic Journal* 67: 591–624.

Kaldor, N. (1972) 'The irrelevance of equilibrium economics', *Economic Journal*, Vol. 52, pp. 1237–1255.

Kaldor, N. (1985) *Economics without Equilibrium*, Cardiff Press, UK.

Köhler, J. (2005) 'Making (Kondratiev) waves: simulating long run technical change' in: Dewick, P., Miozzo, M. and Green, K. (eds) *Technology, Knowledge and the Firm: Implications for Strategy and Industrial Change*, Edward Elgar, pp. 404–426.

Lachenmaier, S. and Wößmann, L. (2005) 'Does innovation cause exports? Evidence from exogenous innovation impulses and obstacles using German micro data', Oxford Economic Papers, forthcoming.

León-Ledesma, (2000) 'R&D Spillovers and Export Performance: Evidence from the OECD Countries', Studies in Economics 0014, Department of Economics, University of Kent.

Maddison, A. (2001) *The World Economy A Millenial Perspective*, OECD, Paris.

Magnier, A. and Toujas-Bernate, J. (1994) 'Technology and Trade: Empirical Evidence for the Five Major European Countries', *Weltwirtschaftliches Archive*, 130, 494–520.

Manne, A. and Richels, R. (2004) 'The impact of learning-by-doing on the timing and costs of CO_2 abatement', *Energy Economics* 26: 603–619.

Scott, M. (1989) *A New View of Economic Growth*, Clarendon Press, Oxford.

McCombie, J. M. and Thirlwall, A.P. (1994) *Economic Growth and the Balance of Payments Constraint,* Macmillan.

McCombie, J.M. and Thirlwall, A.P. (2004) *Essays on Balance of Payments Constrained Growth: Theory and Evidence*, Routledge Press.

McDonald, A. and Schrattenholzer, L. (2001) Learning Rates for Energy Technologies. *Energy Policy* 29: 255–261.

McFarland, J.R. et al. (2004) Representing energy technologies in top-down economic models using bottom-up information. *Energy Economics* 26: 685–707.

Morita, T., Robinson, J., Alcamo, J., Nakicenovic, N., Zhou, D. et al., (2001), *Greenhouse Gas Emission Mitigation Scenarios and*

Implications, Chapter 2 *in* Climate Change 2001: Mitigation, Third Assessment Report, Report of Working Group III of the Intergovernmental Panel on Climate Change. Metz, B., Davidson, O., Swart, R. and Pan, J. Cambridge, Cambridge University Press, Cambridge.

Nakicenovic, N. and Riahi, K. (2003) *Model runs with MESSAGE in the Context of the Further Development of the Kyoto Protocol*, WGBU, Berlin. www.wgbu.de

Pacala, S. and Socolow, R. (2004) 'Stabilization Wedges: Solving the Climate Problem for the Next 50 Years with Current Technologies.' *Science,* Vol. 305, August 13, pp. 968–972.

Palley, T., I. (2003) 'Pitfalls in the theory of growth: an application to the balance of payments constrained growth model', *Review of Political Economy*, Vol. 15, No. 1, pp. 75–84.

Posner, M. (1961) 'International trade and technical change', *Oxford Economic Papers,* 13, pp. 323–341.

Romer, P. (1986) 'Increasing returns and long-run growth', *Journal of Political Economy* 94(5), 1002–37.

Romer, P. (1990) 'Endogenous technological change' *Journal of Political Economy* 98 (5): S71–102.

Roper, S. and Love, J. (2001) 'Innovation and export performance: evidence from UK and German manufacturing plants' Working Paper no. 62, Northern Ireland Economic Research Centre.

Ruttan, V.W. (2002) 'Sources of technical change: induced innovation, evolutionary theory, and path dependence', pp. 9–39 in Grübler, A., Nakicenovic, N. and Nordhaus, W. D. (eds) (2002) *Technological Change and the Environment*, Resources for the Furure, Washington, D.C.

Scherer, F. M. (1982) 'Demand pull and technological inventions: Schmookler revisited', *Journal of Industrial Economics,* Vol. 30, pp. 225–37.

Schmookler, J. (1966) *Invention and Economic Growth*, Harvard University Press.

Setterfield, M. (1997) '"History versus equilibrium" and the theory of economic growth' *Cambridge Journal of Economics*, Vol. 21, pp. 365–378.

Setterfield, M. (2002) (ed) *The Economics of Demand-led Growth – Challenging the Supply-side Vision of the Long Run*, Edward Elgar, Cheltenham, UK.

Solow, R.(1956) 'A Contribution to the Theory of Economic Growth', *Quarterly Journal of Economics*, February 1956.

Solow, R. (1957) 'Technical Change and the Aggregate Production Function', *Review of Economics and Statistics*, August 1957.

Van Vuuren, D.P., de Vries B., Eickhout, B. and Kram, T. (2004) Responses to technology and taxes in a simulated world. *Energy Economics* 26, 579–601.

Wakelin, K. (1998) 'The role of innovation in bilateral OECD trade performance', *Applied Economics*, 30, 10, 1335–1346

Watson, et al. (2001) *IPCC Third Assessment Report, Climate Change 2001: Mitigation, Summary for Policymakers*, IPCC/WMO, Geneva.

Weyant, John P. (2004) 'Introduction and overview: *Energy Economics Special Issue* EMF 19 study Technology and Global Climate Change Policies', *Energy Economics*, 26, 501–515.

Wigley, T.M.L., Richels, R. and Edmonds, J. (1996) 'Economic and environmental choices in the stabilisation of atmospheric CO_2 concentrations', *Nature* 379 (6562) 240–243. (not cited)

Wigley, T.M.L. and Raper, S.C.B. (1997), *Model for the Assessment of Greenhouse-Gas Induced Climate Change (MAGICC Version 2.3)*, The Climate Research Unit, University of East Anglia, UK.

Wolff, E. (1994a) 'Technology, Capital Accumulation, and Long Run Growth', in Fagerberg, J., Verspagen, B. and von Tunzelmann, N. eds., *The Dynamics of Technology, Trade, and Growth*, Edward Elgar Publishing Ltd, 1994, 53–74.

Wolff, E. (1994b) 'Productivity Growth and Capital Intensity on the Sector and Industry Level: Specialization among OECD Countries, 1970–1988', in Gerald Silverberg and Luc Soete eds, *The Economics of Growth and Technical Change: Technologies, Nations, Agents*, Edward Elgar Publishing Ltd.

CHAPTER 39

Carbon Cycle Management with Biotic Fixation and Long-term Sinks

Peter Read
Department of Applied and International Economics, Massey University, New Zealand

39.1 Introduction: Bioenergy in a Two Stage Strategy to Address Abrupt Climate Change

This paper reviews some policy implications of potential abrupt climate change (ACC) on the basis of an Expert Workshop convened in Paris in September–October 2004 by this writer with the support of the Better World Fund of the United Nations Foundation (visit www.accstrategy.org for the prospectus and programme and www.accstrategy.org/simiti for selected peer-reviewed papers forthcoming in a Special Issue of Mitigation and Adaptation Strategies for Global Change).

No scenario was presented in which this threat was imminent, but it was recognised that climate models are more stable than the paleo-climatic record, that a failure of the thermohaline circulation (THC) might prove to be irreversible, and that processes of terrestrial ice loss, sea-ice variability, methane release and climate induced reduction of biotic carbon fixation could precipitate imminent ACC, for which the thresholds and triggers are currently poorly understood. The avoidance of ACC linked to the collapse of land based ice masses may impose a limit to the heat burden that it is safe to inject into the oceans, which would mean that there is a time limit to the elevation of GHG levels above pre-industrial. For such a limit to be met, carbon management may need to be capable of significant annual net absorption of CO_2 from the atmosphere.

Workshop presentations [1, 2] drew attention to risks of serious or irreversible damage that could follow if potential ACC became imminent (or could be realised if ACC became actual) and analysed the path for net greenhouse emissions that would be needed if one of these risks (collapse of the THC) is to be avoided [3]. Those presentations are not reviewed here as other papers in this volume have covered that ground in greater detail. The workshop took the implication of those presentations to be that a need could arise to reduce greenhouse gas levels much more sharply than has been envisaged in scenarios reviewed by the IPCC [4, 5].

The workshop was stimulated by publications [6, 7, 8] which suggest that such sharp reductions are feasible through the large scale adoption of BECS (Bio-Energy with Carbon Storage). This is because BECS is a negative emissions energy system that removes CO_2 from the atmosphere and puts it below the earth's surface, whilst providing useful energy services. It is believed that no other technologically-feasible way exists to achieve that result[1]. Zero-emissions energy technologies, such as may be stimulated by policy to reduce emissions, can at best prevent more CO_2 entering the atmosphere, relying on natural processes to remove it into natural reservoirs such as the ocean[2], resulting in an asymptotic approach to the enhanced level in the natural sinks. Accordingly the workshop came to a policy prescriptive conclusion, conditional on the assumption that policy-makers would be guided by Article 3.3 of the Climate Change Convention, which requires the Parties to take cost-effective precautionary action in response to threats of serious or irreversible damage, without delay on account of a lack of full scientific certainty.

This conditional conclusion was that policy-makers should be urged to stimulate the growth of a global bioenergy market, with world trade (mainly South-North trade) in liquid bio-fuels such as ethanol and synthetic (e.g. Fischer Tropsch) bio-diesel. This action was seen as the long lead-time, low-cost first stage of a two-stage strategy [8] to implement BECS, with the second stage being the retro-fitting of CO_2 Capture and Sequestration (CCS) technology to all large point sources of CO_2 emissions, whether fired by fossil fuel or bio-fuel. This second, high-cost, stage would be stimulated by precursor signals of imminent abrupt climate change and have a much shorter lead-time since it does not involve large scale changes in land use practices. Thus, in the event that scientific progress over the years ahead reveals, say in 2020, that a trigger for ACC is imminent, the existence by then of a large-scale global bioenergy industry would put the world in much better shape to respond effectively. If then linked to CCS under policy urgency prompted by imminent abrupt climate change, BECS can, under strong assumptions regarding biomass productivity, rate of land use change and rate of deployment of CCS, and with parallel efforts with energy efficiency, etc., get the CO_2 level to peak much earlier than projected in the scenarios surveyed in the IPCC Special Report on Emissions Scenarios [4] and IPCC Third Assessment Report [5] and even down to the pre-industrial

---

[1] Theoretic analysis [9] suggests that CO_2 could be removed directly from the atmosphere, but no operational examples exist of this system, which would consume rather than produce energy services.

[2] However, such reliance may lead to unacceptable risks to marine life and ecosystems due to increased ocean acidity [10] – see note 3.

level of 280 ppm by around 2060 [8[3]]. This partial analysis clearly needs to be substantiated through extensive integrated assessment, as noted under bullet 9 in the Conclusion section.

Read and Lermit [8] paper models the inter-related markets for fuel, woody fibre and land, on a global scale over a 70-year horizon, to simulate the effect (relative to a range of reference scenarios – IS92 business as usual [13], Tellus Institute's fossil free energy scenario [13], and a notional 'Kyoto' midway between) of using additional land on a large scale for either conventional forestry or for short rotation energy crops, in two cases, first continued expectation of gradual climate change, and second, bad news of imminent ACC in ~2020. Under the first case, low cost CO_2 stabilisation (possibly negative cost if peak oil induces rapidly rising fossil fuel prices) is achieved at a level below 420 ppm, with an initial long rotation storing carbon for a few decades, while energy sector capital stock adjusts to achieve a smooth transition to using bio-energy. Under the second case, with conditions of ACC-driven urgency, the long rotation land is turned over to energy cropping and a 'maximal' programme of land use change is implemented, along with the (likely costly) linking of all point source emitters, both fossil and bio-fuel, to carbon capture and sequestration. With urgency-driven accelerated technological progress, this results in the substantial dominance of bio-energy by mid-century, whilst CO_2 decreases rapidly to the pre-industrial level of 280 ppm [8].

The view that the first stage would be low cost arises, *prima facie*, from the consideration that energy-related emissions are just over 5% of CO_2 flows into and out of the atmosphere due to terrestrial biotic activity. This suggests that mitigation investment in the heavily-capitalised energy sector is likely to be less cost-effective than investment designed to increase biotic carbon fixation on under-capitalised land. On plausible technological assumptions and oil price projections, CO_2 reductions got from better use of land, and of products of the land, can have low or even negative cost, taking account of added value from the co-production of food or fibre along with biomass for bio-energy raw material. For instance, drivers of dual fuelled vehicles in Brazil select sugar cane derived ethanol in preference to gasoline when the price of crude oil rises above $35 per barrel, with high value food demands met with co-produced refined sugar.

The remaining sections of this paper deal with the social dimensions of the workshop's conclusion, with technical aspects of the two key technology types that were seen to be needed for effective management of the level of CO_2 in the atmosphere, i.e. enhanced biotic fixation of CO_2 and its safe storage out of the atmosphere, and, in conclusion, suggest directions for further research.

39.2 Negotiability – Social and Political Economy Aspects

Additional linkage between abrupt climate change and BECS (which could apply also to other technology-based mitigation strategies, if they can be shown to be relevant to ACC) arises from a legal perspective. This is that actions under Article 3.3 of the Convention would be regarded as different from, and complementary to, action under Article 4.2 which, under the Berlin Mandate, is already addressed by the Kyoto Protocol. Such technology-based actions can therefore become the subject of a separate protocol addressed to the threat of abrupt climate change, without need to revisit the provisions of the Kyoto Protocol[4] or for Parties to the Convention that have not become party to the Kyoto Protocol to alter their position on that matter.

That political will to address potential abrupt climate change in this way could develop may be apparent from the numerous side benefits of the low-cost first stage that were noted at the workshop. These include:

1. Post 2012: more ambitious CO_2 reductions are available at modest cost, with stabilisation by 2025 at 420 ppm CO_2 practicable [8].
2. Transportation emissions: bio-ethanol and bio-diesel are here-and-now technologies that can make an immediate impact on the otherwise intractable problem of vehicle emissions [15]. If the expectation is that crude oil is unlikely to trade at under $30 in the future, bio-fuels can provide the backstop technology that limits price levels in the long term., regardless of greenhouse gas concerns.
3. Energy security: domestic supplies from advanced fermentation technology using cellulosic feedstock can provide the USA and Europe with sufficient transportation fuels to meet essential needs [16]. Diversification of industrialised country imports to include ethanol and bio-diesel from developing counties could break OPEC's residual market power [17].
4. Farm support: a new source of income from biomass supply as bio-energy feedstock, co-produced with food by farmers in developed countries threatened by World

---

[3] Two counter-acting factors were taken into account in papers presented subsequently to the 2005 International Energy Workshop, Kyoto, July 2005. A more sophisticated model of the flows of CO_2 between atmosphere and pools in the ocean and earth [11] allows for the need, if CO_2 levels decline, for CO_2 to be disposed of not only from the atmosphere but also from absorbers such as the ocean surface layer that are close to equilibrium with the atmosphere. Secondly [12] two processes discussed below – *terra preta* soil improvement and co-production of food and bio-fuel from switch-grass feedstock – were not taken into account in [8] which focuses on only the co-production of bio-energy with conventional forest products.

[4] It is not legal practice to have two sets of provisions addressed to the same objective. For instance, actions to protect the ozone layer are the subject of the 1985 Vienna Convention, with substances that deplete the ozone layer controlled under its 1987 Montreal Protocol and excluded from the provenance of the Climate Change Convention.

Trade Organisation decisions on export subsidies, along with woody feedstock co-produced with timber on non-arable (e.g. steep) land [18].

5. Development: carbon credits to developing countries based on absorption of CO_2 in the production of bio-mass feedstock can support the development of bio-fuel production, hence cutting balance of payments costs of imported oil and eventually leading to sustainable growth, led by exports of ethanol and bio-diesel [19]. Carbon credit funded investments can co-produce food and fuel, as in Brazil, where sugar cane ethanol is widely used in lieu of gasoline [17].

6. Problems of Africa: weak institutions and ill-defined land tenure entail a community based approach with tens of thousands of projects and major capacity building needed to train project leaders. Relief of debt needs to be supplemented by the development of sustainable economic activity to enable African people to earn livings consistently with the UN Millennium Development Goals [20]. Exigent communities demand a short-term pay-off such as can be achieved by techniques that raise soil fertility and water retention whilst sequestering carbon in the soil, as well as relieving energy poverty and yielding sustainable rural development [21]. There is no 'silver bullet' and a long-term effort is needed build capacity and to channel funding derived from energy consumers in the 'North' into appropriate local solutions [22]. In this way, addressing Africa's problems and responding to the threat of ACC can go hand in hand.

7. Integrity of the emissions cap: where it is desired both to impose an emissions cap and to drive a technology based response strategy, this can be achieved by requiring emissions permits to be issued initially in exchange for project based certificates that quantify the uptake of policy-desired technology (rather than by auction or grandfathering). If, for instance, the policy commitment is to reduce emissions by one third relative to unrestrained demand, then two emissions permits would be issued in exchange for one project based credit. Under that arrangement for 'allocating permits usefully' [23], project-based credits no longer serve as an offset against emissions permit requirements, so that low quality projects do not result in excess emissions relative to commitments but in a shortage of permits in relation to the demand to emit, with consequent higher permit prices and the choking off of demand. A similar result is achieved with an emissions intensity policy, such as a renewable portfolio standard that requires one third of output to be sourced renewably, providing the policy is enforced rather than a voluntary target[5]. Where the policy is intended to be more technology-specific in response to threats of abrupt

climate change, then the certificate-for-permits exchange can require a minimum proportion of the certificated projects to involve bio-energy and/or CCS technology.

In view of these considerations, it could be felt that a 'win-win' option is available for reducing greenhouse gas levels and that climate change policy has been misled by a plausible fallacy – that the energy sector's problem is best cured in the energy sector. *Prima facie*, it appears that (along with the industrial practicability and here-and-now availability of bio-energy systems) the GHG mitigating potential of land use change to enhance biotic carbon fixation has been overlooked by a policy-making community that has focused mainly on capping energy sector emissions.

39.3 Co-production of Food Fibre and Bio-energy

Managing land so as to substantially increase the total amount of terrestrial biotic carbon fixation, and hence the supply of biomass, raises concerns related to the human 'ecological footprint'. However, ecosystems evolve through resilience to external factors from which intelligent management can provide some degree of protection, thus enabling the sustainable productivity of the land where they have evolved to be increased[6], as with the bulk of farming activity in Western Europe, conducted on land that was formerly in equilibrium climax forest cover. Such simple investments as stock-proof fencing to prevent wild and stray animals destroying crops and plantations at the seedling stage can improve on nature. Carbon fixing soil amendment yields environmental benefits, including enhanced fertility and water retention.

Efficient management of part of the land, in lieu of widespread unsustainable traditional land management, can enable natural bio-diversity to flourish in remaining conservation areas. Such management can yield food and forestry products co-produced with biomass on existing cropland. Alternatively (and hence additionally) estimates of land requirements to effectively mitigate all current anthropogenic emissions of CO_2 fall well short of the ~ 2.3 Gha of potential arable land that is not in use [8, 24]. The workshop concluded there is not a shortage of land but of investment in land. Remedying this can yield additional biomass for use in lieu of fossil fuel as the basis for bio-fuels and bio-electricity; also its carbon content, in the form of bio-char ('charcoal') can be used for soil amendment with ~ 5K-yr half life; or it can be used as wood, or advanced materials such as carbon fibre, in long-lived artefacts.

A review of the variety of existing commercial bio-energy technologies for meeting demands for heat

---

[5] Providing demand is accurately forecast – the effect of an emissions cap is to relieve the policy maker of responsibility for accurate forecasting, transferring the risk to the market through the costs of price volatility.

[6] This is despite such unfornate precedents as the Oklahoma dust bowl, due to ill-informed over-farming, and the introduction of rabbits to Australia to provide sporting opportunities for the early settlers, without regard to the eventual implications for land productivity.

electricity and transportation fuels, their near-term expected development and long-term prospects [25] confirmed the technological assumptions that underlie the land use change implications of the conclusion of the workshop, and provides the first discussion of their suitability for linking to carbon capture and sequestration systems in the event of imminent ACC. The potential for dual-fired fossil and bio-fuelled systems in a low-risk, transitional approach that can establish biomass raw material supply chains, and enable harvesting and farm/forestry management systems to develop was noted. These include the mixing of bio-mass raw material such as wood chips with coal for power station fuel [as is done, for instance, in some parts of the USA to meet local emissions regulations] and the mixing of ethanol with gasoline up to the maximum proportion acceptable to the vehicle fleet.

Novel integrated systems involving near-term achievable advances in technology deployed on existing US cropland to produce biomass for energy co-production, with current food production maintained, have been analysed [16]. 'Consolidated bio-processing' of the cellulosic fraction of crops – in which hydrolysis and fermentation occur in a single process step – is applied to high productivity switch-grass feedstock. One of the most promising options extracts protein, ferments ethanol, and co-produces electricity. Among the scenarios analyzed, this one yields the lowest-cost transport fuel at \$0.56/USgal of gasoline equivalent, with co-produced protein valued equal to the long term average for soy meal protein of \$0.20 per pound and power valued at \$0.04/kWh. This assumes a 20 kton per day plant and technological progress to 2025, when the USDoE forecast gasoline price is in the range \$0.48–1.03/gal around a base case of \$0.79/gal. As with [8], co-production is key to the potential cost-competitiveness of bio-energy systems. In aggregate, about half of current US demand for transportation fuels could, with foreseeable advances in technology, be met in 2025 from biomass produced on existing US cropland, along with the same amount of food as is currently produced. With improved vehicle efficiency and 'smart' urban growth, demand for gasoline imports could be reduced to zero, but even without such developments, sufficient could be supplied to meet emergency needs under conditions of interrupted international trade.

The potential of tropical and sub-tropical cropland for the production of sugar-cane and its conversion to ethanol and electric power at an intensity equivalent to the author's home region in Brazil, i.e. ~10 per cent of all such cropland was evaluated [17]. With the use of only 143 mHa, and with rapid technological advances, 164 EJ of primary energy and 90 EJ of ethanol and electricity can be produced from 4000 units of a scale similar to the largest existing in Brazil (with a capacity to handle 5 Mt of harvested cane annually) with the creation of 'millions of direct and indirect jobs'. This implies that the utilization of 30 per cent of tropical potential cropland in this way would meet all current global commercial energy demands. This is additive to production from energy forest plantations,

involving less fertile land, envisaged in [8] and suggests the export potential of advanced developing countries.

A study of the development of bio-energy on three small island states [22] adopted a 'no one size fits all' philosophy, reviewing issues of scale, human capacity, community involvement, appropriate technology and critical mass. This showed the need for an integrated, multi-disciplinary, cross-sectoral – i.e. whole systems – approach in least developed regions. Strong and consistent policy signals and support systems can enable the potential of bio-energy to be realised in the subsistence economies to be found in the rural sectors of the LDCs. The existence of a global market for bio-fuels would provide eventual linkage to the market economy but initial deployment of appropriate technologies like anaerobic digestion and small-scale bio-diesel is needed to meet local needs.

39.4 Carbon Disposal and Soil Improvement

A pioneering study of the potential for linking bio-energy to CCS [26] noted that suitable sediment sequences of saline aquifers exist in all hydrocarbon-producing areas, are volumetrically much larger than exploited oil and gas fields, and hold the potential to easily store all worldwide emissions until 2050. Geological principles are established to assess the entire continental landmass for candidate sites of CO_2 storage. This shows that opportunity for linking CCS to bio-energy may be widespread, but needs more specific local investigations. Onshore sub-Saharan Africa is considered the most problematic region – but even here there are sediment sequences. No demonstration projects exist using small-scale onshore facilities. A simple estimate, assuming CO_2 value of \$20/ton, suggests that single boreholes onshore may be viable with supply rates of 0.1 Mt/year. It is concluded that, in principle, atmospheric CO_2 could be captured in a country far distant from the original fossil fuel CO_2 emission site by cultivating biomass with the CO_2 emissions from its use as biofuel stored in deep aquifers.

A second pioneer essay [27] makes the first attempt to assess the macro-potential of bio-char soil amendment. Several cases based on existing practice are noted: a change from slash and burn to slash and char, improved 'charcoal' production processes, agricultural waste recycling, and current bio-energy respectively yielding .2, .02, .16 and .2 GtC/yr sequestered. However, the expansion of modern bio-energy in line with a range of scenarios could yield 5.5–9.5 GtC/yr, sufficient to sequester all current emissions.

Practical experience of three examples of bio-char sequestration projects was reported [21]. They involve disposing of urban and rural wastes (in Japan); soil improvement for mallee eucalypt agro-forestry to suppress saline water intrusion (in Australia – a JI project); and managing pulp mill wastes (in Indonesia – a CDM project). Improved soil productivity was reported along with other socio-economic benefits from these projects, which provide an initial basis for quantitative assessment of '*terra preta*' technology.

An advanced (patent pending) process for producing a novel nitrogen-enriched slow release carbon-sequestering fertilizer was also described [8]. Biomass wastes from forestry and agriculture have been pyrolyzed on a laboratory scale, with reforming of volatile fractions to hydrogen. Temperature was controlled to yield a char with affinity for CO_2, SOx and Nox, which were absorbed from fossil fuel flue gases, such as from coal fired power stations, into the pore structure of the char to form a slow release fertilizer. Apart from fertilization, the material provides a substrate congenial to the microbial and fungal activity that facilitates the transport of soil nutrients to rootlets, whilst the slow release feature reduces leaching and polluted run-off. From a systems perspective, the CO_2 impact relative to fossil fuels is $-112\,kg/GJ$ compared with $+48\,kg/GJ$ for natural gas and $+80\,kg/GJ$ for coal. Applied to current global bio-energy usage of $\sim55\,EJ$, this process would absorb $\sim6\,Gt\,CO_2$ without taking account of the dynamic hold-up of carbon in the larger volume of biomass growing on the fertilized soils.

39.5 Conclusion

Scientific certainty does not exist in relation to the potential extremes of climate change addressed by the workshop, and neither is it required in relation to an agreement to act under Article 3.3 of the Convention. Despite the win-win possibilities from the first stage of a response that is effectively precautionary against such extremes, the scale of operations, its organization and the capacity building involved in a land use change programme to underpin large-scale bio-energy is daunting. But it was clear to the workshop that a great many ends could be served by doing as much with bio-energy, and world trade in bio-fuels, as may be. *Prima facie* this direction appeared eminently negotiable, with greater energy security and lowered farm support costs in the North, and with sustainable development and ended energy poverty for many land rich but otherwise impoverished countries in the South.

The environmental and, in an era of high oil prices, the socio-economic benefits from this approach to managing the carbon cycle, additional to the precautionary element in relation to potential abrupt climate change, may turn out to be such that bio-energy comes to dominate the market. Ambient energy technologies (wind, wave and non-photosynthetic solar) may have a smaller role than widely envisaged as a consequence of treating the climate change problem as one of reducing energy sector emissions. However, the workshop did not conclude that this possibility should lead to the putting of all policy eggs into the bio-energy basket: resistance to rational land use may turn out to be very great; technological progress with biomass production and conversion may prove disappointing;

population trends may put greater pressure on land than is currently projected; etc.

Resolution of those issues can only come from research designed to complement the role of large scale bio-energy in a precautionary strategy – some of it action research to enable learning by doing with modern bio-energy, CCS and *terra preta* soil improvement technologies that will provide feedback for the needed integrated assessment of the two-stage precautionary strategy. Addressing the threat of abrupt climate change thus entails a range of activities besides promoting a global bio-energy market, including:

1. doing the climate science needed to specify the precursors of abrupt climate change;
2. setting up the observation systems needed to detect the specified precursors;
3. developing and demonstrating CCS technology, and learning by doing through its deployment;
4. prospecting for disposal sites for carbon dioxide captured from the existing fossil fuelled system and from the newly developing bio-energy system;
5. pressing forward with research and development on advanced bio-energy technologies, and with soil improvement technologies involving long-lived fixation of carbon, such as '*terra preta*' soil amendment with bio-char;
6. developing and ground-truthing detailed inventories at the national (policy design) level, and at the local and community (policy implementing) levels, of the potential for sustainable rural development, based upon carbon-fixing soil improvement and bio-energy, both in terms of carbon cycle management (as the initial financial driver) and improved livelihoods (as the sustaining motivation): – matching crop selection, agronomic practice and product processing to local needs and aspirations under different scenarios for regional and world market conditions;
7. stimulating field-to-factory biomass supply systems, through the early deployment of currently available bio-energy technologies under demands generated by bio-energy oriented renewable portfolio standards;
8. curriculum and delivery methodology development for the capacity building programme – most likely funded through the GEF – needed to support local participation in bio-energy and land use change projects by communities living on the great land banks of the least-developed and developing countries;
9. detailed spatially and temporally differentiated modelling of the roles of the different bio-energy and land use technologies in the evolution of the energy, agriculture and forestry sectors, for the purposes of integrated assessment based on a multi-criterion analysis on a vector of desiderata – including climatic, energetic, socio-economic, and environmental – of the precautionary strategy both globally to yield model closure and internal consistency and locally disaggregated to enable country-by-country projections of its impacts and benefits.

It is not the purpose of this review to pre-judge the results of these studies – which requires a large-scale, co-ordinated, world-wide research effort – but merely to establish the *prima facie* likelihood that it may lead to a low-cost and effective greenhouse gas management strategy. And also to note that initiating a strategy in the context of Article 3.3 of the Convention need not await full scientific certainty of its detailed end-state.

REFERENCES

1. Schlesinger, M. 'Uncoupled ocean GCM and coupled atmosphere/ocean GCM simulation results: theoretical analysis and understanding of the slowdown/shutdown of the THC'. Presentation to the Workshop[†].

2. Nisbet, E. 'Non-THC related potential surprise in the coupled atmosphere-ocean-cryosphere-biosphere system'. Presentation to the Workshop[†].

3. Bruckner, T., and Zickfeld, K. 'Low Risk Emissions Corridors for Safeguarding the Atlantic Thermohaline Circulation' Presentation to the Workshop[†].

4. Nakicenovic, N., Alcamo, J., Davis, G., de Vries, B., Fenhann, J., Gaffin, S., Gregory, K., and Grübler, A. et al., 2000. Special Report on Emissions Scenarios, 2000, Intergovernmental Panel on Climate Change, Cambridge University Press, Cambridge.

5. Morita, T., Robinson, J., Alcamo, J., Nakicenovic, N., and Zhou, D. et al., 2001. *Greenhouse Gas Emission Mitigation Scenarios and Implications*, Chapter 2 *in* Climate Change 2001: Mitigation, Third Assessment Report, *Report of Working Group III of the Intergovernmental Panel on Climate Change*. B. Metz, O. Davidson, R. Swart, and J. Pan. Cambridge, Cambridge University Press, Cambridge.

6. Obersteiner , M., Azar, C., Kauppi, P., Mollerstern, M., Moreira, J., Nilsson, S., Read, P., Riahi, K., Schlamadinger, B., Yamagata, Y., Yan, J., and van Ypersele, J.-P., 2001. 'Managing Climate Risk', Science 294, (5543): 786b.

7. Azar, C., Lindgren, K., Larson, E., Möllersten, K., and Yan, J., 2003. 'Carbon capture and storage from fossil fuels and biomass – Costs and potential role in stabilizing the atmosphere.' In press.

8. Read, P., and Lermit, J., 2005. 'Bio-Energy with Carbon Storage (BECS): a Sequential Decision Approach to the threat of Abrupt Climate Change' Energy 30.

9. Keith, D., and Ha-Duong, M., 2003. 'CO$_2$ Capture from the Air: Technology Assessment and Implications for Climate Policy'. *Proceedings of the 6th Greenhouse Gas Control Conference, Kyoto Japan.* J. Gale and Y. Kaya eds., Pergamon, Oxford UK, p. 187–197.

10. Turley, C., 2005. 'Reviewing the Impact of Increased Atmospheric CO$_2$ on Oceanic pH and the Marine Ecosystem', this volume.

11. Parshotam, A., Enting, I., and Read, P., 2005. 'Negative Emissions Energy and CO$_2$ Levels' International Energy Workshop, July, Kyoto.

12. Read, P., 2005. 'Building a global bio-energy market'. International Energy Workshop, July, Kyoto.

13. Leggett, J., Pepper, W.J., and Swart, R.J., 1992. 'Emission Scenarios for the IPCC: and Update', Chapter A3 in J.T. Houghton, B.A. Callander, and S.K. Varney, (eds) 'Climate Change 1992: the Supplementary Report to the IPCC Scientific Assessment', CUP, Cambridge.

14. Lazarus, M., 1993. 'Towards a Fossil Free Energy Future: the Next Energy Transition': Technical Analysis for Greenpeace International, Stockholm Environment Institute, Boston Center, Boston, MA.

15. Sheppard, J. 'Review of Macro-engineering approaches Symposium [Cambridge, Jan 04]' Presentation to the Workshop[†].

16. Greene, N., Celik, F.E., Dale, B., Jackson, M., Jayawardhana, K., Jin, H., Larson, E., Laser, M., Lynd, L., MacKenzie, D., Jason, M., McBride, J., McLaughlin, S., and Saccardi, D., 2004. NRDC Report 'Growing Energy: how biofuels can help end America's oil dependence' (December).

17. Moreira, J.R., 2004. 'Global biomass energy potential', Mitigation and Adaptation Strategies for Global Change[*].

18. Parris, K. 'Lessons from the OECD expert workshop on biomass and agriculture' Presentation to the Workshop[†].

19. Hooda, N., and Rawat, V.R.S. 'Role of Bio-Energy plantations for carbon-dioxide mitigation with special reference to India', Mitigation and Adaptation Strategies for Global Change[*].

20. J. Sachs, (ed.), 2005. UN Millennium Project, United Nations Development Programme.

21. Ogawa, M., Okimori, Y., and Takahashi, F. 'Carbon sequestration of biomass and forestation: three case studies', Mitigation and Adaptation Strategies for Global Change[*].

22. Woods, J., Hemstock, S., and Burnyeat, W., 2005. 'Bio-Energy systems at the community level in the South Pacific: impacts and monitoring', Mitigation and Adaptation Strategies for Global Change[*].

23. Read, P., 2005. 'Reconciling emissions trading with a technology-based response to potential abrupt climate change', Mitigation and Adaptation Strategies for Global Change[*].

24. Bot, A.J., Nachtergaele, F.O., and Young, A., 2000. 'Land Resource Potential and Constraints at Regional and Country Levels', Land and Water Division, FAO, Rome.

25. Faaij, A.P.C., 2005. 'Modern biomass conversion technologies' Mitigation and Adaptation Strategies for Global Change[*].

26. Haszeldine, R.S., 2005 'Deep geological CO$_2$ storage: principles, and prospecting for bio-energy disposal sites', Mitigation and Adaptation Strategies for Global Change[*].

27. Lehmann, J., Gaunt, J., and Rondon, M. 'Bio-Char sequestration in terrestrial ecosystems – a review', Mitigation and Adaptation Strategies for Global Change[*].

28. Day, D., Evans, R.J., Lee, J.W., and Reicosky, D., 2005. 'Economical CO$_2$, SOx and NOx capture from fossil fuel utilization with combined renewable hydrogen production and large scale carbon sequestration' **Energy** (paper to International Energy Workshop, 2003, revised and forthcoming: www.sciencedirect.com, ref EGY 1407).

[†] Expert Workshop on 'Greenhouse Gas Emissions and Abrupt Climate Change', Paris, Sept. 30th and Oct 1st , 2004 to 'address the policy implications of potential abrupt climate change' – visit www.accstrategy.org.

[*] Special Issue, forthcoming – www.accstrategy.org/simiti has texts of peer-reviewed articles from the Expert Workshop in author-created format prior to typesetting and proof-reading. Also viewable at http://www.springeronline.com/journal/11027.

CHAPTER 40

Scope for Future CO_2 Emission Reductions from Electricity Generation through the Deployment of Carbon Capture and Storage Technologies

Jon Gibbins[1], Stuart Haszeldine[2], Sam Holloway and Jonathan Pearce[3], John Oakey[4], Simon Shackley[5] and Carol Turley[6]

[1] *Imperial College London*
[2] *University of Edinburgh*
[3] *British Geological Survey*
[4] *Cranfield University*
[5] *University of Manchester*
[6] *Plymouth Marine Laboratory*

ABSTRACT: Ongoing work on the potential for carbon (dioxide) capture and storage (CCS) from fossil fuel power stations in the UK suggests that power plants using this family of technologies may be capable of supplying significant amounts of low-emission electricity within one or two decades. Increases in renewable generation are also planned over similar timescales and there is the additional possibility of nuclear replacements being built. If political justification for significant carbon dioxide (CO_2) emission reductions in the UK emerges from global post-Kyoto negotiations, large ($\sim$45%) reductions in CO_2 emissions from the UK electricity generation could be achieved as early as 2020. Both the technical and the political aspects are, however, changing rapidly, and the first clear pointer for the future may only come with the conclusion of the post-Kyoto negotiations. CCS technologies also have considerable potential for future emission reductions worldwide, especially in regions where large numbers of new fossil fuel power plants are being built within $\sim$500 km of sedimentary basins.

40.1 Introduction

In recent years, emissions of carbon dioxide (CO_2) from the UK electricity generation sector have stayed constant or increased slightly. Values from recent UK Department of Trade and Industry (DTI) updated energy projections (UEP) [1] show a decrease over the next two decades, but at a reduced rate compared with the 1990s.

It should be noted, however, that the observed and projected values for CO_2 emissions in Figure 40.1 represent electricity supplies with no (historically) or low (UEP) UK CO_2 reduction targets. A 60% CO_2 emission reduction by 2050 was recommended by the Royal Commission on Environmental Pollution [2] and subsequently endorsed by the Energy White Paper [3]. As Figure 40.1 shows, if overall UK energy use were to match the DTI UEP for 2020, the UK would not be even close to a linear reduction path to 60% reductions by 2050, with the 2020 interim value exceeded by about 110 megatons of carbon dioxide ($MtCO_2$) per year. It is beyond the

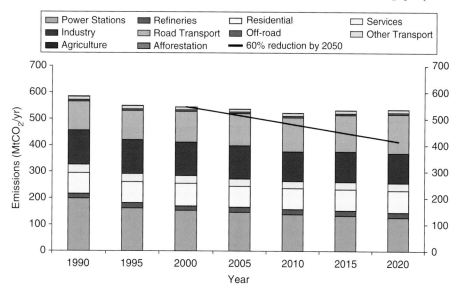

Figure 40.1 UK carbon emissions by sector from *Updated UK Energy Projections* [1] The line corresponds to a linear path to the 60% reduction target for 2050 recommended by the Royal Commission on Environmental Pollution [2].

scope of this paper to discuss the extent to which the sectoral emissions shown in Figure 40.1 might be affected by demand-side measures; however, as discussed later in this paper, even for the same electricity output, a reduction in CO$_2$ emissions from the electricity generation sector of 50–55 MtCO$_2$/yr by 2020 is probably technically feasible, through combinations of increased fuel switching, greater renewable generation, new nuclear and carbon (dioxide) capture and storage (CCS). However, the 'commercial viability' of some or all of these measures for deployment in 2020 depends entirely on final UK carbon emission targets and the ability of alternative options to deliver at a lower price. Additional costs for all the 'decarbonised electricity' options, including CCS, are probably in the range of 1–3 pence per kilowatt-hour (p/kWh).

40.2 Carbon Capture and Dtorage in The UK Electricity Generation Sector

With negotiations on post-Kyoto emission targets only just beginning, it is not possible to provide meaningful projections for UK national CO$_2$ emission caps for the latter part of the next decade and subsequent decades.

Irrespective of the absolute targets to be achieved, it is evident, however, that achieving significant CO$_2$ emission reductions in the UK is likely to involve early and large reductions in CO$_2$ emissions from the electricity sector (e.g. as shown in recent UK MARKAL studies [4]). This reflects lower technical, economic and social barriers to emission reductions from electricity generation; even so, the required changes will be challenging. UEP electricity generation mix figures for 2000–2020 and some alternative scenarios for 2020 are shown in Table 40.1.

The purpose of these alternative, purely illustrative, scenarios is to indicate electricity sector CO$_2$ emissions and reliance on gas for three alternative generation mixes, representing the following scenarios:

A. The maximum emission reductions that can be achieved using fuel switching (coal to gas) to the greatest possible extent (i.e. zero coal), but no other measures;
B. A scenario aiming to achieve emission reductions of about 55 MtCO$_2$/yr but retaining the UEP 2020 coal projection. The emission reductions are assumed to be achieved partly through 2 gigawatts (GW) of new nuclear build, reflecting constraints on the building rate for such new plants, and the remainder through CCS

Table 40.1 UEP electricity generation mix [1] and illustrative alternative scenarios for 2020.

Fuel	Original UEP values					2020 scenarios		
	Electricity Generation, TWh/yr					A No coal, 20% renewables	B 9 GW CCS, 2 GW new nuclear	C Less coal, 13 GW gas CCS
	2000	2005	2010	2015	2020	2020	2020	2020
Coal	111.9	113	106	89	57	0	7	20
Coal + CCS						0	7	20
Oil	2.1	2	2	2	2	2	2	2
Gas	127	116	132	159	225	264	174	144
Gas + CCS						0	17	100
Nuclear	78.3	84	61	41	27	27	43	27
Renewables	10.1	15	39	58	58	76	76	76
Imports	14.3	9	10	10	10	10	10	10
Pumped storage	2.6	3	3	3	3	3	3	3
Total	346.3	344	353	362	382	382	382	382
MtCO$_2$/yr	153.7	147.1	138.9	134.0	126.9	88.1	71.3	72.2
Mean kgCO$_2$/kWh including 8% transmission losses	0.479	0.462	0.425	0.400	0.359	0.249	0.201	0.204
MtCO$_2$ to storage						0	54	31
Low emission power	30%	32%	32%	31%	26%	30%	52%	57%
% gas	37%	34%	37%	44%	59%	69%	50%	64%

Assumptions:

	Coal	Oil	Gas
2020 plant kg CO$_2$/kWh generated	0.903*	0.660	0.329
.... with CCS	0.108	–	0.054
Fraction of CO$_2$ captured	90%	–	85%
Additional fuel for CCS plant	20%	–	10%

* The same UEP value has been used for all coal plants for consistency, although this would be pessimistic for new or upgraded coal plant with CCS.

applied to arbitrary allocations of 6.5 GW of coal generation and 2.5 GW of gas generation;

C. A scenario also aiming to achieve emission reductions of about 15 MtCO$_2$/yr, but without the use of coal with CCS (which is assumed to be less competitive than gas with CCS in this scenario). A modest amount of coal generation is retained for peak load situations. No new nuclear build occurs.

These scenarios are not an attempt to compare the costs or other merits of the different low-emission generation options. To do so would require projections for future fuel prices and equipment costs, which are beyond the scope of this paper, and which in any case would be subject to considerable uncertainty. Policy and regulation yet to be determined will also affect CCS technology development; for example, how retrofit and 'new-build' CCS plants are treated in the European Union's Emission Trading Scheme, and how any additional incentives specifically for CCS are framed to accommodate the fundamental differences between coal and gas CCS plants, enhanced oil recovery (EOR) and aquifer storage, etc. Some features of Table 40.1 are, however, reasonably certain. In all three scenarios it has been assumed that the UK government's aspiration of 20% of electricity from renewables by 2020 has been achieved. It does not appear likely at present that this figure will be significantly exceeded. Scenario A shows that a combination of the existing policies of more renewables and fuel switching, even with the latter carried out to its maximum possible extent, cannot achieve electricity sector emission reductions of more than about 40 MtCO$_2$/yr.

Scenario B shows that a reasonably active nuclear new build programme that brought 2 GW of new nuclear on line by 2020 would not of itself simultaneously achieve a 50 MtCO$_2$/yr reduction and reduced reliance on gas, but that this could be achieved if approximately 90% of the projected 2020 coal generation were from CCS plants. This appears to be a reasonable target, provided that some existing power plants can be retrofitted with CCS technology. This would possibly involve upgrading some existing pulverised coal plants from sub-critical to super-critical steam conditions and adding post-combustion CO$_2$ 'scrubbers'. It is also likely, however, that some new integrated gasifier combined cycle (IGCC) plants, with the carbon monoxide in the gas shifted to hydrogen and CO$_2$ for pre-combustion capture, would be built; several such schemes are already being planned. In the longer term, additional existing coal power plants may be upgraded to oxyfuel operation or be repowered with gasifiers. Gasifiers with CCS might also be used to supply hydrogen-rich gas from coal, instead of natural gas, for existing combined cycle plants.

Scenario C examines a situation with minimal coal and no new nuclear, and thus increased reliance on gas. The somewhat arbitrary ($\sim$50% of 'reasonable' 2020 emission cuts) target of 55 MtCO$_2$/yr requires 13 GW of gas

generation to be fitted with CCS, roughly about three quarters of the new natural gas combined cycle (NGCC) build between now (the 2005 value) and 2020. The relative ease of configuring new plants to include capture technology also appears to make this level of CCS a feasible technical approach.

NGCC plants may also utilise post-combustion CO$_2$ capture technology or, as in the recently-announced Peterhead/Miller project led by BP and Scottish and Southern Energy, pre-combustion conversion of natural gas to hydrogen for combustion in a gas turbine, and to CO$_2$ for storage. Natural gas is likely to offer relatively low-cost CO$_2$ capture so long as gas prices remain low, particularly for new NGCC plants designed for capture from the outset. This highlights the need for all new UK power plants to be 'capture-ready', even if capture equipment is not installed when they are built. Depending on future natural gas supply conditions, some existing NGCC plants may be alternatively modified to operate on gas from new coal gasifiers; these would also need to be suitable for hydrogen production and CO$_2$ capture, either when built or subsequently.

The UK has significant CO$_2$ storage opportunities offshore, with probably the greatest absolute capacity of any European country after Norway and the best combination of CO$_2$ sources relatively close to potential CO$_2$ storage sites. Storage capacity for UK oilfields as a result of enhanced oil recovery has been estimated at approximately 700 MtCO$_2$ [5]. Following an established methodology [6], the CO$_2$ storage capacity of many of the UK gas fields has been estimated on the basis that 90% of the pore space occupied by the recoverable reserves of depletion drive fields and 65% of the pore space occupied by the recoverable reserves of water drive fields could become available for CO$_2$ storage. Storage capacity in UK gas fields in the Southern North Sea Basin has been estimated at approximately 3.7 gigatons (Gt) CO$_2$ [7]. Storage capacity in gas fields in the East Irish Sea Basin has been estimated at approximately 1 GtCO$_2$ [8]. On the above basis, the CO$_2$ storage capacity in the gas and gas/condensate fields of the UK sector of the Northern and Central North Sea Basin is at least 0.8 GtCO$_2$. Therefore, the total CO$_2$ storage capacity of the UK oil and gas fields alone may be in excess of 6 GtCO$_2$

Storage capacity in saline aquifers is more difficult to estimate with confidence owing to the uncertainties surrounding the relatively poorly characterised aquifers. Storage capacity in saline aquifers has been estimated to be up to 14.25 GtCO$_2$ in the Southern North Sea Basin [7] and up to 0.63 GtCO$_2$ in the East Irish Sea Basin [8]. No detailed estimates have yet been made of the aquifer storage capacity of the Northern and Central North Sea Basin or the other sedimentary basins surrounding the UK. Thus, the total CO$_2$ storage capacity of the UK continental shelf is likely to comfortably exceed 20 GtCO$_2$.

The abundance of CCS options in the UK also brings challenges. A range of stakeholders needs to participate in developing effective strategies, and there is a risk of excessive diversification and dissipation of effort. As a result, new integrated research projects have been proposed to study the issues involved in getting the best value for the UK out of CCS applications and to make sure that maximum benefits are achieved through international collaboration on technology development. The DTI Carbon Abatement Strategy (CAT) and the Research Councils' programme Towards a Sustainable Energy Economy (TSEC) are both planned to address CCS issues in depth, to place them in an integrated UK energy system context and to consider the social, environmental, economic, technological and other aspects. Environmental and health and safety issues surrounding CCS on a range of temporal and spatial scales require a focused and coordinated research activity. In the longer term, it is hoped that a UK Carbon Dioxide Capture and Storage Authority will be established by the UK government to take overall responsibility for the regulation of this new industry and

eventually to provide long-term stewardship for the CO$_2$ stored underground.

40.3 Global Applications for Carbon Capture and Storage Technologies

The UK energy economy has the potential to develop and demonstrate CCS technologies that could find applications in many other countries. The UK has the opportunity to make a leading contribution in this field, because of

- its industrial expertise in a number of key areas;
- the need for new UK power plant capacity over the next two decades;
- a window of opportunity in the next decade for enhanced oil recovery in the North Sea;
- a range of additional geological storage options to give extended capacity with, in some cases, access potentially facilitated by existing infrastructure; and

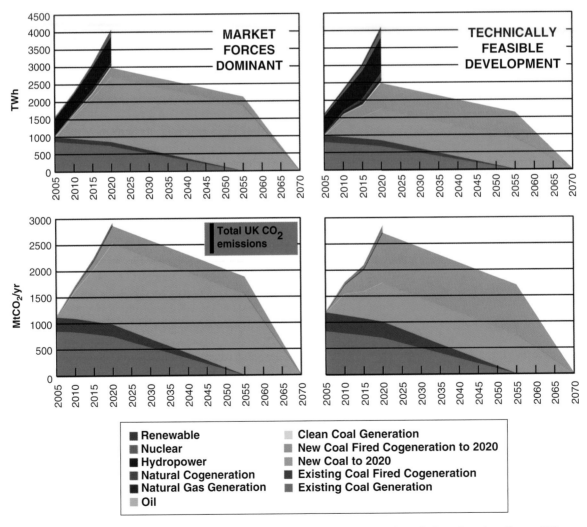

Figure 40.2 Estimates for future Chinese electricity generation and associated CO$_2$ emissions (based on Guo and Zhou [10]). Note the black bar whose length corresponds to total current UK CO$_2$ emissions at the same scale – these are potentially dwarfed by emissions from the new coal plants planned to be built in China up to 2020 alone.

- national aspirational targets for CO_2 emissions that could justify the deep reductions that CCS technologies can give.

CCS is likely also to see early use in other countries over the next two decades and, even where immediate deployment is not justified, it is important to ensure that new power plants are designed and built to be 'capture-ready'. As a minimum, this requires that a feasibility study of how capture will be added later be conducted and that space and essential access requirements be included in the original plant to allow capture-related equipment to be retrofitted. This can generally be done for conventional pulverised coal and NGCC plants at minimal cost and for new IGCC stations at possibly greater expense. It would then be possible to add CO_2 capture rapidly and without any significant additional costs (i.e. costs beyond those for installing capture on an equivalent new plant) whenever political and economic conditions develop to justify it. The capability to achieve rapid and cost-effective deployment of CCS technology, as part of a portfolio of demand- and supply-side options to manage carbon emissions, is also likely to encourage a positive approach to atmospheric CO_2 concentration stabilisation.

Making new power plants at least capture-ready, if not actually built to capture CO_2 from the outset, is particularly important in economies where large numbers of new power plants are being built. China is currently a prime example: as Figure 40.2 shows, new coal plants planned to be built in China up to 2020 offer scope for significant reductions in CO_2 emissions if capture technology can be added in the future. CO_2 emissions from just these new plants are likely to exceed total current UK emissions (black bar) by perhaps a factor of three in the absence of abatement measures. The clean development mechanism (CDM) of the Kyoto Protocol or a similar future mechanism may offer a ready-made route to finance and incentivise such capture retrofits. However,

large amounts of new power plant capacity will also be required in Europe and the USA relatively soon to replace ageing generation capacity built in the latter half of the 20th century. Estimates from the International Energy Agency [9] show that Organisation for Economic Co-operation and Development countries are likely to represent approximately 30% of the demand for new coal and gas generation capacity over the next three decades. CCS, and the initial requirement for capture-ready new power plants as a standard, may therefore be needed soon in all major economies to contribute to avoiding dangerous climate change.

REFERENCES

1 UEP, (2004) *Updated UK Energy Projections*, DTI Working Paper, May 2004. (http://www.dti.gov.uk/energy/sepn/uep.pdf)

2 Royal Commission on Environmental Pollution (2000) *Energy – The Changing Climate*, June 2000. (http://www.rcep.org.uk/energy.htm)

3 DTI (2003) *Our Energy Future – Creating A Low Carbon Economy*, White Paper, HMSO, February 2003.

4 DTI (2004) *Options for a Low Carbon Future*, DTI Economics Paper No. 4, June 2003. (http://www.dti.gov.uk/economics/opt_lowcarbonfut_rep41.pdf)

5 Balbinski, E. (2002) *Potential UKCS CO₂ Retention Capacity from IOR Projects*, SHARP IOR eNewsletter, Issue 3, September 2002. (http://ior.rml.co.uk/issue3/co2/ecl_retention/page1.htm)

6 Bachu, S. & Shaw, J. (2003) Evaluation of the CO₂ sequestration capacity in Alberta's oil and gas reservoirs at depletion and the effect of underlying aquifers, *Journal of Canadian Petroleum Technology*, 42(9), 51–61.

7 Holloway, S., Vincent, C.J., Bentham, M.S. & Kirk, K.L. (in press) Reconciling top-down and bottom-up estimates of CO₂ storage capacity in the UK sector of the Southern North Sea Basin.

8 Kirk, K.L. (2005) *Potential for Storage of Carbon Dioxide in the Rocks beneath the East Irish Sea*. BGS Report No. CR/05/127N, British Geological Survey, Keyworth, Nottingham, UK.

9 IEA (2004) *World Energy Outlook*. International Energy Agency, Paris, France.

10 Guo Y. & Zhou D. (2004) *Low Emission Options in China's Electric Power Generation Sector*, ZETS Conference, Brisbane, February 2004 (and personal communication, Guo Yuan, Energy Research Institute, Beijing, April 2004).

CHAPTER 41

The Technology of Two Degrees

Jae Edmonds and Steven J. Smith

Pacific Northwest National Laboratory, Joint Global Change Research Institute, at the University of Maryland, College Park, MD

ABSTRACT: This paper examines some of the energy technology implications of limiting the change in mean global surface temperature (GMST) to two degrees Celsius (2°C) relative to pre-industrial temperatures. Understanding the implications of this goal is clouded by uncertainty in key physical science parameters, particularly the climate sensitivity. If the climate sensitivity is 2.5°C then stabilization implies stabilization of CO_2 concentrations at less than 500 parts per million (ppm) with a peak in global CO_2 emissions occurring in the next 15 years and with a decline in emissions to approximately 3 petagrams of carbon per year by 2095. Under such circumstances the value of technology improvements beyond those assumed in the reference case is found to be exceptionally high, denominated in trillions of 1990 USD. The role of non-CO_2 greenhouse gases is important. Aerosols could produce significant feedbacks, though uncertainty is significant. If the climate sensitivity is 3.5°C or greater, it may be impossible to hold GMST change below 2°C. On the other hand if the climate sensitivity is 1.5°C, limiting GMST change to 2°C may be a trivial matter requiring little deviation from a reference emissions path until after the middle of the 21st century.

41.1 Introduction

The European Union has set as its goal, the limiting of the change in global mean surface temperature (GMST) to two degrees Celsius (2°C) relative to pre-industrial. This paper makes no attempt to assess the merits of this goal. The purpose of this paper is to ask, what are the implications for energy technology of establishing such a goal? And what implications do energy technologies have for the economic cost of pursuing this goal? This paper utilizes an integrated assessment model, MiniCAM, to examine the implications for energy technology of limiting GMST change to 2°C.

The MiniCAM is an integrated assessment model that considers the sources of emissions of a suite of greenhouse gases[1] (GHG's) emitted in 14 globally disaggregated regions[2], the fate of emissions to the atmosphere, and the consequences of changing concentrations of greenhouse related gases for climate change[3]. The MiniCAM begins with a representation of demographic and economic developments in each of the 14 regions and combines these with assumptions about technology development to describe an internally consistent representation of energy, agriculture, land-use, and economic developments that in turn shape global emissions and concentrations of GHGs. GHG concentrations in turn determine radiative forcing and climate change. The MiniCAM was one of the models employed to develop the IPCC emissions scenarios described in the Special Report on Emissions Scenarios (SRES) [4]. In this paper we update the MiniCAM B2 scenario contained in [4] to include a number of model enhancements [5], and the inclusion of a full suite of greenhouse gases, aerosols, and criteria air pollutants [6].

Briefly, the reference case is one in which global population grows to 9 billion people by the year 2050 and reaches 9.5 billion by the year 2095. Global GDP increases from 30 trillion 2002 USD in 1990 to more than 250 trillion USD[4] in the year 2095. The global energy system increases in scale from about 375 exajoules per year (EJ/y) in 1990 to more than 1200 EJ/y by 2095. Fossil fuel CO_2 emissions increase from approximately 6 PgC/y in 1990 to approximately 20 PgC/y in 2095. Emissions of other CO_2 from industrial applications increase steadily throughout the century. Emissions of high GWP gases also increase significantly on a percentage basis. Other gases such as CH_4 and N_2O increase during the first half of the century but exhibit limited growth thereafter. Emissions of carbonaceous aerosols and sulfur dioxide decline over the course of the scenario.

Reference case emissions trajectories for anthropogenic greenhouse related gases, concentrations for CO_2, CH_4,

---

[1] MiniCAM tracks emissions of 15 greenhouse related gases: CO_2, CH_4, N_2O, NO_x, VOCs, CO, SO_2, carbonaceous aerosols, HFCs, PFCs, and SF_6. Each is associated with multiple human activities that are tracked in MiniCAM.

[2] The United States, Canada, Latin America, Western Europe, Eastern Europe, the Former Soviet Union, the Middle East, Africa, India, China, Other South and East Asia, Australia and New Zealand, Japan and Korea.

[3] The equation structure of the MiniCAM model is described in [1]. Its energy-economy roots can be traced back to [2]. The model has been continuously revised and updated to include an expanded set of processes, such as endogenous agriculture and land-use determination, which is in turn linked to changes in natural system stocks and transient emissions fluxes. Its natural system representation utilizes MAGICC, which has its origins in [3].

[4] Measured using purchasing power parity exchange rates.

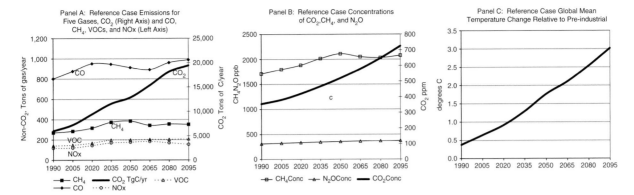

Figure 41.1 Reference Case Emissions (Panel A), Concentrations of CO_2, CH_4, and N_2O (Panel B) and GMST for a Climate Sensitivity of 2.5°C (Panel C).

Table 41.1 Technology assumptions in the reference and advanced technology cases.

Technology	Units	1990	2095 (Ref. Case)	2095 (Adv. Tech)
Electric power generation (fuel + non-fuel cost)[5]				
Solar	1990 USD/kWh	0.610[6]	0.060	0.040
Wind	1990 USD/kWh	0.080	0.040	0.040
CO_2 Capture and storage[7]				
Coal—power output loss	Percent	25	Unavailable	15
Coal—added capital cost	% non-capture capital cost	88	Unavailable	63
Gas—power output loss	Percent	13	Unavailable	10
Gas—added capital cost	% non-capture capital cost	89	Unavailable	72
CO_2 capture efficiency	Percent	90	Unavailable	90
CO_2 storage cost	1990 USD/tC	37	Unavailable	37
Transportation	Service per unit final energy (productivity index, 1990 = 1.0)	1	2	2.6
Agricultural biomass				
Biofuels crop	Average annual productivity growth rate (%.y)	0.70	0.70	1.10
Conventional crops	Average annual productivity growth rate (%.y)	0.70	0.70	1.10
Hydrogen production				
Natural gas to H_2	Percent efficiency	70	Unavailable	80
Natural gas to H_2 + CCS	Percent efficiency	58	Unavailable	71
Coal to H_2	Percent efficiency	62	Unavailable	66
Coal to H_2 + CCS	Percent efficiency	52	Unavailable	58
Electrolysis	Percent efficiency	87	Unavailable	94
Biofuels	Percent efficiency	60	Unavailable	80

[5] Nuclear technology is also subject to technological improvement which we have not explored in this study. Gen III and Gen IV reactor technologies could dramatically reduce costs and enhance performance. The economic implications of advanced reactor and fuel cycle designs will be explored in future work.

[6] Solar power is represented as one aggregate technology in this model version. PV costs are used as a marker in the table, recognizing that costs vary significantly for different solar technologies. Work explicitly addressing solar technologies is underway.

[7] Note that incremental capital costs do not include the cost of CO_2 storage, which is included as a separate cost entry.

and N_2O, and GMST are recorded in Figure 41.1, panels A, B, and C respectively.

41.2 The Technology of the Reference Case

Technology assumptions play a central role in shaping emissions. While fossil fuels are the backbone of the present global energy system, it is energy services that consumers ultimately desire. Fossil fuels are presently the most cost-effective way of delivering energy services to society under most circumstances. Reference case emissions trajectories presume that even in the absence of explicit policies to limit GHG emissions, technologies will evolve substantially. For example, the scenario, IS92a [7], assumed that by the end of the 21st century electric power production

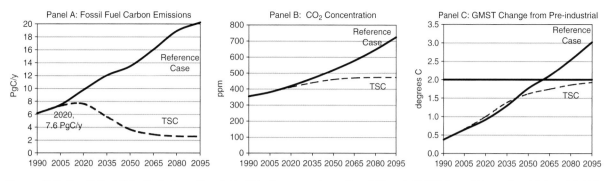

Figure 41.2 Fossil Fuel CO_2 Emissions (Panel A), CO_2 Concentrations, (Panel B) and GMST for a Climate Sensitivity of 2.5°C (Panel C).

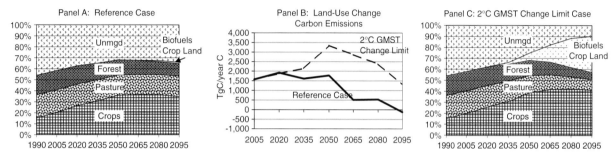

Figure 41.3 Land Use in the Reference Case (Panel A), Land Use Change Emissions (Panel B) and Land Use in the 2°C GMST Change Limit Case (Panel C).

would already be more than 75% non-emitting, and that a commercial biomass industry would develop that was as large as the present global oil and gas enterprise. This study is also characterized by assumed advances in technology over the course of the study given in Table 41.1.

The assumptions given in Table 41.1 are not intended to be predictions about the future or future technology developments. The changes hypothesized are used for illustrative purposes. They are intended to show the sensitivity of the system to technology assumptions. Therefore it is the qualitative response of cost and other variables to technology that is of primary interest. The system is sensitive to assumptions not articulated in Table 41.1 as well. For example, later in this paper we demonstrate the sensitivity of results to end-use energy technology assumptions. The intent of this paper is to initiate consideration of the role of technology in stabilizing climate change, and future work will continue and expand the investigations we have begun here.

41.3 Stabilizing Temperature

Limiting GMST change so it does not exceed 2°C, the Temperature Stabilization Case (TSC), means limiting emissions of the suite of greenhouse gases. Note that stabilizing GMST differs from stabilizing GHG concentrations, the goal of the UNFCCC [8], due to dynamic physical processes such as ocean thermal lag. As a consequence, stabilization of temperature can require that

GHG concentrations peak and then decline, sometimes called concentration 'overshoot' scenarios. GMST is stabilized in the MiniCAM by imposing a common global emissions tax rate denominated in USD/tC equivalent[8] so as to minimize cost. As all countries participate from the start, this trajectory is unrealistic. Any real-world trajectory would encounter greater costs than those reported here owing to heterogeneous timing of regional and national emissions limitations and to the employment of potentially inefficient policy instruments. The estimates of the value of technology improvements are therefore likely biased low.

The CO_2 emissions pathway in the TSC departs dramatically from that of the reference case, Figure 41.2 Panel A. TSC industrial carbon emissions peak in 2020 at 7.6 PgC/y and decline rapidly thereafter. By 2095 industrial emissions have declined to 2.6 PgC/y. The emissions mitigation is accomplished in the reference case by creating a massive commercial biomass industry (more than 400 EJ/y production), deploying additional nuclear power, improving energy efficiency and reducing the use of all fossil fuels. Commercial biomass production plays a major role in providing energy in the TSC. In the reference case the bulk of biomass energy production is traditional fuels and modern fuels derived from waste products, Figure 41.3 Panel A. In the TSC vast tracts of land are devoted to

---

[8] In this study we employ 100-year Global Warming Potentials [9] to convert carbon prices to prices for other GHG's.

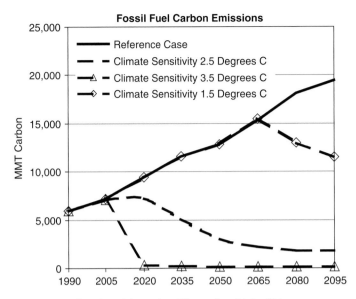

Figure 41.4 Carbon Emissions Pathways for Three Alternative Climate Sensitivity Values.

energy crop production, Figure 41.3 Panel C. This in turn leads to dramatic increases in food prices, transfer of resources to the agriculture sector, transfer of financial resources to producing regions, largely in the south, and a dramatic increase in deforestation rates and associated carbon emissions, Figure 41.3 Panel B.

Cumulative emissions mitigation in the period 2005 to 2095 is more than 850 PgC. It must be noted that emissions mitigation would be still larger were it not for advances in technology assumed to occur in the reference case. While emissions mitigation in the period to 2020 is significant, more than 15 PgC, the bulk of emissions mitigation, approximately 650 PgC, occurs after the year 2050 as opposed to 200 PgC before 2050. Thus, near-term emissions mitigation is merely an overture to the dramatic reductions required in the second half of the 21st century. This in turn implies dramatic value to the improvement of technologies even if the bulk of their deployment occurs after the middle of the century.

The rapid reductions in carbon emissions limit CO_2 concentrations to less than 475 ppm, Figure 41.2 Panel B. Reductions in emissions of fossil fuel CO_2 simultaneously reduce emissions of aerosols and thereby increase GMST prior to 2050, Figure 41.2 Panel C. This effect coupled with ocean thermal inertia is potentially large enough to make achieving the TSC impossible for climate sensitivities of 4.5°C, because the simultaneous reductions in aerosol emissions with carbon emission reductions produces a realized GMST increase relative to pre-industrial that exceeds 2°C.

The assumed climate sensitivity plays a major role in determining the emissions path. Climate sensitivity values ranging from 1.5°C to 4.5°C have long been cited as encompassing a significant portion of overall uncertainty in climate sensitivity[9]. We examine three climate sensitivities: 1.5°C, 2.5°C[10], and 3.5°C[11]. The associated pathways are displayed in Figure 41.4. For a climate sensitivity of

3.5°C emissions must decline to virtually zero by the year 2020. For a climate sensitivity of 1.5°C substantial emissions reductions begin only after 2050. Thus, policy decisions must be taken today in the context of profound uncertainty, a feature which highlights the usefulness of framing the problem in terms of risk management. From this perspective the presence of uncertainty is not a reason for inaction, but rather shapes the nature of near-term actions and recommends policies that provide flexibility in future actions. The uncertainty virtually guarantees that today's decisions will eventually be deemed inappropriate in retrospect, but it is impossible to determine before the facts whether their inadequacy will be in too aggressively preserving other socially desirable resources at the expense of climate, or climate at the expense of other socially desirable resources.

Under the assumed suite of technologies (Table 41.1), the assumed cost-minimizing behavior of human societies and values for physical parameters (e.g. the 2.5°C climate sensitivity), the present discounted economic cost[12] of holding GMST change below 2°C is approximately 18 trillion 1990 constant USD. Costs are calculated for a fixed suite of technologies and profile of technology developments. No attempt has been made to impose a particular model of induced technological change[13]. We do this so as to allow a comparison between alternative technology regimes and thereby to value technology

---

[9] See, for example [9], section 9.3.4.1.

[10] This is the same climate sensitivity employed in Figures 41.1 and 41.2.

[11] It is impossible to prevent GMST from exceeding 2°C for a climate sensitivity of 4.5°C. Thus, we do not show results for the full range of climate sensitivities ranging between 1.5°C and 4.5°C.

[12] Discounted over the period 2005 to 2095 at a discount rate of 5% per year.

[13] While it is popular to impose some simple model such as an 'experience curve' (often mislabeled as a 'learning curve') the long history of

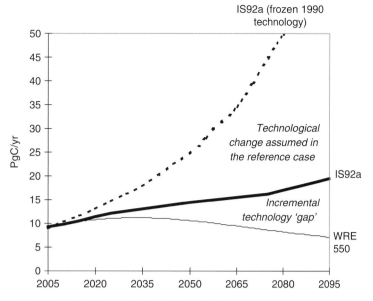

Figure 41.5 Carbon Emissions Trajectories Illustrating the Effect of Reference Case Technological Change.

developments in terms of their contribution to meeting the particular temperature change limit. The value of technology is independent of the source of technological change, and yields a metric for the value of potential technological progress on a variety of fronts.

41.4 The Technology 'Gap'

The imposition of a constraint on climate change implies the deployment of different technologies than in a case without the constraint. Edmonds [12] showed that while it may be convenient to think of the required technological change to meet the environmental constraint as merely the change in technology deployment relative to the reference case, that framing of the issue ignores the potentially huge contribution of technological change incorporated by assumption into the reference case. Figure 41.5 shows a comparison between assumed emissions associated with the IPCC emissions scenario, IS92a [13], and the same scenario run with energy technologies frozen at 1990 levels to illustrate the degree of presumed technological change embedded in the reference scenario[14]. But, the reference case technological improvements are a matter of assumption. They presume the successful execution of a variety of public and private sector long-term energy R&D programs around the world. That success is not guaranteed.

When considering improvements in technology and the change in technology deployment necessary to shift global emissions away from a reference case and onto a climate stabilization trajectory, both the incremental technology 'gap' and the technological change assumed in the reference case must be jointly considered. A failure to develop emissions mitigating technology in the reference case implies a larger incremental technology 'gap'. Conversely, assuming that technologies will be developed and deployed that reduce emissions in the reference case may shrink the expected incremental technology 'gap', but that reduced incremental technology 'gap' is accompanied by an obligation to deliver the technologies in the reference case technology suite. The concept of technology 'gaps' and the role of technological development and deployment have been explored in detail in Battelle [14]. The notion of an incremental 'gap' is therefore a contingent concept.

41.5 The Value of Technology in Limiting GMST Change to 2°C

To illustrate the economic value associated with achieving reference case technology performance, we consider two simple sensitivities – we change the rate of improvement in end-use energy technologies by ±0.25 percent per year. Reference case emissions change by approximately ±5 PgC/y by 2095 with CO_2 concentrations changing by approximately ±75 ppm and GMST changing by approximately ±0.25°C. The value of technological change in stabilizing GMST to less than 2°C (the TSC) is computed as the difference in the cost of the TSC using reference case technology and the cost for each sensitivity case. The benefit of improved end-use energy efficiency

---

research in the field of induced technological change finds no simple model to provide a satisfying explanation of the process by which technologies are created and transformed. See Weyant and Olavson [10] and Clarke and Weyant [11] for reviews of this literature.

[14]The intent of this calculation is not to construct a plausible alternative scenario, but rather to illustrate the degree to which technological change is embedded in the reference scenario and thereby taken for granted.

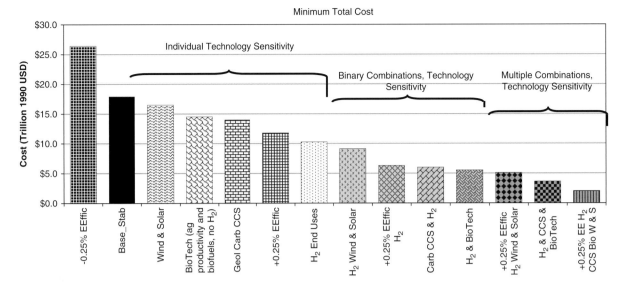

Figure 41.6 Minimum Total Cost of Limiting MGST Change to 2°C Relative to Pre-industrial.

in meeting the climate change limitation is approximately 6 trillion 1990 USD present discounted value[12] (PDV). The economic cost of a lower assumed rate of end-use energy efficiency improvement exceeds 8.5 trillion 1990 USD PDV.

We have hypothesized a variety of alternative technology changes, Table 41.1. While all of the technologies listed in Table 41.1 exist, in that some version of the technology is currently available somewhere in some form[15], mere existence is insufficient to ensure significant global market penetration, even when greenhouse gas emissions have economic value. The principal virtue of technology is not that it makes it possible to achieve a particular environmental goal. That can be accomplished trivially if costs are irrelevant. The great virtue of technology is that it holds the potential for reducing the cost of achieving an environmental goal. We use cost here not in its financial sense, but in the sense of requiring a lesser diversion of scarce social resources from other pursuits.

To explore the value of technology improvement in managing the cost of the TSC, we re-solve the model constrained to hold GMST change below 2°C with each technology change listed in Table 41.1 and record the cost. We then return to original values in Table 1 and move on to the next technology. After examining the sensitivity of cost to hypothesized individual technology changes we next explore the same sensitivity with combinations of two technologies, and then with more than two technologies at a time. Results are reported in Figure 41.6. The technology changes hypothesized in Table 41.1's '2095 (Adv. Tech.)' column are significant, but not beyond the realm of possibility. Some appear closer to deployment than others at the moment. But, the history of technology development is nothing if not a

lesson in forecaster humility. Technologies that were expected to develop have proved more difficult than expected, and technologies that were never envisioned have evolved to play a central role in the economy. We therefore approach the problem of technology analysis not as a forecasting exercise, but rather as a value of technology exercise. We ask the far more tractable question: what is the value of achieving specific cost and performance changes in terms of meeting a specific environmental constraint, imposed in an economically efficient manner, against a prescribed reference case background and a specific set of natural system parameters? Results are therefore highly contingent. Changes hypothesized in Table 41.1's '2095 (Adv. Tech.)' column explore only a subset of potential technology options and improvements. Many important options are left for future work to explore including the potential role of nuclear power. The partial nature of the suite of technology improvements does not negate the qualitative insights regarding the role of technology in managing the economic and environmental risks of climate change. Model representations can also play a role. The impact of wind and solar technologies, for example, are likely understated in this version of the model.

First, the more technologies that are successfully advanced, the lower the cost of implementing the TSC. But, all technologies are not equal. Technologies interact in important ways. Some are substitutes for one another. That is, the lower the cost of one technology, the smaller the deployment of the second. Hydrogen (H_2) and end-use energy efficiency improvements have this property in this analysis. Some technologies are complements and therefore lowering the cost of one technology leads to increased deployment of a second. Carbon dioxide capture and storage (CCS) and cost-competitive H_2 technologies (fuel cells as well as H_2 production, transportation, and storage technologies) are complementary technologies. The availability of CCS implies that H_2 derived from

---

[15] This is the sense employed in [15] and [16].

electricity generated by fossil powered utilities as well as H_2 derived from fossil fuels can reduce emissions more effectively. CCS could also complement commercial biomass crops, implying an ability to produce fuels with negative emissions per unit of end-use energy supply.

Second, important interactions in the global energy-economy system must be considered. For example, crop productivity, biofuels, and land-use-change emissions interact strongly. The package of assumptions associated with biotechnology include increased productivity growth rates for food and fiber as well as energy crops. Furthermore, a hypothetical technology is assumed that employs a biological process to transform input streams such as, for example, waste energy and water, to produce H_2 on insignificant (or currently unused) land areas. A technology with performance characteristics outlined in Table 41.1 reduces the human footprint on land, lowers prices for crops, livestock, and forest products, reduces land-use change emissions to levels below those in the reference case, and reduces the cost of the TSC by between 3.4 and 14.3 trillion 1990 USD, depending upon complementary technology availability (e.g. CCS and H_2). Realized economic value and impact on emissions will depend on technology performance and availability.

Similarly, CCS interacts strongly with other technologies. CCS can be deployed in electric power generation, but capture is incomplete, and the process diverts significant power resources to the capture process and requires significant capital investments. The captured CO_2 must be transported and stored (both temporarily and long-term). CCS technology can also be associated with the conversion of hydrocarbons from one form to another, for example coal to liquids, coal to gas, gas, coal, or bioenergy crops to H_2, or bioenergy crops to gas or liquids. Volumes associated with the successful development and deployment of CCS technology are potentially huge. In this analysis up to 45 PgC were captured and stored in the period between 2005 and 2050 but up to an additional 200 PgC were captured and stored in the period between 2050 and 2095[16]. While annual global capture and storage

did not exceed 0.5 PgC/y in 2020, this rate could be as high as 2.5 PgC/y by 2050 and to more than 6 PgC/y by 2095. The period between 2005 and 2050 is a preparatory period in which technologies that eventually become core components of the global energy system develop and experience initial deployment. The scale of the enterprise changes dramatically in the post-2050 period. A similar story can be told for other technologies such as biotechnology, hydrogen, wind, nuclear, solar or energy efficiency. The requirements of the post-2050 period cast a long shadow back to present technology research and development decision-making.

41.6 Concluding Remarks

In this paper we have explored some of the technology implications of limiting GMST change to 2°C relative to pre-industrial. Climate change is a century scale problem. In this paper analysis has been carried out to the year 2095. Analysis at this time scale yields important insights unavailable in an examination of shorter scope. While the technical and economic challenges of the emissions trajectory between 2005 and 2050 are daunting under the assumption of a 2.5°C climate sensitivity, they are far more modest than the challenges of the 2050 to 2095 period.

We also examined the implications of uncertainty in climate sensitivity for policy and technology development and deployment. In light of the profound uncertainty implied by variation in the climate sensitivity parameter, options which provide flexibility in managing both economic and environmental risks are attractive. One of the attractions of technology development is the flexibility it provides in managing both types of risk. If the climate sensitivity is 2.5°C then limiting GMST to 2°C implies stabilization of CO_2 concentrations at less than 500 ppm with a peak in global CO_2 emissions occurring in the next 15 years and with a decline in emissions to 3 PgC/y by 2095. Under such circumstances the value of technology improvements beyond those assumed in the reference case is denominated in trillions of 1990 USD. The role of non-CO_2 greenhouse gases is important. Aerosols could produce significant feedbacks in addition to their impacts as local pollutants, though uncertainty is significant. If the climate sensitivity is 4.5°C or greater, it may be impossible to limit climate GMST change to 2°C. On the other hand if the climate sensitivity is 1.5°C, limiting GMST change to 2°C may be a trivial matter requiring little deviation from a reference emissions path until after the middle of the 21st century.

---

[16] The question of storage is an important one. Present understanding of potential reservoir capacity suggests that the global storage resource is more than adequate for the volumes suggested by this analysis. Edmonds, Freund and Dooley [17] estimated storage volumes as follows: Deep Saline Reservoirs, 87 to 2,727 PgC; Depleted Gas Reservoirs, 136 to 300 PgC; Depleted Oil Reservoirs, 41 to 191 PgC; and Unminable Coal Seams, >20 PgC, based on Herzog et al. [18] and Freund and Ormerod [19]. The more recent estimate by Dooley and Friedman [20] indicate that resource volumes are: Coal Basins, 48; Depleted Oil Plays, 31 PgC; Gas Basins, 190; Deep Saline Formation On-shore, 1,608; Deep Saline Formation Off-shore, 1,374. Edmonds et al. [21] explore the implication of the heterogeneous regional distribution of storage resources for technology deployment and find that resource distribution has little impact on aggregate global technology deployment. Permanence of storage is another important issue. We have assumed no significant losses from reservoirs. Any real world system will be imperfect. Characterization of real world systems is an important research priority and is a prerequisite to large-scale deployment of CCS technology.

REFERENCES

1. Edmonds, J., Clarke, K., Dooley, J., Kim, S.H., Smith, S.J. 2004. Stabilization of CO_2 in a B2 world: insights on the roles of carbon capture and storage, hydrogen, and transportation technologies, *Energy Economics*, 26(2004): 517–537.

2. Edmonds, J. and Reilly, J., 1985. *Global Energy: Assessing the Future*, Oxford University Press, Oxford, United Kingdom.

3. Hulme, M., Raper, S.C.B. and Wigley, T.M.L., 1995: An integrated framework to address climate change (ESCAPE) and further developments of the global and regional climate modules (MAGICC). *Energy Policy*, 23, 347–355.

4. Nakicenovic, N., and Swart, R. (Eds), 2000. *Special Report on Emissions Scenarios*. Cambridge University Press, Cambridge, United Kingdom.

5. Smith, Steven J. and Wigley, T.M.L., (2005) Multi-Gas Forcing Stabilization with the MiniCAM. Submitted to the ***Energy Journal***.

6. Smith, Steven J., Pitcher, H. and Wigley, T.M.L., (2005) Future Sulfur Dioxide Emissions. ***Climatic Change*** 73(3), pp. 267–318.

7. Edmonds, J., Joos, F., Nakicenovic, N., Richels, R. and Sarmiento, J., 2004. 'Scenarios, Targets, Gaps, and Costs', *The Global Carbon Cycle: Integrating Humans, Climate, and the Natural World. Scope 62*. Island Press pp. 77–102.

8. United Nations. 1992. *Framework Convention on Climate Change*. United Nations, New York.

9. Intergovernmental Panel on Climate Change. 2001. *Climate Change 2001: The Scientific Basis. The Contribution of Working Group I to the Third Assessment Report of the Intergovernmental Panel on Climate Change*. Houghton, J. T., Ding, Y., Griggs, D.J., Noguer, M., van der Linden, P. J. and Xiaosu, D. (Eds.). Cambridge University Press, Cambridge, UK. pp. 944.

10. Weyant, J.P. and Olavson, T., 1999. 'Issues in Modeling Induced Technological Change in Energy, Environment, and Climate Policy,' *Environmental Modeling and Assessment*, 4(2,3): 67–86.

11. Clarke, L. and Weyant, J., 2002. 'Modeling Induced Technological Change: An Overview', In Grübler, A., Nakicenovic, N. and Nordhaus, W. (Eds.), *Technological Change and the Environment*. Resources for the Future, Washington, DC.

12. Edmonds, J. 1999. 'Beyond Kyoto: Toward A Technology Greenhouse Strategy', *Consequences*, 5(1):17–28.

13. Leggett, J., Pepper, W.J., Swart, R.J., Edmonds, J., Meira Filho, L.G., Mintzer, I., Wang, M.X. and Wasson, J., 1992. 'Emissions Scenarios for the IPCC: An Update.' in *Climate Change 1992: The Supplementary Report to the IPCC Scientific Assessment*, University Press, Cambridge, UK.

14. Battelle Memorial Institute. 2001. *A Global Energy Technology Strategy Project Addressing Climate Change: An Initial Report an International Public-Private Collaboration*. Joint Global Change Research Institute, College Park, MD.

15. IPCC (Intergovernmental Panel on Climate Change), 2001. *Climate Change 2001: Mitigation. The Contribution of Working Group III to the Third Assessment Report of the Intergovernmental Panel on Climate Change*. Metz, B., Davidson, O., Swart, R. and Pan, J. (eds.). Cambridge University Press, Cambridge, UK.

16. Pacala, S. and Socolow, R., 2004. Stabilization Wedges: Solving the Climate Problem for the Next 50 Years with Current Technologies. *Science*, 305:968–972.

17. Edmonds, J.A., Freund, P.F. and Dooley, J.J., 2002. 'The Role of Carbon Management Technologies in Addressing Atmospheric Stabilization of Greenhouse Gases.' In *Greenhouse Gas Control. Proceedings of the Fifth International Conference on Greenhouse Gas Control Technologies, GHCT-5*, ed. David Williams, Bob Durie, et. al., pp. 46–51. CSIRO Publishing, Collingwood, Australia.

18. Herzog, H., Drake, E. and Adams, E., 1997. CO_2 Capture, Reuse, and Storage Technologies for Mitigating Global Climate Change. Energy Laboratory, Massachusetts Institute of Technology, Cambridge, MA.

19. Freund, P. and Ormerod, W.G., 1997. 'Progress Toward Storage of Carbon Dioxide.' *Energy Conversion and Management*, 38:S199–S204.

20. Dooley, J.J. and Friedman, S.J., 2004. *A Regionally disaggregated global accounting of CO_2 storage capacity: data and assumptions*. Battelle Pacific Northwest Division Technical Report Number PNWD-3431.

21. Edmonds, J., Dooley, J., Kim, S., Friedman, S. and Wise, M., 2004., Technology in an Integrated Assessment Model: The Potential Regional Deployment of Carbon Capture and Storage in the Context of Global CO_2 Stabilization. Submitted to *Human-Induced Climate Change: An Interdisciplinary Perspective*, Cambridge University Press.

THE EVANS LIBRARY
FULTON-MONTGOMERY COMMUNITY COLLEGE
2805 STATE HIGHWAY 67
JOHNSTOWN, NEW YORK 12095-3790